MECHANICS OF STRUCTURES AND MATERIALS

PROCEEDINGS OF THE FIFTEENTH AUSTRALASIAN CONFERENCE
ON THE MECHANICS OF STRUCTURES AND MATERIALS
MELBOURNE / VICTORIA / AUSTRALIA / 8-10 DECEMBER 1997

Mechanics of Structures and Materials

Edited by

RAPHAEL H.GRZEBIETA & RIADH AL–MAHAIDI
Department of Civil Engineering, Monash University, Clayton, Victoria, Australia

JOHN L.WILSON
Department of Civil and Environmental Engineering, University of Melbourne, Victoria, Australia

A.A.BALKEMA / ROTTERDAM / BROOKFIELD / 1997

Cover photo: Victorian Arts Centre, Melbourne, Victoria, Australia; Glenn Gibson, Blitz Pictures

Published by
A.A. Balkema, P.O. Box 1675, 3000 BR Rotterdam, Netherlands(Fax: +31.10.413.5947)
A.A. Balkema Publishers, Old Post Road, Brookfield, VT 05036-9704, USA
(Fax: 802.276.3837)

ISBN 90 5410 900 9
© 1997 A.A. Balkema, Rotterdam
Printed in the Netherlands

The Mechanics of Structures and Materials, Grzebieta, Al-Mahaidi & Wilson (eds)
© *1997 Balkema, Rotterdam, ISBN 90 5410 900 9*

Table of contents

Composites, materials and biomechanics (bio)

Concrete and steel composite structures

Concrete technology and design

Finite element methods, optimisation and stability

General structures and design

Reliability and geomechanics (geo)

Reliability and geomechanics (rel)

Steel design and welding

Timber and masonry (masonry)

Timber and masonry (timber)

The Mechanics of Structures and Materials, Grzebieta, Al-Mahaidi & Wilson (eds)
© *1997 Balkema, Rotterdam, ISBN 90 5410 900 9*

Preface

The Australasian Conference on the Mechanics of Structures and Materials (ACMSM) is the fifteenth in a series commencing in 1967. It has become a popular biannual event for both researchers and practitioners in the fields of structures, foundations and materials. The conference is usually held at a university campus in different locations around Australia and on this occasion it was held at The University of Melbourne. The conference was jointly run and organised by Civil Engineering staff members from The University of Melbourne, Monash University and the Royal Melbourne Institute of Technology. The conference provides a forum for the exchange of ideas and research findings between structural and geomechanics research scholars, their postgraduate students and practising engineers.

Three keynote speakers were invited: Associate Professor Vistasp Karbhari from University of California San Diego, Dr John Nutt from Ove Arup and Partners and Daryl Jackson from Daryl Jackson Architects Pty Ltd. Their respective addresses covered the wide ranging themes of fibre reinforced composite materials in structures, research collaboration between industry and universities, and architectural design challenges. Papers from the first two addresses listed above are included in these proceedings. The conference also held a professional issues forum chaired by Professors Rob Melchers from the University of Newcastle and Paul Grundy from Monash University. Topical professional issues relating to the new Structural College of the Institution of Engineers Australia, National Register for Structural Engineers (NPER3), international recognition of competency, university/industry collaboration and laboratories and funding were all discussed.

These proceedings are a selection of the broad range of research topics currently being investigated by engineers in Australasia. One of the conference's main features is that it encourages Australasian research students to present their work. Each paper was subjected to a critical review process by at least two experts from the field in which the material was written. The editors would like to thank all of the reviewers listed for their assistance in maintaining the high standard of assessment of each paper.

The organising committee would also like to thank the sponsors for their kind support.

Editorial Committee

Dr Raphael Grzebieta
Dr Riadh Al-Mahaidi
John Wilson

The Mechanics of Structures and Materials, Grzebieta, Al-Mahaidi & Wilson (eds)
© *1997 Balkema, Rotterdam, ISBN 90 5410 900 9*

Organisation

The 15th ACMSM Conference Committee

John Wilson (Chairman)
Department of Civil and Environmental Engineering
The University of Melbourne

Dr Raphael Grzebieta (Editor-in-Chief)
Department of Civil Engineering
Monash University

Barbara Butler (Conference Co-ordinator)
Department of Civil and Environmental Engineering
The University of Melbourne

Dr Riadh Al-Mahaidi
Department of Civil Engineering
Monash University

Dr Kanapathipillai Thirugnanasundralingam
Department of Civil and Geological Engineering
Royal Melbourne Institute of Technology – RMIT

Co-sponsors
Cement & Concrete Association of Australia
The Institution of Engineers, Australia
Gutteridge Haskins and Davey

Cement and Concrete Association of Australia

VICTORIA DIVISION

The Mechanics of Structures and Materials, Grzebieta, Al-Mahaidi & Wilson (eds)
© *1997 Balkema, Rotterdam, ISBN 90 5410 900 9*

Acknowledgements

The Organising Committee gratefully acknowledges the support given by

Sponsors
Australian Premixed Concrete Association
Australian Institute of Steel Construction
Monash University
The University of Melbourne
Royal Melbourne Institute of Technology

Australian Pre • Mixed Concrete Association

RMİT

Reviewers

The conference committee is grateful for the work contributed
by the following reviewers:

Dr Kazem Abhary
Dr Faris Albermani
Dr Riadh Al-Mahaidi
Dr Mario Attard
Prof. Lawrence Rae Baker
Dr Michael Bannister
Dr Fariborz Barzegar-Jamshidi
David Lawrence Beal
Dr Allan Beasley
Assoc. Prof. Geoffrey Boughton
Dr Malek Bouazza
Prof. Russell Bridge
Dr Norbert Burman
Dr W. K. Chiu
Paul Clancy
Frank Collins
Prof. John Corderoy
Brendan Corcoran
Damien Crozier
Dr Hussein Dia
Assoc. Prof. Ian Donald
Dr Stephen Foster
Dr Ala Giero
Prof. Raymond Gilbert
Dr Helen Goldsworthy
Dr Michael Griffith
Prof. Paul Grundy
Dr Raphael H. Grzebieta
Assoc. Prof. Chris Haberfield
Roger Hadgraft
Dr Muhammad Hadi
Prof. Greg Hancock
Assoc. Prof. Nick Haritos
Dean Hewitt
Assoc. Prof. Israel Herszberg
Assoc. Prof. John Hindwood
Dr Alan Holgate
Dr John Howell
Dr Len Koss
Dr Yee Cheong Lam

Dr Brett Lemass
Dr Barry Li
Jeff Lingard
Dr Guoxing Lu
Assoc. Prof. Mahen Mahendran
John Marco
Howard McTier
Prof. Robert Melchers
Dr Pryan Mendis
Assoc. Prof. Bob Milner
Prof. Noel Murray
Basil Naji
Dr Ninh Nguyen
Dr Deric John Oehlers
Prof. Adrian Page
Dr Mark Patrick
Dr Lam Pham
Prof. Vijaya Rangan
Dr R. Ravindrarajah
George Rechnitzer
Dr Ali Saleh
Assoc. Prof. Bijan Samali
Dr G. Sanjayan
Dr Julian Seidel
Dr Bijan Shirinzadeh
Prof. Grant Steven
Dr Mark Stewart
Geoffry Taplin
Dr Peter Thomson
Dr Murray Townsend
Dr Brian Uy
Prof. Bob Warner
Dr Trevor Waechter
Assoc. Prof. John Williams
John Wilson
Dr Bill Wong
Dr Mike Xie
Dr John Qiao Zhang
Dr Xiao Ling Zhao
Dr Roger Zou

Keynote papers

The Mechanics of Structures and Materials, Grzebieta, Al-Mahaidi & Wilson (eds)
© 1997 Balkema, Rotterdam, ISBN 90 5410 900 9

Fibre reinforced composites for civil engineering and infrastructure applications

V. M. Karbhari
Division of Structural Engineering, University of California San Diego, La Jolla, Calif., USA

ABSTRACT: Fibre reinforced composite materials originally developed for the aerospace, defense, and high performance applications areas show great potential for use in civil engineering and infrastructure areas. Based on their high strength- and stiffness-to-weight ratios, corrosion resistance, environmental durability, and inherent tailorability, fibre-reinforced polymer matrix composites are increasingly being considered for use both in repair and retrofit of existing infrastructure, and for the construction of new structures. Applications of this material range from its use as reinforcement in concrete, as patches for the external strengthening of concrete structures, for seismic retrofit of structural elements, to the construction of modular and long span bridges and aesthetically pleasing structural elements. This paper discusses the use of fibre reinforced polymeric composite materials as an addition to the palette of materials for civil engineers, and presents examples of its applications. It also outlines new concepts, development needs, and the potential for this material in the twenty-first century.

1 INTRODUCTION

As we prepare to enter the twenty-first century, there are a number of challenges and opportunities facing the civil engineer, not the least of them being the renewal of infrastructure and transportation lifelines. Over the last few decades, engineering and technological advances have made significant progress in civil infrastructure ranging from construction methods to the development of designs for long-span bridges, and high-rise buildings at a scale that would have been unheard of fifty years ago. However, the deterioration and functional deficiency of existing civil infrastructure represents one of the most significant challenges facing the nations of the world as we move into the 21st century. Further, our traffic needs have increased dramatically over the past two decades with transportation of goods and services being conducted on a global, rather than local basis, resulting in a growing need for the widening of our highway systems to accommodate more lanes, and for the renewal of existing structures to carry heavier loads at higher speeds. It is increasingly becoming apparent that deteriorating transportation related systems such as roads and bridges have a tremendous impact on society in terms of socio-economic losses resulting from delays and accidents. Conventional materials such as steel, concrete and wood have a number of advantages, not the least of which is the relatively low cost of materials and construction. However, it is clear that conventional materials and technologies, although suitable in many cases and with a history of good applicability, lack in longevity in some cases, and in others are susceptible to rapid deterioration, emphasizing the need for better grades of these materials or newer technologies to supplement the conventional ones used. It should also be noted that in a number of cases, design alternatives are constrained by the current limitations of materials used, e.g. in the length of the clear span of a bridge due to weight constraints, or the size of a column due to restrictions on design and minimum cover needed. In a similar

3

manner, often the use of conventional materials is either not possible in cases of retrofit, or may be deemed as ineffective in terms of functionality. In other case restraints such as dead load, restrict the widening of current structures, or the carriage of higher amounts of traffic over existing lifelines. In all such (and other) cases, there is a critical need for the use of new and emerging materials and technologies, with the end goal of facilitating functionality and efficiency.

Since the beginning of mankind the human race has attempted to create new material systems with enhanced properties for the construction of structural systems. The use of combinations of materials to provide both ease of use and enhanced performance, as in the use of straw reinforcement in mud by the ancient Israelites (800 BC), or in the combination of different orientations of plies of wood, has a long history. The concept of combining materials to create a new system having some of the advantages of each of the constituents can be seen in reinforced concrete (steel, aggregate, sand and cement) and to an extent in the use of special compounds and chemicals in alloys of steel. Fibre reinforced composites take this concept one step further through the synergistic combination of fibres in an polymeric resin matrix to form Polymer Matrix Composites (PMCs), wherein the fibre reinforcements carry load in predesigned directions (thereby taking advantage of anisotropy) and the resin acts as a medium to transfer stresses between adjoining fibres through adhesion and also provides protection for the material. This potentially gives the designer a wide palette of materials choices to fit the specific requirements of the structure.

Fibre reinforced composite materials, consisting of stiff and strong reinforcing fibres (such a aramid, carbon and glass), held together by tough and environmentally durable resin systems show immense potential to add to the current palette of materials being used in civil infrastructure. Some of the most important advantages are listed and discussed below.

a) High Specific Stiffness
Fibre reinforced composites generally show very high stiffness-to-weight ratios as compared to most metals and alloys. This obviously is a boon to designers, enabling them to develop designs at lower weights and thicknesses, as well as enabling them to consider new design concepts that would be limited by the ratios of other materials (Karbhari and Wilkins, 1993).

b) High Specific Strength
Fibre reinforced composites also show dramatic improvements in strength-to-weight ratios thereby allowing the designer to develop lighter designs, and when coupled with the high specific stiffness enables the designer to develop designs for bridges that include longer free spans, for example. In the Civil Infrastructure area the weight savings could result in enhancement in seismic resistance, increased speed of erection and a dramatic reduction in time for fabrication of large structures

c) Enhanced Fatigue Life
Most composites are considered to be resistant to fatigue to the extent that fatigue may be neglected at the materials level in a number of structures, leading to design flexibility.

d) Corrosion Resistance
Unlike metals, composites do not rust, making them attractive in application where corrosion is a concern. This has lead to an intense interest in composite reinforcing bars and grids, as well as cables for pre- and post-tensioning, as well as for use in cable stays.

e) Controllable thermal properties
In a number of applications dimensional stability under varying temperature conditions is critical whereas in others temperature gradients and temperature induced expansion and contractions result in thermal strains that may be of extreme concern to the designer. The coefficients of thermal expansion of most reinforcing fibres used in composites are extremely low, resulting in low values for the composites. Also the performance of composites depends on the orientation of fibres, enabling one to "design" composites with $\alpha = o$, leading to immense potential in the design of large or dimensionally critical structural components.

4

f) Parts Integration

With most materials, large, complex components have to be made in sections requiring secondary joining operations that not only increase costs, but also decrease overall systems reliability. An example of this is a large truss or frame structure wherein the assembly of a large number of components can result in dramatic cost increases over the hypothetical cost were it possible to fabricate the structure in one operation. The question related to the reliability and integrity of the field welding associated with stiffeners for girders or with a steel shell/casing used in seismic column retrofit is similar. With composites, it is possible to fabricate large, complex parts in one operation reducing the number of joining operations

g) Tailored Properties

Metals and most building materials (with the exception of wood) are isotropic which means that we design using properties that are equal in all directions, irrespective of whether there is a need for similar properties in all directions. With composites it is possible to design the material such that it responds differently in different directions enabling tailoring of response based on load direction, thereby optimizing material usage. For example, the seismic retrofit of concrete columns requires that the shell/casing provide additional hoop reinforcement in order to develop confinement. The use of steel results in additional strength and stiffness both in the hoop and axial directions, whereas with composites it is possible to tailor properties to comply only in the directions required, thereby increasing efficiency and economy.

h) Non Magnetic Properties

Steel members and reinforcements are often a problem in structures such as hospital operating theaters, areas where radar is operative/housed, and areas which house antennae and other sensitive electronic equipment due to their propensity to cause interference due to electromagnetic waves. The use of glass fibre reinforced composite components is often a solution to these problems.

i) Lower Life-Cycle Costs

Due to the corrosion resistance and enhanced resistance to solvents and the environment, composite structures will potentially require less maintenance, resulting in lower overall life-cycle costs.

It must however be mentioned that composites do suffer from some disadvantages, primary among them being: (a) higher initial materials cost, (b) lack of familiarity in most areas (outside aerospace), (c) lack of comprehensive standards and design guidelines at present, and (d) need for an integrated materials-process-design structure in product development, which entails a critical change in paradigm. The key advantages of fibre reinforced composites, such as free-form and tailored design characteristics, strength/weight and stiffness/weight ratios which significantly exceed those of conventional civil engineering materials, high fatigue resistance, and a high degree of inertness to chemical and environmental factors, are often lost in high materials and manufacturing costs, particularly in direct comparison with conventional structural materials such as steel and concrete. However, the recent downturn in defense spending and the resulting need for new markets has spurred renewed efforts in reducing the costs of both raw materials and manufacturing processes, making it highly feasible to use composites in civil infrastructure. Within the vast applications area in civil infrastructure, this paper will concentrate on a few specific examples of studies conducted at the University of California, San Diego (UCSD) with focus on both rehabilitation and renewal. However, it should be mentioned that there is immense applicability in other areas such as tunnels, buildings, pipelines etc., with significant research in these areas having already been conducted by the Japanese (Karbhari, 1997a).

2 APPLICATION TO THE REHABILITATION OF STRUCTURES

The rehabilitation and retrofit of existing concrete structures with polymer matrix composites can be generically accomplished in one of two ways (i) application of composite overlays or strips, and (ii) external post-tensioning using composite cables, tendons or bars. In this paper we will focus on the first only. The use of composite

5

overlays in the form of complete coatings or in the forms of strips or patches is applicable to (a) seismic retrofit and repair of bridge columns, (b) strengthening of bridge superstructure (deck soffits and girders) for increased capacity, and (c) seismic retrofit and strengthening of reinforced and unreinforced concrete and masonry walls.

2.1 Seismic Retrofit and Repair of Bridge Columns

The confinement of concrete through the use of an external jacket has been proven to result in enhanced system ductility and subsequent performance under seismic events. This is because the external jacket provides constraint to the dilation of concrete in the absence of sufficient hoop steel. Amongst other conventional seismic retrofit strategies, steel jackets have been extensively used in both Japan and the United States. However this method of retrofit necessitates the welding of jacket sections in the field which is a concern as related to effectiveness and overall quality control. Further the use of steel in some areas carries with it the potential for rapid corrosion and degradation. The use of composites as wraps/jackets on columns to replace steel jackets in seismic retrofit has achieved a high level of interest both in Japan and the United States, with the generic methods used to achieve the confinement of concrete including (a) Wet winding of tows, (b) Winding of prepreg tow/tape, (c) Wet layup of fabric, (d) Layup of tape, (e) Adhesive bonding of prefabricated shells, (f) Insitu resin infusion of jackets, and (g) Use of composite cables wrapped around concrete core. A discussion on the development of design guidelines, as well as a comparison of the different methods based on materials choice and method of fabrication is given in (Seible and Karbhari, 1997). As an example of development, tests on 40% scale bridge columns at UCSD have shown that the use of carbon fibre reinforced jackets applied through the continuous winding of prepreg tow which is cured after completion of the jacket, can be as effective as steel jackets for retrofits in flexural lap splice and shear applications (Seible et al., 1997a). The advantages of such a method are (i) the fibre is continuously placed in the direction where it is the most efficient, i.e. the hoop direction and, (ii) the method of application lends itself to overall efficiency with significantly lower thickness of jacketing required, and (iii) the use of an automated method of construction leads to a high degree of quality control at a rapid rate of completion. The use of composites, especially as reinforced with carbon fibres in the hoop direction (providing directed confinement) thus provides a layup that is tailored for the specifics of the application, providing materials efficiencies that were not possible with conventional materials. Extensive testing has been conducted using this technology with both 40% scale and full-scale validation tests. In addition successful field demonstration applications have been conducted on the I-10 Santa Monica Viaduct in Los Angeles and a comprehensive set of design guidelines have been developed (Seible et al, 1995, Seible et al, 1997a). The application of the technique to a rectangular column is shown in Figure 1, wherein the entire jacket was coated with a layer that served both as protection and as an aesthetic coating. The use of composites in such an application has been shown to be cost-competitive with steel retrofits while providing the potential for faster fabrication with less lane closure and potentially far greater durability over the life-time of use. Further details related to testing and materials selection are reported in (Karbhari, 1997b).

Figure 1: Retrofitted Rectangular Column

Table 1: Comparison of jacket thicknesses

Material	Properties* Vf = 60%	Relative Jacket Thickness		
		Shear Strengthening	Plastic Hinge Confinement	Lap Splice Clamping
E-glass/Epoxy	E = 45 GPa σ = 1020 MPa ε = 2.3%	1	1	1
S-glass/Epoxy	E = 55 GPa σ = 1620 MPa ε = 2.9%	0.82	0.50	0.82
Kevlar 49/Epoxy	E = 76 GPa σ = 1380 MPa ε = 1.6%	0.59	1.06	0.59
Graphite/Epoxy	E = 160 GPa σ = 1725 MPa ε = 0.9%	0.28	1.51	0.28
Boron/Epoxy	E = 210 GPa σ = 1240 MPa ε = 0.6%	0.21	3.15	0.21

* Values are representative averages without application of reduction coefficients for aging and environmental durability

For purposes of illustration of materials efficiency, the relative thicknesses of jackets fabricated using E-glass/epoxy, S-glass epoxy, Graphite/epoxy, Kevlar/epoxy and Boron epoxy at 60% fibre volume fractions are shown in Table 1 as a function of the retrofit mechanism.

2.2 External Strengthening of Deck Soffits and Girders

The use of techniques associated with the external attachment of composite plates to the soffit of decks and the underside of beams for purposes of strengthening and retrofit is attractive due to factors related to ease of access and decreased need for extensive changes to the existing structure. Although the bonding of steel plates has been used extensively for over two decades, this method suffers from a number of disadvantages ranging from difficulty in placement, to concerns related to overall durability. Composite plates, on the other hand, do not suffer from most of these deficiencies, due to the high stiffness- and strength-to-weight ratios, corrosion resistance and light weight, as shown schematically in Figure 2.

Beginning with the repair of the Ibach bridge in Switzerland, a number of retrofit and strengthening projects have been completed in Europe and Japan, along with a few demonstration projects in the United States. In recent years, this method has been used for the upgrading of bridge decks to enable the use of higher load levels such as in the case of the Hiyoshigura Viaduct in Chiba, Japan where two layers of carbon fibre unidirectional fabric were bonded to the underside of a deck section in order to increase capacity from a

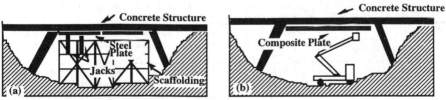

Figure 2: Schematic Comparison of Placement of External Reinforcement (a) Steel Plates, (b) Composite Plates

7

20 ton level to a 25 ton level (Karbhari, 1997a). The use of composites in such applications provides yet another example of how light weight tailored materials can be used effectively to enhance and increase the life of existing civil infrastructure. Details related to use and testing of such applications can be found in (McKenna and Erki, 1994; Karbhari et al, 1997a). The "plate" (or external reinforcement) can itself be fabricated in three generic ways as listed in Table 2. Of these, the premanufactured alternative shows the highest degree of uniformity and quality control for the reinforcement strip, since it is fabricated under controlled conditions. Application is still predicated by the use of an appropriate adhesive and through the achievement of a good bond between the concrete substrate and the composite adherent. Care must be taken to ensure that the adhesive is chosen so as to match both the concrete and the composite and provide an interlayer to reduce mismatch induced stresses. Bonding can be assisted in this case through the temporary use of external clamps or a vacuum bag to provide compaction pressure. The wet layup process is perhaps the most used and gives the maximum flexibility for field application, and is probably also the cheapest alternative. However, it presents the most variability and necessitates the use of excessive resin, and could result in wrinkling of the fabric used, as well as entrapment of air. The in-site infusion method is a fairly new variant and is capable of achieving uniformity and good fabric compaction, while making it easier for the reinforcement to be made to fit the exact contours of the component to be strengthened. In the latter two systems the adhesive function is taken by the resin itself with the bond to the concrete substrate being formed simultaneously with the processing of the composite. This is both an advantage and a disadvantage, since the elimination of a third phase, the adhesive, results in fewer interfaces at which failure could occur, but also eliminates the use of a more compliant interlayer.

2.3 Retrofit of Walls

The seismic repair and retrofitting of reinforced and unreinforced concrete and masonry walls has been shown to be possible through the use of thin composite overlays (Seible, 1995; Priestly and Seible, 1995) with fibres oriented horizontally. Tests have shown that the application of one to two layers of unidirectional fabric can result in significant enhancements in performance with reductions in shear deformation, and increases in ductile flexural in-plane behavior. The application of this technology to deficient wall systems, and to systems where cracking due to uneven settlement is also possible.

3 REPLACEMENT BRIDGE DECK SYSTEMS

Of all elements in a bridge superstructure, bridge decks may perhaps require the maximum maintenance, for reasons ranging from the deterioration of the wearing surface to the degradation of the deck system itself. Added to the problems of deterioration are the issues related to the need for higher load ratings (HS15 to HS20, for example) and increased

Table 2: Methods of Application of External Composite Reinforcement

Procedure	Description
Adhesive Bonding	Composite strip/panel is premanufactured and cured (using wet layup, pultrusion or autoclave cure) and then bonded onto the concrete substrate using an adhesive under pressure
Wet Layup	Resin is applied to the concrete substrate and layers of fabric are then impregnated in place using rollers and squeegees. The composite and bond are formed at the same time.
Resin Infusion	Reinforcing fabric is placed at the spot to be retrofit and the entire area is encapsulated in a vacuum bag. Resin is then infused into the assembly with compaction taking place under vacuum pressure. Unlike the wet layup process this is a closed process and the infusing resin can fill cracks and voids as well.

number of lanes to accommodate the ever increasing traffic flow on major arteries. Beyond the costs and visible consequences associated with continuous retrofit and repair of such structural components are the real consequences related to losses in productivity and overall economies related to time and resources caused by delays and detours. Reasons such as those listed above provide significant impetus for the development of new bridge decks out of materials that are durable, light and easy to install. Besides the potentially lower overall life-cycle costs (due to decreased maintenance requirements), decks fabricated from fibre reinforced composites would be significantly lighter, thereby affecting savings in substructure costs, enabling the use of higher live load levels in the case of replacement decks, and bringing forth the potential of longer unsupported spans and enhanced seismic resistance. There has been considerable activity over the past decade in the area of composite reinforcement for concrete bridges with the reinforcement ranging from composite rebar and grids to composite cables for external and internal post-tensioning. The current application, however, emphasizes the use of fibre reinforced composites for the entire deck or as part of the actual superstructure system itself. Such a concept, in general, is not new, having been used previously in the Miyun bridge in China and the Aberfeldy Footbridge in Scotland, among others. The focus of this section will be on developments in the area of replacement bridge decks capable of being placed on pre-existing concrete and steel girders, as well as being used in new bridge systems.

A systematic test program is being undertaken at the University of California, San Diego (UCSD) with funding from the Federal Highway Administration (FHWA), the Advanced Research Projects Agency (ARPA), and the California Department of Transportation (Caltrans) as part of a University-Industry Consortium, and is aimed at the development of lightweight, degradation resistant fibre reinforced composite decks for primary use in replacement. The overall criteria that were used to guide the development of these decks included: (i) the development of stiffnesses through the appropriate use of face sheets and internal core configurations that would fall in the range between the uncracked and cracked stiffness of existing reinforced concrete decks, (ii) the development

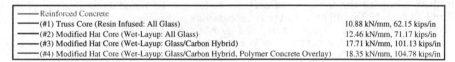

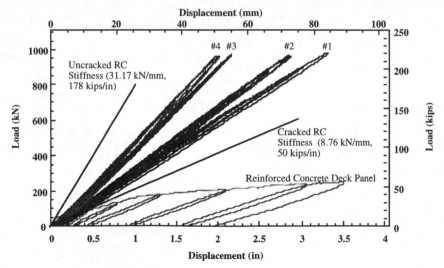

Figure 3: Comparison of Response of Large Scale Deck Panels

9

Figure 4: Test of a Large-Scale Deck Section (Length = 457 cm, Width = 228.6 mm, Depth = 22.9 cm)

Figure 5: Placement of a Replacement Composite Test Panel in a Test Site

of equivalent energy levels at acceptable displacement levels as a means of building in a safety factor due to the elastic behavior of composite sections, and (iii) the development of processing methods that would be cost-effective and which ensured repeatability and uniformity. A building block approach was used in this program with tests being conducted at the subcomponent, component and field-size levels so as to evaluate structural response, effect of fabric architecture and the effect of different core configurations, in components subject to shear and bending as appropriate. Components were fabricated by a number of industrial partners using Wet-Layup, Pultrusion, Resin-Infusion, and processes that were combinations thereof for manufacturing. Details related to these processes and the types of sections tested can be found in (Karbhari, 1996; Karbhari et al, 1997b). In order to verify repeatability and uniformity between samples, a number of components were fabricated and tested to the same specifications as well. It was generally seen that structural response and failure modes could be replicated fairly uniformly. Deck sections were seen to provide almost linear elastic response to failure, with load levels significantly higher than those achieved by reinforced concrete deck panels, but with comparable stiffnesses (Figure 3), at weights one-fourth of those of the reinforced concrete incumbents.

Figure 4 shows a section being tested, whereas Figure 5 depicts the placement of a full size section in a test-bed. The significantly lower weight of the composite deck sections enables very rapid placement in the field.

4 CONCEPTS FOR NEW STRUCTURAL SYSTEMS

Although the areas of rehabilitation and retrofit offer the maximum potential for immediate application of composites in civil infrastructure, the development of new systems that combine the directionality and high performance levels of composites with the dominant characteristics of conventional materials (such as with concrete in compression) show great potential for advances in the design and construction of new civil engineering structures. One such concept is that of composite shell systems for columns, wherein prefabricated composite tubes serve the dual purposes of form-work and reinforcement, thereby replacing the reinforcing steel while enabling faster construction. In this system, the column is constructed by placing the hollow composite shell in place, and then filling it with concrete. Construction details are such that it can be directly incorporated with conventional construction methods. The composite shell system has been successfully tested using two concepts: (a) wherein starter bars from the footing extended into the column providing a conventional anchorage mechanism into the footing and/or bridge

10

superstructure, and (b) wherein steel is completely eliminated with anchorage force mechanisms consisting of the reaction moment due to the compression force couple generated inside the footing and the frictional stresses between the composite shell and concrete (Seible et al, 1997b). The initial success of this system has led to the development of a complete bridge system, wherein the girders and columns can be fabricated from composite shells. An example is shown in Figure 6 for a space truss bridge. The space truss system is comprised of a lower chord suspended from the deck system by means of two inclined Warren-type trusses. The deck itself could either be of precast concrete panels or of composites.

The concept of using composite elements for the deck, girders and supporting columns can also be applied to modular short-span systems wherein composite girders would span between abutments. Preliminary calculations have shown that the typical span to depth ratios for the simply supported concrete filled composite shell bridge systems as described in (Seible et al, 1997b) range from 17 to 20 including the depth of a 150 mm bridge deck. Longer spans may be possible through the use of post-tensioning of light-weight concrete in the girders, thereby enhancing capacity through the combined action of post-tensioning and confinement.

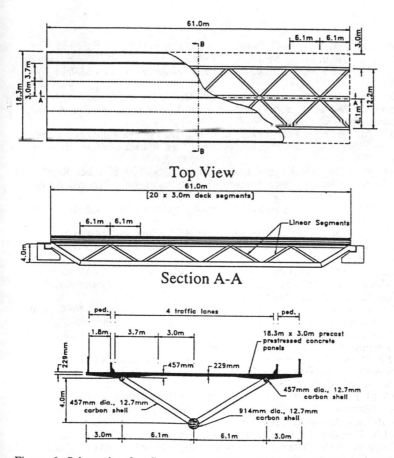

Figure 6: Schematic of a Space Truss Bridge With Superstructure Constructed From Composite Elements

11

5 SUMMARY

Fibre reinforced composites provide an immense palette of opportunity for the civil engineer to combine form and function in a manner not possible to date, and hence open up new vistas for the development of methods for renewal, including those involving new structural systems. These materials offer the potential for design and construction of technologically advanced and aesthetically pleasing structures that combine overall increased durability, with the capability for substantially enhanced performance. The development of new building systems, and of longer-span bridges thus becomes a distinct possibility, with designs constrained to a much lesser extent by materials than before. The extent of these applications will however, depend in large part on the resolution of outstanding critical issues that include (a) durability and fire protection, (b) reparability of composite structural elements, (c) development of validated codes, standards and guidelines of use to the civil engineering community, (d) development of cost-effective design approaches and manufacturing methods, and (e) provision of an appropriate level of quality assurance and control both during manufacturing and installation using unskilled construction labor.

REFERENCES

Karbhari, V.M. and D.J. Wilkins 1993. Development of Composite Materials and Technology for Use in Bridge Structures," *Proceedings of the NSF Symposium on Practical Solutions for Bridge Strengthening and Rehabilitation*, Des Moines, IA, 211-219.

Karbhari, V.M. 1996. Fiber Reinforced Composite Decks for Infrastructure Renewal, *Proceedings of the 2nd International Conference on Advanced Composites in Bridges and Structures*, Montreal, Quebec, 759-766.

Karbhari, V.M., M. Engineer, D.A. Eckel, II 1997a. On the Durability of Composite Rehabilitation Schemes for Concrete: Use of a Peel Test, *Journal of Materials Science*, Vol. 32, 147-156.

Karbhari, V.M., F. Seible, G. Hegemier, G. and L. Zhao 1997b. Fiber Reinforced Composite Decks for Infrastructure Renewal - Results and Issues, *Proceedings of the 1997 International Composites Expo*, Nashville, 3C/1-3C/6.

Karbhari V.M. 1997a. Use of Composites in the Japanese Construction Industry, *NSF/WTEC Report*.

Karbhari, V.M. 1997b. On the Use of Composites for Bridge Renewal: Materials, Manufacturing and Durability, *Proceedings of the 42nd International SAMPE Symposium*, 915-926.

McKenna, J.K. and M.A. Erki 1994. Strengthening of Reinforced Concrete Flexural Members Using Externally Applied Steel Plates and Fibre Composite Sheets - A Survey, *Canadian Society of Civil Engineering*, Vol. 21[1], 16-24.

Priestley, M.J.N. and F. Seible, 1995. Design of Seismic Retrofit Measures for Concrete and Masonry Structures, *Construction and Building Research*, 9[6], 365-377.

Seible, F. 1995. Structural Rehabilitation With Advanced Composites, *Proceedings of the IABSE Symposium*, San Francisco, 391-398.

Seible, F., M.J.N. Priestly and D. Innamorato 1995. Earthquake Retrofit of Bridge Columns With Continuous Carbon Fibre Jackets - Volume II, Design Guidelines, *Report No. ACTT-95/08*, UCSD.

Seible, F. and V.M. Karbhari 1997. Seismic Retrofit of Bridge Columns Using Advanced Composite Materials, *Proceedings of the National Seminar on Advanced Composite Material Bridges*, Arlington, VA, 29 pp.

Seible, F., M.J.N. Priestley, G. Hegemier, and D. Innamorato 1997a. "Seismic Retrofit of RC Columns With Continuous Carbon Fiber Jackets," *ASCE Journal of Composites in Construction* (in press).

Seible, F., G. Hegemier, V. Karbhari, R. Burgueno, and A. Davol 1997b. The Carbon Shell System for Modular Short and Medium Span Bridges," *Proceedings of the 1997 International Composites Expo*, Nashville, pp. 3D/1-3D/6.

The Mechanics of Structures and Materials, Grzebieta, Al-Mahaidi & Wilson (eds)
© 1997 Balkema, Rotterdam, ISBN 90 5410 900 9

Innovation, research and engineering

J. Nutt
Ove Arup and Partners, Sydney, N.S.W., Australia

ABSTRACT: This paper addresses the parameters which will shape the future of engineering research from the perspective of a designer in the construction industry. Topics covered include engineering science, design, innovation, risk, export, the role of research institutions, intellectual property and research funding. A case study is presented which demonstrates how one sector of the construction industry supported an ambitous and generic research programme where there were no obvious beneficiaries apart from the community at large.

INTRODUCTION

The Australasian Conference on the Mechanics of Structures and Materials is an important event in the engineering calender, successfully serving the needs of the engineering community for over a quarter of a century. The theme of these conferences is engineering science, and it attracts papers on structural engineering analysis and design, materials, dynamics, geotechnics, and in recent years, risk analysis, fire engineering, and biomechanics. This address is intended to put the science into the wider context and discuss some of the parameters which will shape the future of engineering research from the background of a designer in the construction industry. Some of these comments will relate to the broad issues as they effect structural engineering, some will identify relationships which have to be enhanced, some will set the Australian and New Zealand scene in an international context. All will relate to activities which are important for the future.

Engineering research in this country is undertaken in universities, government research establishments, occasionally in joint industry research organisations, and sometimes by companies active in the manufacturing and construction industry. The construction industry is characterised by an extremely low level of research funding. Most financial support comes from the government. That source of funding is diminishing.

Engineering research has an important role in the industrial affairs of a nation. It underpins the activities of commercial organisations at home and when they export. There is both anecdotal and research evidence which demonstrates the benefits which accrue to an economy. The changes in the funding mechanisms which have occurred, not only in Australia, but in other countries, requires that the research organisations themselves have to articulate the case for financial support for their work, and actively seek it from sources who will benefit from it. To do so, they must have a understanding of fund raising procedures and the needs of industry so that the benefits of their work is recognised.

This paper's intent is to relate engineering science to the industry it serves, particularly the construction industry, and give, by way of a case study, a successful example of how one sector of the industry supported an ambitious research programme when no obvious or captive sponsor appeared available.

ENGINEERING SCIENCE

Engineering and technology is founded on science which incorporates theory, knowledge,

13

observation and experiment. As the tools of science become more powerful and experimental data accumulates year by year, so does the knowledge which can be passed on to others. Science can be built upon.

Engineering, using scientific laws, has to be created afresh by each individual practitioner. There is a large font of conventional wisdom which can be drawn upon, but creativity and innovation requires additional inputs which do not come from experience and reason alone.

Engineering products are for use by the community. Engineering design which precedes construction or manufacture, is the planning of the form and processes of production - an intellectual activity which incorporates imagination and efficiency.

Engineering Design is not a science. Science studies particular events to find general laws. Engineering design makes use of these laws. There are many solutions. It is an art or craft. It is a creative activity involving imagination, intuition, and deliberate choice......Ove Arup

There are two aspects to engineering which are outside the boundaries of science. The first is that the community, which sees the products of engineering in its environment, assesses these products against criteria unrelated to science, - function, quality of life, visual delight, pleasure and wellbeing. The second is that imagination is part of the creative process. Imagination is beyond reason, outside the scope of our normal rational thought.

DESIGN

"Total Design" is the integration of the design and construction process, and the interdependence of all the professionals involved; it is creative and innovative; and it has social purposes......Ove Arup.

Structural engineers work in teams. It is important that the broad issues are understood and shared. There has been in this country a changing attitude of the community towards engineers. Once it was thought by the community that science and engineering were all powerful because their methods were based on reason, their judgements unquestioned, and their products perfect. That is no longer the case. The community distrusts engineers because they are strong, capable but isolated and incommunicative. This has to change. There is little point in the refinement of a bridge or building through research if the community does not accept the decision to construct it in the first place. The structural engineer must be exposed and understand the community concerns, and develop solutions acceptable to the public.

I referred to bridges. A bridge is the epitome of structural engineering - a structure of such functional simplicity that the engineer was king. Not any more. The most recent competition for a bridge in Melbourne was won by an architect, one who worked closely with his engineer. The selection criteria was not based upon cost and engineering efficiency alone - appearance, the environment, how it related to the public and the culture of the site were governing elements. This occurs not only in Melbourne. In London, the next bridge across the Thames will be from a team which is led by an architect, and in Sweden and North America , architectural/engineering teams are engaged.

Architectural engineering can be the most stimulating of endeavours. The rational deterministic process which the engineer applies to a design is overturned by the unexpected and novel standards of the architect. The engineer is too greatly influenced by analysis. The ability to understand stresses and strains conditions the engineer to believe that skilful analysis leads to the best solutions. That is not so. Design required only an adequate solution. This is recognised in engineering practice and written into all our codes. Take limit state for example. Each limit state has to be satisfied in a structural design. When it can be demonstrated that they are, the design is acceptable. One load path in a high rise building, and only one, must have sufficient strength to support the applied ultimate loads. All other load paths are unnecessary, and provided the servicibility limit states of the whole building are satisfactory, the true load path is irrelevant in the design.

INNOVATION

The process of design is fundamentally different to that of the science on which it is based. Arup called it intuition and choice. As my colleague Richard Hough said: *"Intuition is the immediate apprehension by the mind, without reasoning. At the beginning of design, it is how our mind first starts to select and reject from the infinite range of possibilities. It helps us to pass through the barriers we build with reason, habit, and language, and so to notice parallels, relationships and analogies".*

From that stems innovation - that injection of creative ideas in a framework of careful risk minimisation which is the essential ingredient of excellence. Each successful designer has a personal formula for achieving excellence, but invariably it comprises two parts, a creative part and a systematic procedural part. Without both, a design is unlikely to be successful.

Innovation, by definition, brings new and at times, untried, approaches to a project. This is not without risk, and that risk must be controlled and understood. However, risk cannot be minimised without judgement, judgement cannot be gained without experience, and experience requires the knowledge gained from working on projects and research. Separate the technical issues from the decisions, and little innovation will be achieved.

RISK

The control of risk is fundamental to innovation. The community is intolerant to failure. Litigation and the judgements of the courts have supported the consumer. Project delivery systems for construction have changed, changes brought about through the client's requirement to have a single point of responsibility and the competitiveness which financial alternatives can bring. In this climate, risk definition and risk control is integrated into the design and construction process. Fear of litigation is the greatest deterrent to innovative ideas. That is why it is so important to have available that depth understanding of materials and structures which can only come from research.

EXPORT

Innovation and practice go hand in hand, Industry and research establishments have to be closely linked. If an Australian firm is to **export** from this country, it must be better than the competition. Export is tough. The competition is more intense, the culture is different, the intelligence network is not as effective as at home. More difficult still, the **flow of capital** is **into** this country, not out of it. Australia imports capital to finance its infrastructure, and its manufacturing industry. That import of capital places Australian industry at a tremendous disadvantage. Decisions in a global marketplace are taken close to the sources of capital. In New York for example, a capital and technology hub of the world, foreign projects readily come in from Asia, the Middle East, from Europe. Australian firms have to assiduously seek and hunt for projects on which to apply their engineering skill. It is the same skill, but New York is closer to the centres of finance where many of the investment decisions are taken, and where long associations engender familiarity and trust. To overcome that handicap, Australian firms have to be better, more innovative, more entrepreneurial.

THE ROLE OF RESEARCH INSTITUTIONS

Universities and research institutions have a special role to play in this innovative process. They are repositories of specialist knowledge which must support a national industry. That knowledge must be continually updated, extended, and used. It must be relevant. The demand for the application of that knowledge by industry is a test of its relevance and currency. It is an essential part of success offshore in the absence of long established networks or markets. If Australia is to succeed in the export market, our engineers must be willing to take the initiative and develop their ideas and designs into business opportunities. They must be entrepreneurial, not only the purveyors of technology for others to exploit. If they are only that, the benefits will go to the providers of capital, and since Australia is a net importer of capital, will end up in the hands of overseas investors.

The greatest infrastructure programmes in the world are now underway in Asia. The infrastructure projects are enormous and technically challenging. As a result, the great engineering experiences are being gained in Asia by those countries and firms that are working there. Currently the region is an importer of capital and skills because it is not possible to satisfy the demand from local sources. But this will not always be the case.

15

Asian universities are producing highly skilled and clever graduates who are receiving technical and financial training on major projects which they can build upon for the remainder of their lives. Australians have to capture that experience and they have the ability to do so. But it must be done systematically, not merely as employees of offshore corporations. Australia should not be a mere "body" shop of talent.

INTELLECTUAL PROPERTY

Engineering innovation relies on intellectual property which a research establishment can provide. The intellectual property which a designer brings to a project is frequently in the public domain. However, the export of technology produces a flow-on effect to other parts of the construction industry. It is in these fields that protection of intellectual property is required to maintain a competitive edge. The engineer designer enhances export by the specification of familiar building products, by arranging ongoing maintenance and operations contracts, by networking with foreign counterparts. Intellectual property becomes a tool in export enhancement.

Intellectual property must be continually developed by the research establishments and universities if it is to be state of the art knowledge. Those skilled experts, while not exporting directly themselves, underpin the engineering consultant and help win overseas contracts through innovation. Enhancement of intellectual property requires two things - a challenging environment, and a home market in which innovation is encouraged, and ideas developed.

FUNDING OF RESEARCH

The construction industry spends only 0.13% on research and development to improve its products and productivity. This is not the case in manufacturing or the resource industries. Why is this so? In construction, it is difficult for the various sectors of the construction industry to capture the benefits - builders work to defined documents and receive little for technical improvements; consultants who design and specify products also do not; owners are remote from the construction process and few have an understanding of the issues. In the absence of public pressure, there is little specific incentive for governments to give research high priority. The public, which is the ultimate beneficiary, is too far removed from the decision making process.

A recent study reported by the CSIRO has shown that construction industry savings have greater effect on the Gross Domestic Product than any other industry sector, more than, for example, business services, public administration, or road and rail transport combined. The flow on effect magnifies the savings in the construction industry to achieve a benefit for the whole community of about two and a half times that saving.

The funding situation will get steadily worse. The large government departments, which in the past had responsibility for the construction and operation of infrastructure, are now corporatised or privatised. Many delivery systems in the construction industry are selected to provide financial competition and minimum cost projects. Project teams come together and disband. There is no logic to invest in long term innovation research. There has been widespread privatisation of infrastructure. Here, the emphasis will be on financial return over a defined period. Frequently the projects will be one off. Often the investors will be overseas based, sometimes the overseas participant will provide the technology. There will be little need to develop innovative technology in Australia. In the circumstances, it is not surprising that management cuts back on longer term expenditures which do not conform to the short term political goals.

In the absence of government support for research, funding for research must come from other sources. The research establishments themselves must seek it from sponsors. Active marketing of the products of research is necessary. To be successful, they will have to articulate the benefits which the research can bring. That can only be done from a thorough understanding of industry needs. This knowledge can be obtained by seeking work from industry - a programme of consulting activities, service to industry, and strategic research will lead to fundamental research in the field of activity in which the needs are greatest, and, in the case of universities, balance and complement teaching and training.

16

A CASE STUDY OF RESEARCH SPONSORSHIP

The example in this case study shows how it is possible to fund a research programme when there are no obvious beneficiaries apart from the community at large. Many engineering research activities come within this definition. Frequently they deal with catastrophic events which occur rarely but on which the community relies on well conducted research to define safety and risk, or in the absence of such research, pays a significant penalty through flexible and costly regulations.

Fire safety is the control of risk to life, and sometimes to property. For a decade it has been apparent that the skills existed in Australia to develop a set of properly engineered and scientifically based fire codes and regulations that would be a great improvement over the existing fire regulations. As the design and operation of buildings have changed through the introduction of modern materials and equipment, the prescriptive nature of the regulations imposed limits on planning and construction which added unnecessary costs. Without an underpinning science, both the designers and approvers were venturing into the unknown.

Australia has a good record in building fire-safety. It was agreed by all that this must be maintained. However it is impossible to quantify the risk, or compare risk associated with alternatives. Risk methodology is not new. It is widely used in industry to analyse situations where past experience does not define the commercial and safety risk associated with hazards. It has been successfully adopted in regulations for other catastrophic events, such as earthquakes and extreme winds. The present Australian Standard Wind Loading Code and Earthquake Code which are called up by the Building Code of Australia are based on risk methodology. It is clear that fire risk should be predicted on a similar basic. There is no other way of approaching the analysis of rare catastrophic events in a logical manner.

Industry identified fire regulations as the largest single technical restraint in the cost effective planning of new buildings and the renovation of old buildings. Industry was looking to government, and governments to industry to commission the necessary work. The restraint to successful improvement was scope and scale. However, industry was an ineffective lobbying body. There was therefore no independent champion of the reform process. It was to overcome this problem and delay that the Fire Code Reform Centre was established.

The strategy behind the foundation and operation of the Fire Code Reform Centre (FCRC) is to bring together all major participants in the fire industry in Australia, and through a cooperative effort, undertake and manage the research to underpin codes and regulations for the industry. Researchers, regulators, practitioners, industry groups, fire services, and building owners have taken part in formulating the research program, raising the funds, directing the contracts, and interacting with industry.

Four separate but closely related events occurred between 1989 and 1991 came to fruition in 1994 when the centre was established. A project on 'Fire Safety and Engineering' was undertaken by the Warren Centre of the University of Sydney in 1989. The Warren Centre is a non-profit think tank independently funded within the University. That project drew together some sixty Australian executives representing organisations having an interest in building fire safety. The project leader was a distinguished Australian researcher in fire. Their report came to the attention of the Building Regulation Review Task Force which had been commissioned by all Australian governments to improve the building regulations of which fire safety is the most important technical part. It secured government funding to build on that work and produce a draft code of practice which outlined a possible framework for the research to be used by designers and authorities.

At that stage, the data was not available to make it a useable document, but the systematic concepts were outlined, and it has been used as a draft by other countries, notably the British, and by working groups of the International Organisation of Standards (ISO).

By that time, available funding had been expended. The researchers, comprising a consortium of government and semi/government research organisations and some industry participants applied for funding under the Cooperative Research Centre Scheme. In spite of the application being the only short listed group from the construction industry, it was not selected. One of the reasons for failure was the inability to demonstrate how the benefits could be captured commercially.

After this setback, the group reorganised. Deciding that there would be no funding available in the foreseeable future, the group elected to promote itself and its activities. It commissioned a business plan with the small amount of funds which it had available. It set out in detail its research programme. It developed a work plan, and costed the research.

The research organisations committed themselves to forming a consortium so as to offer the best available resources in Australian in a non-competitive arrangement. It costed the input from the researchers at discounted rates so that sponsors were seeing a commercial benefit not readily achievable by other means. It coopted industry identities on an honorary basis.

The approach matched the procedures adopted by industry for undertaking the feasibility and implementation of its projects.

The consortium received some financial support and pledges from industry on startup, but what was required was a sponsor of substance who would underpin the programme and permit the establishment of a corporate vehicle. That came from the newly established Australian Building Codes Board (ABCB). Established in 1992 to implement reforms to the Building Code of Australia, it had a limited amount of funding available which could be directed towards fire reform. Its Chair came from industry. He was instrumental in obtaining ABCB support on the condition that substantial support was given by industry. With that pledge, a fund raising campaign was launched.

The Fire Code Reform Centre Ltd has been established as a company limited by guarantee. It is controlled by an honorary Board of Directors comprising representatives of Nominating Sponsors, an independent Chairman, and an invited representative of the fire services. Nominating Sponsors are those who contribute $100,000 annually to the FCRC. In addition, when the aggregate sponsorship of other membership categories reach $100,000 p.a, that category may elect a representative to the Board. The first board member from other such categories was elected in August 1996.

The FCRC is an Approved Research Institute under Australian tax law. It undertakes no research itself but places research contracts with a consortium of Australian research organisations who are leaders in their field. These contracts are administered by a Research Supervisory Committee comprising sponsors representatives and others who provide superintendence skills. A broadly based Industry Advisory Committee drawn from all sectors of the fire safety industry contribute their time on a voluntary basis. Membership of the FCRC is open to all organisations and individuals involved in building fire safety who contribute to the support of the programme. Management and administrative costs are kept low and a Business Manager has been employed under contract. His role is to run the centre, raise funds, and administer the research contracts under the direction of the Board and the Research Committee.

The FCRC Research Program commenced in November 1994 and research contracts have been commissioned as funds become available. Because the FCRC Ltd is a company requiring conformance to Australian Corporate Law, it can only commission research when it has funds, i.e. after the cash has been received from its sponsors.

At the 30 June 1997, a total of $ 3.3 million since inception has been raised. The Government, through the Australian Building Codes Board (ABCB) has contributed $ 1.5 million. The FCRC has raised the remainder from other sources, mostly from industry. To date, this has resulted in industry funding of $1 for every $1 contributed by government. Support has taken forms other than cash contributions. The research organisations with whom the contracts have been placed are undertaking the work at discounted rates, because they are universities, other government organisations. One industry research organisation is contributing significantly in kind in addition to its cash contributions. The cash value of the total research program, excluding in kind contributions, is budgeted at $5.7 million. In addition, the value of the in kind and discounted rate contributions at the completion of the program will be nearly $2 million.

The Centre is unique. As far as we are aware, it represents the only comprehensive, integrated program of applied research in the world specifically directed to the improvement of a nation's prescriptive building regulations and the development of verification procedures of an alternative performance based approach.

The lasting benefit to Australia of doing the work here is that it maintains researchers in universities and research establishments who are at the forefront of fire engineering research, and are able to support Australian consultants and contractors in the design of innovative buildings at home and overseas. Ultimately, that will flow through to new products which can be manufactured in Australia for export.

ACKNOWLEDGMENTS

The support of the sponsors or the Fire Code Reform Centre Pty Ltd, and the research consortium is gratefully acknowledged. Without their vision and contributions, the fire engineering research programme would not have taken place. The principal sponsors have been: the Australian Building Control Board, the National Association of Forest Industries, Standards Australia, the Cement and Concrete Research Association, and the Forest and Wood Products R & D Corporation. Other significant sponsors have been the ANZ Bank, AMP Investments, BHP Steel, Building Control Commission Victoria, Civil and Civic, James Hardie, Leighton Contractors, NSW Public Works, St Martins Properties, Steel Reinforcing Institute, Tyco Ltd, and Westfield Design and Construction.

The research consortium comprises : BHP Research, Melbourne; CSIRO Division of Building,Constructionand Engineeing; Scientific Services Laboratory, ACS; University of Technology Sydney; and Victoria University of Technology, under the leadership of Professor Vaughan Beck.

Professional issues

The Mechanics of Structures and Materials, Grzebieta, Al-Mahaidi & Wilson (eds)
© *1997 Balkema, Rotterdam, ISBN 90 5410 900 9*

Work product principles

H. R. Milner
Department of Civil Engineering, Monash University, Melbourne, Vic., Australia

ABSTRACT: Work product principles are misrepresented by a significant number of modern engineering text book writers who link it with the principle of conservation of energy. It is shown herein that the work product principle is totally unrelated to the principle of conservation of energy and, although it is used in Newtonian mechanics, is not associated with Newtonian work and therefore the principle of conservation of energy.

1 INTRODUCTION

Many writers describe virtual work principles as, *inter alia*, having links with Newtonian work, the principle of conservation of energy and constitutive relationships. There is also a widespread, but misguided, impression that a special form of the principle applies to rigid as distinct from deformable bodies.

Herein, a new view is presented of work product methods in which:

(a) in the principle of virtual displacements, products are formed between forces which occur in the system of interest or the "real" system, and geometry changes in a related "conjugate" system, and,

(b) in the principle of virtual forces, products are formed between geometry changes which occur in the system of interest or the "real" system, and forces in a related "conjugate" system.

"Work products" formed according to WP = (forces/stresses in a "real" system) x (displacements/strains in a "conjugate" system) clearly lie outside the realm of Newtonian mechanics.

2 WORK PRODUCTS AND FUNCTION PRODUCTS

In Newtonian mechanics, work, by definition, is the product of force and distance or displacement of the force along the direction in which the force acts, in vector notation $W = \bar{F} \cdot \bar{d}$ (vector dot product). The displacement is regarded as the actual movement of the force. Thus, if a car is pushed along a level road by a constant force, then the work done is simply force x distance. It is possible to imagine multiplying the force to push the car by a dissociated distance such as the movement of the moon in the same time interval that the force acted. This would result in a work measure but it is not Newtonian work. In this paper, such work type quantities will be referenced as "work products" (WP).

In general, a WP = force x (displacement of an associated or a dissociated object). It can be viewed as a generalisation of the work concept. Herein, the dissociation will be restricted to "parallel domains", ie, domains of identical shape with the force system acting on one and the displacements occurring in the other. However, it must be possible to unambigiously map corresponding points one to one in the two domains.

It is also possible to form products of (force x any old thing) and (displacement x any old thing). These products will be referred to as function products; Milner [1966].

3 PROBLEMS WITH ENGINEERING TEXTS AND OUR VIEWS OF VIRTUAL WORK

Charlton [1982] is adamant that virtual work is associated with the principle of conservation of energy. In discussing the principle of virtual work he states that the *"highly significant feature is Calpeyron's use of actual displacements as virtual displacements, for the purpose of deriving the theorem (Calpeyron's Theorem) by the principle of virtual work, a principle which implies conservation of energy!"*

The underlined phrase takes a wrong minded but common view of virtual work. Examples are given below in which a convolution of dissociated forces and displacements reproduce all of the results which are supposedly obtained from conservation of energy principles. The examples make it patently clear why work product principles apply to both conservative and non-conservative systems - Newtonian work or energy simply has nothing to do with the principles. Worse, the principles of virtual work are incompatible with the principle of conservation of energy - a point that is usually avoided by the "energy conservationists".

4 VIRTUAL WORK IN KINEMATICS 1 - A SINGLE LINK SYSTEM CHANGING SHAPE (INVOLVING THE SO-CALLED PRINCIPLE OF VIRTUAL FORCES)

Take the single, pin-ended link (not a structure capable of supporting load) of Fig 1 which rotates about the left end. The analysis is concerned with the geometry changes taking place in the linkage; the focus is upon the relationship of the linkage internal displacements to the external geometry changes.

Purely by geometry, the deflection of the right end is given by $\delta = \theta L$ but a second method will be sought by which the value can be computed by the principle of virtual forces. Note that the link cannot store strain energy since it does not strain.

Suppose that the link internal displacements are multiplied by the internal stresses (bending moment at the left end) of the cantilever shown immediately below; these constitute the dissociated forces referenced in Section 2. Using conventional terminology this product is called "internal" work product $= \theta \times L$. By observation or otherwise, it is noted that $\theta \times L = 1 \times \delta =$ "external" work product.

If the link hinge was rusty it would heat up and dissipate energy but this does not affect the principle. Conversely, if the hinge was frictionless, no "real" energy whatsoever would be involved in the linkage system movement.

5 VIRTUAL WORK IN KINEMATICS 2 - A DOUBLE LINK SYSTEM CHANGING SHAPE

Was the result of Section 4 above a pure fluke? Consider the double link system of Fig 2; δ can be obtained by geometry as $\delta = \theta L + \phi L/2$. The internal link displacements are θ, ϕ and the corresponding cantilever bending moments (the dissociated forces) are $M = L$, $m = L/2$. Again, equate the "internal" and "external" work products.

"Internal" work product $= L \times \theta + (L/2)\phi$, "external" work product $= 1 \times \delta$. If it assumed that external work product = internal work product, then the result $\delta = \theta L + \phi L/2$ is obtained.

Again, the result is independent of energy dissipation anywhere and, indeed, serious thought should be given as to whether the principle of "virtual work" is aptly named. It is obvious that the products are not what James Clerk Maxwell, Calpeyron and others had in mind as "work" in the context of structural analysis. To describe the principle of virtual work accurately, a new vocabulary may be required. The notion that useful principles can be derived from the concept that deductions derived from a principle involved from (forces in object A x displacements in object B) can lead to useful results has not previously been stated.

24

"REAL" SYSTEM WHERE
GEOMETRY CHANGES OCCUR

"CONJUGATE" SYSTEM
REPRESENTING
EQUILIBRIUM FORCE SET

M = L

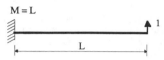

Fig 1:   Displacement of a single link system.

"REAL" SYSTEM WHERE
GEOMETRY CHANGES OCCUR

"REAL" SYSTEM WHERE
GEOMETRY CHANGES OCCUR

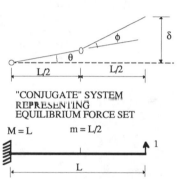

"CONJUGATE" SYSTEM
REPRESENTING
EQUILIBRIUM FORCE SET

M = L   m = L/2

"CONJUGATE" SYSTEM
REPRESENTING
EQUILIBRIUM FORCE SET

M = 0   m = L/2

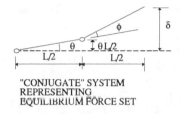

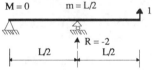

Fig 2: Displacement of a double link
system - force system 1 in conjugate
object.

Fig 3:   Displacement of a double link
system - force system 2 in conjugate object.

6   VIRTUAL WORK IN KINEMATICS 3 - A DOUBLE LINK SYSTEM CHANGING SHAPE WITH ANOTHER VIRTUAL FORCE SYSTEM

Consider the same displacement system but the force system of Fig 3; δ is still given by geometry as $\delta = \theta L + \phi L/2$. The internal link displacements are θ, ϕ but the conjugate force system involves a force at the centre support which must be multiplied by its counterpart displacement in the real system; this has to be determined by geometric analysis. Again, equate the "internal" and "external" work products.

"Internal" work product $= 0 \times \theta \; + (L/2)\phi = (L/2)\phi$, "external" work product $= 1 \times \delta$ $+(-2) \times \theta L/2 \; = 1 \times \delta \; -\theta L$ and, assuming that, external work product = internal work product, and rearranging, $\delta = \theta L + \phi L/2$.

Hence, if the alternative force system is used, additional geometric analysis of the link system is required. It is equivalent, in a beam deflection computation, to choosing an inappropriate virtual force system.

25

7 DUALITY

In the linkage systems studied above, a dual arises. The cantilever can be regarded as the "real" system and the linkage as the "conjugate" system. This role reversal simply transfers us from the principle of virtual forces to the principle of virtual displacements. In the former, kinematical relationships are analysed and, in the latter, statical relationships. A detailed discussion of this point has been provided elsewhere by Milner [1996].

8 BEAMS

Suppose that the sharp angle changes in the linkages are replaced by distributed angle changes in the form of curvatures. These angle changes/curvatures could be moulded into a beam system and involve no straining, could involve a linkage / mechanism system or could occur in a non-conservative system as arises with a beam bending with irreversible plasticity or visco-elasticity. Imagine a second or conjugate beam loaded with an equilibrium force set given by $\dfrac{d^2M}{dx^2} - q = 0$ and obtain the convolution of the two by integration over the length twice by parts. To focus on something specific, assume that the curvatures $\phi = \dfrac{d^2v}{dx^2}$ occur in the "real" system and the equilibrium force set occurs in the "conjugate" system.

$$\int_0^L M \frac{d^2v}{dx^2} dx = \int_0^L M \frac{d\left(\frac{dv}{dx}\right)}{dx} dx = \left[M \frac{dv}{dx} \right]_0^L - \int_0^L \frac{dM}{dx} \frac{dv}{dx} dx$$

$$= [M\theta]_0^L - [Vv]_0^L + \int_0^L \frac{d^2M}{dx^2} v dx = [M\theta]_0^L - [Vv]_0^L + \int_0^L q v dx$$

$$= M_2\theta_2 - M_1\theta_1 - V_2v_2 + V_1v_1 + \int_0^L q v dx \tag{1}$$

In applying this result to obtain deflections, let $M_1, M_2 = 0$ (the "conjugate" beam is always taken as simply supported for convenience in calculating beam deflections) and choose $v_1, v_2 = 0$ (the method is applied to beam segments meeting this requirement, ie, to segments between rigid supports). The loading q takes the form of a unit concentrated load applied at the point where the deflection is required (a distributed load represented by a Dirac function, $\int_0^L q v dx = v$) so that the result is stated in the form $\int_0^L M \dfrac{d^2v}{dx^2} dx = v$.

In the case where the curvature is caused by bending moment, m, in a linearly elastic material, $\dfrac{d^2v}{dx^2} = \dfrac{m}{EI}$ but, if there is plastic flow occurring, then the curvature will be given by some constitutive expression / flow rule even if its evaluation involves tracing the loading history or whatever else is necessary to obtain the local curvature.

9 BEAM EXAMPLE 1

It has been known for some time that the computation of deflections in a "real" continuous beam can involve the use of another "conjugate" beam which is simply supported. As stated above, to make equation 1 work satisfactorily in such cases it is essential that M_1, $M_2 = 0$, ie, always take the conjugate beam as simply supported. Thus, if in the beam of Fig 4, the deflection at mid-span of span AB is sought then it is easiest for the conjugate beam to be simply supported and for equation 1 to be applied to the beam segment AB. At the supports B and C, v_1, $v_2 = 0$ which gives a reference base line for deflection measurement.

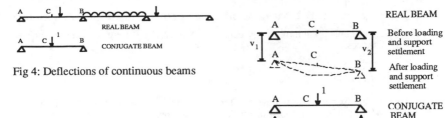

Fig 4: Deflections of continuous beams

Fig 5: "Rigid body" displacement analysis

10 BEAM EXAMPLE 2

Where the supports themselves settle, it is not possible to meet the v_1, $v_2 = 0$ condition and a modified form of the beam deflection formula is required having the form, see Fig 5,

$$\int_0^L M \frac{d^2 v}{dx^2} dx + V_2 v_2 - V_1 v_1 = v \qquad (2)$$

If the beam is perfectly straight, rather than bent as in Fig 5, then $\frac{d^2 v}{dx^2} = 0$ and $V_2 v_2 - V_1 v_1 = v$ which is a form of virtual work statement some authors reference as being applicable only to "rigid" members. The rigidity of the beam is not at issue. This is a common error in texts which unnecessarily try to distinguish virtual work principles for rigid bodies from a seemingly alternative principle for deformable bodies. It is difficult to determine why this distinction is made in so many books.

11 THREE DIMENSIONAL ELASTICITY

The full proof of the theorem of virtual displacements requires use of the divergence theorem of Gauss. As with the beam examples, it can be completed by separating the systems, the "real" one on/in which the surface tractions/stresses occur, and the "conjugate" one on/in which the displacements/strains occur. T_i = surface tractions acting on the "real" system, n_j = direction cosine of surface normal, u_i = displacement along axis i in the "real" system, f_i = body force in the "real" system. The "real" system, which is being analysed for equilibrium, has the same shape as the "conjugate" system after the deformation has taken place.

$$W_{ext} = \int_S T_{ji} n_j \delta u_i dS + \int_V f_i \delta u_i dV$$

The Einstein summation convention is assumed to operate. Convert the surface integral to a volume integral by Gauss' theorem.

$$W_{ext} = \int_V \left[T_{ji} \frac{\partial(\delta u_i)}{\partial x_j} + \left(\frac{\partial T_{ji}}{\partial x_j} + f_i \right) \delta u_i \right] dV$$

The second term vanishes in view of the virtual work requirement that we always deal with equilibrium forces sets. The internal geometry changes (strains) are obtained from the

27

equation $\dfrac{\partial(\delta u_i)}{\partial x_j} = \delta\varepsilon_{ij} + \delta\omega_{ij}$, where $\delta\varepsilon_{ij}$ = strain tensor element, $\delta\omega_{ij}$ = spin tensor element. Because T_{ji} is symmetric and $\delta\omega_{ij}$ skew symmetric ($\delta\omega_{ij} = -\delta\omega_{ji}$, $\delta\omega_{ii} = 0$) then $W_{ext} = \int T_{ij}\delta\varepsilon_{ij}dV = W_{int}$. Again the principle is arrived at without reference to a constitutive law or conservation of energy concepts. The "real - conjugate" system concept works in all instances.

12 CONCLUSIONS

1 The demonstrations above show how an alternative view of work product principles may be taken. While this does not lead to further applications, it does provide an alternative perspective and flies in the face of much conventional wisdom.

2 Within the constraints of small displacement geometry, statements of the type "the principle of virtual displacements as <u>applied to deformable bodies</u>" the under-lined qualification is totally unnecessary. None of the results presented above would be invalidated if small geometry changes took place in the "real" system provided the equation of equilibrium remains valid.

3 The principle of virtual work has nothing to do with Newtonian work and the principle of conservation of energy. This is in conflict with all other writers but especially Jean Bernoulli and Fourier.

4 In the principle of virtual displacements virtual displacement patterns need not be constrained to meet specified geometric constraints placed on the structure under analysis. This assertion is in conflict with, for instance, Timoshenko and Goodier 1970, West 1989. However, the internal and external displacement measures of the conjugate structure must be compatible in the sense of one being derivable from the other, ie, the internal strains must be derivable from the displacement field by differentiation. The strains and displacements cannot be dissociated. Incidentally, this also preserves the original meaning of the term *compatibility* which had nothing to do with geometric admissibility, ie, member continuity and compliance with the geometric conditions inferred by the nature of external supports.

13 REFERENCES

Charlton, T. M. 1982. A history of theory of structures in the nineteenth century, *Cambridge University Press*.

Milner, H. R. 1996. Work and product function product principles, *Australian Civil /Structural Engineering Transactions*, IE Aust, vol CE38, n1, pp49-53.

Timoshenko, S. P., and Goodier, J. N. 1970 Theory of elasticity, New York, *McGraw-Hill*.

West, H. H. 1989 Analysis of structures - an integration of classical and modern techniques, New York, *Wiley*.

The Mechanics of Structures and Materials, Grzebieta, Al-Mahaidi & Wilson (eds)
© *1997 Balkema, Rotterdam, ISBN 90 5410 900 9*

Decayed heritage structures – A professional issue

J.L.van der Molen & N.Grigg
Department of Civil and Environmental Engineering, The University of Melbourne, Vic., Australia

P.F.B.Alsop
Geelong Historial Society, Vic., Australia

ABSTRACT: The paper examines the requirements of condition surveys and structural assessments of structures listed in a Heritage Register, which have sustained some form of decay or damage. It is explained that the structural investigation of existing structures should go well beyond performing a check against an existing Code. A case study serves to emphasise the resulting loss of amenity if these investigations are not properly carried out.

1 INTRODUCTION

A frequently occurring structural engineering task is that of assessing the adequacy of structures which were built some time in the past. Work is generally initiated by the present owners, who engage a structural consulting engineer to advise them on such particulars as the feasibility of changes in occupancy (loading), alterations and/or additions, assessment of deterioration, remedial work, maintenance schedules, etc.

If the question raised is one of structural adequacy and the structure is relatively new the engineer can conduct a site inspection to ascertain that the critical parts of the structure still have adequate load capacity, and perform a check computation using the appropriate Design Code, allowing for any deterioration which has taken place. The engineer may then prescribe such maintenance, strengthening, or replacement as may be required to restore factors of safety to their appropriate levels.

The matter becomes more complex if the structure is older, and has been designed in accordance with an obsolete Design Code. If, in addition, the structure is registered in a Heritage Register, which generally provides it with statutory protection against demolition, alteration, and neglect, this provides an overriding constraint on any action contemplated. This paper deals with the manner in which the engineer and the owner need to tackle structures of the latter kind.

The matter is raised as a professional issue because there is considerable anecdotal evidence suggesting that the manner in which these problems are being dealt with at present often is inimical to our retaining a stock of significant buildings which illustrate our historical development, and are, as such, part of our culture. In addition, older buildings often have considerable aesthetic merit, and serve to enhance the appearance of our built environment.

2 STRUCTURAL ASSESSMENT

Structural assessment of heritage structures presents a unique problem, for the following reasons:
• The structure may have sustained an indeterminate amount of decay, settlement, damage and/or wear,

- The structure is often old, and has been designed in accordance with an obsolete Code, or even in accordance with "common usage" in the absence, at the time, of a recognised Design Code.

The first point needs to be resolved before one can embark upon any useful engineering or economic analysis.

The second point needs special consideration. Obviously, there is no point in checking a structure against a Code which did not exist when it was designed. It would be even more foolhardy to condemn a structure because it does not meet this criterion.

What the engineer ought to determine is whether or not the structure will continue to stand in its present condition, when subjected to the now prevailing dead loads and live loads, and will do so without undue deformations. In practice the engineer would therefore wish to determine the factors of safety against collapse under the various appropriate load combinations (ultimate limit state), and the magnitude of critical deformations (serviceability limit state). We will show that this analysis is substantially different from the normal design, and design checking computations, although the terminology is similar..

2.1 Condition Survey

Obviously, the matter of the initial condition of the structure is of paramount importance in order to answer a number of vital questions which all have an impact on the decision-making process regarding its future. The result of the condition survey should show whether it is technically possible to restore the structure to a level of integrity which will preserve its heritage value, while at the same time allowing for its possible future use.

There is a considerable amount of literature and guidance available on this subject (cf. American Society of Civil Engineers 1981; Building Research Establishment 1991).

Logically, we can only make informed decisions regarding repairs, restoration, ultimate purpose, and their feasibility, when we know in some detail what the present condition of the structure is.

2.2 Collapse Load Analysis

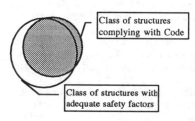

Figure 1 - Venn Diagram, classifying Structures

The above reasoning leads us away from a structural check against some structural Code or other, and towards the determination of the collapse load of the structure. These two approaches are essentially different in nature, as the Venn diagram, Figure 1, shows. The blank circle contains the class of structures having an adequate safety factor, the shaded circle contains the class of structures which comply with their relevant Design Codes.

The Figure indicates the existence of a sub-class of structures which have adequate safety factors, but do not comply with the relevant design Codes. Likewise, it indicates the existence of at least some structures which comply with these Codes, but do not have adequate safety factors.

Figure 1 serves to illustrate the point that compliance or otherwise with a current Design Code is not a valid criterion on which to judge structural competence where structures with a heritage significance are concerned.

2.3 Upper and Lower Bound Methods

The consequence of the necessity of determining the collapse load is that a plastic, or limit, method of analysis has to be employed, which may be either static (lower-bound), or kinematic (upper-bound). (Prager & Hodge 1951). In addition, for the plastic analysis to be valid, the structure must exhibit ductile behaviour, i.e. plastic hinges must form within the structure under increasing load, sufficient in number for the structure to be transformed into a mechanism, thus making use of the alternative load paths available in the structural system. Referring to Figure 2, some interesting conclusions flow from the above:

30

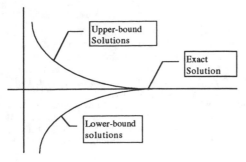

Figure 2 - Upper- and lower-bound solutions

• Without performing a limit-state analysis it is not possible to make any informed pronouncement regarding the magnitude of the collapse load of a structure. Performing a check analysis in accordance with any design Code merely yields information regar-ding the conformance, or otherwise, of the structure to the Code against which it is being checked.

• If the collapse load is not accurately known, it follows that neither is the actual safety factor with respect to any loading condition. Therefore, without prior determination of the collapse load, pronouncements regarding "the safety" of the structure are meaningless, or, depending on the context of such pronouncements, misleading, tendentious, and/or mischievous.

3 CASE STUDY, BARWON RIVER AQUEDUCT

Figure 3 - Barwon river Aqueduct (note foorway above ovoid pipe)

This aqueduct, shown in Figure 3, was built in 1913-1916, as a link in the outfall sewer of the City of Geelong to Black Rock on the South coast of Victoria. It straddles the Barwon river flood plain at Breakwater, just South of Geelong, and has a total length of 750 m, in 14 spans. It was designed and con-structed by Stone & Siddeley. Of this firm, Edward G. Stone was the designing engineer, a somewhat enigmatic figure, who was obviously inspired by the great advances being made in the theory, design and construction in the new medium: reinforced concrete (Willingham 1991).

Two of Stone's major works, the Dennys Lascelles woolstore in Geelong and the aqueduct were inspired by the design philosophy developed by Armand Considère (1841-1914), who designed a number of reinforced concrete truss bridges and roof structures. (Lewis, Alsop & Turnbull 1989). Considère made a definitive contribution to our understanding of the nature of the inter-action between the concrete and the embedded reinforcing steel, and thus of the strength of the composite material (Marsh & Dunn 1906).

Both Stone's structures were listed in the Historic Buildings Register. The woolstore, with its reinforced concrete bowstring trusses spanning 54 m, was demolished following unprece-dented interference by the then Premier of Victoria, Mr. John Cain, and the executive arm of the government of Victoria, following a grossly inaccurate and sensationalist engineering report (van der Molen & Huddle 1990).

3.1 The Aqueduct

The aqueduct is the property of Barwon Water, the "corporatised" successor to the Geelong Waterworks and Sewerage Trust. It was commissioned in 1916, and decommissioned in 1992, its function being taken over by a newly constructed pumped siphon located some 20 metres upstream. It was entered in the Historic Buildings Register in 1981, and is thus protected under the Historic Buildings Act, 1981, and its successor, the Heritage Act, 1995. Under this Act its owner is responsible for the upkeep of the structure.

After considerable public discussion, during which a voluntary, unfunded organisation, the Geelong Aqueduct Committee was formed, Barwon Water lodged with Heritage Victoria an Application for a Permit to Demolish, which it later retracted. In July 1995, Barwon Water commissioned its consulting engineers to report on the structural safety of the aqueduct.

This report was completed in August, 1995, and concludes that the analysis carried out "…..fails to provide any assurance that the structure has any margin of safety under current loading conditions". This conclusion was reached after carrying out a linear elastic analysis, and singling out the first compression vertical (shown bold in the sketch, Figure 4) as the critical member. This member was then analysed using some current design Codes. The conclusions continue: "Were this the case (the absence of a safety factor), the structure would be relying on the development of some alternative load path to avoid collapse". As the possible contribution to the overall strength of these alternative load paths was not investigated, the report in fact failed to address its central objective, which is stated as being "to advise on the current safety of .... the aqueduct". The position was later aggravated by a fax message, dated 7.9.95, from the consulting engineers to the effect that: "it is considered prudent for the Board to assume that collapse of the ovoid sewer aqueduct could occur at any time" and that collapse, were it to occur would happen "with little prior warning".

In November, 1995, the Minister for Planning, the Hon. R.R.C. Maclellan, called a meeting of interested parties, and subsequently commissioned an Independent Panel of Inquiry. During the public hearings of this Inquiry in February, 1996, the two opposing views of Barwon Water (that the aqueduct, being unsafe, should be demolished), and of the Geelong Aqueduct Committee (that it could, and should be restored) were vigorously debated. The Inquiry found (Moles 1996) that the aqueduct should not be demolished.

3.2 Collapse Load Analysis

Investigation of the strut in question revealed that the consulting engineers had calculated the effective buckling length of the member as the centre-to-centre distance of its supporting members, with an effective length factor of 1.0. Taking the effective length as the distance face-to-face of supporting members and accurately calculating the effective length factor one finds that the consulting engineers over-estimated the effective length by a factor of 2.1, thus under-estimating the buckling strength by a factor of 4.6. In addition, the consultants used the assumption suggested to them by Barwon Water's Board member, Professor L. Baker, that corrosion of reinforcement affects tension and compression members to the same extent. This has the effect of further downgrading the computed strength of compression members by a factor of about 1.5, giving a total hidden load factor on compression members of approximately 7. It is obvious that the conclusions communicated to Barwon Water, and widely advertised by them, could not possibly follow from the investigations conducted.

We conducted a collapse load analysis by using the same computer program (Spacegass) as the consultants, but employing an incremental loading technique, and applying a corrosion model to account for the degraded state of the structure. The result of this investigation is shown in Figure 4.

It may be seen that the structure appears capable of sustaining 1.9 x Dead Load. Furthermore, the investigation shows that the critical truss members are the tension chord members at the end of the cantilever, and the end diagonal. These members have comparatively light reinforcement, so that their strength is more sensitive to corrosion. The failure mode is seen to be ductile.

3.3 Load Factors

Translating the results into Dead Load and Pedestrian Live Load over the area of the foot bridge has the following result:

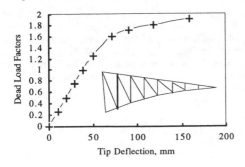

Figure 4 - Aqueduct Truss, Load-Deflection

The overall collapse load is 953 kN. Using the conventional load factors of 1.25 x DL + 1.5 x LL, and applying an overall capacity reduction factor of 0.8, one arrives at a required ultimate design load capacity of 956 kN. This shows that the structure is capable of safely carrying its own weight, together with a design pedestrian load of 5 kPa on its footbridge. Arguably, the latter load is in excess, by a factor of 3.0, of any pedestrian load which could ever be applied. Attention is drawn to the notice affixed to the pylon, shown in Figure 3.

4 CONCLUSIONS AND RECOMMENDATIONS

4.1 Conclusions

The appraisal of decaying heritage structures ("places", using Heritage Act terminology) is a complex engineering task. The first priority is the completion of a condition survey. This survey may have to be divided into progressively more detailed parts. The planning of the survey, and the modifications to these plans as initial steps reveal possible requirements for additional investigations, is of the utmost importance. It must result in the recording of all details which have a bearing on future performance, and cost of repairs or renovations.

Assuming that the place in question has a significant structural component, its structural competence must now be measured, taking into account the effect of all debilitating details which were revealed during the condition survey. In all cases, an estimate of the collapse load of the structure will need to be made, so that available load factors may be realistically assessed, and areas of weakness, either potential or existing, may be defined.

Ideally, the final report of these investigations will provide the owners of the place and the Heritage Authority with a number of alternative management plans, costed in some detail, which can be used as a guide to the performance of any remedial work required, and to alternatives of future use.

The evaluation of existing structures with a view to determining their likely future structural performance is a specialised task. Yao (1981) states:

"In the current practice, relatively few highly-experienced engineers can successfully perform damage assessment and reliability evaluation of existing structures. Even when these experts are willing, their specialised knowledge cannot be transmitted to younger engineers in a direct and systematical way without many years of working together."

As reported in the case study, the aqueduct has so far escaped demolition, mainly through the efforts of the Geelong Aqueduct Committee, whose many representations to Heritage Victoria, and to the Independent Panel of Inquiry commissioned by the Minister for Planning convinced the latter that the aqueduct was structurally viable.

If a proper collapse load analysis had been performed at the outset of these deliberations, this would have shown that the aqueduct was not in danger of collapse, and that, therefore, demolition by virtue of its structural inadequacy was not a consideration.

4.2 Recommendations

It will often be the case that the perceived financial interests of the owner run counter to the provisions of the Heritage Act, 1995. This is a potential source of conflict, for which resolution mechanisms are available through Heritage Victoria. The outcomes of structural assessments should be a scientific input into the resolution process. They should not be manipulated to distort the output of this process.

The crucial importance of structural assessments in the decision-making process regarding the future of heritage structures cannot be over-estimated. It is, therefore, imperative that their quality is impeccable. We are, however, of the opinion, that the quality of at least one of the structural assessments of the Dennys Lascelles wool store, and of the aqueduct were deficient. The effects of this situation were both immediate and far-reaching: the former structure was demolished, while the latter could only be saved from demolition by months of voluntary work, and an expensive inquiry process.

We do not consider that this situation is at all satisfactory, and would recommend:

> that Heritage Victoria seek amendments to the Heritage Act to the effect that condition surveys and structural evaluations of Heritage structures only carry validity if commissioned jointly by the owners and Heritage Victoria, and that a register of properly qualified and experienced engineers should be compiled and used for the purpose.

and

> that any structural evaluation carried out in accordance with the above shall be subject to the checking procedures normally required in the field of engineering design, before it can be publicly quoted, or acted upon by any interested party.

5 REFERENCES

American Society of Civil Engineers, 1981. Guideline for structural condition assessment of existing buildings, *Monograph ANSI/ASCE 11-90*, New York

American Society of Civil Engineers 1986. Forensic engineering: learning from failures, *Proc. Symp. by ASCE Technical Council of Forensic Engineering,*, New York

Building Research Establishment 1991. Structural appraisal of existing buildings for change of use, *BRE Digest No. 366*, Watford, UK

Currie, R.J. 1990. Towards more realistic structural evaluation, *The Structural Engineer, Vol 68, No 12 (1990)*, pp. 223-228

Institution of Structural Engineers 1955. Report on structural safety, *The Structural Engineer Vol 33(1955)*, pp. 141-149

Institution of Structural Engineers 1980. *Appraisal of existing structures*, The Institution of Structural Engineers, London

Johansen, K.W. 1962. *Yield-line theory*. Translated from Danish PhD Thesis, Cement & Concrete Association, London.

Lewis, M.B., Alsop, P.F.B. & Turnbull, M. 1989. *Dennys Lascelles woolstores, Geelong, Australia*, Submission for World Heritage listing, National Trust, Australia (Victoria)

Marsh, C.F. & Dunn, W. 1906. *Reinforced Concrete*, Archibald Constable & Co. Ltd, London

Moles, J.A. 1996. *Barwon sewer aqueduct*, Report by independent panel of inquiry, Heritage Victoria

Prager, W. & Hodge, G 1951. *The theory of perfectly plastic solids*. John Wiley & Sons, New York

Pugsley, Sir Alfred 1970. Practical limits to loadings and strength and their effect on safety, *The Structural Engineer Vol 48, No 12 (1970)*, pp. 489-491

van der Molen, J.L. & Huddle, L. 1990. The death of a building. *The Age*, 15.6.90

Willingham, A. 1991. *The ovoid sewer aqueduct at Breakwater, Geelong, Victoria, Assessment of cultural significance*, A conservation plan for the Geelong & District Water Board

Yao, James T.P. 1981. Safety and reliability of existing structures, *Developments in Civil Engineering 4, Structural Safety and Reliability, Proc, 3rd Int. Conf. on Structural Safety and Reliability*, pp. 283-293, Elsevier Scientific Publishing Co, Amsterdam

The Mechanics of Structures and Materials, Grzebieta, Al-Mahaidi & Wilson (eds)
© *1997 Balkema, Rotterdam, ISBN 90 5410 900 9*

The brothers Michell: Their contributions to engineering theory and practice

P.G. Lowe
Department of Civil and Resource Engineering, University of Auckland, New Zealand

ABSTRACT: The aim of this paper is to outline and discuss the contributions made to engineering, both theoretical and practical, by the Melbourne brothers J.H. and A.G.M. Michell. To appreciate their contributions, during what were the formative years of much of present day engineering, it is necessary to widen the horizon and discuss influences with which they were in contact, and to delve a little into the social history of the time.

1 INTRODUCTION

The brothers John Henry (1863-1940) and Anthony George Maldon (1870-1959) Michell lived and worked for the greater parts of their lives in Melbourne. Their engineering-related achievements were very substantial. Our aim is to consider what they achieved and, particularly, what motivated them in the original ideas they developed.

John, at least in his early career, was a high profile professional academic mathematician with strong engineering/technology leanings: George was a consulting civil and mechanical engineer. They were the two sons in a family of five children of an immigrant English couple from Devon: George was youngest and John may have been the eldest. George took one of his names from the mining district in Victoria where the family settled for a time, although he was actually born in London: the other children were born in Australia. There was a long family connection with mining, in Cornwall, Devon and elsewhere. A characteristic of this family appears to be that until the children were in their late teens, they travelled, probably at great expense and disruption to some family members, to achieve various desired goals. The most obvious data to support this view is the decision taken to move the entire family to England so enabling John to study for the Mathematical Tripos at Cambridge University, after he had shown clear talent in his mathematical studies at the then almost new and quite tiny University and Faculty in Melbourne. This was in 1884, nearly thirty years after the parents had emigrated to Victoria.

First some background on the study of mathematics in Cambridge at that time. Recall that the tertiary educational scene then in the U.K. consisted primarily of Cambridge and Oxford, ancient foundations and establishment dominated, together with a few new centres such as Owens College now the University of Manchester, Belfast, Cardiff, four ancient Scottish Universities of St. Andrews, Glasgow, Edinburgh and Aberdeen, Cork and Dublin. Mathematics was a primary study area, with roots extending back to the Renaissance and classical learning. There were strong mathematical traditions particularly at Cambridge, Glasgow and Edinburgh. But the taught mathematics was generally old fashioned, even for the time. In Cambridge it was dominated by Euclidean geometry and mechanics in the Newtonian mould, with limited applications of the calculus and more modern developments, which were largely European in origin. In Cambridge the actual teaching environment was dominated not by faculty staff members as we know them today, but by self-employed coaches. It was they who charged (large) fees for students to be schooled (coached) for the competitive examinations known as Triposes, of which Mathematics was the most dominant. There were some lectures available, usually given by professors holding named chairs, but commonly students attended no lectures at all, even top students, because their coach deemed that he could better organise their time, the better to prepare for the gruelling problem-based examination which emphasised particularly manipulative skill. The pressure to compete well in the Tripos was considerable, in some measure because career opportunities were very much geared to the result. The top group in the mathematics examination were called Wranglers, and were listed in order of merit, with the Senior Wrangler first followed by the Second Wrangler and so on. Competition was intense and essentially nation-wide, since entry to the University was either through the Public (that is private) School system, or by College conducted entry examination. The most able students undertook this intensive, may be even ruthless, course. After much agitation the Mathematical Tripos was reformed in the early 1900's and the order of merit abolished. It is not clear in what circumstances John Michell gained entry. The most distinguished mathematical coach operating in Cambridge at that time was Edward John Routh (1831 - 1907). He had achieved early notice when he edged James Clerk Maxwell (1831 - 1879) out of the Senior Wrangler position in 1854. Routh was Michell's coach.

The Michell mission was singularly successful in the sense that John was bracketed (shared) Senior Wrangler in 1887, probably the only Australian ever to achieve this distinction. He went on to complete the double of the equally coveted Smith Prize (shared again) two years later. George was at this time a schoolboy at the Perse School in Cambridge and attended some of the lectures by academic celebrities of the time, such as J.J.Thomson .

Much of the available personal detail about the brothers derives from the Royal Society published Biographical essays (Michell 1941,1962). John was elected F.R.S. in 1902 and George in 1934. George wrote his brother's essay (Michell 1941) but then, unusually, went on to prewrite much of his own (Michell 1962). They are both excellent essays, full of interesting and pertinent detail. George notes (Michell 1941,p365), regarding the Tripos, 'It is probable that the strain which he imposed on himself caused permanent injury to his health and spirits'. This is a very telling comment. In 1890 the family returned to Melbourne, John to a newly created lectureship on the University Mathematical Staff and George to study for his B.C.E.. John could probably have commanded a chair somewhere in the Dominions, but he opted to return to a relatively junior post in Melbourne. Their father, John , died in 1891, aged c. 65.

George thought that his mathematical talents were not comparable to those of his brother and so chose engineering as a career. While still a student he became associated with Bernhard A. Smith who both lectured on course and had a professional practice. He later was a partner with Smith before becoming independent as a consultant in about 1903. Thereafter he built up a considerable practice dealing particularly with irrigation works on the Murray and hydro-electric developments in Tasmania.

2 ORIGINAL WORK OF THE BROTHERS

John quickly produced a body of published work starting in 1890 and published 23 papers by 1902. The first four papers appeared in *Messenger of Mathematics*,a respectable but not a top-shelf journal. His next paper is a classic, *'On the theory of free stream lines'* (Michell 1890), and was published in the Phil. Trans. of the Royal Society. In this work he applied, for the first time, what has become known as the Schwarz-Christoffel transformation to this class of problem in classical fluid mechanics. This was one of the first fruits of his independent research of the European mathematics, as an antidote to the Mathematical Tripos which had left little time for such study. In this paper John used complex variable techniques to considerable advantage and in a largely pioneering mode. Another great paper was his *'On the direct determination of stress in an elastic solid, with application to the theory of plates'* (Michell,J.H. 1899). This work is still referred to in current literature. We note a third paper,which clearly gave him much trouble to complete and which was a considerable disappointment to him from the lack of response (Michell 1898). It dealt with an important practical engineering problem of *'The wave resistance of a ship'*. A majority of his papers deal with solid mechanics. He rather specialised in two dimensional solid mechanics and the partial differential equations which result, including the biharmonic. He was clearly a master of the then comparatively new Fourier Series and associated integrals. The paper (Michell 1892) has only come to light recently with the publication of the collected papers (Niedenfuhr 1964). At the time it was published only in Abstract. There he discusses the buckling of the circular plate under in-plane thrust. It seems reasonable to suppose that this paper sparked the first of the classic papers by George. Apart from the 1937-published textbook which John wrote with a colleague after retirement, he did not publish anything after 1902. Thirty two years after joining the staff John became Professor of Mathematics after the 72 year old Nanson resigned in 1922, see later. He reached retiring age in 1928.

The opportunity was not open to him in Cambridge in 1884 but he would probably have benefitted more in career and personal terms if he had undertaken research, if not in Cambridge then may be in Göttingen, rather than subject himself to the Tripos. Ten years later Rutherford from N.Z. went to Cambridge in a research capacity supported by the 1851 Exhibition and did not get embroiled in any Tripos.

Turning now to younger brother George, in quick time he published three papers, all of which are classics and all three are in totally different fields. The first (Michell A.G.M. 1899) is his lateral instability paper which predates the better known work by Prandtl. Next was his highly original *'The limits of economy of material in framed structures'* (Michell 1904) which was a pioneer discussion of structural optimization. It builds on a little known Maxwell Theorem (or Lemma as George calls it). The third is his best known and most widely applied paper, *'The lubrication of plane surfaces'* (Michell 1905). Little notice seems to have been taken of the first paper, published when he was 29. Remember too that he was earning a living as a working consulting engineer. The 'Limits' paper was also ignored for forty years, although it has had a substantial vogue in the past fifty years. With the 'Lubrication' paper he hit the jackpot. He was 35. He essentially solved in closed form the practical problem of a finite size of lubricated plane bearing. Previous work had been more academic, with the basic early papers of Reynolds and then Sommerfeld, both high profile professors. Their solutions dealt with infinitely long, two dimensional, bearings whereas the sideways leakage which George introduced proved to be the key. George's result was confirmed by experiment and lead very quickly to a patent on his thrust bearing which he was able to sell the rights for to a consortium of U.K. industrial companies, including the then mighty Vickers Ltd., and from which he made a substantial fortune.

But this is far from the end of his series of engineering achievements. The consulting work he undertook required that he specify many pump installations. He had a close relationship with the Melbourne manufacturers George Weymouth Pty. Ltd., and incorporation of thrust bearings were some of the improvements he was able to make. There was a rival U.S. patent by Kingsbury which was also successful. It is not clear when first, but certainly by the early twenties George had turned his attention to another mechanical device to which he gave the name of the *'Crankless engine'*, and for many years this device held his attention and absorbed probably virtually all of his patent revenues. It is a reciprocating piston device which can either be operated as an engine or as a pump. A late example of the hardware is on display in the Mechanical Engineering Department at Melbourne University. It is an ingenious, compact, relatively light and potentially powerful and mechanically well behaved engine which depends critically on the successful functioning of thrust bearings of the Michell type. Hence it is easy to see why he was attracted to such a device. A very interesting recent and independent contemporary account of these engines has been published (Irving,1992). There is described the machine-shop of Michell Bearings Ltd. in Greeves Street, Fitzroy where the engines were completed from castings made at the Castlemaine works of John Thompson Pty. Ltd.. George spent the period 1925 - 1933 in the U.S.A. and Europe promoting the crankless engine, but the outcome must have been a bitter disappointment to him. No worthwhile interest was generated. Curiously almost all the completed engines and all of the road going vehicles in which they were installed have disappeared. Quite where and who George visited and almost all aspects of this period are at best sketchy. He was however in close business association with Richard G. Casey (1890-1976) who later became well known in politics and as Lord Casey served as Governor General of Australia. Casey took sample engines to the U.S. to promote them to Ford and General Motors, but with no success.

3 SOURCES OF CREATIVITY AND IDEAS

Identifying the factors which contributed to the creativity and seeking the sources of ideas which these two brothers were able to evoke is the motivation for the paper. There is a common mathematical theme to the original work of the brothers, and in George's case if the thought processes were not deeply mathematical they were certainly geometrical. The creative process has been written about extensively (Hadamard 1945). In what follows a number of facts and relationships are noted but these do not add up to anything like a convincing picture of how the brothers were motivated.

In the period up until about 1900 it seems likely that George was primarily stimulated by the work of his brother. John himself seems to have been fired if not inspired in many of his papers from his study of the major themes in European mathematical and mechanics literature. Most references in his papers are to St. Venant, Kirchhoff etc. But then in 1902 came the virtual end to John's productive career. The reasons for this sudden and sustained decline in formal productivity are probably complex and may relate in part to the success of others seen through his eyes.

One group of circumstances which it is suggested is relevant relates to some of John's contemporaries in the academic world. The most relevant person in this group is Augustus Edward Hough Love (1863 - 1940). This is Love, author of the great 'A treatise on the mathematical theory of elasticity'. The two men paralleled one another in a number of ways, and each wrote important early papers in fluid and solid mechanics. Love was just seven months older than John, and outlived him by only four months. They were both bachelors, as was George. Love took the Tripos in 1885 and was Second Wrangler. John was two years later entering the Cambridge scene. Love's early papers were distinguished but not as distinguished as John's, but he maintained a steady output throughout most of his career and had early success with his (initially two volume first edition) elasticity book in 1892 - 93. In 1911 he was awarded the Adams Prize which was the most prestigious award of its type in the gift of the University of Cambridge. The prize submission is in the form of an extended essay on a nominated topic, in this case on seismology. In his essay 'Some problems of geodynamics', Love discusses surface waves for the first time. One can only surmise, but it would seem reasonable that Michell and Love probably thought of themselves as rivals. Love was elected to the Royal Society nearly ten years before John Michell. There is a slight twist to this association in that Love's elder brother Ernest emigrated to Melbourne in 1888 and had a modest career in the Natural Philosophy Department at Melbourne: he died in 1929.

Englishman Edward John Nanson (1850 - 1936), the professor of mathematics at Melbourne when John was a student and for the bulk of his subsequent career, was Second Wrangler in 1873, was only a moderately productive mathematician and seems never to have really felt at ease in the Melbourne scene. His was also a lifetime appointment as professor. Another relevant personality was another Englishman and major contributor to mechanics, (Sir) Horace Lamb (1849 - 1934), best known for his highly regarded *'Hydrodynamics'*. He was Second Wrangler in 1872, and from 1875 to 1885 was foundation professor of mathematics at Adelaide, from where he wrote the first edition of his treatise under the title *'A treatise on the mathematical theory of the motion of fluids'* in 1879. Lamb and Michell may have met since Lamb became a close friend of Nanson. Michell was a keen student of French and German and once he started publishing he makes reference to the important European works of the day, such as St.Venant's much annotated 1883 French edition of Clebsch's *'Theorie der elasticität fester körper'*, which may have been the model for Love's book. A possible conclusion to be drawn from these points is that John Michell had regrets about his career to that time, aged 39 in 1902, and these may have become dominant in his thinking and slowed his personal ambition. George's Tripos related fears, expressed late in life and quoted earlier, are relevant.

The sources of inspiration and motivation for George's three papers can only be guessed at. The Lateral Instability topic probably arose from his brother's solution of the circular plate instability problem, coupled with what were concerns at the time of actual failures of compression chords in trusses. His 'Limits' paper, which in the second half of this century has been a source paper for structural optimization studies, just seems to come from nowhere. He probably did not know about Maxwell's Lemma from any course he had attended. Few of the textbooks of the time refer to it: Routh in his 'Statics' of 1896 gives it a few lines. Indeed hardly any textbooks today do, despite the simplicity and usefulness of the result. In George's hands the Lemma is judiciously exploited to enable him to put a lower limit on the *total* weight of the structure, where Maxwell essentially finds an invariant for the *difference* in weight between the material which is in tension and in compression. He also seems to have had a sense of place which lead him to publish his result in the same journal in which Maxwell had published the Lemma, the Philoshophical Magazine.

It would be comforting to think that discussion with other engineers at the time played a significant part in sorting ideas and obtaining stimulation, but George emphasised the point in the autobiographical notes that he avoided joining professional engineering organisations. In particular he did not join in the Institution of Engineers Australia, which came into being (quite late) in 1919.

A notable difference between most of the Michell's papers and more modern papers is the very few references to other work. This is not to suggest that they were uninformed or ignored the works of others but rather is a measure of the pioneering nature of many of their papers. Sir Thomas M. Cherry (1898-1966), professor of mathematics following Michell, who was a close friend of both brothers and edited the essay (Michell 1962), draws attention to the economy with which they each wrote their papers. The three George Michell papers in the Collected Mathematical Papers (Niedenfuhr 1964) are particularly pithy but are highly readable and complete, even including experimental detail. They amount to just 36 pages in all. None of George's papers or other writings relating to his engineering innovations are included in the collected papers. It is also worth observing that neither brother collaborated with any other author in their original papers. The same is true of Love.

It seems very likely that George derived primary intellectual stimulation from his elder brother, certainly until his career path was set and he became involved with his patents, fabrication facilities and business commitments. John's ceasing any outwardly visible conventional publication after 1902 is a puzzle. There is good evidence that he expended great effort on his classes although he would appear not to have had research students.

Certainly he showed interest and offered encouragement to younger independent mathematicians who came to Melbourne. An example is Charles Ernest Weatherburn (1884 - 1974), a Sydney graduate who followed a path rather similar to John Michell to Cambridge. He was described by Horatio Scott Carslaw (1870 - 1954), the Scot 4th wrangler in 1894, who directed Sydney mathematics for more than a generation, as 'the best student to pass through this department'. Weatherburn was a pioneer in the application of vector methods in mechanics and applied mathematics and dedicates his first volume of five well known works to 'John Henry Michell.....' . The book was written while he was on the staff of Ormond College prior to taking up the Chair of Mathematics and Natural Philosophy at Canterbury College in New Zealand in 1923. Nanson's unease in his post, which was probably aggravated by the reasonably rough treatment he was dealt out by the Universitites Royal Commission in the early years of this century, and being in the decidedly senior role to Michell for so long, probably also took a toll on John Michell's enthusiasm. Another relevant personality who spent many years in an influential University post in Melbourne was Thomas Howell Laby (1880 - 1946) who was elected FRS in 1931 despite the lack of even a first degree. Victorian born, brought up in rural NSW, he largely through his own efforts became a researcher under J.J. Thomson in Cambridge, then to the Chair of Physics at Victoria University College, Wellington before moving to Melbourne. He was an abrasive and thrusting individual, best known for the Kaye and Laby *Physical and Chemical Tables*, first published in 1911, and for his collaboration with research students and staff in making precise measurements of the mechanical equivalent of heat and other quantities. He was Nanson and later Michell's counterpart in Natural Philosophy from 1913 and in many respects was a major personality contrast to John Michell.

4 CONCLUSION

The Michell brothers were a remarkable pair of Australian contributors to the science and practice of engineering during the formative stage of the modern era. Some of their contributions were hailed at the time, others were ignored and yet others have surfaced many years later as important pioneer contributions.

John, after an outstanding academic career in early life, faded from the limelight. After returning to Melbourne in 1890 he never travelled abroad again and probably rarely outside of Melbourne. George showed similar brilliance in an academic sense and went on to develop professional and business interests, but at the end of a long career it probably seemed to him that he had had more than his share of major disappointments. Both of them were early environmentalists and spent much effort preserving the natural vegetation.

For the top group of our students today there may be lessons to be learnt from the Michell experience. Our courses are heavily loaded, probably overloaded, with non-essential detail and this absorbs valuable on-course time. We academics first, and the profession later, should be making greater efforts to identify exciting possible fields for development in most of our established technologies in the medium term which the students can be introduced to at an earlier stage in their careers. We are still in the 'Tripos' phase as John Michell encountered it over a hundred years ago, and we should be beyond it. The educational environment being created by many present day governments is corrosive of intellectual development and far too accepting of what Industry is doing as being the best possible. History is likely to show that the beneficial changes and best ideas do not arise or blossom in the Industry environment, but in the more humane environment of our educational institutions.

REFERENCES

Hadamard, J.   1945 *The psychology of invention in the mathematical field*, Princeton.
Irving ,P.E.   1992 *Phil Irving - an autobiography*. Sydney: Turton and Armstrong.
Michell,A.G.M.   1899 Elastic stability of long beams under transverse forces, *Phil.Mag.* (5), **48**,298
                  1904 The limits of economy of material in frame structures. *Phil. Mag.*(6), **8**,589
                  1905 The lubrication of plane surfaces. *Zeit. f. Math. u Phys.* 52 Band
                  1941 *Obituary notices of fellows of the royal society*, 9 ,**3** , 363.
                  1962 *Biographical memoirs of fellows of the royal society*, **8**, 91. Cherry, T.M. (ed)
Michell, J.H.   1890 On the theory of free stream lines, *Phil. Trans. A*, **181** , 389
                  1892 On the bulging of flat plates, *Australasian Assoc. for the Adv. Of Science*. (Abstract only).
                  1898 The wave resistance of a ship. *Phil. Mag.* (5),**45**, 106
                  1899 On the direct determination of the stress in an elastic solid, with application to the theory of plates. *Proc. Lond. Math. Soc.*,**31**, 124.
Niedenfuhr, F.W. 1964 *The collected mathematical papers of J.H. and A.G.M. Michell.*
& J.R.M.Radok,   Groningen: Noordhoff,.

Loads on buildings – Variation within a political region

L. Pham
CSIRO, Building, Construction and Engineering, Melbourne, Vic., Australia

D. S. Mansell
International Development Technologies Centre, University of Melbourne, Vic., Australia

ABSTRACT:

Australia's political linkages have, in the past, influenced profoundly the setting of regulations, standards and codes used by designers and builders. New situations in the international political arena and significantly changed commercial relationships now call for a reappraisal of the technological expressions of those relationships. One form of those expressions is the writing of regulations and codes for design. Progress has been made in Australasia because of the policy of Closer Economic Relations (CER) so there is scope for greater harmony in the APEC (Asia Pacific Economic Cooperation) region.

1 INTRODUCTION

Many of the contributors to the technical sessions of the Australasian Conference on the Mechanics of Structures and Materials, since its inception, have been those whose studies have been influential in the drafting of Australian Standards. Other conferences, such as those sponsored by the Institution of Engineers, Australia and by the industry organisations that have a commercial interest in the use of a particular material, are also a valuable source of information on the background to changes in Standards, but ACMSM has historically provided a forum for discussion of more basic issues of mechanics, some of which appear later in the deliberations of committees drafting Standards.

Those who have been involved in standards committees know that the discussions in the committee meetings are not purely scientific in nature - the drafting process tries to find negotiated answers to technical and economic questions such that a legally enforceable document can be written. The committee tries to optimise the outcome to the satisfaction of all of the parties to any contract which will later incorporate the Standard. A very obvious example of such optimisation, or compromise, is the choice of the preferred sizes of standard components where the manufacturer's view of the desirable number in the range of sizes is likely to differ from that of the designer or the owner.

In this paper the writers are inviting the conference to consider this matter of compromise on a regional rather than on a national scale, drawing on the outcomes of meetings in Melbourne in November 1996 on regional harmonisation of Standards for loadings and general design requirements for buildings. The context for the work undertaken in those meetings, and therefore for this contribution to ACMSM, is the drive for liberalisation of trade between the APEC countries. There have been other meetings in other places, bearing on the same general issue.

2 THE "FOREIGN RELATIONS" OF STANDARDISATION

The historical dependence of Australian[1] Standards on British precedents lasted well into the period following World War II. The political, technological, commercial and economic changes that were triggered or accelerated by the trauma of that war eventually helped loosen the ties between Australia and Britain, and that loosening is reflected in the work of many institutions, including those involved in creating and up-dating Standards.

There are two international compacts to which the Australian and New Zealand governments have assented and which are likely to have a profound impact on the practice of engineering here. The first is the accord called CER (Closer Economic Relations) between Australia and New Zealand, which has forced the pace of harmonisation of Standards in the two countries. Outcomes of that increased collaboration are already evident, with common documents now published and joint drafting committees and working groups in place.

The second is the APEC agreement (Asia Pacific Economic Cooperation) which is now driving further negotiations in the region covered by that agreement. APEC arose partly out of an awareness of the trade opportunities developing around the rim of the Pacific Ocean and partly out of an awareness of the gains to be made by abandoning a passive stance which allows all the leadership to come from Western Europe and North America. It also fitted well with an ideology of liberalised trade, and much of the expression of the APEC accord has been found in high-level, much-publicised discussions on trade barriers.[2] Many Australian engineers have been mere observers of these important diplomatic initiatives without realising that all forms of trade will be caught up in the resulting changes, including their own. They appear to be unaware of the fact that their work will be influenced by a political agenda of regional and then possibly global conformance to international standards to facilitate free trade. The relevance of that political agenda to this Conference is that an important sector of our national and international trade is engineering goods and services.

Design companies, for example, that were at one stage branch offices of major consultancies in Britain and USA have grown in size and reputation and now have their own branches in Asia, or have formal agreements on cooperation with similar Asian companies. Therefore it is possible for a company based in one country using a particular set of Standards to be carrying out work at the same time in other countries using different Standards. Clearly it would be helpful if an engineer could design structures for various countries while using, as part of a tool-kit of design software, design Standards that are largely common to those other countries. The trend that will eventually lead to that internationalisation has started, but we are at present in a transition phase which provides the genesis of this paper.

3 STANDARDS - COMMON, HARMONISED OR ALIGNED

Reference was made in the previous section to "foreign relations" because the major tool in constructing a successful international cooperation is diplomacy. One should not underestimate the importance of diplomacy in obtaining cooperation in the drafting and adoption of engineering Standards. The difficulties in finding the compromises referred to earlier are familiar to members of national drafting committees, and the profession has been vociferous in its reactions to such changes as the move from Working Stress Design to Limit States Design. How much greater will be the difficulties when we try to exercise abroad the kind of diplomacy that has already proved to be difficult at home?

---

[1] This paper is written from an Australian perspective although much of it applies equally to New Zealand.
[2] It is of interest to note that, at the time of writing, a similar move has been started, to encourage cooperation between countries around the rim of the Indian Ocean.

The easy way out, of course, is to adopt a passive position, allowing the most powerful economies to dictate the philosophy, the format and the content of Standards. We are now launched on the more demanding alternative - the diplomatic task of engagement - not because of the assertiveness of Australian engineering, but because political and economic policies such as those stemming from CER and APEC leave us little choice.

In the case of Australia and New Zealand (which we might regard as "sub-regional") there is bi-lateral government policy in place requiring the preparation of common Standards written to cover all the special needs of both countries. That has proved to be not a trivial task, but considerable progress has been made.

When we expand our horizon to include countries to our north in approximately the same time-zone, the diplomatic task can be expected to increase in complexity. Consider the main economies involved: Indonesia, Singapore, Malaysia, Thailand, Philippines, China, Korea and Japan, to which we might soon need to add Vietnam and Papua New Guinea. The colonial powers whose influence remains observable include Britain, Netherlands, France, Russia, USA and, to some extent, Germany. The range of climatic, social, political, legal and economic characteristics to be found when contrasting the former colonies or dependencies is wide, and sensitivities to outside influence remain. Therefore any attempt to negotiate a common Standard is indeed a daunting task. Perhaps there is scope for "harmonising" Standards, by negotiating to amend aspects of Standards that are in conflict (i.e. removing discord). At least that task is consistent with the Asian tradition which sees the avoidance of discord and the promotion of harmony as the mark of civilised behaviour. The term "harmonisation" applied to Standards has been defined by the International Organisation for Standardisation as "Standards on the same subject approved by different standardising bodies, that establish interchangeability of products, processes or services, or mutual understanding of test results or information provided according to these Standards". This definition is written in terms of outcomes rather than content of the standard itself. Having agreed outcomes contributes to the lowering of barriers to trade in products, and presumably enables the user of the Standard to work internationally with documents which are familiar at least in philosophy, and possibly in certain detail.

"Alignment" of standards is another term in use. It seems to imply documents that differ from each other but that have much in common. The term "alignment" is used by the APEC subcommittee on Standards and Conformance (SCSC), to mean that national Standards are brought into alignment with an international model.

It is not our purpose to become distracted by the semantics of these terms, as long as there is the prospect that a reasonable and achievable goal is to have all of the cooperating countries shift the direction of their standardisation policies so that the underlying philosophy of related Standards is common to all, and its expression in terms of rules, nomenclature and units is consistent.

4 ILLUSTRATION OF THE SCOPE FOR ALIGNMENT

A series of meetings over three days took place at the University of Melbourne in November 1996 arranged by CSIRO Building, Construction & Engineering with AusAID support. The first day was devoted to a public seminar on "Loadings and General Design Requirements for Buildings in Asia", and a reading of the ten papers (comprising 80 pages) gives some insight into the situation in the following participating countries : Australia, China, Indonesia, Japan, Malaysia, New Zealand, the Philippines, Singapore and Thailand. Some of the issues which were brought to light for later consideration were as follows :

(a)   There is a variety of types of documents which are influential to a greater or lesser degree in the same way that Standards are influential. They range from legislation, through government-backed Standards and technical memoranda from disinterested research oganisations to technical notes and manuals on practice prepared by companies or industry associations. They range from legislation, through mandatory requirements empowered by legislative regulation to informative advice. Design decisions which in one country might be based on informative material only might, in another country, be determined by a prescriptive law. Consider the information used by designers to determine loads on buildings. In Western countries the loadings are usually specified in a document or documents created by a national standards body or its equivalent. Such documents are then invoked by the building regulations and are thus incorporated into contracts. The situation in Japan is different: the loads are prescribed directly in the Building Standard Law of Japan, which can be amended only by the Japanese parliament.

(b)   The bodies preparing Standards have various relationships with government, government instrumentalities and industry representative bodies. Even within one country the Building Code might be legislated by the national legislature while the Standards used for technical design are prepared by a non-government organisation such as a professional body.

(c)   The political structure of each country influences the extent of devolution of responsibility for building control. A federal system such as found in Australia, with its three levels of government, might be expected to create a distribution of authority and responsibility different from that found in a country with only two tiers. Similarly countries might designate certain areas as "special zones" with special requirements or special privileges. An example of the latter, in Australia, is the convention that State building control authority does not apply, except by consent, over construction on Commonwealth property. Capital cities are similarly often designated as special zones, such as the Capital City Special Distinction of Jakarta in Indonesia.

(d)   Terminology creates problems - even within monolingual Australia it is doubtful whether many engineers in construction could set down confidently the precise meanings of "regulation", "standard", "code", "norm", "technical note", "manual", "code of practice", "specification", "rule" etc. When translating between English and another language the scope for misunderstanding is considerable.[3]

(e)   There are important differences between the formal structures for standardisation notionally in place in a particular country and the actual way of working. The discrepancy is almost inevitable if the system of Standards is not complete so that the designer is forced to go outside it in any event. It is also more likely where the considerable expenditure of effort to keep Standards up-to-date cannot be supported. A third factor undermining the authority of indigenous Standards is the hiring of foreign designers who simply ignore local requirements and advice and use the documents with which they are familiar (British, German, Russian, French, Japanese, American, Canadian, Australian, New Zealand or some other country's documentation). Of course, some countries recognise the inevitability of that and formally approve it, at a cost of considerable inconsistency and probable lack of optimal performance.

(f)   There are intriguing variations in the sophistication, complexity and "user-friendliness" of documents in different countries. Are complex codes necessarily more advanced? Should alignment necessarily occur through adoption of the most scientifically sophisticated content?

(g)   There are important differences in philosophy - notably in a varying rate of progress from allowable stress design to limit states design.

---

[3] In preparing for the one-day seminar on which this paper is based there was some difficulty in reassuring foreign participants that the "workshops" following the seminar were opportunities to work together as colleagues without committing their home governments. The term "Workshop" differs in its implications in different cultures, carrying quite different obligations on workshop participants.

(h)   Units of measurement are inconsistent - in fact confusion abounds, with documents derived from different sources using SI, kg-m-s, British Imperial measures and others.

(i)   Methods of testing, standards and criteria for validation of the performance of products are not consistent. This means that a product approved for use in the one country is not necessarily approved, or even readily validated for approval in another country.

5 CAN PROGRESS BE ACHIEVED?

The participants in the technical Workshops that followed immediately on the November Seminar in Melbourne were largely unknown to each other. Yet, when presented with the information they all brought to the meetings, there was an astonishing degree of consensus on the technical feasibility of improving the present state of practice. A total of 28 engineers agreed to a communiqué from the Workshops in the final plenary session and put their signatures to it. It is reproduced as an Appendix.

To achieve agreement across the diversity of economies they represent will require documentation that has not only a high degree of commonality but also the flexibility to allow users to choose the level of service and reliability they consider affordable. In some instances there is a need for that flexibility even within one economy because of uneven development or because of more technical reasons such as climatic variation. There was a strong conviction that it is feasible to produce Standards that accommodate those varying needs.

6 CONCLUSION

The political pressure to incorporate Australia and New Zealand more fully into global economic markets is being applied to those sectors of the Australian economies which engage engineers as designers, manufacturers or contractors. Part of that move to global exchange of goods, services and intellectual property is a concerted effort to make the APEC region a significant partner in the negotiation of international standardisation. The APEC countries in roughly the same time-zone as Australia and New Zealand share with those two countries a vested interest in ensuring that Standards relating to building construction reflect not only the best available models of the mechanics of structures and materials but other local factors such as climate, political structure and the state of industrialisation. Therefore, work has started on the diplomatic and technical task of developing Standards which are aligned to a shared regional model document; the degree of alignment should be sufficient to free the exchange of design services, building products and construction services, while retaining in the hands of each individual economy and culture autonomy over performance levels.

For engineers interested in modelling loading conditions, performance requirements and structural reliability, a new challenge remains - to achieve international accord in Standards that allow proprietors to choose the level of service and reliability that they consider affordable.

APPENDIX

Harmonisation of Structural Loading Standards for Building Design & Construction in APEC Countries

20-22 November 1996, Melbourne, Australia

A three day meeting was held with the support of AusAID in Melbourne with participants from Australia, China, Indonesia, Japan, the Philippines, Singapore, New Zealand. Written submissions were received from Malaysia and Thailand.

We reviewed the current status of the structural loading specifications of the above countries (including wind and earthquake loadings) and explored the ways to align these specifications with International Standards (ISO) and with each other.

We noted that "Standards" for general design and structural design loading (including earthquake and wind loading) has been accepted by the APEC Standards and Conformance Sub-committee meeting in Manila 13-14 October as an "additional priority area for alignment of Member Economies Standards with International Standards".

We have reached the following consensus view :
- Alignment of the loading standards in our respective countries with each other is technically feasible.
- This alignment should be done in close collaboration with countries in our region.
- The alignment is expected to bring substantial benefits in terms of trade liberalization and professional engineering development.
- A plan for the implementation of this alignment should be developed as follows :
  1. A draft framework and a guideline for producing harmonised loading standards should be developed within the next twelve months. An example of the use of this guideline can function as a model code.
  2. Regular consultation should occur during the drafting process.
  3. Each draft should be circulated to all participating countries for comments until a consensus is achieved.
  4. Each country can then develop its own loading standard in accordance with the proposed structure and guideline.
- Support from all governments of the region is necessary for the above task.
- We believe the matter is urgent and we resolve to seek support from our respective governments to carry out the above work.

The above points represent the view of the participants as engineering professionals and do not necessarily represent the view of their respective governments.

SIGNATORIES

Australia
Dr. G. Foliente
Dr. J. Holmes
Dr. P. Kleeman
Assoc. Prof. K. Kwok
Assoc. Prof. D. Mansell
Dr. L. Pham
Prof. L. Schmidt
Prof. Em. L. Stevens

Japan
Prof. Yukio Tamura
Prof. Jun Kanda
Dr. Hisashi Okada

The Philippines
Prof. Jose M. de Castro
Mr. Carlos Villaraza
Dr. Benito Pacheco

China
Prof. Hu Dexin
Prof. Wei Lian
Prof. Jin Xinyang

Indonesia
Prof. Dr. Ir. Wiratman
  Wangsadinata
Dr. Ir. Indra Djati Sidi

Dr. Ir. Widiadnyana
  Merati
Dr. Ir. Adang Surahman
Ir. H.R. Sidjabat

New Zealand
Mr. Andrew King

Singapore
Assoc. Prof. T. Balendra
Dr. C.G. Koh
Dr. K.H. Tan
Mr. K.K. Lim
Mr. Ong See Ho

Composites, materials and biomechanics

The Mechanics of Structures and Materials, Grzebieta, Al-Mahaidi & Wilson (eds)
© *1997 Balkema, Rotterdam, ISBN 90 5410 900 9*

Mechanical properties and micromechanics of porous materials

G. Lu
*School of Mechanical and Manufacturing Engineering, Swinburne University of Technology,
Hawthorn, Vic., Australia*

Z. M. Xiao
School of Mechanical and Production Engineering, Nanyang Technological University, Singapore

ABSTRACT: This paper studies the mechanical properties of a few common porous materials: carbon rods, ceramics, polymeric foams and bricks. The characterisation of pore structures was performed using a Mercury Porosimeter. Detailed information was obtained on the density, porosity, surface area and pore size distribution. A large number of experiments on either bending or compression was conducted in order to obtain macro-mechanical properties such as Young's modulus, hardness and strength. Based on the experimental observations, theoretical models were employed to predict the macroscopic properties from the micromechanics viewpoint. By studying the deformation of pores the global behaviour was calculated. The study has provided insights into the mechanical properties of porous materials.

1. INTRODUCTION

With increasing demand for porous materials in various industries, more and more such materials are produced using powder technology in order to achieve desired mechanical properties. The understanding of the relationship between the microstructural properties such as porosity and the global mechanical behaviour is, however, far from satisfactory. Research in this area has received attention recently (Lowell and Shields 1991, Unger and Rouquerol 1988), which includes the properties of carbon (Patrick et al. 1989) and foams (Daniel and Kurt 1990).

However, there seems to be a gap in the current research between materials scientists and mechanicians and this paper attempts to bridge the gap by applying mechanics knowledge to explaining material properties. In the paper, the mechanical properties of some common porous materials: (carbon rods, ceramics, polymeric foams and bricks,) are investigated both experimentally and theoretically.. A large number of experiments was conducted in order to obtain the macro-mechanical properties such as Young's modulus, hardness and strength. Based on the experimental observations, a theoretical model was proposed to predict the macro-properties from the micromechanics viewpoint. Based on the theoretical considerations, two simplified formulae are presented applicable for the whole range of porosity values.

2. EXPERIMENTAL TECHNIQUES AND RESULTS

A mercury porosimeter (Micromeritics Poresizer 9320) was used to characterise porous materials in terms of porosity and surface area. A porosimeter of this type is capable of generating a pressure up to 410MPa detecting a range from meso- to macro- pores, but is not suitable for detecting micropores. It has four low pressure ports and two high pressure chambers, which are controlled by a PC. During each test, mercury was forced into the porous samples by hydraulic pressure and the volume of mercury penetrating the pores was measured directly as a function of applied pressure. Mercury would intrude into big pores initially, and then into smaller pores as the pressure increases. Typical samples were prepared to the size 3x3x1mm. All samples were dried in an oven to remove the water vapour. For high pressure runs the equilibrium time was set at 10-15 seconds.

An Ultrapycnometer (Quantachrome Model 1000) was used to measure the true solid volume of the specimen and, therefore, density of the specimen. The displacing gas was helium, which can penetrate the finest pores. The pressure was set at 2.1MPa and the purge time was set between 15 and 25 minutes. About 20 runs were performed for each specimen and the mean value was taken.

Carbon rods were tested for their flexural strength by means of three point bending with a span of 80mm. From the maximum force attained the strength was calculated based on the elementary bending theory. For foams, compression tests were performed on an Instron machine in order to deduce their mechanical properties. A Vickers hardness tester was used to evaluate the hardness values for cermets while a Rockwell test was used for bricks. Foam cell walls are usually weak compared with , for example, cermets and bricks. They cannot sustain the high pressure of mercury intrusion in a mercury porosimeter. A Scanning Electron Microscope (SEM) with an image analyser was therefore employed to determine the porosity of foams. The SEM magnified the cell structures of foams and the image so obtained was processed by an image analyser, which identified the cell walls and determines the area fraction.

Four groups of porous materials were tested and analysed. They were baked and green carbon rods of diameter 8mm, cermets, bricks and foams. Within each group, materials of different porosity were obtained by changing processing parameters slightly.

Table 1 Test results for carbon rods (G: green; B: baked)

Samples	Porosity ϕ (%)	Surface Area (m^2/g)	Pore Diameter (μm)	E (GN/m^2)	σ (MN/m^2)
Carbon 1 (G)	8.42	17.047	0.0098	10.7	47.7
Carbon 2 (G)	5.82	13.252	0.0250	11.6	53.4
Carbon 5 (B)	18.60	11.255	1.5313	18.1	95.0
Carbon 6 (B)	22.75	11.179	1.6538	16.2	101.8
Carbon B (B)	26.31	10.990	1.5496	15.6	93.4
Carbon C (B)	21.66	14.052	1.3971	16.5	99.1

Table 2 Test results for foams

Samples	Porosity ϕ (%)	E (MN/m^2)
Foam (pack)	79.97	1.515
Foam (float)	70.00	4.940
Polystyrene	65.33	9.230
Polyurethane	82.20	2.160

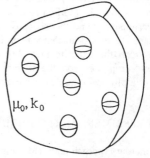

Fig 1 Micromechanics modelling of porous materials.

Because of limited space, only some selected experimental results are listed in this paper. Table 1 contains the data for carbon rods in terms of their porosity, ϕ, surface area, median pore diameter, Young's modulus, E, and strength. The values of porosity and Young's modulus for foams are given in Table 2.

3. THEORETICAL ANALYSIS

As elastic modulus, E, is a key material property, we restrict our theoretical analysis to the relationship between the porosity and the elastic modulus of porous materials. The existing theoretical models are suitable for limited cases and some are mathematically complicated. Therefore, a new micromechnics model is proposed here to explain the present experimental results for materials of porosity less than say 30%. One existing theory is employed for foams which has a porosity value larger than 30%.

3.1 A micromechanics model

When the porosity is relatively low (less than 30%), the theory of micromechanics can be applied to obtain a relationship between the porosity and the elastic moduli. Fig 1 shows a schematic of the porous material being modelled as an isotropic material (the matrix) containing n spherical voids of different sizes. We evaluate the average shear strain $\bar{\gamma}$ when an average shear stress is given as $\sigma_{12}^0 = \bar{\sigma}$. When the relation between $\bar{\gamma}$ and $\bar{\sigma}$ is obtained, the average shear modulus can be determined.

The stress in the porous material varies in space. Here it is denoted by $\sigma_{12}^0 + \sigma_{12}$, where σ_{12} is the disturbance due to voids. Eshelby's equivalent inclusion method (Mura 1987) is employed. All the voids are simulated by a homogeneous material having the same moduli as those of the matrix with eigenstrain ε_1^* , we then have

$$\sigma_{12}^0 + \sigma_{12} = 2\mu_0(\varepsilon_{12}^0 + \varepsilon_{12} - \varepsilon_{12}^*) = 0 \qquad (1)$$

where $\varepsilon_{12} = 2S_{1212}\varepsilon_{12}^*$, $\sigma_{12}^0 = 2\mu_0\varepsilon_{12}^0$, and S_{1212} is one component of the Eshelby tensor. In this case, S_{1212} is only dependent upon v_0 and is equal to $(4 - 5v_0)/15(1 - v_0)$. μ_0, v_0 are the shear modulus and Poisson's ratio of the matrix (when the porosity is zero). From the above equation the following results are obtained

$$\varepsilon_{12}^* = \frac{\varepsilon_{12}^0}{1 - 2S_{1212}}, \quad \varepsilon_{12} = \frac{2S_{1212}\varepsilon_{12}^0}{1 - 2S_{1212}}, \quad \sigma_{12} = \frac{2(2S_{1212} - 1)\mu_0\varepsilon_{12}^0}{1 - 2S_{1212}}. \qquad (2)$$

According to Tanaka-Mori's theorem (1970), we have

$$\int_{D-V} \sigma_{12} dD = 0, \quad \int_D \sigma_{12} dD = 0, \qquad (3)$$

in which V is the volume occupied by the voids, D is the whole volume of the porous material. It is easily found that

$$\int_D \sigma_{12} dD = \frac{\phi(2S_{1212} - 1)}{1 - 2S_{1212}} \sigma_{12}^0 \neq 0,$$

where ϕ is the porosity. Therefore a uniform stress

53

$$-\frac{\phi(2S_{1212} - 1)}{1 - 2S_{1212}}\sigma_{12}^0$$

is added to σ_{12} in Eq.(2). As a result, a uniform strain

$$-\frac{\phi(2S_{1212} - 1)}{1 - 2S_{1212}}\frac{\sigma_{12}^0}{2\mu_0}$$

is added to $\varepsilon_{12}^0 + \varepsilon_{12}$, and the average strain is thus given by

$$\bar{\gamma} = 2\int_D (\varepsilon_{12}^0 + \varepsilon_{12})dD \tag{4}$$

Combining Eqs.(2), (3) and (4), the shear modulus of the porous material μ which defined as $\mu = \bar{\sigma}/\bar{\gamma}$, is given by

$$\frac{\mu_0}{\mu} = 1 + \frac{15(1 - \nu_0)}{7 - 5\nu_0}\phi, \tag{5}$$

and similarly, the bulk modulus, k, of the porous material can be obtained as

$$\frac{k_0}{k} = 1 + \frac{3(1 - \nu_0)}{2(1 - 2\nu_0)}\phi. \tag{6}$$

In the above two equations, μ_0, ν_0 and k_0 are the elastic moduli of the material when the porosity is zero, and ϕ is the porosity. The Young's modulus E is related to μ and k by

$$E = \frac{9k\mu}{3k + \mu}. \tag{7}$$

As the parameters μ_0, ν_0, E_0 and k_0 are inter-dependent. we may say that ultimately the elastic modulus E is dependent upon the initial value of elastic modulus (when porosity ϕ equals to zero) E_0 and the initial value of the Poisson's ratio ν_0. The value of the Poisson's ratio in general is about 0.3 for carbon rods and ceramics. Numerical calculations show that the results are insensitive to the value of ν_0 in this range. Therefore, the elastic modulus mainly depends on the value of E_0. Calculated results based on this theory are shown in Fig.2 together with the experimental data. It can be seen that the agreement between the theory and the experiments is reasonable.

3.2 Theory for foams

When the porosity is high (larger than 30%), the above micromechanics model is invalid. The deformation mechanism at the micro-level is generally bending of cell walls. There are various theories for foams. Here we describe a simple one based on the scaling argument (Gibson & Ashby 1988).

A general open cell may be idealised into its simplest form, a cubic array of members of length l and square cross-section of side t, as shown in Fig 3. The relative density of the cell ρ / ρ_0 is related to t and l by

$$\frac{\rho}{\rho_0} \propto (\frac{t}{l})^2 . \tag{8}$$

and the second moment of area of a member, I, is

$$I \propto t^4 \tag{9}$$

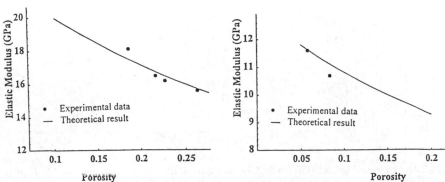

Fig 2 Plot of elastic modulus vs porosity for (a) baked carbon rods; (E_0=24 GPa, ν_0=0.3) (b) green carbon rods (E_0=13 GPa, ν_0=0.3).

When a load F is applied at the midpoint of a member, the deflection is proportional to Fl^3/E_0I where E_0 is the elastic modulus for the material of the beam. The remote stress σ is related to this force by $F \propto \sigma\, l^2$ and the remote strain ε is related to the displacement by $\varepsilon \propto \delta/l$. It follows that the elastic modulus for the foam is

$$E = \frac{\sigma}{\varepsilon} \propto \frac{E_0 I}{l^4} \propto E_0 (\frac{\rho}{\rho_0})^2 \tag{10}$$

or $\quad E \propto E_0(1-\phi)^2 \tag{11}$

This relationship is used to fit the experimental data. Fig 4 shows the comparison between the test results and Eq (11). The agreement is reasonable in spite of the simplicity of the model.

4. DISCUSSION AND CONCLUSION

In this paper, we have performed a large number of experiments in order to investigate the effect of porosity on the mechanical properties. As expected, as the porosity increases, the strength, elastic modulus and hardness decrease substantially.

We have presented theoretical analyses of the behaviour of elastic modulus using micromechanics models. This approach has produced reasonable results. It would be worthwhile to establish similar models for the change in hardness and strength.

55

It may be interesting to note that several previous workers have attempted to fit the test data using functional relationships such as exponential or polynomial, based on the test results only. This approach is purely empirical and is useful in the early stage of studies. However, the approach is unable to explain the mechanics involved. The present theoretical analysis has offered a fundamental way of studying the effect of porosity on the elastic modulus. For example, in the case of foams, equation [11] suggests that when we first analyse the experimental data, we should plot E against $(1-\phi)$ on a logarithmic scale, which would produce a straight line. The slope of the line would be close to the value of 2 and the intercept would indicate the elastic modulus of the cell walls (E_0). Semi-empirical formulae from such analyses would be more mechanics-based and would offer more insight into the physics involved.

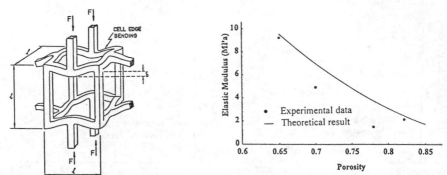

Fig 3 An idealised foam cell (Gibson & Ashby 1988).   Fig 4 Plot of E vs porosity for foams.

In the case of materials of low porosity, the calculations in the model may seem less straightforward. However, we may make simplifications. For most of the porous materials such as carbon, bricks and ceramics, the value of Poisson's ratio is close to 0.3. Thus, if we substitute this value in Eqns [5-7] and neglect the high order terms for ϕ, we obtain the following approximate relation:

$$E = E_0(1 - 2\phi)(1 + 4\phi^2)$$
(12)

The error introduced by neglecting the higher order terms is about 12% for $\phi=0.3$, and 2.0 % for $\phi=0.2$. Eqns [11] and [12] provide simple guidelines for assessing the elastic modulus of a range of porous materials.

REFERENCES

Daniel K. and Kurt C., *Handbook of Polymeric Foams and Foam Technology*. (Hanser Publishers, Nwe York, 1990).
Gibson L. and Ashby M.F., *Cellular Solids: structures and properties*. (Pergamon , 1988).
Lowell S. and Shields J.E., *Powder Surface Area and Porosity*, 3rd edition, (Chapman & Hall, 1991).
Marsh H., *Introduction to Carbon Science*. (Butterworth, 1989).
Mura T, *Micromechanics of defects in solids,* Sec. Edn., Martinus Nijhoff Publishers, 1987.
Patrick J.W., Sorlie M. and Walker A., *Carbon* **27** (1989) pp 469-74.
Tanaka K. and Mori T., The hardening of crystals by non-deforming particles and fibres. *Act. Metall.,* **18** (1970) pp.931-941.
Unger K.K. and Rouquerol J., *Characterization of Porous Solids*, (Elsevier Science Publishers, 1988).

The Mechanics of Structures and Materials, Grzebieta, Al-Mahaidi & Wilson (eds)
© 1997 Balkema, Rotterdam, ISBN 90 5410 900 9

Test requirements for elastomeric sealing pipe joints

J.P.Lu & L.S.Burn
CSIRO Building, Construction and Engineering, Melbourne, Vic., Australia

ABSTRACT: It has been known for some time that inadequate jointing performance of pipe joints can lead to problems such as root intrusion, and infiltration and exfiltration through pipe joints, and that rectification of these problems absorbs a significant part of water authorities' maintenance budgets. This paper reports on a literature search of previous and current research on the performance and evaluation of elastomeric joints. The existing test requirements for elastomeric pipe joints given in Australian and overseas standards are reviewed and the inconsistencies across the different pipe material types are discussed. Initial results of an ongoing test on root penetration are also presented. Finally, the performance requirements for elastomeric pipe joints for pressure and non-pressure pipeline system components are proposed.

1 INTRODUCTION

It has been known for some time that inadequate jointing performance of pipe joints can lead to problems such as root intrusion, and infiltration and exfiltration through pipe joints, and that rectification of these problems absorbs a significant part of water authorities' maintenance budgets. Root intrusion through pipe joints has always been a major concern throughout most of Australia, and results in significant maintenance costs for sewerage systems. Infiltration through pipe joints results in silting of pipelines, contamination of potable water and increases load on sewage treatment plants. Exfiltration through pipe joints from sewerage systems can lead to contamination of ground water and waterways, and public health hazard.

Although pipe joints have undergone development and improvement over the years, the performance requirements needed for pipe joints have not necessarily been fully researched or comprehensively defined within Australian Standards, and the current performance requirements given in Standards have a number of inconsistencies across the different pipe material types. Certainly, the widespread adoption of rubber seals for sewer pipe joints in the 1970s dramatically reduced leakage and root invasion through the joints. However, root systems are still found to penetrate sewer pipe joints, and there is limited research into the long-term sealing performance of elastomeric seals in both sewage and water pipelines.

This paper reports on a literature search of previous and current research on the performance and evaluation of elastomeric joints. The existing test requirements for elastomeric pipe joints given in Australian and overseas standards are reviewed and the inconsistencies across the different pipe material types are discussed. Initial results of an ongoing test on root penetration are also presented. Finally, the performance requirements for elastomeric pipe joints for pressure and non-pressure pipeline system components are proposed.

2 CURRENT PERFORMANCE REQUIREMENTS

The performance requirements of pipes and fittings are controlled in the long-term by the performance of the elastomeric joint systems which have been used to define the requirements for both pipes and fittings. The major factor that appears to have been of concern in current standards is pressure testing, with smaller emphasis being placed on other performance requirements.

2.1 *Pipes and fittings*

Pipes and fittings classified for pressure use (AS 1477, 2280, 2544, 2977, 1579) have a required pressure resistance within the range 0.45–8.5 MPa.

Pipes and fittings classified for dual pressure/non-pressure use (AS 3571, 4058) have a required pressure resistance within the range 0.9–7.4 MPa.

Pipes and fittings classified for non-pressure use only (AS 1254, 1260, 1415, 1273, 1631, 1741, 4139) have a required pressure resistance within the range 0.03–0.350 MPa.

2.2 *Joints*

The performance requirements for joints can be divided into three categories.

(a) Joints that are subject to an application of pressure only:
  - internal pressure (ISO/DIS 44225 and AS 2544, 2977, 1477, 1254, 1415, 1273, 1631, 4139, 1579);
  - external pressure (ISO/DIS 4422-5 and AS 1260) – PVC sewer will change to pressure differential; and
  - internal vacuum (AS 2977, 1477).

(b) Joints that are subject to both pressures and the combination or one of the following additional mechanical load conditions (AS 2280, 4058, 3571, 1741):
  - a deflection test in which two lengths of pipe, axially aligned through a joint, are deflected by a particular angle about a fulcrum that is located within the joint;
  - a straight draw test in which two lengths of pipe, axially aligned through a joint and fully engaged, are separated longitudinally by a particular displacement; and
  - a shear resistance test in which two lengths of pipe, axially aligned through a joint, are subject to a shear load perpendicular to the axis at the joint.

(c) Joints that have a performance requirement of the resistance to tree root penetration (AS 1260, 1741). The objective of this performance requirement is to produce a standard average interface pressure between the rubber ring and the pipe surface, so that the same quantity can be estimated at 50 years after jointing, taking into account stress relaxation, creep and joint movement.

The first two categories fall into the class of positive internal pressures. There are, however, two additional variations in the method of pressure application. They are external pressure and an internal vacuum of 0.09 MPa respectively.

Joints classified for pressure use, dual pressure/non-pressure use (AS 3571, 4058) and non-pressure use, have a required pressure resistance within the range 0.08–7.2 MPa, 0.005–7.5 MPa and 0.03–0.35 MPa, respectively.

2.3 *Remarks*

It is seen that the current performance requirements of pipes, fittings and joints given in the standards have the following inconsistencies across the different pipe material types:
  - performance requirements are not uniform; and
  - pressure ranges for hydrostatic pressure tests vary significantly.

With regard to the requirement of draw (AS 1741, 3571), CSIRO experience shows that large diameter pipes are not easily joined, even without the additional requirement for draw. A large number of terracotta pipes have also been joined and they are extremely difficult to separate once joined. It is therefore suggested that the requirement of draw be omitted.

3 ROOT PENETRATION TEST

For rubber ring joints to resist plant root penetration, it has been reported and promulgated into Australian Standards that a sufficiently high pressure at the contact between the rubber and pipe must be maintained. However, the evaluation of the appropriate interfacial pressure to be used has been under discussion for some time and different values have been used in different standards (see Table 1). This matter has been the subject of considerable controversy, and decisions on the interface pressure used will affect the performance and ease of insertion of a rubber ring joint.

Table 1. Contact pressure values in current codes.

Standard title	Initial contact pressure (MPa)	Contact length (mm)
AS 1462.13 and AS 1741	0.55	7
NZS 7649:1988	0.4	4
ASTM C425	0.21	–

To clarify the appropriate interfacial pressure value to be used, interface pressures of 14 joints (DN 100) have been assessed in accordance with NZS 7649:1988 – 1 mm Hole Test. Currently these joints for PVC, VC and FRC are being tested for a 12-month period by 'accelerated horizontal root ingression' in Adelaide Botanic Gardens. To examine the effect of the porosity of pipe materials on root intrusion into the joint system, the pipe surface of VC and FRC was coated with two-part epoxy resin (40% diluted with solvent; Intergard Ez Epoxy (4): En-cap Ac Sealer (1)). The initial interfacial pressure results are summarised in Table 2.

Table 2. Test results* for interfacial sealing pressure assessment (IFSP).

Pipe type	Joint type	Contact length (mm) when IFSP >0.4 MPa	Interfacial pressure (MPa)	
			Min.	Max.
uPVC	Slip coupling (SWV)	<1	0.37	0.54
	NZ joint	7.54	0.723	0.82
	Australian sewer joint	4.5	0.44	0.85
VC	Maximum collar	7.84	0.7	>0.8
	Minimum collar	9.46	>0.8	>0.8
VC (coated)	Maximum collar	7.84	0.7	>0.8
	Minimum collar	9.46	>0.8	>0.8
FRC	Maximum collar	16	0.73	0.73
	Minimum collar	14	0.855	0.915
FRC (coated)	Maximum collar	16.25	0.735	0.77
	Minimum collar	13.9	0.845	0.9

* The values in the table are the average of two samples for FRC and VC joints, and five samples for PVC joints, and the contact length is the sum of all the contacts between the rubber and pipe.

To measure the effects of creep of the pipe material and relaxation of the rubber ring, another test on long-term interface pressure evaluation of assembled rubber ring joints is also currently being performed.

4 RECOMMENDATION FOR UNIFORM TEST REQUIREMENTS FOR DIFFERENT WATER/SEWER PIPES

From the studies outlined previously, the recommended performance requirements for elastomeric pipe joints for pressure and non-pressure pipeline system components, in terms of two pipe type categories – rigid and flexible materials – are given in Table 3. The test method for assessing the elastomeric joint performance at the recommended maximum deflection, as given in Table 4, is based on the ASTM D3212 method and the apparatus is illustrated in Figure 1. The test procedures suggested by the authors are summarised in the following.

Table 3. Test requirements for different water/sewer pipes.

Test	Pressure pipe application		Non-pressure pipe application	
	Test pressure	Time	Test pressure	Time
Flexible/semi-flexible material				
Hydrostatic pressure & diameter distortion	$\beta\,Pw$	1 hour	85 kPa	1 hour
	$\alpha\,Pw$	1000 hours		
Negative pressure & diameter distortion	−85 kPa	2 hours	−85 kPa	1 hour
Hydrostatic pressure & angular deflection	$\beta\,Pw$	1 hour	85 kPa	1 hour
	$\alpha\,Pw$	1000 hours		
Negative pressure & angular deflection	−85 kPa	2 hours	−85 kPa	1 hour
Rubber ring initial interface pressure	Not required		0.4 MPa	0.5 hour
Rigid material				
Hydrostatic pressure & angular deflection	$\beta\,Pw$	1 hour	85 kPa	1 hour
	$\alpha\,Pw$	1000 hours		
Negative pressure & angular deflection	−85 kPa	2 hours	−85 kPa	1 hour
Rubber ring initial interface pressure	Not required		0.4 MPa	0.5 hour

Note: Pw is the working pressure specified by the water authorities or taken from relevant standards. The coefficients α and β are the multipliers for test pressure which are defined in Appendix A.

Table 4. Joint deflections

Pipe materials	Pressure application		Non-pressure application	
	Angular deflection ϕ deg. (°)	Diameter distortion γ (% of OD)	Angular deflection ϕ deg. (°)	Diameter distortion γ (% of OD)
Rigid (VC, concrete)	At manufacturer's maximum recommended deflection	NA	At manufacturer's maximum recommended deflection	NA
Semi-flexible (plastics)	At manufacturer's maximum recommended deflection	3	At manufacturer's maximum recommended deflection	7.5
Flexible (steel, ductile iron)	At manufacturer's maximum recommended deflection	3	At manufacturer's maximum recommended deflection	7.5

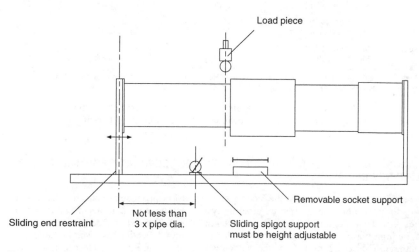

Figure 1. Test apparatus.

(a) Assemble the components of the test assembly to the manufacturer's recommendations.
(b) Perform the required tests as detailed in Table 3 on two properly jointed assemblies.
(c) Apply a distorting load so as to cause a γ% deflection reduction of the original outside diameter when measured at the end of the beam remote from the socket face (*diameter distortion*) or an angular deflection ϕ (degree, °) measured from component to component appropriate to nominal outside diameter (*angular deflection*) as given in Table 4. The test method for assessing the elastomeric joint performance at the recommended maximum deflection is based on the ASTM D3212 method and the apparatus is illustrated in Figure 1.
(d) While maintaining the distortion or deflection of the joint, subject the test assembly to the internal pressure and time requirements as given in Table 3.
(e) While maintaining the distortion or deflection of the joint, subject the test assembly to the internal vacuum and time requirements as given in Table 3.
(f) No leakage shall occur in either the internal pressure or internal vacuum test.
(g) Determine the rubber ring initial interfacial pressure in accordance with AS/NZS 1462.13.

5   CONCLUSION

The current performance requirements given in the standards are not uniform and the pressure ranges for hydrostatic pressure tests vary significantly across the different pipe material types.

Different values of interfacial pressure and contact length required to evaluate the resistance of rubber ring joints to plant root penetration have been used in different standards (Table 1), and a research program initiated by the authors aiming at clarifying this value is currently underway.

A proposal for uniform test requirements for elastomeric pipe joints for pressure and non-pressure pipeline system components has been presented.

ACKNOWLEDGMENT

The authors are grateful to the financial support of Melbourne Water, Quality Systems Team Asset Management Division, Water Services, and the Urban Water Research Association of Australia.

REFERENCES

ASTM 1991. ASTM C425, Standard specification for compression joints for vitrified clay pipe and fittings. Philadelphia: American Society for Testing and Materials.
Burn, S. 1995. Specific comment on draft on PVC pipes and fittings for drain, waste and vent – elastomeric joint tests.
ISO 1992. ISO/DIS 4422–5, *System standard for water supply – unplasticized poly(vinyl chloride) (PVC-U): Part 5 – Fitness for purpose of the system*, International Organization for Standardization.
Standards Australia 1982. AS 2544, *Grey iron pressure pipes and fittings*. Sydney: Standards Australia.
Standards Australia 1992. AS 4058, *Precast concrete pipes (pressure and non-pressure)*. Sydney: SAA.
Standards Australia 1984. AS 1260, *Unplasticized PVC (UPVC) pipes and fittings for sewerage applications*. Sydney: SAA.
Standards Australia 1984. AS 1415, *Unplasticized PVC (UPVC) pipes and fittings for soil, waste and vent (SWV) applications*. Sydney: SAA.
Standards Australia 1988. AS 1477, *Unplasticized PVC (UPVC) pipes and fittings for pressure applications*. Sydney: SAA.
Standards Australia 1988. AS 2977, *Unplasticized PVC (UPVC) pipe for pressure applications – compatible with cast iron outside diameters*. Sydney: SAA.
Standards Australia 1989. AS 3571, *Glass filament reinforced thermosetting plastics (GRP) pipes – polyester based – water supply, sewerage and drainage applications*. Sydney: SAA.
Standards Australia 1991. AS 1254, *Unplasticized PVC (UPVC) pipes and fittings for storm and surface water applications*. Sydney: SAA.
Standards Australia 1991. AS 1273, *Unplasticized PVC (UPVC) downpipe and fittings for rainwater*. Sydney: SAA.
Standards Australia 1991. AS 1741, *Vitrified clay pipes and fittings with flexible joints – sewer quality*. Sydney: SAA.
Standards Australia 1991. AS 2280, *Ductile iron pressure pipes and fittings*. Sydney: Standards Australia.
Standards Australia 1993. AS 1579, *Arc welded steel pipes and fittings for water and waste water*. Sydney: SAA.
Standards Australia 1993. AS 4139, *Fibre-reinforced concrete pipes and fittings*. Sydney: Standards Australia.
Standards Australia 1994. AS 1631, *Cast grey and ductile iron non-pressure pipes and fittings*. Sydney: SAA.
Standards New Zealand 1988. NZS 7649, *Unplasticized PVC sewer and drain pipe and fittings*. Wellington: Standards New Zealand.

APPENDIX A

The coefficients α and β are the multipliers for test pressure which are defined in such a way that an adequate performance of the assembled rubber ring joints can be ensured during the service life (50 years). That is, the test is carried out under such conditions that the expected rubber stress relaxation and the creep of the pipe material after 50 years are reached in 1000 hours. Based on either the rate of compression stress relaxation of rubber ring (defined in Table 3.2 in AS 1646–1992; denoted as R_r here) or the creep properties of pipe materials, α is defined as the larger of $(1 - 3\,R_r)/(1 - 5.64\,R_r)$ which is $(\sigma_{1000}/\sigma_{50year})_r$ derived from eq. (A1), and $(\sigma_{1000}/\sigma_S)_p$, and β is defined as the larger of $1/(1 - 5.64\,R_r)$ which is $(\sigma_0/\sigma_{50year})_r$, and $(\sigma_0/\sigma_S)_r$, where t is the time in hours, σ_0 is the initial rubber or pipe stress (taken as the stress at 1 hour), σ_s is the design stress of pipe material, and σ_{1000} and σ_{50year} are the rubber or pipe stresses at 1000 hours and 50 years, respectively. The subscripts 'r' and 'p' stand for rubber and pipe, respectively. These factors should be rounded to the next highest 0.05.

$(\sigma_0/\sigma_S)_p$ and $(\sigma_{1000}/\sigma_S)_p$ are determined based on the approach given in ISO/DIS 4422–5, where the long-term behaviour of the pipe material is incorporated by considering the creep strain at the operating stress, applied over the design life. The strain in the pipe material according to the design stress (σ_S) in the system for 50 years can be determined from isochronous stress/strain diagrams for the service temperature.

Instead of a safety factor for the test, an additional test strain of 0.5 times the calculated strain is added. The test stress for 1 hour (σ_0) and 1000 hours (σ_{1000}) are again determined from the isochronous stress/strain diagram.

Rubber:
$$\sigma = \sigma_0 \,(1 - R_r \log t) \tag{A1}$$

A.1 Calculation Example for Rubber Ring Joined PVC Pipe

Rubber   Take R_r as given in AS 1646–1992, i.e. 6%, $\alpha = (1 - 3\,R_r)/(1 - 5.64\,R_r) = 1.24$, and $\beta = 1/(1 - 5.64\,R_r) = 1.51$.

PVC pipe   design stress σ_S: 11.0 MPa (AS 2566.1)
strain (ε) at an induced stress equal to σ_S at 50 years: 0.483
additional strain for test (ε_A): 0.242
strain value for test $(\varepsilon_T = \varepsilon + \varepsilon_A)$: 0.725
test stresses (σ_0) and (σ_{1000}) related to: 22.75 MPa and 18.3 MPa
$\alpha = \sigma_{1000}/\sigma_S = 1.7$ and $\beta = \sigma_0/\sigma_S = 2.10$

As the values of α and β calculated are larger for the PVC pipe than those for rubber, the multipliers for test pressure shall be those from PVC, i.e. $\alpha = \mathbf{1.7}$ and $\beta = \mathbf{2.10}$.

The Mechanics of Structures and Materials, Grzebieta, Al-Mahaidi & Wilson (eds)
© 1997 Balkema, Rotterdam, ISBN 90 5410 900 9

Lifetime analysis of uPVC pressure pipes constrained by a tapping band

L.S. Burn & J.P. Lu
CSIRO Building, Construction and Engineering, Melbourne, Vic., Australia

ABSTRACT: Finite element analysis and strain gauging have been conducted on a uPVC pipe installed with a tapping band to allow determination of the imposed stress levels in a pipe that had failed in the field. These stresses were used in a fracture mechanics approach to determine if the applied stresses were the critical factor controlling pipe failure. It is shown that finite element modelling can be used to effectively predict the operating stress levels in pipelines constrained by tapping bands and it can pinpoint stress concentration areas which are likely sites for crack initiation and propagation. Combined with a fracture mechanics approach, the lifetimes of plastic pipes constrained by tapping bands can be predicted.

1 INTRODUCTION

Unplasticised PVC (uPVC) pipes have proved to have advantages over more traditional pipe materials in terms of cost, corrosion resistance and ease of installation, and have been used extensively worldwide in pressure applications. In order to provide service connections between water service pipes and water mains, tapping bands have been widely used in PVC pressure pipeline applications. However, there have been a number of brittle failures which are associated with pressure tapping for service connections.

Brittle failure of materials results from a combination of stress, flaws and a reduction of the pipes material characteristics with time. In regions of high stress, such as those at the edge of the hole under a tapping point, conditions are ideal for crack initiation, propagation and early pipe failure. To assess the likelihood of failure, static stress analysis in conjunction with fracture mechanics models are essential tools to allow failure investigation and lifetime prediction of plastics pipelines.

In this paper, finite element analysis and strain gauging results are discussed for a 200 mm Class 20 uPVC pipe subjected to operating stress levels and the application of a tapping band. Utilising these results, lifetime prediction models are developed to determine if the stress level imposed by the tapping band is the factor controlling failure.

2 STRAIN GAUGING

To assess the stresses imposed by the tapping band in a pipe (without a hole under the tapping point) and confirm the accuracy of the subsequent finite element analysis, hoop strains at different locations inside the pipe were measured by using SHINKOH strain gauges with a gauge length of 5. An internal pressure of 0.7 MPa was applied to simulate the maximum water head which was about 66 m. While maintaining the above pressure, the tapping band was tightened onto the pressurised pipe.

Figure 1 shows details of the pipe sample and the locations of the strain gauges. The output of strain gauges was automatically logged using a DT 505 datataker.

3 FINITE ELEMENT ANALYSIS

An ANSYS finite element package which can model both material and geometric non-linearity was employed to study the stresses imposed by the tapping band in the PVC pipe. In this analysis, because of the symmetry, only one-quarter of the sample needed to be modelled. To ensure that accurate stresses in the structure could be calculated and the concentration close to the point of tapping could be modelled in sufficient detail, three-dimensional solid model brick elements were used for mesh generation to allow efficient model definition. Greater mesh refinement was adopted near the tapping point and coarser mesh was used away from the tapping point. Two to three layers of elements through the wall thickness were used to increase the accuracy. ANSYS's error estimation technique was employed to ensure that the amount of solution error due to mesh discretisation was small in the region of high stress.

Because the steel tapping band was much stiffer than the pipe material, a condition of fixed, rotationless and immovable constraint was applied along the contact points between the pipe and the tapping band. The constraint force caused by the tapping band was approximated by assigning a measured displacement to the rubber ring.

Symmetry boundary conditions with zero out-of-plane translations and in-plane rotations were applied to all nodes on the symmetry planes. The boundary conditions imposed on the model were fully built-in (zero translations and rotations) at the ends of the pipes. A uniform normal pressure of 0.7 MPa was applied along the internal surface of the model, together with a node displacement of $D_y = -3.5$ mm at the top of the rubber ring to simulate the constraint force caused by the tapping band on the rubber ring.

As rubber is hyper-elastic, it requires the finite element analysis (FEA) code to model the complex material properties such as near incompressibility, large strains and displacements, and generalised strain energy functions. In this study, the three-dimensional element HYPER 86 and two-term Mooney–Rivlin strain energy function are used. The Mooney–Rivlin constants, taken from Raos (1993), are C1 = 0.333 MPa and C2 = 0.0141 MPa.

Table 1 shows the hoop strains corresponding to the locations shown in Figure 1 obtained from both finite element analysis and strain gauging for the pipe without a hole drilled under the tapping band. The reasonably good agreement between the two methods gives us confidence to apply the FEA to the situation of a pipe with a tapping band and hole, as experienced in practice. The Von Mises contour displays of the pipe without a drilled hole is shown in Figure 2 to show how the equivalent stresses vary over the model. It is seen that a very high stress con-

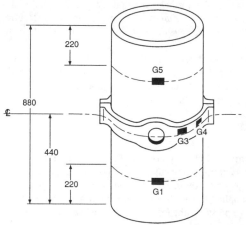

Figure 1. Detail of pipe sample showing location of strain gauges.

centration of up to 8.98 MPa occurs in the internal pipe under the tapping band. If a hole is then drilled in the pipe, as shown in Figure 3, the maximum stress is increased to 19.05 MPa for the same loading condition. If the internal pressure is increased from 0.7 MPa to 1.2 MPa, as could occur with a pressure surge, then the maximum stress rises to 28.14 MPa at the edge of the hole under the tapping band.

Table 1. Hoop strain obtained from FEA and strain gauging

Location	Hoop strain (FEA) (με)	Hoop strain (gauging) (με)
G1	1364	1916
G3	−808	−984
G4	1707	1916

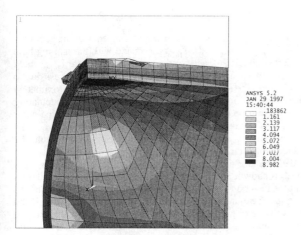

Figure 2. Von Mises stress contour plot – pipe without hole.

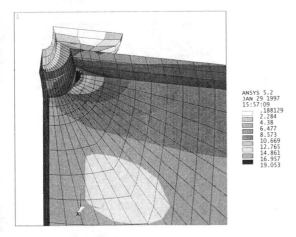

Figure 3. Von Mises stress contour plot – pipe with hole.

65

4 LIFETIME ANALYSIS

The lifetime expected from a pipe in use, is determined by the interaction between the levels of flaws or installation damage in the pipe, by the crack initiation time, the rate at which the crack propagates through the pipe wall, the material properties of the pipe and how these change with time, and the stress conditions under which the pipe operates.

The initiation and propagation of cracks is controlled by the fracture toughness of the material and the pipeline operating conditions. Under static pressures, the fracture toughness (K_{IC}) of the material decreases with time, as shown in Figure 4, where the fracture toughness of the material was measured according to BS 3505 (1986). Under constant stress the fracture toughness decreases, until a point is reached where the stress associated with a flaw equals the fracture toughness, at which point a crack begins to grow.

If the equation for the fracture toughness (K_{IC}) is known, then the size of flaw needed to cause failure or the stress to cause failure associated with a certain size flaw can be calculated from the following equation:

$$K_{IC} = Y_{flaw}\sigma\sqrt{\pi a}$$

where σ is the stress at a flaw of size a and Y_{flaw} is a geometric factor associated with the geometry of the specimen. For a flaw close to the pipe inner wall, Y_{flaw} associated with the tapping hole can be determined from Rooke and Cartwright (1976), and for the case examined here Y_{flaw} would be approximately 1.0. Because the pipe inner wall is subjected to a residual tensile stress due to manufacture of approximately 1.0 MPa, and the stress changes to compression of approximately 3.5 MPa at the pipe outer wall, these stresses needs to be taken into account when calculating the flaw size that would cause the crack to propagate from the pipe inner wall to the pipe outer wall, as discussed by Burn (1994). Also to be taken into account are the stresses imposed on the pipe wall due to the deformations associated with the installation of the tapping band and the stresses imposed on the pipe wall due to the internal hydrostatic pressure.

In Figure 5 the total stress needed to cause a crack to propagate from a range of flaw sizes at the pipe inner wall to the outer wall is shown, and it can be seen that the required stress levels needed to cause failure decrease in time in accordance with the reduction in fracture toughness.

The operating conditions of the pipe correlate to an operating hoop stress of 3.9 MPa. If the residual stresses existing in the pipe wall due to production conditions are taken into account, the hoop stress is 4.9 MPa at the inner wall and 0.4 MPa at the outer wall. Stresses associated with the tapping band and drilled hole raise the stress level at the pipe inner wall to 20.0 MPa for an internal hydrostatic pressure of 0.7 MPa. At this level of operating stress, the stresses in the pipe

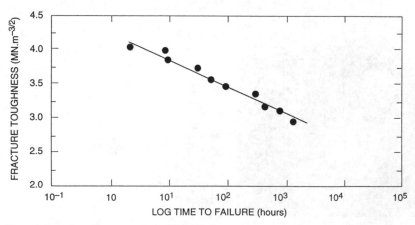

Figure 4. Reduction of fracture toughness (K_{IC}) with time.

66

wall are below the levels needed to cause a crack to propagate from a flaw and cause failure at the one-year, post-installation, time frame encountered in practice, unless a very large flaw exceeding 2.0 mm in size exists. As seen in Figure 6, a flaw of this size does not exist at the crack initiation point.

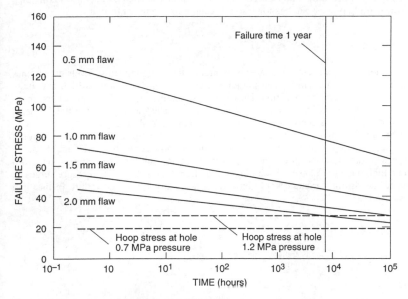

Figure 5. Failure stress versus time curve for different flaw sizes.

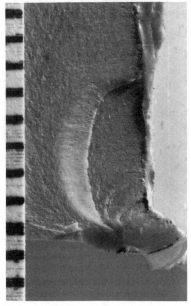

Figure 6. Failure initiation point at drilled hole on pipe inner wall.

It is only if the pipe is placed under additional levels of stress that crack propagation and failure would be likely. There are two likely causes for this additional stress: higher operating pressures, such as water hammer associated with the tapping procedure; or stress imposed on the pipe wall due to the drilling procedure. If the internal pressure of the pipe was raised to 1.2 MPa, as shown in Figure 5, then a 2.0 mm flaw could cause failure. The stresses imposed on the pipe wall due to the drilling process are unknown so this cannot be discussed further.

5 CONCLUSIONS

Finite element modelling can be used to effectively predict the operating stress levels in pipelines constrained by tapping bands and it can pinpoint stress concentration areas which are likely sites for crack initiation and propagation. Utilising finite element techniques, high stress concentration of up to 19.0 and 28.1 MPa are predicted to occur at the edge of the hole under a tapping band in a uPVC pipe operating at pressures of 0.7 and 1.2 MPa respectively.

Using a fracture mechanics approach, the lifetimes of plastic pipes constrained by tapping bands can be predicted, provided that the operating conditions and material properties are known.

For the case examined here, the stresses associated with applying the tapping band are unlikely to be high enough to have caused failure. Additional factors such as pressure surges or stresses due to the drilling process would have been necessary to initiate failure in the time frame experienced.

ACKNOWLEDGMENT

The authors wish to thank Sydney Water for help in funding this work.

REFERENCES

British Standards Institution 1986. BS 3505 *Unplasticized polyvinyl chloride (PVC-U) pressure pipes for cold potable water.*
Burn, L.S. 1994. Lifetime prediction of uPVC pipes – experimental and theoretical comparisons. *Plastics, Rubber & Composite Processing & Applications* 21: 99–108.
Raos, P. 1993. Modelling of elastic behaviour of rubber and its application in FEA. *Plastics, Rubber & Composites Processing & Applications* 19: 293–303.
Rooke, D.P. & Cartwright, D.J. 1976. *Compendium of stress intensity factors.* London: HMSO.

The Mechanics of Structures and Materials, Grzebieta, Al-Mahaidi & Wilson (eds)
© *1997 Balkema, Rotterdam, ISBN 90 5410 900 9*

Cyclic deformation behaviours of some non-wood natural fibre-reinforced composites

T.D.Ip, B.C.Tobias, J.He & T.Liu
Department of Mechanical Engineering, Victoria University of Technology, Melbourne, Vic., Australia

ABSTRACT: A series of experimental studies on the mechanical properties of non-wood natural fibre-reinforced composites has been carried out using Abaca, banana and rice hull fibres embedded randomly in various thermoset resin matrice. The strain softening characteristics of these composites were evaluated under quasi-static load cycles with combined modes of tension and compression using a servohydraulic universal tester. The aim of these studies is to observe the ability of different natural-fibre composites to arrest fracture and delay crack growth. This progressive damage process is manifested by the strain softening characteristics and observed under programmed strain-controlled load cycles which can be accurately illustrated by monitoring the values of elastic modulus under different strains.

1 INTRODUCTION

In this time of diminishing resources and population explosion, efforts can be made to minimize wastes to the environment and maximize uses of natural resources. Natural fibres are available in abundance in Australia and many parts of the world where agriculture and forests are their major natural resources. Wood fibres have been heavily used throughout the history of paper manufacturing. Non-wood plant fibres have also been used in papermaking, but to a much less extent due to the higher costs associated with the production processes and uneven fibre supplies in bulk volumes. Only a few countries which apply stringent government import restrictions on wood pulp produce non-wood based paper in large quantities. China and India alone produce nearly half of the 10 million tonnes of annual world production of non-wood based paper. Applications of non-wood fibres other than papermaking have been limited. It was estimated that non-wood plant fibres, e.g. plant straw, sugarcane bagasse, bast fibres (jute, etc.), corn, sorghum and cotton stalks, papyrus and leaf fibres, account for an annual world production (1982 estimate) of some 2,270 million tonnes (Atchison 1983). Most of these fibres are wasted and burnt after harvest.

One of the growing interest in the application of non-wood fibres has been in fibre-reinforced composites. Natural fibres, known to have lower elastic modulus than most man-made fibres and less effective in taking up transferred loads from the composite matrix, can be significant in bridging microcracks thus delaying critical failure in cementitious materials (Cotterell & Mai 1996) and perhaps in other materials. In this study, the strain softening characteristics of selected non-wood based fibre composites were evaluated under quasi-static loading using an Instron 8501 servohydraulic universal tester with DT–Vee control platform. The aim of this study is to observe the ability of different types of fibre samples (banana, rice hull and abaca) with similar geometric configurations by their force-displacement

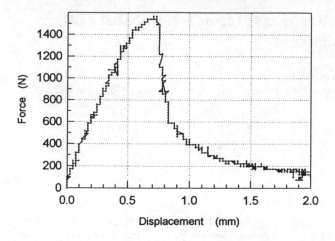

Figure 1. Force-displacement relationships under tension for banana-fibre composite

relationships (Figure 1), and the ability of these fibres to arrest fracture and delay crack growth, before and after reaching the critical stress, respectively. This progressive damage process is manifested by the strain softening characteristics and observed under programmed strain-controlled load cycles which can be accurately illustrated by monitoring the elastic modulus values under different strains, using an approach for strain-softening characterisation of hard rock (norite) and relatively soft sandstone samples (Bieniawski 1971) by cyclic tension-compression uniaxial load tests (Figure 2). Based on the relationships between the effective elastic moduli of each composite at various strains, it can be shown whether failure is dominated by macrocracks (as the case of most metals) or by different modes such as fibre-breaking, debonding, or matrix cracking, and whether the sample has undergone catastrophic or progressive failure.

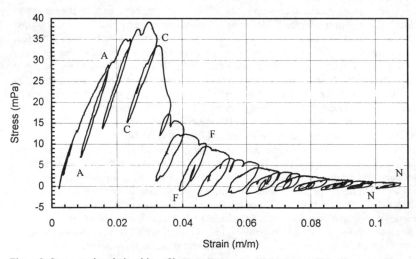

Figure 2. Stress-strain relationships of banana-fibre composite under cyclic loading

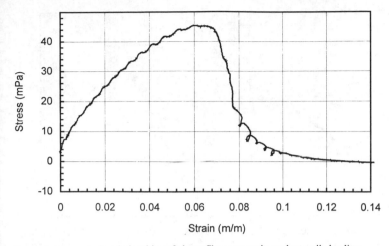

Figure 3. Stress-strain relationships of abaca-fibre composite under tensile loading

2 DESIGN OF EXPERIMENT

The experiment comprises two parts: quasi-static tension loading and cyclic tension-compression loading (e.g. Figure 2). Three types of natural based composites were tested: banana, rice hull and abaca fibre-based samples. They were prepared by embedding the fibres randomly in hybrid amino-phenolic based (rice hull and banana) or polyester based (abaca) resin matrix before pressed to required thickness (3 mm for rice hull and banana fibre samples and 5 mm for abaca samples) by a hydraulic platen, with an objective density of 0.8 grams per cubic centimeters for banana and abaca composites, and 1.2 grams per cubic centimeters for rice hull composites. Test specimens were prepared by cutting to sizes of 50 mm by 10 mm for rice hull and abaca specimens and 50 mm by 12 mm for banana specimens. Test lengths between the upper and lower grips of the Instron tester were set at 20 mm for all specimens.

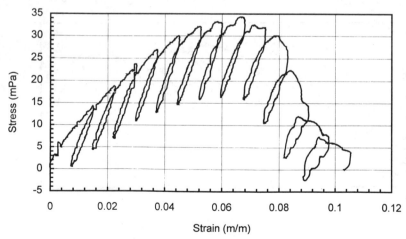

Figure 4. Stress-strain relationships of abaca-fibre composite under cyclic loading

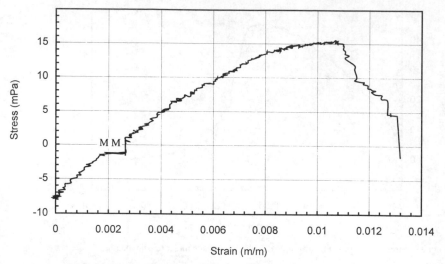

Figure 5. Stress-strain relationships of rice hull-fibre composite under tensile loading

All tension-only tests were performed with a constant strain rate of 5 x 10^{-4} /sec by setting the tension speed at 0.01 mm per second. For each cycle during cyclic load tests, abaca and banana specimens were initially stretched under tension for 0.3 mm, followed immediately by a compression of 0.15 mm, at an advancement to retreat ratio of 2. In other words, net elongation during each cycle is 0.15 mm. The tension/compression speeds were unchanged at 0.01 mm per second. Because of the inherent weakness of structural strength and brittleness for rice hull specimens, it was decided to reset the deformation under tension and compression to 0.15 mm and 0.075 mm, respectively, for each cycle, but maintaining the same tension/compression speeds at 0.01 mm per second, and the same advancement to retreat ratio of 2. Also, the initial grip was set slightly under compression for all rice hull sample tests, in order to extend the range of stress-strain curves.

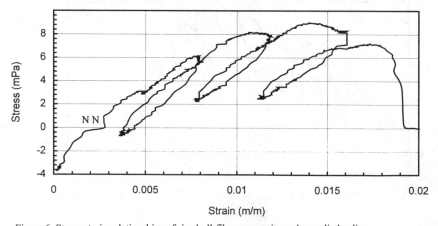

Figure 6. Stress-strain relationships of rice hull-fibre composite under cyclic loading

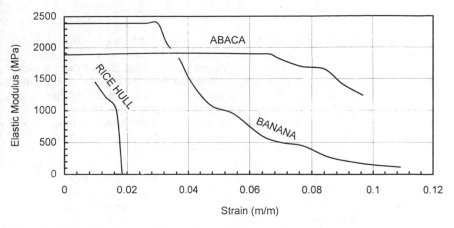

Figure 7. Relationships between effective elastic moduli and engineering strains (cyclic tests)

3 RESULTS AND DISCUSSION

Results from the six test modes were obtained in the form of force-deformation curves, such as Figure 1 for banana-fibre specimen under tension-only test. They were then transformed into stress-engineering strain curves, as shown in Figures 2 to 6, having exactly the same shapes as their corresponding force-deformation curves since changes in dimensions were negligible. Necking was only observed in the case of abaca sample tests, but the change is small enough to be negligible.

3.1 *Stress-strain relationships*

Stress-strain curves of typical tension-only (TO) runs are shown in Figure 3 (abaca) and Figure 5 (rice hull) and compared with those of cyclic (C) runs in Figure 2 (banana), Figure 4 (abaca) and Figure 6 (rice hull).

The TO stress-strain curve for banana fibre is not shown. It reaches a tensile strength of 41 MPa before failure. There were no significant change in shape nor stress-strain values between TO and C (reaching a tensile strength of 39 MPa before failure) runs for banana fibres, which is the least affected by cyclic loading. It displays almost linear stiffening effect up to failure for both the TO and C runs, failing at an engineering strain of 3.0 %, comparable to typical low modulus epoxy composites (Daniel & Ishai 1994). However, it suffers catastrophic failure, dropping 60% of its strength from an increase in strain from 3.0 % to 3.2 % (Figure 2).

For the two rice hull runs, a slight instability was recorded in each case ("MM" and "NN" in Figures 5 and 6, respectively), which is the result of unstable control signal input during the initial tension loading, as the grips were pre-set at slight compression. For the TO case (Figure 5), it failed at a strain of 1.1 %, after a linear stiffening, reaching an tensile strength of 15 MPa and completely fractured. For the C run, it failed at a strain of 1.4 % (Figure 6), reaching a tensile strength of 9 MPa.

For the abaca runs, the comparatively stronger capability of abaca fibre to arrest failure is shown (Figures 3 and 4), reaching a tensile strength of 45 MPa (at 6.8 % strain) for TO run and 34 MPa (at 6.7 % strain) for C run. It is interesting to note that, for TO run, it loses 78% of its strength in 1.6% of strain after failure. But for C run, it loses 88% of its strength in a much larger strain (3.6%) after reaching its tensile strength, failing by more progressive mode.

3.2 Effective elastic moduli

The effective (bulk and shear) moduli of a composite as obtained from experimental results, whether it is isotropic or heterogeneous in nature, is able to describe its overall, macro-scale behaviour (Kohn 1988). From the stress-strain curves of cyclic tests, the fact that the cyclic loops are narrow permits the conclusion that their slope represents the effective moduli at that particular strain. Thus, a relationship between the effective modulus (bulk modulus in our case) and strain throughout the damage process can be established. As the fibres in a composite normally take up most of the load from the matrix, it may be safe to suggest that a drastic decrease in effective modulus beyond the tensile strenth signifies the start of fibre breaking. This can be illustrated by the cyclic banana run (Figure 2). The parallel slopes from AA to CC suggest that the failure is primarily due to matrix cracking or debonding, and to fibre breaking beyond CC with drastic (but not catastrophic) failure, followed by a progressive failure mode due to all three mechanisms. We recall that the shape of the stress-strain curves for both the tension-only and cyclic runs are very much the same, suggesting that both cases underwent similar failure mechanisms.

For the rice hull cyclic run (Figure 6), the effective moduli started to change even before tensile strength is reached, suggesting that the matrix micro-cracking started upon loading and resulting in a macrocrack (fracture) with catastrophic failure.

For the cyclic abaca run (Figure 4), there is no substantial change in the slopes either before or after the tensile strength was reached, suggesting that the failure mechanisms were primarily due to debonding and/or matrix cracking only. This was confirmed by observation on the specimen after the test, with no noticable breakage in all the fibres. By comparing the tension-only and cyclic test results, it is clear that the significant drop in strength for the case of cyclic abaca tests is primarily due to debonding during the advance-retreat deformations.

4 SUMMARY AND CONCLUSIONS

A series of effective elastic moduli for abaca-, banana- and rice hull-based composites was generated throughout their damage processes by using cyclic tension-compression load tests. These cyclic test results (Figure 7), together with those obtained from tension-only tests, can be used to identify different failure mechanisms. A convex slope (abaca) in Figure 7 suggests a gradual drop due to debonding and/or matrix cracking, and any sharp drop which follows may signify the start of fibre breaking, while a concave slope (banana) suggests a drastic drop as a result of an initial dominating fibre breakage, followed by a more progressive damage. Any catastrophic failure will be shown by a vertical drop in strength, as in the case of rice hull specimen.

REFERENCES

Atchison, J.E. 1983. Data on non-wood plant fibres. In M.J. Kocurek & C.F.B. Stevens (eds), *Pulp and paper manufacture, vol.1: Properties of fibrous raw materials and their preparation for pulping*. Atlanta: Tappi Press.

Bieniawski, Z.T. 1971. Deformational behavior of fractured rock under multiaxial compression. In M. Teeni (ed), *Structures, solid mechanics and engineering materials*: 589-598. New York: John Wiley.

Cotterell, B. & Mai, Y.W. 1996. *Fracture mechanics of cementitious materials*. London: Blackie Academic.

Daniel, I.M. & Ishai, O. 1994. *Engineering mechanics of composite materials*: 31. New York: OUP.

Kohn, R.V. 1988. Recent progress in the mathematical modeling of composite materials. In G.C. Sih, G.F. Smith, I. H. Marshall & J.J. Wu (eds), *Composite material response: constitutive relations and damage mechanisms*: 155-177. London: Elsevier.

The Mechanics of Structures and Materials, Grzebieta, Al-Mahaidi & Wilson (eds)
© *1997 Balkema, Rotterdam, ISBN 90 5410 900 9*

Fibre reinforced plastics – Development and properties

X.W.Zou, N.Gowripalan & R.I.Gilbert
School of Civil Engineering, The University of New South Wales, Sydney, N.S.W., Australia

ABSTRACT: Corrosion of steel reinforcement in concrete is a major problem in terms of durability of concrete structures. One of the most promising developments is the use of Fibre Reinforced Plastics (FRP) as reinforcement or prestressing tendon in concrete structures. FRP reinforcement consist of aligned continuous fibres, mainly carbon, aramid or glass, embedded in a resin matrix such as epoxy, polyester or vinyl ester. FRP possess high tensile strength, light weight, and are non-magnetic and non-corrodible. These properties can lead to maintenance free durable concrete structures.

This paper briefly describes the historical development, the production processes and the advantages of FRP. It also summarises the short-term and long-term mechanical properties of the common types of FRP reinforcement, together with some experimental results obtained by the authors. An equation for predicting the long-term creep coefficient of FRP tendons containing aramid fibres with epoxy resin matrix is proposed.

1 INTRODUCTION

Concrete has been extensively used in the construction industry for over a century, because it is cheap, strong in compression and relatively durable. Concrete elements or structures are commonly reinforced or prestressed with steel to carry the tensile forces. However, steel bars and tendons are vulnerable to corrosion problems. It has been estimated that in the USA, some 160,000 bridges have been seriously affected by corrosion of steel requiring a major repair program which has been estimated at US$20 billion. It has also been estimated that the annual cost of corrosion repairs was about one billion pounds in Europe (Clarke, 1993).

The corrosion problem of steel in concrete has led to several research projects to minimise or eliminate it. Some possible approaches involve the use of high quality concrete cover, stainless steel, epoxy coated steel reinforcement and cathodic protection. One of the most promising recent developments is the use of FRP reinforcement or tendons. Over the past twenty years, research and development on FRP reinforcement for concrete structures has been progressing rapidly. Japan, United States, Canada, Belgium, U.K., Germany, Netherlands and Switzerland have assumed a major role in terms development, research and applications while Australia has also been active in this area since 1993. Currently there are a number of manufacturers producing more than 20 different types of FRP as alternatives to the conventional steel reinforcements or tendons. To date more than 700 structures and bridges have been either constructed or repaired (Nanni, 1995) with FRP reinforcements or tendons.

This paper describes the historical development, the advantages of FRP over steel reinforcement, the production process, and the mechanical properties of FRP made of carbon (CFRP), aramid (AFRP), and glass (GFRP) fibres. Some preliminary experimental results on the mechanical properties obtained by the authors are discussed. The flexural behaviour of concrete beams prestressed with FRP tendons is discussed in a subsequent paper (Zou, Gowripalan and Gilbert, 1997).

2 HISTORICAL DEVELOPMENT

FRP materials have been one of the main focuses of the engineering community since the development of light weight, high strength, high stiffness fibres. Since the 1950's glass fibre was considered as a potential substitute for steel in reinforced or prestressed concrete structures. However, early studies noted problems with anchorage, surface protection of bars and bonding of glass fibres to concrete (Dolan, 1993).

There was a period of some 20 years where research in this field remained dormant. In the 1970's, corrosion of steel in concrete structures, particularly bridge decks, led to a renewed interest into design strategies and choice of materials that reduce structural susceptibility to corrosive environments (Dolan, 1993). By the 1980's, the corrosion resistant properties of FRP bars and tendons also took on a renewed interest. Many of the initial shortcomings have been overcome, resulting in renewed interest in the use of these materials in civil engineering.

Since the early 1990's many practical applications have emerged in various types of concrete structures, e.g. prestressed concrete bridges with prestressing tendons of FRPs such as the bridges in Canada (Rizkalla and Tadros, 1994) and the Ulenbergstrasse bridge in Germany (Wolff and Miesseler, 1993), architectural curtain walls and tunnel lining walls with FRP grid bars, prestressed concrete piers and pontoons for marine structures. FRP materials have also been used to repair and strengthen existing concrete structures.

3 ADVANTAGES AND SUITABLE APPLICATIONS

There are many advantages of FRP over steel as reinforcement or tendons. It has a high specific strength (ratio of strength to density). The high strength and low density of FRPs result in a specific strength that is 10 to 15 times higher than that of steel. Carbon and Aramid fibre based FRP (CFRP and AFRP) have good fatigue strength, as much as three times that of steel. They also have good corrosion resistance which is an important requirement for aggressive environments and reduces the maintenance cost of the structure. Unlike steel, FRP needs no protective coating. Its low thermal expansion gives another advantage in widely varying climatic conditions. Other advantages are its electromagnetic neutrality, high resistance against abrasion and excellent chemical resistance. FRP is easy to handle in construction due to its light weight.

Hence, FRP is suitable for exposed concrete structures in the presence of de-icing salts, in splash zones, in salt-containing concrete, when an accelerator (e.g. $CaCl_2$) is used, in non-magnetic and non-conductive elements and in other aggressive environments. FRP is also suitable as permanent grout anchors and for structures where fatigue resistance is important. The good fatigue strengths of CFRP and AFRP make them suitable for prestressing bridge girders or beams or reinforcing bridge decks.

However, at present the cost of FRP products is relatively high when compared with steel due to the high cost of production and small quantities produced. It is believed that the cost will reduce with time when it is used more often in practical applications.

4 PRODUCTION

FRP reinforcements are composite rods made of a high strength and high modulus fibre and a resin matrix. Fibres used for this purpose include carbon, aramid and glass. The resins include epoxy, polyester and vinyl ester. The FRP rods are produced mainly by a pultrusion method.

Commercially available fibres include carbon, aramid, glass, polypropylene, nylon, polyethylene, acrylic and polyester. Among these, carbon, aramid and glass fibres are used to produce FRP reinforcement because of their superior strength (higher than steel) and comparable stiffness. For example, carbon fibres have high stiffness and resistance to chemical attack. Aramid fibres are less dense than carbon fibres, have a toughness similar to that of glass fibres and have a modulus half of that of steel. Glass fibres tend to become weaker in the presence of an alkaline environment, but they are quite tough.

The primary role of the matrix in FRPs is to provide lateral support to the fibres and to protect the fibres from physical and chemical attack due to the surroundings. The matrix may also be used to impart desired physical properties to the FRP. Some of the important material characteristics to consider in selecting a matrix for a structural FRP are stiffness, strength, fracture toughness, thermal and electrical conductivity, upper service use temperature, coefficient of thermal expansion, processing temperature, chemical shrinkage during processing, ability to impregnate and bond to fibres, flame resistance and sensitivity to environmental factors such as moisture, chemicals, or ultraviolet radiation. One important factor to consider in the selection of a matrix for an FRP application is the relative mismatch in shrinkage or expansion between the fibre and the matrix that can occur during processing. One source of such mismatch is unequal coefficients of thermal expansion in the presence of a temperature change, and another is chemical shrinkage of the matrix during curing (in thermosets) or crystallite growth (in semi-crystalline thermoplastics) during processing (Bakis, 1993).

In the pultrusion method, the fibres are bonded together, drawn through a resin mix and then pulled through a shaping die. As the rods emerge from the shaping die, they pass through a curing chamber where the resin is allowed to harden. The pultrusion process is a viable technique with all the high modulus fibres and with a wide variety of resins and it allows considerable latitude in the selection of a structural shape (Dolan, 1993).

In the construction industry, much attention has been focused on pultruded rods with "deformed" or sand-coated surfaces which promote better mechanical bonding with concrete. Hybrid fibre FRPs can also be easily made by the pultrusion process since any mix of fibres can be selected at the time of manufacture. While circular rods are most common, flat bars and other shapes have also been manufactured. The Pulforming method is also used in Australia to produce glass fibre reinforced plastic (GFRP) rock anchors (AROA,1995).

5 MECHANICAL PROPERTIES

The short-term and long-term properties of FRP are presented and discussed in this section with some experimental results obtained by the authors.

5.1 Short-term properties

Typical stress-strain curves of CFRP rod (Leadline, 8 mm dia.), AFRP rod (Arapree, 7.9 mm dia.) and steel strand (9.3 mm dia.) obtained experimentally, are compared in Figure 1. It can be seen that FRP reinforcements have linear stress-strain relationships up to failure and the strains at failure are less than that of steel tendons (3 times or more). It can also be seen that FRP reinforcements have comparable tensile strengths to steel tendons while the moduli of elasticity are lower. The lower modulus of elasticity of a FRP reinforcement may lead to larger deflection of a concrete member after cracking. On the other hand, in prestressed beams, it will lead to a lower loss of prestress due to shrinkage and creep of concrete when used as tendons.

The tensile strength of the rod is based almost exclusively on the fibre content since the resin strength is an order of magnitude lower than the fibres. The other factors which influence the strength of an FRP reinforcement are shear transfer through the matrix, fibre alignment within the matrix and surface finishing of the reinforcement. The diameter of an FRP rod also affects the tensile strength. The strength can decrease as the diameter of the rod increases. The principal reason for the strength reduction with increasing rod size is the shear lag as the surface bond stresses are transferred to the core of the rod. Excessive transverse

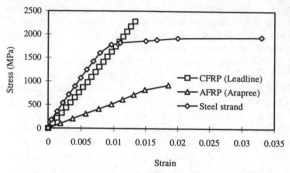

Figure 1. Stress-strain curves of CFRP, AFRP and steel tendons

shear forces can initiate a progressive failure of the individual fibres. This behaviour has generally led to the development of small diameter prestressing reinforcement (Dolan, 1993).

While the pultrusion process is intended to provide a highly uniform product, often some variation in the fibre alignment and initial stress occur. These misalignments can lead to a variation in the strength of the rods. Due to these reasons, the tensile strength guaranteed by the manufacturer is based on a larger margin of safety. The larger margin of 3σ is recommended so that the probability of failure is minimised to about 0.1%.

Guaranteed strength = Mean strength - 3σ $\hfill (1)$

In a pultruded rod, the modulus of elasticity of the composites may be determined by the rule of mixture, i.e., each fibres and resin contributes in proportion to its volume in the composite. However, this method has limitations. Current research (Figure 2) showed that under the static loading, the modulus of elasticity increased as the load increased in the case of CFRP (from 150 GPa at 300 MPa of stress to 172 GPa at 2300 MPa stress), while AFRP only had marginal increase. This stiffening occurs as the fibres align to a straight configuration within the composite material. Research by Dolan (1993) also drew a similar conclusion.

The short-term properties of some products are summarised in Table 1. Due to the difference in mechanical properties and stress-strain relationship between FRP and steel, it is necessary for designers to have a thorough understanding of the properties of these new materials.

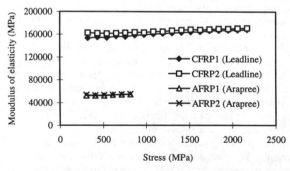

Figure 2. Modulus of elasticity versus stress level for CFRP (Leadline) and AFRP (Arapree)

Table 1. Characteristics of FRP reinforcement [+]

Trade name	Supplier	Tensile strength (MPa)	Elastic modulus (GPa)	Strain at failure (%)
Carbon fibre reinforced plastic (CFRP)				
Leadline	Mitsubishi Chemical, Japan	1800	147	1.6
CFCC	Tokyo Rope, Japan	1800	137	1.6
Jitec (1 to 25 mm diameter)	Cousin Frere, France	1500	125	N/A
Bri-Ten (Carbon-HS)	British Ropes, U.K.	1480	136	1.1
LCR (AS4) (Carbon)	Hercules Co. USA	1860		5-7
Torray (T800 HB-12000-)	Japan	1320	118	1.15
Aramid fibre reinforced plastic (AFRP)				
Arapree	AKZO&HBG, Holland	3000[++]	125[++]	2.3[++]
Parafil Ropes (Type A)	ICI Linear Comp. U.K	620	12	5.3
(Type F)	As above	1930	77.7	2.5
(Type G)	As above	1930	126.5	1.5
Phillystran 8mm(3/8 in.)	Unit. Ropework U.S.A	1900	124	N/A
13mm (1/2 in.)	As above	1810	124	N/A
15mm (5/8 in.)	As above	1770	110	N/A
N/A	Teijin, Japan	1900	54	3.7
FiBRA Rod (8 mm)	Mitsui, Japan	1390	64.8	2.1
(12 mm)	As above	1200	64.8	2.1
Aramid(ZhSVM-300)	Russia	1120	55	3.2
(16 mm)	As above	1210	64.8	2.1
Glass fibre reinforced plastic (GFRP)				
Polystal (E-Glass)	Stabag AG and Bayer AG, Germany	1670	51	3.3
(S-Glass)		2090	63.8	3.3
Isorod	Pultall Inc., Canada	700	44.8	1.8
IMCO	IMCO Reinforced Plastics Inc., USA	1034	41.4	1.8
Jitec (1 to 25 mm)	Cousin Frere, France	1000-1600	35 to 50	4.0
LCR (S2-449) Glass	3M Co. U.S.A	1580		10-12
Kodiak (10 to 25 mm)	IGI Inter., USA	690	50	N/A
Plalloy	Asahi Glass, Japan	590 to 690	24 to 39	N/A
GFRP rock Anchor	AROA	477	56.7	N/A
Hybrid fibres reinforced plastic (HFRP)				
1/3Carbon+2/3Aramid	Chalmers Uni of Tech., Sweden	1050	67	2.36
2/3Carbon+1/3Aramid	As above	1310	100	1.62
1/3C+1/3A+1/3Polypropyl.	As above	420	38	1.24
1/2Carbon+1/2Aramid	As above	1620	101	1.7
1/2Carbon+1/2Glass	As above	1500	88	1.85
1/4Carbon+3/4Glass	As above	1170	63	1.9

[+] The table published Erki and Rizkalla (1993) is modified by the authors to include more products.
[++] based on fibres only.

5.2 *Long-term properties*

The development of FRP reinforcement has yielded a range of products with very attractive and favourable properties. However, for reliable applications more information is required about the long-term characteristics. In particular, assessments of relaxation, creep and stress-rupture, in environments aggressive to the tendon (e.g. alkaline) are necessary. It seems that relaxation as well as creep of AFRP are independent of the stress-level applied within normal prestress levels (up to 50 to 60% of ultimate strength) but are influenced by environmental moisture. They have also been proved not to be sensitive to temperature up to 60°C (Gerritse and Uijl, 1995) when an epoxy matrix is used. With other matrix materials, temperature change also can affect creep and relaxation characteristics.

In a relaxation test, the elongation of the specimen is held constant and the force reduction is measured as a function of time. The expected 100 year relaxation extrapolated from tests up to 50,000 hours for AFRP is in the range of 10 to 15% in normal temperature (in air) (Dolan, 1993). Gerritse and Den Uijl (1995) reported that the relaxation of AFRP in an alkaline liquid is about 40% more than that in air. A final relaxation is estimated to be about 15% in air and 20 to 25% in alkaline liquid (after 100 years).

The relaxation processes observed with GFRP bars are mainly caused by defects in manufacturing (e.g. deviations from the strict axial orientation of the fibres within a bar). As with prestressing steel, stress losses resulting from relaxation show linear behaviour if plotted against a logarithmic time axis. A relaxation loss of 3.2% was recorded for a time period of 5000 hours and from this 100 year values can be estimated. A significant change in this behaviour is not expected even at higher temperatures (Wolff and Miesseler, 1993).

On the other hand, CFRP reinforcement has a low relaxation loss of about 3% after 100 years (estimated) which is comparable to low relaxation steel tendons. A series of relaxation and creep test on CFRP are currently being conducted by the authors.

By extending sustained loading tests until failure occurs, at various stress levels, information on both creep and stress-rupture can be obtained. The creep coefficient, which is defined as the ratio of creep strain to elastic strain under a constant sustained stress, of one particular type of AFRP (Arapree) at different ages, obtained experimentally, is shown in Figure 3. The tests were carried out at about 75±5% humidity and at a temperature of 20±2°C. According to the regression analysis, the creep coefficient ϕ_{tt}, can be expressed as a function of time t (in hours) as follows:

$$\phi_{tt} = 0.054 \, t^{0.082} \qquad\qquad (R^2 = 0.9675) \qquad\qquad (2)$$

According to this formula, the predicted 100 year creep coefficient is 16.5%. This value is comparable to the results reported by Dolan (1993) or Gerritse and Den Uijl (1995), which are 10 - 15% and 15% respectively, within the same temperature and humidity range.

While the creep coefficient of AFRP at 60°C did not significantly differ from that at 20°C, high temperatures showed accelerated ageing effects, indicating that the progressive (e.g. chemical) damage to the fibre-structure was enhanced. The time to failure (due to rupture) was 10 to 15 times shorter at 60°C than at 20°C according to Gerritse and Den Uijl (1995). They suggested a lower bound value of the load that can be carried by AFRP at 20°C for at least 100 years is 52% of the characteristic failure load. Current practice also limits the prestressing level to 60% of f_u to avoid sustained load failure.

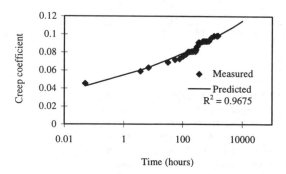

Figure 3. Creep coefficient of 8 mm AFRP (Arapree) tendon at 20°C

6 SUMMARY

From the above information, the following conclusions can be drawn:

- The use of FRP reinforcement or tendons in concrete structural members has emerged as one of the most promising developments in applications of FRP when corrosion of reinforcement is a potential problem.
- The tensile strength guaranteed by the manufacturer for FRP reinforcement is based on the mean strength minus 3 times the standard deviation. This larger margin of 3σ is recommended so that the probability of failure is minimised to about 0.1%. This is to allow for variations due to manufacturing processes.
- The fully linear stress-strain relationship and the relatively small strain at failure of FRP indicates that the ductility of concrete members with FRP reinforcement needs to be carefully assessed.
- The effects of the long-term properties of FRP such as relaxation/creep and stress-rupture on the behaviour of prestressing members needs to be further investigated.

REFERENCES

Applied Research of Australia Pty Ltd (AROA) 1995. *Technical information on GFRP rock anchors,* Adelaide, Australia.

Bakis, C.E. 1993. FRP reinforcement: materials and manufacturing. *Fibre-reinforced-plastic reinforcement for concrete structures: Properties and Applications.* Nanni, A. (Ed): 13-58. Elsevier.

Clarke, J. L. 1993. The need for durable reinforcement. *Alternative materials for the reinforcement and prestressing of concrete.* Clarke, J.L. (Ed): 1-33. Blackie Academic and Profession.

Dolan, C. 1993. FRP development in the United States. *Fibre-reinforced-plastic reinforcement for concrete structures: Properties and Applications*; Nanni, A.; (Ed): 129-163. Elsevier.

Erki, M.A. and Rizkalla, S.H. 1993. A sample of international production, FRP reinforcement for concrete structures; *Concrete international,* 15 (6): 48-53.

Gerritse, A. and Den Uijl, J.A. 1995. Long-term behaviour of Arapree; *Non-metallic (FRP) reinforcement for concrete structures, Proceedings of the second international RILEM symposium on FRP reinforcement for concrete structures:*57-66. Belgium, E & FN Spon.

Mitsubishi chemical corporation, Japan. 1996. *CFRP rod--Leadline product information.* 1-4.

Naani, A. 1995. Concrete repair with externally bonded FRP reinforcement. *Concrete international,* 17 (6): 22-26.

Rizkalla, S.H. and Tadros,G. 1994. A smart highway bridge in Canada. *Concrete International,* 16(6): 42-44

Scheibe, M. and Rostasy, F.S. 1995. Engineering model of stress-rupture of AFRP in concrete; *Proceedings, 2nd international symposium on FRP reinforcement for concrete structures:*74-81. Belgium, E & FN Spon.

Wolff, R, and Miesseler, H.J. 1993. Glass Fibre prestressing System, *Fibre-reinforced-plastic (FRP) reinforcement for concrete structures: Properties and Applications*; Nanni, A. (Ed): 305-332. Elsevier.

Zou, X.W., Gowripalan, N. and Gilbert, R.I. 1997. Flexural behaviour and ductility of HSC beams prestressed with AFRP tendons, *(to be presented) 15th Australian Conference on the Mechanics of Structures and materials;* Melbourne, Australia.

The Mechanics of Structures and Materials, Grzebieta, Al-Mahaidi & Wilson (eds)
© *1997 Balkema, Rotterdam, ISBN 90 5410 900 9*

Interactive buckling in fibrous pultruded I-section composite beams

M.Z. Kabir
Department of Civil Engineering, AmirKabir University of Technology, Tehran, Iran

ABSTRACT: Interactive buckling of I-section fibrous composite beams is discussed in this paper. The flanges are considered as beam elements when the member bends lateraly under the prebuckling stress distribution, and as plate elements for normal bending and twisting; the web is analyzed as a plate element. A detailed parametric study demonstrates that improved designs can be suggested which show superior performance of the flange in both pre and postbuckling, while web distortion caused by interactive buckling requires improvement of the bending stiffness in a vertical direction.

1 INTRODUCTION

The beam can buckle either locally, or laterally, or with a combination of local and lateral modes. If neither local nor lateral modes are suppressed in the buckling analysis, then the beam may buckle in an interactive mode as a result of a linear combination of local and lateral modes In interactive buckling modes, due to local buckling of compression elements of the cross section, the stiffness of the member is reduced. As a result, lateral buckling takes place at a lower load than the member would carry in the absence of local buckling. This is especially critical for intermediate length beams with large width-to-thickness ratios of the component plates of the beam section.

2 ANALYTICAL FORMULATION

The Rayleigh-Ritz energy method presented by Bradford and Waters (1988) is extended to the laminated composite beam. As the member buckles, the total stored potential energy, Π, is

$$\Pi = \Pi_{FT} + \Pi_{FB} + \Pi_{W} - \frac{1}{2}\int_{V}(\epsilon_{l}^{T}D\epsilon_{l} + 2\sigma^{T}\epsilon_{n})dV \tag{1}$$

ϵ_l and ϵ_n represent the linear and quadratic membrane strain vectors, respectively, while σ is the corresponding associated component of the Cauchy stress tensor and D is the Hookean constant relating material properties. V is the volume of the member.

2.1 Total Potential Energy of Flanges

Flanges are considered as beam elements when the member bends laterally under a prebuckling stress distribution, and as plate elements for its bending and twisting. Thus, the total potential energy in the flanges is produced by their beam and plate actions.

$$\Pi_F = \Pi_F)^b + \Pi_F)^p \tag{2}$$

Since the isolated flanges are essentially rectangular beam elements, the components of the strain tensor for ϵ_{xx}, ϵ_{yy} and ϵ_{xy} can be ignored (Wang *et al* 1991). Thus, the total potential energy of the flange due to linear membrane strains modeled as a beam element is:

$$\Pi_F)^b_l = \frac{1}{2} \int_0^L \int_0^s (N^f_{zz} \epsilon^f_{zz} + N^f_{zz} \epsilon^f_{zz}) \, ds \, dz \tag{3}$$

and the total potential energy due to the quadratic membrane strain tensor is related to the applied loads

$$\Pi_F)^b_n = \int_0^L \int_0^s (N^f_{zz} \epsilon_{zz} + M^f_z k_z) \, ds \, dz \tag{4}$$

Here, N^f_{zz} and M^f_z denote the flange in-plane stress and bending moment, respectively. Classical lamination theory, based on Kirchhoff's hypothesis, gives force (N) and moment (M) resultants from the following constitutive relations for symmetric plies with $B_{ij}=0$ (Jones 1975):

$$\begin{Bmatrix} N \\ M \end{Bmatrix} = \begin{bmatrix} A_{ij} & 0 \\ 0 & D_{ij} \end{bmatrix} \begin{Bmatrix} \epsilon \\ k \end{Bmatrix} \quad i,j = 1..3 \tag{5}$$

2.2 Total Potential Energy of Web

The web is analyzed as a plate (Robert and Jhita 1983) which has thickness t_w, and depth, d_w. Considered as a two-dimensional plane stress element, its membrane stress tensor components N_{xx}, N_{xz}, N_{xy} and strain components, ϵ_{xy} and ϵ_{xz} are equal to zero. Moreover, the strain component ϵ_{xx} can be neglected, (Wang *et al* 1991). If N_{zz}, N_{yy} and N_{zy} are the membrane forces per unit length and M_{zz} and M_{yy} and M_{zy} are the bending and twisting moments per unit length, the total potential energy of the web, Π_W, can be expressed as

$$\Pi_W = \frac{1}{2} \int_0^L \int_0^s (N_{zz} \epsilon_{zz} + N_{yy} \epsilon_{yy} + N_{zy} \epsilon_{zy} + M_z k_z + M_y k_y + 2M_{zy} k_{zy}) \, ds \, dz \tag{6}$$

3 POST BUCKLING STIFFNESS

As was mentioned earlier, lateral distortional buckling takes place due to interaction of local buckling of the compressive elements of the beam cross-section and overall lateral buckling.

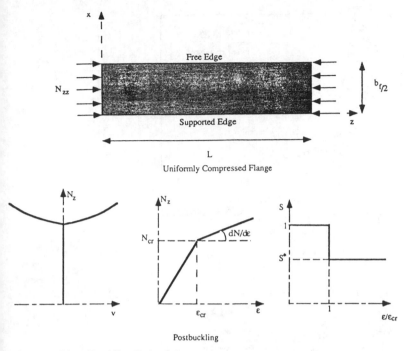

Figure 1. Plate Buckling Behaviour

In this phenomenon, lateral bending increases the curvatures and, therefore, the maximum stresses on the compressive side of the member which has already experienced in-plane critical load, resulting in a corresponding lower in-plane stiffness. This region of the beam section will thus be continuously weakened as to its effective properties as the lateral deflection increases. Since this study investigates only incipient lateral distortional buckling at local buckling of the beam, the initial postbuckling stiffness, indicated by the slope of the load-end shortening relation at bifurcation, Fig. 1, is utilized as an indicator of postbuckling behaviour. It is proposed to adopt a reduced cross-section concept to account for the post-buckling strength of the buckled plate components. In this study, the relative stiffness of the compressive top-flange in post-local-buckling, E^*/E, for I-section beams can be adopted from the results calculated by (Rhodes 1986) for SSFS and SSFC boundary conditions as 4/9 and 0.556, respectively.

4 NUMERICAL RESULTS

A wide flange pultruded I-section beam with the geometry of (8"×8"×3/8") for Glass/Polyester materials with the following elastic constants is used to analyze the combined local-lateral (distortional) buckling of a beam.

Elastic Modulus: E_f=68.9E+9 N/m², E_m=3.45E+9 N/m², Poisson Ratio: v_f=0.2 , v_m=0.3 and Fibre Volume Fraction: 50%. A parametric study is conducted to investigate

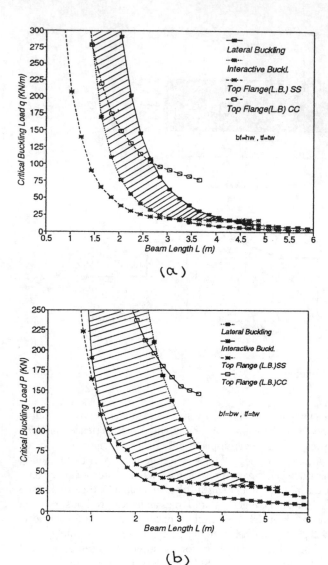

Figure 2. Beam Failure Modes Under UDL and CCL (Local and Overall Effects)

the importance of the local-lateral interaction phenomenon. In practice, structural components are very often subjected to stress gradient and most studies so far have been carried out only for beams subjected to uniform terminal moments. In a departure from traditional work on this subject and to treat the problem realistically, two common loading systems (UDL) and (CCL), representing linear and non-linear stress gradients are applied The calculated critical load-beam length characteristics for the four typical buckling modes, (lateral buckling, interactive buckling, local buckling of compressed top flange which is simply supported to the web and local buckling of the compressed top flange when is assumed to be completely fixed to the web) are shown in Fig. 2 (a,b). Contrary to traditional studies of beam local buckling, which assume the plate subjected to constant compressive stress, the local buckling results presented for the limiting cases of boundary conditions are sensitive to beam length. Premature plate buckling under stress gradient has an overpowering effect on load carrying ability, causing large reductions in the collapse loads. The shaded area represents, in general way, the effect of interaction between local and overall buckling on the location and shape of the failure curves for thin-walled anisotropic beams. Retaining the transverse shear forces in the calculations leads to a substantial reduction in the collapse load, more specificly for the CCL case in which the shear distribution is constant along the beam. For both cases, in pultruded sections, web local buckling under bending gradient stress was found to occur at a much higher load than flange local buckling. Therefore, the interaction of web (local) buckling and overall (lateral-torsional) buckling was not taken into account.

Attention is focused on examining the optimal fibre orientation in the web for interactive local-lateral buckling of pultruded beams. In a previous section, it was mentioned that the postbuckling stiffness of the flange is often maximum for fibre angles in the vicinity of $0°$. This direction provides the maximum prebuckling flange stiffness in the lateral-stability of beams. In lateral-torsional buckling it is already known that, due to the rotation of the total section, the web torsional stiffness, D^w_{33}, has an important effect on the load-carrying capacity, hence placing the web fibre angle at $±45°$ leads to considerable improvement in the collapse load. In the coupled local-lateral buckling mode, it is assumed that the web is distorted and the bending stiffnesses D^w_{11}, D^w_{22}, D^w_{12} and also the coupling terms D^w_{13} and D^w_{23} become important. These terms are sensitive to the fibre orientation. It is, therefore, logical to think of designs that combine the optimum value of several terms with inherent superior performance in the distortional buckling mode for the web. A significant increase of the buckling load can be observed by placing the fibres in the optimal directions reported in Table (1).

Table 1: Optimal Web Fibre Angle in Distortional Failure Mode

Beam Length (L)	1.5 (m)	3.0 (m)	4.5 (m)
CCL	±75°	±70°	±65°
UDL	±50°	±60°	±60°

The percentage increase in failure load for optimal web fibre angle relative to $0°$ is given in Table (2)

Table 2: Improvement of Collapse Load in Optimal Design

Beam Length (L)	1.5 (m)	3.0 (m)	4.5 (m)
CCL	49%	80%	83%
UDL	4%	16%	25%

5. CONCLUSIONS

It is realized that, for a slender beam section (deep beam) under transverse loads, the failure modes approach interactive buckling resulting in elastic postcritical behaviour with a sudden drop in load carrying capacity. It is shown that optimal flange fibre angles in prebuckling offer considerable improvement in postbuckling stiffness. An extended parametric study, in order to find the optimal web fibre orientation, reveals that web distortion caused by interactive buckling requires improvement of the bending stiffness in a vertical direction. This phenomenon, representing combined action among the bending, torsional and coupling terms, shifts the optimal fibre angle , which is $\pm 45°$ in prebuckling, towards $90°$.

REFERENCES

Bradford, M.A. and Waters, S.W. (1988). "Distortional instability of fabricated monosymmetric I-beam", *Computer and Structures*, 29(40), pp. 715-724

Jones, R.M. (1975). *Mechanics of Composite Materials*, McGraw Hill Co. NY.

Rhodes, J. (1986), *Behaviour of Thin-walled Structures*, Applied science, London, UK, Chapter 4: Effective widths in plate buckling.

Roberts, T.M. and Jhita, P.S. (1983), "Lateral local and distortional buckling of I-beams, *Thin-Walled Struct.*, 1, pp. 289-308

Wang, C.M., Chin, C.K. and Kitipornchai, S. (1991). "Parametric study on distortional buckling of monosymmetric beam-columns", *J. Construct Steel Research*, 18, pp. 89-110.

The Mechanics of Structures and Materials, Grzebieta, Al-Mahaidi & Wilson (eds)
© *1997 Balkema, Rotterdam, ISBN 90 5410 900 9*

Fibre reinforced polymer composite short span bridges: Are they viable?

A. M. C. Sonnenberg
Connell Wagner Pty Ltd, Melbourne, Vic., Australia

R. H. Grzebieta
Department of Civil Engineering, Monash University, Clayton, Vic., Australia

ABSTRACT: This paper summarises an investigation into the cost of designing and constructing a short span rural bridge from polymer composite (PC) material. Actual cost of a super T beam concrete bridge, including design, construction and life-time maintenance, is compared with the estimated price of an equivalent PC bridge. Results show material and labour constitute the majority of cost and was an order of magnitude higher than for the concrete structure. Advantages of using polymer composites in bridge construction were found to be corrosion resistance, low density and potential savings in installation time. Disadvantages were found to be cost of construction, depth of sections required, uncertainty in design life and lack of design load, material and section property tables. The paper proposes that if polymer composites are to ever permeate civil engineering infrastructure, plastics manufacturers must begin directing research funds towards significantly reducing material costs and developing structural design guides.

1 INTRODUCTION

Polymer composites are rarely used in civil engineering structures. Most structures rely heavily on materials such as steel and concrete to meet their construction needs. Nevertheless it is possible to use polymer composites in such structures. An impressive 30.5 m diameter circular market building at Argenteuil, outside Paris, used glass-fibre reinforced polyester beams for its primary structural members (Broutman and Krock, 1974). Moreover there has been keen interest in the use of PC in bridge construction. With major composite bridges like the Aberfeldy footbridge in Scotland (Composites Institute, 1995), Lockheed Martins experimental highway bridge (Popular Mechanics, 1996) and the Miyun bridge in Beijing, China (Hollaway, 1993) to name a few. However, such structures are usually of an experimental nature, designed and constructed using government research funds and considerable "in-kind" contributions from devoted scholars and composites disciples.

This paper presents a study of whether it is economically feasible to use polymer composites to construct a short span bridge. A number of different bridge types are currently being investigated by the second author - a short span rural bridge, a medium span freeway overpass bridge, a long span cable stayed bridge, and a very long span suspension bridge. A typical short span rural bridge was chosen for the first case study. Case studies of longer spans will be presented elsewhere. The feasibility study compared the cost, weight, construction times, design methodology, maintenance and dimensions of a polymer composite bridge to a reinforced prestressed concrete bridge.

2 CASE STUDY

2.1 *Concrete Bridge*

The bridge chosen for the case study is located north of Melbourne in the Rural City of Wangaratta. This bridge was designed in May 1996 and is named "Frasers Bridge". Frasers bridge is essentially composed of precast pretension super-T concrete slabs. The bridge has a span of 14.69 meters between abutments which are skewed at an angle of 30°. Figure 1-1 shows sketches of the bridge plan and cross-section.

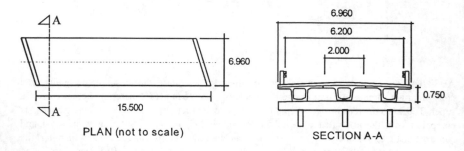

Figure 1 Plan view and cross-section of Fraser's bridge

2.2 *Alternative plastic composite*

Sketches of an alternative bridge design constructed from plastic composite is shown in Figure 2. The design of the bridge was based on Lockheed Martins demonstration bridge (Composites News International, 1995). It is comprised of four 1.4 m deep hand lay-up composite beams fabricated from Glass Polyester. Attached to the beams are five sandwich deck panels (Figure 3). The sandwich panels are comprised of two hand layed plates with 102 mm Pultruded square hollow sections in between.

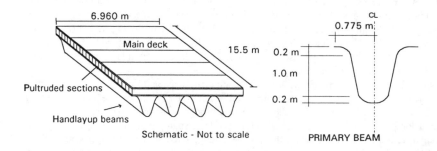

Figure 2 Schematic diagram of conceptual plastic composite bridge design.

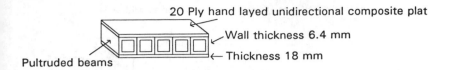

Figure 3 Schematic diagram of deck panels

3 COMPARISON OF BRIDGE TYPES

The following sections compare the key elements of each bridge design. All prices are quoted in Australian dollars.

3.1 *Design procedure*

Both bridges where designed in accordance with the Austroads bridge design code (Austroads, 1992). Due to the anisotropic material properties of the plastic composite bridge finite element analysis was required. On the other hand, the capacity and deflection of the reinforced concrete bridge beams subjected to the design loads may be calculated relatively easily by hand or by use of computer programs.

3.2 *Bridge maintenance and repairs*

After discussions with Vicroads, Australia representatives the following maintenance cost were estimated for Frasers bridge.

A concrete bridge such a Fraser's bridge would cost approximately $100 a year to maintain for the first 30-40 years. After this period it would be expected that the concrete may start to spall. Maintenance would then cost an additional $2,000 every five years (on top of the $100 a year for inspection). Hence for a design life of 100 years it is expected that $30,000 of maintenance may be required.

If Frasers bridge was close to the sea and subject to sea spray then additional maintenance cost would be incurred after 25-30 years. These cost would be in the vicinity of $8,000 every five years (four times greater). In such a case the total maintenance cost over the life of the structure could be $130,000.

The estimated cost above can be compared with maintenance costs for a plastic composite bridge. It is estimated that for a 100 year design life of a PC bridge would be $17,000.

Hence it is evident that a composite bridge the size of Frasers bridge could potentially save a road authority approximately $110,000 in maintenance costs over the life of the structure in a corrosive environment. In a more temperate environment savings would be closer to $15,000.

3.3 *Cost comparison of bridge designs*

A comparison of the approximate cost involved in both concrete and PC bridge design is shown in Figure 4. The cost of the concrete bridge is based on actual cost obtain from the Rural City of Wangaratta, T.B. Gallagher & Co. (design engineers) and Herring Construction. The cost of the composite bridge was obtained from a quotation provided by GR Plastics, Australia.

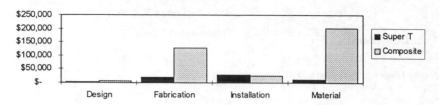

Comparison of cost - Composites v's Concrete bridge

Figure 4 Comparison of the cost of plastic composite bridge design to concrete bridge design

Clearly the material costs are greater for the PC design. The fabrication/labour costs are also much greater for the PC bridge design. The total initial cost of the PC bridge is estimated to be $362,000 compared with the concrete bridge design which costs $62,000. Hence the composite bridge is 5.8 times the cost of the concrete bridge.

If the 100 year maintenance costs (calculated in section 3.2) are added to the initial cost of the bridges the concrete bridge would cost $92,000 while the PC bridge would cost $379,000. Hence it is estimated that the PC bridge would be 4.1 times the cost. Moreover if the PC bridge was located in a coastal environment with only routine maintenance required the PC bridge would still be 1.9 times the cost of an equivalent concrete bridge. Hence it is concluded that PC short span bridges are not economically viable under present conditions.

The cost benefit analysis did not include a discount rate. Inclusion of discount rates to the analysis would have favoured a concrete bridge due to its lower initial cost.

Material costs for the PC bridge were approximately 20 times greater than for the concrete bridge. If the material cost of the PC bridge design could be reduced to that of the concrete design the initial cost of a composite bridge would be 2.7 times that of the concrete bridge. By including 100 years of maintenance costs for a corrosive environment the bridges would then be of a similar cost.

Potentially fabrication cost could be reduced by either automating the production process with new technology or by fabricating bridge components in a country where the cost of labour is low. If this was able to be achieved in conjunction with large material cost savings then the PC bridge would be more attractive. The most costly material component in the bridge's primary beams was the general purpose resin attributing to 49% of the material cost of beams. In the bridge decking the Pultruded beams were the most costly material component at 80% of the material cost. It would therefore be of most advantage to reduce the cost of these components.

3.4 *Weight of bridge decks*

The concrete bridge weighed approximately 95 tonnes compared with the PC bridge design of approximately 22 tonnes. Hence the PC bridge is 23% of the weight of the concrete bridge. If the weight of bitumen surfacing is included in the total weight of the bridge the PC bridge construction is then 36% of the weight. Hence it is seen that the weight of PC bridges is substantially less than conventional concrete bridges.

Fraser's bridge has a design dead load of approximately 40% of the total design load. This design dead load can be reduced by 30% due to the use of polymer composites. Hence the overall design loads may be reduced by 12%. The reduction in design load could lead to some savings in foundation, transportation and installation cost. Again these saving alone will not make polymer composites attractive.

The reduction in design load also means that the design capacity of bridge beams required is reduced. Unfortunately the cost of materials required for a given moment capacity is higher for polymer composite materials, compared with conventional materials such as steel. For example polymer composite Pultruded beams are approximately five times more costly than steel beams based on equivalent section moment capacity. For polymer composites produced by the hand lay-up technique the cost for a given moment capacity may be less than for Pultruded beams but still appears to be higher than for conventional materials. Hence for short span bridges it is unlikely to be an advantage to use polymer composites to reduce initial construction costs.

3.5 *Dimensions*

The depth of the main PC bridge beams was 1400 mm compared to the Super-T beams with a depth of 750 mm. The increase depth of the composite beams may be aesthetically unacceptable and may cause clearance problems. The increase depth is required to maintain acceptable deflections.

3.6 *Design, Fabrication and Installation Times*

A comparison of design, fabrication and installation times for the PC bridge design versus the concrete bridge design is shown in Figure 5.

From Figure 5 it is clear that both design and fabrication times are likely to be longer for PC bridges while installation times are likely to be less. This is because the concrete in the Super-T bridge deck must be allowed to cure.

4 CONCLUSIONS

The following conclusions were drawn from the case study:

- Polymer composites have the potential to be durable over a long period of time and hence maintenance costs may be reduced.

- Polymer composites light weight can allow complete bridge sections to be moved in one lift.

Comparison of estimated design, fabrication and installation times

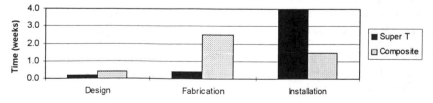

Figure 5 Comparison of estimated design, fabrication and installation times

- Foundation design loads are reduced.

- Installation times are reduced.

- Polymer composites may be more resistant to earthquakes due to the low density.

- Bridges and beams manufactured from polymer composites require larger depths of section.

- The initial cost of the PC material is too high.

- Polymer composites are not common in bridge construction

- Long term durability more than 30-50 years for polymer composite bridges is uncertain at present.

- Manufacture of PC bridges can be labour intensive and involved.

5 ACKNOWLEDGMENTS

The author wishes to acknowledge the assistance provided by the following people in relation to costing and design issues.

Bernie - T.B. Gallagher & Co. , Australia.
Darrin Boxall and Grant Corrin - GRPlastics, Australia.
Russell Bittner, Ken Mc Gregor and Graeme Walter - Vicroads, Australia.
Max Coles - Lockheed Martin Corporation, Australia.
Gary Edwards - Rural City of Wangaratta, Australia.
Steven - Herring Construction, Australia.
Hank van Herk - Technical Director Pacific Composites , Australia.
Dennis Southam - Composite Design engineer, Adelaide, Australia.

6 REFERENCES

Australian Composites Institute, 1995. *Industry Guide 1995 - 1998*, Australia.

Austroads, 1992. Bridge Design Code - Section 2: Design Loads. Austroads publication.

Broutman L.J. and Krock R.H. 1974. *Composite materials,* Volume 3 Engineering Applications of Composites, ed B. R. Noton, New York, London, Academic Press, pp 278-279.

Composites News International, 1995. Solana Beach, California, No. 33, October.

Popular Mechanics. 1996. *A bridge to last*, Vol 173, No. 3, March, p 22.

Holloway L. 1993. *Polymer Composites for Civil and Structural Engineering*, London, New York, Blackie Academic, 1993.

Composites, materials and biomechanics (bio)

The Mechanics of Structures and Materials, Grzebieta, Al-Mahaidi & Wilson (eds)
© *1997 Balkema, Rotterdam, ISBN 90 5410 900 9*

Study of shoulder injuries on swinging motion of arm

T. Nishimura, M. Itoh & S. Yanagi
*Department of Mechanical Engineering, College of Science and Technology, Nihon University,
Tokyo, Japan*

Y. Wada
Department of General Phy., Junior College, Nihon University, Chiba Pref., Japan

ABSTRACT: Volleyball spikers and baseball pitchers often injure the end of long head tendon.
The tendon is attached to the upper side of the scapula. Because a shoulder joint is a very
shallow dimple, the restriction of the lateral movement is performed by mainly the tendons
around the joint. Hence it is imagined that the lateral force may cause the shoulder injuries.
    In this study, the volleyball spiking motion and baseball throwing motion are recorded by high
speed video cameras, and the motions of the arm are analyzed as the mechanical links system. In
order to calculate the generated force at the shoulder joint, the acceleration is derived form the
obtained video image. The generated force is calculated and is resolved into the longitudinal
(direction from the shoulder to the throat) and the lateral components. An affection by the lateral
force at the joint is discussed in this paper

1. INTRODUCTION

Sports players often suffer from many kinds of the sports injuries. An injury is a very serious
problem for top or professional players because players are sometimes forced to retire from top
or active players by the sports injuries. Although many doctors try to treat the sports injuries, the
prevention of the injuries is more important than the treatment. For the prevention, we must study
the mechanism of the sports injuries. As the sports injuries are brought by the physiological
and/or mechanical matters, it is necessary to study the mechanism of the sports injuries from the
medical (physiological) and mechanical approaches. In order to discuss the mechanism of the
sports injuries, the generated force and moment on the motion must be estimated at the injured
portion by the mechanical knowledge. The generated force and moment can be obtained by
analyzing an actual human motion dynamically. It is effective for the prevention of sports
injuries to discuss a possibility of the injuries caused by the force and moment with a support of
the medical knowledge. It is possible to find an ideal motion free from the sports injuries on the
actual motion. In this paper, the sports injuries is discussed only from the view of mechanical
field.
    It is reported that many volleyball spikers and baseball pitchers suffer from the shoulder
injuries by the excessive motion of the arm. The injuries are mainly the dislocation or tendinitis
on the end of long head tendon. Because a shoulder joint is shaped like a shallow dimple, the
lateral movement on joint is restricted by the mainly muscle and tendon. Although the muscle and
tendon have enough strength against the longitudinal force, they are weak against the lateral
force. Therefore, if the tendon is subjected to a strong lateral force repeatedly, it is imagined that
the lateral force causes the shoulder injuries.
    In this study, the actual volleyball spiking motion and baseball throwing motion are analyzed

with assuming the arm as the mechanical links system in order to resolve the mechanism of the shoulder injuries. The actual swinging motions are taken by high speed video cameras. The displacement of the each body element is derived by the obtained video image, then the acceleration at the joints and the generated force on shoulder joint can be calculated. Since the long head tendon, which is often

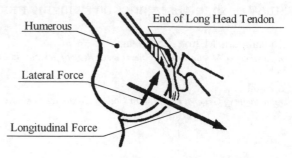

Figure 1. Schematic View of the Shoulder Joint

injured by the lateral force, is attached to the upper side of the scapula, the generated force is resolved into the longitudinal component (the direction from the shoulder to the throat) and lateral component which is in the normal plane to the longitudinal axis (Figure 1). Because it is imagined that the injury of the tendon is caused by the lateral force, it is important to examine whether the lateral force affects the tendon. Hence, we discuss a possibility that the lateral force could cause the shoulder injuries with superimposing the lateral force on the normal plane at the shoulder joint. Our final goal of this study is to suggest the effective form reducing the large lateral force for the prevention of the shoulder injuries.

2. ANALYSIS OF SPIKING MOTION

For resolving the mechanism of the shoulder injuries, it is necessary to know the generated force at the shoulder joint. Since it is difficult to measure the force directly, we attempt to obtain the force by analyzing the video image of the actual swinging motion. The motion is analyzed with assuming the arm as two links system, the first link is the upper arm and the second is forearm including the hand. The links system has three nodal points, that is the shoulder joint, the elbow and the wrist.

Let us introduce a right-handed orthogonal coordinate system. The positive direction of Y-axis is the spiking or throwing direction. Z-axis is the vertical upper direction. X-axis is defined in accordance with a right-handed coordinate rule.

In this study, the motion of the volleyball spiker, who is a female professional volleyball player (lefty, height=1.84 [m], weight=66 [kg]), and the motion of the baseball pitcher, who is a male non-professional player (righty, height=1.72 [m], weight=66 [kg]), are analyzed(Skata 1997). We estimate the mass of each link and mass center of the individual body element with referring to the published paper (Hai-peng 1994). For measuring the nodal point easily, four marks are painted at the center of each joints (that is, the shoulder joint, elbow, wrist and throat). The actual swinging motions are taken by two high speed video cameras (1/200 [sec/frame]) which are orthogonally placed to observe the three dimensional motion.

The obtained video image data are transformed digitally into numerical nodal point data (displacement) by using computers. On these procedures, the obtained displacement contains the errors inevitably because of the restriction of the dimension per pixel on a computer display.

For calculating the generated force at the shoulder joint, it is required to derive the acceleration of arm by the second derivative operation of the displacement. However, since the displacement includes the errors, the errors are amplified by the divided difference operation. As the velocity given by the first divided difference includes enormous error, the acceleration can not be calculated without filtering operation. Hence, the previous paper (Itoh 1995) suggested that the Fourier Series is effective to explain the velocity which is derived by the first divided difference of the displacement. And the integration and the differentiation of the velocity Fourier expansion are introduced as the displacement and the acceleration, respectively.

It is important to determine what number of Fourier terms should be employed on this procedure. The least number of Fourier terms should be determined to express the motion well.

The vertical displacement, velocity and acceleration of the hand on the spiking motion as an example are shown in Figures. 2, 3 and 4. Here, the initial time (t=0) is defined the moment when the subject hits or releases the ball. The circles in Figure 2 show the measured vertical displacement. The vertical velocity obtained by the first divided difference of the displacement is shown by the circles in Figure 3. The velocity by the Fourier expansion is given by the solid line in Figure 3. The displacement and acceleration, which are obtained by the integration and differentiation of the velocity Fourier expansion, are shown by the solid line in Figures 2 and 4.

3. LATERAL FORCE ON SWINGING MOTION

Volleyball spikers and baseball pitchers often injure the long head tendon, which is attached to the upper side of the scapula. Because the shoulder joint is one of the unstable joint and the lateral movement on joint is restricted by the mainly long head tendon, it is imagined that the tendon is injured by the lateral force. If the long head tendon is subjected to the large lateral force

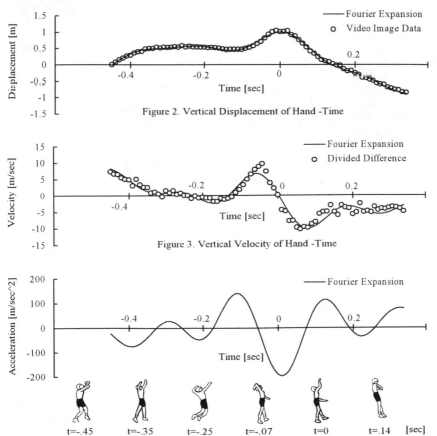

Figure 2. Vertical Displacement of Hand -Time

Figure 3. Vertical Velocity of Hand -Time

Figure 4. Vertical Acceleration of Hand -Time

repeatedly, there is a possibility that the tendon is injured by the lateral force. In this section, the method to derive the lateral force on the swinging motion of the left arm is discussed.

The position vector of the left shoulder joint is denoted as $\mathbf{X}_0$. The relative vector from the shoulder joint to the elbow and from the elbow to the hand are denoted by $\mathbf{r}_1$ and $\mathbf{r}_2$, respectively. It is already known from the previous study (Itoh 1996) that the generated force $\mathbf{F}_0$ at the shoulder joint is given by

$$\mathbf{F}_0 = \left(\ddot{\mathbf{X}}_0 + \lambda_{G1}\ddot{\mathbf{r}}_1\right)m_1 + \left(\ddot{\mathbf{X}}_0 + \ddot{\mathbf{r}}_1 + \lambda_{G2}\ddot{\mathbf{r}}_2\right)m_2 \qquad (1)$$

where m and λ_G denote the mass of the link and the ratio of mass center to the length of the link (Subscript 1: first link, 2: second link), respectively.

Since the injured tendon is connected to the upper side of the scapula, the generated force is resolved into the longitudinal component (direction from the left shoulder to the throat $\mathbf{r}_3$) and normal plane to $\mathbf{r}_3$. Here, let us introduce the right-handed orthogonal local coordinate system $\mathbf{H}_0$ embedded in the left shoulder joint as shown in Figure 5. Each axis is

    x-axis : Unit vector from the shoulder to the throat $\mathbf{r}_3$

    y-axis : Outer product of the unit vector of $\mathbf{r}_3$ and unit vector from the left shoulder to the
          navel $\mathbf{r}_4$

    z-axis : Outer product of the vector of x-axis and y-axis

The local coordinate system $\mathbf{H}_0$ is

$$\mathbf{H}_0 = \begin{bmatrix} \dfrac{\mathbf{i}_x}{|\mathbf{i}_x|} & \dfrac{\mathbf{i}_y}{|\mathbf{i}_y|} & \dfrac{\mathbf{i}_z}{|\mathbf{i}_z|} \end{bmatrix} \qquad (2)$$

where the components of H_0 are

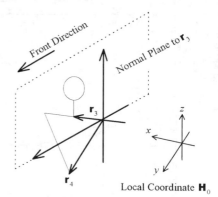

$$\mathbf{i}_x = \frac{\mathbf{r}_3}{|\mathbf{r}_3|}, \quad \mathbf{i}_y = \frac{\mathbf{r}_3}{|\mathbf{r}_3|} \times \frac{\mathbf{r}_4}{|\mathbf{r}_4|}, \quad \mathbf{i}_z = \mathbf{i}_x \times \mathbf{i}_y$$

The generated force on the local coordinate system $\mathbf{H}_0$ is resolved into the longitudinal and lateral components as

$$\begin{bmatrix} F_x \\ F_y \\ F_z \end{bmatrix} = \mathbf{H}_0^T \, \mathbf{F}_0 \qquad (3)$$

Figure 5. Local Coordinate System Embedded in the Left Shoulder Joint

where F_x is the longitudinal force. The resultant of forces F_y and F_z is the lateral force.

4. DISCUSSION

The generated force at the shoulder joint can be derived by substituting the obtained acceleration of each nodal point into Eq.(1). The calculated force is resolved into the lateral and longitudinal forces by substituting the displacement into Eq.(3). In order to discuss a possibility of the injury at the long head tendon, it is necessary to evaluate the magnitude and the direction of the acting lateral force.

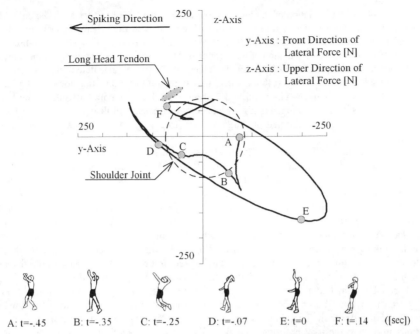

Figure 6. Lateral Force on the Left Shoulder Joint (Spiking Motion)

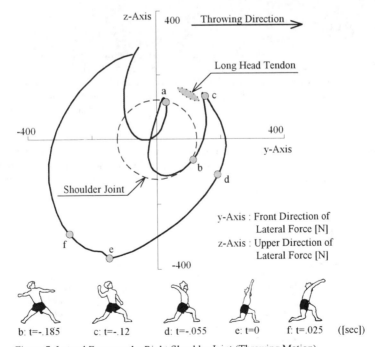

Figure 7. Lateral Force on the Right Shoulder Joint (Throwing Motion)

Let us obtain the variation of the lateral force against the long head tendon. The calculated lateral force on the spiking motion is shown in Figure 6 with the schematic view of the left shoulder joint marked the attached portion of the long head tendon. Figure 7 also shows the lateral force on the throwing motion of the right arm. The normal plane at the right shoulder joint is defined in similar procedure of the left shoulder joint. Since the calculated force is generated by the muscle contraction, the reaction force, that is the opposite direction force of the lateral force, acts against shoulder. Figures 6 and 7 show that almost maximum value of the lateral force toward the long head tendon that is produced at the instance of the spiking or releasing the ball. Therefore, it is supposed that the large reaction force toward the long head tendon causes the shoulder injuries.

The spiking motion E in Figure 6 (t= 0.025 [sec]), just after spiking the ball, generated the maximum lateral force, about 295 [N], toward the tendon. Therefore the reaction (295 [N]) to this lateral force acts against the shoulder. This reaction force must be supported by the muscle and the tendon. However, since the muscle and the tendon are weak to bear the lateral force, the attached portion of the long head tendon may be subjected to the almost lateral force. On the other hand, just after releasing the ball (Figure 7, t = 0.025 [sec], form No.f), the lateral force has the maximum value. It is about 400 [N]. Here, as compared with the maximum lateral force on the spiking motion, the lateral force of the throwing motion is generated by about 100 [N]. Because, on the throwing motion, the foot catches the ground and the reaction force from the ground is expected, the generated acceleration is larger than the spiking motion.

It can be seem that the reaction force acts against the long head tendon varies strongly through the motion. With considering that the shoulder joint is very shallow dimple and the long head tendon restricts the lateral movement around the shoulder joint, if the lateral force gives a strong impact to the long head tendon repeatedly, there is a possibility that the lateral force causes the shoulder injuries.

5. CONCLUSION

The lateral force on the shoulder joint can be derived by the suggested procedure. Generally, it has been considered that the shoulder injuries are occurred by the centrifugal force. But, as the tendon is weak against the lateral force and mainly the long head tendon restricts the lateral movement around the joint, there is a high possibility that the shoulder injuries are caused by the lateral force if the lateral force (the shearing force) acts against the long head tendon repeatedly and strongly.

For preventing the shoulder injuries, we must find the ideal form reducing the lateral force with keeping the result of activity. It is important to investigate the ideal form with a support of medical doctors. The suggested procedure could contribute to the prevention of the sports injuries with a support of medical doctors. The mechanical engineers and medical doctors should cooperate with each other for the prevention of sports injuries.

REFERENCE

Hai-pang Tang, et al., 1994. Estimation of inertia properties of the body segments in Chinese athletes. *Japan J. Phy. Edu.* 38 :487-499
Itoh, M. et al., 1995. Study the mechanism of shoulder injuries by rotational motion of arm, *Pacific-asia conference on mechanical engineering* (PACNE'95) :127-132
Itoh, M. et al., 1996. Basic study of sports injuries, *Asian-pacific conference on strength of material structures* (APCSMS'96) :333-338
Sakata, T. et al., 1997. On expression of human body motion, *15th Australasian conference on the mechanics of structures and materials* (15th ACMSM)

The Mechanics of Structures and Materials, Grzebieta, Al-Mahaidi & Wilson (eds)
© 1997 Balkema, Rotterdam, ISBN 90 5410 900 9

Elasto-plastic behavior of the structure on multi-axial loading

S.Takeda, T.Nishimura, S.Bai & H.Shiraishi
Department of Mechanical Engineering, College of Science and Technology, Nihon University,
Tokyo, Japan

ABSTRACT : In the limit analysis of a structure, the Interaction Curve (abbreviated to I.C.) defined by the generalized internal forces is usually treated as similar to a yield curve defined by the conventional stress. On the analysis, the plastic deformation is believed to proceed in the normal direction on an I.C.; that is defined as the normal flow rule. Although the normal flow rule, suggested by R.Hill or D.C.Drucker for the conventional plastic flow, is applied to the generalized plastic deformation, our previous studies assert that the radial flow rule is reasonable rather than the normal flow rule for the plastic deformation. On the radial flow rule, the plastic deformation proceeds in the radial direction on an I.C.

In this research, the plastic behavior of the right-angle bent bar structure (introduced by Heyman) is studied. The collapse load according to the radial flow rule is compared with that according to the normal flow rule by investigating the deformational behavior. Moreover, the collapse load derived by the upper and lower bound theorems is discussed here.

1.INTRODUCTION

Because the limit analysis is convenient for estimating the collapse load of a structure, it is usually used for structural design. In limit analysis, the collapse load is derived by the upper and lower bound theorems without the very lengthy calculations of elastic-plastic behavior. The upper and lower bound theorems are based on the normal flow rule for the plastic deformation and the normal flow rule comes from the plastic flow rule for conventional plastic strain, suggested by R.Hill and D.C.Drucker. However, there is little experimental research to show that the conventional plastic flow rule is valid. Hence, it should be proved whether the normal flow rule is correct or not for plastic deformation. Usually in limit analysis, an interaction curve defined by the generalized internal force is utilized as similar to a yield curve defined by the conventional stress, nevertheless there is a little difference between an I.C. and a yield curve. The properties of each curve is different in nature. For example, the inside region of a yield curve is pure by elastic, but that of an I.C. is composed of a pure elastic and an elastic-plastic region. Moreover, the variation of the stress along a yield curve is given by loading or neutral loading, but the variation of the internal force along an I.C. can not be defined.

It has been shown by previous studies that the plastic deformation of a bar member under the combined loading of the bending and twisting moments proceed not in the normal direction but in the radial direction on an I.C. Hence, the collapse load derived by the upper and lower bound theorems is suspicious, because the radial flow rule is reasonable rather than the normal flow rule.

In this paper, a right-angle bent bar structure is investigated. The internal forces, bending and twisting moments, are produced on each member by a concentrated lateral force. As an external lateral force is increased, the internal force at a certain cross section attains to the ultimate value on an I.C. This section becomes the plastic hinge. However, the plastic hinge does not deform abruptly because of the structural constraint, and the only deformation at the hinge proceeds with remaining the internal force on the I.C. A second plastic hinge is produced at another cross section as the external force increases. Finally, after several plastic hinges are produced, the unstatically states force the structure to collapse. It is expected that the collapse load given by the radial flow rule is different from that of the normal flow rule.

The collapse load of a right-angle bent bar structure is investigated experimentally, and the collapse load derived by the upper and lower theorems is discussed.

2.EXPERIMENTAL STUDY ON PLASTIC FLOW RULE

Although the normal flow rule is believed to be reasonable for plastic flow, there are few experimental verification. The behavior of the conventional plastic strain is studied preliminarily. An annealed mild steel is adopted in the experiment. The yield plateau of an annealed mild steel is about eight times as long as the elastic limit strain. Therefore, a mild steel is regarded as an elastic perfectly plastic material within the deformation on a yield plateau.

The experimental results on a thin walled circular tube (mean diameter; 21.5mm, wall thickness; 1.5mm), which is loaded with a tensile force and a twisting moment at the same time, are shown Figs.1(a) and 1(b). The abscissa denotes the nondimensional deviatoric tensile stress and strain, and the ordinate denotes the nondimensional shearing stress and strain. The ellipse in the figures is Tresca yield curve. Fig.1(a) shows the deformation under proportional loading. Fig.1(b) shows the deformation under a traction with a constant shearing stress. From these figures, the only plastic strain after the stress reaches Tresca yield curve proceeds in the outward radial direction, while the stress remains on the yield curve. The experimental results lead us to conclude that mild steel is a Tresca material, and the radial flow rule is reasonable rather than the normal flow rule for plastic flow.

Because the generalized internal force and deformation are treated directly in limit analysis, it is also necessary to study elastic-plastic deformation by the internal force. Fig.2 shows the deformation of an annealed mild steel circular bar which is loaded by bending and twisting moments simultaneously. The abscissa denotes the nondimensional generalized bending moment and curvature, and the ordinate denotes the nondimennsional generalized twisting moment and rotational angle. The unit circle in the figures is the I.C. based on the Tresca criterion. Figs.2(a) and 2(b) show respectively the deformation under proportional loading and by a bending moment

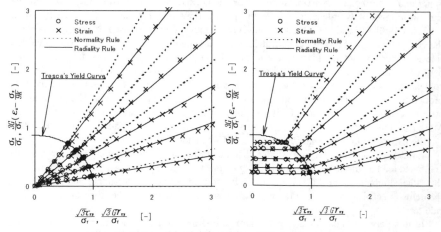

Figs.1(a)(b) combined deformation of the tension and shear

104

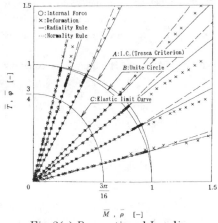

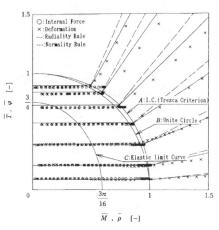

Fig.2(a) Proportional Loading

Fig.2(b) Bending under A Constant
Twisting

at the constant twist. The generalized deformation varies with the generalized internal force within the inside region of an I.C. After arriving at an I.C., the deformation proceeds in the radial direction not in the normal direction while the generalized internal force remains on the I.C. As a result, the whole inside region of an I.C. can be regarded as a pure elastic region. And the relationship between the generalized internal force and deformation can be treated similarly to the conventional stress-strain relationship. Moreover, it is verified that the plastic behavior under a generalized deformation obeys the radial flow rule rather than the normal flow rule.

3.EXPERIMENT ON A RIGHT-ANGLE BENT BAR STRUCTURE

The experimental right-angle bent bar structure is shown in Fig.3. Two bar members of circuler cross section are connected rigidly at C and the opposite ends are fixed rigidly at the walls, A and D. As a concentrated load is applied at the section B, twisting and bending moments are induced on both members. An arbitrary combination of moments is provided by changing the loading position.

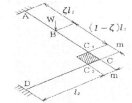

Fig.3 Right-Angle Structure

4.ELASTIC-PLASTIC BEHAVIOR OF A RIGHT-ANGLE BENT BAR STRUCTURE

Because the structural members mentioned in section 3 have a linearly distributed bending and twisting moments, it is assumed that the plastic hinges are built at the critical sections $A,B,C_1,C_2,$or D. When the first section yields to from a plastic hinge, the curvature at the section increases rapidly, but the growth of the curvature will be restricted by strain hardening and structural constraint. The actual plastic hinge is not formed at a section with no length but over a certain finite length. A plastic hinge is also produced at the adjacent section. Let us denote the length of the plastic hinge as L_p (the equivalent length of the plastic hinge). The equivalent plastic hinge can be obtained experimentally by four points bending and three points bending.

The generalized deformations at the critical section, after the generalized internal forces arrived at an I.C., are defined as

$$\bar\rho=\frac{\rho^*+\theta/L_p}{\rho_P}, \qquad \bar\varphi=\frac{\varphi^*+\psi/L_p}{\varphi_P} \tag{1}$$

105

where ρ^* and φ^* are respectively the elastic curvature and the rotational angle corresponding to an ultimate internal force. θ and φ are respectively the angles by a bending and a twisting moments rotate around the plastic hinge. θ/L_ρ and ψ/L_ρ are equivalent curvatures at the plastic hinge.

The relation between $d\theta$ and $d\psi$ define the plastic deformation. The appropriate equations are given by the plastic flow rule.

$$\frac{d\psi}{d\theta} = \frac{EI}{GI_p}\frac{T}{M} \qquad \text{(Radial flow rule)} \qquad (2)$$

$$\frac{d\psi}{d\theta} = \left(\frac{M_P}{T_P}\right)^2 \frac{T}{M} \qquad \text{(Normal flow rule)} \qquad (3)$$

where EI and GI_p are respectively the flexural and the torsional rigidity.

5. EXPERIMENTAL RESULTS

Figs.4, 5(a) (b) and (c) show the experimental results under the condition of $l_1/l_2=1$, $EI_1/EI_2=1$ and $\zeta=0.6$ (refer to Fig.3). Fig.4 shows the relation between the deflection at the point C and the load at B. The abscissa and the ordinate denote respectively the nondimensional deflection and

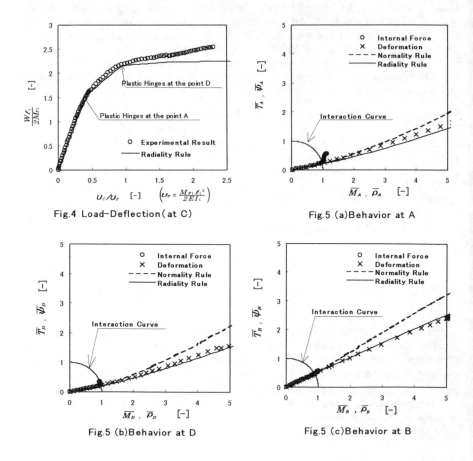

Fig.4 Load-Deflection(at C)

Fig.5 (a)Behavior at A

Fig.5 (b)Behavior at D

Fig.5 (c)Behavior at B

the nondimensional load. The experimental results are represented by the symbol "○", and calculation under the radial flow rule is shown by the solid line. Under the above conditions, the elastic-plastic numerical calculations by both the radial and the normal flow rules lead to the result that plastic hinges are produced serially at the sections A,D, and B, in order, and the structure finally collapses. It is found from Fig.4 that the experimental results are similar to the results of calculation. Figs.5 (a), (b),and (c) show respectively the path of the internal forces and the deformations at the critical sections A, D, and B. In these figures, the abscissa denotes a generalized bending moment $\overline{M}$ and a generalized curvature $\overline{\rho}$, the ordinate denotes a generalized twisting moment $\overline{T}$ and a generalized angular rotational rate $\overline{\varphi}$. The experimental results of the generalized internal forces and deformations are represented by the symbols "○" and "×", and calculations according to the radial flow and the normal flow rules are shown by solid and dashed lines, respectively. The figures show that the experimental results within an I.C. agree with the calculation, and after the internal forces reach an I.C., the internal forces move along the I.C., but the deformations proceed to the outward radial direction from the I.C. Moreover, it is confirmed that the behavior in generalized plastic deformation agrees with the calculation by the radial flow rule rather than the normal flow rule.

6.DISCUSSION OF THE BOUNDARY THEOREM

As mentioned above, it was recognized experimentally that plastic deformation proceeds in the outward radial direction. Although the collapse load of a structure is usually resolved by the method of the boundary theorem which is based on the normal flow rule, the correctness of the boundary theorem should be studied.

The structure discussed here is expected six types of the collapse mechanism. The plastic hinges are mainly developed at A and D. As the first example, let consider the case in which the plastic hinges occur at the sections A, D, and B. If we suppose that the sections A and D yield as the plastic hinges, we have three equilibrium equations and two yield conditions for four unknowns at A and D. Therefore we can resolve a load multiplier for given arbitrary forces T_A and T_D. The calculated result is shown in Fig.6. Since the internal force at B exceeds the I.C. on the curved surface shown by dotted lines, the remaining surface shown by solid lines gives the statically admissible load multiplier. The symbol "○" designates the result calculated by the radial flow rule. The statically admissible load multiplier is determined uniquely independence of the plastic flow rule.

The collapse load given by the radial flow rule therefore cannot exceed the collapse load given by the normal flow rule. Figs.7 and 8 show the kinematically admissible load multiplier obtained respectively by the normal and radial flow rules. Because they depend on the kinematic mechanism, two different load multiplier curves are derived from each plastic flow rule. It can be seen from Fig.7 for the normal flow rule that the upper bound surface contacts the lower bound surface at the maximum and minimum points. On the other hand, for the radial flow rule, although the lower and upper surfaces do not contact at the maximum and minimum points, the two curves contact at a single point. In the example cited, the difference in the load multiplier is very small, because the lower bound surface is not steep near the maximum point. However, because the measured deformation and internal force in the experiment agree with the calculation by the radiality rule(rather than that by

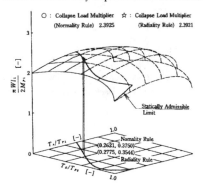

Fig.6 Lower Bound Surface

the normal flow rule, the collapse load obtained by the radial flow rule is certainly smaller than that by the normal flow rule.

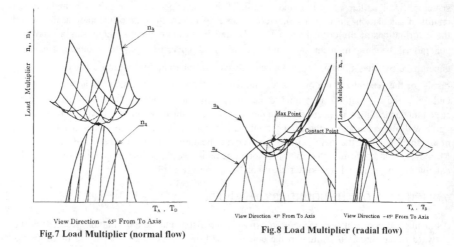

Fig.7 Load Multiplier (normal flow)

Fig.8 Load Multiplier (radial flow)

7.CONCLUSION

The elastic-plastic behavior of a structure is discussed in this paper. The mild steel considered is assumed to be an elastic-perfectly plastic material within our range of interest because of the length of the yield plateau. Hence, a beam under pure bending behaves similarly to a perfectly plastic material. This fact makes to expect that, in plastic analysis or limit design, the internal force and the associated deformation can be treated as similar to conventional stress and strain in the mathematical theory of plasticity.

We can conclude as follows;
1) the radial flow rule is reasonable rather than the normality rule.
2) the upper and lower bound theorems derived from the normal flow rule is suspicious, because the validity of the normal flow rule is not likely.
3) the collapse load derived by the radial flow rule is surely smaller than that by the normal flow rule.

REFERENCE

Nishimura,T. & Bai,S.,1996. Elasto-plastic behavior of a generalized deformation in a multiaxtial loading Proc.APCSHS'96. 7-9 Oct.1996. Beijing.
Nishimura,T. & Nemoto,O & Yanase,H.,1996 Elasto-plastic analysis of a right-angle bent bar structure Proc.PVPC'96(ASME). 23-25 July 1996. Montreal.

The Mechanics of Structures and Materials, Grzebieta, Al-Mahaidi & Wilson (eds)
© *1997 Balkema, Rotterdam, ISBN 90 5410 900 9*

On expression of human body motion

T. Sakata, M. Itoh, T. Nishimura & Y. Wada
Department of Mechanical Engineering, College of Science and Technology, Nihon University, Tokyo, Japan

ABSTRACT: In order to study a human body motion, a human body is sometimes replaced by a mechanical links system composed of human body elements. Each link is supposed to be connected by a spherical joint, if we neglect a small deformation or distortion at a human joint. Since it is assumed that a complex human body motion is performed by the superposition of the simple relative rotational motion of body elements on many human joints, it is necessary to resolve a body motion into the relative rotational motion for a study of a human body motion. Here, we discuss the method of resolving a relative rotational motion at a human joint and try to apply it to an actual swinging motion of arm.

1. INTRODUCTION

A study of human has been done for a long time from the view of many fields, that is, physiological, psychological, and sociological field. Recently, requirements for biomechanical study have increase for the promotion of sports performance and as protection against sports injuries. It is so difficult to win a gold medal at the Olympic Games even for the world-wide top players without the support of medical and mechanical knowledge. The analysis of human body motion demands the contribution especially by mechanical approach. Medical doctors or sports trainers have a lot of ideas and knowledge in their own fields, but they generally lack mechanical knowledge. However few mechanical researchers are interested in biomechanics.

In order to study a human body motion, a human body is sometimes replaced by a mechanical links system composed of human body elements. Although a human joint has a small deformation or distortion on the actual motion, if we intend to investigate a rough motion of a human body, we should neglect a small deformation or distortion at a human joint for saving enormous calculations. Hence a human joint is supposed to be a spherical joint. Since it is assumed that a complex human body motion is performed by the superposition of the simple relative rotational motion of body element on each joint, it is necessary to obtain the relative rotational motion at each joint in order to study an actual human motion. As the actual human motion can be observed by a high speed video camera, it is required to develop the method of resolving the relative rotational motion at the joint from the obtained image data. If we get the information of the relative motion at each joint, we can estimate the muscle force causing the motion and it may contribute to rectifying the motion. Here, we discuss the method of resolving a relative rotational motion at a human joint, and apply the method to an actual swinging motion of arm.

We focus our discussion on a swinging motion of an arm as the first step. Since an upper arm and a forearm are serially connected to the shoulder, we study the motion of a serial connected links system, in this paper.

2. FORMULATION OF SERIAL LINK SYSTEM

Rotational motion of a link element attached on a spherical joint generally has three degrees of freedom, that is, a spin motion of link system about the self-axis and spatial swinging motions. In order to resolve the relative rotational motion, we introduce the local right handed coordinate system, for which z-axis is embedded in the link element. As the length and the local coordinate of j-th element are denoted by L_j and H_j, respectively, the relative vector of j-th element is shown by

$$r_j = L_j \, H_j \, i_0,$$ (1)

where the matrix H_j is composed of column vectors x_j, y_j and z_j, and i_0 is a unit vector, the components of which are zero except z-component. The position vector of the top of j-th element at the current state can be written below (Nishimura 1993),

$$R_j = X_0 + r_1 + r_2 + \cdots + r_j = X_0 + L_1 \, H_1 i_0 + L_2 \, H_2 i_0 + \cdots + L_j \, H_j i_0$$
$$= X_0 + H_0 \left\{ H_0^T H_1 \left\{ L_1 \, E + H_1^T H_2 \left(L_2 \, E + \cdots + H_{j-2}^T H_{j-1} \left| L_{j-1} \, E + L_j \, H_{j-1}^T H_j \right| \right) \right\} \right\} i_0$$ (2)

where X_0 is a position vector at the root of the first link, and H_0 is the coordinate system embedded in base position X_0. If the base position is a shoulder joint, H_0 explains the rotated state of a shoulder. The initial states are distinguished from the current states by adding the subscript '0' to each subscript. For example, H_{j0} denotes the initial state of H_j. Let us suppose that the j-th joint has three degrees of freedom. The element vector r_j at the current state moves from the initial state r_{j0} as shown in Fig.1. r_{j0} rotates by θ_j about the ϕ_j-vector which is placed in x_{j0}-y_{j0} plane and inclined by ϕ_j from the x_{j0} axis, and is given the spin ω_j. The matrix $H_{(j-1)0}^T H_{j0}$ describes the local coordinate system

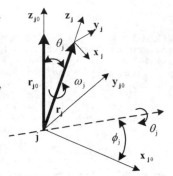

Fig.1 Motion of Element Vector

of H_{j0} observed on $H_{(j-1)0}$. As the element vector r_{j0} swings by the angle θ_j, and has the spin ω_j, the local coordinate H_{j0} embedded in r_{j0} is translated into H_j. The current coordinate H_j embedded in r_j can be derived as follow,

$$H_j = H_{j-1} H_{(j-1)0}^T H_{j0} \Phi_j \Theta_j \Phi_j^T \Omega_j,$$ (3)

where H_{j-1} explains the current state of $H_{(j-1)0}$, and Φ_j, Θ_j and Ω_j are defined by

$$\Phi_j = \begin{bmatrix} \cos\phi_j & -\sin\phi_j & 0 \\ \sin\phi_j & \cos\phi_j & 0 \\ 0 & 0 & 1 \end{bmatrix}, \quad \Theta_j = \begin{bmatrix} 1 & 0 & 0 \\ 0 & \cos\theta_j & -\sin\theta_j \\ 0 & \sin\theta_j & \cos\theta_j \end{bmatrix}, \quad \Omega_j = \begin{bmatrix} \cos\omega_j & -\sin\omega_j & 0 \\ \sin\omega_j & \cos\omega_j & 0 \\ 0 & 0 & 1 \end{bmatrix}.$$ (4)

In Eq.3, $H_{(j-1)0}^T H_{j0} \Phi_j \Theta_j \Phi_j^T \Omega_j$ denotes the relative translation of H_{j0} observed on H_{j-1}.

3. MOTION OF ARM

The equations introduced in previous section are applied for a motion of a right arm in this section. The first and the second link are the upper arm and the forearm, respectively. X_0 is the position vector of the shoulder joint and H_0 is determined by the orientation of the shoulder. The initial or standard state of human body is usually defined by the configuration of upright standing position with the palm of hand facing to the front. The x- and z-axis of the space right handed coordinate are placed to the front and upper direction, respectively. The z-axis of the local coordinate H_0 is directed from the neck to the shoulder joint and the x-axis is normal to the plane

110

made by the vector product of $\mathbf{r}_t \times \mathbf{r}_s$ (in Fig.2). The local coordinates of shoulder joint ,upper arm and forearm in the initial state are given below,

$$
\mathbf{H}_{00} = \begin{bmatrix} 1 & 0 & 0 \\ 0 & 0 & -1 \\ 0 & 1 & 0 \end{bmatrix} \quad
\mathbf{H}_{10} = \begin{bmatrix} 1 & 0 & 0 \\ 0 & -1 & 0 \\ 0 & 0 & -1 \end{bmatrix} \quad
\mathbf{H}_{20} = \begin{bmatrix} 1 & 0 & 0 \\ 0 & -1 & 0 \\ 0 & 0 & -1 \end{bmatrix} \tag{5}
$$

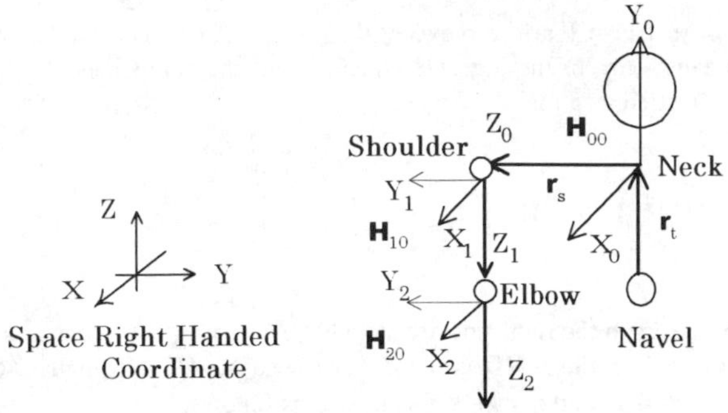

Fig.2 Coordinate system

It is supposed that the shoulder joint has three degrees of freedom in motion, and the elbow joint can rotate only about the y-axis of $\mathbf{H}_{20}$, that is, the elbow joint has one degree of freedom. Therefore the spin of forearm ω_2 is zero, and the direction angle ϕ_2 is fixed at 90 degrees. Then the matrices representing relative motion Φ_2 and Ω_2 are as follows,

$$
\Phi_2 = \begin{bmatrix} 0 & -1 & 0 \\ 1 & 0 & 0 \\ 0 & 0 & 1 \end{bmatrix} \quad
\Omega_2 = \begin{bmatrix} 1 & 0 & 0 \\ 0 & 1 & 0 \\ 0 & 0 & 1 \end{bmatrix}. \tag{6}
$$

Each human joint is restricted in motion. Since the motion of a right arm is discussed, the maximum movable range on a shoulder joint and an elbow are shown in Table 1. Here the clockwise rotation is positive.

Table 1. Movable Range on a Joint

Joint	Degree of Freedom	θ(deg.)	ϕ(deg.)	ω(deg.)
Shoulder	3	0~180	30~270	-90~90
Elbow	1	0~160	90	0

4. RESOLVING RELATIVE ROTATIONAL MOTIONS

A complex human body motion is performed by a superposition of relative rotational motion of body element at each joint with a different phase. Hence, it is more effective to know the relative motion at each joint for rectifying the motion or the estimation of muscle force. As the actual motion can be observed by a high speed video camera, it is required to derive the relative motions at joints from the obtained image data. In this section the procedure for resolving of relative motion is mentioned. The image data are transformed into the digital data by a computer, and the position data of each joint are obtained from a computer display. The information of the position at joints gives the relative vectors of body elements. The position of the shoulder joint $\mathbf{X}_0$ and the local coordinate of shoulder $\mathbf{H}_0$ are also calculated from the position data of the shoulder, neck and navel directly. From Eq.(4) the relative vector of upper arm in the initial and current state are shown below,

111

$$r_{10} = L_1 H_{10} i_0, \qquad r_1 = L_1 H_0 H_{00}^T H_{10} \Phi_1 \Theta_1 \Phi_1^T \Omega_1 i_0 \tag{7}$$

As all components in the unit vector i_0 and the third column of Ω_1 are zero except z-component, the vector r_1 is not affected by Ω_1. Therefore Ω_1 can be omitted from Eq.(7). With rewriting Eqs.(7), the following formulation is derived,

$$L_1^2 \Phi_1 \Theta_1 \Phi_1^T i_0 i_0^T = H_{10}^T H_{00} H_0^T r_1 r_{10}^T H_{10}. \tag{8}$$

As the components of matrix $i_0 i_0^T$ are zero except the (3, 3) component, the last column of the matrix $\Phi_1 \Theta_1 \Phi_1^T$ remains only in the left side equation and the components of other columns vanish. Hence Eq.(8) is reduced to

$$L_1^2 \begin{bmatrix} \sin\phi_1 \cos\theta_1 \\ -\cos\phi_1 \sin\theta_1 \\ \cos\theta_1 \end{bmatrix} = H_{10}^T H_{00} H_0^T r_1 r_{10}^T H_{10} i_0 \tag{9}$$

As the matrices and vectors in the right side are already known, Eq.(9) gives ϕ_1 and θ_1.

Although the local coordinate H_1 is not defined because the spin matrix Ω_1 is not yet derived, the element vectors r_1 and r_2 can be formulated as follows,

$$r_1 = L_1 H_1 i_0 \qquad r_2 = L_2 H_2 i_0 = L_2 H_1 H_{10}^T H_{20} \Phi_2 \Theta_2 \Phi_2^T \Omega_2 i_0. \tag{10}$$

The spin Ω_2 can be removed from the above equation because of no affection in motion to r_2. The both equations (10) are combined into the following equation.

$$L_1 L_2 i_0^T H_{10}^T H_{20} \Phi_2 \Theta_2 \Phi_2^T i_0 = r_1^T r_2 \tag{11}$$

Substituting Eqs.(5) and (6) in (11),

$$L_1 L_2 \cos\theta_2 = r_1^T r_2 \tag{12}$$

is derived, the angular rotation θ_2 is resolved.

Here the new local coordinate H_1^* is introduced, then the local coordinate of the first element is $H_1 = H_1^* \Omega_1$. The element vector r_1 and r_2 are

$$r_1 = L_1 H_1^* i_0 \qquad r_2 = L_2 H_1^* \Omega_1 H_{10}^T H_{20} \Phi_2 \Theta_2 \Phi_2^T \Omega_2 i_0 = L_2 H_1^* \Omega_1 \Phi_2 \Theta_2 \Phi_2^T i_0 \tag{13}$$

From Eqs.(13), $r_2 r_1^T$ is derived as follows.

$$r_2 r_1^T = H_1^* \Omega_1 \Phi_2 \Theta_2 \Phi_2^T i_0 i_0^T H_1^{*T} \tag{14}$$

Rewriting Eq.(14),

$$L_1 L_2 \begin{bmatrix} \cos\omega_1 & -\sin\omega_1 & 0 \\ \sin\omega_1 & \cos\omega_1 & 0 \\ 0 & 0 & 1 \end{bmatrix} \begin{bmatrix} \sin\theta_2 \\ 0 \\ \cos\theta_2 \end{bmatrix} = H_1^{*T} r_2 r_1^T H_1^* i_0 \tag{15}$$

is obtained. Above linear equation gives the spin angle ω_1.

The relative rotational motion at the joints can be resolved from the obtained image data by the above mentioned procedure.

112

5. CALCULATION OF RELATIVE MOTION

In this section, the actual spiking motion of volleyball is resolved into the relative angular rotational motion. The spiking motion by the female professional volleyball lefty player (Nisimura 1997) is observed by high speed video cameras, and the obtained image data are translated into the digital data. The three dimensional position data of each joint are taken from the computer display. Then the relative vector of each element are already known. The procedure mentioned in the previous section gives the relative rotational angle at each joint.

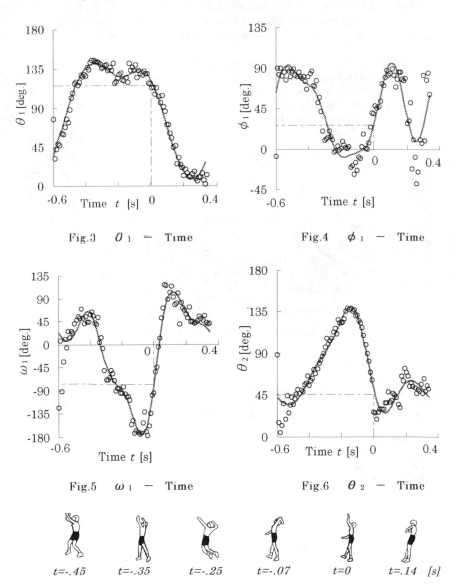

Fig.3 θ_1 — Time

Fig.4 ϕ_1 — Time

Fig.5 ω_1 — Time

Fig.6 θ_2 — Time

$t=-.45$ $t=-.35$ $t=-.25$ $t=-.07$ $t=0$ $t=.14$ [s]

Since the obtained position data have inevitable error because of pixel size and unclearness of image data, a certain filtering operation is required on the process of calculation. In this research, the Fourier expansion is employed for smoothing the discrete data. It is known from the previous study (Itoh 1996) that four or five terms of Fourier series are enough to explain a human motion well. Fig.3 to 6 show the calculated results of the relative rotational angle on spiking motion of volleyball. The circles represent the results calculated by the discrete data directly, and the solid line is given by the Fourier series with the first four terms. The time zero on the abscissa is arranged at the instance just hitting a ball.

The angular rotation θ_1 of the upper arm at the shoulder joint are shown in Fig.3. $\theta_1 = 120$ [deg.] and $\phi_1 = 30$ [deg.] express that the upper arm is forced to swing upward for hitting. At that time, the spin of the upper arm is about -80[deg.]. and the elbow is almost stretched (Fig.5 and 6).

The swinging motion of arm can be explained inversely by using the obtained relative angular rotation at each joint. The displacement calculated from the relative angular variation is compare with data directly obtained from the video image. The solid line shows calculated results from the angular variation and the dotted line shows the displacement from the video image data in Fig.7. Both displacement is calculated by adopting the Fourier Series.

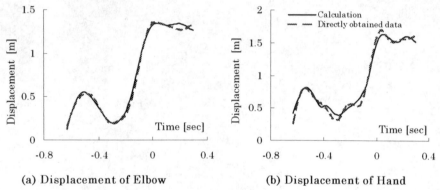

(a) Displacement of Elbow (b) Displacement of Hand

Fig.7 Displacement Along Spiking Direction - Time

6.   CONCLUSION

The relative rotational motions at the joints can be resolved from the video image of the actual human body motion by the procedure mentioned in this paper. As an example, an actual spiking motion of volleyball player is studied. The suggested procedure is effective and the obtained relative angular motion explains the actual human motion well.

If it is required to rectify the motion for preventing a sports injury or for peak performance, the relative angular motion may be useful. As a human motion is performed by the superposition of relative motion at each joint, the same relative motion can be added with shifting the phase for rectifying the motion.

REFERENCE

Itoh, M. et al., 1996. Basic study of sports injuries, *Asian-pacific conference on strength of material structures* (APCSMS'96) :333-338
Nishimura, T. & Wada, Y. 1993.Formulation of human body motion, *International Society of Biomechanics 14th Congress* 952-953; Paris.
Nishimura, T. et al., 1997. Study of shoulder injuries on swinging motion form, *15th Australsian Conference on the Mechanics of Structures and Materials* (15th ACMSM), Melbourne.

Concrete and steel composite structures

The Mechanics of Structures and Materials, Grzebieta, Al-Mahaidi & Wilson (eds)
© *1997 Balkema, Rotterdam, ISBN 90 5410 900 9*

Beneficial effect of friction on the fatigue life of shear connectors

R. Seracino, D.J.Oehlers & M.F.Yeo
The University of Adelaide, S.A., Australia

ABSTRACT: In the design of steel-concrete composite bridge beams, it is assumed that the shear connectors carry all of the longitudinal shear force developed at the steel-concrete interface. In reality, however, friction at the interface reduces the longitudinal shear force acting on the connectors. A finite element program has been developed to model the effects of friction at the steel-concrete interface of composite bridge beams subjected to either static or vehicular fatigue loading. The analyses have shown that friction reduces the longitudinal shear force relatively uniformly along the length of a simply supported beam. The implication of this is that the design life of stud shear connectors may be greater than that currently assumed.

1 INTRODUCTION

In today's aging infrastructure, many composite bridge beams are approaching the end of their anticipated design lives. Assessments must be made in order to determine the remaining life of the structures. The difficulty in assessing the condition of stud shear connectors, unlike other bridge components, is that they are not visible. In order to predict the forces acting on the connectors, it is necessary to model their behaviour.

A finite element program has been developed to model the behaviour of the connectors allowing for the effects of friction at the steel-concrete interface. It has been found that friction along the interface increases the fatigue life of the stud shear connectors. Subsequent development of the program will allow for other phenomena including incremental set (Oehlers & Bradford 1995).

The composite bridge beam and loading used in the analyses are briefly described. This is then followed by a description of the finite element program, and an outline of the technique used to account for friction. A summary of the results is then presented, followed by a discussion of the significance of the results obtained.

2 DESIGN OF COMPOSITE BRIDGE BEAM

The composite bridge beam used in this investigation was designed using hand calculation techniques and assuming typical fatigue vehicle loading. It was assumed that, with the exception of the stud shear connectors, the bridge will remain within the linear elastic range, and effects such as local buckling are not present at the strength limit state. The type of loading which was of interest involved a large number of vehicle traversals within the serviceability limits of the structure.

The length of the simply supported span was chosen to be 19.8m, to suit the dimensions of the standard fatigue vehicle (SFV) and the width of the finite elements used in the analyses. The SFV used is shown in Figure 1, and the design load for the strength limit state of the structure

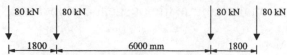

Figure 1. Standard fatigue vehicle.

was the traversal of a vehicle that was nine times the weight of the SFV. It was assumed that the bridge design life was 100 years, with 2 million traversals per year.

The cross-section of the composite bridge beam designed is shown in Figure 2. A concrete compressive strength (f_c) of 35MPa, and a steel yield stress (f_y) of 250MPa was used in the design. Stud shear connectors with a shank diameter (d_{sh}) of 19mm and a tensile strength (f_u) of 450MPa were used in the design.

3 FINITE ELEMENT PROGRAM

The steel and concrete components of the composite bridge beam are modelled using standard 4-noded isoparametric elements (Cheung & Yeo 1979), containing two translational degrees of freedom per node.

The steel and concrete stiffnesses, E_s and E_c respectively, are defined in the input data and remain constant throughout the analysis.

3.1 *Stud shear connectors*

The stud shear connectors at the steel-concrete interface are modelled by two orthogonal spring elements, illustrated in Figure 3.

The vertical spring represents the axial stiffness and the horizontal spring models the shear stiffness of the stud shear connectors. The springs at each nodal point along the steel-concrete interface represent the group of connectors located within that node's tributary length.

The axial spring stiffness is made large relative to the shear stiffness, and has a minimal effect on the flexural behaviour of the beam. The shear stiffness, when taking into account the effects of friction, is determined using an iterative secant stiffness approach described in the following section.

3.2 *Shear stiffness*

The shear stiffness of a spring, representing a group of connectors, is determined according to the type of analysis being performed and the axial force acting on the spring. The two types of

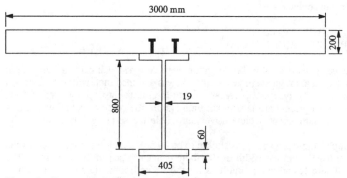

Figure 2. Cross-section of composite beam.

118

Figure 3. Spring element.

analysis procedures are the full interaction analysis (FIA) and the partial interaction analysis (PIA).

A FIA was used in the design phase to determine the distribution of shear connectors required. This type of analysis assumes that there is no relative displacement, or slip, between the steel and concrete components. Consequently, the shear stiffness of the springs is made sufficiently large in order to minimise the slip.

The PIA is one where slip is permitted at the steel-concrete interface, and is usually performed with the connector distribution obtained from the FIA. The initial stiffness of a connector is calculated by the equations given below (Oehlers & Bradford 1995):

$$D_{max} = 4.3 A_{sh} f_u^{0.65} f_c^{0.35} \left(\frac{E_c}{E_s} \right)^{0.40} \tag{1}$$

$$K_{si} = \frac{D_{max}}{d_{sh}(0.16 - 0.0017 f_c)} \tag{2}$$

where D_{max} = the shear strength of a stud shear connector in a composite beam, K_{si} = the initial shear stiffness of a stud shear connector (see Figure 4), and A_{sh} = the cross-sectional area of the stud shear connector. All units are in Newtons (N) and millimetres (mm).

The non-linear behaviour of stud shear connectors is taken into account in a PIA, and it is with this procedure that the effect of friction on the connectors is modelled. The role friction plays on the shear stiffness is dependent on the magnitude and direction of the normal force acting across the interface.

If the normal force (F_n) across the interface is tensile, there is no frictional resistance and the initial spring stiffness is maintained.

However, when F_n is compressive, it must be determined if the frictional resistance (F_r), given by Equation 3, is greater than the shear force (F_v) at that point.

$$F_r = \mu F_n \tag{3}$$

where μ = the coefficient of friction along the steel-concrete interface.

If F_r is greater than or equal to F_v, slip would be prevented and none of the shear force would be carried by the stud shear connector. In the finite element program, this is modelled by increasing the spring shear stiffness to the same order of magnitude as that used in a FIA where interfacial slip is prevented. If, however, F_r is less than F_v, then the connector would be required to resist the shear force that is in excess of F_r, as calculated in the following equation:

$$F_{dwl} = F_v - F_r \tag{4}$$

where F_{dwl} = the shear force resisted by the dowel action of the stud shear connector.

The current secant stiffness at each spring ($Ksec_{old}$) is used to perform an analysis. From the results of the analysis, the slip (δ) and F_r (at each spring) are determined and the new secant stiffness of each spring ($Ksec_{new}$) is calculated, as shown graphically in Figure 4. The new spring secant stiffness is defined as the slope of the line that passes through the origin and the point along 'line a' where the slip is equal to δ. The slope of 'line a' is equal to K_{si}, and the

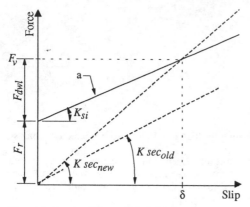

Figure 4. Secant stiffness of stud shear connectors.

vertical offset is equal to F_r. The analysis is repeated once all of the spring secant stiffnesses are updated. Iterations continue until the largest difference in the secant stiffness of the springs, in successive analyses, has met the convergence limit prescribed by the user.

4 RESULTS

A segment of the finite element mesh used to perform the analyses is shown in Figure 5. The elements were distributed uniformly along the length of the beam. The mesh is relatively coarse, but was used at this stage of the investigation to reduce the time required to perform an analysis. A comparison of results with a finer mesh showed that although the magnitude of the results changed slightly, the relative proportions are the same. Therefore, the results are suitable for the comparisons being made. However, the reader is advised to bear this in mind when considering the shear and axial flows of Figure 6.

The loads were applied to the structure at the nodal points of the top surface of the concrete elements. A total of 50 loads stages were required to model the behaviour after a complete traversal of the SFV (Figure 1). To move the SFV along the beam, each nodal load was moved from the current node to the next node on the right.

A friction coefficient of 0.8 was used in all of the analyses. This is an approximately average value based on experimental results obtained from cyclically loaded specimens (Singleton 1985).

4.1 Static load case

Figure 6 shows the shear and axial flow forces acting along the steel-concrete interface for a typical analysis. The shear flow force resisted by the stud shear connectors is reduced relatively uniformly along the span when friction is taken into account (line b) compared to when friction is ignored (line c).

This indicates that slip occurs despite the frictional resistance being very high under the highly concentrated compressive forces (line d). It would initially be expected that the connectors do not resist any of the shear force since the frictional resistance is so high. However, when examining the total shear flow force taking into account friction (line a), the mechanism in place is revealed. The increased stiffness of the region subjected to the high compressive forces attracts the shear flow force from the adjacent, more flexible, regions. The result being that a state of equilibrium is reached when slip occurs along the entire span of the beam without any sharp transitions.

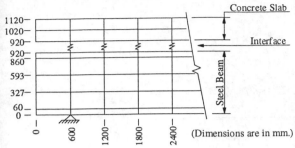

Figure 5. The finite element mesh.

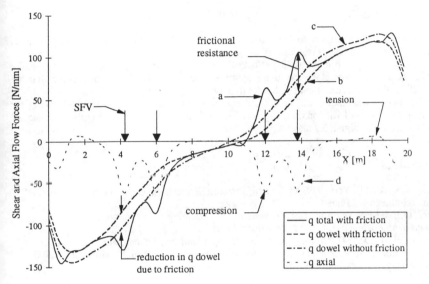

Figure 6. Shear and axial flow forces.

4.2 Moving load case

When the SFV is moved across the beam, the shear flow force influence line diagram at each spring location, or design point, can be determined from which the maximum range is calculated. The fatigue damage of the stud shear connectors is dependent on the total range of load applied to them (Oehlers & Bradford 1995). It was found that the magnitude of the maximum range, when taking into account the effects of friction, is less than the magnitude of the maximum range when friction is ignored. This occurred at all of the spring locations. Figure 7 shows the percentage reduction in the maximum range along the span of the beam when friction is taken into account.

The reduction in the range of load translates into a reduction in the fatigue damage. At midspan, where the range is reduced by only 3%, the fatigue damage is reduced by 13%. The significance of this being that the bridge may be able to safely continue to carry load once the original design life is surpassed. For example, at the critical location (midspan), assuming that the load conditions have not and will not change beyond the design life of the bridge, the stud shear connectors should be able to resist an additional 29.8 million cycles, or last another 14.9 more years.

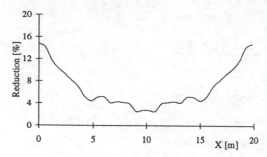

Figure 7. Percentage reduction in maximum range.

5 CONCLUSIONS

A finite element program has been developed to model the behaviour of composite bridge beams, taking into account the effects of friction.

The results show that friction has the effect of reducing the shear flow forces to be resisted by the stud shear connectors along the entire length of the beam, the significance being that the fatigue life of the connectors would be longer than originally anticipated.

Further parametric studies will be carried out in an attempt to develop a simple model for predicting the effects of friction on composite bridge beams. After confirming the accuracy of the model, a set of design rules may be established.

REFERENCES

Cheung, Y.K. & M.F. Yeo 1979. *A practical introduction to finite element analysis.* London: Pitman Publishing Limited.
Oehlers, D.J. & M.A. Bradford 1995. *Composite steel and concrete structural members: Fundamental behaviour.* Oxford: Pergamon Press.
Singleton, W.M. 1985. *The transfer of shear in simply supported composite beams subjected to fatigue loading.* M.Sc. Thesis submitted to the University of Ireland, Department of Civil Engineering, University College. Cork.

The Mechanics of Structures and Materials, Grzebieta, Al-Mahaidi & Wilson (eds)
© *1997 Balkema, Rotterdam, ISBN 90 5410 900 9*

Preventing debonding through shear of tension face plated concrete beams

M.S.Mohamed Ali & D.J.Oehlers
Department of Civil Engineering, University of Adelaide, S.A., Australia

ABSTRACT: The technique of increasing the flexural strength of reinforced concrete beams by gluing plates to their tension faces has been popular for more than two decades. However, the load carrying capacity of such strengthened beams can be heavily curtailed by premature peeling of the plates at their ends. This is because of the formation of diagonal shear cracks due to increased vertical shear loads. A study has been undertaken to improve the shear peeling strength of such soffit plated beams by gluing additional plates to the sides. Predictive equations have been developed to account for the increase in shear peeling strength due to the glued side plates.

1 INTRODUCTION

The technique of strengthening reinforced concrete (RC) beams by gluing mild steel plates to their tension faces using strong epoxy resins has been adopted widely in many countries to improve the load carrying capacity of concrete structures. Such plated beams are designed for flexure using conventional procedures for RC beams that are prescribed by relevant national codes of practice on the assumption that full bond will be developed between the concrete and the plate. However, the premature failure of the strengthened beams in the vicinity of plate ends due to abnormal debonding stresses may not permit the beams to realise their full flexural carrying capacity. Under such circumstances, external shear reinforcements in the form of plates bonded to the sides of the beams have to be provided to prevent debonding, in order to carry the increased flexural loads. This paper mainly addresses the issues pertaining to this technique.

2 FAILURE MECHANISM IN BEAMS WITH GLUED SOFFIT PLATES

The earlier studies conducted by McDonald (1982) and Swamy et.al (1987) established the general structural feasibility of strengthening RC beams by glued soffit plates. These studies primarily attempted to provide the designer with some simple design guidelines like restricting the width to thickness ratio of the plates and the neutral axis depth of plated concrete sections, both to maintain ductility and to avoid premature debonding of the soffit plates. This sort of conservative approach may not always hold good, as it is generally based on shear flow debonding which occurs rarely. Even though there was a general appreciation of the premature failure of plated beams, the failure mechanism behind it was not rationally explained.

Later, Oehlers (1989,1992) conducted detailed studies on the failure mechanism of steel plated beams and the various debonding failure modes were categorised rationally. The three identified failure mechanisms are shear peeling, flexural peeling, and a combination of both shear and flexural peeling. According to this study, flexural cracks, shear cracks, the bond stress along the bottom reinforcement, and aggregate interlock forces across the peeling cracks, all play a potentially major part in the failure mechanism. The physical characteristics of the peeling

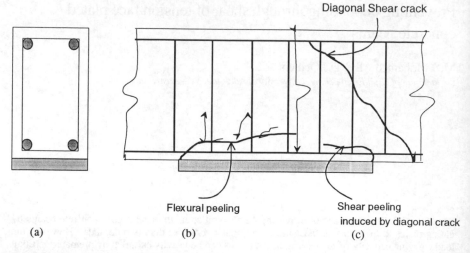

Diagonal Shear crack

Flexural peeling

Shear peeling
induced by diagonal crack

(a)   (b)   (c)

Fig.(1).Failure modes in a soffit plated beam

failure can vary from a straight debonding of the plate leaving the cover intact, to cases where the cover peels off with the plate exposing the reinforcement. The different failure modes are briefly described below.

2.1. *Flexural peeling :* Flexural peeling is described as that induced by increasing curvature and which is associated with gradual separation of plate as shown in Fig.1(b). This form of debonding was studied in beams in which the soffit plate was terminated in a region of constant bending moment and hence zero vertical shear forces. Based on experimental studies conducted on 57 plated simply supported RC beams that covered a wide range of geometrical and material properties, Oehlers and Moran (1990) suggested the following formula to calculate the moment to cause flexural peeling M_{up}:

$$M_{up}=(EI)_{cp}f_b/(0.474E_st_p) \tag{1}$$

where $(EI)_{cp}$ is the flexural rigidity of the cracked plated section derived assuming elastic material properties and ignoring the tensile strength of the concrete; f_b is the Brazilian tensile strength of the concrete; E_s is the modulus of elasticity for steel and t_p is the thickness of plate steel.

2.2. *Shear peeling*: Shear peeling is caused by the formation of diagonal shear cracks as shown in Fig.1(c) and is associated with rapid separation of the plate. This form of peeling occurs in beams where the plates are terminated close to the supports in a region of high shear force and low bending moment. Diagonal cracks first form in the concrete beam adjacent to the end of the plate and these are followed by the horizontal peeling crack at the level of bottom reinforcement. This sequence of crack formation continues along the length of the plate, causing the composite plate and concrete element, below the bottom reinforcement, to fall away over the full shear span. Experimental results (Oehlers 1990) have shown that the inclusion of shear stirrups does not prevent shear peeling as it is caused by the formation of diagonal shear cracks adjacent to the plate end and it is not associated with the shear failure of the reinforced flexural member. Therefore, it can be inferred that shear peeling is not controlled by the shear flow along the steel /concrete interface, and that the vertical shear force on the beam to cause shear peeling V_{peel} occurs when the shear strength of the beam without shear stirrups V_{uc} is reached, i.e.

$$V_{peel}=V_{uc} \tag{2}$$

2.3 *Interaction between flexural and shear peeling*: In a related work, Oehlers (1992) studied the combined effects of shear and flexural peeling when the plate end is terminated in a region of significant flexure and shear. The study showed a strong interaction between the two modes of failure. An empirical failure envelope that employs results obtained from tests on simply supported reinforced concrete beams subjected to point loads, with plates terminated in regions with varying M/V ratios, where M and V are the moment and shear at the end of the plate respectively, was developed and the recommended equation was as follows:

$$(M_p/M_{up}) + (V_p/V_{uc}) \leq 1.17 \tag{3}$$

where $M_p \leq M_{up}$ and $V_p \leq V_{uc}$ and M_p is the design moment at the end of the plate when peeling occurs, M_{up} is the design peeling moment at the end of the plate when V=0 as given by Eqn.1, V_p is the design shear force at the end of the plate when peeling occurs, and V_{uc} is the shear strength of the reinforced concrete beam without stirrups.

3 SOFFIT PLATED RC BEAMS WITH GLUED SIDE PLATES

Among the various failure modes observed, shear peeling of the soffit plate is the most critical as this leads to loss of ductility and a sudden catastrophic failure. As the shear peeling strength is entirely controlled by the shear strength of the concrete element, it is important to provide additional shear reinforcement in the form of plates glued to the sides of the beam in the critical region, i.e. in the vicinity of ends of the soffit plates as shown in Fig.2. This section describes the equations developed for quantifying the increase in shear peeling strength due to side plates. The validity of the equations is established by comparing with the results obtained from tests conducted at the University of Adelaide on a number of RC beams strengthened by gluing plates both to their sides and soffits.

3.1 *Predictive equations*

As there are no standard equations available to quantify the increase in shear peeling strength due to side plates, the conventional equations for calculating shear capacity are adopted as a basis and suitably modified. As previously mentioned the presence of internal shear stirrups plays no significant role in preventing shear peeling of soffit plated beams. The development of the analytical procedure is briefly explained below.

Step1: Ignoring the presence of shear stirrups, the shear peeling strength of an RC beam with only a soffit plate (V_p) equals the shear capacity of the concrete element and can be written as (Zsutty 1968)

$$V_p \propto (b_v d_o)^{2/3} (A_{st})^{1/3} \tag{4}$$

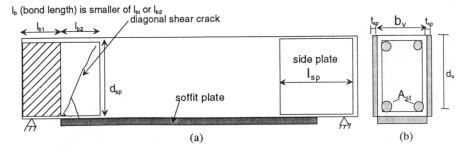

Fig.(2) Beam strengthened with soffit and side plates

where b_v is the effective width of the web, d_o is the distance from the extreme compressive fibre of concrete to the centroid of the outermost tensile reinforcement and A_{st} is the area of fully anchored tensile reinforcement, refer Fig 2(b).

Step 2: The addition of glued side plates to a soffit plated beam provides two additional benefits, viz.: the side plate acts like the conventional tensile reinforcement in improving the dowel action; and it increases the transformed area of concrete web cross section in resisting the shear stresses. The full area of the cross section of the side plates may not always be available for resisting the shear peeling forces due to the dual role played by the side plates. Hence, the concept of the effective area of the cross section of the side plate ($A_{sp,eff}$) is introduced as there is a possibility of the side plate debonding before it attains its full yield strength. Therefore, the shear peeling strength of a soffit plated beam with glued side plates can be written as

$$V_{p,sp} \propto (b_v d_o + m A_{sp,eff})^{2/3} (A_{st} + A_{sp,eff})^{1/3} \qquad (5)$$

where m=modular ratio $=E_s/E_c$. E_s and E_c are the modulus of elasticity of the side plate steel and the concrete, respectively.

Step 3: From Eqns. 4 and 5, the proportional increase in shear peeling strength (P_v) can be written as

$$P_v = V_{p,sp}/V_p = (1 + (A_{sp,eff}/A_{st}))^{1/3} * (1 + (m A_{sp,eff}/b_v d_o))^{2/3} \qquad (6)$$

Now, the only unknown factor in eqn.(3) is $A_{sp,eff}$ and it can be calculated as follows:

Case-1: When the plate is fully bonded, that is,when bonding force ($l_b.d_{sp}. f_{bt}$) is greater than the yielding force ($A_{sp}.f_{yp}$), then the whole area of the side plate is effective so that

$$A_{sp,eff} = A_{sp} \qquad (7)$$

where l_b = bond length provided (Refer fig.2(b)), d_{sp} = depth of side plate, f_{bt} = allowable bond stress in concrete, (Assumed to be 80 % 0f Brazilian strength which is determined by actual tests), A_{sp} = area of cross section of side plate, and f_{yp} = yield stress of side plates.

Case-2 : When the plates are not fully bonded, that is $l_b.d_{sp}.f_{bt} < A_{sp}.f_{yp}$, then the maximum force is $l_b.d_{sp}.f_{bt}$ and hence the plate acts as a fully yielded section of reduced area $A_{sp,eff}$ such that

$$l_b.d_{sp}.f_{bt} = A_{sp,eff}.f_{yp}$$

then $A_{sp,eff} = l_b.d_{sp}.f_{bt} / f_{yp}$ $\qquad (8)$

3.2 Correlation with test results

Wemin Lou (1993) conducted tests on RC beams with steel plates glued to both the soffit and sides as in Fig.2 to study the effect of side plates on the shear peeling strength. All the beams were provided with soffit steel plates 5 mm thick and 1590 mm long. The main parameter varied was the length of the side plates.(l_{sp}) (Refer Fig.2). Both shear spans of a beam were provided with side plates of depth 180 mm and thickness 4 mm. Two shear spans had no side plates and they were used as reference specimens. The results obtained from these tests are compared with theoretical predictions from Eqn.6.

Figure 3 compares the results obtained from the tests with that calculated from the prediction Eqn.6.

It can be observed that a good correlation exists between the theoretical and experimental results as the mean value of $P_{v,test}/P_{v,theory}$ is 0.954 with a standard deviation value of 0.067.

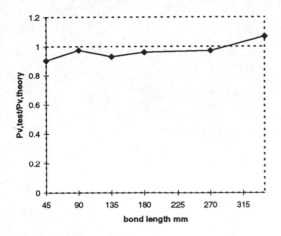

Fig(3). Comparison between theoretical and test shear peeling strength

4 SUMMARY AND CONCLUSION

Shear peeling failure in reinforced concrete beams with plates bonded to the tension face leads to rapid separation of the plates and causes premature failure of the strengthened beam. A study conducted on such beams which were provided with additional shear reinforcement in the form of steel plates bonded to their sides proved the efficiency of the technique in improving the shear peeling strength. It was shown experimentally that a properly designed and detailed side plate can double the shear peeling strength of beams with soffit plates.

An analytical procedure was developed to determine the increase in shear peeling strength due to side plates. It was derived by suitably modifying the conventional approach used for calculating the shear capacity of a reinforced concrete beam without internal stirrups and it shows good correlation with the test results.

REFERENCES

Zsutty, T.C. 1968. Beam shear strength prediction analysis of existing data. Journal ACI, Vol.65,Novomber: 943-951
Macdonald, M.D 1982. The flexural performance of 3.5m concrete beams with various bonded external reinforcements. Supplementary Report 728, TRRL(UK): 41 pp.
Swamy, R.N., Jones, R., & Bloxham, J.W. 1987 . Structural behaviour of reinforced concrete beams strengthened by epoxy-bonded steel plates. Structural Engineer, Vol.65A, No.2, Feb.1987: 59-68.
Oehlers, D.J.1989. Discussion on Roberts, T.M., & Haji-Kazemi.1989. Proc.Instn.Civ.Engrs., Part 2, Vol.87, Dec.1989: 651-653.
Oehlers , D.J. & Moran, J.P.1990. Premature failure of externally plated reinforced concrete beams. Journal of Structural Engg.,(ASCE). Vol.116, No.4, Apr.1990: 2033-2038..
Oehlers , D.J. 1992. Reinforced concrete beams with plates glued to their soffits. Journal of Structural Engg.,(ASCE). Vol.118, No.8, Aug.1992: 2033-2038.
Wemin-Lou. 1993. Strengthening of post-tensioned and reinforced concrete beams strengthened by steel plates. Thesis Submitted to the University of Adelaide, Jan.1993.

The Mechanics of Structures and Materials, Grzebieta, Al-Mahaidi & Wilson (eds)
© *1997 Balkema, Rotterdam, ISBN 90 5410 900 9*

Behaviour and design of concrete filled fabricated steel box columns

B. Uy & S. Das
Department of Civil and Mining Engineering, University of Wollongong, N.S.W., Australia

ABSTRACT: This paper describes the behaviour and design aspects of concrete filled box columns fabricated with very thin steel plate in multistorey buildings. The construction and service load aspects of these members and the implications to design are discussed. Furthermore, the behaviour and design of these members for ultimate loading is considered and a numerical model is presented and compared with a set of experiments undertaken.

1 INTRODUCTION

Concrete filled steel box columns have seen a renaissance during the last decade in Australia with the use of thin-walled fabricated steel columns in the Forrest Centre and Exchange Plaza in Perth, Casselden Place in Melbourne, Myer Centre in Adelaide, Market City in Sydney and in the recently completed Brisbane International Airport, (Watson & O'Brien 1990).

 The use of concrete filled fabricated steel columns provides an efficient and cost effective method for constructing steel framed buildings as they eliminate the need for formwork and reinforcement. Furthermore, vibration or compaction of the concrete is eliminated as the concrete is pumped in from below once several levels of floor construction have been completed. The use of this method requires close consideration of each of the construction, service and ultimate load stages due to the thin-walled nature of the steel column. These loading stages will be described. Experiments and design procedures will be discussed for the ultimate loading stage. Typical cross-sections of the column which have been used in previous projects are illustrated in Fig. 1 where a rectangular, tubular or triangular geometry is shown. The use of these types of geometries in fabricated steel columns provide architects with a great variety of shapes which were not previously available for hot rolled sections.

Figure 1. Concrete filled steel box cross-sections

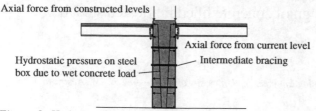

Figure 2. Hydrostatic pressure and axial forces during construction on a steel box

2 CONSTRUCTION LOADING

The construction procedure of concrete filled steel columns requires concrete to be pumped from below after the steel has been erected and used to support several levels of floor construction. This procedure imposes both axial stress and hydrostatic pressure on the column as shown in Fig. 2. Austmeier (1996) described this behaviour in the construction of the tallest building in Europe, the Commerzbank building in Frankfurt. He described the initial difficulties which were experienced during construction with bulging of the columns and outlined methods which were adopted to restrain these by intermediate bracing. Uy and Das (1997c) have considered this behaviour firstly by determining appropriate plate slenderness values to satisfy acceptable deflection limits. Furthermore a study for appropriate intermediate bracing strategies has been carried out to determine the optimum plate slenderness limits and the optimum bracing strategies to be adopted for multistorey building construction, (Uy and Das 1997 a). A diagram showing a typical bracing strategy is given in Fig. 2.

3 SERVICE LOADING

Once a concrete filled steel column has been pumped full of concrete, the concrete cures and begins to shrink. The creep and shrinkage effects inside concrete filled steel columns have been measured by researchers previously, (Morino et al. 1996, Terrey et al. 1994 and Nakai et al. 1991) and these results have been incorporated in an analysis of tall buildings by Uy and Das (1997b). The results have shown that the effects of creep and shrinkage can be quite substantial. Figure 3 shows that the time dependent component of shortening is equivalent to the elastic shortening of the concrete filled column. This was based on a typical column in a 60 storey building where the time dependent shortening was 80 mm which is half of the total shortening of the column and therefore suggests the importance of considering creep and shrinkage in design.

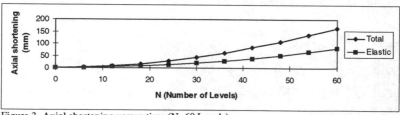

Figure 3. Axial shortening versus time (N=60 Levels)

130

4 ULTIMATE LOADING

4.1 *Experiments*

Two series of concrete filled fabricated steel box columns have been tested at the University of Wollongong to determine the ultimate strength in bending and compression. The specimens were constructed from mild structural steel and filled with normal strength concrete. The details of each test series including geometrical and material properties measured are given in Tables 1 and 2 respectively where C denotes a column and B denotes a beam specimen. It should be noted that the properties have been rounded for initial calibration of the numerical model described in this paper. Each specimen was initially tack welded and then a 3 mm longitudinal fillet weld was passed along the four edges of the column. The residual stresses were measured and shown to be approximately 15 % of the yield stress. Geometrical imperfections were not measured as these would change after pumping the concrete inside the column.

Tables 1 and 2 outline the maximum load of each of the columns and the beam specimens, where N_u is the maximum applied load and the maximum moment M_u is calculated as $M_u=N_u e$ for the column specimens, where e is the applied eccentricity for combined bending and compression of the specimens. The beams were loaded using a symmetric two point load arrangement with a distance L from the support to a point load. The maximum support reaction was $N_u/2$ and the maximum moment was calculated as $M_u=N_u.L/2$.

Table 1. Series 1 - Normal Strength Concrete Filled Box Columns (186 mm)

SpecimenNo.	b (mm)	t (mm)	f_y (MPa)	f_c (MPa)	e (mm)	N_u (kN)	M_u (kNm)
NS1 (C)	186	3	300	32	0	1555	0
NS2 (C)	186	3	300	32	37	1069	39.6
NS3 (C)	186	3	300	32	56	1133	63.4
NS4 (C)	186	3	300	32	84	895	75.2
NS6 (B)	186	3	300	32	L=950	131	62.6

Table 2. Series 2 - Normal Strength Concrete Filled Box Columns (246 mm)

SpecimenNo.	b (mm)	t (mm)	f_y (MPa)	f_c (MPa)	e (mm)	N_u (kN)	M_u (kNm)
NS7 (C)	246	3	300	37	0	3095	0
NS8 (C)	246	3	300	37	48	2255	108.2
NS9 (C)	246	3	300	37	74	1900	140.6
NS10 (C)	246	3	300	37	100	1279	127.9
NS12(B)	246	3	300	37	L=1210	171	103.5

4.2 *Moment-curvature response*

In order to model the results of the experiments and to calibrate a model which may be useful in design, a thrust-moment-curvature analysis has been developed by Uy (1996) which uses the non-linear material properties of concrete and steel. This model idealises the cross-section as a series of finite slices throughout the depth of the cross-section. The strain in each slice is idealised as a function of the depth and a stress is determined according to the material and the constitutive relationship. The forces are established and

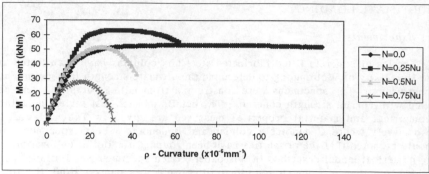

Figure 4. Thrust-moment-curvature response of Series 1

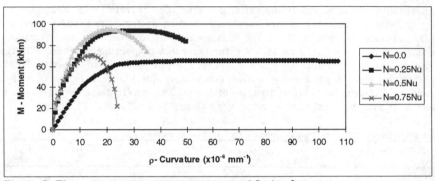

Figure 5. Thrust-moment-curvature response of Series 2

horizontal axial force equilibrium is enforced for steps of increased curvature where the cross-section is subjected to both bending and compression. The results for each series are given in Figs 4 and 5, respectively. The axial thrust is represented as a proportion of the ultimate axial strength N_u which is calculated as $N_u=0.85f_cA_c+f_yA_s$ where f_c,A_c are the concrete strength and area, respectively and f_y, A_s are the steel yield strength and area, respectively.

4.3 Strength interaction diagrams and comparisons

The results from the peaks of the thrust-moment-curvature curves can be used to determine a strength interaction diagram for combined bending and compression. The results from the experiments are compared with the strength interaction envelopes in Figs 6 and 7 respectively. The numerical model is shown to be conservative in the prediction of ultimate strength for both series of tests. It is worth noting that the ultimate axial strength N_u for each series was very close in the prediction of ultimate strength. For other values of combined bending and axial force the experimental values were greater than the model. The model would therefore be useful for design as it gives a lower bound result and is therefore safe. Confinement was ignored in the analysis of this paper and more accurate calibration may be achieved by modifying the concrete stress-strain law of the model to allow for a small degree of

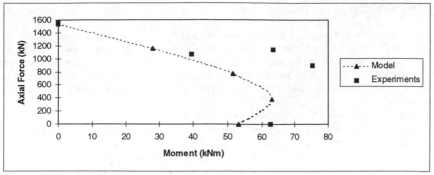

Figure 6. Strength interaction diagram for Series 1

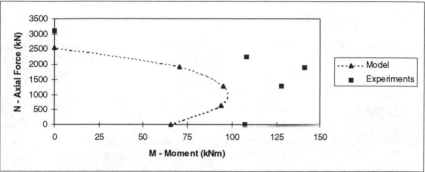

Figure 7. Strength interaction diagram for Series 2

confinement. The confinement may occur for columns where crushing of the concrete occurs prior to local buckling of the steel and this would generally be for columns where the plate slenderness limit is fairly small.

4.4 *Design Recommendations*

The strength interaction diagrams developed in Figs 6 and 7 can be used in the determination of the ultimate strength of the cross-sections. For limit states design it is necessary to factor the strength by a capacity reduction factor. For composite cross-sections where the cross-section is comprised of both concrete and steel the capacity reduction factors need to be quantified so that an appropriate method for design can be developed.

 The percentage of steel area as a proportion of the total concrete area for Series, 1 and 2 was 7 and 5 % respectively. Further testing is necessary for varying degrees of steel percentages and this may then enable the development of an empirical strength interaction diagram. The development of a suitable strength interaction diagram and capacity reduction factors will enable the design of cross-sections for strength.

5 CONCLUSIONS

This paper has described the construction, service and ultimate load stages of concrete filled steel box columns in tall buildings. A series of experiments has been conducted and this has been used to calibrate a numerical model developed elsewhere. The strength interaction envelope has been developed for various cross-sections and this has been compared with the experimental results.

Research has been conducted here for thin walled steel box columns constructed with mild structural steel and filled with normal strength concrete. For the application in tall buildings considerable structural efficiencies can be achieved by the use of high strength materials including high strength concrete and high strength steels. Further experiments will be conducted in this area and important design relationships will be further developed.

6 ACKNOWLEDGEMENTS

The authors would like to thank the Australian Research Council for the financial assistance of this project and BHP Flat Products Division, Port Kembla for the supply of the steel plate used to fabricate the columns.

7 REFERENCES

Austmeier, H.F. 1996 Extension of the Commerzbank Headquarters Frankfurt/Main, *Comp. Constr. III, Proc. of an Engg Foundn Conf.*, (In press).

Morino, S. Kawaguchi, J. and Cao, Z.S. 1996 Creep behavior of concrete-filled steel tubular members, *Comp. Constr. III, Proc. of an Engg Foundn Conf.* (In press).

Nakai, H. Kurita, A. and Ichinose, L.H. 1991 An experimental study on creep of concrete filled steel pipes, *Proc. of 3rd Int. Conf. on Steel-Concrete Composite Structures*: 55-60.

Terrey, P.J., Bradford, M.A., and Gilbert, R.I. 1994 Creep and shrinkage in concrete filled steel tubes, *Tubular Structures VI, Proc. of the 6th Int. Symp. on Tubular Structures*: 293-298.

Uy, B. 1996 Strength and ductility of fabricated steel-concrete filled box columns. *Comp. Constr. III, Proc. of an Engg Foundn Conf.*, (In press).

Uy, B., and Das, S. 1997a Bracing of thin walled steel box columns during pumping of wet concrete in tall buildings, *Submitted for publication.*

Uy, B., and Das, S. 1997b Time effects in concrete filled steel box columns in tall buildings. *The Int. J. of The Struct. Des. of Tall Bldgs,* 6 (1): 1-22.

Uy, B., and Das, S. 1997c Wet concrete loading of thin-walled steel box columns during construction of a tall building, *To appear in J. of Constr Steel Res., An Int. Journal,* (In press)

Watson, K.B. and O'Brien, L.J. 1990 Tubular composite columns and their development in Australia, *The Instit. of Engrs Aust., Struct. Engg Conf., Australia*: 186-190.

The Mechanics of Structures and Materials, Grzebieta, Al-Mahaidi & Wilson (eds)
© *1997 Balkema, Rotterdam, ISBN 90 5410 900 9*

Flexural behaviour and ductility of HSC beams prestressed with AFRP tendons

X.W.Zou, N.Gowripalan & R.I.Gilbert
School of Civil Engineering, The University of New South Wales, Sydney, N.S.W., Australia

ABSTRACT: This paper presents experimental results on the flexural behaviour and ductility of High Strength Concrete (HSC) beams prestressed by Aramid Fibre Reinforced Plastic (AFRP) tendons. The study showed that the load-deflection relationship of HSC beams pretensioned with AFRP tendons, varied linearly before cracking and almost linearly after cracking with a reduced stiffness. An increase of concrete strength led to an increase in cracking moment, but did not necessarily result in an increase of the ultimate moment capacity. The sustained service load had little effect on the ultimate flexural strength. The study also showed that at failure, the deflections of AFRP beams were less than those with steel strands. The deflection at ultimate of beams with AFRP decreased with the increase of concrete strength. On the other hand, the HSC beams absorbed more energy than Normal Strength Concrete (NSC) ones. A new definition of an overall factor expressed as a combination of the strength factor and energy factor, appears to be a suitable indicator of ductility of beams prestressed with AFRP tendons.

1 INTRODUCTION

In order to evaluate the performance of Fibre Reinforced Plastic (FRP) tendons as prestressing elements in concrete structures, a research project is being carried out at the University of New South Wales. Eight simply-supported beams prestressed with AFRP tendons and steel tendons were tested. The results of the beams prestressed by AFRP, including the load-deflection relationships, the cracking behaviour and the ultimate flexural strength, were compared with the results of the beams prestressed with steel strands.

One of the main concerns when using either HSC or AFRP tendons in concrete structural members is the perceived reduction in ductility. Conventional definitions of ductility indices which are based on NSC and steel reinforcement, are inappropriate for the evaluation of the ductility of HSC beams prestressed with AFRP tendons due to the relative brittleness of the concrete, and the linear elastic stress-strain relationship of AFRP. In this paper, a new definition of ductility index expressed as an overall factor is proposed and verified by the experimental results.

2 EXPERIMENTAL PROGRAM

Beam details: Eight beams having the same rectangular cross section, (150 mm wide and 300 mm deep), and the same span (3000 mm) were cast. The beams were fabricated with 6 mm diameter steel stirrups spaced at 150 mm throughout the length of the beam. Four beams were constructed using commercial concrete mixes with a 28 day characteristic strength of 40 MPa (two with AFRP and two with steel tendons) and another four with 80 MPa concrete. The tendons used were 8 mm in diameter. A constant eccentricity of prestress of 85 mm were maintained for all beams. All beams had identical levels of initial prestress. For the designation of the beams, the following sequence was used. First letter stands for tendon type: A for AFRP, S for steel strand; the following two-digit number indicates the concrete strength and the last letter gives the loading history: L - long term sustained loading and S - testing to failure without any sustained load. e.g., A40-L means 40 MPa concrete beam with AFRP tendon, subjected to long term service load before ultimate flexural strength testing. The properties of concrete, AFRP and steel used in the beams are listed in Table 1.

Table 1 Material properties

Tendons	AFRP (Arapree)	Steel strand	Steel bar
No. of tendons	4	2	2 at top, 2 at bottom
Area (mm^2)	196 / 88*	74.4	56.5
Tensile strength (MPa)	1298 / 3000*	1890	400
Young's modulus (GPa) ·	53.5 / 125*	200	200
Ultimate strain (%)	2.3	>4	>8
Poisson's ratio	0.38	0.3	0.3
Concrete (Nominal f'c)	40 MPa	80 MPa	
Strength at transfer (7 days)	27.3	64.4	
Strength at testing(28 days)	33.5	72	
Strength at testing (200 days)	39.5	85.7	

* specified by the manufacturer.

Beam construction: The beams were cast in a 8 m long prestressing bed. A load cell was used to monitor the force in each tendon during jacking and up to the transfer of prestress. The two sides of the moulds of the beams were removed at the age of 5 days. Demec points were then attached to the side surface of the beam at midspan across the depth of the beam for the measurement of concrete strains. At the age of 7 days, the prestressing force was transferred to the concrete by cutting the tendons suddenly at one end. The beams were simply supported immediately after the transfer of prestress. Cylinders having 100 and 150 mm diameter were also cast to evaluate the compressive strength.

Beam testing: At the age of 28 or 35 days, four beams, i.e. one of each type, were subjected to one cycle of loading, unloading and reloading under two equal point loads applied at third points and finally loaded to failure. The other four beams were subjected to a sustained uniformly distributed service load (UDL) for 200 days, and then, subjected to four cycles of loading, unloading and reloading. The unloading part of the cycle was carried out at first cracking, span/200, span/120 and span/75 deflection levels, under two equal point loads applied at third points and finally loaded to failure. It should be noted that the latter four beams were cracked under the sustained UDL. Deflections at midspan and at third points were measured using LVDT (Linear Variable Differential Transformer) transducers linked to a computer. The concrete strains at midspan were measured using a Demec gauge. The slippage of the tendons was monitored by placing dial gauges at the end of the tendons.

3. RESULTS AND DISCUSSION

The results of the beams tested at the age of 28 or 35 days have been reported in detail elsewhere (Gowripalan, Zou and Gilbert, 1996), only some main results are recalled here for comparison. The time dependent behaviour of the beams under sustained service load are present elsewhere (Zou, Gowripalan and Gilbert, 1997). Short term test results, including load-deflection relationship, ultimate flexural strength, energy absorption, ductility and mode of failure are presented in this paper.

3.1 Load-deflection relationship

Precracking behaviour: All beams had similar stiffness with a linear load-deflection relationship before cracking. The load-deflection behaviour can be modeled using the second moment of area of the gross cross-section (I_g).

At cracking: The deflections of the beams previously subjected to a UDL had larger values at cracking than those without any previous sustained load, as much as twice (Table 2). This is due to the higher loss of prestress, the creep and shrinkage induced curvature and due to the cracks in the beams caused by the UDL.

After cracking: The beams with AFRP showed a reduced stiffness (I_{eff1}) but still followed a linear load-deflection relationship up to a certain stage (i.e. within service load range) and then, a further drop in section stiffness to I_{eff2}, but still apparently linear up to failure. This is because the beams were under reinforced and the stress-strain relationship of the AFRP tendon is linear. The beams with steel tendons behaved in a non-linear manner as deflections became large, with relatively small increase in load after cracking, as expected.

At ultimate: Near failure, the beams A40-L and A40-S deflected approximately 68 mm, the beams A80-L and A80-S deflected about 50 mm. The beams with steel tendons deflected more than 80 mm for both 40 MPa and 80 MPa concrete. The beams prestressed by AFRP tendons gave less deflections than the beams with steel strands. For the beams with AFRP, the deflection at ultimate reduced when the concrete strength increased, while for the beams with steel strands the ultimate deflection was independent of the concrete strength.

Residual deflections: The residual deflections when unloaded after first cracking can be ignored for all the beams. After cracking, the beams with AFRP showed a better recovery when unloaded than the beams with steel strands, especially when initially loaded to high deflection levels.

Typical load-deflection curves of 40 and 80 MPa concrete beams with AFRP tendons are shown in Figures 1 and 2. Abdelraman and Rizkalla (1995) reported that the prestressed beams with CFRP showed a bilinear load-deflection relationship while Dolan (1993) reported a trilinear relationship. Current research showed the load-deflection behaviour could be simplified as a trilinear relationship for beams prestressed with AFRP tendons. This behaviour can be modeled by establishing a relationship between I_g, I_{eff1}, and I_{eff2}. This study showed I_{eff1} was about 10 to 20% of I_g and I_{eff2} was less than 10% of I_g. This is due to the low elastic modulus of AFRP.

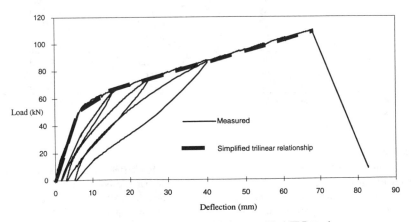

Figure 1 Load-midspan deflection of a 40 MPa beam with AFRP tendons

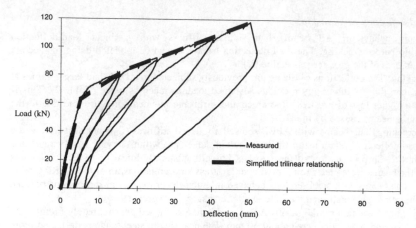

Figure 2 Load-midspan deflection of a 80 MPa beam with AFRP tendons

3.2 Cracking and ultimate load

Table 2 summarises the measured cracking load and ultimate load. The beams had similar cracking loads because the prestressing forces were identical but the ultimate loads for the beams with AFRP and steel tendons were different for both 40 and 80 MPa concrete because of the different tensile capacities of the AFRP and steel tendons. The cracking load increased with an increase of concrete strength, but the ultimate loads were only marginally affected by the change in concrete strength (Gowripalan, Zou and Gilbert 1996). The ultimate load of the beams with AFRP was 30% higher than that of steel beams in both 40 and 80 MPa beams. It can be seen that the sustained service load has little effect of the ultimate flexural strength on both types of beams.

Table 2 Test results

Beam No.	P_{cr}	P_u	Δ_{cr}	Δ_u	ε_{cu}	P_u/P_{cr}	Δ_u/Δ_{cr}
A40-L	50	109.4	6.33	68.04	0.00297	2.19	10.75
A40-S	40	100	3.88	67.77	0.00311	2.50	17.47
S40-L	45	80.1	3.57	84.38	0.00315	1.78	23.53
S40-S	40	75	2.65	83.99	0.00324	1.88	31.66
A80-L	62	118.1	5.24	50.11	0.00236	1.90	9.56
A80-S	50	105	3.40	53.97	0.00249	2.10	15.87
S80-L	65.2	85.6	6.94	93.21	0.00253	1.31	13.43
S80-S	50	73.2	3.32	94.28	0.00235	1.46	28.40

P_{cr}: cracking load (kN); P_u: ultimate load (kN); Δ_{cr}: deflection at cracking; Δ_u: deflection at ultimate, ε_{cu}: Concrete strain at top fibre at ultimate.

3.3 Ductility index

The conventional ductility index which is based on the yield of steel reinforcement is not suitable for beams prestressed with FRP tendons since these bars do not have a yield point. Also, the conventional ductility index does not take into account the energy (the area under the load-deflection curve). The new ductility index proposed by Gowripalan and Zou (1997) is applied to assess the ductility of these beams. The ductility index is expressed as the overall factor which combines the strength factor and energy factor:

$$Overall\ factor = Strength\ factor \times Energy\ factor \dotfill (1)$$

where the strength factor is the ratio of the ultimate load (P_u) to the cracking load (P_{cr}) and the energy factor is the ratio of the total energy (E_{tot}) to the elastic energy (E_{ela}) at any stage of loading of the beam. This ductility index takes into account not only deformation, but also the strength and the energy. This is a suitable indicator of ductility in HSC beams pretensioned with AFRP tendons. Because it takes the brittleness of both the HSC and the AFRP tendons, as well as the linear elastic stress-strain relationship up to failure of AFRP, into account. Table 3 summarises the ductility factors at three deflection levels. These test results show that the pretensioned HSC beams underwent less deformation at ultimate, but absorbed more energy than NSC ones at the same deflection level (Figure 3b). All beams showed similar ductility in terms of the overall factor.

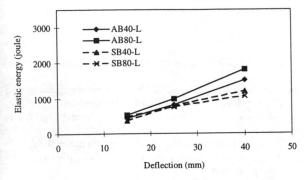

Figure 3a Elastic energy vs deflection of the beams tested

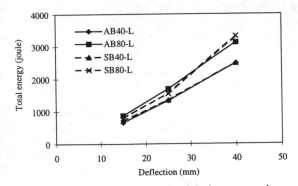

Figure 3b Total energy vs deflection of the beams tested

Table 3 Ductility indices

Beam No.	P_u/P_{cr}	E_{tot}/E_{ela} at different deflection level			Overall factor at different deflection level		
		15 mm	25 mm	40 mm	15 mm	25 mm	40 mm
A40-L	2.19	1.50	1.61	1.64	3.29	3.53	3.59
A80-L	1.90	1.63	1.72	1.73	3.10	3.27	3.29
S40-L	1.78	1.92	1.71	2.07	3.42	3.04	3.68
S80-L	1.31	1.64	2.04	3.15	2.15	2.67	4.13

3.4 Cracking behaviour and Mode of failure

The beams cracked at the expected loading level (calculated cracking load level) in the pure flexural region. The cracks extended with the increase of load in terms of height and number of cracks. At about 70% of ultimate load, in addition to the flexural cracks, some fine horizontal cracks formed at the level of the AFRP tendons for both the 40 and 80 MPa concrete. The horizontal cracks occurred due to slippage of the tendon. It was noted that some cracks formed outside the pure flexural region at high loading levels. For the beams prestressed with steel tendons, flexural cracks formed in a regular pattern with a space of about 150 mm which is the space of the stirrups in the pure flexural region , and at the ultimate loading stage, a major crack developed leading to the failure of the beams.

The observed mode of failure in this investigation was the crushing of the concrete for 40 MPa beams and the rupture of tendons for 80 MPa beams irrespective of the type of tendon. This was confirmed by investigating the concrete strain at the top fibre at ultimate(Table 2).

4. CONCLUSIONS

Based on this experimental investigation, the following conclusions can be reached:

(1) The load-deflection of the beams prestressed by AFRP is linear before cracking, and bi-linear after cracking with a reduced section stiffness. The reduced (effective) section stiffnesses after cracking, I_{eff1}, I_{eff2} can be modelled as a function of I_g, both in the service load range and up to failure.

(2) The use of HSC up to 80 MPa has a significant effect on the cracking moment but little effect in ultimate flexural strength of prestressed beams. Hence HSC can be used with advantage where cracking is not permitted.

(3) The sustained service load has little effect on the ultimate flexural strength or deflection at ultimate.

(4) AFRP tendons can be used in both normal and high strength concrete beams to produce ductile prestressed concrete elements with energy absorption at failure similar to those with steel tendons.

REFERENCES

Abdelrahman, A.A. and Rizkalla, S.H. 1995. Serviceability of concrete beams prestressed by carbon fibre plastic rods, *Non-metallic (FRP) reinforcement for concrete structures. Edited by L. Taerwe.* 403-412.

Dolan, C. 1993. FRP development in the United State. *Fibre-reinforced-plastic (FRP) reinforcement for concrete structures: Properties and Applications*; Nanni, A. (Ed); Elsevier.

Gowripalan, N., Zou, X. W. and Gilbert, R. I. 1996. Flexural behaviour of prestressed beams using AFRP tendons and high strength concrete, *2nd Conference of advanced composite materials in bridges and structures*, Montreal, Canada, 325-333.

Gowripalan, N. and Zou, X. W. 1997. Flexural behaviour and ductility of prestressed HSC beams, to be published in ACI conference, Malaysia.

Zou, X.W., Gowripalan, N. and Gilbert, R.I. 1997. Time dependent behaviour of prestressed beams using AFRP tendons and high strength concrete, *(to be published)*.

The Mechanics of Structures and Materials, Grzebieta, Al-Mahaidi & Wilson (eds)
© 1997 Balkema, Rotterdam, ISBN 90 5410 900 9

Load moment interaction curves for concrete filled tubes

R.Q. Bridge & M.D.O'Shea
Civil Engineering, University of Western Sydney, Nepean, N.S.W., Australia

J.Q. Zhang
Construction and Building Sciences, University of Western Sydney, Hawkesbury, N.S.W., Australia

ABSTRACT: Load-moment interaction curves for concrete filled tubes can be generated using a variety of methods including: accurate numerical methods using equilibrium, strain compatibility and appropriate material stress-strain relationships; approximate numerical methods in which the non-linear material stress-strain relationships are simplified; or empirical methods in which equations are fitted to either results from tests or results from numerical methods. Each of these methods are examined and compared with reference to current design codes. It is found that: discretisation of the cross-section influenced the accuracy of numerical methods; the stress-strain relationships for the materials must take account of the insitu conditions of the actual member; care should be taken if empirical or semi-empirical methods are to be extrapolated beyond their calibrated range; and methods in design codes are not necessarily conservative over the full load-moment range.

1 INTRODUCTION

Load-moment interaction curves for cross-sections are used in the design of members comprising one or more materials. The interaction curves can be explicit in the case of reinforced concrete or implicit in the case of bare steel. For composite steel-concrete members, both implicit and explicit formulations have been used.

Accurate curves can be derived using analyses based on considerations of equilibrium, strain compatibility, and stress-strain relationships that accurately model the stress-strain characteristics of the materials including strain hardening or strain softening. Modelling generally requires the determination of the full load-moment-curvature relationship (surface) where the load-moment interaction curve is a bound to this surface. Techniques are generally iterative in nature.

Approximate curves have been developed for a variety of cross-sections and materials. These are deterministic and generally make use of a limiting ultimate strain, and modified stress blocks to simplify integration (AS3600 1994). In some simple cases, some design standards have used a simple fully plastic approach with reduction factors (Eurocode 4 1992). Further simplifications include representing the curve by a few straight lines to key points (Eurocode 4 1992).

The approximate methods are compared with the accurate approaches. Some of the simplified curves have been found to be unconservative, with even negative values of moment and higher loads than the theoretical squash load for some particular cross-sections. Simple linear approximations are not necessarily a good representation, particularly for thin-walled tubes filled with high strength concrete. The stress-strain relationship for the concrete, particularly in the post-ultimate region, is affected by the confinement provided by the steel tube and needs to be carefully considered.

2 MODELS

2.1 Accurate numerical models

A number of accurate numerical methods have been developed (Pfrang et al 1964, Bode 1976, Chen and Atsuta 1976, 1977, Bridge and Roderick 1978, Wheeler and Bridge 1993 and Hajjar and Gourley 1996). The strip method of Bridge and Roderick (1978) and the fibre method of Wheeler and Bridge (1993) have been used in this paper as typical examples. In the strip method, the cross-section is discretised into a series of finite strips across the section, parallel to the axis of bending. In the fibre method, a grid of fibre monitoring points over the section is established. The assumptions are: plane sections remain plane; no slip between concrete and steel; no local buckling of the steel; a single stress-strain relationship for all strips/fibres of the same material irrespective of strain gradient; and zero concrete tensile strength.

A typical load-moment-curvature relationship (surface) for a thick-walled circular steel tube filled with medium strength concrete is shown in Figure 1 with contours of constant non-dimensionalised curvature. In Figure 1, the load-moment interaction curve (heavy solid line) is the envelope to the surface of maximum axial load N and moment M values with individual curvature contours asymptoting to the envelope.

In this study on composite tubes, the steel was assumed to be elastic perfectly plastic with an elastic modulus of 200000 MPa and a yield stress f_y of 350 MPa. Three different unconfined concrete cylinder strengths f'_c were considered, 40 MPa, 80 MPa and 120 MPa. The actual strength f_c of the concrete in the tubes was assumed to 0.85 f'_c as recommended in AS3600 (1994). The CEB (1973) stress-strain curve was used for the 40 MPa concrete. The C&CA (1992) curve was used for the 120 MPa concrete. The effects of concrete confinement were not included in these two curves shown in Figure 2. For the 80 MPa concrete, the possible effects of concrete confinement by a square steel tube are shown in Figure 2 using the recommendations of Hajjar and Gourley (1996). It can be seen that the ductility of the concrete, compared to the equivalent CEB curve, has been significantly improved.

2.2 Approximate numerical models

In models such as that in AS3600 (1994), the following simplifications are generally made: a rectangular stress block for ease of integration of the non-linear stresses; the point of maximum moment corresponding to a given load occurs at particular value of extreme fibre strain, usually 0.003; and the steel is yielded although elasticity of the steel can be taken into account (Bradford 1991). In models such as that in Eurocode 4 (1992), both the concrete and steel are stressed fully to their maximum strength right up to the neutral axis ie. a fully plastic model. In

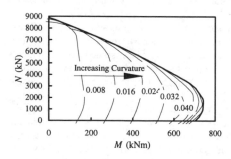
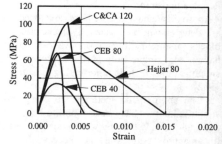

Figure 1. Load-moment-curvature diagram   Figure 2. Concrete stress-strain curves

this latter model, a correction factor to the moment capacity (eg. 0.9) is applied to account for an overestimate of the moment capacity resulting from this simplification.

2.3 *Empirical Methods*

These are generally equations fitted to the results of tests and/or numerical models. In Eurocode 4 (1992), exact equations for five critical points A, B, C, D, and E on the load-moment interaction curve have been derived for rectangular concrete filled tubes assuming full plasticity of both concrete and steel. Linear interpolation between the five points is assumed. These equations are modified for circular tubes. Bradford (1991) has derived equations for the complete interaction curve of rectangular tubes based on curve fitting to approximate numerical methods. Hajjar and Gourley (1996) have also derived equations for rectangular tubes based on accurate numerical results and calibration with available test results.

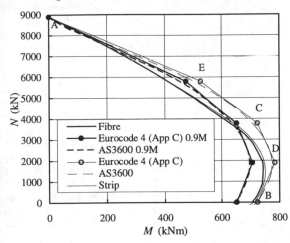

Figure 3. Interaction curve for circular tube, $t = 12$ mm, $f'_c = 40$ MPa

3 COMPARISON OF INTERACTION CURVES

3.1 *Circular concrete filled steel tubes*

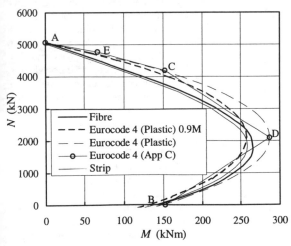

Figure 4. Interaction curve for circular tube, $t = 2$ mm, $f'_c = 40$ MPa

In this study, the following material and geometrical properties were examined: a diameter of 400 mm; a yield stress of 350 MPa, wall thicknesses of 2 mm and 12 mm; and concrete strengths of 40 MPa and 120 MPa.

The interaction curves for a thick walled tube (12 mm) filled with medium strength concrete (40 MPa) are shown in Figure 3. The following observations can be made: the accurate fibre and strip numerical methods give similar results; the equations of Appendix C of Eurocode 4 (points A, B, C, D and E)

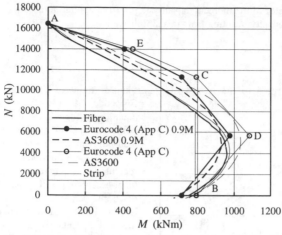

Figure 5. Interaction curve for circular tube, $t = 12$ mm, $f'_c = 120$ MPa

overestimates the moment capacity as expected; the 0.9 moment reduction factor in Eurocode 4 is conservative for low axial force N but still slightly unconservative for high axial force; and the approximate numerical model of AS3600 (1994) assuming the steel to be fully yielded gives similar results to the fully plastic method.

The interaction curves for a thin walled tube (2mm) filled with medium strength concrete (40 MPa) are shown in Figure 4. The following observations can be made: the strip method gives slightly lower values than the fibre method due the strips being wider than the wall thickness; the equations of Appendix C of Eurocode 4 (points A, B, C, D and E) do not accurately reflect the shape of the interaction curve; the fully plastic numerical method of Eurocode 4 gives the required curve shape but overestimates the moment capacity as expected; the 0.9 moment reduction factor in Eurocode 4 is conservative for low axial force N but still slightly unconservative for high axial force.

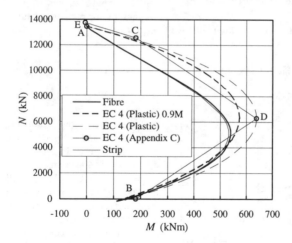

Figure 6. Interaction curve for circular tube, $t = 2$ mm, $f'_c = 120$ MPa

The interaction curves for a thick walled tube (12 mm) filled with high strength concrete (120 MPa) are shown in Figure 5. The following observations can be made: the accurate fibre and strip numerical methods give similar results; the equations (fully plastic) of Appendix C of Eurocode 4 (points A, B, C, D and E) overestimate the moment capacity; the 0.9 moment reduction factor in Eurocode 4 is conservative for low axial force N and is quite unconservative for high axial force; and the rectangular stress block in the AS3600 method is not accurate for high strength concrete.

The interaction curves for a thin walled tube (2mm) filled with high strength concrete (120 MPa) are shown in Figure 6. The following observations can be made: the equations of Appendix C of Eurocode 4 (points A, B, C, D and E) do not accurately reflect the shape of the interaction curve, particularly point E; the fully plastic numerical method of Eurocode 4 gives the required curve shape but overestimates the moment capacity as expected; the 0.9 moment reduction factor in Eurocode 4 is accurate for low axial force N but very unconservative for high axial force; the steel tube provides only a small proportion of the capacity and if the moment capacity is limited to the pure bending capacity at point B, as recommended in Eurocode 4, the effect of large increase in moment capacity with axial load can not be utilised.

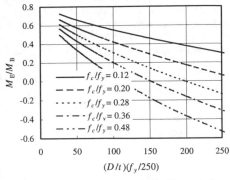

Figure 7. Variation of moment at point E of Eurocode 4

It can be seen from Figure 6 that the Eurocode 4 equation for point E is totally inaccurate (negative moment values and a load higher than the axial squash load) but the fully plastic numerical method gives reasonable results. The reason for this is that the equation for circular tubes was an approximation to the correct equation for rectangular tubes. In addition, the wall thickness and concrete strength used for the results in Fig. 6 were outside the limits of Eurocode 4. A parametric study of the effect of material and geometric properties on the value of the moment capacity at point E is shown in Figure 7. The errors are greatest for thin-walled tubes with very high strength concrete.

3.2 *Rectangular tubes*

The interaction curves for a 400 mm square steel tube 6 mm thick filled with high strength concrete (80 MPa) are shown in Figure 8. The following observations can be made: the effect

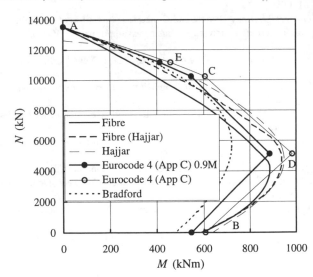

Figure 8. Interaction curve for square tube, $t = 6$ mm, $f'_c = 80$ MPa

of the shape of the concrete stress-strain curve as influenced by confinement can be seen by comparing results of the accurate fibre analysis for the unconfined CEB curve and the more ductile curve used by Hajjar and Gourley (Figure 2); the equation by Bradford (1991) is generally conservative and it should be noted that it was derived for concrete strengths not exceeding 40 MPa; the equations derived by Hajjar and Gourley (1996) are a reasonable approximation to the fibre analysis using the ductile concrete stress-strain curve; and the equations in Eurocode 4 for points A, B, C, D and E are unconservative when compared to the fibre analysis using the CEB concrete curve but are reasonable when the more ductile concrete stress-strain curve is used. Comparisons with test results (Hajjar and Gourley 1996) would indicate that the use of the more ductile concrete stress-strain curve is warranted.

145

4 CONCLUSIONS

From comparison with accurate fibre or strip analyses, the approximate fully plastic numerical method in Eurocode 4 and the corresponding equations for five points on the interaction diagram can not be extrapolated beyond the present limits on tube diameter to wall thickness ratio ($D/t = 80$) and concrete strength ($f'_c = 50$ MPa). Even within these limits, the method is still unconservative at higher values of axial force. Also, the five points with linear interpolation are not a good representation for thin-walled tubes filled with high strength concrete.

For concrete filled tubes, the influence of the tube confinement on the stress-strain response of the concrete, particularly on the ductility of the post-ultimate response, is significant. At present, this phenomena has been modelled mainly from tests on axially loaded tubes with zero strain gradient (Tomii and Sakino 1979, Hajjar and Gourley 1996). The influence of strain gradient on the response has been considered by Sugupta (1994) but needs experimental verification.

5 REFERENCES

AS3600 1994. *Concrete structures.* Sydney: Standards Australia.

Bode, H. 1976. Columns of steel tubular sections filled with concrete - design and application, Acier-Stahl-Steel, 11(12):388-393

Bradford, M.A. 1991. Design of short concrete filled RHS sections, *IEAust Civil Engg Transactions,* CE33(3):189-194

Bridge, R.Q. and Roderick, J.W. 1978. Behaviour of built-up composite columns, *ASCE Jnl Struct. Div.,* 104(ST7):1141-1155.

C&CA 1992. *High strength concrete,* Sydney:Cement and Concrete Association of Australia.

CEB 1973. Deformability of concrete structures - basic assumptions, *CEB Bulletin d'Information, No. 90, Part 2.*

Chen, W.F. and Atsuta, T. 1976. *Theory of beam-columns, vols. 1 & 2,* New York: McGraw Hill

Eurocode 4 1992, *ENV 1994-1-1 Eurocode 4, Design of composite steel and concrete structures,* Brussels: European Committee for Standardisation.

Hajjar, J.F. and Gourley, B.C. 1996. Representation of concrete filled steel tube cross-section strength, *ASCE Jnl Struct. Div.,* 122(11):1327-1336

Pfrang, E.O., Siess, C.P. and Sozen, M.A. 1964. Load-moment-curvature characteristics of reinforced concrete cross-sections, *ACI Journal,* 61(7):763-778.

Sugupta, D.P.G. 1994. Composite members normal and high strength concrete filled circular steel tube column design, *Seminar Proceedings, High Performance Concrete, Dept. Civil Env. Engg, Uni. Melbourne, Australia*: 135-158.

Tomii, M. and Sakino, K. 1979. Elasto-plastic behaviour of concrete filled square steel tubular beam-columns, *Trans. Arch. Inst. Japan,* 280:111-120.

Wheeler, A.T. and Bridge, R.Q. 1993. Analysis of cross-sections in composite materials, *Proc. 13th Aust. Conf. Mech. Struct. Matls., Wollongong, Australia*: 929-936

The Mechanics of Structures and Materials, Grzebieta, Al-Mahaidi & Wilson (eds)
© *1997 Balkema, Rotterdam, ISBN 90 5410 900 9*

Simplified design of continuous composite slabs including moment redistribution and crack control

D.J.Proe, M.Patrick & C.C.Goh
BHP Research, Melbourne Laboratories, Clayton, Vic., Australia

ABSTRACT: Efficient design of continuous composite slabs in steel-frame or masonry wall construction is discussed, with particular attention being paid to three aspects, viz. moment redistribution, crack control and the calculation of positive moment capacity. Simplified equations are given which enable composite slabs to be designed by hand calculation.

1 INTRODUCTION

A composite slab is formed by pouring concrete onto steel sheeting. For a sheeting profile which can develop strong and ductile mechanical resistance in conjunction with the concrete, the sheeting acts as effective bottom-face reinforcement. Because the sheeting typically has substantial cross-sectional area and a relatively high yield stress, the resulting slab generally has substantial intrinsic positive moment capacity. For efficient use of this capacity in continuous slabs, use in design of the maximum permissible amount of redistribution of moments from negative to positive moment regions is desirable.

The redistribution of moments from negative to positive moment regions reduces the required amount of conventional top-face reinforcement (steel bars or welded-wire fabric) in negative moment regions. The reduced reinforcement area is subjected to a greater stress under serviceability loading, which increases flexural crack widths. Where crack control is a relevant criterion, the designer must ensure that the required amount of top-face reinforcement is provided.

The positive moment capacity of a composite slab is affected by the degree of shear connection between the sheeting and the concrete, which increases with the distance from the end of the sheeting and depends also on the support reaction passing through the sheeting. The partial shear region of a composite slab is analogous to the development length of a reinforcing bar.

Simplified sets of equations are presented in this paper to enable the designer to efficiently handle the above three aspects of the design of continuous composite slabs in steel-frame or masonry wall construction. No Australian Standard currently exists for the design of composite slabs, and the design provisions described in this paper are being considered for inclusion in the relevant part of AS 2327. For negative moment regions, the design of composite slabs incorporating sheeting with relatively narrow steel ribs is similar to that for one-way reinforced-concrete slabs and reference will therefore be made to AS 3600 [SA 1994].

2 NEGATIVE MOMENT REGIONS – MOMENT REDISTRIBUTION

It will be assumed that moment redistribution is permitted only from negative to positive moment regions. The moment redistribution parameter at a support, η, will be defined as follows:

$$M_r^* = (1 - \eta)M_e^* \tag{1}$$

where M_e^* is the elastically-determined design negative bending moment and M_r^* is the design negative bending moment after redistribution.

Clause 7.6.8 of AS 3600 defines a limit on η which reduces with increasing values of the neutral axis parameter, k_u. The parameter k_u is directly proportional to the amount of reinforcement and is defined as follows:

$$k_u = \frac{p f_{sy.r}}{0.85\gamma f'_c} \quad \text{where} \quad p = \frac{A_r}{bd} \tag{2}$$

and A_r is the area of reinforcement over width, b, with effective depth, d, f'_c is the characteristic compressive cylinder strength of concrete at 28 days, $f_{sy.r}$ is the reinforcement minimum yield stress and γ is the stress-block parameter for concrete as defined in AS 3600, Clause 8.1.2.

According to Clause 7.6.8 of AS 3600, the moment redistribution limit at a support depends on the maximum value of k_u at any peak moment region in the adjacent spans. These peak moment regions include the peak positive moment locations in the two adjoining spans and the negative moment location being considered. This rule is considered to be more restrictive than necessary. It is anticipated that in any span of a composite slab the positive moment hinge will always be the last to form, since redistribution is permitted only from negative to positive moment regions and since the positive moment capacity is normally well in excess of the positive moment. Hence the k_u of the positive moment regions need not normally be considered, and the redistribution permitted at any support cross-section becomes simply a function of the k_u at that cross-section only, as follows:

$$\eta = \begin{cases} 0.3 & \text{for } k_u \leq 0.2 \\ 0.3 - 0.75k_u & \text{for } 0.2 < k_u \leq 0.4 \end{cases} \tag{3}$$

For convenience, all bending moments, M, will be expressed as normalised moments, m, by dividing by bd^2. Based on the rectangular stress-block approximation for a singly-reinforced concrete cross-section with $k_u \leq 0.4$, the normalised nominal moment capacity, m_u, is given by equation 4. The design equation relating the normalised design moment capacity, ϕm_u, and the normalised elastic design moment, m_e^*, is as shown in equation 5.

$$m_u = 0.85\gamma k_u (1 - 0.5\gamma k_u) f'_c \tag{4}$$

$$\phi m_u \geq (1 - \eta)m_e^* \tag{5}$$

For design including redistribution to the AS 3600 limit, equations 3, 4 and 5 can be solved to produce the expression for k_u shown in equation 6. This expression enables the required amount of negative reinforcement to be determined directly without iteration, and the amount of moment redistribution permitted can then be calculated by substitution into equation 3.

$$k_u \geq \left(a_1 - \sqrt{a_1^2 - a_2}\right) / \gamma \tag{6}$$

where $a_1 = 1$ for $k_u \leq 0.2$; $a_1 = 1 - \dfrac{0.75}{0.85\phi\gamma}\dfrac{m_e^*}{f'_c}$ for $0.2 < k_u \leq 0.4$; and $a_2 = \dfrac{1.4}{0.85\phi}\dfrac{m_e^*}{f'_c}$

An alternative approach to designing for the above redistribution limit is to design the continuous slab as a series of simply-supported spans, corresponding to 100% redistribution. Note that redistribution between these two regimes is not permissible, since the resulting cross-sections may not have sufficient rotation capacity to sustain the assumed negative moment capacity. The 100% redistribution case would be expected to produce large crack widths, but this can be avoided by providing sufficient top-face reinforcement, as described below.

148

3 NEGATIVE MOMENT REGIONS – CRACK CONTROL

The application of vertical loads to continuous slabs will generally result in cracking over the supports due to flexure. This cracking will normally occur at load levels well within the service loading range of the member and cannot be completely avoided, but design calculations can be done to ensure that crack widths are limited to an acceptable value. Top-surface cracking in slabs is particularly important where this surface will be visible, and inclusion of crack control in the design process is generally desirable.

Design for crack control is not explicitly mentioned in AS 3600. Instead flexural cracking is simply deemed to be controlled by satisfying a bar spacing limit. Parametric studies on continuous slabs have been conducted and have shown that this approach does not provide for satisfactory crack control, and a more rigorous approach is recommended. Clause 3.2.4 of BS 8110.2 [BSI 1985] suggests a crack width limit of 0.3 mm, and a method for calculating crack widths at the concrete surface is given in BS 8110.2, Clause 3.8.3. This method is based on extensive theoretical and experimental work on reinforced-concrete beams and slabs [Beeby 1979], and it is assumed to apply also to composite slabs provided they have relatively narrow steel ribs such that the slabs are effectively solid. It will be assumed that the long-term serviceability loading case to AS 1170.1 [SA 1989] is appropriate for incorporation of the British method of crack width calculation into design to Australian Standards.

The method for calculating crack width as defined in BS 8110 is unfortunately cumbersome to use in design, and a parametric study has therefore been conducted to develop a simpler method. A singly-reinforced negative moment cross-section has been considered, with the presence of the steel sheeting being ignored. Using a crack width limit of 0.3 mm, it has been found that the normalised service load moment at which this limit occurs, $m_{s.fc}$, can be related to the steel reinforcement ratio, p, the concrete cover, c, and the bar spacing, s, by the following relationship, provided c does not exceed $D/3$, where D is the slab depth [Proe and Patrick 1995b]:

$$m_{s.fc} = 1100 \left(\frac{p}{c} \right)^{3/4} \left(\frac{s}{200} \right)^{-1/8} \tag{7}$$

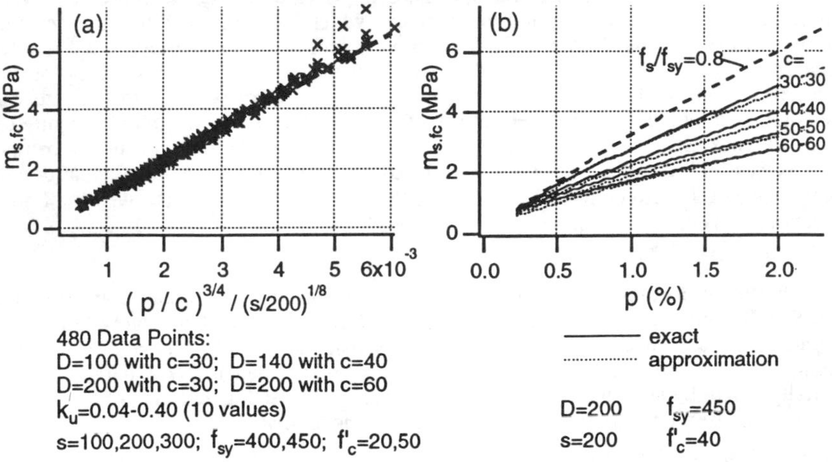

480 Data Points:
D=100 with c=30; D=140 with c=40
D=200 with c=30; D=200 with c=60
k_u=0.04-0.40 (10 values)
s=100,200,300; f_{sy}=400,450; f'_c=20,50

D=200 f_{sy}=450
s=200 f'_c=40

——— exact
·············· approximation

Figure 1. Crack control - comparison of simplified calculation with calculation to BS 8110

A comparison of this relationship with the full calculation of crack width is shown in Fig. 1(a). This graph shows data points calculated for 480 combinations of D, c, k_u and s, as listed in the figure, excluding cases which violated the AS 3600 first-cracking requirement ($M_u \geq 1.2 M_{cr}$, where M_{cr} is the first-cracking moment). The approximate relationship gives errors which are generally less than ±5%. A further comparison of this approximation with the full calculation is shown in Fig. 1(b), in which $m_{s.fc}$ is plotted against p and the influence of the various parameters is illustrated.

For design to the flexural cracking limit, the normalised design bending moment under long-term serviceability conditions, m_{sl}, must not exceed $m_{s.fc}$. Transforming equation 7 and putting $m_{sl} \leq m_{s.fc}$ gives:

$$p \geq \left(\frac{m_{sl}}{1100}\right)^{4/3} \left(\frac{s}{200}\right)^{1/6} c \tag{8}$$

Equation 8 should only be used within the ranges of parameters listed in Fig. 1(a). A modified equation covering slab depths, D, up to 400 mm has also been developed and reported elsewhere [Proe and Patrick 1995b]. Equation 8 clearly shows the strong dependence of the required quantity of reinforcement on m_{sl} and c and the relative insensitivity to s. For example, if the value of s is doubled then the required quantity of reinforcement is increased by only 12%. This result is in contrast to the provision for crack control reinforcement in AS 3600, in which s is the only parameter considered and no upper limit on c is specified.

The BS 8110 method for calculating crack width which has been used to derive equation 8 is only valid when the steel stress (f_s) does not exceed $0.8 f_{sy}$ (see Fig. 1(b)). The AS 3600 first-cracking requirement ($M_u \geq 1.2 M_{cr}$) must also be satisfied, Again, equations have been developed which enable the required amount of negative reinforcement for these criteria to be calculated directly without iteration, and these equations are reported elsewhere [Proe and Patrick 1995a, 1995b].

4 POSITIVE MOMENT REGIONS – MOMENT CAPACITY

For a sheeting profile which develops ductile mechanical resistance in conjunction with the concrete, the positive moment capacity at any cross-section may be calculated by solving the equations of equilibrium and compatibility as for a reinforced-concrete cross-section. This approach can account for the contribution of both the steel sheeting and any conventional tensile bottom-face reinforcement. The main difference from a reinforced-concrete cross-section is that account may need to be taken of the degree of composite connection between the sheeting and the concrete, which increases with the distance from either end of the sheeting. Partial shear connection is accounted for by limiting the resultant tensile force in the sheeting, T_{sh}, to that which can be developed at the cross-section considered. When the distance from an end of the sheeting increases such that T_{sh} reaches either that corresponding to the yield stress of the sheeting or the magnitude required to balance the maximum possible compressive force in the concrete, then no further increase in T_{sh} is possible and the cross-section is in complete shear connection thereafter. The limiting value of T_{sh} at complete shear connection is termed T_{csc}, and the degree of shear connection, β, is defined as follows:

$$\beta = T_{sh} / T_{csc}, \ 0 \leq \beta \leq 1 \tag{9}$$

The sheeting properties which determine its shear connection performance are measured in the Slip-Block™ Test [SA 1996]. In this test, adhesion bond is deliberately broken and two parameters are measured, a mechanical resistance factor, H_r, and a coefficient of friction, μ. The sheeting performance is termed ductile when these parameters are essentially independent of the amount of slip between the sheeting and the concrete. These parameters can be related to the sheeting tensile force as follows:

$$T_{sh} = \min\left(H_r bx + \mu R^*, \ T_{csc}\right) \tag{10}$$

$$T_{csc} = \min\left(0.85 f'_c (D - h_r)b - T_{y.r}, \ T_{y.sh}\right) \tag{11}$$

$$T_{y.sh} = A_{sh} f_{sy.sh} \tag{12}$$

where x is the distance from the end of the sheeting considered, R^* is the support reaction passing through the sheeting and A_{sh}, $f_{sy.sh}$ and h_r are the cross-sectional area, yield stress and rib height of the sheeting respectively, while $T_{y.r}$ is defined in equation 15 below.

Calculations of positive nominal moment capacity have been performed for composite slabs incorporating one proprietary brand of steel sheeting [BHP Building Products 1991]. Three levels of cross-section analysis have been used, with the first being the most rigorous and the third being the most simplified:

- moment-curvature analysis (CEB), in which the stress in the concrete is obtained from the CEB-FIP stress-strain curve and account is taken of strain compatibility over the depth of the cross-section;
- rectangular stress-block theory (RSBT), in which the stress in the concrete is obtained from the Whitney rectangular stress-block, strain compatibility is ignored and conventional reinforcement is always assumed to be at its yield stress when it lies on the tensile side of the plastic neutral axis and is ignored otherwise; and
- rectangular stress-block theory as above, except the sheeting is lumped at the height of its centroid above the soffit, $y_{c.sh}$, which varies with the degree of shear connection.

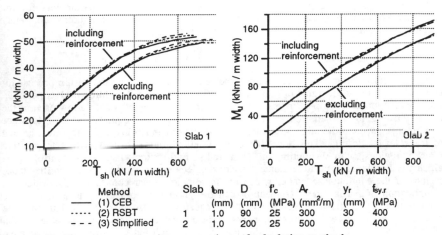

Method	Slab	bm	D	f'c	Ar	yr	fsy.r
—— (1) CEB		(mm)	(mm)	(MPa)	(mm²/m)	(mm)	(MPa)
····· (2) RSBT	1	1.0	90	25	300	30	400
– – – (3) Simplified	2	1.0	200	25	500	60	400

Figure 2. Positive moment capacity - comparison of calculation methods

These three approaches have been found to be in excellent agreement for the typical range of parameters used for these slabs (see Fig. 2), and the third method is therefore proposed for design. The equations for calculating the design moment capacity, ϕM_u, are as follows:

$$\phi M_u = \phi T \left(D - y_c - \frac{0.5T}{0.85 f'_c b} \right) + \phi M_{u.sh} (1 - \beta_1^2) \tag{13}$$

$$T = T_{sh} + T_{y.r} \tag{14}$$

$$T_{y.r} = A_r f_{sy.r} \tag{15}$$

$$y_c = \frac{T_{sh} y_{c.sh} + T_{y.r} y_r}{T} \tag{16}$$

$$\beta_1 = T_{sh} / T_{y.sh} \tag{17}$$

where $M_{u.sh}$ is the nominal moment capacity of the sheeting alone, and A_r, $f_{sy.r}$ and y_r are the cross-sectional area, yield stress and centroidal height above the slab soffit of the conventional reinforcement respectively.

151

In this formulation, the conventional reinforcement is assumed to be at its yield stress. Equation 18 must be satisfied to ensure that this assumption is valid, otherwise the contribution of the conventional reinforcement should be ignored.

$$y_r \leq D - \frac{\left(1 + f_{sy.r} / 600\right)\left(T_{y.sh} + T_{y.r}\right)}{0.85 \gamma f'_c b} \tag{18}$$

This formulation applies both to partial shear and complete shear connection regions. As a further simplification, it is useful for the designer to ascertain those regions which are guaranteed to have complete shear connection. Equation 19 is obtained from equations 10, 11 and 12 by putting $T_{sh} = T_{y.sh}$ and $\mu R^* = 0$. Any cross-section which is further than x_{csc} from the nearer end of the sheeting will always be in complete shear connection. The distance, x_{csc}, typically ranges from 1.0 m to 1.6 m.

$$x_{csc} = \frac{T_{y.sh}}{H_r b} \tag{19}$$

In order to apply the above equations, values of the following parameters for the particular narrow-rib sheeting profile must be obtained from the manufacturer's proprietary specifications and test data: A_{sh}, $f_{sy.sh}$, H_r, μ, $y_{c.sh}$ and $M_{u.sh}$.

5 CONCLUSION

The design formulations proposed in this paper are aimed at making the design process for composite slabs in steel-frame or masonry wall construction as straightforward as possible. For negative moment regions, independent equations are presented which allow each of the design criteria to be satisfied by direct calculation of the required area of top-face reinforcement without iteration, after which the designer can readily select the most critical of the relevant criteria. This approach enables redistribution from negative to positive moment regions, with the option of 100% redistribution, and also crack control to be efficiently handled. For positive moment regions, an equation is given which identifies those regions of the continuous member which are in complete shear connection, for which the complication of partial shear connection can be ignored. In practice, these regions are commonly most of the member. A simplified equation is also given for calculating the design moment capacity in complete or partial shear connection, including the contribution of any conventional bottom-face reinforcement. All simplified equations presented have been verified by rigorous methods of analysis and are suitable for use with any narrow-rib steel sheeting profile for which the shear connection performance is ductile and suitable proprietary test data is available.

6 REFERENCES

Beeby, A W (1979). The Prediction of Crack Widths in Hardened Concrete, *The Structural Engineer*, 57A, 9-17.
BHP Building Products (1991). *Bondek II*. Reference No B. Dist.
British Standards Institution (BSI) (1985). *Structural Use of Concrete, Part 2 - Code of Practice for Special Circumstances*, BS 8110 : Part 2 : 1985.
Proe, D J and Patrick, M (1995a). Design for Control of Flexural Cracking in Support Regions of Continuous One-Way Slabs, *4th Pacific Structural Steel Conf, Singapore*.
Proe, D J and Patrick, M (1995b). Design of the Support Regions of Continuous One-Way Slabs, with Particular Reference to Crack Control, BHP Research – Melbourne Laboratories Report No BHPR/SM/R/015
Standards Australia (SA) (1996). *Methods of Test for Elements of Composite Construction, Method 1 : Slip-Block™ Test*, Committee Draft, Document No BD/32/4/96-2.
Standards Australia (SA) (1994). *Concrete Structures*, AS 3600 - 1994.
Standards Australia (SA) (1989). *SAA Loading Code, Part 1: Dead and Live Loads and Load Combinations*, AS1170.1–1989.

The Mechanics of Structures and Materials, Grzebieta, Al-Mahaidi & Wilson (eds)
© *1997 Balkema, Rotterdam, ISBN 90 5410 900 9*

Load-slip characteristics of composite slab specimens

G.Taplin
Department of Civil Engineering, Monash University, Melbourne, Vic., Australia

ABSTRACT: The load-slip characteristics of two Australian composite slab decks - Bondek II and Condeck HP - are presented based upon push-off test results. The tests were conducted with a known and constant value of clamping force applied across the steel-concrete interface. The effect of embossments on the behaviour of Bondek II is shown.

1 INTRODUCTION

Composite slabs rely for their structural integrity on adequate transfer of horizontal forces across the steel/concrete interface. The interface shear strength comprises three components (Patrick 1990):

- adhesion
- mechanical interlock
- friction

Adhesion is lost once slip happens, so it is usually disregarded in an assessment of the interface shear strength. The purpose of the tests reported in this paper was to see how the interface shear strength due to mechanical interlock plus friction varies with slip. Load-slip test results are given for two steel decking types that are commonly used in Australia - Bondek II and Condeck HP, as well as for the superseded Condeck profile. The tests were conducted on small push-off specimens, which were subjected to a constant clamping force across the steel-concrete interface.

The use of a constant clamping force does not enable any information to be obtained on the separate contributions from mechanical interlock and friction, and therefore the results obtained from the tests reported here cannot be used for the prediction of the strength of composite slabs. The SLIP-BLOCK™ test (Patrick 1993) enables mechanical interlock and friction values to be separately obtained, and this test should be used to obtain values for composite slab design.

2 TEST SET UP

The push-off tests were conducted in a shear box in the Department of Civil Engineering Laboratories at Monash University (Figure 1). The shear box has servo controlled hydraulic actuators for both vertical and horizontal load.

All of the tests described in this paper were tested at a vertical clamping force of 25 kN, which was maintained under load control. This allowed upwards displacement of the concrete slab to occur, if required, as the concrete moved over embossments on the steel decking. Horizontal (push off) displacement was applied under displacement control at the rate of 0.2

Table 1: Details of the specimens tested

No.[1]	Decking	Nominal decking base metal thickness (mm)	Nominal specimen thickness (mm)	Concrete strength (MPa)	Comments	load at first slip (kN)	peak load (kN)
AM1	Condeck	0.75	125	39		78	78
AM2	Condeck	0.75	125	39		68	68
AM3	Condeck	0.75	125	39		63	63
AM4	Condeck	0.75	125	39		73	73
AM5	Condeck	0.75	125	39		56	56
AM6	Condeck	0.75	125	39		55	55
AM14	Condeck HP	0.75	125	51		56	56
AM15	Condeck HP	0.75	125	51		63	63
AM16	Condeck HP	0.75	125	51		45	45
AM20	Condeck HP	0.75	125	51		68	68
AM21	Condeck HP	0.75	125	51		71	71
AM22	Condeck HP	0.75	125	51		64	64
P1A	Condeck HP	0.75	100	41		67	67
P1B	Condeck HP	0.75	100	41		54	54
P1C	Condeck HP	0.75	100	41		57	57
P1D	Condeck HP	0.75	100	41		71	71
P2A	Bondek II lap rib	1.00	100	41	against trailing edge	51	64
P2B	Bondek II lap rib	1.00	100	41	against trailing edge	76	76
P2C	Bondek II lap rib	1.00	100	41	against trailing edge	55	74
P2D	Bondek II lap rib	1.00	100	41	against trailing edge	69	87
KA2	Bondek II lap rib	1.00	100	38	against trailing edge	68	85
KA31	Bondek II lap rib	1.00	100	35	against trailing edge	63	84
KB1	Bondek II lap rib	1.00	100	38	against leading edge	62	83
KB2	Bondek II lap rib	1.00	100	38	against leading edge	67	85
KB3	Bondek II lap rib	1.00	100	38	against leading edge	83	102
KC1	Bondek II dovetail	1.00	100	31	against trailing edge	17	30
KC2	Bondek II dovetail	1.00	100	31	against trailing edge	26	37
KC3	Bondek II dovetail	1.00	100	31	against trailing edge	28	51
P3A	Bondek II dovetail	1.00	100	28	against leading edge	17	30
P3B	Bondek II dovetail	1.00	100	28	against leading edge	29	41
P3C	Bondek II dovetail	1.00	100	28	against leading edge	30	41

1 AM series tests were carried out by Abdel-Malek (1994)
   P series tests were carried out by Pavlovic (1994)
   K series tests were carried out by Konstanty (1995)

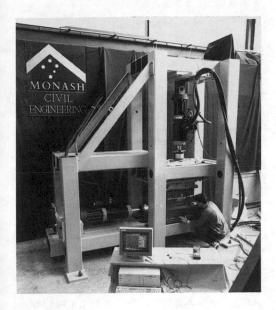

Figure 1. Test rig

mm/minute. Horizontal and vertical load, and slip between the concrete and the steel decking were recorded.

3 SPECIMEN DETAILS

Details of the specimens tested are given in Table 1. The surface of the decking was cleaned in each case to ensure that there was no grease on the surface to interfere with bond between the concrete and the steel. Each specimen was reinforced with one piece of steel mesh comprising 7 mm bars at 200 mm centres. The specimens were all 300 x 300 mm in plan dimension and 100 or 125 mm thick (nominal dimensions). Each specimen contained one decking rib cast into the centre of the specimen, as illustrated in Figure 2 below.

3.1 *Condeck specimens*

Condeck has nominal dimensions of 55 mm x 23 mm L shaped ribs at 300 mm centres (Stramit Industries 1989). Each L rib is formed by overlapping male and female ribs from adjacent sheets. Six Condeck specimens were tested.

3.2 *Condeck HP specimens*

The Condeck HP profile has nominal dimensions of 55 mm x 30 mm L shaped ribs at 300 mm centres (Stramit Industries 1992). Like Condeck, each L-shaped rib is formed by overlapping male and female ribs from adjacent sheets, however unlike Condeck, the ribs have formed, rather than flat, surfaces. Ten Condeck HP specimens were tested.

155

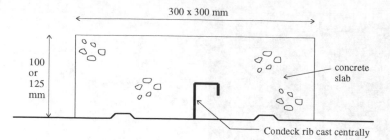

Figure 2. End view of "Condeck" specimen showing cast in rib

3.3 Bondek II specimens

The Bondek II profile comprises two dovetail ribs, and one L shaped rib, at nominally 200 mm centres (Lysaght Building Industries 1991). The L-shaped rib is formed where the sheets overlap. As the specimens were only 300 mm wide, the dovetail ribs and the L shaped rib were tested separately. The interface shear strength of the sheeting containing both ribs can be determined by taking a "weighted average" of each type of rib. Bondek II also has embossments rolled into the top surface of the ribs. Due to the action of the roll-forming process these are not symmetric - the dimensions are given in Figure 3. In the testing, distinction was therefore made between slip occurring against the trailing edge of the embossment, and slip occurring against the leading edge of the embossment (as defined by the relative concrete movement indicated in Figure 3). Six Bondek II specimens were tested.

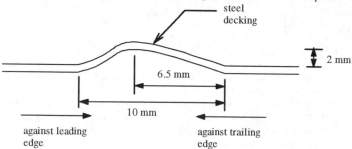

Figure 3. *Embossment on Bondek II rib (not to scale)*

4 EXPERIMENTAL RESULTS

Typical load slip plots for Condeck and Condeck HP are presented in Figures 4 and 5 below. All specimens tested exhibited a similar response. The horizontal load dropped after first slip occurred, and then stabilised at approximately 70-80% of the peak value over a large slip range. Typical load-slip results for Bondek II are presented in Figure 6. All specimens tested exhibited a similar response. The dovetail rib had a weaker interface shear strength than the L rib. This is because the inside faces of the dovetail are not restrained, and can bend inwards, away from the concrete.

Bondek II exhibited an increase in the horizontal load after the advent of slip. To investigate this, large slip displacements were applied. These displacements are much greater than would occur at failure in composite slabs, where the range of interest is slip values up to about 10 mm,

156

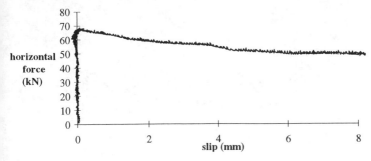

Figure 4. *Typical Condeck load-slip response at a vertical clamping force of 25 kN*

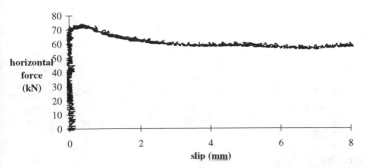

Figure 5. *Typical Condeck HP load-slip response at a vertical clamping force of 25 kN*

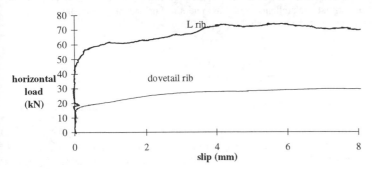

Figure 6. *Typical Bondek II load-slip response at a vertical clamping force of 25 kN*

but the tests served to illustrate the importance of the embossments in the performance of Bondek II. Figure 7 shows the horizontal force oscillating in a periodic manner. The peaks in the horizontal force occur every 25-26 mm of slip. This corresponds to the spacing of the embossments on the rib of the Bondek II.

The load at first slip, and the peak load results are summarised in Table 1. It must be remembered that these results are all for a clamping force of 25 kN, so no general conclusions can be drawn about the interface shear strength properties of the composite slab specimens at other values of clamping force, nor can direct comparisons be made between the two products, Bondek II and Condeck HP.

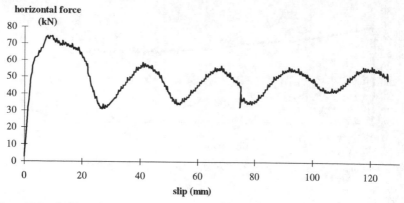

Figure 7. *Bondek II load-slip response at large values of slip (L-shaped rib)*

5 CONCLUSIONS

These tests, carried out with a clamping force of 25 kN, show that the interface shear strength of Condeck and Condeck HP decreases by 20-30% after slip commences. Over the range of slips that is of practical interest in composite slabs (up to 10 mm say), Bondek II showed an increase in interface shear strength above that when slip commenced. By testing Bondek II over a large slip range, the effect of the embossments was illustrated.

6 ACKNOWLEDGMENTS

The author would like to acknowledge the assistance of former postgraduate student Nabil Abdel-Malek, and undergraduate students Nick Pavlovic and Susan Konstanty who carried out the experimental work that formed the basis of this paper.

REFERENCES

Abdel-Malek, N. 1994 Interface shear strength of composite slab blocks. *MEngSc minor thesis* Monash University

Konstanty, S. 1995 Slip behaviour of Bondek II composite slabs. *CIV4210 Project A* Monash University

Lysaght Building Industries 1991 Bondek II Composite Slabs BDII-2 Design Manual for Masonry Wall & Steel-Frame Construction Part A Composite Design Manual

Patrick, M. 1990 A new partial shear connection strength model for composite slabs *AISC Steel Construction* Vol 24 No 3

Patrick, M. 1993 Testing and design of Bondek II composite slabs for vertical shear *AISC Steel Construction* Vol 27 No 2

Pavlovic, N. 1994 Comparative performance of Condeck HP and Bondek II composite slabs. *CIV4210 Project A* Monash University

Stramit Industries 1989 Condeck Structural Steel Deck Technical Design Manual

Stramit Industries 1992 Condeck HP Structural Steel Deck Technical Design Manual

The Mechanics of Structures and Materials, Grzebieta, Al-Mahaidi & Wilson (eds)
© 1997 Balkema, Rotterdam, ISBN 90 5410 900 9

Design of composite beams with large steel web penetrations

C.G.Chick, P.H.Dayawansa & M.Patrick
BHP Research, Melbourne Laboratories, Clayton, Vic., Australia

ABSTRACT: Methods for designing simply-supported composite beams incorporating large steel web penetrations for both strength and deflection are proposed which suit Australian practice. A penetration may be either circular or rectangular, unreinforced or reinforced, and possibly eccentric to the steel beam centroid. The methods can also be used to design the steel beam prior to composite action. Four methods of strength design have been reviewed with regard to their accuracy and simplicity of use, and one of these methods has been suitably modified to conform with the latest Australian composite beam design requirements. Two methods for calculating deflection have also been reviewed, and a method suitable for use in a wide range of practical situations is proposed. Both design methods are validated using test results from overseas and Australia.

1 INTRODUCTION

Large rectangular or circular penetrations are often made in the steel web of composite beams for the passage of horizontal building services. They may be up to 70 per cent of the depth of the steel beam, with a length-to-depth ratio of up to two. Penetrations of this size weaken the beam locally and reduce its overall flexural stiffness. This can partly be overcome by welding narrow steel plates along their horizontal edges, although the economics of reinforced penetrations needs to be carefully considered. The steel beam may also have to be designed for the construction stages prior to the development of composite action. No specific information is provided in an Australian Standard for either of these design situations. This is also the case in overseas design Standards.

Four methods of strength design for composite beams incorporating web penetrations have been reviewed, viz.: Redwood and Cho (1993), ASCE Task Committee (1992), Lawson (1987) and Oehlers and Bradford (1995). The method specified by the ASCE Task Committee is based mainly on the research of Darwin and his co-workers. In this paper, these design methods are referred to as those by Redwood, Darwin, Lawson and Oehlers, respectively. All of these methods can also be used to design bare steel beams with web penetrations, which is a situation that occurs during construction of a composite beam prior to the concrete hardening.

Two methods of deflection design for composite beams incorporating web penetrations proposed by Lawson (1987) and Tse and Dayawansa (1992) have also been reviewed.

Large experimental programs were undertaken during the development of the Redwood and Darwin methods. The composite beams that were tested either had solid concrete slabs or composite slabs formed using trapezoidal profiled steel sheeting with wide steel ribs. The new Australian Standard for simply-supported composite beams AS 2327.1 (SAA 1996) requires that the profiled steel sheeting has narrow steel ribs (not more than 20% void area). Trapezoidal profiles like those tested by Redwood and Darwin were excluded at the time because they are prone to rib shear failure (Patrick et al. 1995). The results of tests performed by the authors on composite beams incorporating Australian decks conforming to AS 2327.1 have been used

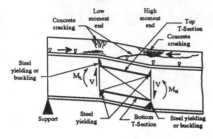

(a) Behaviour at ultimate load

(b) Australian test beam

Figure 1. Composite beam with web penetration

while reviewing the design methods, and are considered more relevant than the tests involving trapezoidal profiles.

2 BEHAVIOUR OF WEB PENETRATION REGION AT FAILURE

The behaviour of the web penetration region at failure is briefly described in relation to Figure 1(a). One of the Australian test beams referred to later in the paper is shown in Figure 1(b).

In the absence of shear force, a penetration in the steel web of a composite beam reduces the moment capacity of cross-sections within the width of the opening due to the reduction in steel cross-sectional area. Vierendeel action occurs over the penetration length when shear force is present. This causes additional secondary moments to develop in the top and bottom T-sections. The shear force in the region is shared by the concrete slab and the webs of these T-sections.

Tests show that failure may occur when plastic hinges form in the T-sections at the high and low moment ends of the penetration due to the combined effects of flexure, shear and Vierendeel action. Failure may be accompanied by a major diagonal crack which forms suddenly in the concrete slab above the penetration and extends into the span. Factors significantly affecting its formation are the geometry of the profiled steel sheeting and the type and amount of reinforcement in the slab. A separate investigation is being undertaken to find practical means of controlling this type of cracking.

Owing to the complexity of the behaviour illustrated in Figure 1(a), a semi-empirical approach has been taken for strength design which is based on some simplifying assumptions.

3 REVIEW OF STRENGTH DESIGN METHODS

The four strength design methods referred to in Section 1 are briefly reviewed in this section.

The methods all use rectangular stress block theory to calculate the design moment capacity (ϕM_b) of overall cross-sections ignoring the effects of shear and Vierendeel action. Account is taken of the degree of shear connection at a cross-section.

The strength design methods by Redwood and Darwin share the same moment-shear interaction curve given by Equation (1) which has been calibrated against the results of a large number of tests. In this equation, the design shear capacity (ϕV_u) of the beam section at the penetration is calculated assuming a contribution from the concrete slab in addition to that from the steel webs of the T-sections. Account is taken of the presence of flexural stresses in the T-sections caused by Vierendeel action when calculating the nominal shear capacity of these steel sections. The action effects M^* and V^* are calculated at the mid-length of the penetration.

$$\left(\frac{M^*}{\phi M_b}\right)^3 + \left(\frac{V^*}{\phi V_u}\right)^3 = 1 \tag{1}$$

In Darwin's method, a significant simplifying assumption made with regard to the calculations required is that the neutral axes of the top and bottom T-sections lie in their

160

respective steel flanges. However, this assumption has little effect on the accuracy of the method.

The methods by Redwood and Darwin essentially only differ in the way the nominal shear capacity V_u is calculated. Although the values calculated are similar, Darwin's approach is preferred because the calculations are simpler and therefore are more suitable for design.

Lawson accounts for moment-shear interaction using a simplified model rather than an empirically-based curve like Equation 1. The shear capacity (ϕV_u) is calculated ignoring the contribution of the bottom T-section. A similar amount of calculation is required using this method compared with either Redwood's or Darwin's methods, and it is also less accurate. Therefore, its use is not recommended.

Oehlers method is also based on the moment-shear interaction relationship given in Equation 1. However, ϕV_u is calculated differently using an iterative procedure to satisfy von Mises yield criterion more exactly than the other methods, and any contribution from the concrete slab is ignored. This iterative procedure can have difficulty converging. Also, ignoring the shear capacity of the concrete is very conservative. Therefore, this method is also not favoured.

4 RECOMMENDED STRENGTH DESIGN METHOD

Darwin's strength design method has been adapted to suit Australian practice and is briefly outlined below. A value of $\phi = 0.9$ has been chosen for use in Equation (1) which is the same value adopted for bending and vertical shear in AS 2327.1.

The design moment capacity (ϕM_b) of cross-sections is determined using rectangular stress block theory in accordance with the principles set down in AS 2327.1, but ignoring the presence of shear force. Equations of the type needed to calculate ϕM_b have been derived by Patrick and Poon (1989). Darwin requires that for a reinforced penetration, ϕM_b must not exceed the design moment capacity of the composite beam without a web penetration (for the same degree of shear connection, and also ignoring shear force).

At the web penetration, V_u is given as:

$$V_u = V_t + V_b \tag{2}$$

where,

$$V_t = \frac{\sqrt{6} + \mu_t}{v_t + \sqrt{3}} V_{pt} \le V_{pt} + 0.29\sqrt{f_c'} A_{vc} \tag{3}$$

$$V_b = \frac{\sqrt{6} + \mu_b}{v_b + \sqrt{3}} V_{pb} \le V_{pb}$$

$$V_{pt} = 0.6 f_{yw} s_t t_w \;;\; V_{pb} = 0.6 f_{yw} s_b t_w$$

$$\mu_t = \frac{2F_r d_r + F_{c.H} d_{c.H} - F_{c.L} d_{c.L}}{V_{pt} s_t} \;;\; \mu_b = \frac{2F_r d_r}{V_{pb} s_b} \;;\; v_t = \frac{L}{s_t} \text{ and } v_b = \frac{L}{s_b}$$

$F_{c.H}$ equals F_{cc} or F_{cp} depending on whether there is complete or partial shear connection at the high-moment end, and it follows that:

$$F_{c.L} = F_{c.H} - (n_H - n_L) f_{ds}$$

Also, $d_{c.H} = D_c - \dfrac{F_{c.H}}{1.7 f_c' b_{cf}}$, $d_{c.L} = \dfrac{F_{c.L}}{1.7 f_c' b_{cf}}$ for solid slabs and $d_{c.L} = h_r + \dfrac{F_{c.L}}{1.7 f_c' b_{cf}}$ for composite slabs with sheeting ribs perpendicular to the steel beam.

If $\dfrac{\sqrt{6} + \mu_t}{v_t + \sqrt{3}} > 1.0$ in Equation (3), then a further condition is that $F_{c.H} \le f_{yf}(t_f(b_f - t_w) + A_r)$

and μ_t should be recalculated, and then applied in the following equation:

$$V_t = \frac{\mu_t}{v_t} V_{pt} \ge V_{pt} \;;\; V_t \le V_{pt} + 0.29\sqrt{f_c'} A_{vc}$$

The accuracy of this method is examined in Section 7.1 using the results obtained from overseas and Australian testing of simply-supported composite beams.

5 REVIEW OF DEFLECTION CALCULATION METHODS

Two methods for calculating the deflection of simply-supported composite beams are discussed. They are applicable provided the behaviour under service loads remains elastic. It is reasonable to suggest that this check should be made on the basis of overall bending of the beam with the penetrations, but ignoring Vierendeel action which may cause limited yielding around the corners of the penetrations.

Lawson (1987) presents a method for calculating the mid-span deflection of a simply-supported composite beam with two, symmetrically-placed web penetrations. Two simple equations are given to directly calculate the additional deflection components due to bending and shear deformations in the region of the web penetrations. The loading cases considered are either uniform loading or a single point load applied at mid-span.

Tse and Dayawansa (1992) present a more general method than Lawson, which is also applicable to bare steel beams. It is shown in Section 7.2 that the method can accurately predict the deflected shape of simply-supported composite beams, even with asymmetrically-placed penetrations.

6 RECOMMENDED DEFLECTION CALCULATION METHOD

The method of Tse and Dayawansa can be used to calculate the total deflection at any cross-section of a simply-supported composite beam arising from both short- and long-term loading effects. The approach taken is to calculate three separate components and to add them together to give the total deflection. These components are: (1) the deflection with no penetration calculated in accordance with the simplified method in AS 2327.1, (2) the additional bending deflection due to the penetration, and (3) the additional shear deflection due to the penetration. The simplified method in AS 2327.1 accounts for the effects of long-term loading and partial shear connection. These effects are not considered when calculating the latter two components.

The additional deflection components are calculated assuming that away from the penetrations the beam is rigid, and that these rigid portions are connected together top and bottom by flexible T-sections. A differential design moment (M_d^*) is assumed to act across the penetration and is calculated as follows:

$$M_d^* = M_H^* - M_L^* \tag{4}$$

The additional bending deflection across the penetration (δ_H) is a function of the rigid arm end rotations (θ_L and θ_H) and the secondary moment induced by Vierendeel action at the penetration $\left(M_{se}^* \right)$ and is given as follows for a concentric penetration

$$\delta_H = \frac{L}{3}(\theta_L - 2\theta_H) - \left(\frac{M_{se}^* L^2}{6EI_T} \right)$$

where,

$$\theta_L = \left(\frac{M_{se}^* I_0 \left(L^2 - 2L(3b - 2L) \right) - M_d^* I_T L(3b + 2L)}{6EI_0 I_T S} \right); \quad \theta_H = -\left(\frac{\left(M_d^* I_T + 2M_{se}^* I_0 \right) L}{2EI_0 I_T} \right) - \theta_L$$

$$M_{se}^* = -\frac{V^* L}{2}$$

The additional shear deflection across a penetration δ_v' for the same situation is calculated as:

$$\delta_v' = \delta_s' + \frac{L}{3}(\theta_L' - 2\theta_H'), \text{ where } \delta_s' = \frac{kV^* L}{Gs_t t_w}, \quad \theta_L' = \frac{2L\delta_s'}{3bS} \text{ and } \theta_H' = \left(\frac{\delta_s'}{b} \right) - \theta_L'$$

Reference should be made to Tse and Dayawansa (1992) to take account of penetration eccentricity.

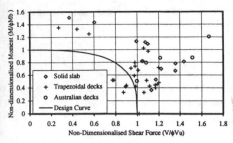

Figure 2: Experimental vs. Design Strengths

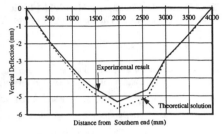

Figure 3: Experimental vs. Theoretical Deflections

7 COMPARISON WITH TEST RESULTS

7.1 Strength Design Method

The method was applied to five Australian tests conducted by the authors and thirty seven overseas tests found in the literature, and the results are shown in Figure 2. A strength reduction factor $\phi = 0.9$ was used. Although several test with trapezoidal profiled steel sheeting fall inside the curve, the results show the conservative nature of the proposed design method for Australian practice.

7.2 Deflection Calculation Method

The experimental and theoretical deflected shape of a beam tested as part of the current investigation is shown in Figure 3. The deflection predicted by the method agrees well with the experimental results.

8 CONCLUSIONS

Four strength design methods for composite beams with web penetrations have been reviewed, and one of these methods has been modified to conform with AS 2327.1. A method has also been recommended for calculating deflections. The accuracy of these methods has been examined using test results, and both are considered suitable for design use in Australia.

9 REFERENCES

ASCE Task Committee on Design Criteria for Composite Structures in Steel and Concrete 1992. Proposed Specification for Structural Steel Beams with Web Openings. *Journal of Structural Engineering*. 118(12):3315-3349.

Darwin, D. and Donahey, R.C. 1988. LRFD for Composite Beams with Unreinforced Web Openings. *Journal of Structural Engineering*. 114(3):535-552.

Lawson, R.M. 1987. Design for Openings in the Webs of Composite Beams. *CIRIA Special Publication 51*. CIRIA/SCI, London.

Oehlers, D. and Bradford, M. 1995. Composite Steel and Concrete Structural Members: Fundamental Behaviour, *Pergamon*.

Patrick, M., Dayawansa, P.H. and Watson, K.B. 1995. A Reinforcing Component for Preventing Longitudinal Shear Failure of Composite Edge Beams. *Proc. 4th Pacific Structural Steel Conference*, Singapore.

Patrick, M. and Poon, S.L. 1989. Composite Beam Design and Safe Load Tables. *Australian Institute of Steel Construction*.

Redwood, R.G. and Cho, S.H. 1993. Design of Steel and Composite Beams with Web Openings. *Journal of Constructional Steel Research*. 25:23-41.

Standards Association of Australia (SAA). 1996. AS 2327.1—1996 Composite structures, Part 1: Simply supported beams.

Tse, D. and Dayawansa, P.H. 1992, Elastic Deflection of Steel and Composite Beams with Web Penetrations. *The Structural Engineer*. 70(21):372-376.

10 NOTATION

a, b	= distance from the high- and low-moment ends, respectively, to the supports
A_r	= cross-sectional area of steel plate reinforcement along top or bottom edge of the penetration
A_{vc}	$= 3D_c(D_c - h_r)$, effective area of concrete slab for shear
b_{cf}	= effective width of the slab of the composite beam cross-section
b_f	= width of the steel top flange
$d_{c.H}, d_{c.L}$	= depth from outside edge of top flange to centroid of concrete force at high- and low-moment ends, respectively
d_r	= distance from the outside edge of flange to centroid of steel plate reinforcement
D_c	= overall depth of slab
f_c'	= 28-day characteristic compressive cylinder strength of concrete
f_{ds}	= design shear capacity of a shear connector corresponding to n_M connectors
f_{yf}, f_{yw}, f_{yr}	= yield stresses of the flange, web and steel plate reinforcement, respectively
F_{cc} and F_{cp}	= see Paragraphs D2.3.2 and D2.3.3, AS 2327.1 for definitions
$F_{c.H}, F_{c.L}$	= forces in the concrete slab at high- and low-moment ends of the penetration, respectively
F_r	= force in steel plate reinforcement along one edge of the penetration = $f_{yr}A_r \le (f_{yw}t_w L)/2\sqrt{3}$
G	= shear modulus of the material (for steel, $G = 80000$ MPa)
h_r	= height of the ribs of the profiled steel sheeting
I_0	= second moment of area of the gross cross-section at the penetration
I_T	= second moment of area of the top T-section
k	= shear coefficient ($k = 1$ for the current study)
L	= length of the penetration
M^*	= design moment acting at the mid-length of a penetration
M_b	= nominal moment capacity of a composite beam in the region of a penetration ignoring the presence of shear force, calculated with a degree of shear connection corresponding to that at the mid-length of the penetration
M_H^*, M_L^*	= design moment at high- and low-moment ends, respectively for serviceability limit state
n_H, n_L	= number of shear connectors between end of the beam and high- and low-moment ends, respectively
n_M	= number of shear connectors between end of the beam and mid-length of the penetration
s_t, s_b	= depth of the web of the top and bottom steel T-sections, respectively
S	= span of the beam
t_f, t_w	= thickness of the flange and web of the steel beam, respectively
V^*	= design shear force acting at the mid-length of a penetration
V_c	= nominal shear capacity of the concrete slab
V_{pt}, V_{pb}	= plastic shear capacity of the web of top and bottom steel T-sections, respectively
V_t, V_b	= nominal vertical shear capacities of the top and bottom steel T-sections, respectively
V_u	= nominal shear capacity of a composite beam in the region of a penetration
δ_s'	= additional shear deflection neglecting geometric continuity
δ_v'	= differential shear deflection across the penetration

The Mechanics of Structures and Materials, Grzebieta, Al-Mahaidi & Wilson (eds)
© *1997 Balkema, Rotterdam, ISBN 90 5410 900 9*

Behaviour of concrete filled steel tubular columns subjected to repeated loading

K.Thirugnanasundralingam, P.Thayalan & I.Patnaikuni
Department of Civil and Geological Engineering, RMIT University, Melbourne, Vic., Australia

ABSTRACT: Concrete filled steel tubular columns (CFST) have been used in structures such as high rise buildings, bridges and offshore structures for sometime. These structures are often subjected to *variable repeated loading* (VRL). When the VRL exceeds certain limit it causes excessive inelastic deformation which grows with repetition of the load and eventually leads to incremental collapse. An experimental investigation carried out on concrete filled steel tubular columns subjected to static and variable repeated loadings has been reported in this paper. Variables considered in this study are, concrete strength, end moment and effect of filling. It has been found from the experiments that the incremental collapse limit lies between 70% and 79% of the static collapse load for composite columns.

1 INTRODUCTION

Concrete filled steel tubular columns (composite columns) have been used in structural engineering applications for sometime. Use of this type of columns in high rise buildings is increasing due to its high performance, constructability, architectural feature and other economical reasons (Webb and Peyton, 1990). Composite columns have also been used in bridge piers and in offshore platforms.

Structures such as high rise buildings, offshore platforms and bridges are often subjected to heavy winds, waves and/or earthquake loading. The magnitude and position of such loads are highly variable and repeated a number of times. When the magnitude of such load exceeds certain limit, it causes inelastic deformation in the structure. This inelastic deformation grows with repetition of the load and eventually leads to local or complete collapse of the structure. It may also cause serviceability problems such as excessive deformation, settlement and cracking under seismic load. The failure of the structure in this circumstance is progressive or incremental and termed "Incremental Collapse"(IC), rather than static collapse (plastic collapse) (Neal, 1965).

Research on concrete filled steel tubular (circular) columns todate were mainly focused on using high strength concrete and thin walled steel tubes (O'Shea, 1995), effect of confinement (Ruoquan, 1992), behaviour of slender columns (Neogi, 1969 & O'Brien, 1993) etc. These investigations were carried out under static loading. Although investigations have been carried out on frames and joints made of hollow sections subjected to variable repeated loading (VRL) (Grundy, 1994) however, none investigated the behaviour of composite columns under VRL.

This paper reports the experimental investigation carried out on concrete filled steel tubular (circular) columns under static and VRL. Results of the experiments are presented and discussed.

2 TEST PROGRAM

Eight concrete filled steel tubular columns of 114.3 mm outer diameter, 3.2 mm thick steel tube and 1200 mm height were tested under static and VRL. Steel tubes of Grade 350 cold formed black structural grade were filled with concrete of nominal strength of 40 MPa and 80 MPa. Columns were tested under 15 mm and 30 mm eccentricity using a 1000 kN capacity material testing system (MTS). In addition to composite columns, four hollow steel tubular columns of the same size were also tested under static and VRL.

The steel tubes were cut to size and delivered. Sharp edges of the tubes were filed out. The steel tubes were covered at the bottom by a plastic cap and concrete was poured in three layers. Each layer was compacted using a poker vibrator. After casting, column specimens were covered with steel lids at the top to avoid moisture loss.

Two different concrete mixes were used with water/cement ratios of 0.45 and 0.3 to obtain 40 MPa and 80 MPa respectively. Condensed silica fume and superplasticiser were also used in order to obtain higher strength.

Eighteen 75 mm diameter and 150 mm height concrete cylinders were also cast with the same batch of concrete in two layers and vibrated in the vibrating table for two minutes. Twenty four hours after casting the concrete cylinders were demoulded and cured in a water tank until they were tested.

Specially fabricated steel end caps were welded to the long column specimens. Two embossments were provided to the horizontal surface of the end caps which enabled the required eccentric (15 mm and 30 mm) application of the axial load (Figure 1).

The column specimens were instrumented with strain gauges and LVDTs at mid-height. Three strain gauges, two along the longitudinal direction and other along the hoop direction were provided in the bending plane. Three LVDTs were also mounted at mid height of which two at bending plane and one perpendicular to bending plane (Figure 1) to measure lateral deflections. In addition to these load and vertical stroke were also measured by the MTS.

The long column specimen was setup in the compression testing machine at the required eccentricity. A safety belt was then wrapped around the specimen and tied to the loading frame of the machine. The MTS was used to test long specimens and concrete cylinders. The machine is provided with a sophisticated control system, Testar which is capable of loading at any desired rate. A datataker DT 500 series data acquisition system was used to record the data.

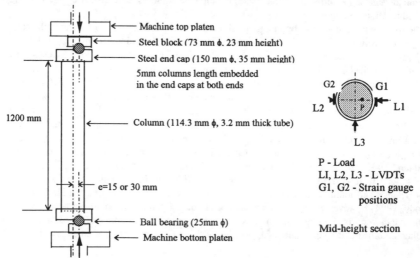

Figure 1. Typical column setup

2.1 *Testing*

Static tests were performed in four composite column specimens and two hollow steel tube specimens to determine their ultimate load carrying capacity. Columns were tested under pin end conditions with 15 and 30 mm eccentricities (Figure 1) by having two steel blocks of 73 mm diameter glued onto the top and bottom platens of the machine. The load was applied under displacement control at a rate of 2.5mm/10minute which enabled the investigation of the complete behaviour of the column. The test was continued well into the post peak range. Axial load, axial shortening, lateral deflection and strain readings were recorded at every second interval.

Four composite column specimens and two hollow steel tube column specimens were tested under variable repeated loading. Pin end conditions with 15 mm and 30 mm eccentricities were maintained as for static tests. In this test, the load was repeated (cycled) between prescribed values. The lower limit was chosen to be 5% of the collapse load for all the tests. The upper limit was set to 20%, 40%, 50%, 60%, 70%, 80% and 90% of the static collapse load. At each load level, the loading was repeated 30 - 50 times depending on the load range. The load was applied at a rate of 100 kN/minute. At the end of each load range the specimen was unloaded to 2 kN and residual measurements were recorded. The specimens were not unloaded to zero load as it required resetting of the machine etc. Axial load, deflection and strain measurements were recorded at 3 seconds intervals using the Datataker data acquisition system. After the repeated load test, the specimens were subjected to static test to determine their remaining capacity.

Concrete cylinders were tested at a loading rate of 2.5mm/10minute. Prior to the tests surfaces of the cylinders were cut using a rotating saw to obtain a smooth surface. At least three cylinders were tested with each column.

2.2 *Test observations*

The MTS control system, Testar displayed on-line plot of axial load vs vertical stroke. This facilitated relating naked eye observations to load and vertical stroke from which other information could be extracted (strains and lateral deflections). Observation for local buckling of the tube was made for each specimen.

A local plastic collapse mechanism formed at the mid-height region in the static collapse test after the peak load was reached. Inward local collapse mechanism was observed in hollow tubes while outward collapse mechanism was seen in composite columns.

In the case of the composite column with 80 MPa concrete and 15 mm eccentricity, the specimen collapsed during the 15th cycle of the 5% to 90% load range. Close observation revealed an outward local buckling (bulging) in the compression side at the mid-height region. It should be noted here that in the other five columns subjected to variable repeated loading local buckling was not observed during the repeated load tests. They were observed after the peak load when subjected to static load.

3 RESULTS AND DISCUSSIONS

3.1 *Static test*

Load vs strain and load vs lateral deflection relationships obtained for the composite column of 40 MPa with 30 mm eccentricity subjected to static test are shown in Figures 2 to 5. Mid-height deflections and strains in the tension(T) and compression(C) sides of the bending plane at the peak load are tabulated in Table 1 for all the columns.

It can be seen from the load - deflection relationship shown in Figure 2 that the behaviour of the column is elastic at the initial stage of loading and becomes inelastic. In the post peak range, the deflection growth is very fast while unloading, this indicates buckling of the column. Similar phenomena were observed in Figures 3, 4 and 5.

It can be seen from Figure 3 that the compressive steel at mid-height has yielded well into the plastic range at the peak load. This is based on the assumption that the yield strain of Grade 350 cold formed steel is 1733 $\mu\varepsilon$ according to the Manufacturer's data (Hancock, 1994). Tensile steel at mid-height is also yielding at peak load (Figure 4). However, the hoop strain at mid-height was well below yield strain (Figure 5). Local buckling could have occurred in the compressive steel when the strain is approximately 15500 $\mu\varepsilon$ according to hollow stub column tests.

Behaviour of the other columns were similar except in the case of the hollow tubes and the composite column of 40 MPa and 15 mm eccentricity. For hollow tubes and the column, the tensile steel at mid-height was in the elastic region at peak load.

3.2 *Repeated load test*

The lateral deflection vs time and longitudinal strain vs time relationships obtained for the composite column of 40 MPa concrete with 30 mm eccentricity subjected to variable repeated

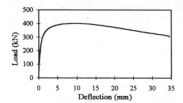

Figure 2. Load vs mid-height deflection

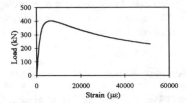

Figure 3. Load vs longitudinal strain (Compression)

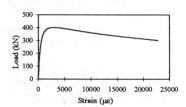

Figure 4. Load vs longitudinal strain (T)

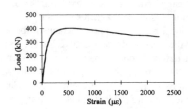

Figure 5. Load vs hoop strain (Tension)

Table 1. Static test

Specimen	Hollow tube		f_c' =40 MPa		f_c' =80 MPa	
Eccentricity (mm)	15	30	15	30	15	30
Concrete strength (MPa)	-	-	44.1	44.1	81.1	81.1
Age of concrete (days)	-	-	125	125	119	119
Static Collapse load (kN)	244	204	527	401	683	469
Vertical stroke at peak load (mm)	6.9	7.9	7.5	7.4	17.8	7.7
Deflection at peak load (mm)	11.7	14.3	11.1	13.5	12.6	11.6
Longitudinal strain at peak load ($\mu\varepsilon$) - T	763	1489	1617	2916	-	2228
Longitudinal strain at peak load ($\mu\varepsilon$) - C	7165	7833	5460	6316	6267	7920
Hoop strain at peak load ($\mu\varepsilon$) - T	355	623	331	524	334	490

168

loading are shown in Figures 6 and 7. For clarity of presentation only maximum and minimum deflections and strains obtained in each cycle are given.

It can be seen from the deflection vs time (Figure 6) and strain vs time (Figure 7) plots that the growth of deflections and strains have settled with repetition of load up to 5% to 70% load ranges and started growing with repetition of load once the load exceeds approximately 70%. The rate of growth is high at 80% and 90%. Incremental collapse could be clearly seen beyond 80% to 90%. Similar phenomena have been observed in other columns including hollow tubes.

Accurate determination of the IC limit experimentally is not an easy task. Instead of that, the IC limit was calculated using two successive load ranges where growth was observed. A hypothesis developed by Tin Loi (1977) has been used to determine the IC limit. The hypothesis states that " cyclic growth rate of deflection is proportional to the load by which IC limit is

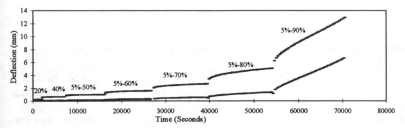

Figure 6. Deflection vs time - Variable repeated loading

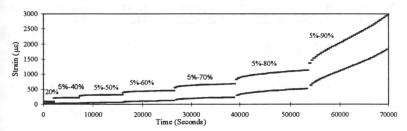

Figure 7. Strain vs Time - Variable repeated loading

Table 2. Repeated load test

Specimen	Hollow tube		f'_c =40 MPa		f'_c =80 MPa	
Eccentricity (mm)	15	30	15	30	15	30
Concrete strength (MPa)	-	-	52	52	109	109
Age of concrete (days)	-	-	144	151	158	151
Incremental collapse limit (%)	80-90	80-90	80-90	80-90	80-90	80-90
Incremental collapse load (kN)	171	147	411	309	540	342
IC load/SC load (%) -IC limit	70	72	78	77	79	73
Remaining capacity (kN)	242.6	203.5	519.4	379.5	615	467.8
Mid - height deflection (mm)	10.2	-	10.6	16.0	-	10.6
Longitudinal strain (µε) - T	-	1550	2323	4059	-	2270
Longitudinal strain (µε) - C	6663	9424	-	9140	-	5561

exceeded ". This has been applied to composite beams (Thirugnanasundralingam, 1991) and slabs (Taplin and Grundy, 1995) and found to be working well. The growth rate of deflection and strains per cycle were obtained by linear regression analysis.

Incremental collapse limit load ranges obtained from the data plots, Incremental collapse load calculated using the Tin Loi's hypothesis, the remaining capacity of all the columns are tabulated in Table 2.

The remaining capacities of columns after undergoing variable repeated loading in Table 2 are much closer to the Static collapse loads of the similar columns in Table 1, except for the composite column of 80 MPa with 15 mm eccentricity. Premature failure of this column could be due to inferior quality of core concrete.

4 CONCLUSIONS

- Failure of hollow tube and composite column specimens under static loading was due to column instability.

- Incremental collapse failure occurs in composite and hollow tube columns when subjected to variable repeated loading.

- The incremental collapse limit lies within 70% to 72% of the static collapse load for hollow steel tube columns and 73% to 79% of the static collapse load for concrete filled steel tubular columns.

- Although the ultimate capacity of hollow tube and composite columns remain the same even after undergoing repeated loading, however the deformation was not. The deformation at IC limit is large enough to render these columns unserviceable for many structural applications.

- More columns including reinforced concrete and other types of columns need to be tested with varying slenderness ratio, eccentricity and degree of confinement to explore the IC limit further, and implications of IC limit to practical structures could be investigated.

REFERENCES

Grundy, P. 1994. Incremental collapse of hollow sections. *Proc. of the 6th Int. Sym. on Tubular Structures, Melbourne, Australia*: 497-503.

Hancock, G. J. 1994. *Design of cold-formed steel structures. 2nd Edition*: Australian Institute of Steel Construction.

Neal, B. G. 1965. *The plastic methods of structural analysis*. London: Chapman & Hall Ltd.

Neogi, P. K. & H. K. Sen 1969. Concrete-filled tubular columns under eccentric loading. *The Structural Engineer*, 47(5): 187-195.

O'Brien, A. D. & B. V Rangan 1993. Tests on slender tubular steel columns filled with high-strength concrete. *Australian Civil Engineering Transactions*, CE35(4): 287-298.

O'Shea, M. D. & R. Q. Bridge 1995. Circular thin walled concrete filled steel tubes. *Proc. of the 4th Pacific Structural Steel Conference, Singapore*: 53-60.

Ruoquan, He. 1992. Behaviour of long concrete filled steel columns. *Composite construction in steel and Concrete II, Proc. of Engineering Foundation Conference* : 728-737: ASCE.

Taplin, G. & P. Grundy 1995. The incremental slip behaviour of stud shear connectors. *Proc. of the 14th Australasian Conference on the Mechanics of Structures and Materials, Hobart, 1995*: 84-87.

Thirugnanasundralingam, K. 1991. Continuous composite beams under moving loads. *Ph. D. Thesis*. Monash University, Australia.

Tin Loi, F. S. K. 1977. Shakedown and deflection prediction of steel frame structures. *Ph. D. Thesis*. Monash University, Australia.

Webb, J. & J. J. Peyton, 1990. Composite concrete filled steel tubular columns. *Proc. of the Structural Engineering Conference, Adelaide, 1990*.

The Mechanics of Structures and Materials, Grzebieta, Al-Mahaidi & Wilson (eds)
© *1997 Balkema, Rotterdam, ISBN 90 5410 900 9*

Workability of high performance steel fibre reinforced concrete

I. Patnaikuni & K. Thirugnanasundralingam
Royal Melbourne Institute of Technology, Melbourne, Vic., Australia

Q. Chunxiang
Southeast University, Nanjing, People's Republic of China

A. K. Patnaik
Curtin University of Technology, Perth, Australia

ABSTRACT: Research information on workability of high performance concrete with Australian steel fibres is lacking. In this paper the research finding on the workability of high performance steel fibre reinforced concrete of strengths between 50 and 100 MPa is presented. American Standard specifies a dynamic test for the slump of steel fibre reinforced normal strength concrete. In this investigation a novel method of utilising both the slump cone test and a dynamic test were used. It was found that a combination of optimum range of slump and time of flow give a better measure of the workability of high performance steel fibre reinforced concrete (HPSFRC). The paper presents the test results and gives the optimum ranges of slump and time of flow.

1 INTRODUCTION

The use of steel fibres in reinforced concrete is not a new phenomena. Steel and other fibres have been used for some time to improve the ductility of normal strength concrete and steel fibre reinforced concrete has been studied for the past few decades. But the concrete used was normal strength concrete. Research carried out on high performance steel fibre reinforced concrete (HPSFRC) is only limited. Information on workability of HPSFRC is in particular limited. The large surface area of fibres tend to restrain flowability and mobility of the concrete mix (Bayasi and Soroushian, 1992). Interparticle friction of fibres and fibres and aggregates is another factor reducing fresh mix workability of steel fibre reinforced concrete. Most of the standards specify a slump cone test for determining the workability of concrete without steel fibres. Australian standard AS 1012, part 3 (1983) specifies a hollow frustum of a cone with bottom and top diameters of 200 mm and 100 mm respectively and a height of 300 mm for the slump test. For the consistency and workability of fibre reinforced concrete, ASTM-C995 (1991) specifies a standard test method for time of flow of fibre-reinforced concrete through inverted slump cone. This test is applicable to ordinary fibre-reinforced concrete. For high performance steel fibre-reinforced concrete the flow test itself may not be adequate to determine the workability of concrete. In this paper the results of combination of slump test and flow test for HPSFRC are presented.

2 EXPERIMENTAL PROGRAM

In this research program three different types of enlarged end steel fibres with rectangular cross-section supplied by BHP were used to produce the HPSFRC. The dimensions of these fibres were 18x0.4x0.3 mm(L/d = 46, Type I), 18x0.6x0.3 mm (L/d = 38, Type II) and 25x0.6x0.4 mm

(L/d = 45, Type III). The aggregate used was Basalt from Kilmore with a maximum size of 10 mm. Silica fume produced in Australia was used. The composition of the silica fume used is given elsewhere (Ting et al, 1992). A silica fume content of 10% was used. Four different kinds of super-plasticisers were used to produce workable fresh concrete mixture with water-binder (W/B) ratio ranging from 0.272 to 0.305. Two of the super plasticisers used (SP1 & SP2) were modified naphthalene polymers. The other two were modified naphthalene sulphonate polymer (SP3) and sulphonated naphthalene formaldehyde condensate (SP4).

The sand to total aggregate ratio for steel fibre reinforced concrete should be increased compared to concrete without fibres to reduce the harshness of fresh concrete. Recommendations of Banthia and Trothier (1995), Ashour and Wafa (1993) and Lim and Hsu (1994) were followed and a sand to total aggregate ratio of 45% was used in this research program. An aggregate to binder ratio of 3.3 as suggested by Ting and Patnaikuni (1992) was used. Steel fibres of one percent (1%) by volume were used in the mixture.

From the experimental program it was found that the W/B ratio for a workable concrete depends on the super-plasticiser and mixing procedure. The properties of super-plasticiser is one of the key factors which affects the production of HPSFRC in terms of workability with the lowest possible W/B ratio. Out of the four kinds of super-plasticisers used, super-plasticiser SP1 seemed to be the best from the point of view of good workability, proper setting time and lowest W/B ratio. The dosage of super-plasticisers is also an important factor, which depends on the chemical composition of the super-plasticiser and the concrete mix. The highest dosage of super-plasticiser SP1 was 2.5% by weight of binder for a workable concrete without segregation.

The mix proportions used were :
cement : silica fume : sand : coarse aggregate = 0.9 : 0.1 : 1.48 : 1.81. Further details of the mixtures are shown in Table 1.

Table 1. Concrete Mixture Details

Mix No.	Fibre Type	SP Type	SP Dosage	W/B Ratio	Mixing Method
M1	I	SP1	2.0	0.272	MM-3
M2	I	SP1	2.5	0.285	MM-2
M3	I	SP1	2.0	0.290	MM-2
M4	I	SP1	3.0	0.290	MM-1
M5	I	SP2	2.5	0.297	MM-3
M6	I	SP2	1.2	0.300	MM-2
M7	I	SP4	2.5	0.300	MM-3
M8	II	SP1	2.0	0.295	MM-2
M9	II	SP1	2.5	0.300	MM-1
M10	II	SP2	1.5	0.295	MM-2
M11	III	SP1	2.5	0.305	MM-1
M12	*	SP1	1.5	0.295	MM-2
M13	*	SP1	2.5	0.300	MM-1

* Note : No steel fibres were used for mixes M12 and M13 for comparison purposes.

Three different mixing methods were used as shown in Table 1. In the first mixing method MM-1, all the water and super-plasticiser were added into the mixer at the same time. In the second mixing method MM-2, about 90% of water along with all the super plasticiser were added to the dry materials in the mixer and the remaining 10% of water was added at regular intervals later on. The materials in the mixer were mixed for about 3 to 4 minutes each time after adding the water. In the mixing method MM-3, about 90% of water along with 80% of super-plasticiser were added to the dry materials and the remaining water and super-plasticiser were added in approximately three equal amounts at regular intervals. The materials were mixed for approximately 3 to 4 minutes each time after the water and super-plasticiser were added.

3 WORKABILITY TESTS

Two types of workability tests were carried out for the HPSFRC. The first test is static slump cone test as per Australian Standard AS 1012 part 3 (1983). This test was carried out using the standard slump cone mould and tamping rod according to the procedure specified in AS 1012.3. The second test was a dynamic method of inverted slump cone test similar to that specified in ASTM C995 (1991). The inverted cone used for the dynamic method is a standard slump cone. The inverted cone was filled in three layers, each approximately equal to one third of the volume of the cone. Compacting the concrete was avoided but each layer was levelled lightly with a scoop or trowel to minimise the entrapment of large voids. The surface of the top layer was struck off by means of a screeding and rolling motion of the tamping rod. Protruding fibres which inhibit screeding were removed by hand. Vibrator and stop watch were started and the element of the vibrator was inserted centrally and vertically into the top of the sample in the cone. It was allowed to descend at a rate such that it touched the bottom of bucket in 3±1 seconds. Vibrating element was maintained vertically and in contact with the bottom of the bucket. Touching the cone was avoided. The stop watch was stopped when the cone became empty which occurred when an opening became visible at the bottom of the cone. Only difference in the dynamic test compared to ASTM was the distance between the small end of the cone and the bottom of the bucket was made larger. In the case of HPSFRC, this distance of 101.6 mm specified by ASTM C995 (1991) was found to be inadequate. So the distance was increased to 200 mm. The increase in the distance has the effect of allowing the concrete to flow through the cone with much ease when the vibrator is started. The authors believe that the increase in distance will decrease the time taken for the flow. In future an investigation will be carried out to determine the relation between the method specified by ASTM and the method used in this research. The test results of the static slump cone test and the dynamic inverted slump cone test are shown in Table 2.

4 RESULTS AND DISCUSSION

From Table 2 it can be seen that Mixes M1, M7, M9 and M11 gave higher compressive strengths in addition to Mix M10. Mix M10 is a very stiff mix and workability obtained was poor. Mixes M1, M7, M9 and M11 were workable mixes. Mixes M4 and M5 have exhibited slight segregation and were not set until the next day. Mix M12 was a very stiff mix and was not workable. Mix M13 has exhibited slight segregation but was set on the next day. A close look at the slump and time of flow from Table 2 indicates that an optimum range exists. From Figure 1 also it can be seen that concretes of higher strengths have shown a slump value between 20-60 mm and time of flow between 20-40 seconds. Figure 2 also shows that a workable slump of 20-60 mm is in the time of flow range of 20 to 40 seconds. For Mix M3 the slump value of 5 mm for a time of flow of 22 seconds obtained may be due to experimental error. A suitable slump for workable HPSFRC seems to be between 20 and 60 mm, and the time of flow seems to be between 20 and 40 seconds. If these two conditions are satisfied, the fresh HPSFRC will neither be stiff (dry) nor will segregate. A combination of these two tests appear to give a good measure of the workability of HPSFRC.

Table 2. Workability Tests Results and Compressive Strength

Mix No.	Slump (mm)	Time of Flow (secs)	Compressive Strength (MPa)
M1	50	28	84.6
M2	15	31	67.8
M3	5	22	60.3
M4	118	11	64.4
M5	115	21	46.3
M6	0	22	67.4
M7	30	30	70.0
M8	5	41	63.5
M9	35	27	80.1
M10	0	117	71.7
M11	55	31	73.8
M12	0	126	64.7
M13	130	12	65.8

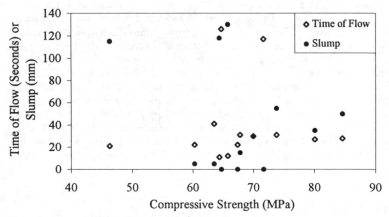

Figure 1 : Time of Flow & Slump vs Compressive Strength

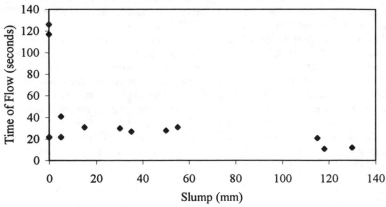

Figure 2 : Time of Flow vs Slump

174

As mentioned earlier the type and dosage of super-plasticiser and the mixing method are important for achieving a good workable HPSFRC. The dosage of super-plasticiser and the W/B ratio of mix M1 were in the lower range and the slump value and the time of flow were in the suitable range. Mix M1 gave the highest compressive strength of 84.6 MPa. The sequence of adding materials and the mixing time also had an affect on the workability and compressive strength of HPSFRC. Among the mixing methods considered in this investigation mixing method MM-3 gave better results compared to the other two.

5 CONCLUSIONS

For HPSFRC there is an optimum range of slump and time of flow of concrete through the inverted cone.
For workable HPSFRC a suitable slump is in the range of 20-60 mm and the time of flow of concrete through the inverted cone is in the range of 20-40 seconds.
Both the conditions of the slump in the range of 20-60 mm and time of flow in the range of 20-40 seconds have to be satisfied for a workable HPSFRC to achieve higher strength.
Type and dosage of super-plasticiser is important to achieve a workable HPSFRC for a given mixture.
Mixing method has an influence on the workability and compressive strength of HPSFRC.

6 ACKNOWLEDGMENT

The authors wish to acknowledge the support given by BHP Steel to this project by supplying the steel fibres required for the project.

7 REFERENCES

AS 1012-Part 3 (1983). Australian standard of methods for the determination of properties related to the consistence of concrete. *Standards Association of Australia* : Sydney.

Ashour, S. A., and Wafa, F. F. (1993). Flexural behaviour of high-strength fiber reinforced concrete beams, *ACI Structural Journal*, Vol. 90, No.3, May-June, pp 279-287.

ASTM C995 (1991). Standard test method for time of flow of fiber-reinforced concrete through inverted slump cone. *ASTM Standard*.

Banthia, N. & Trottier, J. F. (1995). Concrete reinforced with deformed steel fibers, part II : toughness characterisation. *ACI Materials Journal*. Vol. 92. No. 2. March-April : 146-154.

Bayasi, M. Z., & Soroushian, P. 1992. Effect of steel fibre reinforcement on fresh mix properties of concrete, *ACI Materials Journal*. Vol. 89. No. 4. July-August : 369-374.

Lim, S. H. & Hsu, C. T. 1994. Stress-strain behaviour of steel-fiber high-strength concrete under compression. *ACI Structural Journal*. Vol. 91. No.4. July-August : 448-457.

Ting, E. S. K. & Patnaikuni, I. 1992. Influence of mix ingredients on the compressive Strength of very high strength concrete, *Proc. of the 17th conference on Our World in Concrete & Structures*. August. Singapore : 207-215.

Ting, E. S. K., Patnaikuni, I., Pendyala, R. S, & Johansons, H. A. 1992. Effectiveness of silica fume available in Australia to enhance the compressive strength of very high strength concrete. *Proceedings of the International Conference on The Concrete Future*. February, Kuala Lumpur, Malaysia : 257-261.

Concrete technology and design

The Mechanics of Structures and Materials, Grzebieta, Al-Mahaidi & Wilson (eds)
© *1997 Balkema, Rotterdam, ISBN 90 5410 900 9*

High strength reinforcement in concrete structures: Serviceability implications

R.I.Gilbert
School of Civil and Environmental Engineering, University of New South Wales, Kensington, N.S.W., Australia

ABSTRACT: The serviceability of reinforced concrete beams and slabs is investigated for elements designed using 500 Grade (f_{sy}=500 MPa) reinforcement. The short-term and long-term in-service behaviour are considered and the applicability of the serviceability design provisions of AS3600-1994 are examined.

1 INTRODUCTION

Over the last two or three years, reinforcing steels with strengths in excess of 500 MPa have become increasingly available in the market place and are now commonly used in concrete structures. These products typically have significantly less ductility than conventional 400 MPa tempcore reinforcement (Y bars). In structural design, the use of higher strength steels usually results in less steel being required to satisfy the design requirements for the ultimate limit states. However, the reduced strain capacity of the new higher strength reinforcement has implications with regard to the ductility of the structure.

Standards Australia has commissioned several working groups to assess the implications of the use of higher strength steels, particularly in relation to the applicability of existing design provisions in AS3600-1994. To date, much of the attention of the working groups has been directed, quite correctly, at the strength limit states. However, when high strength steels are employed in design, there may be significant implications with regard to the serviceability requirements of the Standard, particularly those related to deflection and crack control.

For many reinforced concrete flexural members, the size of the cross-section (in particular, its depth) is governed by the deflection requirements for the member. This is particularly true for structural elements with relatively large span to depth ratios, such as floor slabs. For such elements, designers frequently use the deemed-to-comply maximum span-to-depth ratios specified in the code to establish the slab thickness and then determine the tensile reinforcement quantities to satisfy the design requirements for the strength limit states. When 500 MPa steel is employed instead of 400 MPa steel, a smaller amount of flexural reinforcement is required to provide adequate strength. This inevitably leads to less stiffness after cracking and, consequently, increased immediate deflections and increased crack widths under service conditions, where the stress in the tensile steel is much less than the yield stress of the material (ie. the steel stress is in the elastic range) and it is the stiffness of the reinforcement, rather than its strength, that affects structural behaviour.

When deflection is critical, a significant reduction in steel area cannot be tolerated, unless it is accompanied by an increase in effective depth. On the other hand, a significant part of the total deflection of any reinforced concrete flexural member occurs over a period of months and years due to the gradual development of creep and shrinkage strains in the concrete. A reduction in the amount of tensile reinforcement results in a reduction in the restraint provided to these time-dependent strains and may result in a reduction in the long-term deflection, particularly that caused by shrinkage. This may in fact balance the increased immediate deflection resulting from the reduction of tensile steel area in the member.

This paper assesses the effects of using 500 MPa reinforcement (instead of 400 MPa reinforcement) on the deflection and crack control of concrete beams and slabs. The serviceability provisions of AS3600-1994, including the deemed-to-comply span to depth ratios and minimum steel quantities for crack control, are also critically examined.

2 IN-SERVICE BEHAVIOUR OF BEAMS AND SLABS

This section examines the short-term and time-dependent behaviour of the singly-reinforced cross-section shown in Figure 1. A comparison is made of the behaviour of two such cross-sections, one containing 400 MPa steel and the other containing 500 MPa steel. For each section, the steel capacity $(A_{st}f_{sy})$ is identical.

2.1 Short-Term Behaviour

The theoretical moment-curvature relationships for the cross-sections of Figure 1 are shown in Figure 2 for the in-service range of moments. The curvature represents the average curvature over a gauge length long enough to include several flexural cracks. The curves were calculated using well established elastic analysis (modular ratio theory), with tension stiffening included using Branson's method, as specified in AS3600-1994. For the range of moments given in Figure 2, the maximum concrete compressive stress in the top fibre is less than $0.5f_c$, so that the assumption of linear-elastic behaviour for concrete in compression is reasonable.

At a typical maximum in-service moment level for these cross-sections (M=150 kNm), the curvature of the beam containing 500 MPa steel ($\kappa = 2.52x10^{-6}mm^{-1}$) was 17% greater than that of the beam containing 400 MPa steel ($\kappa = 2.15x10^{-6}mm^{-1}$). At the same moment, the maximum tensile steel stress at a crack in the 500 Grade steel ($\sigma_{st}= 228 MPa$) is 24% greater than the maximum stress in the 400 Grade steel ($\sigma_{st} = 184 MPa$).

It is clear that by adopting 500 Grade steel, rather than 400 Grade, with a correspondingly reduced steel area, both the initial deflection (which is proportional to curvature) and the flexural crack width (which is proportional to steel stress) will be significantly greater at the same service load.

2.2 Time-Dependent Behaviour

Effects of Creep:

Under sustained stress, concrete strains gradually increase due to the development of creep. On an uncracked and unreinforced concrete section subject to moment, both the compressive and tensile zones creep and the time-dependent change in curvature is similar to the time-dependent change in strain. On an uncracked, singly reinforced section, the tensile reinforcement provides restraint to tensile creep and causes a redistribution of stress between the concrete and the tensile steel, a slight lowering of the neutral axis of strain and a reduced time-dependent curvature.

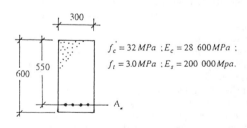

$f_c' = 32 MPa \; ; E_c = 28\ 600 MPa \; ;$
$f_t = 3.0 MPa \; ; E_s = 200\ 000 Mpa.$

500 Grade: $A_s = 1320$ mm² $\quad f_{sy} = 500$ MPa
400 Grade: $A_s = 1650$ mm² $\quad f_{sy} = 400$ MPa

Figure 1   Cross-Sections used in analysis

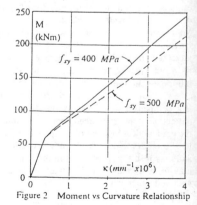

Figure 2   Moment vs Curvature Relationship

On a cracked singly-reinforced section, only the compressive zone creeps (the tension being carried primarily by the steel). As the strains in the compressive zone increase, the neutral axis depth increases, the compressive zone becomes larger and the concrete compressive stress distribution becomes slightly non-linear with a significant decrease in extreme fibre stress. The strain at the level of the tensile steel increases by a relatively small amount, so that the tensile force in the steel increases slightly to compensate for the slightly reduced lever arm between the compressive and tensile stress resultants, thereby maintaining equilibrium. The relative change in curvature with time is much less than on an uncracked cross-section. The inclusion of compressive reinforcement would further reduce the time-dependent curvature by relieving the concrete compressive zone of stress and reducing the compressive creep strains.

The moment induced curvature $\kappa(t)$ on a singly-reinforced concrete cross-section at any time t can be expressed as

$$\kappa(t) = \kappa(o)\left[1 + \frac{\phi(t)}{\alpha}\right] \qquad (1)$$

where $\kappa(o)$ is the initial curvature, $\phi(t)$ is the creep coefficient (which may be obtained from the data in AS3600-1994) and α is a parameter that accounts for the depth of cracking (if any) and the restraining action of the tensile reinforcement. For an uncracked singly-reinforced cross-section α is usually within the range 1.0 to 1.6, with $\alpha = 1.0$ for an unreinforced section and α increasing with increasing amounts of tensile steel. For example, $\alpha = 1.2$ when the reinforcement ratio A_{st}/bd equals about 0.005 and $\alpha = 1.6$ when A_{st}/bd is about 0.02. For a cracked singly-reinforced cross-section, α is usually within the range 3.0 to 8.0, with the higher end of the range being for the less heavily reinforced cross-sections where the depth of cracking and the initial curvature are larger.

Figure 3 shows the initial and the time-dependent strain and stress distributions for the fully cracked cross-section of Figure 1, with $A_{st} = 1650mm^2$; M=150 kNm, $\phi(t) = 3.0$ and shrinkage has not been included. The time-dependent strains were calculated using the age-adjusted effective modulus method of analysis and a computer program developed by the writer. The method has been fully presented elsewhere (Gilbert, 1988).

Table 1 provides details of the effect of varying the quantity of tensile reinforcement on the initial and creep induced strains and curvatures for the same cross-section, applied moment and material properties. Also tabulated are the corresponding values of the parameter α used in Eqn 1.

If 500 MPa steel is used instead of 400 MPa steel, with a corresponding reduction in steel area (ie. a reduction in steel area of 20%), the final moment induced curvature $\kappa(t)$ is increased by

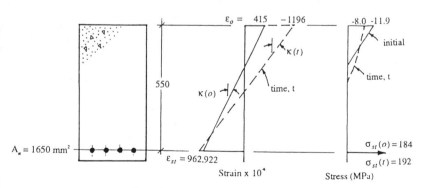

Figure 3: Moment dependent strain and curvature

181

Table 1:　　　Short-term and Final Load-induced Strains and Curvatures.

A_{st} (mm²)	Reinf ratio, A_{st}/bd	Short-term			Final (Initial + Creep)			$\kappa(t)/\kappa(o)$	α
		ε_o $x10^{-6}$	ε_{st} $x10^{-6}$	$\kappa(o)$ $x10^{-6}mm^{-1}$	ε_o $x10^{-6}$	ε_{st} $x10^{-6}$	$\kappa(t)$ $x10^{-6}mm^{-1}$		
990	0.006	-503	1503	3.65	-1408	1556	5.39	1.48	6.28
1320	0.008	-451	1141	2.90	-1281	1186	4.49	1.55	5.46
1650	0.010	-415	922	2.43	-1196	962	3.92	1.61	4.89
1980	0.012	-389	775	2.12	-1134	811	3.54	1.67	4.47
2310	0.014	-369	670	1.89	-1086	702	3.25	1.72	4.17
2640	0.016	-353	590	1.72	-1048	620	3.03	1.77	3.91
3300	0.020	-328	478	1.47	-991	504	2.72	1.86	3.51

between 11% and 15%, with the lower end of the range corresponding to more heavily reinforced cross-sections. For sections containing less than 1% steel, the increase in $\kappa(t)$ is between 14% and 15%. The ratio $\kappa(t)/\kappa(o)$ increases and the parameter α decreases with increasing reinforcement ratio (for a cracked section) and α is typically about 5 when the reinforcement ratio is 1%. For reinforced concrete slabs, where the tensile reinforcement ratio is typically between 0.003 and 0.005, α is usually between 7 and 8.

Effects of Shrinkage:

　　　　Shrinkage in an unsymmetrically reinforced concrete beam can produce deflections of significant magnitude, even if the beam is unloaded. As the concrete shrinks, it compresses the steel reinforcement which, in turn, imposes an equal and opposite tensile force ΔT on the concrete at the level of the steel. This gradually increasing tensile force, acting at some eccentricity to the centroid of the uncracked part of the concrete cross-section, produces curvature (elastic plus creep) and a gradual warping of the beam. The magnitude of ΔT and the resulting curvature depend on the quantity of reinforcement and on whether or not the cross-section is cracked. Shrinkage warping in a cracked beam is significantly greater than in an uncracked beam.

　　　　Compressive reinforcement reduces shrinkage curvature. By providing restraint at both the top and bottom of the section, the eccentricity of the resultant tension in the concrete is reduced and , consequently, so is the shrinkage curvature. An uncracked, symmetrically reinforced section will suffer no shrinkage curvature.

　　　　Unlike creep, shrinkage in concrete structures is invariably detrimental and this is largely related to the problem of cracking. The advent of shrinkage cracks depends on the degree of restraint to shrinkage and the extensibility and strength of concrete. Although the tensile strength of concrete increases with time, so too does the elastic modulus and, therefore, the tensile stresses induced by shrinkage. Furthermore, the relief offered by creep decreases with age. The existence of load induced tensions in uncracked regions accelerates the formation of time-dependent cracking and shrinkage therefore causes a gradual reduction in the tension stiffening effect.

　　　　Table 2 contains the results of analyses performed on the cross-section of Figure 1, with identical data as that use to generate the results in Table 1, except that the shrinkage strain ε_{sh} is -0.0006 (600 microstrain). The shrinkage induced curvature κ_{sh} is obtained by subtracting the final curvature $\kappa(t)$ in Table 1 (obtained with $\varepsilon_{sh} = 0$) from the value of $\kappa(t)$ in Table 2. For this cracked cross-section, κ_{sh} increases by only 8% as the reinforcement ratio increases from 0.006 to 0.02. A reasonable approximation for the shrinkage induced curvature on a cracked section is

$$\kappa_{sh} = \frac{1.15 \ \varepsilon_{sh}}{d} \tag{2}$$

For sections containing compressive steel, A_{sc}, the right hand side of Eqn 2 should be multiplied by $(1-0.5 A_{sc}/A_{st})$. It can be concluded that the shrinkage induced curvature, and hence

Table 2 Initial and Final (Creep+Shrinkage) Strains and Curvatures.

A_{st} (mm²)	Reinf ratio, A_{st}/bd	Short-term			Final (Initial + Creep+ Shrinkage)			κ_{sh} $x10^{-6}mm^{-1}$
		ε_o $x10^{-6}$	ε_{st} $x10^{-6}$	$\kappa(o)$ $x10^{-6}mm^{-1}$	ε_o $x10^{-6}$	ε_{st} $x10^{-6}$	$\kappa(t)$ $x10^{-6}mm^{-1}$	
990	0.006	-503	1503	3.65	-2088	15.49	6.61	1.22
1320	0.008	-451	1141	2.90	-1973	1178	5.73	1.24
1650	0.010	-415	922	2.43	-1897	953	5.18	1.26
1980	0.012	-389	775	2.12	-1843	801	4.81	1.27
2310	0.014	-369	670	1.89	-1802	692	4.54	1.28
2640	0.016	-353	590	1.72	-1771	610	4.33	1.30
3300	0.020	-328	478	1.47	-1726	493	4.04	1.32

deflection of a cracked flexural member, is not significantly affected by using 500 MPa steel rather than 400 MPa steel (with a correspondingly reduced area of tensile steel).

3 MAXIMUM SPAN TO DEPTH RATIOS - AS3600-1994

The deemed to comply span to depth ratios specified in AS3600-1994 for deflection control of beams and slabs were calibrated to be consistent and compatible with deflection calculations. Certain approximations, with regard to the effective moment of inertia of the element, were made to simplify the procedures and a maximum long term to instantaneous deflection ratio of 2.0 was assumed for singly-reinforced cross-sections. In its current form, the maximum span to depth ratios in AS3600 are independent of the quantity of flexural reinforcement and hence independent of the grade of steel used in design. They are also independent of the creep and shrinkage characteristics of the concrete and the flexural tensile strength of the concrete (and hence the extent of cracking in the element). Clearly, the use of 500 Grade steel (instead of 400 Grade steel) will render the procedure less conservative.

Table 3 gives some of the results of a parametric study of the in-service behaviour of singly-reinforced simply-supported one-way slabs. Each row in the table compares the AS3600-1994 maximum span to depth ratio for a maximum total deflection of span/250 to the corresponding value obtained by calculation, assuming that the amount of the tensile steel is the minimum amount required to satisfy the flexural strength requirements of the Standard. The calculations were carried out using a microcomputer using the procedures referred to in the preceding section. The effects of cracking, tension stiffening, creep and shrinkage of the concrete were included in the calculations, including the time-dependent reduction in tension stiffening due primarily to shrinkage.

In Table 3, w_G is the superimposed dead load; w_Q is the live load; ϕ is the final creep coefficient; ε_{sh} is the final shrinkage strain; Δ_s is the short-term deflection caused by the sustained part of the service load; Δ_Q is the short-term live load deflection; Δ_{cs} is the creep and shrinkage induced long-term deflection; and Δ_{ToT} is the total deflection $(\Delta_s + \Delta_Q + \Delta_{cs})$. In all causes, the concrete cover to the flexural reinforcement was taken to be 20mm, the bar diameter was assumed to be 12 mm and the short-term and long-term serviceability live load factors were 0.7 and 0.3, respectively. The concrete was assumed to weigh 24 kN/m³ and the short-term flexural tensile strength of the concrete was taken to be $0.6\sqrt{f_c'}$.

The results shown in Table 3 illustrate that the L/d ratios in AS3600 are generally in close agreement with the calculated deflections for typical values of final creep coefficient and shrinkage strain (ie. when $\phi = 3.0$ and $\varepsilon_{sh} = -0.0006$). The effect on the maximum L/d ratio of using 500 MPa steel instead of 400 MPa steel is relatively small and the code procedure appears to be adequate.

The ratio of long-term deflection to short-term sustained load deflection for all the runs in Table 3 with $\phi = 3.0$ and $\varepsilon_{sh} = -0.0006$ fall within the range 3.5 to 7.0 and this highlights a problem with the code procedure for predicting slab deflection by "simplified calculation". For most practical slabs, the maximum in-service moment is usually not much greater than the short-term cracking moment and the effective moment of inertia is closer to the uncracked than the fully-cracked moment of inertia. The calculated short-term deflection is therefore relatively small. With time shrinkage induced tension causes additional cracking and a gradual reduction in the tension stiffening effect.

183

Table 3 Calculated and AS3600 Span to depth ratios for maximum total deflection of Span/250.

Run	f_c'	f_{sy}	Span L_{ef}	d	A_{st}	Calculated Deflections				L_{ef}/d	
No.	MPa	MPa	mm	mm	mm^2	Δ_s mm	Δ_Q mm	Δ_{cs} mm	Δ_{ToT} mm	Calculated	AS3600 ($\varphi9.3.4$)
$w_G = 1.0$ kPa; $w_Q = 3.0kPa$; $\phi = 3.0$; $\varepsilon_{sh} = -0.0006$:											
1	25	400	4000	139.5	497	2.23	0.46	13.30	16.0	28.7	27.9
2	"	"	5000	188.0	654	3.68	0.63	15.67	20.0	26.6	26.4
3	"	"	6000	241.1	830	5.35	0.77	17.87	24.0	24.9	25.1
4	"	500	4000	141.2	394	2.10	0.43	13.46	16.0	28.3	27.9
5	"	"	5000	190.2	519	3.52	0.60	15.88	20.0	26.3	26.4
6	"	"	6000	244.1	659	5.15	0.74	18.11	24.0	24.6	25.1
7	32	400	4000	133.8	506	1.97	0.41	13.63	16.0	29.9	29.4
8	"	"	5000	178.2	669	3.16	0.56	16.29	20.0	28.1	27.9
9	"	"	6000	228.0	848	4.68	0.70	18.60	24.0	26.3	26.5
10	"	500	4000	136.0	401	1.91	0.40	13.70	16.0	29.4	29.4
11	"	"	5000	180.4	532	2.98	0.52	16.49	20.0	27.7	27.9
12	"	"	6000	230.8	674	4.47	0.66	18.86	24.0	26.0	26.5
$w_G = 1.0$ kPa; $w_Q = 3.0kPa$; $\phi = 2.0$; $\varepsilon_{sh} = -0.0004$:											
13	32	400	4000	114.7	564	3.78	0.86	11.35	16.0	34.9	29.4
14	"	"	5000	156.5	725	5.81	1.11	13.09	20.0	31.9	27.9
15	"	"	6000	202.4	905	7.97	1.30	14.72	24.0	29.6	26.5
16	"	500	4000	115.9	448	3.64	0.82	11.52	16.0	34.5	29.4
17	"	"	5000	158.3	576	5.65	1.07	13.27	20.0	31.6	27.9
18	"	"	6000	204.9	719	7.81	1.26	14.92	24.0	29.3	26.5
$w_G = 2.0$ kPa; $w_Q = 3.0kPa$; $\phi = 3.0$; $\varepsilon_{sh} = -0.0006$:											
19	32	400	4000	140.8	549	2.25	0.39	13.37	16.0	28.4	27.6
20	"	"	5000	188.0	717	3.65	0.55	15.80	20.0	26.6	26.4
21	"	"	6000	239.3	904	5.23	0.68	18.08	24.0	25.1	25.3
22	"	500	4000	142.4	435	2.12	0.37	13.53	16.0	28.1	27.6
23	"	"	5000	190.2	570	3.47	0.52	16.01	20.0	26.3	26.4
24	"	"	6000	242.1	719	5.03	0.65	18.34	24.0	24.8	25.3

The consequent time-dependent reduction in stiffness is not adequately accounted for by the long-term deflection multiplier of 2.0 for singly-reinforced cross-sections.

The span to depth ratios in AS3600 overcome this problem by assuming the member is more extensively cracked to begin with (thereby overestimating short-term deflection) and, in this case, a long-term deflection multiplier of 2.0 provides a quite reasonable approximation of the eventual final deflection. The span to depth ratios are therefore compatible with a "refined" deflection calculation but are not at all compatible with the "simplified" procedure contained in the Standard. To render the "simplified" deflection calculation procedure more accurate (in terms of total deflection only), a reduction in the assumed value of the cracking moment is required. A simplified deflection calculation procedure for slabs which provides a close approximation to the total deflection calculated using the refined method is as follows:

(i) The flexural tensile strength of the concrete is taken as $0.45\sqrt{f_c'}$ in the determination of the in cracking moment (rather than $0.6\sqrt{f_c'}$ specified in AS3600);

(ii) The short-term deflections Δ_s and Δ_Q are calculated using the effective moment of inertia specified in AS3600 (Branson's method);

(iii) The creep induced deflection is obtained from Eqn 3, where for slabs α may be taken as 7.5.

$$\Delta_{cr} = \Delta_s \phi / \alpha \tag{3}$$

184

(iv) The shrinkage induced deflection may be calculated from

$$\Delta_{sh} = \beta. \ \kappa_{sh}. \ L_{ef}^2 \qquad (4)$$

where κ_{sh} is given by Eqn 2 and $\beta = 0.125$ for simply supported spans, 0.0625 for end spans of continuous members and 0.045 for interior spans of continuous member.

In these calculations, the short-term deflections (Δ_s and Δ_Q) will be slightly overestimated, the time-dependent deflection ($\Delta_{cs} = \Delta_{cr} + \Delta_{sh}$) will be slightly underestimated, but total deflection ($\Delta_s + \Delta_Q + \Delta_{cr} + \Delta_{sh}$) will be quite reasonable. For each of the 24 runs in Table 3, the calculated Δ_{ToT} and the corresponding value obtained using the suggested "simplified" calculation procedure differ by less than 13.5% in all cases.

4 CRACK CONTROL FOR SHRINKAGE AND TEMPERATURE EFFECTS

Where a slab is restrained from expanding or contracting in the secondary direction, significant reinforcement is required in that direction to control cracking. For reinforced concrete slabs, AS3600-1994 specifies a minimum amount of reinforcement in the secondary direction as

$$(A_s)_{\min} = \gamma \ bD / f_{sy} \qquad (5)$$

For non-aggressive environments, $\gamma = 0.7$, 1.4 and 2.5 when a minor degree, a moderate degree and a strong degree of control over cracking, respectively, is required. For aggressive environments, $\gamma = 2.5$ in all situations.

While Eqn 5 gives reasonable values for 400 Grade steel bars and 450 Grade welded wire mesh, an increase in f_{sy} will give a smaller minimum area of steel. Smaller quantities of steel will lead to higher steel stresses at each crack and hence wider crack widths. Eqn 5 is therefore inappropriate for steel bars when f_{sy} exceeds 400 MPa. As a replacement for Eqn 5, the following equation is proposed:

$$(A_s)_{\min} = \gamma \ b \ D \qquad (6)$$

where for non-aggressive environments, $\gamma = 0.00175$, 0.0035 and 0.00625 when a minor degree, a moderate degree and a strong degree of control over cracking, respectively, is required. For aggressive environments, $\gamma = 0.00625$. When welded wire mesh is used, the respective values for γ may be reduced to 0.00155, 0.0031 and 0.0055.

5 CONCLUSIONS

The serviceability of reinforced concrete beams and slabs is examined for elements designed using 500 Grade reinforcement. The short-term and long-term deflections are considered and the applicability of the serviceability design provisions of AS3600-1994 are examined. It is concluded that the "simplified" method for deflection calculation for slabs has certain fundamental weaknesses and the use of higher strength reinforcement (with reduced amounts of tensile reinforcement) will exacerbate serviceability problems in deflection sensitive structures. An alternative "simplified" method for deflection calculation is proposed for reinforced concrete slabs which provides close agreement with the total deflection obtained using more refined methods. The deemed to comply maximum L/d ratios in the Standard are still, however, generally applicable and can be used with relative confidence. Suggestions are also made to modify the minimum steel provisions in AS3600 for crack control for shrinkage and temperature effects when higher strengths steels are employed.

6 REFERENCES

AS3600-1994, *Australian Standard for Concrete Structures*, Standards Australia.

Gilbert, R.I. 1988, *Time effects in concrete structures*, Elsevier Science Publishers, Amsterdam.

The Mechanics of Structures and Materials, Grzebieta, Al-Mahaidi & Wilson (eds)
© *1997 Balkema, Rotterdam, ISBN 90 5410 900 9*

Advances in concrete wall design

S. Fragomeni
School of Engineering, Griffith University, Gold Coast, Southport, Qld, Australia

P.A. Mendis
Department of Civil and Environmental Engineering, University of Melbourne, Vic., Australia

ABSTRACT: The AS3600 and ACI318 concrete codes of practice used for the design of load bearing walls fail to recognise the effect on load capacity of high strength concrete and side restraints. It is important to include a more rational design method, in AS3600, incorporating the various support conditions and concrete strengths encountered in practice. This paper initially presents a brief review of the code methods and gives an overview of the relevant studies undertaken on the axial load capacity of concrete walls. Then state-of-the-art wall design formulae are presented. The formulae are the result of an extensive research program undertaken by the authors and can be used for various situations, including when high strength concrete or side restraints are present. The development of the factors used to account for high strength concrete and side restraints are also highlighted.

1 INTRODUCTION

At present, wall design practice in Australia involves the use of the concrete standard - AS3600 (1994), with some consulting firms preferring to use the American concrete code - ACI318 (1989), for guidance. Even though these codes provide for safe designs, they can produce unsuitable results in some cases. The code methods are intended for normal strength concrete walls, supported top and bottom only. They fail to recognise the effect on load capacity of high strength concrete and side restraints. When these conditions are considered thinner walls may result, thus providing huge savings.

It is important, therefore, to have wall design methods available to suit the various design situations encountered. This paper initially presents a brief review of the code methods and gives an overview of the relevant studies undertaken on the axial load capacity of concrete walls. These studies also fail to recognise the effect of high strength concrete and side restraints

The extensive research program recently completed by Fragomeni (1995) involved the testing and investigation of normal and high strength reinforced concrete walls. New state-of-the-art design formulae were proposed to replace the current AS3600 equation. These formulae are presented in this paper. They can be used for various conditions encountered in practice, including when high strength concrete or side restraints are present. The development of the factors used to account for high strength concrete and side restraints are also highlighted.

2 CURRENT CODE WALL DESIGN METHODS

The AS3600 and ACI318 wall design method are generally intended for load bearing walls free from side restraints and for concrete strengths in the range 20 to 50MPa.

2.1 AS3600-1994 Method

The Australian concrete standard - AS3600 (1994), gives two methods for the design of concrete walls. Section 11 of the code specifies a simplified equation which can be used for the design of walls when certain loading and bracing restrictions are met. The code also allows any wall to be designed as a column using the provisions of Section 10.

For the *simplified design method*, the ultimate design axial strength per unit length, N_u, of a braced wall in compression is given by the following formula.

$$\phi N_u = \phi(t - 1.2e - 2e_a)0.6f'_c \qquad (1)$$

where

e = eccentricity of the load measured at right angles to the plane of the wall.
e_a = $(H_{we})^2/2500t$, an additional eccentricity due to deflections in the wall.
f'_c = characteristic compressive strength of concrete.
H_{we} = lesser of kH or kL, the effective height of a wall.
H = unsupported height of the wall.
k = 0.75 restrained against rotation both ends & 1.0 not restrained against rotation at ends.
L = unsupported length of the wall
t = thickness of the wall.
ϕ = 0.6, the strength reduction factor.

This equation applies to walls where the slenderness ratio, $H_{we}/t \leq 30$ (*if the ultimate design axial force, $N^* \leq 0.03 f'_c A_g$ then $H_{we}/t \leq 50$*). A practice sometimes adopted in Australia is to use $H_{we}/t \leq 20$ when large axial loads are encountered. The walls are required to have minimum reinforcement ratios of 0.0015 vertically, ρ_v, and 0.0025 horizontally, ρ_h.

2.2 ACI318-1989 Method

ACI318 (1989) also offers two alternatives for the design of concrete walls, a simplified method and a more accurate method using column design. For the *simplified method*, ACI318 gives an empirical equation for the design axial load strength of a wall as:

$$P_u = 0.55\phi f'_c A_g[1 - (kH/32t)^2] \qquad (2)$$

where

A_g = Lt, gross area of the wall panel section.
k = 0.8 and 1.0 for walls restrained and unrestrained against rotation respectively, 2.0 walls not braced.
ϕ = 0.7 for compression members.

The equation applies to walls where $H/t \leq 25$ or $L/t \leq 25$ whichever is less for loadbearing walls. The minimum allowable thickness is 100mm. The resultant load must be in the 'middle third' of the overall thickness of the wall. This allows for a maximum eccentricity allowance of t/6. The walls are required to have minimum reinforcement ratios of 0.0015 vertically, ρ_v, and 0.0025 horizontally, ρ_h. These values can be reduced to 0.0012 and 0.0020 respectively if bars are less than 16mm diameter or if mesh is used.

3 TESTING DONE BY OTHER RESEARCHERS

Table 1 indicates that the experimental studies undertaken so far on concrete wall panels have been with normal concrete strengths between 20 and 42MPa. No testing has been done on high strength panels. Most of the testing, except for the work of Seddon (1956), relate to the ACI318 equation. The researchers focusing on the ACI method have suggested equations as shown in Table 2. Interestingly, the equations of Zielinski et al (1982) and Saheb & Desayi (1989) have incorporated the effect of reinforcement which was not done previously. A comprehensive comparison of these equations has been made by Fragomeni et al (1994b).

In Table 2, the wall equations indicate that a linear increase in axial load strength occurs for increasing concrete strength. As concluded by Fragomeni (1995), this may be considered appropriate for walls with concrete strengths under 50MPa but may not hold for higher strength concrete walls since no tests have been conducted using higher strengths.

Table 1. Summary of tests and variables used

Research	Number of tests	Steel ratio ρv	Concrete Strength f_c (MPa)	Slenderness ratio H/t	Aspect ratio H/L
Seddon (1956)	not given	0.008 single or 0.004 two-layers	17.5 to 28	18 to 54	1.5
Oberlender/Everard (1977)	54 half-scale	0.0033 or 0.0047 two-layers	28 to 42	8 to 28	1.0 to 3.5
Kripanarayanan (1977)	theoretical analysis	0.0000 or 0.0015 or 0.0100 two-layers	28	0 to 32	0.0 to 2.66
Pillai /Parthasarthy (1977)	18 large scale	0.0015 or 0.0030 one-layer	16 to 31.5	5 to 30	0.5 to 3.0
Zielinski et al (1982)	5 large-scale	relatively larger than others	33 to 37.5	72 *stiffened by side ribs	2.25
Saheb/Desayi (1989)	24 quarter & half- scale	0.00173 to 0.00856	20.2 to 25.7	12 to 27	0.67 to 2.0

Table 2 Design axial strength formulae

Equation	Design axial stress - (P_u/A_g)	
ACI318 Kripanarayanan (1977)	$0.55\phi[1 - (kH/32t)^2]f_c$	
Oberlender /Everard (1977)	$0.60\phi[1 - (H/30t)^2]f_c$	
Pillai/Parthasarathy (1977)	$0.57\phi[1 - (H/50t)^2]f_c$	
Zielinski et al. (1982)	$0.55\phi[1 - (H/40t)^2][1+(f_y/f_c - 1)\rho_v]f_c$	
Saheb/Desayi (1989)	$0.55\phi[1 - (H/32t)^2][1+(f_y/f_c - 1)\rho_v]f_c$	for H/L $\geq$ 2.0
	$0.55\phi[1 - (H/32t)^2][1+(f_y/f_c -1)\rho_v][1.2 - H/10L]f_c$	for H/L < 2.0

4 RECENT DEVELOPMENTS

Recent research by Fragomeni (1995) involved the testing of normal and high concrete strength reinforced concrete walls with various support conditions. For walls supported top and bottom only, ten normal and ten high strength concrete panels were tested. Also, testing was performed on two normal and two high strength concrete panels supported on four sides. The wall panels had minimum reinforcement placed centrally and were axially loaded at an eccentricity of t/6 to match similar testing done by other researchers. Simple supports were simulated by locating steel bars of 22mm diameter between the loading plates. For the walls requiring side restraints, angle sections were used to act as V-grove simple side supports. A comprehensive description of testing and results were reported by Fragomeni et al (1994a, 1995).

The test results along with data from other researchers were used to propose modified wall design formulae. These formulae are divided in the separate categories of one-way action (walls supported top and bottom only) and two-way action (walls supported on four sides). The formulae do not include the contribution of reinforcement and are based on the AS3600 (1994) wall design equation.

4.1 Proposed One-Way Wall Design Equations

The design axial strength per unit length of a braced wall in compression, supported top and bottom only (one-way action), shall be taken as:

$$\phi N_u = \phi(t - 1.2e - 2e_a)0.6f_c \qquad\qquad 20 \le f_c \le 50 \text{ MPa} \qquad (3a)$$

$$\phi N_u = \phi(t - 1.2e - 2e_a)30[1+(f_c - 50)/80] \qquad 50 \le f_c \le 80 \text{ MPa} \qquad (3b)$$

The one-way equations are exactly the same as the current AS3600 wall design formula except that for concrete strengths greater than 50 MPa an adjustment was required. It was found that for an 60% increase in concrete strength (from 50 to 80 MPa) a 37.5% increase in wall strength was appropriate, not 60% as indicated by using the current code equation. Limitations on slenderness and reinforcement restrictions are the same as those given for the AS3600 (1994) equation. Figure 1 gives an indication of the significant impact of the modified equation.

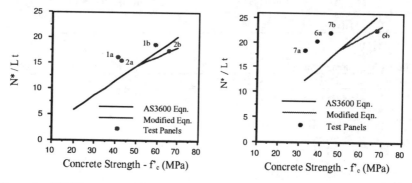

Figure 1. AS3600 versus Modified equation (a) H/t = 20 (b) H/t = 15

Figure 1 compares actual test panel failure loads against predicted results from design equations. The panels marked 'a' are composed of normal strength concrete and those marked 'b' are of high strength concrete. The following observations can be made:

a) The current AS3600 equation overestimated the failure load of two high strength panels,

b) The modified equations safely predict all test panel failure loads,

c) The slope of an imaginary line through panel pairs (ie. 1a and 1b or 7a to 7b, which have identical dimensional and reinforcement properties and only differ in concrete strength) more closely follows the slope of the modified equation line rather than the AS3600 line.

4.2 Proposed Two-way Wall Design Equations

To account for the added side supports, effective length factors from the German code - DIN1045 (1988) are incorporated. For two-way action (walls supported on four sides), the design axial strength per unit length of a braced wall in compression, shall be taken as:

$$\phi N_u = \phi(t - 1.2e - 2e_a)0.7f_c \qquad\qquad 20 \le f_c \le 50 \text{ MPa} \qquad (4a)$$

$$\phi N_u = \phi(t - 1.2e - 2e_a)35[1 + (f_c - 50)/80] \qquad 50 \le f_c \le 80 \text{ MPa} \qquad (4b)$$

where

e_a = $(H_{wc})^2/2500t$, an additional eccentricity due to deflections in the wall.

H_{wc} = βH, the effective height of a wall.

L = the horizontal distance between centres of lateral restraint, or horizontal length of wall.

β = $1/(1 + (H/L)^2)$ if $H \le L$ <u>or</u> L/2H if H > L, factor for walls supported on four sides.

The two-way formulae were devised using the experimental data from the research of Fragomeni (1995) and test results from Saheb & Desayi (1990). The normal strength Eq. (4a) was found to give good prediction of the test results. The high strength Eq (4b) was incorporated as a modified version of Eq. (3b), allowing for high strength concrete.

The equations are for the aspect ratios (H/L) and slenderness ratios (H/t) used in the experiments, ie. $0.67 \le H/L \le 2.0$ and $H/t \le 30$. Also, the walls should have minimum reinforcement ratios of 0.0015 vertically, ρ_v, and 0.0025 horizontally, ρ_h. The full comparison of predicted results with test data were given elsewhere by Fragomeni et al (1995).

4.3 Background of German code effective length factors.

The German code effective length factors used in Equations (4a) & (4b) are based on the plate theory proposed by Timoshenko and Gere (1961). According to the theory, the **Increase in Strength (Iₛ)** of an axially loaded rectangular plate simply supported on four sides when compared to an identical axially loaded plate simply supported top and bottom only is given by:

$$I_s = k_b (H/L)^2/(1-\mu^2) \qquad (5)$$

where

k_b = $(m_bL/H + H/m_bL)^2$, depends on the dimensions of the plate.

m_b = the number of half waves the plate will buckle in.

μ = Poisson's ratio, in the range 0.15 - 0.20.

When $H/L \ge 1.0$ then $k_b = 4$ can be conservatively used. When H/L < 1.0, then $k_b > 4$. If a plate has H/L > 2, then it will generally buckle into half waves such that $H/L = m_b$. If Poisson's ratio is assume negligible, an **effective length factor, b_s,** for a plate simply supported on four sides can be taken approximately as the square root of the reciprocal of Equation (5). That is:

$$\beta_s = 1/(m_b + H^2/m_bL^2) \qquad (6)$$

This factor is the same as the β term given in DIN1045 for $H \le L$ when $m_b = 1$. When H > L the factor seems to differ but when numerical values of H/L are substituted into in each term the results are almost identical, as shown in Table 3.

Table 3. Effective length values (β_s and β)for plates simply supported on four sides

Aspect ratio H/L	Timoshenko and Gere (1961) - effective length factor, β_s	DIN1045 (1988) - effective length factor, β
0.00	1.00	1.00
0.25	0.94	0.94
0.50	0.80	0.80
1.00	0.50	0.50
1.25	0.39	0.40
1.50	0.32	0.33
2.00	0.28	0.29
3.00	0.25	0.25

5 CONCLUSION

New wall design formulae for use in AS3600 have been proposed to cover concrete strengths up to 80 MPa and can be used for walls supported on either two or four sides. An overview of the origins of the factors used for high strength concrete and effective length was given. The formulae are in good agreement with published test results. More testing is desirable to further verify the two-way formulae. Also, an additional factor accounting for reinforcement contribution is desirable.

6 REFERENCES

AS3600-1994. *Concrete Structures*, North Sydeny: Standards Association of Australia.
ACI318-1989. *Building Code Requirements for Reinforced Concrete*, Detroit: ACI.
DIN1045-1988. Reinforced Concrete Strucutres: Design and Construction, Berlin:DIN.
Fragomeni, S. 1995. *Design of normal and high strength reinforced concrete walls*, Ph.D Thesis, The University of Melbourne, Australia.
Fragomeni, S. Mendis, P.A. and Grayson, W.R. 1994b. A Review of Reinforced Concrete Wall design Formulae, *ACI Structural Journal, USA*, Vol. 91 No. 5, Sept-Oct: pp. 521-529.
Fragomeni, S. Mendis, P.A. & Grayson, W.R. 1994a. Axial Load Tests on Normal and High Strength Wall Panels, *Civil Eng. Transactions, IEAust*, Vol. 36 No. 3, Aug: pp 257-263.
Fragomeni, S. Mendis, P.A. and Grayson, W.R. 1995. Axial Load Tests on Concrete Wall Panels Supported on Four Sides, *Proceedings of the 14th ACMSM conference, University of Tasmania, Australia,* Dec: pp. 313-318.
Kripanarayanan, K.M. 1977. Interesting aspect of the Empirical Wall Design Equation, *ACI Structural Journal,*Vol. 74, No. 5, May: pp. 204-207.
Oberlender, G.D., & Everard, N.J. 1977. Investigation of reinforced Concrete Wall Panels, *ACI Structural Journal,* Vol. 74, No. 6, June: pp. 256-263.
Pillai, S.U, & Parthasarathy, C.V. 1977. Ultimate Strength and Design of Concrete Walls, *Building and Environment,* London, Vol. 12: pp.25-29.
Saheb, S.M., & Desayi, P. 1989. Ultimate Strength of R.C. Wall Panels in one-way in-plane action, *Journal of Structural Engineering, ASCE,* Vol. 115, No.10, Oct.: pp. 2617-2630.
Saheb, S.M., & Desayi, P. 1990. Ultimate Strength of R.C. Wall Panels in two-way in-plane action, *Journal of Structural Engineering, ASCE,* Vol. 116, No. 5, May: pp. 1384-1402.
Seddon, A.E. 1956. The Strength of Concrete Walls Under Axial and Eccentric Loads, *Symposium on Strength of Concrete Structures - May,* London: Cement & Concrete Assoc.
Timoshenko, S.P., & Gere, J.M. 1961. *Theory of Elastic Stability*, Sydney: Mcgraw-Hill.
Zielinski, Z.A., Troitsky, M.S., & Christodoulou, H. 1982. Full-scale Bearing Strength Investigation of Thin Wall-Ribbed Reinforced Concrete Panels, *ACI Structural Journa.,* Vol. 79, No. 4, Jul-Aug: pp. 313-321.

The Mechanics of Structures and Materials, Grzebieta, Al-Mahaidi & Wilson (eds)
© 1997 Balkema, Rotterdam, ISBN 90 5410 900 9

A stochastic method to assess life cycle performance of concrete structures

C.Q.Li
Department of Civil Engineering, Monash University, Caulfield, Vic., Australia

ABSTRACT: Since material degradation is inevitable in reinforced concrete structures more attention should be drawn to the maintenance and repairs of the structure. This has, in essence, raised an issue of life cycle performance evaluation of concrete structures. The present paper is concerened with the development of a reliability-based method to assess the life cycle performance of concrete structures, and to predict maintenance and repair times and ultimately the service life of concrete structures. In the paper a stochastic model for structural deterioration is proposed from the investigation of material degradation. Based on the model, reliability analysis of deteriorating structures is presented, and the maintenance and repair times and ultimately the service life of the structures are predicted. The proposed method can be used as a tool for rating and benchmarking of concrete structures, which may assist management in decision-making regarding their maintenance and rehabilitation.

1 INTRODUCTION

Since the failure of concrete structures is usually caused by deterioration of structural strength it is becoming increasingly important that a method be developed to monitor and assess structural performance of these structures. Structural performance is generally understood as behaviour associated with a certain service. It is a measure to which a structure fulfils its functions. In principle, performance can be related to strength, stability, serviceability and aesthetics of the structure. From practical point of view, it is desirable that performance be a quantifiable property of the structure. When maintenance and repairs are considered necessary during the service life of a concrete structure, the whole service life has been divided into a number of "life cycles", which may be defined as the period between two maintenance times. The basic concept of life cycle performance is as shown schematically in Fig. 1 where structural performance is represented by its strength.

Control of the deterioration of concrete structures has gained increasing importance at design and maintenance stages of these structures. Most of the current deterioration problems of concrete structures in, for example, bridges, dams and facades of buildings could have been avoided if a systematic durability design and maintenance strategy had been developed. Such a design, however, requires both an overall methodology and detailed calculation models of actual deterioration processes. Unfortunately, both the methodology and deterioration models are not widely available and need to be developed. Since the factors that affect the deterioration of concrete structures are not only uncertain but also time-variant it is well justified that the assessment of life cycle performance of concrete structures should be based on the theory of time-dependent structural reliability.

Time-dependent structural reliability problems are those in which the applied loads are modelled as stochastic processes, and the structural resistance is changing (typically deteriorating) with time and/or loading or both. To some extent, the state-of-the-art approaches

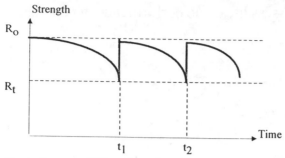

Fig. 1   Concept of life cycle performance

in time-dependent structural reliability theory are quite successful (see, e.g., Li and Melchers, 1993a). However, the application of time-dependent reliability approaches to practical engineering structures, when the structural resistance is deteriorating, has not been accorded much attention by researchers.

The present paper is concerened with the development of a reliability-based method to assess the life cycle performance of concrete structures, and to predict maintenance and repair times and ultimately the service life of concrete structures. In the paper a stochastic model for structural deterioration is proposed from the investigation of material degradation. Based on the model, reliability analysis of deteriorating structures is presented, and the maintenance and repair times and ultimately the service life of the structures are predicted. The proposed method can be used as a tool for rating and benchmarking of concrete structures, which may assist management in decision-making regarding their maintenance and rehabilitation.

2   MODELLING OF STRUCTURAL DETERIORATION

Structural deterioration stems basically from two sources: material degradation and the damage by external laods. It is one of the least attended topic in structural reliability analysis. The most difficulty lies in lack of deterioration models. Existing models for structural deterioration are either too difficult to tackle or too simple to make any practical sense. From the viewpoint of material degradation, only steel corrosion has the capacity to deteriorate the structural strength and hence cause structural collapse. However the majority of the research on steel corrosion is limited more to corrosion mechanisms than to residual strength. As a consequence, there are few stochastic models that could simulate the behaviour of deteriorating structures, a situation which has actually overlooked the random nature of the problem.

In general, the deterioration of structural strength can be modelled by a residual strength, which can be expressed as

$$R(t) = \varphi(t)R_0 \qquad (1)$$

where $\varphi(t)$ denotes the deterioration function and R_0 is the initial strength, both of which are functions of structural parameters, e.g., concrete grade, cover, stiffness, dimensions and so on.

Given the current state of development of structural deterioration modelling, the best ways to model structural material (resistance) deterioration are either to compare it to the known results of others, or to employ Monte Carlo simulation from observations. Since experimental results on structural deterioration, e.g., on residual strength, are not available it seems that Monte Carlo (computer) simulation is the only realistic alternative. Therefore, the deterioration model, to be proposed herein, is based on the simulation results of a case study of structural strength deterioration due to the corrosion of reinforcing steel in concrete structures. A more detailed description of the model is available in Li and Melchers (1993b).

The deterioration function, $\varphi(t)$, can be obtained for a flexural member in reinforced concrete structures, as follows. From reinforced concrete theory (Warner, et al, 1989)

194

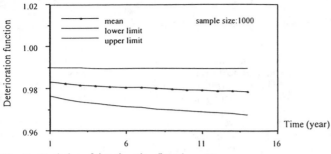

Fig. 2 Statistics of deterioration function φ(t)

$$R(t) = A_s(t)f_y d'$$ (2)

where $A_s(t)$ is the cross-section area of the tensile steel at time t and d' is the lever arm. Using an area reduction parameter $A_r = \dfrac{A_s(0) - A_s(t)}{A_s(0)}$ (in percentage), equation (2) becomes (see, Li and Melchers, 1993b)

$$R(t) = [A_s(0) - 0.01 A_r(t)A_s(0)] \cdot f_y \cdot d'$$

$$= [1 - 0.01 A_r(t)]R_0$$ (3)

So that the deterioration function in this specific case is

$$\varphi(t) = 1 - 0.01 A_r(t)$$ (4)

The statistics of φ(t) may be obtained using a simulation procedure described in Li and Melchers (1993b) the results are shown in Fig. 2 and analytical expressions are as follows

$$\mu_\varphi = 0.983 - 0.0003t$$ (5a)

$$\sigma_\varphi = 0.004t^{0.18}$$ (5b)

The analysis of Li and Melbourne (1993b) indicated that φ(t) is approximately of normal distribution so only the mean and standard deviation functions are of interest.

3 PERFORMANCE ASSESSMENT

3.1 Basic theory

Structural deterioration of a reinforced concrete structure can be assessed using time-dependent reliability approaches. The basic theory of structural reliability is briefly described below. Details of the theory are out of the scope of the paper but are available in text books such as Melchers (1987).

For reliability problems involving stochastic processes, the structural reliability depends on the time that is expected to elapse before the first occurrence of the stochastic processes crossing out of the safe domain D, sometime during the lifetime [0, t_L] of the structure. Equivalently, the probability of the first occurrence of such an excursion is the probability of structural failure $p_f(t)$ during that time period. Under some assumptions (see, e.g., Melchers, 1987), the so-called "first passage probability" can be determined from

$$p_f(t) = p_f(0) + [1 - p_f(0)][1 - e^{-\int_0^t v(\tau)d\tau}] \leq p_f(0) + \int_0^t v(\tau) \tag{6}$$

where $p_f(0)$ is the probability that the structure fails on first loading, i.e., at time $t = 0$. Evidently $p_f(0)$ is time-independent and for structural deterioration problems, it may be assumed to be zero approximately, which means that it is unlikely that the structure deteriorates as soon as it is put in use. The main difficulty in application of equation (6) to realistic problems is the determination of the outcrossing rate $v(t)$.

When the applied load is a scalar stochastic process $X(t)$, the outcrossing problem becomes an upcrossing one so that the outcrossing rate in equation (6) can be replaced by an upcrossing rate, which can be evaluated directly by the well-known Rice formula (Rice, 1944)

$$v = v_a^+ = \int_a^\infty (\dot{x} - \dot{a}) f_{x\dot{x}}(a,\dot{x}) d\dot{x} \tag{7}$$

where v_a^+ is the upcrossing rate of the scalar process $X(t)$ relative to a, the barrier level (threshold) to be upcrossed, $\dot{a}$ is the slope of a with respect to time, $\dot{X}(t)$ is the time derivative process of $X(t)$ and $f_{x\dot{x}}(\,)$ is the joint probability density function for X and $\dot{X}$.

When the threshold a is not a random variable the solution to equation (7) can be expressed analytically as (a detailed derivation is given in Li and Melchers, 1993b)

$$v = \frac{\sigma_{\dot{x}|x}}{\sigma_x} \phi\left(\frac{a - \mu_x}{\sigma_x}\right) \left\{ \phi\left(\frac{\dot{a} - \mu_{\dot{x}|x}}{\sigma_{\dot{x}|x}}\right) - \frac{\dot{a} - \mu_{\dot{x}|x}}{\sigma_{\dot{x}|x}} \Phi\left(\frac{\dot{a} - \mu_{\dot{x}|x}}{\sigma_{\dot{x}|x}}\right) \right\} \tag{8}$$

where $\phi(\,)$ and $\Phi(\,)$ are standard normal density and distribution functions respectively, μ and σ denote the mean and standard deviation of random variables X and $\dot{X}$, represented by subscripts and " | " denotes the conditional probability. For a given stochastic process, all the variables in equation (8) can be obtained as shown in Melchers (1987).

3.2 Practical application

Let the structural deterioration of a reinforced concrete structure be represented by its flexural capacity of a critical section of the structure. Let the applied load $Q(t)$ be a continuous normal stochastic processes with n components (representing a load combination). To evaluate the deterioration of the structure, i.e., the strength deterioration, a criterion for the structural performance needs to be established, which can be defined as a limit state in general

$$S[Q(t)] \leq R(t) \tag{9}$$

where $S(Q)$ is the load effect (bending moment) produced by the load process at the critical section and $R(t)$ is the corresponding structural resistance (flexural capacity). When the structural analysis is elastic (as the general practice in reinforced concrete structures), the load effect $S(Q)$ has a linear form

$$S = c_1 q_1 + c_2 q_2 + \cdots + c_n q_n \tag{10}$$

so that the load effect $S(Q) = CQ$ is a scalar process, where $C = \{c_i, i = 1, ..., n\}$ is the coefficient vector which can be determined from structural analysis. Owing to the linear form of equation (10), the load effect process S is also of normal and its statistics can be obtained using stochastic process theory (Papoulis, 1965).

The deterioration of a structural member is unacceptable when the load effect in the member is greater than its resistance, i.e., $S > R$. Because both of S and R are time-variant equation (11) is readily used to calculate the probability of this unacceptable deterioration, which is defined as a structural failure. From equations (6) to (8), it can be obtained that the probability of the deterioration failure at time t is

$$p_f = \int_0^t v d\tau \qquad (11)$$

where v is upcrossing rate, the solution of which is given by equation (8) with S(t) replacing X(t) and R(t) replacing the threshold a. Since the structural strength (resistance) is also a time-variant random variable, the equation (8) needs to be modified to account for the uncertainty of the strength R, i.e.

$$p_f = \int_0^t \left(\int_0^\infty v \cdot f_R(r) dr \right) d\tau \qquad (12)$$

where $f_R(r)$ is the probability density function of R at the critical section. The easiest way to evaluate equation (12) is through simulation, i.e.,

$$p_f = \int_0^t \left(\frac{1}{k} \sum_{i=1}^k v_i \right) d\tau \qquad (13)$$

where k is the number of samples taken from R in the simulation.

4 MAINTENANCE TIME

Assigning $t = t_L$ in equation (13), an unique relationship between the performance (strength deterioration) and the service life is established (although an analytical solution may not be easy). Therefore, for a given structural performance and an acceptable probability of structural failure p_{fa}, if $p_f(t) < p_{fa}$, the performance is acceptable at time t. At the time, e.g., t_1, that $p_f(t_1)$ is greater than or equal to p_{fa} the maintenance is needed and one life cycle for the structure ends. Alternatively, if no maintenance is to be carried out the service life of the structure is determined, i.e., $t_L = t_1$. When there are a number of maintenance actions the service life of the structure becomes the union of all time intervals between maintenance times, i.e.,

$$t_L = \bigcup_{i=1}^m t_{Mi} \qquad (14)$$

where m is the number of maintenance actions that are expected during the service life of the structure, $\cup$ denotes union of random events and t_{Mi} is the time interval between two maintenance times, i.e., $t_{Mi} = t_i - t_{i-1}$. As has been seen the structural performance is closely associated with the service life of the structure. This is the nature of structural deterioration problems.

As can be seen there is a problem in using equation (14) which is how to determine the acceptable limit for the probability of structural deterioration failure, i.e., p_{fa}. The acceptable limit for the probability of structural failure can be determined in the framework of reliability-based economic optimisation of structures. The basic idea of economic optimisation is to find an optimal balance between the total cost of the structure (including initial construction cost, maintenance cost, rehabilitation cost etc.) and the required safety level, which should be based on a socio-economic criterion. Due to the length limit of the paper this topic is not to be discussed but see Li (1995).

5 WORKED EXAMPLE

As a simple example, a reinforced concrete structural member is subjected to severe steel corrosion. The load effect (bending moment) is modelled as a stationary process, $S(t)$, with a mean $\mu_s = 200$ kNm, a standard deviation $\sigma_s = 45$ kNm and an auto-correlation function $\rho(\tau)$, where $\tau = t'$ - t'' (t' and t'' are two time points in the service life). The original flexural strength of the member (resistance), R_0, has a mean value of 128 kNm and a standard deviation of 12.8 kNm. The deterioration model of the member is of the form of equation (4) with a mean function and a standard deviation function of equation (5). Since the load effect (bending moment), $S(t)$, is a stationary process, the solution (i.e., equation (13)) is reduced, with S replacing X and R replacing a, to

$$v = \frac{\sigma_{\dot{s}}}{\sqrt{2\pi}\sigma_s}\phi\left(\frac{R-\mu_s}{\sigma_s}\right) \tag{15}$$

where $\sigma_{\dot{s}} = -\dfrac{\partial^2\rho(0)}{\partial\tau^2} = -\rho''(0) = 0.2$. The acceptable limit for the deterioration failure is taken as $p_{fa} = 10^{-4}$. It is assumed three major maintenance actions would be carried out during the service life of the structure and each maintenance is independent.

The computation procedure is as follows: (1) at a time t, take a sample from the probability distribution N(128, 12.8) to determine the structural strength at time t, i.e., R in equation (15); (2) calculate the upcrossing rate using equation (15); (3) repeat step (2) for a large number of times. In this example it was 100 times, i.e., the sample size is 100; (4) calculate the probability of deterioration failure using equation (13); (5) if $p_f(t) \geq p_{fa}$ then $t_1 = t$ is the first maintenance time; otherwise repeat the procedure from (1). After the maintenance, it was assumed the strength can be restored to 95% of the original strength. Following the above procedure, the second and third maintenance times can be determined. The preliminary results were: the first maintenance time is $t_1 = 21$ (years); the second and third times are $t_2 = 17$ and $t_3 = 13$. Due to the independence of each maintenance, the service life for the structure is $t_L = 51$ (years). A user-friendly computer program was developed for the computation procedure.

6 CONCLUSION

A reliability-based method has been developed to assess the life cycle performance of concrete structures, and to predict maintenance and repair times and ultimately the service life of concrete structures. A stochastic model for structural deterioration has been proposed from the investigation of material degradation and the reliability analysis of deteriorating structures has been presented. The proposed method can be used as a tool for rating and benchmarking of concrete structures, which may assist management in decision-making regarding their maintenance and rehabilitation.

7 REFERENCES

Melchers, R.E., 1987, Structural Reliability Analysis and Prediction, John Wiley and Sons, Chichester, U.K..

Li, C.Q., 1995, "Optimisation of Reliability-Based Structural Design", *Civil Engrg.Trans.*, **37**, (4), 303 - 308.

Li, C.Q. and Melchers, R.E., 1993a, Outcrossings from Normal Convex Polyhedrons for Nonstationary Processes ", J. Engrg. Mech., ASCE, 119, (9).

Li, C.Q. and Melchers, R.E., 1993b, "Simulation of Corrosion of Reinforcing Steel in Concrete Structures", Proc. 13th Aust. Conf. on Mech. of Struct. and Mater., Wollongong, 517 - 524.

Papoulis, A., 1965, Probability, Random Variables, and Stochastic Processes, McGraw-Hill, New York.

Rice, S.O., (1944), "Mathematical Analysis of Random Noise", Bell. System Tech. J., 23, 282-332, (1945), 24, 46-156.

Warner, R.F., Rangan, B.V. and Hall, A.S., (1989), Reinforced Concrete, Longman Cheshire, (3rd Edition), Melbourne.

The Mechanics of Structures and Materials, Grzebieta, Al-Mahaidi & Wilson (eds)
© *1997 Balkema, Rotterdam, ISBN 90 5410 900 9*

Behaviour of reinforced concrete frames under non-proportional loadings

K.W.Wong & R.F.Warner
Department of Civil and Environmental Engineering, University of Adelaide, S.A., Australia

ABSTRACT: This paper gives details of a study of the overload behaviour and strength of concrete framed structures when subjected to non-proportional loadings. A special analytic procedure is described which simulates the overload behaviour and collapse of a concrete frame when subjected to vertical and horizontal loads which are applied in sequence. Comparisons are made of the effects of applying loads sequentially and simultaneously to a portal frame system. The results indicate that load capacity for sequentially applied loads can be predicted with good accuracy by using an equivalent proportional load system.

1 INTRODUCTION

Concrete structures are designed to resist various combinations of loads which act both vertically and horizontally. To simplify the design calculations it is always assumed that the loads comprising any one combination are applied simultaneously and proportionally, and the load effects are usually evaluated by means of linear elastic frame analysis.

In reality, the loads occur in sequence, for example with the vertical loads due to self weight and dead and live load preceding the horizontal loads due to wind or earthquake. Furthermore, concrete structures behave in a highly non-linear manner, even under working load conditions. Although non-linear analysis packages are gradually becoming available commercially, they usually predict structural behaviour under conditions of a progressively increasing, but proportional, system of loads.

This paper reports on an investigation of concrete structural behaviour under non-proportional loading. The aim of the study was to evaluate the effect of the sequence of loading on the load capacity of concrete frames. The situation investigated is illustrated in Figure 1.

In Figure 1a, a simple frame is considered with a vertical load system V and a horizontal load system H. Three load sequences are considered. In the first sequence, Figure 1b, the vertical load is increased to a value represented by V_{1max} and the horizontal load is then progressively increased until collapse occurs at the value H_{1ult}. In the second sequence, Figure 1c, the horizontal load is first increased to a maximum value of H_{2max}, which is chosen to be equal to H_{1ult}, and the vertical load is then increased until failure occurs at V_{2ult}. In the third case, Figure 1d, the vertical and horizontal loads are applied simultaneously. They are proportionally increased in the ratio $V_{1max}:H_{1ult}$ until failure occurs at loads of V_{3ult} and H_{3ult}. Clearly, the load combinations at collapse are not necessarily the same in these three cases. To date, very little research has been undertaken on the effects of load sequences and non-proportional loading (Kenyon & Warner, 1993), and the purpose of the present study was to obtain quantitative estimates of the effect of load sequence on load carrying capacity.

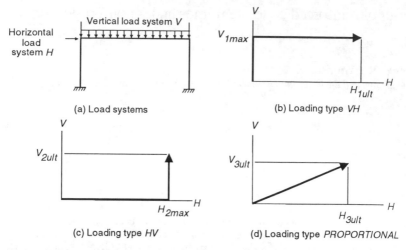

Figure 1. Diagrams showing the various loading paths.

2 METHOD OF ANALYSIS

In order to undertake the investigation a computational procedure was developed to simulate the sequential application of the loads. Only two load systems were considered, in order to limit the computational intensity of the work; however, the method can easily be extended to treat any number of load systems. The first load system is increased to a pre-determined value and maintained. The second load system is then progressively increased on the loaded structure until a condition of collapse is achieved.

The simulation program has been developed as part of an ongoing study of the non-linear behaviour of concrete structures (Warner &Yeo 1984; Wong, Yeo & Warner 1988; Kenyon & Warner 1993; Kawano & Warner 1995). Non-linear analysis can be undertaken using deformation control, load control, or a combined load-deformation control. Load-control methods tend to have convergence problems when used to determine load capacity, although methods have been devised to deal with peak loads and post failure decreasing loads. Deformation control procedures have been developed which use curvature control in a local region (Wong, Yeo & Warner 1988), and deflection control at a suitable point on the frame (Kawano & Warner 1995). An early application of the combined load-deformation control (Crisfield 1983) is very effective although computationally intensive in dealing with peak load behaviour. While the deformation control and load-deformation control methods are both suitable for determining the behaviour of the structure at collapse under proportional loading, the curvature control method was found in the present study to provide a simple and efficient basis for the analysis of non-proportional loading. A previous preliminary study (Kenyon & Warner 1993) used load control for the first load system and the first part of the second load application, and then changed over to defomation control for the later part of the second load application to failure. In the present study, the use of curvature control throughout has greatly simplified the computations.

The computational procedure with curvature control for proportional analysis is described in an earlier paper (Wong, Yeo & Warner 1988). The extension of this procedure to non-proportional analysis will now be explained.

The two load systems are represented by two unit load patterns which are progressively increased by means of scaling factors. In Figure 1a, the rectangular portal frame is subjected to a vertical load system consisting of a uniformly distributed load of w kN/m and two point loads of αwL kN which act on each column. To represent various levels of loading, an initial value of w (eg 1.0 kN/m) is taken, and a progressively increasing scaling factor λ_1 is introduced. For the horizontal load system a unit load H (eg 1.0 kN) is scaled using λ_2.

200

Load pattern 1 is applied in several steps by increasing the load factor λ_1 using curvature control until it reaches λ_{1max}. From this stage on, λ_1 is kept at this value while curvature control is used to increase λ_2 until collapse occurs at λ_{2ult}.

The frame is modelled by dividing it into numerous segments and each of these segments is assumed to have uniform flexural stiffness equal to that of the most critically stressed section within it. The section behaviour is modelled by dividing it into concrete and steel layers. The concrete layers are assumed to have a non-linear stress-strain relation, and the steel layers are assumed to have a linearly elastic-plastic stress-strain relation. These stress-strain relations allow for unloading paths in case of strain reversal. Structural members such as beams and columns are assumed to have equal size segments. The number of segments chosen for each member gives a length to depth ratio of approximately unity for the segments.

An incremental-iterative curvature control solution strategy is used. The full range behaviour is obtained by subjecting a nominated key segment to increasing curvatures. This key segment is selected at the start of the analysis. A basis for the selection is that it has monotonically increasing curvatures as the loading process progresses, initially under the influence of load pattern 1 and subsequently under the additional influence of load pattern 2. At the start of a typical curvature-increment step, the targeted curvature for the key segment is calculated by adding a preset curvature increment to the curvature used in the preceding step. The load factors λ_1 and λ_2 consistent with this targeted curvature in the key segment are then determined by carrying out several iterative cycles of calculations until convergence is achieved.

Computational steps required for a typical iterative cycle within an incremental step which has been assigned a targeted curvature K_{step} are:

1. The frame with the latest full-load stiffnesses EI_i in all the segments is subjected to the unit load pattern 1. If this is the first cycle of iterations, then these stiffnesses are set to those calculated at the end of the previous step, else these stiffnesses are set to those calculated at the end of the previous cycle. A second order elastic analysis routine is used to determine curvatures $k_{11}, k_{21},....k_{key1}, k_{i1},......, k_{nseg1}$ in all the segments. Similarly, the frame with the latest full-load stiffnesses EI_i in all the segments is subjected to the unit load pattern 2. A second order elastic analysis routine is used also to determine curvatures $k_{12}, k_{22},....k_{key2}, k_{i2},......, k_{nseg2}$ in all the segments.

2. The load factor λ_1 to give the key segment a total curvature (the term total curvature means curvature under full loading) of K_{step} is then estimated :

$$\lambda_1 = \frac{K_{step}}{|k_{key1}|} \tag{1}$$

where k_{key1} is the curvature in the key segment under unit load pattern 1.

3. If λ_1 is less than or equal to λ_{1max}, λ_1 is taken to be the value calculated in step 2 above and λ_2 is set to zero. Proceed to Step 5.

4. If λ_1 exceeds λ_{1max} the following steps are required:

   • Estimate the full load curvature of the key segment under load pattern 1 K_{key1} which will give a λ_1 value equal to λ_{1max}:

   $$K_{key1} = \lambda_{1max} \times |k_{key1}| \tag{2}$$

   • Load factor λ_1 is then set to λ_{1max} :

   $$\lambda_1 = \lambda_{1max} \tag{3}$$

   • The required total curvature of the key segment for the frame subjected to load pattern 2 to give a total curvature of K_{step} under the combined application of the two load patterns is calculated:

   $$K_{key2} = K_{step} - K_{key1} \tag{4}$$

- The load factor λ_2 to give the key segment a total curvature of K_{key2} is then calculated:

$$\lambda_2 = \frac{K_{key2}}{|k_{key2}|} \tag{5}$$

where k_{key2} is the curvature of key segment under unit load pattern 2.

5. After having calculated both λ_1 and λ_2, they are used to scale the corresponding curvatures of the segments to give corresponding total curvatures. For a typical segment i, the total curvature K_i is obtained by adding the component curvatures:

$$K_i = \left(\lambda_1 \times |k_{i1}|\right) + \left(\lambda_2 \times |k_{i2}|\right) \tag{6}$$

where k_{i1} is the curvature of the ith segment under unit load pattern 1 and k_{i2} is the curvature of the ith segment under unit load pattern 2. Note that the simultaneous application of λ_1 and λ_2 in the above equation will give a curvature of K_{step} in the key segment.

6. For each segment i with curvature K_i, a section analysis routine which takes account of the material stress-strain relations is used to calculate the corresponding bending moment, M_i. The stiffnesses of all the segments are updated:

$$EI_i = \frac{M_i}{K_i} \tag{7}$$

7. Convergence of the EI_i values is checked at this stage. If convergence is achieved, a solution has been obtained for the present curvature step. The targeted curvature is then set to that of the next step. Proceed to step (1) to start a new set of iterative cycles. If convergence is not achieved then proceed to step (1) maintaining the present targeted curvature for more iterative cycles.

The solution procedure caters for both geometric and material nonlinearities. The geometrical nonlinearity is taken into consideration when the second order elastic analysis is carried out, and the material non-linearity is taken into consideration when the section analysis is used to update the stiffnesses of the segments.

3 PORTAL FRAMES UNDER NON-PROPORTIONAL LOADINGS

To illustrate the method of analysis, the portal frame shown in Figure 2a was analysed for three loading histories. The following properties were assumed: $f'_c = 32$ MPa, $f_{sy} = 400$ MPa with mean values: $f_{cm} = 37.5$ MPa and $f_{sm} = 460$ MPa. The peak stress used for the stress-strain relation of in-situ concrete is assumed to be $0.85\, f_{cm}$. This takes into account of the difference in the strength of concrete in the structure when compared with the cylinder strength obtained from tests.

The portal frame was first subjected to non-proportional load of type VH. The vertical pattern load was applied by increasing α until it reached a predetermined value α_{1max}. When α reached this value, the uniformly distributed load on the beam simultaneously reached a value of $0.5w_{beam}$, where w_{beam} is the failure load of the frame with only the vertical uniformly distributed load w acting along the beam. Limiting the maximum value of α to this value ensured that the frame did not fail prematurely with a local beam failure mechanism. Curvature control was used to increase the horizontal load until it reached its peak value of H_{1ult}.

The frame was then subjected to an equivalent proportional loading (ie with the $H:\alpha$ ratio maintained at $H_{1ult}:\alpha_{1max}$) until failure occurred at $(H_{3ult}:\alpha_{3ult})$. The frame was also subjected to non-proportional loading of type HV. For this loading, system H was increased until it reached a value of H_{2max} (with H_{2max} set equal to H_{1ult}). It was then maintained constant while α was caused to increase until it reached its peak value α_{2ult}. Results for the three loading types for a frame with $h = 4$m are presented in the form of a strength interaction diagram

202

shown in Figure 2b. Note that this interaction diagram shows the effects of both geometric and material non-linearities.

The results indicate that, generally, equivalent proportional loading gives a good estimate of the peak loads for frames subjected to non-proportional loading. Note that states represented by the portion of the interaction diagram to the right of the horizontal load H of 227 kN are not attainable using the loading type HV.

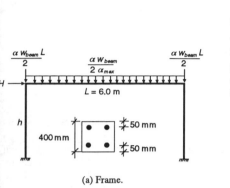

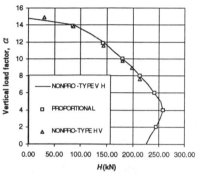

(a) Frame.

(b) Strength interaction diagram.

Figure 2. Portal frame and its strength interaction diagram.

Horizontal load versus sway deflection curves for type VH loading with $\alpha_{max}=10.0$ are shown in Figure 3a. The sway deflection is the horizontal deflection at the top of the left column of the portal frame. Results obtained for the equivalent proportional load are also plotted in the same figure. The results show that while the peak loads for both loading types are almost the same, type VH has a smaller sway stiffness over a large portion of the load range.

Curves of vertical load factor α versus sway deflection are also shown in Figure 3b for the non-proportional load type HV and the proportional loading type. The sway deflection is also taken as the horizontal deflection at the top of the left column of the portal frame. The results show that for the former, the lateral stiffness of the frame increases initially on the application of the vertical load. This behaviour is consistent with previous observations of the stiffening effect of thrust on moment curvature relations of concrete sections.

The use of an equivalent proportional load for loading type HV underestimates the sway deflection of this frame over most of its loading history. At half the peak load level of α, the sway deflection for the equivalent proportional load case is less than half that of non-proportional type HV.

Another two sets of rectangular portal frames were also analysed by subjecting portal frames to the three different loading types described above. These frames are the same as those described above with only the column height being different. The column heights of the second and third set of portal frames are 8m and 12m respectively. Owing to lack of space, results obtained for these frames are not shown in this paper. The results, and results obtained for a free-standing column and a three-storey frame which are available elsewhere (Wong & Warner 1997), all suggest that the equivalent loading is a good approximation for the V-H load pattern.

Results from the analysis carried out show that equivalent proportional pattern loads can generally be used to predict the strength of practical concrete portal frames subjected to non-proportional pattern loads.

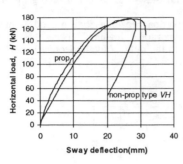 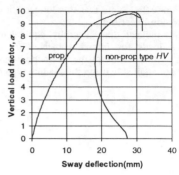

(a) Type *VH* with its equivalent proportional type   (b) Type *HV* with its equivalent proportional type

Figure 3. Load versus sway deflection plots for case $\alpha_{max} = 10$.

4 CONCLUDING REMARKS

A curvature-controlled based procedure has been developed which is efficient for tracing the full-range behaviour of non-linear reinforced concrete frames under non-proportional pattern loads. The superiority of the approach is that throughout the entire range, a key-segment curvature-controlled procedure is used. This approach is complementary to an earlier computational approach developed to study concrete frames under proportional loadings. Both are currently being used to study the non-linear behaviour of concrete frames.

Load versus deflection relations for columns and frames under non-proportional loadings have been obtained. They were compared with those obtained for corresponding frames under equivalent proportional loadings.

Results obtained show that an equivalent proportional pattern load may be used to predict the peak loads for a structure subjected to non-proportional load patterns, and that for unbraced, practical frames designed to carry predominantly vertical loads, the peak loads are virtually independent of the loading paths taken.

5 REFERENCES

Crisfield, M.A. 1983. An arc-length method including line searches and accelerations, *International Journal for Numerical Methods in Engineering*, 19:1269-1289.

Kawano, A. & Warner, R.F. 1995. Nonlinear analysis of the time-dependent behaviour of reinforced concrete frames, *Research Report No R125, Department of Civil and Environmental Engineering, The University of Adelaide*, January, 1-41

Kenyon, J.M. & Warner, R.F. 1993. Refined analysis of non-linear behaviour of concrete structures, *Civil Engineering Transactions, Instn of Engineers Australia*, Vol CE35, No 3, 213-220.

Warner, R.F. & Yeo, M.F. 1984. Ductility requirements for partially prestressed concrete, *Proceedings of the NATO Advanced Research Workshop on 'Partially Prestressing,From Theory to Practice', Paris, France*, June, 315-326.

Wong, K.W., Yeo, M.F.& Warner, R.F. 1988. Non-linear behaviour of reinforced concrete frames, *Civil Engineering Transactions, Instn of Engineers Australia*, Vol CE30, No 2, July, 57-65.

Wong, K.W. & Warner, R.F. 1997. Behaviour of reinforced concrete frames under non-proportional loadings, *Research Report R146, Department of Civil and Environmental Engineering, The University of Adelaide*, February, 1-16.

The Mechanics of Structures and Materials, Grzebieta, Al-Mahaidi & Wilson (eds)
© 1997 Balkema, Rotterdam, ISBN 90 5410 900 9

Use of rice husk ash as a pozzolanic admixture

M.R.Smajila & J.L.van der Molen
Department of Civil and Environmental Engineering, The University of Melbourne, Vic., Australia

ABSTRACT: An experimental study of the effect of rice husk ash (RHA) on concrete and mortar was undertaken. The majority of concrete produced in Australia is low strength, 20-25 MPa at 28 days. In the hope to capitalise on this large market, the focus of the experiments was on the use of RHA in concrete with a compressive strength of 25 MPa at 28 days. The RHA used has a silica content of 85% which is in an amorphous form. The characteristics of concrete made with RHA investigated include water demand, strength growth under different curing regimes and for different mixes, pozzolanic reactivity, permeability and carbonation.

1. INTRODUCTION

Rice hulls make up on average 20%, (range 16-23%), of the weight of paddy rice. The 1995 rice crop produced 1.2 million tonnes of paddy rice. This produced 222,000 tonnes of rice hulls, of which 126,000 tonnes were utilised leaving 96,000 tonnes as waste. The disposal of rice hulls is presenting problems to rice millers throughout the world. Burning the rice husks in a controlled manner produces an ash high in silica content (80% or more) that can be used as a pozzolanic admixture in concrete, thus providing an attractive means of disposing of rice husks and the possibility of reducing the use of cement and improving the characteristics of concrete.

The sponsor of this research, Biocon Pty Ltd, a subsidiary of the NSW Ricegrowers Co-op Ltd (RCL), provided the rice husk ash (RHA) for our experiments. Biocon has developed a method of burning rice husks, producing RHA containing 85% SiO_2 in an amorphous form, making it highly reactive. The RHA has a specific mass of 2040 kg/m^3 with a loss on ignition (LOI) value of 8.8%. The work presented is part of the second section of research studying the RHA produced by Biocon at The University of Melbourne. Due to the implementation of a more economical burning method the material characteristics changed from the first section of research to the second, necessitating some repetition of experimental work. The focus of the current research is to establish the effects of the RHA on 25 MPa concrete.

The only mortar tests carried out were the water requirement test specified in ASTM-C311 to determine the water demand, and the permeability and carbonation tests. To take advantage of the well-documented benefits of pozzolanic materials on the paste-aggregate interface (Mindess 1995, Mitsui 1994), all other tests use concrete mixes. The areas investigated include workability, curing regimes, strength, cement replacement levels, comparison between RHA and fly ash (FA), addition of lime, lime-silica mixes, permeability and carbonation.

2. TEST RESULTS AND DISCUSSION

For the preparation, mixing and testing of all specimens set procedures were followed to help maintain consistency of results. These procedures were based on the corresponding Australian or American Standards, where possible.

Table 1. Cement and Pozzolanic material characteristics

Constituent % weight	RHA	FA	Type GP cement	Particle size μm	RHA % passing	FA % passing	GPcem % passing
SiO_2	85.4	46.4	20.1	75	100	100	100
Al_2O_3	0.1	27.8	5.5	56	100	99	99.8
Fe_2O_3	0.1	15.5	3.4	32	99.5	90.2	89.5
CaO	0.8	3.2	64.5	24	94.7	81.2	72
MgO	0.5	-	1.2	16	77.2	62.4	36.3
Na_2O	0.0	0.2	0.1	12	62.9	48.1	22.5
K_2O	2.19	0.5	0.6	8	51.1	33.7	15.6
SO_3	0.01	-	2.9	6	28.3	18.5	8.4
L.O.I. (carbon)	8.8	2.1	1.5	4	20	7.7	3.2
S.G.	2.04	2.42	3.15	3	13	4.5	1.9
Ave. particle size (μm)	8.8	11.6	18.4	2	0	0.2	0

A 14 mm graded basalt from the Pakenham quarry and a fine sand from Lyndhurst were used in a 2:1 ratio for all mixes. Water/Cementitious ratios were slightly effected by absorption due to the aggregate being used in an air dry state. The cement and mineral admixtures characteristics are listed in Table 1. General purpose Portland type GP cement from Geelong Cement Ltd was used, the FA was provided by Pozzolanic Industries Ltd, SF by Microsilica Pty Ltd and lime by David Mitchell Pty Ltd. The water reducer used was WRDA from Grace W.R. Australia Ltd and the curing compound used was Concure CR from Fosroc Pty Ltd.

Water/cementitious ratios for the mixes without a water reducer were kept at 0.80 except for the FA mix where it was reduced to 0.78 to take advantage of the reduced water demand of FA. For mixes using a water reducer, the water cementitious ratio was reduced to 0.70. Cementitious aggregate ratios were kept within a range of 0.122 to 0.143.

2.1 Workability/water demand

The water requirement test results show that replacing 20% of the cement with RHA will slightly increase the water demand of the mix. The replacement of cement with FA will slightly decrease the water demand while SF will increase the water demand considerably. For the concrete mixes, to maintain a constant slump the amount of aggregate was adjusted while keeping the water/cementitious ratio constant. The slump test results are not as clear. With increase in RHA replacement (causing fines to increase due to specific gravity of RHA being lower than GP cement) the mixture became more cohesive leading to more consistent slump result. The reductions in slump from 15% to 25% replacement and from 30% to 45% replacement (water reducer used) suggest a correlation with the mortar results where the RHA had a similar or slightly increased water demand to that of GP cement. The addition of lime to the mix tended to increase the slump. So a combination of RHA and lime can directly replace an amount of cement without increasing the water demand.

Table 2. Concrete slump test results

Mix	Slump mm	Mix	Slump mm
cem	65,90,105, 90,150	40 rha-wr	85
10 rha	160, 60,90	45 rha-wr	65
15 rha	115	-10cem/+20rha	95
20 rha	90	-15cem/+20rha	80
25 rha	85	20rha 15Lime	125
30 rha-wr	75	20rha 30Lime	110
35 rha-wr	80		

Note : wr = water reducer

Table 3. ASTM-C311 water requirement results

Mix	Water ml	wr	w/c	Flow
GPC	484	100	0.484	79
RHA	515	106	0.515	83
FA	465	96	0.465	82
SF	590	122	0.59	78

Note : wr = water requirement

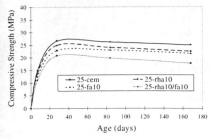

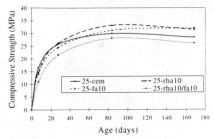

Figure 1. Curing regime 1: Water cured for 7 days and then stored in open air

Figure 2. Curing regime 2 : Water cured for 3 days followed by a curing compound

2.2 *Variation in curing*

Three different curing regimes were tested. In curing regime 1 specimens were placed in lime saturated water until the age of 7 days and then placed in the laboratory store room (T$\cong$20^{0}C, RH$\cong$50%), compressive strength results for these specimens are displayed in Figure 1. For this curing regime the effect of 10% replacement with RHA or FA simply reduced the strength of concrete at all ages. The RHA slightly out-performed the FA, while the RHA/FA combined mix which replaced 20% of the cement producing the lowest strengths at all ages. After 28 days all mixes lost strength at approx. the same rate. A possible explanation for the strength loss when cylinders dry out is proposed by de Larrad & Aitcin (1992).

For curing regime 2 the specimens were stored in lime saturated water until 3 days of age, after which a curing compound was applied. They were then placed in the laboratory store room. The results for this curing regime are shown in Figure 2 and are obviously very different to the previous curing regime used. Clearly, the increase in strength (at 28 and 84 days) in absolute terms (and relative to the control mix), indicate that a pozzolanic reaction has occurred for the three mixes containing pozzolans. The pozzolanic reaction is a secondary reaction, dependent on the primary hydration reaction of cement to produce enough lime for the silica to begin reacting. Hence the pozzolanic reaction takes longer to have an effect on the concrete strength, as seen in Figure 2 where by 84 days of age the RHA and FA mixes are stronger than the control mix. Since the pozzolanic reaction also requires the presence of moisture, a curing regime that keeps the concrete moist for as long as possible is desirable (Salas 1988). From the results in Figure 2 it is evident that 3 days water curing followed by application of a curing compound provides conditions sufficient for the pozzolanic reaction to take place. After 84 days all mixes except the FA mix suffered some strength loss similar to that in the first curing regime tested, suggesting that the cylinders have started to dry out. It is interesting that the FA mix followed the same pattern of strength loss as the other mixes in the first curing regime but not in the second.

For the third curing regime tested specimens were cured in lime water continuously until the day of testing. At this stage of the experimentation the 24 week specimens are not ready for testing. The results available so far are shown in Figure 3. They show a very similar trend to the second curing regime where RHA and control mixes achieved similar strengths at 28 days and the RHA mix being stronger at 84 days. It is expected that strengths will continue to increase due to the constant availability of water. The 3 days water curing followed by the curing compound provides conditions which facilitate the pozzolanic reaction.

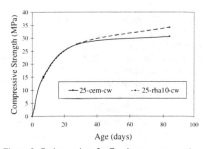

Figure 3. Curing regime 3 : Continuous water curing

Curing regime 2 is capable of application over a broad range of in-situ and precast concrete products. It was therefore used for all remaining tests.

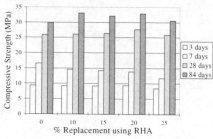

Figure 4. Mixes with no water reducer

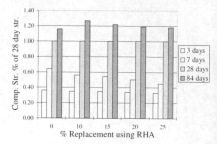

Figure 5. Mixes with no water reducer

2.3 *Strength growth of 25 MPa mixes*

The effect on compressive strength of replacing cement with RHA in varying amounts is shown in Figure 4. From 10 to 20 % replacement, the strength at all ages is approximately the same. At 25% replacement the strengths at all ages start to reduce. The drop off in strength at 25% rather than a steady decrease from 10%, could be due to a lack of available lime at the 25% replacement level.

The compressive strength at different ages relative to the 28 day strength is plotted in Figure 5 for the same mixes. The diagram highlights the change in strength growth as more RHA is used. The main finding is that with increasing concentrations of RHA there is a proportional increase in strength between 7 and 28 days, suggesting that the bulk of the RHA pozzolanic reaction occurs in this period. The tapering off of strength at 84 days could indicate the available lime is slowing declining.

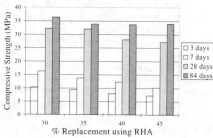

Figure 6. Mixes using a water reducer

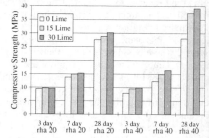

Figure 7. Mixes with added lime

The use of low-range water reducers in concrete is becoming standard practise, so a series of mixes was tested with a combined water reducer and set accelerator. The water reducer was used to lower the water/cementitious ratio from 0.8 to 0.7 so more RHA could be used while still achieving the 25 MPa design strength. From the results in Figure 6, the water reducers allowed up to 45 % replacement with RHA while achieving the target strength. As expected, with more RHA replacement strength values steadily decreased, especially for 3 and 7 days. No higher than 45% replacement was tested.

To help establish when the $Ca(OH)_2$ was being critically depleted, varying amounts of lime were introduced in two RHA mixes. For 20 and 40 % replacement mixes, 15 and 30 % of lime as a proportion of the RHA, was added, the results are shown in Figure 7. The lime had the most effect on the 40 % replacement mix at 28 days. This suggests that there is enough lime produced by the cement for the 20% replacement mix but not enough for the 40% mix. The small increases in strength at 3 and 7 days for all mixes would be most likely due to the introduced lime being available from the commencement of mixing.

2.4 Pozzolanic reaction test

Comparative pozzolanic reactivity was tested by producing mixes of RHA, SF & FA with lime and sand. The pozzolan lime ratio used was 40:60 and, water/cementitious ratio was maintained at 0.5 with the aid of a water reducer. The specimens were kept in a saturated condition at 55-60 degrees Celsius until the age of testing. Results are shown in Table 4. Clearly the RHA mix proved to be the most reactive with the FA producing approximately half the strength of RHA. A possible reason for the high reactivity of rice husk ash was put forward by Mehta (1992), who contributes it to the microporous surface of the ash particles. Mehta suggests that the microporous surface of RHA could be due to its lower temperatures of production compared to the other pozzolanic admixtures. The very low strength attained from the SF mix was surprising, especially since the SF mix hardened the quickest from observations.

Table 4. Lime pozzolan test

mix	7 day comp. (MPa)	28 day comp. (MPa)
RHA-lime	15.9	32.6
SF-lime	4.1	7.5
FA-lime	6.3	18.8

2.5 Permeability and carbonation

The tests reported on above show RHA to have a powerful pozzolanic action. It has all the advantages claimed for it: greater strength development, and lower permeability. Using the customary chemical shorthand:

$$CH + S \rightarrow C\text{-}S\text{-}H$$

Here, the CH ($Ca(OH)_2$) is the lime produced during the hydration of the cement, while the S (SiO_2), is the major constituent of the RHA. While this reaction produces additional solid matter (C-S-H), reducing porosity and increasing strength in doing so, it also consumes CH, which will provide protection of the embedded reinforcing steel against corrosion due to carbonation of the concrete, by maintaining an alkaline environment (pH > 13). The CH resides partly dissolved in the capillary pore water, and partly as small crystals embedded in the mass of C-S-H. CH does not dissolve well, hence this crystalline surplus.

Now, carbonation of concrete is the reaction which takes place between the 0.038% CO_2 available in the air, and the dissolved CH in the pore water. The reaction (in chemical longhand) is as follows:

$$Ca(OH)_2 + CO_2 \rightarrow CaCO_3\downarrow + H_2O$$

The downward arrow denotes that the calcium carbonate, $CaCO_3$, is insoluble, and leaves the solution. Part of the crystalline CH can now enter into solution, resulting in the formation of additional $CaCO_3$ until the total store of CH is exhausted. The carbonation front can now move inward, and the process continues. It follows, that if there is an additional "client" for the consumption of CH, this will accelerate the speed with which the carbonation front moves inward by depleting the total store of CH available to resist the carbonation.

Referring to Table 5, we observe that the introduction of RHA in excess of 20% causes both a decrease in porosity, and a substantial increase in carbonation rate. We can compensate for this effect by also adding CH (lime) in powder form at the time of mixing the concrete. The adverse effect on carbonation of high proportions of RHA is shown in the last line of the table, which indicates that 40% replacement of cement with RHA achieves both the lowest porosity and the highest carbonation rate.

These tests were specially developed, using a Leeds permeability cell (van der Molen 1994). The porosity was measured with respect to nitrogen at 100 kPa pressure on a mortar sample of 50 mm dia. x 50 mm. The carbonation was measured in the same apparatus using a mixture of 10% CO_2 and 90% N_2 at a pressure of 10 kPa for 72 hours.

Table 5. Porosity and Carbonation

Mix	Porosity N_2 cm3/sec	Carbonation depth mm
m25-cem	0.106	3.3
m25-RHA20	0.050	3.2
m25-RHA20-L15	0.088	3.5
m25-RHA20-L30	0.035	2.1
m25-RHA40-wr	0.037	9.6

It should be noted that this testing programme is ongoing, and the above are preliminary results, showing the trend described above. Permeability and carbonation test results show a considerable spread, ideally requiring the averaging of five results. The results in Table 5 are the mean of two tests. The results are included here because they show an important characteristic of RHA addition to concrete. It is recommended at this stage, that if carbonation could prove a long-term problem, lime should be added to the mix in the proportion of 1/3 by weight of the added RHA.

3. CONCLUSIONS

- water demand of RHA is similar or slightly higher than type GP cement, with the addition of lime the water demand can be reduced
- a curing compound or continuous water curing should be used for mixes containing RHA
- 20% cement can be replaced with RHA causing a slight increase in 28 & 84 day strength
- the bulk of the pozzolanic reaction with RHA takes place between 7 and 28 days
- up to 40% cement can be replaced with RHA when a water reducer is used without reducing 28 and 84 day strengths
- for over 20% cement replacement with RHA lime should be added to increase strength
- RHA is highly reactive with lime in the presence of moisture
- RHA addition will reduce permeability
- for over 20% cement replacement with RHA, additional lime should be added to help reduce carbonation (approximately 1/3 by weight of RHA should be added in lime)

4. ACKNOWLEDGMENTS

The contribution of Biocon Pty Ltd. in providing financial assistance, technical knowledge, and samples of RHA is gratefully acknowledged. We wish to thank Dr. P. Mendis who instigated and organised the Australian Postgraduate Award (industry) scholarship associated with this research. We also wish to thank Mr Ken Day from Concrete Advice Pty Ltd. for his technical advice and guidance throughout this research.

5. REFERENCES

Mitsui, K., Z. Li, D.A. Lange & S.P. Shah 1994. Relationship between microstructure and mechanical properties of the paste-aggregate interface. *ACI Materials Journal*: V.91, No.1, pp. 30-39.
Mindess, S. 1995. Mechanical properties of the interfacial transition zone: a review. *Interface Fracture and Bond*: ACI SP-156, pp. 1-9
de Larrard, F. & P. Aitcin 1993. Apparent strength retrogression of silica-fume concrete. *ACI Materials Journal*: V.90, No.6, pp. 581-585
Mehta, P.K. 1992. Rice husk ash - a unique supplementary material. *Advances in Concrete Technology*: CANMET pp. 407-431
Salas, J., M. Alvarez, G. Gomez & J. Veras 1988. Crucial curing of rice husk ash concrete. *The International Journal of Cement Composites and Lightweight Concrete*: V.9, N.3, pp. 177-182
van der Molen, J.L. 1994. Assessment of concrete durability by determination of its intrinsic permeability. *Proceedings of the AUSTROADS 1994 Bridges Conference*: V.1, No.24

The Mechanics of Structures and Materials, Grzebieta, Al-Mahaidi & Wilson (eds)
© *1997 Balkema, Rotterdam, ISBN 90 5410 900 9*

Design of high-strength concrete members

R.S. Pendyala, P.A. Mendis & A.S. Bajaj
Department of Civil and Environmental Engineering, University of Melbourne, Vic., Australia

ABSTRACT: High-Strength Concrete has revolutionised the construction industry in the 1980's and Australian researchers, consultants and contractors have been very active in this area. However, most major codes for design and construction with concrete are applicable only to normal strength concrete(f'c < 50MPa) and are based on research on normal strength concrete (NSC). Even the Australian Code for Concrete Structures, AS3600-1994, does not contain provisions for HSC, mainly due to paucity of research data, in particular with Australian materials. The research at The University of Melbourne and elsewhere has shown that simply extrapolating the constitutive equations for NSC structures to HSC structures could either be unsafe or uneconomical.

This paper discusses design rules suggested by the HSC research group at the University of Melbourne to update AS3600 to cover design areas for HSC, such as ductility of beams and columns, rectangular stress block for HSC, shear design for HSC and bond and anchorage of reinforcement in HSC members.

1 INTRODUCTION

Ever since the extensive use of high-strength concrete(HSC) began in construction of buildings, bridges and pavement work more than two decades ago, a significant amount of research into all aspects of HSC has been conducted and rational and practical design rules are in the process of being established. High-strength concrete is defined as concrete with strengths between 50MPa and 100MPa.

Most design practitioners have simply extrapolated normal strength design rules to HSC. It has to be recognised that HSC is a new or a different material with different fracture mechanics, micro-mechanics, homogeneity of the interfacial zone and less difference in the strengths between high-strength concrete matrix and aggregates, etc. Thus simply extrapolating the design rules meant for normal strength concrete are not appropriate for HSC without due validation.

Due to the popularity of the usage of HSC in construction, research output in HSC has been rising at an exponential rate in the past two decades. This paper briefly reviews the relevant research done in the areas of ductility of beams and columns, rectangular stress block for HSC, shear design for HSC and Bond and anchorage of reinforcement in HSC members. This research at the University of Melbourne clearly shows that there are differences in structural behaviour between NSC and HSC members in the crucial areas mentioned above.

2 RECTANGULAR STRESS BLOCK (RSB)

The RSB is defined by two parameters α, the parameter to determine the intensity of stress in the stress block and β(or γ, as in the case of AS3600 1994), the ratio of the depth of the stress block to the depth of the neutral axis (refer to Fig. 1). In AS3600, the value of α, the factor for stress intensity has been fixed at 0.85, but as many investigators, such as Attard 1995 and Pendyala 1997 have shown there is no justification in keeping the value of α fixed. In fact the values of both α and γ both reduce with increase in f'c.

Many investigators, for example, Pastor 1984, Attard 1995 and Pendyala 1997, have concluded that 0.003 is a reasonable proposition for the ultimate strain for high-strength concrete. Thus referring to clause 8.2.1.2(c) of AS3600 1994, it is suggested that provisions for the equivalent rectangular stress block for HSC shall be deemed to be satisfied by assuming that a uniform compressive stress of αf'c acts on an area bounded by:

(a) edges of the cross-section

(b) a line parallel to the neutral axis at the strength limit state under the loading concerned, and located at a distance $\gamma k_u d$ from the extreme compressive fibre, where

$$\gamma = 0.65\text{-}0.00125(\text{f'c-}60) \text{ for } 60 < f'c \leq 100 MPa \tag{1}$$
$$\alpha = 0.85 - 0.0025(f'c - 60) \text{ for } 60 < f'c \leq 100 MPa \tag{2}$$

For details of derivation of the equations (1) and (2) refer to Pendyala and Mendis 1996a and 1996b.

It may be noted that, even though the present AS3600 provisions are valid up to 50MPa, a comprehensive analysis by Pendyala 1996a has shown that the current rectangular provisions can be safely extended to 60MPa.

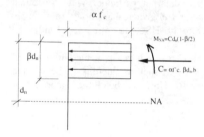

Figure 1. The Rectangular Stress Block

3 DUCTILITY

A number of research projects have been conducted on the ductility of high-strength concrete beams and columns by investigators such as Leslie et. al. 1976 , Smith 1982, Pastor et. al. 1984, Attard and Mendis 1993, Pendyala et. al. 1996c and Shehata and Shehata 1996. All the authors mentioned above have concluded similar trend in the ductility of HSC beams viz.:

1) ductility of members, which are predominantly in flexure have increased ductility when higher strength of concretes are used. This is due to the fact that because of the higher strength of concrete, the neutral axis depth, $k_u d$, is reduced for beams with similar cross-sectional details and this has the effect of improving ductility of under-reinforced beams.

In an experimental and analytical investigation by Pendyala 1997, it has been shown that:

a) For a beam used in the study, the curvature ductility index, μ_c, using a 78MPa concrete was 13.9 as compared to 4.8 using 30MPa concrete. This observation reinforces the enhanced ductility of HSC beams.

b) For the same level of curvature ductility, the decrease in the neutral axis depth, k_u for a HSC beam, of 100MPa as compared to a NSC beam, of 30MPa, was 25%. Looking at this observation from another point of view, for the same level of k_u, there is a 25% increase in ductility for HSC beams.

c) Another observation can be made on the moment curvature capacities of the beams made with HSC and NSC is that, for the same neutral axis depth k_u , the amount of effective tension reinforcement can be increased by 1.9 times and still maintain the same level of ductility (Pendyala 1997). In other words for the same level of ductility and same overall dimensions it is possible to increase the moment capacity of the beam by 1.9 times. This observation is particularly important when, due to architectural or other requirements, when the depth of the beam cannot be increased, HSC has a definite advantage in increasing both the ductility and moment capacity of the beam.

2) Experimental work and analytical studies by Attard and Mendis 1993 and Kovacic 1995 have shown that ductility of HSC columns is lower compared to similar NSC columns. This is contrary to what happens in beams when HSC is used. Defining the appropriate level of ductility in a concrete column may depend on the importance of the structure or the level of earthquake risk in a particular area. The required amount of ductility is achieved by providing closely spaced stirrups. The design of lateral reinforcement in a column is based on the following three criteria-

    i. To prevent longitudinal steel from buckling
    ii. To prevent fracture of lateral steel due to sudden loss of concrete steel
    iii. To provide adequate confinement for ductility.

The recommendation for stirrup spacing for columns are that the spacings of stirrups shall not exceed the smaller of values given in Table 1:

Table 1: Recommendations for Stirrup Spacing for HSC Columns (after Mendis and Kovacic 1997)

f'c < 50 MPa	50<f'c<80MPa (N*/f'cAg <0.3)	80<f'c<100MPa (N*/f'cAg <0.3)
D_c, 15 d_b, 48 d_t	D_c, 12 d_b, 30 d_t	D_c, 8d_b, 24 d_t

* For f'c > 50 MPa and N*/f'cAg > 0.3, the spacings have to be reduced by a factor of 0.3/ (N*/f'cAg).

For f'c > 50 MPa, the recommended spacings are based on 12 mm, f_{yt} = 400 MPa stirrups. If other bar types are used the spacings have to be multiplied by a factor of $\dfrac{f_{yt}}{400} \times \dfrac{d_t}{12}$.

D_c = the smaller column dimension
d_b = the diameter of the smallest longitudinal bar in the column
d_t = the diameter of stirrups
f_{yt} = yield strength of stirrups

4 MINIMUM TENSILE REINFORCEMENT

The minimum reinforcement depends on the cracking moment M_{cr} of the beam. M_{cr} inturn depends on the area of longitudinal steel, width of beam, effective depth of beam, strength of concrete and yield stress of reinforcement steel values. The suggested formula for minimum reinforcement value for longitudinal bars for beams is (Pendyala 1997):

$$\frac{Ast_{min}}{bd} = \frac{0.20\sqrt{f'c}}{f_{sy}} \tag{3}$$

The Eq. 3 is based on the work by Nilson 1985 equating the cracking moment of a beam section to the resisting moment after cracking and limiting the steel stress at cracking to 2/3 of the yield stress.

5 SHEAR STRENGTH OF HSC BEAMS EXCLUDING SHEAR REINFORCEMENT

The shear strength of high-strength concrete does not proportionately increase with f'c. The lack of aggregate interlock mechanism, which resists a substantial portion of the post- crack shear forces, is the factor which reduces the shear strength of high-strength concrete. Unlike normal strength concrete, high-strength concrete fracture surfaces are smooth with fracture passing through aggregates. A comprehensive analytical and research study by Pendyala 1997 has shown that an additional factor β_4 may be used to compute the shear strength of high-strength concrete beams, where

$\beta_4 = 1.25 - 0.005f'c \geq 0.9$, for 50MPa < fc < 100MPa

Therefore, referring to AS3600 1994 cl.8.2.7.1, this clause for high-strength concrete should read(Eq. 4):

$$Vuc = \beta_1\beta_2\beta_3\beta_4 b_v d_o \left[\frac{A_{st}f'_c}{b_v d_o} \right]^{\frac{1}{3}}$$ (4)

The value of β_4 reduces from 1.0 at 50 MPa to 0.9 at 70 MPa.

6 BOND AND ANCHORAGE FOR HSC

Unlike normal strength concrete, where the bearing stresses are uniform over the entire length of the splice, the bond stresses in HSC are non uniform and follow a triangular pattern with peak stresses at the splice and gradually decreasing towards the end (Azizinamini (1993)). The high intensity bearing stresses have the tendency to split the concrete at the tension splice leading to brittle fracture. Inclusion of a minium amount of ligatures (for f'c greater than 70 MPa) over the length of the splice helps to distribute the bearing stresses uniformly and reduce the splitting tendency due to high bearing stresses. As a guide, bars with minimum diameter of 10mm should be used to prevent fracture of the ligatures.

Current design provisions of AS3600 (1994) (Eq. 5) are based on empirical relationships developed from tests on low-strength concrete. The results of a series of tests on high-strength concrete, up to 110 MPa, completed at the University of Minnesota and 101 test results from five other research studies were used by Mendis and French (1997) to review the existing recommendations in AS3600 for design of splices and anchorage of reinforcement. It was shown that AS3600 (1994) formula was conservative for concrete strengths up to 100MPa.

$$L_{sy.t} = \frac{k_1 k_2 f_{sy} A_b}{(2a + d_b)\sqrt{f'_c}} \geq 25k_1 d_b$$ (5)

Therefore Eq. 5 can be retained for high-strength concrete up to 100MPa.

7 SLENDER COLUMNS

For a slender column, any *initial curvature* or deflection, caused by lateral loads or end moments, will be magnified when an axial load is applied. Such a *moment magnification* effect is dealt with in AS3600 by increasing the design ultimate moment for slender columns by a factor which depends on the slenderness ratio.

The value of *flexural-rigidity*, EI, in the equation for N_C will vary along a column because of cracking induced by bending moments, and because of *tension stiffening* between these cracks. A formula is given in AS3600 for the calculation of EI in slender columns.

While the aforementioned formulae can be applied with confidence when dealing with normal strength concrete, more research is required to verify whether they are applicable to high strength concrete columns. Specifically, the topics requiring review are:

214

1) The applicability of the Moment Magnification Method given in AS3600 to high strength concrete.
2) The adequacy of the formula given in AS3600 to calculate the flexural rigidity of columns.
3) The adequacy of the criteria set out by AS3600 to classify short and slender columns.
These aspects are investigated in a project conducted at The University of Melbourne.

8 BIAXIAL BENDING OF HSC COLUMNS

Biaxial bending of high strength concrete columns is a very important engineering aspect of concrete. A preliminary literature survey and other initial work done at the University of Melbourne have shown the inadequacy of the design rules in this area. None of the design codes throughout the world specifies rules for design of biaxially loaded HSC columns. There is a need to formulate design rules for these columns.

In AS 3600, a horizontal plane through the failure surface is approximated by an interaction formula given by:

$$\left(\frac{M_x}{\phi\, M_{ux}}\right)^{\alpha_n} + \left(\frac{M_y}{\phi\, M_{uy}}\right)^{\alpha_n} = 1.0 \tag{6}$$

In Eq (6), the design strengths in uniaxial bending about the principal axes x and y are $\phi\, M_{ux}$ and $\phi\, M_{uy}$, respectively, for a design axial force N^*, while M^*_x and M^*_y are the design moments, and α_n varies with N^* and is given by:

$$\alpha_n = 0.7 + 1.7\left(N^* / 0.6 N_{uo}\right) \text{ with } 1.0 < \alpha_n < 2.0 \tag{7}$$

The format of Eq (6) was originally proposed by Bresler (1960), although he did not specify values for α_n. The value of α_n given by Eq (7) has been adopted from the British Standard BS 8110 (1985). Some preliminary research at the University of Melbourne has shown that simply extrapolating the constitutive equations for normal strength concrete to HSC is unsafe. Therefore, The value of α_n needs to be derived again from the failure surfaces of high-strength concrete columns with different dimensions, reinforcement arrangements, etc. This work is continuing at the University of Melbourne.

9 SUMMARY

Based upon a number of experimental and analytical studies conducted at The University of Melbourne and other research institutions around the world, some design rules are suggested to revise the present code, AS3600 (1994), to cover high-strength concrete. This paper briefly addressed issues relating to ductility of beams and columns, rectangular stress blocks for HSC, shear design for HSC and bond and anchorage of reinforcement in HSC members. More details can be found in the references listed below.

REFERENCES

AS 3600 1994. Australian Standard Concrete Structures Code. *Standard Australia*. NSW.
Attard, M.M. 1995. Rectangular Stress Block Parameters for High-Strength Concrete, *14th Australasian Conference on Mechanics of Structures and Materials:*147-151.Hobart.

Attard, M.M. & Mendis, P.A. 1993. Ductility of High-strength Concrete Columns, *Australian Civil Engineering Transactions, The Institution of Engineers, Australia, Vol CE35, No. 4:* 295-306.

Azizinamini, A, Stark, M., Roller, J. & Ghosh., S.K. 1993. Bond Performance of Reinforcing Bars Embedded in High-Strength Concrete. *ACI Structural Journal, v90 No.5:* 554-561.

Bresler B. 1960. Design Criteria for Reinforced Columns under Axial Load and Biaxial Bending. *Journal Of The American Concrete Institute:* 481-490.

CSA A23.3. 1994. Canadian Concrete Structures Code .

Kovacic, D. 1996. Design of High Strength Concrete Walls, M.Eng. Thesis, *The University of Melbourne.*

Leslie, K.E., Rajagopalan, K.S. & Everard, N.J. 1976. Flexural Behaviour of High-Strength Concrete Beams, *ACI Journal:* 517-521.

Mendis, P. A. & Kovacic, D. 1997. Lateral Reinforcement Spacings for High-Strength Concrete Columns in Ordinary Moment Resisting Frames, Submitted for Review. *Civil Engineering Transactions, The Institution of Engineers, Australia.*

Mendis, P.A & French, C. 1997. Development Lengths of Reinforcement In High-Strength Concrete, Accepted for Publication, *Civil Engineering Transactions, The Institution of Engineers, Australia.*

Nilson A.H. 1986. Design Implications of Current Research on High Strength Concrete. *SP87, ACI Special Publication, American Concrete Institute:* 85-118.

Pastor, J.A. 1984. Behaviour of High-Strength Concrete Beams. *Research Report No. 84-3, School of Civil and Environmental Engineering, Cornell University.*

Pendyala, R.S. 1997. Behaviour of High Strength Concrete Flexural Members. *Ph.D Thesis. Dept. of Civil And Env. Eng., The University of Melbourne.*

Pendyala R.S & Mendis P. A. 1996a. Review of the Rectangular Stress Block for High-Strength Concrete. *Research Report - Dept. of Civil and Environmental Engineering, The University of Melbourne, Melbourne, Australia.*

Pendyala R.S. & Mendis P.A. 1996b. A Rectangular Stress Block For High-Strength Concrete, Submitted for Review. *Civil Engineering Transactions, The Institution of Engineers, Australia.*

Pendyala, R.S, Mendis, P.A. & Patnaikuni, I. 1996c. Full-range Behaviour of HSC Flexural Members. *ACI Structures Journal, Vol. 93, No. 1:* 30-35.

Shehata, I & Shehata, L. 1996. Ductility of HSC Beams in Flexure. *Proc. of Conf. BHP-96:* 945-954. Paris.

Smith, R.G. 1982. Aspects of Structural Behaviour In Very High-Strength Concrete, Proc. of Seminar Harris: High-Strength in Concrete. *Concrete Institute of Australia:* 5.1-5.22. (NSW Branch).

The Mechanics of Structures and Materials, Grzebieta, Al-Mahaidi & Wilson (eds)
© 1997 Balkema, Rotterdam, ISBN 90 5410 900 9

Design rules for bond strength of epoxy-coated bars

P. Mendis
Department of Civil and Environmental Engineering, The University of Melbourne, Vic., Australia

C.W. French
Department of Civil Engineering, University of Minnesota, Minn., USA

ABSTRACT: Epoxy-coated reinforcement has been commonly used to inhibit reinforcement corrosion in severe environments throughout the world. Because of factors such as reduced adhesion associated with the epoxy coating, longer anchorage lengths are required to fully develop epoxy-coated reinforcement compared with uncoated reinforcement. The Concrete Structures Standard of Australia, AS3600 does not specify rules for epoxy-coated reinforcement. This paper provides a basic background in understanding bond between epoxy coated bars and concrete. Modifications to the existing recommendations in AS3600 for design of splices and anchorage of uncoated reinforcement are proposed for epoxy-coated reinforcement. Finally the results of a study conducted at the University of Minnesota to investigate the bond behaviour of epoxy-coated bars in micro-silica concrete are briefly presented.

1 INTRODUCTION

Epoxy-coated reinforcement has been commonly used to inhibit reinforcement corrosion in severe environments. Although the use of epoxy-coated bars has been relatively limited in Australia, engineers and builders are beginning to realise the value of epoxy-coated reinforcement in highway and marine construction. One of the reasons why epoxy-coated reinforcement has not been popular in Australia is due to the lack of provisions in the Concrete Structures Standard, AS3600 (SAA 1994), to address the problem of reduction of bond between steel and concrete due to the epoxy coating. The results of a series of tests completed recently at the University of Minnesota and seven other research studies were used by Mendis and French (1996) to modify the existing recommendations in AS3600 for design of splices and anchorage of uncoated reinforcement to cover the case of epoxy-coated reinforcement. These recommendations are given in Section 3.

2 BOND MECHANISM AND FAILURE MODES

The importance of the bond between the reinforcing bars and concrete on cracking, deflection, ductility and anchorage of steel in reinforced concrete members has been realized for a long time. Bond stresses modify the steel stresses along the length of the bar by transferring load between the bar and the surrounding concrete. The following expression may be derived from equilibrium of the concrete and bar forces:

$$A_b f_s = u \pi d_b l_d \tag{1}$$

where A_b and d_b are the area and diameter of the reinforcing bar, l_d is the bond length of bar, f_s is the stress developed in the bar, and u is the bond stress. The bond stress can be related to the bar diameter, bar stress, and bond length:

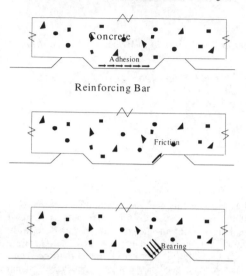

Fig. 1 Bond Mechanisms for Deformed Bars in Concrete

$$u = \frac{f_s d_b}{4 l_d} \qquad\qquad (2)$$

This formula is used to determine the average bond stress developed between the reinforcing bar and concrete.

Fig. 1 illustrates schematically the three mechanisms of bond, adhesion, friction and mechanical bearing. Upon initial loading, the forces are transferred by adhesion created through chemical bonding between the steel and concrete. The adhesion is not a sustained resistance and at low bar stresses the adhesion is lost. After adhesion is lost the bar slips relative to the concrete which enables development of the friction and mechanical bearing mechanisms. The frictional mechanism results from the contact of one material rubbing against the other due to bar slippage. The properties of the interface between the bar and the concrete are important to the friction. For example, the ability of concrete particles to attach to the bar surface will increase the bond due to the greater concrete-on-concrete friction. The bond stress due to mechanical bearing is the horizontal shearing component of the bearing stress, created when the lug pushes against the concrete. The mechanical bearing and friction between the bar and concrete produce a vertical splitting force radiating from the lug.

The types of bond failures that occur can be either a pull-out or splitting type failure. Pull-out type failures are characteristic in specimens with large cover, transverse reinforcement, or both, which provide adequate confinement of the reinforcing bar. Failure occurs when the concrete between the mechanical lugs is sheared off and the bar is pulled from the specimen. The specimen fails by conical cracking of the concrete at the loaded end of the bar. Splitting failures are more common and occur when little confinement from cover or transverse reinforcement is provided around the bar (e.g. slabs or bridge decks). The radial tensile force resulting from the bond and bearing forces splits the concrete cover and the bar slips due to wedging action or crushing of the concrete at the mechanical lugs.

Current ACI (1995) and AASHTO codes (1989) recognize the decreased ultimate bond strength of epoxy coated reinforcement by specifying amplified development lengths for epoxy coated reinforcement. The amplified development lengths are 20 to 50 percent greater than those of uncoated reinforcement (Johnston and Zia 1982, Treece and Jirsa 1989). It is generally

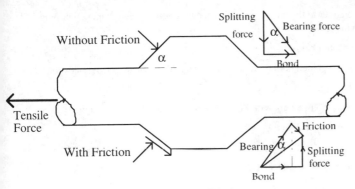

Fig. 2 Forces on Bar Lug, with and without Friction

assumed that the decreased bond capacity of epoxy-coated reinforcement is due to the reduced friction and the lack of chemical adhesion between the bars and the concrete. The reduced friction of epoxy-coated bars increases the radial pressure component which sets up radial tension in the concrete cover (Fig. 2). Therefore the bond strength is controlled by the magnitude of the radial pressure that the concrete cover can resist. Another reason for the decrease in bond strength, found in a study conducted by Grundhoffer (1992) at the University of Minnesota study, was the reduction in related rib area due to the epoxy coating, which then decreases the bond strength and stiffness.

3 AS3600 PROVISIONS AND SUGGESTED MODIFICATIONS

Mendis and French (1996) suggested a modification to the existing recommendations in AS3600 for design of splices and anchorage of uncoated reinforcement to cover epoxy-coated reinforcement by considering a total of 182 test results from the work at the University of Minnesota and seven other research studies conducted in the USA by Treece and Jirsa (1989), Choi et al. (1990), Devries et al. (1991), Hadje-Ghaffari et al. (1991), Cleary and Ramirez (1991), Hester et al. (1991) and Hamad and Jirsa (1993). In all cases, tests were conducted on lapped joints or beam-end specimens of coated and uncoated bars to assess the reduction in bond due to coating. A considerable variety of bar types and diameters, concrete strengths, concrete covers and bars cast top and bottom were represented in the various studies. These details are presented elsewhere (Mendis and French 1996).

It was suggested that development lengths should be multiplied by 1.35 for splitting-type failures with epoxy-coated bars. For pull-out type failures, a reduced factor of 1.15 based on present ACI recommendations was suggested.

The revised AS3600 formula for bond and anchorage is given below.

$$L_s = \frac{k_1 k_2 k_3 f_{sy} A_b}{(2a + d_b)\sqrt{f'_c}} \geq 25 k_1 k_3 d_b \qquad (3)$$

where k_1 = Top bar factor equal to 1.25 for a horizontal bar with more than 300 mm of concrete cast below.

k_2 = 1.7 for bars in slabs and walls with clear distance between bars > 150 mm

= 2.2 for bars in beams and columns with stirrups

= 2.4 for other bars

2a = Twice the cover to the bar or clear distance between adjacent bars developing stress, whichever is less.

k_3 = 1.0 for uncoated bars

= 1.35 for epoxy-coated bars with cover less than $3d_b$ or clear spacing between bars less than 6 d_b

= 1.15 for epoxy-coated bars with cover larger than $3d_b$ and clear spacing between bars larger than 6 d_b

4 THE EFFECT OF MICROSILICA ON BOND

The use of epoxy-coated reinforcing steel and micro silica concrete is becoming popular in the design of reinforced concrete structures. Epoxy coating is used to protect the reinforcing steel from corrosive agents, thus prolonging the life of the structure. Micro-silica is added to the concrete to increase its impermeability, stiffness, and strength.

Some researchers have suggested that the presence of silica fume may actually alter the concrete transition zone at the bar surface (Gjorv et al. 1986, Hwang et al. 1994). However, how the surface is altered and the direct effects of silica fume on bond appear inconclusive, as it has been found to both increase and decrease the bond capacity of concrete. Hwang et al. (1994) suggested that the presence of silica fume decreases the adhesion at the surface of the bar and concrete, invariably decreasing the developed bond stress. Gjorv et al. (1986) suggested that the presence of silica fume increases the adhesion between the concrete and bar surface, invariably increasing the bond stress due to increased adhesion and friction.

In the study by Grundhoffer (1992) at the University of Minnesota the effect of micro-silica on the bond strengths of epoxy-coated bars was investigated. Grundhoffer tested inverted half-beam specimens (Fig. 3) with uncoated and epoxy coated bars. The concrete compressive strengths ranged from 42 to 96 MPa. This study focussed on splitting bond failures.

All the bars were cast in the bottom position. 19 mm, 25 mm and 36 mm bars were used with yield strengths of 428, 428 and 477 MPa respectively. The specimens were all 305 mm wide and 457 mm deep, and the length of the specimens varied from 900 mm to 1200 mm depending on the size of the reinforcing bar tested to accommodate the bar embedment length. The nominal cover for the specimens was 38.1 mm for the 19 mm bar specimens and 63.5 mm for the 25 mm and 36 mm bar specimens, but varied slightly for each specimen. The covers were chosen to insure a bond failure with a realistic embedment length. U-shaped transverse reinforcement was used only as a cage for the auxiliary reinforcement and did not contact the embedment length of the bar tested in bond.

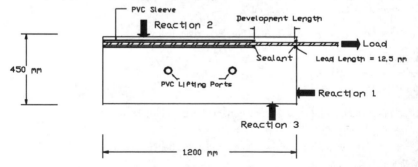

Fig. 3 Inverted Half-beam Specimen

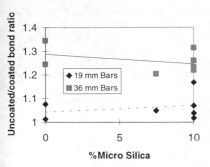

Fig. 4 Bond Ratio versus percentage Micro silica

Fig. 4 shows the best fit lines through the data for 19 mm bars and data for 36 mm bars respectively, for the ultimate bond strength ratios (Uncoated/Coated) versus percentage micro silica relationship. Overall, the limited data suggests that the micro silica does not effect the bond ratio significantly. Therefore the proposed design rules given in Section 3 can be used for micro-silica concrete without any modifications. However it can be seen that there is a considerable amount of scatter in the plot. Therefore more work is required to confirm these findings.

5 CONCLUSIONS

1. The reduced friction of epoxy-coated bars increases the radial pressure component which sets up radial tension in the concrete cover. The reduction in related rib area due to the epoxy coating also decreases the bond strength and stiffness.

2. Based on a comprehensive analysis of test results from the University of Minnesota and seven other research studies, it was recommended to multiply the existing formula given in AS3600 for the calculation of splice lengths and development lengths by 1.35 for epoxy-coated bars. This factor can be reduced to 1.15 for epoxy-coated bars with cover larger than $3d_b$ and clear spacing between bars larger than 6 d_b.

3. The limited data suggests that the micro silica does not affect the bond behaviour of epoxy-coated reinforcement. More work is required to confirm this finding.

REFERENCES

ACI Committee 318: 1995. *Building Code Requirements for Reinforced Concrete*, Detroit. American Concrete Institute.
AS3600: *Concrete Structures Standard*, 1994. Standards Association of Australia.
Choi, O.C., Hadje-Ghaffari, H., Darwin, D. & McCabe, S.L. 1990. Bond: Epoxy-Coated Reinforcement to Concrete: Bar Parameters. SL Report 90-1. University of Kansas Center for Research.
Cleary, D.B. & Ramirez, J.A., 1991. Bond Strength of Epoxy Coated Reinforcement", ACI Materials Journal, Vol. 88, No.2, pp. 146-149.
DeVries, R.A., Moehle, J.P. & Hester, W., 1991. Lap Splice Strength of Plain and Epoxy-Coated Reinforcements - An Experimental Study Considering Concrete Strength, Casting Position, and Anti-Bleeding Additives, Berkeley. University of California. Report UCB/SEMM-91/02.

Gjorv, O.E., Monteiro, P.J.M.& Mehta, P.K., 1986. Effect of Condensed Silica Fume on the Steel-Concrete Bond, Trondheim. Norwegian Institute of Technology. BML Report 86.201.

Grundhoffer, T, 1992. Bond Behavior of Uncoated and Epoxy-Coated Reinforcement in Concrete, M.S. Thesis, The University of Minnesota.

Hadje-Ghaffari, H., Darwin, D. & McCabe, S.L., 1991. Effects of Epoxy-Coating on the Bond of Reinforcing Steel to Concrete, Lawrence. The University of Kansas Center for Research. SM Report No. 28.

Hamad, B.S. & Jirsa, J.O., 1993. Strength of Epoxy-coated Reinforcing Bar Splices Confined with Transverse Reinforcement, ACI Structures Journal, Vol. 90, No. 1, pp. 77-88.

Hester, C.J., Salamizavaregh, S, Darwin, D. & McCabe, S.L., 1991. Bond of Epoxy-coated Reinforcement to Concrete: Splices, SL Report 91-1. University of Kansas Center for Research.

Hwang, S., Lee, Y. & Lee, C., 1994. Effect of Silica Fume on the Splice Strength of Deformed Bars in High performance Concrete, ACI Structural Journal, Vol. 91, No. 3, pp. 294-302.

Johnston, D.W. & Zia, P., 1982. Bond Characteristics of Epoxy Coated Reinforcing Bars, Washington, D.C. Federal Highway Administration, Report No. FHWA-NC-82-002.

Mendis, P.A. & French, C.W., 1996. Bond Strength of Epoxy-Coated Bars, Research Report, Dept. of Civil & Environmental Engineering, University of Melbourne.

Standard Specifications for Highway Bridges, 1989. 14th Edition. Washington D.C. AASHTO.

Treece, R. A. & Jirsa, J. O., 1989. Bond Strength of Epoxy-Coated Reinforcing Bars, ACI Mat. Journal. Vol. 86. No. 2.

The Mechanics of Structures and Materials, Grzebieta, Al-Mahaidi & Wilson (eds)
© *1997 Balkema, Rotterdam, ISBN 90 5410 900 9*

Amplified moments in reinforced concrete core-walls

Mario M. Attard & Stephen J. Foster
School of Civil Engineering, University of New South Wales, Sydney, N.S.W., Australia

Nguyen Dai Minh
Institute for Building Science and Technology, Hanoi, Vietnam

ABSTRACT: A finite element formulation for the out-of-plane buckling analysis of reinforced concrete walls is presented. The formulation is based on an orthotropic constitutive model for concrete. An in-plane analysis is firstly carried out to determine the stresses and tangent moduli. Out-of-plane, the concrete is modelled using non-linear orthotropic 16 DOF plate bending elements; the reinforcing steel using elasto-plastic fibre-type elements. In the plane of the structure stresses are calculated using either 4 or 8 node membrane elements with bar elements used for the reinforcing steel. Examples are presented for different boundary conditions and for walls with out-of-plane imperfections. The magnitude and distribution of amplified moments due to second order effects are discussed.

1 INTRODUCTION

Many multi-storey structures have reinforced concrete cores which are designed to carry either the whole or part of the lateral loading (usually wind shear), as well as the dead load and live loads. The cores normally form the main structural element through which the lifts move. Cores can consist of several walls making a number of closed cells. Some cells can contain a central diaphragm wall, which may only be restrained along its sides by connecting walls. The critical slenderness of these walls, which can run the length of the building, is the horizontal span to thickness ratio. As with columns in the lower storey of multi-storey buildings, economic benefits can be gained by reducing the core wall thickness through the use of high strength concrete. Reducing the wall thickness however, results in walls of greater slenderness requiring an assessment of the amplification effects of buckling on the out-of-plane design moments. Present design methods for walls are only applicable to walls in one-way bending. For core walls, where the axial load eccentricity is small and where any out-of-plane buckling effects involve two-way bending, the present design methods are very conservative.

Reinforced concrete walls differ from columns because of the two-way nature bending within a wall supported on all four sides. Core walls are predominantly loaded by in-plane axial and shear stresses. Out-of-plane bending stresses occur due to the presence of initial imperfections and out-of-plane loading eccentricities This paper briefly summarises a new finite element formulation developed for investigating the effects of small initial imperfections and small loading eccentricities, involving the solution of second-order effects.

2 WALL EQUILIBRIUM EQUATIONS

The displacement functions describing the movement of a thin flat wall with an initial out-of-plane imperfection of the middle surface described by w_o, can be derived using the Kirchhoff hypothesis, and are:

$$u_x = u - z(w_{o,x} + w_{m,x}) \qquad u_y = v - z(w_{o,y} + w_{m,y}) \qquad u_z = w_o + w_m \tag{1}$$

where z is the out-of-plane coordinate measured perpendicular to the middle surface; u_x, u_y, and u_z are the displacements of a point within the wall in the x y z directions, respectively; w_m is the measured displacement caused by the loading; and u, v and w are the total displacements of the middle surface in the x y z directions, respectively. Substituting Eqn 1 into the definition for finite strain, we have:

$$\varepsilon_{xx} = u_{,x} - z\, w_{m,xx} + \frac{1}{2}\,(w_{m,x}^2 + 2\, w_{o,x}\, w_{m,x}) = \varepsilon_{xx}^o - z\, w_{m,xx}$$

$$\varepsilon_{yy} = v_{,y} - z\, w_{m,yy} + \frac{1}{2}(w_{m,y}^2 + 2\, w_{o,y}\, w_{m,y}) = \varepsilon_{yy}^o - z\, w_{m,yy} \qquad (2)$$

$$\varepsilon_{xy} = \frac{1}{2}(u_{,y} + v_{,x}) - z\, w_{m,xy} + (w_{m,x}w_{m,y} + w_{o,x}\, w_{m,y} + w_{o,y}\, w_{m,x}) = \varepsilon_{xy}^o - z\, w_{m,xy}$$

where ε_{xx}^o, ε_{yy}^o and ε_{xy}^o are the strains at the middle surface (z=0). For equilibrium, virtual work must be satisfied and hence:

$$\delta V \equiv \int_A \int_{-\frac{t}{2}}^{\frac{t}{2}} S_{xx}\delta\varepsilon_{xx} + S_{yy}\delta\varepsilon_{yy} + 2S_{xy}\delta\varepsilon_{xy}\ dz\, dA \ - \int_{S_o} T_i\ \delta u_i dS_o = 0 \qquad (3)$$

where S_{xx}, S_{yy} and S_{xy} are Kirchhoff stresses; $\delta\varepsilon_{xx}$, $\delta\varepsilon_{yy}$ and $\delta\varepsilon_{xy}$ are the first variation of the plate finite strains; T_i are the force tractions; δu_i are the first variation of the displacements in the ith direction (i=1,2,3 correspond to the x,y,z axes respectively); t is the plate thickness, A the area measured in the plane of the plate and S_o is the surface over which the tractions act.

Plate internal forces can be written in terms of the resultant of the Kirchhoff stresses. Hence the normal forces N_{xx} and N_{yy} in the x and y directions, respectively, and the shear force resultants N_{xy} (=N_{yx}) are:

$$N_{xx} = \int_{-\frac{t}{2}}^{+\frac{t}{2}} S_{xx}\, dz \qquad N_{yy} = \int_{-\frac{t}{2}}^{+\frac{t}{2}} S_{yy}\, dz \qquad N_{xy} = \int_{-\frac{t}{2}}^{+\frac{t}{2}} S_{xy}\, dz \qquad (4)$$

The bending moments M_{xx} and M_{yy} about the y and x axes, respectively, and the twisting moments M_{xy} (= M_{yx}) are:

$$M_{xx} = \int_{-\frac{t}{2}}^{+\frac{t}{2}} S_{xx}\, z\, dz \qquad M_{yy} = \int_{-\frac{t}{2}}^{+\frac{t}{2}} S_{yy}\, z\, dz \qquad M_{xy} = \int_{-\frac{t}{2}}^{+\frac{t}{2}} S_{xy}\, z\, dz \qquad (5)$$

Using Eqn 3, the expressions for the plate finite strains Eqn 2, as well as the relationship between the Kirchhoff stresses and the internal force resultants Eqns 4 & 5, the following two equilibrium equations are derived (assuming only in-plane loads):

$$\int_A \delta u_{,x} N_{xx} + \delta v_{,y} N_{yy} + \frac{1}{2}\big(\delta u_{,y} + \delta v_{,x}\big) N_{xy}\, dA - \iint T_1\delta u\, dzdy - \iint T_2\delta v dzdx = 0 \qquad (6)$$

$$\int_A (\delta w_{m,xx} M_{xx} + \delta w_{m,yy} M_{yy} + 2\,\delta w_{m,xy} M_{xy} + \begin{bmatrix} \delta w_{m,x} & \delta w_{m,y} \end{bmatrix} \begin{bmatrix} N_{xx} & N_{xy} \\ N_{yx} & N_{yy} \end{bmatrix} \begin{bmatrix} w_{m,x} \\ w_{m,y} \end{bmatrix}$$

$$+ \begin{bmatrix} \delta w_{m,x} & \delta w_{m,y} \end{bmatrix} \begin{bmatrix} N_{xx} & N_{xy} \\ N_{yx} & N_{yy} \end{bmatrix} \begin{bmatrix} w_{o,x} \\ w_{o,y} \end{bmatrix})\, dA - \iint T_1 z\delta w_{m,x} dzdy - \iint T_2 z\delta w_{m,y} dzdx = 0 \qquad (7)$$

Equation (6) represents the in-plane equilibrium equation while Eqn 7 is the out-of-plane equilibrium equation. These equations do not depend on the stress-strain behaviour of the wall material. For linear elastic walls, Eqns (6) and (7) are not coupled, so the in-plane problem and the out-of-plane problem can be solved independently. However, for inelastic concrete walls, Eqns (6) and (7) involve the coupling of in-plane and out-of-plane behaviour. The solution involves both a highly non-linear material and non-linear geometric effects.

224

Since we are only interested here with walls predominantly loaded axially (the out-of-plane bending contribution to the stress is small and arises from second order effects), the following assumption are made: (i) Equations (6) and (7) are uncoupled; (ii) The stress distribution through the wall thickness is approximated by a parabola (the loading eccentricity is small).

3 ORTHOTROPIC CONSTITUTIVE MODEL FOR CONCRETE

One method of analysing reinforced concrete walls with out-of-plane bending is to use a layered shell element, which can fully describe the non-linear stress-strain relationship through the wall thickness. In this study, an approximate solution is presented which eliminates the need for layers. For undamaged concrete, the stress-strain relationship is formulated by using the principle of equivalent uniaxial strains as proposed by Darwin and Perknold (1977). The "equivalent uniaxial strain" model was developed in order to predict the biaxial or in-plane behaviour of concrete by subtracting the dilation effect, thus allowing the use of uniaxial stress-strain relationships. The principal strains at a given depth within the wall ε_1 and ε_2, can be broken into their equivalent uniaxial components as follows:

$$\varepsilon_1 = \varepsilon_{1u} - \nu_1 \varepsilon_{2u} \qquad \varepsilon_2 = \varepsilon_{2u} - \nu_2 \varepsilon_{1u} \tag{8}$$

where ε_{1u} and ε_{2u} are the equivalent uniaxial strains and ν_1 and ν_2 are the dilation constants.
The relationship between the equivalent uniaxial strains and the strains in a general coordinate system is:

$$\begin{bmatrix} \varepsilon_{1u} \\ \varepsilon_{2u} \\ \gamma \end{bmatrix} = \frac{1}{1-\nu_1\nu_2} \begin{bmatrix} 1 & \nu_1 & 0 \\ \nu_2 & 1 & 0 \\ 0 & 0 & 1-\nu_1\nu_2 \end{bmatrix} \begin{bmatrix} \varepsilon_1 \\ \varepsilon_2 \\ \gamma \end{bmatrix} = \mathbf{D} \begin{bmatrix} \varepsilon_1 \\ \varepsilon_2 \\ \gamma \end{bmatrix} = \mathbf{DT} \begin{bmatrix} \varepsilon_{11} \\ \varepsilon_{22} \\ \varepsilon_{12} \end{bmatrix} \tag{9}$$

In Eqn (9), $\mathbf{T}$ is the transformation matrix between the principal coordinate system and the x-y coordinate system. It is assumed that the transformation matrix $\mathbf{T}$ remains constant through the thickness of the plate (the angle θ between the principal coordinate system and the x-y coordinate is taken as approximately constant through the plate thickness).
The corresponding principal stresses σ_1, σ_2 and the maximum shear stress τ are given by

$$\sigma_1 = E_1 \varepsilon_{1u} \qquad \sigma_2 = E_2 \varepsilon_{2u} \qquad \tau = G\gamma \tag{10}$$

where γ is the maximum shear strain and E_1, E_2 are the secant moduli and G is the shear modulus. The relationship between stresses in the x-y coordinates and the principal coordinates can be written as

$$\begin{bmatrix} S_{xx} \\ S_{yy} \\ S_{xy} \end{bmatrix} = \mathbf{T}^T \begin{bmatrix} \sigma_1 \\ \sigma_2 \\ \tau \end{bmatrix} \tag{11}$$

The principal stress σ_i is expanded as a Taylorian series, about the equivalent uniaxial stress at the middle surface of the wall, hence:

$$\sigma_i = \sigma_{i0} + (\varepsilon_{iu} - \varepsilon_{iu0})E_i^t + \frac{1}{2}(\varepsilon_{iu} - \varepsilon_{iu0})^2 \frac{d^2\sigma_i}{d\varepsilon_{iu}^2}_{(\varepsilon_{iu0})} + \frac{1}{6}(\varepsilon_{iu} - \varepsilon_{iu0})^3 \frac{d^3\sigma_i}{d\varepsilon_{iu}^3}_{(\varepsilon_{iu0})} + \dots \tag{12}$$

225

where σ_{i0} is the principal stress in i-th direction at the middle surface, ε_{iu0} is the corresponding equivalent uniaxial strain at the middle surface, and E_i^t is the corresponding tangent modulus of concrete. If the principal stress distribution is taken as parabolic, then $\dfrac{d^2\sigma_i}{d\varepsilon_{iu}^2}$ is a constant and higher derivatives of the principal stress are zero. Hence Eqn 11 becomes:

$$\sigma_i = \sigma_{i0} + (\varepsilon_{iu} - \varepsilon_{iu0})E_i^t + \frac{1}{2}(\varepsilon_{iu} - \varepsilon_{iu0})^2 a_i \tag{13}$$

with a_i being a constant. A similar expression can be derived for the shear stress and is

$$\tau = \tau_0 + (\gamma - \gamma_0)G^t + \frac{1}{2}(\gamma - \gamma_0)^2 a_3 \tag{14}$$

With the plate forces defined in Eqn (5) and Eqns (11), (13) & (14), the bending moments within the wall are therefore:

$$\begin{bmatrix} M_{xx} \\ M_{yy} \\ M_{xy} \end{bmatrix} = \int_{-\frac{t}{2}}^{\frac{t}{2}} z\,\mathbf{T}^T\left(\begin{bmatrix} \sigma_{10} \\ \sigma_{20} \\ \tau_0 \end{bmatrix} - \begin{bmatrix} (\varepsilon_{1u} - \varepsilon_{1u0})E_1^t \\ (\varepsilon_{2u} - \varepsilon_{2u0})E_2^t \\ (\gamma - \gamma_0)G^t \end{bmatrix} + \frac{1}{2}\begin{bmatrix} (\varepsilon_{1u} - \varepsilon_{1u0})^2 a_1 \\ (\varepsilon_{2u} - \varepsilon_{2u0})^2 a_2 \\ (\gamma - \gamma_0)^2 a_3 \end{bmatrix}\right) dz \tag{15}$$

From Eqns (2) and (9) we obtain:

$$\begin{bmatrix} \varepsilon_{1u} - \varepsilon_{1u0} \\ \varepsilon_{2u} - \varepsilon_{2u0} \\ \gamma - \gamma_0 \end{bmatrix} = \mathbf{DT}\begin{bmatrix} -z\,w_{m,xx} \\ -z\,w_{m,yy} \\ -z\,w_{m,xy} \end{bmatrix} \tag{16}$$

Substituting Eqn (16) into Eqn (15) and integrating through the plate thickness we have:

$$\begin{bmatrix} M_{xx} \\ M_{yy} \\ M_{xy} \end{bmatrix} = -\frac{t^3}{12}\,\mathbf{T}^T\,\frac{1}{1-\nu_1\nu_2}\begin{bmatrix} E_1^t & \nu_1 E_1^t & 0 \\ \nu_2 E_2^t & E_2^t & 0 \\ 0 & 0 & 0.5\,(1-\nu_1\nu_2)\,G^t \end{bmatrix}\mathbf{T}\begin{bmatrix} w_{m,xx} \\ w_{m,yy} \\ 2w_{m,xy} \end{bmatrix} \tag{17}$$

or

$$\begin{bmatrix} M_{xx} \\ M_{yy} \\ M_{xy} \end{bmatrix} = -\frac{t^3}{12}\,\mathbf{T}^T\,\mathbf{D_{mt}}\,\mathbf{T}\begin{bmatrix} w_{m,xx} \\ w_{m,yy} \\ 2w_{m,xy} \end{bmatrix} \tag{18}$$

Note because of the symmetry requirement for hyperelastic materials $\nu_1 E_1^t = \nu_2 E_2^t$.

4. FINITE ELEMENT EQUILIBRIUM EQUATIONS

By substituting Eqn (18) into Eqn (7) we can write the equation for the virtual work of the out-of-plane forces. A 16-degrees-of-freedom plate bending element is used for the concrete with the reinforcing steel modelled using elasto-plastic fibre-type elements (see Nguyen 1996 and Attard et al. 1996). It is assumed that

$$w_m(x,y) = \mathbf{a}^T \cdot \mathbf{q}_{em} \qquad w_o(x,y) = \mathbf{a}^T \cdot \mathbf{q}_{eo} \tag{19}$$

where $\mathbf{a}$ is the vector of coefficients of the displacement function, $\mathbf{q}_{em}$ is the vector of degrees-of-freedom of the element, and $\mathbf{q}_{eo}$ is the vector of the initial out-of-plane displacements of the plate element. The resulting out-of-plane virtual work equation for an element is therefore

$$\delta\,\mathbf{q}_{em}^T \cdot \{[\,\mathbf{K}_e + \mathbf{S}_e\,]\,\mathbf{q}_{em} + \mathbf{S}_e\,\mathbf{q}_{eo} + \mathbf{R}_{om}\,\} = 0 \tag{20}$$

where $\mathbf{K}_e$ and $\mathbf{S}_e$ are given in Nguyen (1996) and $\mathbf{R}_{om}$ is the element load vector. Assembling for the whole structure, the displacements are calculated from

$$[\,\mathbf{K}_t + \mathbf{S}\,]\,\mathbf{q}_m + \mathbf{S}\,\mathbf{q}_o + \mathbf{R}_o = 0 \tag{21}$$

226

where K_t is the structure tangential stiffness matrix (determined using the tangent moduli given by the in-plane analysis); S is the stability matrix which is a function of the in-plane plate forces (also given by the in-plane analysis); q_m is the vector of the structure nodal unknown displacements; q_0 is the vector of the structure nodal initial displacements (imperfections) and R_0 is the structure out-of-load load vector.

The in-plane analysis is obtained by solving Eqn (6) with the in-plane stresses calculated using either 4 or 8 node membrane elements and 2 or 3 node bar elements used for the reinforcing steel. The program RECAP developed by Foster and Gilbert (1990) was used to perform the in-plane analysis with slight modification. The steps followed to arrive at a solution are:

- The loading is divided into the load steps;
- At each load step the in-plane analysis is carried out until the *required load level* is reached. The tangent moduli and the in-plane forces are then calculated;
- The element tangential stiffness matrix and the stability matrix are then obtained numerically through a nine point Gaussian integration; The structure tangential stiffness matrix and stability matrix, as well as, the structure out-of-plane load vector are then assembled;
- Equation (21) is solved for the displacements;
- The wall moments are calculated using Eqn (18). These are the amplified or second-order moments.

5 EXAMPLE - SIMPLY SUPPORTED WALL

The following example is similar to the one in Attard (1995). Consider a simply supported wall of width b=8.0 m, thickness t=300 mm and variable length a. The aspect ratio of the wall is taken as a/b=0.5, 1, 1.5, 2, 3 and 5. The origin is taken at the top left corner while the x axis is across the width and the y axis follows the length. The concrete strength is 100 MPa. A uniform compressive load of 9000 N/mm is applied with an out-of-plane eccentricity of 0.05*t=15 mm at each end of the wall. The axial loading is in the y direction. The wall is reinforced with a minimum steel ratio $\rho_{min} = 0.4\%$.

An analysis was carried out using the present finite element model. Out-of-plane displacements and bending moments were obtained both with and without second-order effects. A summary of the results of the analysis with different aspect ratios is given in Table 1. Results for an aspect ratio of 2 are shown in Fig. 1. The amplification factors are taken as the ratio of quantities (such as the bending moments and out-of-plane displacement) with second order effects to the quantities without second order effects.

The amplification of the out-of-plane displacement and maximum moment M_{xx} is fairly constant ranging between 1.15 to 1.31. Although the out-of-plane displacement is amplified it has little effect on the maximum M_{yy} moment whose distribution quickly reduces from a

Table 1 Summary of the Results of the Analysis of Walls with Different Aspect Ratios

Aspect ratio a/b	Max w (mm)	δ_w	max M_{xx} (kNm/m)	δ_{Mxx}	max M_{yy} (kNm/m)	δ_{Myy}
0.5	2.25	1.25	47	1.15	135	1
1.0	3.66	1.31	63	1.26	135	1
1.5	2.74	1.27	48 at (y=0.47b)	1.23	135	1
2.0	2.34	1.26	43 at (y=0.375b)	1.19	135	1
3.0	2.21	1.27	42 at (y=0.375b)	1.20	135	1
5.0	2.21	1.28	42 at (y=0.375b)	1.22	135	1

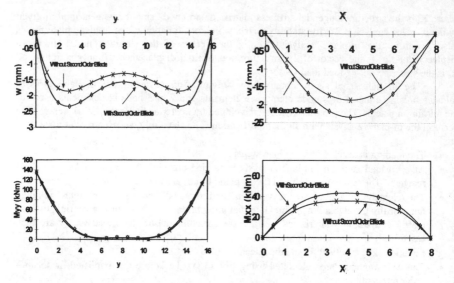

Figure 1 a) Displacement w and Moment M_{yy} at x = b/2. (Aspect Ratio a/b = 2.0)

Figure 1 b) Displacement w and Moment M_{xx} at y = 0.375 b. (Aspect Ratio a/b = 2.0)

maximum at the ends. Since the end moment M_{yy} is unaffected by the out-of-plane displacement, the amplification factor is 1 for M_{yy}. This is very different to the behaviour of columns where the applied ends moments are amplified if the column is in single curvature.

6 REFERENCES

Attard, M. M. (1995) "Buckling of Reinforced Concrete Walls". *Civil Engineering Transaction*, Australia, Vol. CE37 No. 4, PP. 271-278.
Attard, M. M., Nguyen Dai Minh and Foster, S. J. (1996) "Finite Element Analysis of the Out-of-Plane Buckling of Reinforced Concrete Walls". *International Journal of Computers and Structures*, Vol. 61. No. 6, pp. 1037-1042
Darwin, D. and Pecknold, D. A. (1977) "Nonlinear Biaxial Stress-Strain Law for Concrete." *Journal of Eng. Mech.* Division, ASCE, Vol. 103, No. EM2, April, pp. 229-241.
Foster, S. J. and Gilbert, R. I. (1990) "Non-Linear Finite Element Model for Reinforced Concrete Deep Beams and Panels." The University of New South Wales, December, *UNICIV Report*, No. R-275.
Nguyen Dai Minh (1996), "Buckling of Reinforced Concrete Walls by the Finite Element Method", ME thesis, School of Civil Engineering, The University of New South Wales, Feb.

The Mechanics of Structures and Materials, Grzebieta, Al-Mahaidi & Wilson (eds)
© 1997 Balkema, Rotterdam, ISBN 90 5410 900 9

Tendency to cracking in hardening concrete under different degrees of restraint

N.Vitharana
Major Works Division, Department of Water Resources, Parramatta, N.S.W., Australia

ABSTRACT: In the current study, the relationship between the degree of restraint and thermal stresses has been established for typical concrete wall sections subject to heat-of-hydration, based on a rational analysis methodology. This relationship was shown to be non-linear due to creep effects and material non-linearity. Similar analyses are being carried out for other structural elements under different exposure conditions.

1 INTRODUCTION

Concrete structures are subjected to thermal strains during the hardening of concrete due to the heat-of-hydration of cement. If these strains are restrained, thermal stresses are generated. When a thermal stress exceeds the current tensile strength of hardening concrete, cracking of concrete would occur. The mechanisms of temperature and stress developments vary significantly from structure to structure or within the same structure under different conditions. The prediction of temperature and thermal stresses developed in young concrete is complex due to their dependency on hydration, thermal, material, structural, and environmental factors. In order to avoid the thermal cracking of concrete structures at early age, the conventional guidelines [e.g., BS 8007 1987] specify a limit on the maximum temperature rise as the sole criterion. However, thermal cracking is determined by the tensile thermal stresses developed and not by the temperature rise. Consequently, the conventional guidelines are cursory and would give inaccurate predictions being either conservative or unconservative under different conditions.

A rational analysis methodology was developed earlier (Vitharana 1994) to predict the early-age stress development in concrete structures and has been applied successfully to several practical situations. This methodology is based on the ratio of thermal stress to tensile strength as the major criterion for evaluating the tendency to cracking and incorporates the following aspects, summarily:
- Hydration characteristics of different cement types and concretes: Present-day cement types are different from the past cement types on which the conventional recommendations were based. The differences would be in chemical and physical (e.g., fineness) aspects.
- Correct heat-of-hydration modeling: Hydration is a thermally-activated process. Therefore, the effects of the reaction temperature on hydration should be incorporated by coupling the hydration process and heat-transfer between concrete and the ambient, i.e., adiabatic and varying-temperature conditions.
- Environmental interaction: This would be significant as heat would be lost or gained from the ambient in the form of convection, solar radiation etc. Usually, the environmental interaction is either ignored or considered irrationally. The type of formwork and their removal time would modify the development of temperature and thermal stresses significantly.
- Maturity of concrete: The material (e.g., Young's modulus of elasticity) and strength (e.g., tensile strength) properties mature rapidly with age (time since casting concrete). These properties are also dependent of the temperature-moisture-time histories as the hydration is particularly a thermally-activated process.
- Creep of concrete: Strain-induced stresses are relaxed significantly by the creep of concrete, by about 40 - 50% typically (Vitharana 1995). At early age, creep of concrete is higher as concrete is

in a plastic state. Early-age creep depends on the temperature-time histories, moisture variations, transient effects due to elevated temperatures occurring after loading etc.

- Construction stages: Hydration-induced stresses occur while the construction progresses. Heat-transfer and hydration are highly-transient processes and therefore the construction stages should be considered. It was shown (Vitharana 1995) that the multi-stage and single-stage concreting for large foundations have the same tendency to cracking if the construction stages were considered in a rational analysis.

- Autogeneous shrinkage: At present, high-performance concretes, having water cement ratios as low as 0.25 by mass, are introduced for various structures subject to aggressive loading and exposure conditions. However, they undergo early-age shrinkage due to the consumption of water in the hydration process. Autogeneous shrinkage occurs simultaneously with the heat-of-hydration effects, and the compounding effect is severe enough to cause excessive cracking.

- Stress redistributions: Strain-induced stresses, such as thermal stresses, are stiffness-dependent. The cracking of concrete, resulting in reduced stiffness, would therefore relax thermal stresses. The differences in stiffness would redistribute the thermal stresses among the structural elements.

In this paper, the thermal stress generation in concrete members with different degrees of restraint, representative of walls in particular, is presented. The correlation between thermal stress and degree of restraint is established for typical cases. The findings will be useful to structural engineers and constructors concerned with the early-age thermal crack occurrence.

2.0 EARLY-AGE STRESS GENERATION AND CRACKING

Figure 1 shows a schematic representation of the mechanism of thermal stress generation due to heat-of-hydration. The incremental thermal stress $\Delta\sigma_t$ developed within time step Δt, at age t, is given by: $\Delta\sigma_t = E_t \, \alpha \, \Delta T_t$ where α is the coefficient of thermal expansion of concrete, E_t is Young's modulus of elasticity of concrete (E_c) at age t, and ΔT_t is the temperature rise within the time step. During the temperature rising stage, typically within the first day or two, concrete is in a plastic stage having a very low E_c value. Consequently, the induced stresses are low compressive stresses on the order of 1 - 2 MPa. Once the hydration retards, the temperature begins to drop at a rate which is governed by the heat-transfer characteristics between the member and the ambient. This temperature drop takes place under an increased E_c due to the maturity of concrete, and consequently the induced incremental stresses are tensile stresses of higher magnitude. The total stress σ_t at an age t is given by the algebraic sum of the preceding stress increments $\Delta\sigma_t$. The net stress towards the end of hydration is therefore tensile. These stresses are however relaxed by the early-age creep. If the induced tensile stress exceeds the available tensile strength, cracking of concrete would occur.

Shown also is the development of stresses in mature concrete having a constant E_c subjected to the same temperature cycle. With constant E_c, the thermal stress at the end of the temperature cycle returns to a zero value.

The above example illustrates the thermal stress generation in a simplified form by assuming that the induced strain (= $\alpha \, \Delta T_t$) is fully-restrained and no stress redistribution takes place. Thermal stresses can be generated in concrete sections even if they are unrestrained externally or by structural indeterminacy. The temperature distribution (also corresponding free thermal strains) across a given cross section is non-linear generally. As a result, even in the absence of any external restraints, stresses are generated in structural members whose behaviour is governed by flexural and axial actions having a final linear strain distribution according to the plane sections remain plane hypotheses. These stresses under unrestrained conditions are self-equilibrating on the cross-section and generally known as primary stresses.

Figure 2 shows a concrete section subject to a non-linear but symmetrical temperature distribution (Fig. 2a) which is a typical situation occurring in concrete walls. The unrestrained primary stress at point i can be calculated by considering the equilibrium of the axial force. Similarly, unsymmetrical temperature distributions will result in a flexural deformation (i.e., curvature across the thickness) which can be obtained by considering the equilibrium of the flexural moment. Unrestrained primary stress (tension positive) σ_i at point i is then given by (Fig. 2b):

$$\sigma_i = -E_i\alpha\left[T_i - T_e\right] \qquad (1)$$

230

and the equilibrium unrestrained uniform temperature T_e is given by:

$$T_e = \frac{\sum T_i E_i \Delta y_i}{\sum E_i \Delta y_i} \qquad (2)$$

Under fully-restrained conditions, the thermal stress at point i is given by (Fig. 2c):

$$\sigma_i = -E_i \alpha T_i \qquad (3)$$

where E_i is Young's modulus of elasticity of concrete in layer i, α is the coefficient of thermal expansion of concrete, T_i is the temperature increase, and Δy_i is the thickness of the layer represented by point i.

In young concrete, material properties change with age and temperature. The above simplified procedure therefore becomes invalid and an incremental analysis in time-steps is needed, incorporating creep effects of concrete. In an incremental analysis, it is first necessary to obtain the free incremental strains (tensile as positive) $\Delta \varepsilon_i$ at point i in the cross-section within a given time-step Δt at time t:

$$\Delta \varepsilon_i = \alpha \left[T_{(t+\Delta t)} - T_{(t)} \right] + D\varepsilon_{cr} \qquad (4)$$

where $T_{(t)}$ is the temperature at the beginning of the time-step, $T_{(t+\Delta t)}$ is the temperature at the end of the time-step, and $D\varepsilon_{cr}$ is the incremental creep strain at point i due to the previously-generated thermal stresses. The evaluation of the creep effects is complicated in young concrete due to the involvement of numerous interactive parameters and details can be found in (Vitharana 1994).

As the temperature within the section varies during the hydration process, each point within the section has an unique creep behaviour. Visco-elastic creep characteristics depend on various parameters such as age at loading, concrete composition, volume/surface ratio, moisture-temperature-time histories etc., and therefore the accurate estimate of creep effects is somewhat difficult at present due to the limited knowledge on the creep behaviour of young concrete.

The primary and fully-restrained stresses can be calculated by using $\Delta \varepsilon_i$ value in the place of T_i value in Eqs 1 to 3 appropriately.

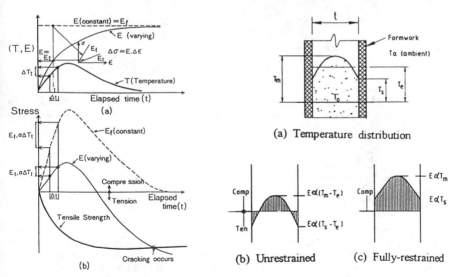

Fig. 1   Cracking mechanism of concrete at early-age.   Fig. 2   Typical conditions on a wall section.
(a) Temperature and Young's modulus
(b) Stress and strength

3.0 DIFFERENT DEGREES OF AXIAL AND FLEXURAL RESTRAINTS

The degree of restraint in a hardening concrete mass varies significantly between the elements. In an accurate analysis methodology, it is necessary to introduce the free strains in a structural analysis of the complete system. However, due to the variation of material properties with time, an incremental analysis in time-steps is required. The visco-elastic creep of concrete has a memory of the applied loading, and consequently the effect of each incremental loading, over the time period considered, needs to be stored during the analysis. In hydrating concrete, Young's modulus of elasticity varies with time, and thus the stiffness ratio when monolithically connected with a flexible body such as a rock foundation. Therefore, the analysis of the complete system would be time-consuming and expensive in comparison with only a single analysis required in mature concrete generally. Alternatively, if the stress distribution is known for a particular structure or a region under a given temperature change (e.g., the stress distribution in a concrete wall rigidly-fixed along the base when undergoing a temperature change), then the tendency to cracking in different regions can be predicted. The degree of restraint can be defined between 0% for unrestrained and 100% for fully-restrained conditions respectively.

Figure 3 shows the distribution of the thermal stress coefficient (which represents the degree of restraint) on the wall of a square (in plan) concrete reservoir with a wall aspect ratio of 1:2 (i.e., length = 2 x height) when subjected to a uniform temperature change across the wall thickness (Vitharana 1991). The stress coefficients C_σ are given along the vertical corner and center of the wall. The tensile thermal stress σ at a point is given by: $\sigma = - C_\sigma E_c \alpha \Delta T$ where ΔT is the temperature change (an increase is assumed positive). As can be seen, the degree of restraint varies significantly over the wall and is more than unity close to the bottom corner of the reservoir, on the order of 3 in this particular case.

Thermal stresses induced by the heat-of-hydration effects would not be linearly proportional to the degree of restraint due to the concrete creep and material non-linearities dependent of the stress level. Therefore, it is important to establish the effect of the degree of restraint on the induced stresses.

The effect of the degree of external restraint on the hydration-induced stresses on concrete wall sections is presented by adopting separate restraint factors with respect to axial and flexural actions. The total stress (primary stress + secondary stress due to external restraints) at a point i within a given section can be calculated by: (i) fully restraining the free thermal strain at each layer (point) individually, (ii) releasing the restrained stresses as equal and opposite stresses on the section and carrying out a section analysis, and (iii) total stress is then given by the algebraic sum of fully-restrained and resulting released stresses. Therefore, in an incremental analysis in time-steps, the total incremental stress $D\sigma_{(i)}$ at point i is given by:

$$D\sigma_{(i)} = -E_i\alpha\Delta\varepsilon_i + (1 - RF)\frac{\sum_{i=1}^{i=n}E_i\alpha\Delta\varepsilon_i\Delta y_i}{\sum_{i=1}^{i=n}E_i\Delta y_i}E_i + (1 - RM)\frac{\sum_{i=1}^{i=n}E_i\alpha\Delta\varepsilon_i\Delta y_i(y_i - y_g)}{\sum_{i=1}^{i=n}E_i\Delta y_i(y_i - y_g)^2}y_iE_i \quad (9)$$

where RF and RM are the restraint factors against axial and flexural actions respectively, $\Delta\varepsilon_i$ is given by Eq 4, Δy_i is the layer thickness, y_i is the distance to layer i from a reference axis, and y_g is the distance to the elastic centroid of the section from the reference axis:

$$y_g = \frac{\sum_{i=1}^{i=n}E_i\Delta y_i y_i}{\sum_{i=1}^{n}E_i\Delta y_i} \quad (10)$$

Some of the typical values for RF and RM are: (a) for horizontal direction stresses in long walls rigidly-fixed at the base; RF = RM = 1, and (b) for a raft foundation founded on soil; RF = 0.0 and RM = 1.0. A common practical situation, where the hydration-induced thermal crack occurrence is a major concern, is the mass concreting on a rock foundation, e.g., RCC dams. In (JCI 1986), axial and flexural restraint factors are given for a range of stiffness ratios (ratio of Young's moduli between concrete E_c and foundation material E_r) and aspect ratios of the concrete mass (length/height). These values are based on extensive finite element analyses assuming the foundation to be an elastic-half-space, and no tensile stress

232

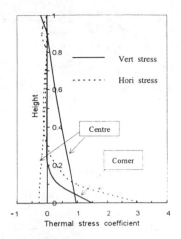

Fig. 3   Stress distribution coefficients C_σ
on the wall of a reservoir.

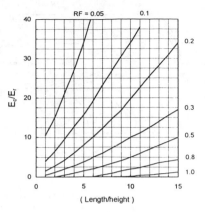

Fig. 4   Axial restraint factors (RF) for a
concrete mass (JCI 1986).

development between the concrete mass and the foundation. Figure 4 shows the axial restraint factors for a hardening concrete mass (JCI 1986). The stress generation can therefore be determined in time-steps in hydrating concrete with the use of the appropriate E_c values and corresponding RF values from Fig. 4.

4.0   THERMAL STRESSES IN WALL SECTIONS

In this Section, typical analysis results are given to highlight the significance of the degree of restraint on the thermal stress prediction in concrete sections. The details of the heat-transfer analysis and stress calculations can be found in (Vitharana 1994). The following values and conditions were assumed for the major parameters:

Concrete: Concrete is ordinary Portland cement. Young's modulus of elasticity at 28 days $E_{c(28)}$ under the standard temperature of 20 °C is 30,000 MPa. The development of E_c with age and temperature are incorporated based on a maturity function and the activation energy of hydration (Vitharana 1994). Unit cement content S is 500 kg/m³.

Heat-transfer: There is no solar radiation on the wall, and the ambient temperature remains constant. Wood formwork is provided on both wall surfaces upto 2 days since casting concrete (formwork thickness on each face is 30 mm). The value of the surface heat-transfer coefficient is 10 J/m²/s/°C, corresponding with still-air conditions. The concrete placing temperature T_o is 20 °C. A multi-layer heat transfer analysis was carried out incorporating both concrete and formwork, and the formwork interaction was neglected in the stress calculations.

Figure 5a shows the temperature developments at the center and the mid-thickness of a 750 mm thick wall. As can be seen, the formwork removal at 48 hrs has suddenly increased the temperature differential between the surface and the mid-thickness while dissipating the overall temperature rapidly. Also shown in Fig. 5a are the corresponding curves if steel formwork, having a higher heat conductivity, is used. The development of primary (self-equilibrating) stresses, which are responsible for the surface cracking in concrete walls, is shown in **Fig. 5b**. The surface stresses change from compressive to tensile as the hydration retards. Consequently, the surface cracking of the wall, if occurs, would close at the end of the hydration. These stresses are compared with a typical tensile strength development, having a 28-day value of 2.5 MPa. As can be seen, the tendency to cracking is very small in this particular case.

Figure 6 shows the development of total stresses at the 750 mm thick wall center and surface respectively under different degrees of restraint. They vary from initial compressive stresses in the heating phase to higher tensile stresses during the cooling phase. As can be seen, the effect of the degree of restraint is significant, and the stress development is not linearly proportional to the degree of restraint. Figure 7 shows the dependency of thermal stresses at 7 days on the degree of restraint. When the degree of restraint increases from 100 to 200% (i.e., 100% increase), the increase in thermal stress is only 60%.

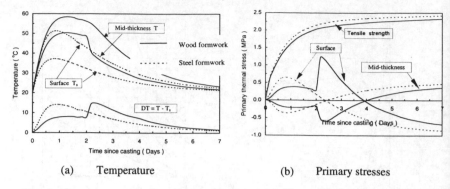

(a)   Temperature   (b)   Primary stresses

Fig. 5 Temperature and primary stress developments with wood and steel formworks.

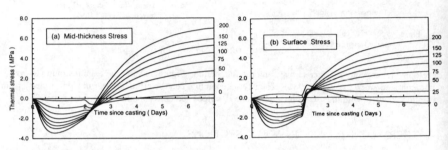

Fig. 6 Effect of degree of restraint on thermal stress development.

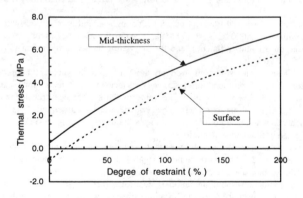

Fig. 7 Stress vs degree of restraint at 7 days since casting.

Therefore, is important to consider the non-linear stress vs degree of restraint relationship if an accurate prediction of the thermal stresses is required.

5.0 CONCLUDING REMARKS

1).   It is important to rationally evaluate the thermal stress development at early age in order to avoid thermal crack occurrence. The conventional methods of limiting the maximum temperature rise was shown to be cursory as cracking of concrete is caused by the thermal stresses which exceed the concrete tensile strength.

2).   A non-linear incremental analysis is required for predicting early-age thermal stresses. The creep of concrete would relax the thermal stresses significantly. Environmental interaction modifies the temperature and stress developments. The type and removal time of formwork should be considered in the heat-transfer analysis. The proposed methodology incorporates these factors.

3).   The effect of the degree of restraint on the thermal stress development is non-linear. The typical calculations for wall sections showed this non-linearity. A parametric study is currently being carried out covering a range of cement types, concretes, and exposure conditions.

6.0 REFERENCES

BS 5337 1987. *Design of concrete structures for retaining aqueous liquids*. British Standards Institute. London.

Vitharana, N. D. 1994. *Evaluation of early-age behaviour of concrete wall sections: hydration and shrinkage effects*. Report No. 94/I. Hokkaido Development Bureau, Sapporo, Japan.

Vitharana, N.D. and Sakai, K. 1995. *Single and multi-stage concrete placements for large foundations: hydration-induced thermal stresses*. 5th East-Asia Pacific Conference on Structural Engineering and Construction. Gold Coast. Australia: 2311-2316.

Vitharana, N.D., Priestley, M.J.N. and Dean, J.A. 1991. *Strain-induced stressing of concrete storage tanks*. Report 91-8, University of Canterbury, New Zealand.

Japan Concrete Institute. 1986. *Standard specifications for design and construction of concrete structures. Part 2 (construction)*. Tokyo. Japan.

Self-stress development of expansive concrete under axisymmetric confinement

C.M. Haberfield
Department of Civil Engineering, Monash University, Melbourne, Vic., Australia

ABSTRACT: Uses for expansive cements have extended beyond off setting the drying shrinkage of concrete to include improved rock anchor and pile performance, as an alternative connection in tubular frames, and to replace blasting in open pit mining operations. Unfortunately, design rules for estimating the appropriate dosage of expansive cement are not available. This paper presents analytical solutions for estimating the expansion and self-stress developed in radially confined expansive concrete cylinders. Expansion is modelled using a thermal analogy which requires only one parameter to define behaviour. A simple laboratory procedure for determining this parameter is outlined. Predictions from the model are compared with laboratory experiments.

INTRODUCTION

After cement grout (or concrete) sets and is allowed to dry, it undergoes the sometimes-destructive process of drying shrinkage. Expansive cements were basically developed to minimise drying shrinkage and hence their use has generally been restricted to shrinkage compensation rather than developing self-stress. The difference between the two is essentially the amount of expansion that can take place. A shrinkage-compensating cement will expand no more than the magnitude of the drying shrinkage of the grout. A self-stressing cement will expand more (if not restricted), so that the net volume after expansion and shrinkage is greater than the initial volume.

Recently, the possible benefits that can be obtained by utilising self-stressing cements have been realised. These include enhancing the performance of anchors and piles in rock (Haberfield et al., 1994) and as a more efficient and reliable method of connecting tubular members in off-shore structures (Foo, 1990). Both applications essentially utilise the expansive cement in the same way.

When an anchor or pile is constructed with expansive cement, the expansion process causes the concrete of the pile or anchor to expand against the stiffness of the surrounding rock (Fig. 1a and 1b). Since the rock resists this expansion, normal stresses are generated across the grout/rock interface. This increase in normal stresses increases the frictional resistance of the pile or anchor, thereby leading to enhanced performance. Haberfield et al. (1994) showed that increases in pile and anchor capacity in excess of 200 % can be achieved through using expansive cements.

Foo (1990) proposed that expansive cement could be utilised in tubular steel connections. He showed that a strong joint could be obtained by placing one tube inside another and filling the annulus between the two tubes with expansive cement grout (Fig. 1c). The grout expands against the stiffness of the two tubes, thereby generating very large normal stresses which lock the two tubes together.

In both applications the expansion of the grout in a confined space causes normal stresses or self-stress to be generated, which in turn leads to enhanced performance. The degree of self-stress depends on the level of the confinement and the concentration of expansive additive. If confinement is low, then the expansive grout will expand freely and result in severe loss of grout strength, stiffness and integrity. However, for relatively high levels of confinement, the expansion of the grout is severely restricted, resulting in high levels of self-stress and increased grout strength.

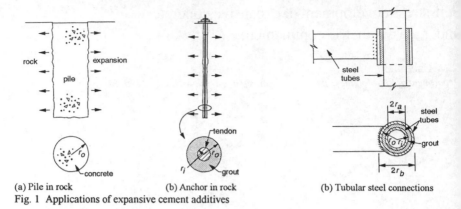

(a) Pile in rock   (b) Anchor in rock   (b) Tubular steel connections

Fig. 1 Applications of expansive cement additives

The difficulty arises in determining the self-stress that can be generated, and if the confinement is adequate to ensure integrity of the grout. By using a thermal analogy to model the expansion process, this paper develops theoretical solutions for the performance of expansive grouts that are confined radially by an elastic boundary. These solutions require only one input parameter (analogous to a thermal coefficient of expansion) which can be readily determined from a simple laboratory test.

ELASTIC CAVITY EXPANSION

The applications mentioned above all involve the forced expansion of a cylindrical cavity. Cross-sections of a pile, an anchor and a tubular joint are shown in Fig. 1. For the case of a pile in rock (Fig 1a), the expansive grout forms a long cylindrical body with its outer boundary surrounded by a rock mass of infinite extent. The rock bolt (Fig. 1b) is similar to the pile, but has also an inner boundary formed by the steel bolt or tendon. The tubular joint (Fig. 1c) has inner and outer boundaries formed by steel tubes of finite wall thickness. In each case, the grout is restrained from expanding by these inner and outer boundaries at radii of r_i (= 0 for Fig. 1a) and r_o respectively. For purposes of analysis, the inner and outer boundaries are replaced by normal stiffnesses K_i and K_o respectively. It is assumed that the expansion does not cause yielding at the boundaries and therefore K_i and K_o are constant.

Expansion of the grout must force a change in the position of the inner and outer boundaries. The axisymmetric displacement of these boundaries implies that a reaction stress will be generated to off-set the expansion. The reaction stresses are normal to the surface of the grout and are denoted p_i and p_o for inner and outer surfaces of the grout respectively. Note that both p_i and p_o are assumed to be compressive. Since the thickness of the grout in each application is small compared to the out of plane direction (i.e. the length of the pile, anchor or joint), plane strain conditions have been assumed.

If it is assumed that the grout remains elastic and that it expands evenly in all in-plane directions then the total expansion of the grout in each of the orthogonal directions r, θ and z can be determined from

$$\varepsilon_{r,\theta,z}^{T} = \varepsilon_{r,\theta,z}^{S} + \varepsilon^{E} \tag{1}$$

where $\varepsilon_{r,\theta,z}^{T}$ are the total strains in each direction, ε^{E} is the strain due to the chemical expansion process and $\varepsilon_{r,\theta,z}^{S}$ are the strains in each direction due to any stresses that are applied. From Hookes law, and enforcing the plane strain condition ($\varepsilon_{z}^{T} = 0$), leads to

$$\varepsilon_{r,\theta}^{T} = \frac{(1-v_g)(1+v_g)}{E_g}\left(\sigma_{r,\theta} - \frac{v_g}{1-v_g}\sigma_{\theta,r}\right) + (1+v_g)\varepsilon^{E} \tag{2}$$

where $\sigma_{r,\theta,z}$ are the stresses in the r, θ and z directions respectively and E_g and v_g are the elastic Young's modulus and Poisson's ratio of the grout. Solving for σ_r and σ_θ, and enforcing the

238

equilibrium and strain displacement relationships for small strain leads to Eq. 3 involving radial displacement U_r at radius r. Eq. 3 can be solved to give the general solution of Eq. 4.

$$\frac{d^2 U_r}{dr^2} + \frac{1}{r}\frac{dU_r}{dr} - \frac{U_r}{r^2} - \left(\frac{1+v}{1-v}\right)\frac{d\varepsilon^E}{dr} = 0 \tag{3}$$

$$U_r = A_g r + \frac{B_g}{r} + \left(\frac{1+v_g}{1-v_g}\right)\frac{1}{r}\int_{r_i}^{r}\varepsilon^E r\,dr \tag{4}$$

where A_g and B_g are constants of integration. Applying the boundary conditions at $r = r_i$, $\sigma_r = -p_i$ and $r = r_o$, $\sigma_r = -p_o$ (assuming tensile stresses positive) leads to

$$A_g = \left(1-2v_g\right)\frac{B_g}{r_i^2} - \frac{p_i}{\lambda_g} \qquad B_g = \frac{r_i^2 r_o^2}{r_o^2 - r_i^2}\left[\frac{p_i - p_o}{\lambda_g\left(1-2v_g\right)} + \left(\frac{1+v_g}{1-v_g}\right)\frac{I_{r_i}^{r_o}}{r_o^2}\right] \tag{5}$$

where $\lambda_g = \dfrac{E_g}{(1-2v_g)(1+v_g)}$ and $I_{r_i}^{r_o} = \displaystyle\int_{r_i}^{r_o}\varepsilon^E r\,dr$ (6)

For the special case of a solid grout cylinder (eg. pile) where $r_i = 0$,

$$B_g = 0 \quad \text{and} \quad A_g = \frac{(1+v_g)(1-2v_g)}{(1-v_g)}\frac{I_0^{r_o}}{r_o^2} - \frac{p_o}{\lambda_g} \tag{7}$$

Finally, back substituting for strains leads to Eqs. 8 & 9 for displacements and stresses at radius r within the expanding grout. The integral $I_{r_i}^{r}$ is dependent on the quantity and type of expansive cement used, the stresses developed and the total volume of grout per unit length.

$$U_r = A_g r + \frac{B_g}{r} + \left(\frac{1+v_g}{1-v_g}\right)\frac{1}{r}I_{r_i}^{r} \tag{8}$$

$$\sigma_{r,\theta} = \lambda_g\left[A_g \mp \frac{B_g}{r^2}(1-2v_g) - \frac{I_{r_i}^{r}}{r^2}\frac{(1+v_g)(1-2v_g)}{(1-v_g)}\right] \tag{9}$$

DETERMINATION OF CONFINING STRESSES AND STIFFNESSES

The stresses on the inner and outer boundaries, p_i and p_o depend on the stiffnesses K_i and K_o of the surrounding elastic material. The inner boundary can involve either a solid cylinder (eg rock anchor) or a hollow tube (eg tubular connection). The situation is analogous to the expanding grout annulus described above. By substituting appropriate tube dimensions and properties and setting $p_i = 0$ and $\varepsilon^E = 0$, the following equations, which model the elastic behaviour of the inner tube, can be derived.

$$U_r = A_i r + \frac{B_i}{r} \qquad\qquad \sigma_{r,\theta} = \lambda_i\left[A_i \mp \frac{B_i}{r^2}(1-2v_i)\right] \tag{10}$$

239

where $A_i = -\dfrac{r_i^2}{r_i^2 - r_a^2}\dfrac{p_i}{\lambda_i}$ $\qquad B_i = -\dfrac{r_i^2 r_a^2}{r_i^2 - r_a^2}\dfrac{p_i}{\lambda_i(1-2v_i)}$ $\qquad \lambda_i = \dfrac{E_i}{(1-2v_i)(1+v_i)}$ (11)

E_i and v_i are elastic Young's modulus and Poisson's ratio for the tube material and r_a is the internal radius of the tube. The elastic, normal stiffness of the tube at $r = r_i$ can then be defined by

$$K_i = \frac{p_i}{U_{r=r_i}} = -\frac{\lambda_i(1-2v_i)(r_i^2 - r_a^2)}{r_i(r_i^2(1-2v_i)+r_a^2)}$$ (12)

For a solid inner tube (eg rock bolt), $r_a = 0$ and Eq. (12) reduces to $K_i = -\lambda_i/r_i$.

Similar equations are also appropriate for an elastic outer tube of inner radius r_o, outer radius r_b, Young's modulus, E_o, Poisson's ratio v_o, . The normal stiffness of the outer tube is given by

$$K_o = \frac{p_o}{U_{r=r_o}} = \frac{\lambda_o(1-2v_o)(r_b^2 - r_o^2)}{r_o(r_o^2(1-2v_o)+r_b^2)}$$ (13)

For an outer tube of infinite extent, eg a pile or anchor in a rock mass with an insitu stress of p_{og}, $r_b \rightarrow \infty$ and Eq. 13 reduces to $K_o = \lambda_o(1-2v)/r_o$. By enforcing compatibility of radial displacements at both boundaries; ie. at $r = r_i$ and $r = r_o$, equations for p_i and p_o can be determined :

$$p_i = \frac{2(1+v_g)r_i K_i I_{r_i}^{r_o}}{r_o K_o\left[\dfrac{r_i K_i - \lambda_g(1-2v_g)}{r_o K_o - \lambda_g(1-2v_g)}\right]\left[\dfrac{r_o^2 K_o + r_o\lambda_g}{\lambda_g K_o}\right] - r_i^2\left[\dfrac{r_i K_i + \lambda_g}{\lambda_g}\right]}$$ (14)

$$p_o = \frac{2(1+v_g)r_o K_o I_{r_i}^{r_o}\left[\dfrac{r_i K_i - \lambda_g(1-2v)}{r_o K_o - \lambda_g(1-2v)}\right]}{r_o K_o\left[\dfrac{r_i K_i - \lambda_g(1-2v_g)}{r_o K_o - \lambda_g(1-2v_g)}\right]\left[\dfrac{r_o^2 K_o + r_o\lambda_g}{\lambda_g K_o}\right] - r_i^2\left[\dfrac{r_i K_i + \lambda_g}{\lambda_g}\right]}$$ (15)

For a pile in rock, there is no inner boundary (ie. $r_i = r_a = 0$) and Eq. 15 reduces to

$$p_o = \frac{2(1+v_g)\lambda_g K_o I_0^{r_o}}{r_o(\lambda_g + K_o r_o)}$$ (16)

DETERMINATION OF EXPANSIVE POTENTIAL

The above equations require an estimate of the expansive strain, ε^E. Unfortunately, ε^E is not constant but depends on expansive cement type and content, mix design, stress level and curing conditions. However, it is a relatively simple process to determine the variation of ε^E with the above parameters, by utilising a standard laboratory oedometer test usually used for determining the consolidation properties of soil. Baycan (1996) describes a series of such tests carried out on pastes made from Denka CSA, a Class C expansive cement which generates expansion through the formation of ettringite crystals. Four expansive paste mixes were tested. In each mix, a specified amount of CSA was added to normal cement and then mixed with water. The four mixes all used a water/total cement ratio of 0.45 and had CSA to total cement ratios (CSA') of either 0, 0.069, 0.138 or 0.207.

In these tests, fluid cement paste was placed into a standard steel oedometer ring (r_i = 75 mm, r_o = 80 mm, height = 20 mm). Filter paper and porous stones were then placed on the top and bottom of samples and the sample placed into the oedometer rig. A specified constant preload (of between 100 and 2500 kPa) was applied to the sample, and measurements of vertical displacement taken at regular intervals for 30 days. At all times, the sample was immersed in water; the porous stones and filter paper providing a conduit for water to the top and bottom of the samples.

Since the oedometer ring is relatively stiff (but finite), most expansion is forced to occur in the vertical direction. Analysis of the test set-up using analytical solutions similar to those given earlier, show that the finite stiffness of the oedometer ring has a negligible effect on the vertical expansion of the grout and can be ignored. Typical plots of vertical expansion versus time obtained from the oedometer tests are shown in Fig. 2. Fig 2(a) compares expansion vs time plots for the four different mixes all at a preload of 1500 kPa, while Fig 2(b) compares curves for a CSA' = 0.207 mix at several different preloads. Fig. 2 clearly shows the dependence of expansion on preload and CSA content. It is worth noting that all expansion ceases after an initial period of 10 to 20 days, but samples containing greater CSA contents and higher normal stresses require more time to complete the expansion process.

The maximum expansion from each test (corrected for elastic displacements) and expressed in terms of strain, are plotted against applied normal stress in Fig. 3. Using standard curve fitting procedures, the following empirical equation to describe the variation of expansive strain, ε^E, with expansive cement content, CSA', and normal stress ratio, σ_n has been determined :

$$\varepsilon^E = \left(\frac{CSA'}{0.0072\overline{\sigma}_n + 0.3283} \right)^{3.774} \tag{17}$$

where $\overline{\sigma}_n = \sigma_n / p_a$ and σ_n is the applied normal stress and p_a is atmospheric pressure (100 kPa). Fig. 3 shows that Eq. 17 fits the experimental data well. Note that Eq. 17 has been determined for cement pastes made from Denka CSA, normal cement and water and cured under near ideal conditions. Any variation from these conditions will result in a different relationship. Nevertheless, this procedure can be used to establish a similar relationship for any type of expansive cement and for any mix design. Research in this area is continuing. The amount of expansion also depends on the curing conditions. In particular, the amount of free water available for the cement hydration process governs the amount of ettringite formed. Hence, full expansion potential may not be realised if there is not enough free water available. In such cases, Eq. 17 will overpredict the amount of expansion.

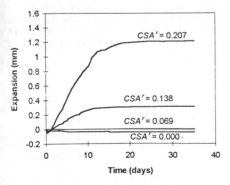

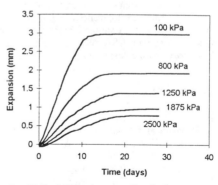

(a) 1500 kPa pre-load for a range of CSA' (b) $CSA' = 0.207$ and a range of pre-loads

Fig. 2   Expansion vs time responses for expansion tests using soil oedometer rigs

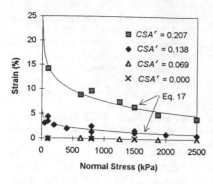

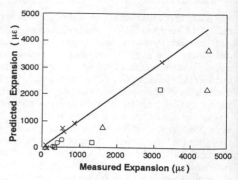

Fig. 3   Variation of expansion with CSA content and normal stress

Fig. 4   Predicted vs measured strains for confined expansion tests

COMPARISON WITH LABORATORY CONFINED EXPANSION TESTS

The above model was tested in the laboratory by confined expansion tests. These tests involved casting cement paste of 4 different CSA' values into tubes and measuring expansion (using strain gauges) with time. The tubes remained completely immersed in water at 20°C throughout testing. All confining tubes had length to diameter ratios of 2:1 and ranged in stiffness from 52 MPa/mm to 1310 MPa/mm. The range in stiffnesses was made possible by using tubes of different diameter and wall thickness and made from different materials. The tubes showed similar expansion time responses to those observed in the oedometer tests. Full details of these tests are provided by Baycan (1996).

Fig. 4 compares predicted and measured values of peak circumferential strain. The outliers correspond to tests in which the confining tube yielded and hence the above model is not applicable. Although reasonable predictions have been obtained, it appears that the theoretical model slightly under-estimates expansion. Friction between the expanding grout and the oedometer ring, which could lead to lower expansions being measured in the vertical direction, may be the reason for this underestimate.

SUMMARY

Theoretical equations describing the behaviour of expansive cement grout under radially confined conditions have been derived. Such equations are useful for determining the degree of expansion and self-stress development in piles and rock anchors and tubular connections that utilise expansive cement. A simple laboratory test for determining the expansion potential of expansive grouts containing CSA has been described. Good agreement has been obtained between laboratory confined expansion tests and theoretical predictions.

REFERENCES

Baycan, S. 1996. *Field Performance of Expansive Anchors and Piles in Rock*. PHD Thesis, Department of Civil Engineering, Monash University

Haberfield, C.M., Baycan, S. & Seidel, J.P. 1994. Improving the Capacity of Piles in Rock Through the Use of Expansive Cement Additives. *5th Intl. Conf. and Exhibition on Piling and Deep Foundations*, Belgium, June 1994 : 1.2.1-1.2.7.

Foo, J.E.K. 1990. *Prestressed Grouted Tubular Connections*. PHD Thesis, Department of Civil Engineering, Monash University

The Mechanics of Structures and Materials, Grzebieta, Al-Mahaidi & Wilson (eds)
© *1997 Balkema, Rotterdam, ISBN 90 5410 900 9*

Optimal application of high-strength concrete in cantilever structural walls

B.G. Bong
Victoria University of Technology, Melbourne, Vic., Australia

A.R. Hira
Department of Civil and Building Engineering, Victoria University of Technology, Melbourne, Vic., Australia

P.A. Mendis
Department of Civil and Environmental Engineering, The University of Melbourne, Vic., Australia

ABSTRACT: The availability of very high-strength concretes in the market place has been a major factor in the development of very tall concrete buildings worldwide. Although concrete strengths of up to 120 MPa is commercially available in Australia and overseas, the costs of production associated with higher strengths and the lack of awareness of the advantages of using this versatile material effectively have discouraged many designers from specifying it for tall building applications. This paper presents the cost benefits associated with optimal use of high-strength concrete in construction of walls in tall buildings. It is shown that the increased lettable space due to thinner walls, makes high strength concrete application extremely cost effective for tall buildings.

1 INTRODUCTION

In the last decade, record setting tall concrete buildings are being built around the world. This includes the world's tallest building, the 98-storey twin Petronas Towers in Kuala Lumpur with an overall height of 450 m. The external columns and central core in the lower levels of the towers were constructed with 80 MPa concrete. Another striking example, where high strength concrete has been extensively utilised is the 88 storey Jin Mao Building, currently under construction in Shanghai, which will become the third tallest building in the world. A significantly important factor contributing to the viability of achieving such heights is the development of efficient structural systems and the availability of high strength concrete which was used for the shear core elements and the principal column elements.

The biggest advantages of HSC that makes its use attractive in high-rise buildings are the larger strength/unit cost, strength/unit weight and stiffness/unit cost compared to normal strength concrete and other structural materials. The resulting lower mass reduces inertial load providing a further incentive for its application in tall buildings located in high seismic regions. The benefits of HSC applications in tall building structural systems are explored in a project conducted at Victoria University of Technology. The preliminary results of this project are presented in this paper.

2 STRUCTURAL MODEL - CASE STUDY

A 40-storey high single cantilever structural wall with a 4.0 m inter-storey high is considered for this case study (Fig. 1). The dimensions are selected on the basis of satisfying an acceptable lateral drift of 0.002 of the building height when subjected to derived seismic loads based on

the Indonesian earthquake code (Departemen Pekerjaan Umum, 1983). The seismic mass i
based on a vertical floor loading (dead and live) of 5.75 kPa with a tributary area of 160 m².
The elastic response spectrum and the resulting dynamic storey inertia load distribution
generated by the program ETABS (Habibullah, 1989) is shown in Fig. 2. Using a concrete
strength of 40 MPa for the wall section, the calculated first mode period and maximum inter
storey drift ratio is 4.5 seconds and 0.003 respectively.

3 WALL OPTIMISATION

An optimisation process by use of Displacement Participation Factors and Sensitivity Index
based on the principle of virtual work (Charney, 1991) is utilised. The objective of the process
is to obtain the optimum solution corresponding to the use of minimum volume of material
while attaining a target stiffness in terms of the lateral displacement at the top of the structure.

The resulting optimised section properties and the equivalent thicknesses are shown in
Fig. 3a and 3b respectively. V is the concrete volume and EI the flexural rigidity of the section
As expected the wall thickness profile up the height of the structure closely follows a parabolic
profile. However in practice the number of transitions in wall thickness are kept to a minimum
on both practical and economical grounds with regards to constructability.

The computer program 'OPTIC' is developed to calculate the optimum solutions for
cantilever structural walls with different concrete strengths and designated number of wall
thickness transitions. For a given deflection limit and a minimum wall thickness the objective
of the program is to locate the positions of the transitions and to determine the optimal wall
thicknesses that produce minimum overall volume. The procedure involves a systematic sizing-
analysing and resizing process within the given limits for the parameters. The process is
continued until the minimum volume criterion is reached.

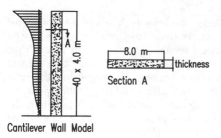

Cantilever Wall Model

Fig. 1 Elevation and plan of wall model

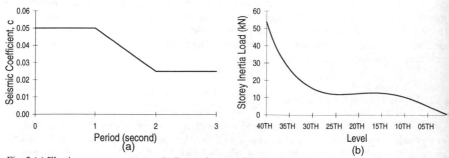

Fig. 2 (a) Elastic response spectrum (b) Dynamic storey inertia load distribution

244

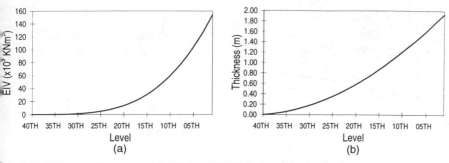

Fig. 3 (a) Optimum section property distribution, EIV (b) Optimum theoretical wall thickness distribution

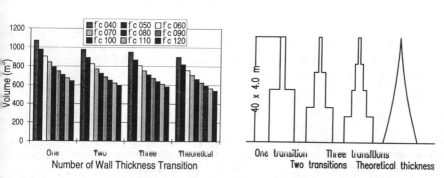

Fig. 4 Optimum volumes with different number of transitions and different concrete strengths

Table 1 Concrete cost

Concrete strength (MPa)	40	50	60	70 ·	80	90	100	110	120
Cost ($/m3)	138	149	163	180	200	230	260	290	320

4 BENEFITS OF USING HIGH-STRENGTH CONCRETE

Uniform concrete grade for full height. Fig. 4 illustrates the optimal concrete volumes for the case of a cantilever wall using the program OPTIC, for different concrete grades with differing number of wall thickness transitions. For comparison purposes, results for a cantilever wall with a parabolic wall thickness profile giving an ideal optimal solution are also presented. The principal observation is the significance of the volume reduction that can be obtained by increasing the concrete strength and the insignificance of the effect of number of transitions on the total volume. When one considers the substantial cost increase for introducing each wall thickness transition in terms of construction delay it is clear that use of higher strength concrete is an effective means of minimising concrete volume.

A cost benefit analysis is also carried out. The analysis explores the effect of concrete cost and the additional financial return due to the extra lettable floor space made available by reduction in wall thickness. The cost of concrete used for the cost analysis is given in Table 1.

Fig. 5 summarises the results of the cost benefit analysis at different capitalised rates for the extra lettable area gained by the use of higher concrete strengths resulting in thinner walls. The zero rate considers concrete costs only. The cost of 40 MPa wall is used as the base for cost comparison purposes.

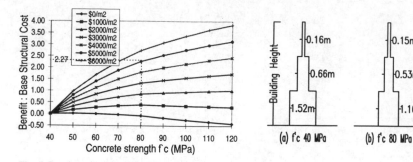

Fig. 5 Cost benefits of using higher strength concretes
(a) Vol.=978 m³; Concrete cost =$134964
(b) Vol.=726 m³; Concrete cost =$145200; Area saving =63.25m2; Cost Saving =$316250; Benefit : Initial cost =2.27

The results clearly illustrate that despite the reduction in concrete volume the cost increases with concrete strength on the basis of material cost only. The benefit of using high-strength concrete only becomes apparent if the capitalised value of the extra lettable space gained from the use of higher strength concrete is taken into consideration. Fig. 5 also indicates that the cost benefit of using very high-strength concrete for buildings attracting low financial returns, diminishes. For example for buildings yielding less than $2000/m². there is no benefit using concrete strengths exceeding 80 MPa. However for the higher yielding buildings, corresponding to prestigious buildings located in central business districts of most cities, there is a tremendous financial benefit in using very high concrete strength, with benefits increasing with increasing concrete strengths. For example, for a building yielding $5000/m², utilising 80 MPa concrete for the wall will gain a cost benefit corresponding to 3.27 times the structural cost of the 40 MPa wall, a significant amount.

Locations of wall thickness transitions for optimum solutions are shown in Fig. 6. The results show that to achieve optimum solution for a one transition wall the transition point is approximately at mid height of the building and is independent of the concrete grade. A similar trend is also observed in optimum solution for walls with two transitions where the three zones are approximately equal in height.

Concrete strength (MPa)						Concrete strength (MPa)				
Level	f c 040	f c 060	f c 080	f c 100	f c 120	Level	f c 040	f c 060	f c 080	f c 100	f c 120
40th	0.29	0.26	0.25	0.18	0.19	40th	0.16	0.16	0.15	0.15	0.15
38th	0.29	0.26	0.25	0.18	0.19	38th	0.16	0.16	0.15	0.15	0.15
36th	0.29	0.26	0.25	0.18	0.19	36th	0.16	0.16	0.15	0.15	0.15
34th	0.29	0.26	0.25	0.18	0.19	34th	0.16	0.16	0.15	0.15	0.15
32nd	0.29	0.26	Transition Levels			32nd	0.16	0.16	0.15	0.15	0.15
30th	0.29	0.26				30th	0.16	0.16	0.15	0.15	0.15
28th	0.29	0.26	0.25	0.18	0.19	28th	0.16	0.16	0.15	0.15	0.15
26th	0.29	0.26	0.25	0.18	0.19	26th	0.66	0.16	0.15	0.15	0.15
24th	0.29	0.26	0.25	0.18	0.19	24th	0.66	0.62	0.53	0.46	0.15
22nd	0.29	0.26	0.25	0.18	0.19	22nd	0.66	0.62	0.53	0.46	0.51
20th	0.29	0.26	0.25	0.93	0.19	20th	0.66	0.62	0.53	0.46	0.51
18th	1.44	1.20	1.07	0.93	0.85	18th	0.66	0.62	0.53	0.46	0.51
16th	1.44	1.20	1.07	0.93	0.85	16th	0.66	0.62	0.53	0.46	0.51
14th	1.44	1.20	1.07	0.93	0.85	14th	0.66	0.62	0.53	0.46	0.51
12th	1.44	1.20	1.07	0.93	0.85	12th	1.52	1.32	1.16	1.04	0.51
10th	1.44	1.20	1.07	0.93	0.85	10th	1.52	1.32	1.16	1.04	0.98
08th	1.44	1.20	1.07	0.93	0.85	08th	1.52	1.32	1.16	1.04	0.98
06th	1.44	1.20	1.07	0.93	0.85	06th	1.52	1.32	1.16	1.04	0.98
04th	1.44	1.20	1.07	0.93	0.85	04th	1.52	1.32	1.16	1.04	0.98
02nd	1.44	1.20	1.07	0.93	0.85	02nd	1.52	1.32	1.16	1.04	0.98
Volume	1070	904	792	710	644	Volume	979	827	726	653	596

(a) One wall thickness transiton   (b.1) Two wall thickness transitons   (b.2) Diagram

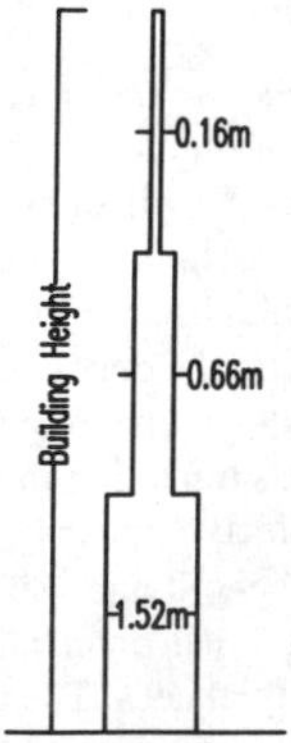

Fig. 6 Locations of wall thickness transitions at optimum solutions

246

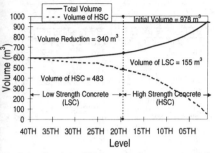
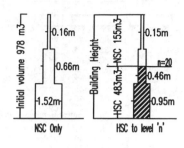

Fig.7 Application of high-strength concrete and the associated reduction of concrete volume

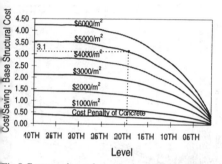

Fig.8 Cost penalty and cost saving of using higher strength concrete.

Base structural volume	= 978m³
Base structural cost	= 978m³ x $138/m³ = $134964
Cost penalty	= (483m³ x $320/m³ + 155m³ x $138/m³) - $134964 = $40986
Cost saving	= 340m³ : 4m¹ x $5000/m² = $425000
Cost Penalty : Base cost	= 0.30; Cost saving : Base cost = 3.15

Varying concrete grade with height. The study is repeated to assess the cost benefit of using higher concrete strengths for a more realistic case where concrete strength varies with height. For the purposes of this study the cantilever wall with two concrete strengths is considered, a higher strength concrete for the lower levels of the wall and lower strength concrete for the upper levels for a one and two transition walls. Fig. 7 shows the resulting volume reductions achieved by introducing 120 MPa high-strength concrete up to a particular floor with the remainder of the wall being 40MPa. in a two transition wall. A wall with concrete strength of 40 MPa throughout is used as the base for volume and cost comparisons As an example Fig 7 shows a volume reduction of 340 m³ when 120 MPa concrete is used for the first 20 floors.

The total cost benefit analysis of the above wall, taking into account the effects of capitalising the gain in net lettable area and the concrete cost, is carried out with the results shown in Fig. 8. As expected there is a penalty in concrete cost associated with replacing the 40MPa. concrete in the lower levels of the wall with 120MPa. strength concrete, however the benefits from the extra lettable area far exceed the cost penalty of the material. An interesting feature of results reveal that there is no significant gain in cost benefits by extending the higher strength concrete much beyond the mid-level (20th level) particularly for buildings yielding low rentals. The main gain in cost benefits is achieved by introducing higher strength concrete in the lower 30% to 40% of the wall height.

247

| Capitalised value | | | | | | | | | | | | | | | |
| | $0/m2 | | | | $2000/m2 | | | | $4000/m2 | | | | $6000/m2 | | | |
Level	f'c 060	f'c 080	f'c 100	f'c 120	f'c 060	f'c 080	f'c 100	f'c 120	f'c 060	f'c 080	f'c 100	f'c 120	f'c 060	f'c 080	f'c 100	f'c 120
40th	0	-10	-35	-56	76	116	128	135	152	243	291	326	228	369	454	517
35th	1	-9	-32	-51	77	118	131	140	153	244	294	331	228	371	457	522
30th	1	-8	-30	-48	76	117	131	141	151	241	292	331	226	366	454	520
25th	0	-9	-31	-49	71	109	122	133	141	227	276	314	211	346	429	495
24th	1	-8	-29	-47	71	109	122	132	140	226	274	311	210	342	425	490
23rd	1	-7	-27	-45	70	108	122	132	140	224	272	309	209	339	422	487
22nd	1	-7	-28	-45	70	108	120	131	139	222	268	306	207	337	416	482
21st	1	-6	-27	-43	69	107	119	131	137	220	265	304	204	333	412	477
20th	2	-6	-26	-41	68	105	118	129	135	216	261	300	202	327	405	470
19th	2	-6	-25	-41	67	103	117	126	132	212	259	293	198	321	400	460
18th	2	-5	-25	-40	65	101	113	124	129	208	251	287	193	314	389	451
17th	2	-6	-25	-39	64	97	109	120	126	200	243	280	188	303	378	439
16th	2	-6	-25	-38	61	95	105	117	121	195	235	271	181	295	366	426
15th	1	-5	-22	-38	59	91	104	111	117	188	229	260	174	284	355	409
10th	2	-3	-16	-27	46	73	84	91	91	150	183	209	135	226	283	328
05th	2	0	-8	-15	30	47	53	58	57	94	114	130	85	141	175	203

Notes: Concrete strengths in MPa, Benefit in $x1000

Fig. 9 Cut off levels for 90% of maximum cost benefit

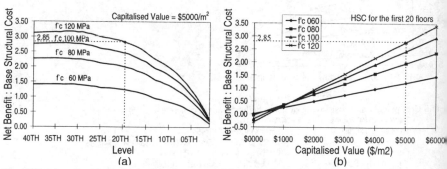

Fig. 10 Cost benefit of using high-strength concrete (a) at capitalised value of $5000/m² (b) level n = 20

Figure 9 provides tabulated values of cost benefits for varying concrete grades and varying capitalisation rates. The table also identifies the level designated as 90% cost benefit. Extending of the use of high-strength concrete above this level will reap a maximum increase in the cost benefit of less than 10%. It can be seen that the line is consistently located at the mid-height of the wall. This suggests that a convenient transition for concrete grade is approximately mid-height of the wall as there is no significant cost benefit in using higher strength concrete in top half of the wall.

Fig. 10a shows the benefit of using different concrete strengths compared with the structural cost of 40 MPa wall. The benefit is calculated for the value of lettable area capitalised to $5000/m². Fig. 10b indicates that a linear relationship can be assumed between the net benefits and the values of lettable area.

5 CONCLUSIONS

This paper presents some of the early results of assessing the benefits of using high-strength concrete in tall buildings. Results reveal that the capitalisation value of the additional lettable area gained through thinner structural walls by use of higher strength concrete provides significant cost savings for the building owners.

248

REFERENCES

Charney, F.A. 1991. The use of displacement participation factors in the optimization of wind drift controlled buildings. Proceeding of the second conference on tall buildings in seismic regions. Los Angeles: Council on Tall Buildings and Urban Habitat

Charney, F.A. 1994. DISPAR for ETABS ver. 5.4, users manual. Colorado: Advance Structural Concept.

Departemen Pekerjaan Umum. 1983. Buku Pedoman Perencanaan untuk Struktur Beton Bertulang Biasa dan Struktur Tembok Bertulang untuk Gedung. Bandung: Pusat Penelitian dan Pengembangan Pemukiman.

Habibullah, A. 1989. ETABS ver. 5.10, users manual. California: Computers & Structures, Inc.

The Mechanics of Structures and Materials, Grzebieta, Al-Mahaidi & Wilson (eds)
© 1997 Balkema, Rotterdam, ISBN 90 5410 900 9

Expert system methodology for deterioration of reinforced concrete structures

A. R. Sabouni
Civil Engineering Department, United Arab Emirates University, U.A.E.

ABSTRACT: Structural engineers involved in the repair and maintenance of reinforced concrete structures need to diagnose the types of deterioration encountered in reinforced concrete in order to provide appropriate decisions regarding remedial measures. Knowledge-based expert systems established means by which one can use computers to solve ill-structured problems, such as the assessment of distressed reinforced concrete structures. This paper describes the development methodology of a prototype knowledge-based expert system for providing advice related to common distress problems in reinforced concrete structures. System development is presented, and followed by appropriate conclusions.

1 INTRODUCTION

Recent global trends on environmental awareness accompanied with economic recession in many countries have placed more emphasis on the preservation of existing facilities and infrastructure as opposed to spending on new establishments. The problem of deterioration of structures is becoming more critical than any time before. This problem has strongly affected the building industry in many countries, especially those suffering from harsh weather (Alshamsi et al 1994). Recently, extensive research efforts have been spent on dealing with the complicated problems of diagnosis, assessment and repair of distressed reinforced concrete buildings (Sherif et al 1996, Sabouni 1994).

The durability of reinforced concrete and the factors involved in various aspects of maintenance and repair are of concern in many parts of the world (Campbell-Allen & Roper 1991). These matters are of particular importance in the Arabian Gulf region because of the difficulties encountered in the use of reinforced concrete in the hot, arid and salt-laden environment of the Gulf coast of the Arabian Peninsula. In such conditions, poor quality concrete deteriorates rapidly.

Precautions based on standard recommendations for good concreting practices, and general guidelines for improved concrete durability can substantially improve the situation. Nevertheless, there are always some practices that are uniquely governed by local conditions and need to be based on local expertise (Sabouni 1994). Developments in the current research are intended to contribute to the efforts of documentation and dissemination of knowledge and experience concerned with improving durability of concrete structures. This paper is aimed at defining the main considerations need to be taken into account in the development of a prototype knowledge-based expert system advisor for the deterioration of reinforced concrete structures (DETEREX).

2 DETERIORATION OF REINFORCED CONCRETE STRUCTURES

Historically, codes of practice in structural concrete have put much greater emphasis on the mechanics of the design process (which is usually based on material strength), as opposed to durability-related issues (Campbell-Allen & Roper 1991). It is only very recently that codes of practice have paid any serious attention to aspects of durability. ACI 318-89 code included a new chapter on durability requirements (ACI-318 1989). Although no one expects that mechanical systems of a building for example will serve indefinitely without regular inspection and maintenance, there is still a widespread misconception among practicing engineers and construction professionals that concrete structures may last forever without any attention.

Evaluation of reinforced concrete buildings showing distress requires a great deal of engineering judgment (Emmons 1993). Procedures that can be used to evaluate concrete structures are outlined in the ACI committee 364 report for evaluation of concrete structures prior to rehabilitation (ACI-364 1993). The report presented a series of recommended guidelines, based on experience drawn from "existing resources" and "past investigations". However, the procedures are not intended to replace engineering judgment. It is extremely difficult to standardize the evaluation process into well defined steps because the number and type of steps vary depending on the type and physical condition of the structure, the strength and quality of the materials, the completeness of the available design and construction documents, and the purpose of the investigation. Based on the evaluation of the structure, repair strategies are determined (Sabouni 1994). In order to establish the knowledge need to be incorporated in the expert system development, the defects in deteriorated reinforced structures have to be defined and classified. The main types of defects in reinforced concrete structures that are included in the prototype knowledge base follow the classification provided by ACI special publications (ACI-201 1984, ACI 1996).

These defects are grouped in three categories. The first group is the cracking defects. This group includes defects that represent incomplete separation in the body of the concrete structure with or without gaps. Cracks are classified according to appearance, direction, width, depth, pattern, spacing, and randomness. The textural defects group includes localized channels, streaks and voids in the concrete texture due to bleeding and separation of the constituents of the concrete mix. Defects in reinforced concrete other than cracking and textural defects are grouped under deterioration defects. This group includes a wide spectrum of defects reflecting adverse changes of normal mechanical, physical and chemical properties either on the surface or in the whole body of concrete generally through separation of its components.

3 CONDITION ASSESSMENT OF THE STRUCTURE

Frequent and well planned inspections are the most effective means to reducing maintenance costs. Such inspections monitor any defects, and help in making decisions on when repair becomes economical (ACI-364 1993). But how frequent do routine inspections need to be carried out? The answer depends on the type of structure and the environmental conditions. Inspections of reinforced concrete buildings can be divided into two types 'routine' inspections, and 'extended' inspections (Campbell-Allen & Roper 1991).

To achieve an effective and lasting repair, the nature and extent of the problem must first be thoroughly investigated. A structured program of testing is the most reliable approach of ensuring proper diagnosis. In most cases, it is recommended to have a suitably qualified and independent testing authority to apply the appropriate tests. The tests should be capable of handling and determining the extents of *External behavior* (which includes cracks, deformation, and bleeding), and *Internal behavior* (which includes delaminations, steel reinforcement, strength tests, carbonation depth, and chloride content).

4 EXPERT SYSTEMS TECHNIQUES

Knowledge Based Expert Systems (KBES) are computer programs that employ artificial intelligence principles in trying to emulate to a certain extent human expertise in a specific domain of knowledge (Durkin 1994). Many expert systems have been reported in the literature for various structural engineering applications (Topping 1991, Sabouni 1995, Sabouni & Al-Mourad 1997). Early developments of expert systems in structural engineering employed standard programming languages for system implementation, while more recent systems employed domain-independent programming tools for expert systems development, referred to as expert system shells.

4.1 Main components of a KBES

The basic architecture of an expert system consists of the main components shown in Figure 1. These components are: a) a knowledge base, b) an inference mechanism, c) a user interface, and sometimes d) a data base. The knowledge base consists of a set of rules and facts in the field of specialization of the program being developed. The inference mechanism (or inference engine) is considered as the brain of the expert system since it controls the procedure in which the rules of the knowledge-base are examined and the search process for a conclusion is directed. Knowledge base rules in the prototype expert system advisor DETEREX is based mainly on textual knowledge published in the literature, in addition to a collection of heuristics and rules of thumbs accumulated from previous experience and contacts with experts and practitioners.

4.2 The expert system shell

The expert system shell used in the development of the program DETEREX is called EXSYS Professional (EXSYS 1993). This shell has been chosen for its versatility and ability to develop complicated expert systems. It works under various operating systems including VAX/VMS, UNIX, Macintosh, DOS, and Windows. It provides convenient developer environment and allows for graphical user interface. Reasoning is carried out through backward chaining, forward chaining, or a combination of both. There are five different ways of assigning confidence values in the knowledge base rules (EXSYS 1993).

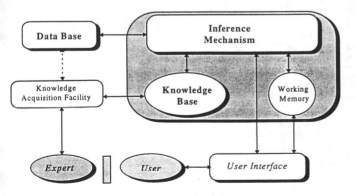

Figure 1. Components of a knowledge-based expert system

5 DEVELOPMENT OF THE PROTOTYPE

Experts rely mostly on "instinct" in providing judgment needed for decision making without consciously referring to the knowledge they have accumulated. Knowledge-based expert systems play an important role in the documentation of such experienced knowledge and ultimately imparting the expertise to junior engineers. To some extent, expert systems may substitute the lack of human expertise in a particular domain of knowledge.

Based on classification of the deterioration types in reinforced concrete structures the list of choices that constitutes the main output of the prototype advisory expert system DETEREX are established.

Development of the current prototype expert system is carried out in three phases: the system analysis and knowledge acquisition phase, the phase of building the knowledge base, and the system implementation and verification phase. Each phase consists of various steps intended to achieve specific goals as explained below. Fig. 2 illustrates the general development scheme of the expert system DETEREX.

5.1. System Analysis and knowledge acquisition

This is the fundamental phase for development of any knowledge-based expert system regardless of the developing media. The importance of this step is that it represents the basis for creation of the decisions that the expert system should make. In the development of the expert system DETEREX, the objective is to provide expert advice for the diagnosis of deterioration in reinforced concrete structures.

The value of any expert system comes from its knowledge base. The main sources of knowledge for DETEREX are published literature (textual) and human experts. Acquiring

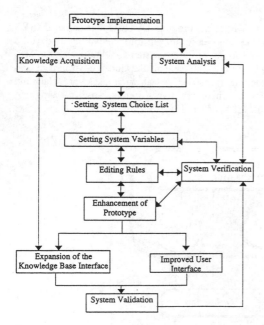

Figure 2. Development scheme for the expert system DETEREX.

textual knowledge is easier than acquiring knowledge from human expert. The main method of knowledge acquisition from human experts is consulting sessions on how to diagnose various defects and what are the factors affecting the diagnosis. The aim of system analysis is to establish the reasoning process of the system that best represents human knowledge and gives most accurate results. This task is the core of the knowledge base and is accomplished through knowledge acquisition from various sources. In this step it is essential to establish the relations between factors and decisions and to define the importance (weight) of every relation and/or factor.

5.2 Building the knowledge base

Selection of the development tool is the first step in this phase, and it controls the subsequent implementation process.

5.2.1 Selection of development tool

There are two main programming options for development of expert systems: programming languages and development shells. Programming languages are generally more flexible and powerful but knowing and mastering the language is a must.

Expert system development shells are easier in development since they provide the developer with tools such as knowledge acquisition facility and tools to enhance the user interface and interaction such as explanation facility. Because of the nature of deterioration diagnosis problems, they have been implemented in DETEREX through rules in the knowledge base representing the relations and importance of these problems.

5.2.2. Building up the rules

This step is bounded by the abilities and limitations of the development tool which controls also the procedure followed in development. In the expert system shell EXSYS the first step is to identify the decisions or choices that the system should make. Diagnosis of defects is based on identifying the types of the distress. These types are set as the main decisions the expert system should make.

The next step is to setup the variables that represent factors affecting and leading to the decisions. When dealing with the diagnosis of deterioration there are many factors to be considered such as the type of the structural member having the defect, location of the defect, time of formation, age of structure, and visual features. The relations between these factors and the weight of each relation is set through development of the rules. However it is very important to select the weighting system (confidence mode) prior to editing the rules, (EXSYS 1993)

5.3. System Implementation and Verification

The aim of this development phase is to check the prototype logic and components for any missing information, and to pinpoint any possible bugs in the prototype. It is very important in the development of expert systems that each module is checked thoroughly before proceeding to the next one. The decision making logic may rapidly become so complicated and involved for debugging. After editing each group of rules they can be checked for correctness and completeness.

6 SUMMARY AND CONCLUSIONS

In order to ensure serviceability of a concrete structure and to enhance its useful life, systematic inspection and maintenance plans are indispensable. The heuristic nature of the problem indicates that knowledge-based expert systems are very valuable to the quality of judgment to be used in selecting appropriate maintenance and repair schemes. Expert systems offer the potential for automating decision support systems in some structural engineering applications which rely heavily on accumulated knowledge, established rules of thumb, and professional judgment. This paper has described the operation of a proposed expert system for the diagnosis of distressed reinforced concrete buildings. Important parameters affecting the durability of reinforced concrete structures were presented, classified, correlated and transformed into knowledge base rules.

ACKNOWLEDGMENT: This paper is part of a research project funded by a grant from the UAE University Research Council.

REFERENCES

ACI 1996. *Manual of Concrete Practice.* American Concrete Institute. USA.

ACI-201 1984. *Guide to Making Condition Survey of Concrete in Service.* (ACI-201.1R-68 Committee report, revised 1984). American Concrete Institute. USA.

ACI-318 1989. *Building Code Requirements for Structural Concrete.* (ACI 318-98) and Commentary (ACI 318R-98). American Concrete Institute. USA.

ACI-364 1993. ACI-364.1R-93 Committee Report: How to Evaluate Concrete Structures, *Concrete International,* Vol. 15, No. 9, September, pp. 56-57.

Alshamsi, A. M., Sabouni, A. R., Alhosani, K. I., and Bushlaibi, A. H. (ed.) 1994. *Reinforced Concrete Materials in Hot Climates,* Proceedings of the First International Conference on Reinforced Concrete Materials in Hot Climates. Al-Ain, UAE. UAE University and the American Concrete Institute (ACI).

Campbell-Allen, D., and Roper, H. 1991. *Concrete Structures: Materials, Maintenance, and Repair.* UK. Longman Scientific & Technical.

Durkin, J. 1994. *Expert Systems, Design and Development.* UK: Printice Hall.

Emmons, P. H. 1993. *Concrete Repair and Maintenance Illustrated,* R. S. Means Company Inc., USA, 295 pp.

EXSYS 1993. *Expert System Development Software, Reference Manual.* EXSYS Inc., Albuqureque, NM, 1993, USA.

Sabouni, A. R. 1994. Proposed Operation of an Expert System for Diagnosis and Repair of Cracking in Reinforced Concrete Buildings in Hot Climates, *Proceedings of the First International Conference on Reinforced Concrete Materials in Hot Climates,* Vol. 2, UAE University and the American Concrete Institute (ACI), Al-Ain, UAE, April, pp. 721-730.

Sabouni, A. R. 1995. An Expert System for the Preliminary Design of Earthquake Resistant Buildings, *Proceedings of the Second Congress on Computing in Civil Engineering,* Atlanta, Georgia, USA, June 5-7, pp. 1188-1195.

Sabouni, A. R., and Al-Mourad, O. M. 1997. A Quantitative Knowledge Based Approach for the Preliminary Design of Tall Buildings, *Journal of Artificial Intelligence in Engineering,* Elsevier Applied Sciences, UK, Vol. 11, No. 4, April, pp. 143-154.

Sherif, A. M., Sabouni, A. R., Berrais, A., and Alshamsi, A. M. 1996. Advisory Graphical System for Crack Diagnosis and Assessment, *Proceedings of the Second International Conference in Civil Engineering on Computer Applications, Research and Practice,* Bahrain University, *Practice,* April 6-8, Bahrain. Bahrain University. pp. 31-40.

Topping, B. H. V. (Editor) 1991. *Artificial Intelligence Techniques and Applications for Civil and Structural Engineering,* CIVIL-COMP Press, U. K., 319 pp. 47-56.

Environmental loading, testing and response (dynamics)

The Mechanics of Structures and Materials, Grzebieta, Al-Mahaidi & Wilson (eds)
© 1997 Balkema, Rotterdam, ISBN 90 5410 900 9

Reduced analysis of pin-supported beams under blast loading

M. Boutros
Department of Civil Engineering, University of Western Australia, Perth, W.A., Australia

ABSTRACT: An elastic-plastic formulation of simply supported beams with end membrane restraints subjected to uniformly distributed loading is developed. Flexibility and mass are distributed along the beam axis. Instantaneous plastification of the section at midspan due to bending and axial actions is considered. The deflected shape of the beam satisfies the boundary conditions at the supports and midspan. The equations of motion are derived by virtual work. These equations are reduced by relating the variations of the elastic and plastic deformations. The analysis using Lagrange equation of motion is compared with this analysis showing that the standard form of the Lagrange equation is energy dissipative for displacement dependent shape functions. This model is better suited for this analysis than earlier models as it takes into account elastic deformations and their effect on the stretch force and the distribution of inertia forces.

1 INTRODUCTION

The effect of short duration pressure loading on beams was analysed using several approximations. Symonds and Jones (1972) developed models for the behaviour of beams by considering the interaction of bending and membrane actions on rigid-perfectly plastic beams. They justified neglecting the elastic deformations by the fact that the magnitude of permanent deformations was large compared to the depth of the beam and thus to the elastic deflections. They also identified the effects of strain sensitivity on the results of their models. Yankelevsky and Boymel (1984) included elastic deformations in a lumped form by analysing beams composed of rigid segments connected by elastic-plastic hinges at the plastic points.

In this paper, a simply supported beam with horizontal support restraints and distributed flexibility is analysed. Elastic and plastic strain rate sensitivities are considered. The results of this model show that the deformations of the elastic region of the beam are significant enough to affect the distribution of inertia forces. These, in turn, determine the response of the beam.

In the following section, the analytical model is presented. Section 3 is an assessment of the accuracy of the time integration of the equations of motion on the basis of conservation of energy and an application of the model. Finally, some general conclusions are drawn.

2 ANALYTICAL MODEL

The case studied here is shown in figure 1. The beam is uniformly loaded by the blast pressure wave. In the horizontal direction, it is restrained at the pinned ends. The transverse deflection

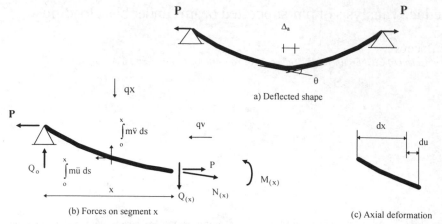

a) Deflected shape

(b) Forces on segment x

(c) Axial deformation

Figure 1: Membrane force and deformation

of the axis of the beam was derived in terms of the generalised variables Δ_e, θ and Δ_a where Δ_e is an elasticly varying midspan deflection, θ is the plastic kink at midspan and Δ_a is the plastic axial deformation at midspan (Figure 1).

2.1 Internal actions and deformations

The longitudinal movement, u, and the support reaction, P, in the x-direction were considered to be in phase with transverse deflection. They were determined by the equilibrium of forces for a given configuration Δ_e, θ and Δ_a as shown in figure 1.

In the case of blast loading, the pressure q is of such a short duration that it fades out before any significant deformations develop. Therefore, its follower (horizontal) component was neglected. For moderate deformations, the slope of the axis is small and the inertia forces in the horizontal direction are negligible compared to those in the vertical direction. Hence, the former were neglected in this analysis. From equilibrium of an element dx (Figure 1(c)), the horizontal component P, the vertical component Q of the internal forces and the axial force N along the deflected axis are:

$$P = \text{constant}, \quad Q_{(x)} = Pv' + \frac{dM}{dx} \quad \text{and} \quad N_{(x)} = \frac{P + Q_{(x)}v'}{\sqrt{1 + v'^2}} \tag{1}$$

where a prime denotes a derivative w.r.to x. The movement in the x-direction is (Figure 1(c)):

$$\frac{du}{dx} = \frac{1}{\sqrt{1 + v'^2}}\left(1 + \frac{N_{(x)}}{EA}\right) - 1 \tag{2}$$

P and $u_{(x)}$ were determined by applying the boundary condition:

$$\int_0^{L/2} \frac{du}{dx}dx = -\frac{\Delta_a}{2} \tag{3}$$

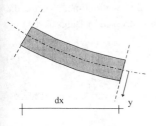

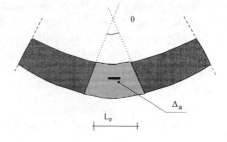

a) Elastic deformations   (b) Plastic deformations

Figure 2. Strains along the beam axis

The strains were expressed as elastic and plastic strains, ε_e and ε_p, from figure 2. as:

$$\varepsilon_e = \frac{N}{EA} - yv'' \qquad \text{and} \qquad \varepsilon_p = \frac{1}{L_p}[\Delta_a + \theta y] \qquad (4)$$

where L_p, the plastic length, was taken as $L/20$.

2.2 Deflection function and equation of motion

The deflected shape of the beam axis was defined in terms of the three degrees of freedom Δ_e, θ and Δ_a for $0 \le x \le \frac{1}{2}$ by:

$$v = \operatorname{Sin}\frac{\pi x}{L}\Delta_e + \frac{\operatorname{Sinh}(kx)}{2k\operatorname{Cosh}\left(\dfrac{kL}{2}\right)}\theta \qquad (5)$$

where $k = \sqrt{\frac{P}{EI}}$. The first term is the small deflection natural shape of a beam without axial restraint (Biggs, 1964). The second term satisfies the support conditions and symmetry whereby the vertical component of the internal forces, Q in equation 1, vanishes at midspan.

The yield function, $G_{(P,M)}=0$ (where P and M are the axial force and bending moment at midspan, respectively, Symonds and Jones, 1972), defines the relation between the variations of Δ_e and θ; and the associative flow rule (Hill, 1956) defines the relation between the variations of θ and Δ_a:

$$\delta\theta = \rho\,\delta\Delta_e = -\frac{G_{,\Delta_e}}{G_{,\theta}}\delta\Delta_e \qquad \text{and} \qquad \delta\Delta_a = \lambda\,\delta\theta = \frac{G_{,P}}{G_{,M}}\delta\theta \qquad (6)$$

where the coma denotes a partial derivative w.r.to the following variable. In the elastic stage, both ρ and λ are zero. Using the common operator:

$$\frac{\partial}{\partial q}\delta q = \left[\frac{\partial}{\partial\Delta_e} + \rho\left(\frac{\partial}{\partial\theta} + \lambda\frac{\partial}{\partial\Delta_a}\right)\right]\delta\Delta_e \qquad (7)$$

261

where q is a generalised coordinate, the particle velocity and acceleration are expressed in terms of Δ_e, θ, Δ_a, $\dot{\Delta}_e$ and $\ddot{\Delta}_e$ where a dot denotes a derivative w.r.to time. Using the principle of virtual work to find the equilibrium configuration as expressed by Biggs (1964),

$$\oint m\ddot{v}\frac{\partial v}{\partial q}d\Gamma + \oint \sigma \, \delta\varepsilon \, d\Gamma = \frac{\partial W_{nc}}{\partial q} \tag{8}$$

where m is the mass per unit volume, W_{nc} is the work done by non-conservative external forces, Γ is the volume and σ is the axial stress which is defined as:

$$\sigma = E\varepsilon_e + C_e\dot{\varepsilon}_e + C_p\dot{\varepsilon}_p \tag{9}$$

where E is the modulus of elasticity, $\dot{\varepsilon}_e$ and $\dot{\varepsilon}_p$ are the elastic and plastic strain rates and C_e and C_p are the elastic and plastic strain rate moduli. They represent the material viscous damping (Clough and Penzien, 1975). In this analysis, C_e and C_p were taken as proportions of the elastic small deflection critical damping. The material was considered to be elastic-perfectly plastic.

In the special case where:

$$\frac{\partial v}{\partial q} = \frac{\partial \dot{v}}{\partial \dot{q}} \tag{10}$$

equation of motion 10 reduces to Lagrange's equation (Biggs, 1964):

$$\frac{\partial}{\partial t}\left(\frac{\partial T}{\partial \dot{q}}\right) - \frac{\partial T}{\partial q} + \frac{\partial V}{\partial q} = \frac{\partial W_{nc}}{\partial q} \tag{11}$$

where T is the kinetic energy and V is the potential energy. The condition in equation 10 is not satisfied in the case studied here because the shape function $\varphi_{(x,P)}$ is dependant on the stretch force P which is dependant on the displacements. Then,

$$\dot{\varphi}_{(x,P)} = \varphi_{,P}\,\dot{P} \neq 0 \tag{12}$$

From equation 8, a single equation of motion was expressed in terms of the three degrees of freedom in the form:

$$M\ddot{\Delta}_e + \left(C + T_1\dot{\Delta}_e\right)\dot{\Delta}_e + K_1\Delta_e + K_2\theta = f \tag{13}$$

The term T_1 results from the differentiation of v w.r.to time which yields the acceleration in the form:

$$\ddot{v} = A_a\ddot{\Delta}_e + B_a\dot{\Delta}_e^2 \tag{14}$$

where A_a and B_a are dependant on the displacements and their rates of variation. By linearising the yield constraint in the plastic stage to the form:

$$G = G_1 + G_{,\Delta_e} \Delta_e + G_{,\theta}\, \theta = 0 \tag{15}$$

the constrained problem was expressed in the form:

$$
\begin{bmatrix} M \\ 0 \end{bmatrix} \ddot{\Delta}_e + \begin{bmatrix} C + T_1\dot{\Delta}_e \\ 0 \end{bmatrix} \dot{\Delta}_e + \begin{bmatrix} K_1 & K_2 \\ G_{,\Delta_e} & G_{,\theta} \end{bmatrix} \begin{Bmatrix} \Delta_e \\ \theta \end{Bmatrix} = \begin{Bmatrix} f \\ -G_1 \end{Bmatrix} \tag{16}
$$

Equation 16 is a reduced form of the equations expressed in terms of the two variables Δ_e and θ and the yield constraint applied using Lagrange multiplier (Boutros, 1996), where the multiplier is eliminated by substitution.

Using Newmark's integration method with constant average acceleration (α=0.5 and β=0.25, Bathe, 1982) the velocity and acceleration were expressed in terms of the displacement Δ_e. Then, the set of equations were solved for Δ_e and θ.

3 ANALYSIS

In this section, sample cases are analysed. First, the accuracy of the governing equation is investigated for an undamped beam case. Then, a full analysis is performed.

A case of an undamped beam subjected to a triangular pulse was analysed using the two equations of motion (Equations 8 and 11) for comparison. The steel beam was 1000mm long, 10mm deep and 100mm wide. Its natural bending small deflection period was 0.0426sec. The pulse varied linearly from 0 to 50N/mm at 0.001 sec and back to 0 at 0.002sec. The total energy was evaluated by summing the kinetic and the strain energy components. The former was calculated explicitly. The latter was numerically integrated over the loading path for both strain and strain rate components of stresses. The integration time steps were 10^{-5} sec in the elastic stage and 10^{-7} sec in the plastic stage. The analysis was performed until the stretch P vanished as equation 5 is valid for tension only. Initial crookedness values of θ=10^{-7}radian and Δ_a=-10^{-7}mm and a yield stress of 100MPa were used in all cases.

The responses resulting from these analyses are shown in figure 3. Figure 3(a) shows that the total energy using equation 8 was unchanged while using equation 11 it showed a gradual drop as the midspan deflection increased. This was accentuated by yielding. An increase in θ alters the deflected shape significantly such that the error of the approximation in equation 11 is

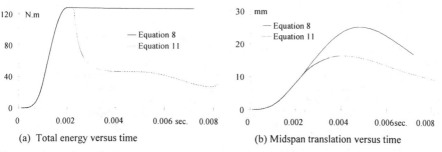

(a) Total energy versus time   (b) Midspan translation versus time

Figure 3. Comparison of responses using equations of motion 8 and 11 for an undamped beam.

263

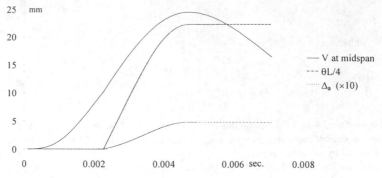

Figure 4. Response of a damped beam.

large. A total energy loss of about 75% appears in this analysis. Consequently, the elastic and plastic deformations were underestimated as shown in figure 3(b).

The analysis was performed for the same conditions as above with strain rate moduli C_e and C_p of 0.1% of the elastic small deflection critical damping. The results of the analysis are shown in figure 4. This last case was also stable and energy conservative. The drop in total energy over the response period was less than 0.4%.

4 CONCLUSIONS

A pinned beam with membrane restraint at the supports and a plastic hinge at midspan was analysed. The shape of the axis was derived by satisfying equilibrium and deformation conditions. The equations of motion in the plastic stage were reduced to a single equation. It was shown that the standard form of Lagrange's equation of motion did not yield accurate (energy conservative) results and that the equation of motion derived from equilibrium of forces by virtual work was an accurate form in this case where the shape of the axis was displacement dependant. A full analysis was also performed showing that the solution was rapidly converging and stable. Compared to previous models where the elastic deformations were neglected or lumped at the plastic-hinge location, this model is more accurate because it takes into account the effects of elastic deformations on the stretch and inertia forces.

REFERENCES

Bathe, K.-J. 1982. *Finite element procedures in engineering analysis.* New Jersey. Prentice-Hall.
Biggs, J.M. 1964. *Introduction to structural dynamics*, New York: McGraw-Hill.
Boutros, M.K. 1996. Elastic-plastic response of beams to blast loading. *Proceedings of the First Australasian Congress on Applied Mechanics.* The Institution of Engineers Australia. 1:29-34.
Clough, R.W. and Penzien, J. 1975. *Dynamics of structures.* New York. McGraw-Hill.
Hill, R. 1956. *The mathematical theory of plasticity.* Oxford at the Clarendon Press.
Symonds, P.S. and Jones, N. 1972. Impulsive Loading of Fully Clamped Beams with Finite Plastic Deflections and Strain Rate Sensitivity. *International Journal of Mechanical Science*, 14:49-69.
Yankelevsky, D.Z. and Boymel, A. 1984. Dynamic Elasto-Plastic Response of Beams - A New Model. *International Journal of Impact Engineering.* 2(4):285-298.

The Mechanics of Structures and Materials, Grzebieta, Al-Mahaidi & Wilson (eds)
© *1997 Balkema, Rotterdam, ISBN 90 5410 900 9*

Ambient vibration testing of bridge superstructures

N. Haritos
Department of Civil and Environmental Engineering, University of Melbourne, Parkville, Vic.,
Australia

D. N. Mai
Development Technologies Unit, University of Melbourne, Parkville, Vic., Australia

ABSTRACT: A span over the Goulburn River floodplains of McCoy's Bridge near Nathalia Victoria was dynamically tested using a shaker and natural excitation from traffic. Traffic records were grouped into "low" and "higher" levels of excitation (car and truck records) for analysis via the RANDEC method and a simplified NEXT interpretation of the Experimental Modal Analysis testing technique. Results for the estimates of natural frequency, damping level and corresponding mode shape from this analysis were then compared with those from the shaker testing and the predictions from a finite element model. The agreement obtained was considered to be reasonably good suggesting that the use of natural traffic excitation can be a cost-effective unobtrusive method for estimating the modal properties of short span bridges.

1 INTRODUCTION

Non-destructive evaluation (NDE) techniques for performing structural identification, integrity assessment and/or monitoring is receiving accelerated international interest from authorities concerned with maintaining aging infrastructure. VicRoads, the State of Victoria road and bridge authority has for some time now been collaboratively involved with the Department of Civil and Environmental Engineering at The University of Melbourne in the assessment of the relative merits of various dynamic testing techniques as structural identification tools as part of their corporate sponsored R & D project: "Load Capacity of In-service Bridges", (Haritos 1996).

Several bridges have now been tested as part of this program using various implementations of The University of Melbourne Experimental Modal Analysis (EMA) testing system. This system features a +/- 10 tonne servo-controlled actuator powered by a 120 litre/min hydraulic power pack. The actuator can operate in either direct "force" or "shaker/seismic" modes.

Whilst this system has demonstrated a capacity to perform controlled vibration testing measurable to a high degree of accuracy and yielding corresponding highly accurate estimates of the natural modes of vibration (mode shapes, natural frequencies and damping levels) of the principal modes of the bridges tested in the 0-50 Hz bandwidth, the system itself is rather bulky and obtrusive. Use of the system necessitates closure of the bridge under test for its assembly/ disassembly and during the actual performance of the dynamic testing and also when relocating and setting up the response measurement transducers (usually accelerometers) over the grid of measurements chosen. In situations where access is available from the underside of the bridge, it has been appreciated that *natural ambient traffic excitation* may offer an alternative *unobtrusive* means of performing a *simplified form of EMA testing* (generally referred to in the literature as a NEXT technique, (Farrar et. al. 1994)) in which *all* measurements of the dynamic response of the bridge superstructure are made from the underside of the bridge.

The first opportunity for using such a method was afforded to The University of Melbourne team on the La Trobe River Bridge near Rosedale, where a simplified interpretation of the Relative Response Functions (RRFs) in combination with the Random Decrement (RANDEC) method was used with natural traffic excitation to successfully obtain the estimates of several modes of the bridge span tested in the 0-50 Hz bandwidth, (Aberle 1996). The results obtained, although not as reliable as those from the use of controlled excitation, (performed on the same bridge span some two years earlier), were nonetheless encouraging and prompted the performance of similar testing on another bridge span, (McCoy's Bridge over the Goulburn River near Nathalia) in February, 1996, which was also tested using controlled excitation via the shaker.

This paper describes the dynamic testing performed on McCoy's Bridge and highlights the results obtained using the simplified NEXT approach comparing them with the results from the forced excitation testing series and Finite Element Modelling (FEM) predictions.

2 DESCRIPTION OF EMA TESTING

Experimental Modal Analysis (EMA) is the process by which the natural modes (mode shapes, frequencies and damping) of the bridge structure under test are identified from performance of a Modal Experiment. The Modal Experiment itself usually involves the simultaneous measurement of the dynamic response of the bridge over a predetermined grid of points to a controlled (measured) force input. The time traces of response and force measurement are transformed in the frequency domain to obtain estimates of the Frequency Response Functions (FRFs), $h_{jk}(\omega)$, via:

$$\tilde{h}_{jk}(\omega) = \frac{X_j(\omega)}{F_k(\omega)} \tag{1}$$

where $X_j(\omega)$ and $F_k(\omega)$ are the Fourier Transforms of the displacement response at point "j", given by $x_j(t)$, and the excitation force from the exciter located at position "k" and given by $f_k(t)$, respectively, (Ewins 1985).

The theoretical form of the FRF for the situation of eqn(1), $h_{jk}(\omega)$, is given by:

$$h_{jk}(\omega) = \sum_{n=1}^{N} \left(\frac{\varphi_{jn}\,\varphi_{kn}}{(i\omega - \lambda_n)} + \frac{\varphi^*_{jn}\varphi^*_{kn}}{(i\omega - \lambda^*_n)} \right) \tag{2}$$

in which φ_{jn} and φ_{kn} represent the jth and k^{th} elements of the complex eigenvector for the nth mode of vibration and λ_n is the complex eigenvalue for this mode. The complex eigenvalue λ_n in eqn. (2) is related to the damped and undamped circular frequencies for the nth mode, ω_{dn} and ω_d respectively, and the ratio to critical damping for that mode, ζ_n, via:

$$\lambda_n = -\zeta_n\omega_n + i\omega_{dn} \tag{3}$$

"Ensemble averaging" of several realisations of $h_{jk}(\omega)$ from repeated tests reduces the effect of measurement "noise" for subsequent processing by suitable EMA algorithms. The Direct Simultaneous Modal Analysis (DSMA) algorithm, developed by the University of Melbourne team (Chalko et al 1996), achieves "best" estimation of φ_n the complex eigenvectors (mode shapes) and λ_n the eigenvalues (natural frequencies and damping ratios) by minimising on the square of the difference between all of the measured forms of FRF and their corresponding theoretical counterparts via a non-linear least squares fitting procedure.

In practice, the number of grid points when testing far exceeds the number of transducers available for measuring the response, so that these transducers need be re-located a sufficient number of times to "cover" the required grid and the excitation repeated as necessary.

3 THE SIMPLIFIED NEXT TECHNIQUE

In this interpretation of the EMA testing technique, forcing occurs from "natural" sources (eg traffic) over the entire grid of points in which case the response measured at point "q" becomes:

$$x_q(\omega) = \sum_{k=1}^{N} h_{qk}(\omega)f_k(\omega) \tag{4}$$

Consequently, under the assumption that at a circular frequency ω corresponding to one of the natural frequencies ω_i, the Relative Response Function RRF, $R_{qo}(\omega_i)$, is dominated by the modal contribution of mode "i", then the amplitude of the mode at point "q" relative to reference point "o", becomes, (Haritos & Aberle 1985):

$$R_{qo}(\omega_i) = \frac{x_q(\omega_i)}{x_o(\omega_i)} \approx \frac{\Phi_{qi}}{\Phi_{oi}} \tag{5}$$

4 THE "RANDEC" METHOD

The Random Decrement (RANDEC) method can be applied to time domain traces of response measurements in order to obtain estimates of the natural frequency and damping value associated with the level of response chosen. The method essentially superposes a sequence of record portions of band-pass filtered response records taken from commencement of an upcrossing with corresponding lengths of these record portions associated with a downcrossing at the same level for a large number of such observations. The resultant trace can be shown to possess the properties of a typical "pluck" test, ie a decaying cosine wave with a frequency associated with the natural frequency of the mode that has specifically been band-pass filtered for this purpose and an amplitude decay rate associated with the applicable level of damping, (Aberle 1996). By choosing different levels of response, y*, for application of the RANDEC method, an investigation of the amplitude dependence (if any) in the damping can be made to ascertain to what extent the structural system under investigation exhibits non-linear features reflected in its damping levels.

5 EMA TESTING OF McCOY'S BRIDGE

Two spans (one over the floodplains and another over the river itself) of McCoy's Bridge over the Goulburn River on The Murray Valley Highway near Nathalia were dynamically tested in February 1996 using the University of Melbourne EMA testing system, (Haritos & Chalko 1996). The span over the floodplains (a simply supported reinforced concrete deck acting compositely with three steel support beams) was also investigated for its modal characteristics using natural traffic excitation as the source. A regular grid of 7 x 7 = 49 acceleration response points was chosen necessitating 4 location sequences of the 15 accelerometers used for performing response measurements three of which (identified as M, N and O) were fixed in position for all measurements with the remaining 12 (A to L) "roving".

A total of 289 sequences of records (approx. 70 per accelerometer group setting) of all 15 measurements (channels) consisting of 2048 data points per channel over a 16 second sampling period were captured. These acceleration response traces were then sorted into "car" and "truck" records, according to response level, and the relatively small number of records containing minor "faults", discarded from subsequent processing (NEXT and RANDEC analyses).

6 PRESENTATION OF RESULTS

A typical example of a car ($y* < 0.5$ m/s^2) and truck ($y* > 0.5$ m/s^2) accelerometer response trace is depicted in Fig. 1.

The autospectra of the acceleration response at each grid point were computed and averaged over all records separately for the car and truck record classification. These spectra were then summed over all grid points to obtain the "global" autospectrum under each classification.

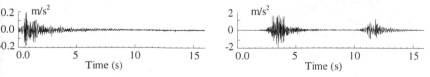

Figure 1. Typical traces for a single car and a truck (accelerometer "B")

(Figure 2 depicts the form of these autospectra in the case of the car and truck records separately). The locations of "peaks" in these spectra are considered to be indicative of locations of natural frequencies (modal frequencies) excited by these vehicle types. Whilst there appears to be several well defined peaks "common" to both types of vehicular excitation in the spectra of Fig. 2, beyond approximately 25 Hz this distinction is observed to be less clear.

Modal frequencies obtained from the car and truck record autospectra are compared with those obtained from the shaker test and from FEM modelling in Table 1. (An reduced effective thickness in flexure was assumed for the concrete deck in the FEM model). Modal Index values (equivalent to R^2 correlation) were calculated to compare the "match" between mode shapes obtained from the Simplified NEXT technique using car records and truck records separately. These values and associated comments thereon are presented in the last two columns of Table 1.

Modeshapes obtained from the Simplified NEXT technique for two selected modes are presented for comparison with the modeshapes obtained for corresponding modes from the shaker test and FEM modelling in Figures 3 and 4. It is clear from simple visual inspection that these modeshapes are highly correlated.

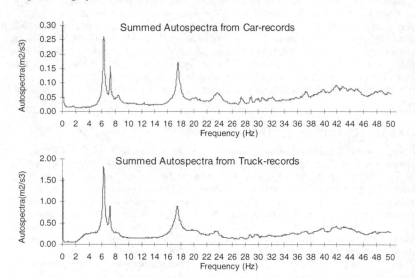

Figure 2. Assemblage of all autospectra over the measurement grid (car and truck records)

Table 1. Modal natural frequencies and comparison of modal parameters.
Modal frequencies (Hz)

Mode	FEM	Shaker	Traffic		Modal	Comments
			Car	Truck	Index	on "match"
1	6.12	6.11	6.312	6.248	0.998	excellent
2	6.76	6.84	7.312	7.251	0.945	very good
		7.76	8.388	8.445	0.968	excellent
		8.65				detected by shaker only
			12.377			detected by cars only
3	11.8	16.9	17.623	17.505	0.997	excellent
4	21.5	20.3	20.315	19.875	0.782	fair
5	21.6	18.7				detected by shaker only
6	24.4	23.3	23.625	23.685	0.953	excellent
7	26.4					predicted by FEM only
8	32.9	25.6	25.815	26.125	0.264	not acceptable
9	33.7	26.5	26.625	26.565	0.534	barely acceptable
		27.1	27.737	27.495	0.785	fair

RANDEC traces corresponding to mode#1, #4 and # 6 were evaluated for estimating damping corresponding to these modes at a number of response levels and for re-estimation of the modal frequencies. All 170 car records for accelerometer channel O and 120 truck records for accelerometer channel N were used to obtain the associated RANDEC traces for this purpose. Examples of these RANDEC traces for mode#1 corresponding to car (left) and truck records (right) are presented in Figure 5. The characteristic form for this trace commencing at the response level value chosen for each record type is clearly distinguished in these example records.

Values for the damping ratio and frequency of mode#1 estimated from the RANDEC traces at 3 different levels of response in each of the vehicular record types are provided in Table 2. A mild dependence of damping value with amplitude of response is clearly evident in the results presented suggesting that the bridge exhibits mildly non-linear characteristics in its response.

RANDEC traces for the three modes most strongly excited by traffic were used to estimate damping and natural frequencies. These estimates are compared with the results of the shaker testing (Haritos and Chalko, 1996) in Table 3. It is observed from this table, that the RANDEC estimates for damping were consistently lower than those observed in the shaker testing whereas the agreement between frequency estimates from these different testing techniques was much closer.

Table 2. Estimates for modal frequencies and damping for mode 1 at different response levels.

	Car			Truck		
Level (m/s^2)	0.1	0.2	0.35	0.4	0.8	0.11
Frequencies (Hz)	6.251	6.25	6.246	6.173	6.21	6.23
Damping %	0.98	1.24	1.38	1.43	1.53	1.57

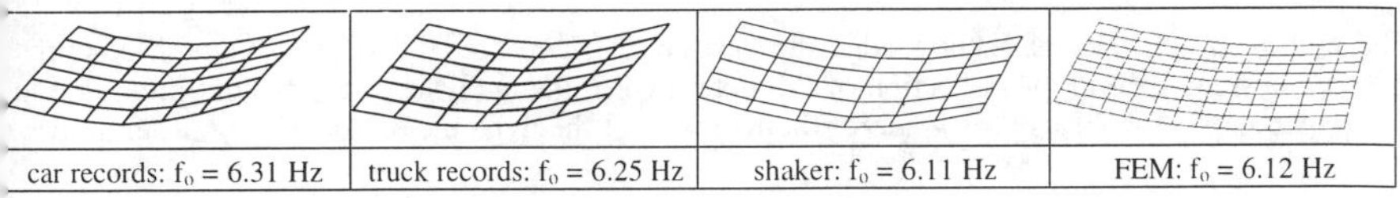

car records: f$_0$ = 6.31 Hz	truck records: f$_0$ = 6.25 Hz	shaker: f$_0$ = 6.11 Hz	FEM: f$_0$ = 6.12 Hz

Figure 3. Mode 1: NEXT: car and truck (left pair), shaker and FEM (right pair)

car records: f$_0$ = 23.6 Hz	truck records: f$_0$ = 23.7 Hz	shaker: f$_0$ = 23.3 Hz	FEM: f$_0$ = 24.4 Hz

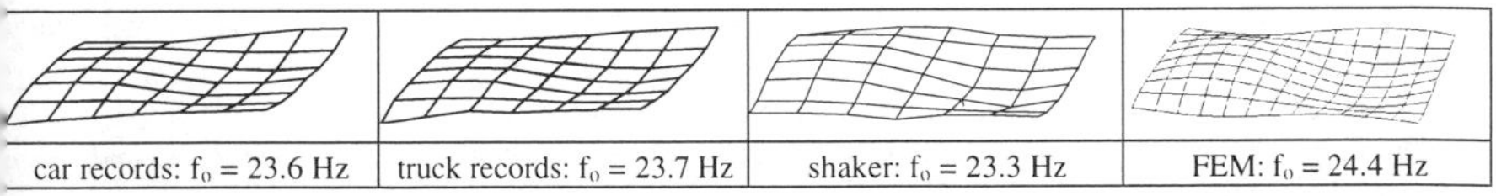

Figure 4. Mode 6: NEXT: car and truck (left pair), shaker and FEM (right pair)

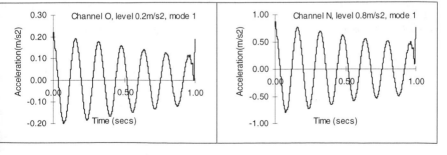

Figure 5. RANDEC traces derived from car (left) and truck records (right) for mode 1.

269

Table 3. Comparison between shaker and traffic tests

		Shaker	Traffic Car @0.2 m/s^2	Truck @0.8 m/s^2
Mode 1	Frequencies (Hz)	6.11	6.25	6.21
	Damping %	3.17	1.24	1.53
Mode 2	Frequencies (Hz)	16.88	17.7	17.5
	Damping %	2.87	1.42	2.31
Mode 3	Frequencies (Hz)	23.29	23.15	22.7
	Damping %	3.14	1.48	1.15

7 CONCLUDING REMARKS

A Simplified EMA approach using natural traffic excitation on a span of McCoy's Bridge over the Goulburn River floodplains has been implemented on acceleration response data records captured over a regular 7x7 grid of points on the underside of the bridge deck. Records were classified according to response level into two categories: "cars" and "trucks".

The Simplified EMA results obtained for mode shapes and natural frequencies from this data were largely in agreement with the "shaker" test results of traditional EMA testing obtained independently for the same test span. An RANDEC analysis at several response levels of both the car and truck records suggested a "mild" amplitude dependence of damping at a significantly lower level than obtained from traditional EMA testing in a number of modes. This would suggest a "mild" non-linearity in the composite action of the bridge deck under cyclic action and that the shaker itself may enhance the value of implied damping when it is used in EMA testing.

ACKNOWLEDGMENTS

The authors would like to acknowledge the assistance offered by the University of Melbourne dynamic testing team in the collection of the data used in the analysis of the results presented in this paper and to VicRoads personnel for their support of the dynamic bridge testing programme.

REFERENCES

Aberle, M. 1996. *Using traffic excitation to establish modal properties of bridges*, M.Eng.Sci. thesis, University of Melbourne.

Chalko, T., V. Gershkovich & N. Haritos 1996. The Direct Simultaneous Modal Approximation method, *Proc. 14th Intl. Modal Analysis Conf.*, Dearbourn, Michigan, February, 1130-1136, SEM.

Ewins, D. J. 1985. *Modal Testing: Theory and Practice*, New York: John Wiley.

Farrar, et. al. 1994. *Dynamic characterization and damage detection in the I-40 Bridge over the Rio Grande*, Los Alamos, New Mexico: Los Alamos National Lab.

Haritos, N. 1996. Application of EMA testing to the identification of the in-service condition of bridge superstructures, *Proceedings of Asia-Pacific Symposium on Bridge Loading and Fatigue*, 91-100, Dec., Monash University.

Haritos, N. & M. Aberle 1995. Using traffic excitation to establish modal properties of bridges, Proc. MODSIM '95 Conf., Nov., Newcastle, Aust., 1:243-248.

Haritos, N. & T. Chalko 1996. *Dynamic Testing of Three Bridges*, VicRoads Report: UNIMELB.

The Mechanics of Structures and Materials, Grzebieta, Al-Mahaidi & Wilson (eds)
© *1997 Balkema, Rotterdam, ISBN 90 5410 900 9*

The in-line and transverse force characteristics of slender surface-piercing cylinders

N. Haritos & D. Smith
Department of Civil and Environmental Engineering, University of Melbourne, Vic., Australia

ABSTRACT: Wave tank studies have been performed on the influence of proximity and placement orientation relative to wave direction on the forcing experienced by each member of a two cylinder group of nominally identical bottom-pivoted surface-piercing test cylinders under a variety of incident wave conditions. Results for the in-line force characteristics of these influences have been reported via the observed variation in the Morison force coefficients fitted to the measured cylinder tip restraint forces. Variations on cylinder forcing in the transverse direction have also been noted. Results obtained indicate that the influence on wave force characteristics is particularly significant when the relative cylinder spacing is less than approximately two cylinder diameters and that pattern orientation is also important in determining these force characteristics.

1 INTRODUCTION

The in-line force per unit length at position z from the Mean Water Level (MWL), $f(z,t)$, experienced by an isolated rigid vertical surface-piercing cylinder under separated flow conditions from uni-directional gravity waves can be described by the Morison equation, (Morison et al 1950), viz:

$$f(z,t) = C_M f_I(z,t) + C_D f_D(z,t) \tag{1}$$

where:
$$f_I(z,t) = \frac{\pi}{4}\rho D^2 \dot{u}(z,t) \tag{2}$$

and
$$f_D(z,t) = \frac{1}{2}\rho D|u(z,t)|u(z,t) \tag{3}$$

in which ρ represents fluid density, D the diameter of the cylinder and $u(z,t)$ and its time derivative, $\dot{u}(z,t)$, represent the alongwave water particle velocity and acceleration at the cylinder centreline position produced by the wave concerned in the absence of the cylinder. The so-called Morison force coefficients in eqn (1) for inertia and drag, C_M and C_D respectively, are found to be dependent upon the flow conditions as described by the Keulegan Carpenter number, KC, (= $u_m T/D$, where u_m is the peak alongwave water particle velocity for the wave with period T), and the traditional Reynold's number, R_N, (Sarpkaya & Isaacson 1981).

The situation of the forcing experienced by isolated vertical surface-piercing cylinders in the transverse direction under the action of uni-directional waves can be a great deal more complex. Flow separation can lead to the formation of a number of vortices which impart transverse force to the cylinder concerned as they "shed". The peak value of this force transverse force, $f_L(z)_m$, is described using the empirically determined lift force coefficient, C_L, via the equation

$$f_L(z)_m = \frac{1}{2}\rho C_L Du_m^2(z,t) \tag{4}$$

In the case of vertical surface-piercing cylinder groups, ie when one or more additional such cylinders are located relatively closely one to the other to form a pattern (such as to be found in riser systems in offshore oil rig applications), the in-line and transverse force characteristics of each cylinder in the group will be influenced to some degree by its neighbouring cylinders, and this influence can be gauged by determining the empirical force coefficients, C_M, C_D and C_L for these conditions and comparing them against corresponding values for the same wave conditions for the isolated cylinder case.

The above approach has been adopted by the authors in their wave tank studies performed in the Michell laboratory of The University of Melbourne on the simplest grouping condition, that of the two cylinder group, concentrating on the in-line, transverse and 45° cylinder orientations relative to incident wave direction, and using a variety of regular and irregular wave conditions, (Haritos & Smith 1995, 1996; Smith & Haritos 1996).

This paper summarises some of the key results from this extensive experimental program and highlights some of the relative features observed in the coincident transverse and in-line force characteristics of the cylinder pairs posed by the "group" effect.

2 DESCRIPTION OF EXPERIMENTS

2.1 General

A pilot study in the "inertia dominant" to onset of the "drag force significant" Keulegan-Carpenter number range (KC~5 to KC~10), followed by a main series of tests extending from the onset of the "troublesome" regime to well within "drag force significant" conditions (KC~5 to KC~20) were performed on model two-cylinder group pairs in the Michell wave tank, (Smith 1997). Two sizes of test cylinder (external diameters of 16.0mm and 25.0mm respectively), were utilised in the program, both pairs being constructed from aluminium tube which was powder coated to retain a smooth exterior surface. Values for the inter-cylinder spacing condition parameter S (= centre-to-centre spacing relative to cylinder diameter) chosen for testing were: S = 1.125, 1.25, 1.5, 1.75, 2.0, 2.5, 3.0 and ~17.0, the last value of which, for all intensive purposes, corresponds to "isolated" cylinder conditions. Most emphasis was placed in the testing upon the orientation of the cylinder group in-line and transverse to the direction of motion of the generated waves. (These settings extended the conditions tested by He (1993) in earlier studies).

Regular waves were generated in the main series of tests for a range of wave amplitudes at frequencies of 0.4Hz to 1.8Hz in 0.1Hz increments and at peak frequencies of 0.25Hz, 0.3Hz, 0.35Hz and 0.4Hz for the irregular waves. Due to the large difference in the magnitude of the forces exerted on the cylinders over the range of wave heights utilised in the main test series, experimentation was conducted in "high" (>70mm) and "low" (<70mm) amplitude wave groupings with cylinder tip restraint force transducers suitably ranged and calibrated to suit each.

2.2 Experimental Rig - description and measurement principle

Figure 1 depicts a schematic of the experimental rig designed for the test program illustrating the measurement principle adopted for each cylinder in the two cylinder group configuration tested in the series. The cylinder bottom-pivot joint was especially designed to achieve very low friction. Cantilever force transducers with sensitivities designed to accommodate the force range from each wave amplitude grouping were oriented to provide support at the tip of each cylinder in the pair in orthogonal directions via rigid connecting rods. Wave probes were located adjacent to each test cylinder to measure the surface elevation traces describing each wave test.

The total alongwave cylinder restraint force for the bottom-pivoted cylinder, $F_R(t)$, as depicted in Fig. 1, can be obtained from integration of the Morison force equation for wave surface elevation profile $\eta(t)$ measured alongside the cylinder and is given by:

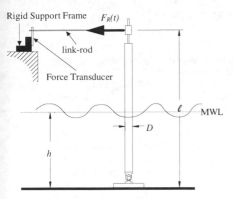

Figure 1 Schematic of experimental rig

$$F_R(t) \approx \int_{-h}^{\eta(t)} f(z,t)\left(\frac{h+z}{\ell}\right)dz \qquad\qquad (5)$$

$$= \int_{-h}^{\eta(t)} \left[C_M f_I(z,t) + C_D f_D(z,t)\right]\left(\frac{h+z}{\ell}\right)dz \qquad (6)$$

$$= C_M F_I(t) + C_D F_D(t) \qquad\qquad (7)$$

where $F_I(t)$ and $F_D(t)$ are interpreted as the inertia and drag force contributions towards $F_R(t)$ per unit C_M and C_D respectively. Throughout these experiments, $h = 1$m and $\ell = 2.3$m.

The wave particle kinematics $u(z,t)$ and $\dot{u}(z,t)$ necessary to the evaluation of $F_I(t)$ and $F_D(t)$ in eqn (7) via use of eqns (2) and (3) need be inferred from the water surface elevation profile $\eta(t)$ and a suitable wave theory. Linear (or Airy) wave theory has been found to give reasonable results for laboratory situations is easy to apply and hence widely used. However, Fenton's stream function theory (Rienecker & Fenton 1981) is an alternative, non-linear theory that is more accurate and directly applicable to situations when the wave amplitude to water depth ratio, a/h, is significant or for steep waves and shallow wave conditions. Both theories were adopted in the performance of the studies being reported herein.

2.3 Data Acquisition and Analysis

Digital recording and control of wave generation allowed reproduction of wave records and simultaneous measurement of the wave surface elevation and cylinder forces via an XT-IBM compatible computer with an internal data card with 1 channel digital-to-analog output and 16 channel analog-to-digital input capabilities. The data acquisition and control software in this system uses pre-recorded digital wave records to produce the analog electrical output to control the Salter duck style wave generator at an operating frequency of 100 Hz whilst capturing data over all measurement channels at a frequency of 20 Hz, digitally.

Test waves were generated for 50 seconds to exclude the effect of their reflections from the beach at the far end of the wave tank. However, data acquisition of all measurements began immediately at the start of wave generation and for 100 seconds thereafter. This allowed observation of the signals from all measurement transducers for the still conditions prior to hydrodynamic forcing of the cylinders and capture of the effects of reflected waves after wave generation had ceased. Software especially developed for the purpose, then extracted the useable 50 second imbedded portions of data for subsequent analysis from these records.

273

3 PRESENTATION AND DISCUSSION OF RESULTS

3.1 *Summary of method of analysis*

A suite of Fortran programs were purpose written to undertake processing and analysis of the experimental data. These programs performed three functions:

i. PRE-PROCESSING - locate and extract the 50 second long record of wave activity from the 100 second long data file after band-pass filtering (between 0.2 Hz and 6 Hz) and removal of any "spikes" and convert each data channel to engineering units.

ii. ANALYSIS - perform a wave-by-wave capture on the measured in-line force and water surface elevation data, based on the latter. Derive the period and wave height of each individual wavelet of the wave record as an input to the determination of the Airy and Fenton's stream function water particle kinematics over each wavelet cycle. Perform a "least squares" fit of the Morison force coefficients in eqn (6), for each wavelet in the entire wave record, after performing numerical integration for evaluation of $F_I(t)$ and $F_D(t)$ in this same equation utilising the predicted water particle kinematics and the observed cylinder restraint force for $F_R(t)$, (Haritos & Smith, 1994). Perform a half-wave cycle by half-wave cycle fit to C_L in eqn (4) after numerical integration for the tip restraint force measured in the transverse direction over the entire record of each data series.

iii. POST-PROCESSING - summarise and collate results against KC number, cylinder spacing parameter, S, and other parameters as necessary for the presentation of results.

In the case of the regular wave tests, the individual results from the wave-by-wave analysis (KC number, C_M and C_D) were averaged over all waves in the test record concerned for the purposes of reporting these values.

3.2 *Results for in-line cylinder orientation*

Figures 2 (a) and (b) provide graphs of the C_M and C_D values obtained from the regular wave tests grouped for KC number and plotted against spacing parameter S (the ratio of centre-centre spacing to cylinder diameter) distinguishing between the "upstream" and "downstream" cylinders in this orientation, respectively. The "closely spaced" region (S < 2~3) is clearly distinguished from the "separated" region (S > 2~3) in these plots as the effects on the C_M and C_D values appear to be significant only when S < 2~3.

In the case of the downstream cylinder, it can be seen from these plots that the influence on both the C_M and C_D values attributed to cylinder proximity effects is more pronounced than for the upstream cylinder. At very close spacings, (S=1.125) C_M seems to increase whilst there appears to be a local "dip" at around a value of S=1.25 and a local "peak" at around S=1.75 over the broad KC range tested for the downstream cylinder. In the case of the C_D variation for this cylinder, the closely spaced region appears to have a marked affect in decreasing C_D with decreasing value in spacing parameter S over the broad KC range tested. (Similar features in the C_M and C_D variation with S and KC number were observed in the results for the irregular waves).

Figures 3 (a) and (b) provide plots of the locus of the orthogonal cylinder tip restraint force measurements (F_x, F_y) for different spacing ratio values in the case of a regular wave with amplitude of 70mm and frequency of 0.5 Hz distinguishing the upstream and downstream cylinders variations respectively. It can be observed in these plots that acrosswave forcing (vortex separation) effects are significant over a broad range of cylinder spacings but appear to be "suppressed" at the closest cylinder spacing (S=1.125) and again at S=1.75, coinciding with the local "dip" and "peak" respectively in the C_M variation observed for the downstream cylinder in Fig. 2 (a). In general, acrosswave forcing was observed to be less than ~half the in-line forcing.

3.3 *Results for transverse cylinder orientation*

Figure 4 provides graphs of the C_M and C_D values obtained from the regular wave tests grouped for KC number and plotted against spacing parameter S (the ratio of centre-centre spacing to

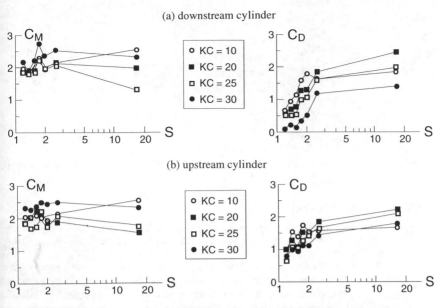

Figure 2 C_M and C_D variation with spacing for in-line cylinder configuration and regular waves

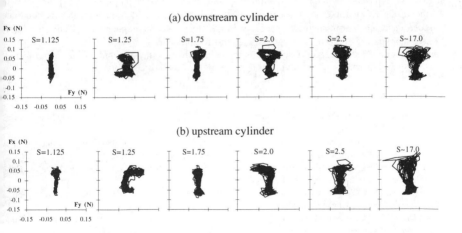

Figure 3 Locus of (F_x, F_y) for different spacing condition and in-line cylinder configuration

cylinder diameter) for the transverse cylinder orientation, using the results for one of the cylinders from the pair. (Results for the other adjacent cylinder were found to be near identical, as would be expected of this configuration).

Again, the "closely spaced" region (S < 2~3) is clearly distinguished from the "separated" region (S > 2~3) in these plots as the effects on the C_M and C_D values appear to be significant only when S < 2~3. C_M values are observed to increase markedly from the classical value of ~2.0 (expected of the "separated" condition) with decrease in cylinder spacing in the closely spaced region, whereas the C_D variation is almost invariant with spacing but decreases with increasing KC number (i.e. as drag force assumes greater significance compared to inertia force).

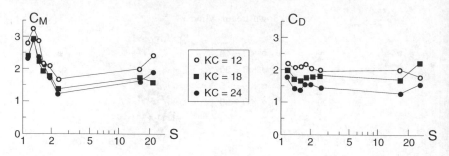

Figure 4 C_M and C_D variation with spacing for transverse cylinder orientation and regular waves

4 CONCLUDING REMARKS

This experimental programme of the influence on the Morison force coefficients of members in two cylinder groups attributed to cylinder spacing and orientation relative to incident wave direction has determined that:

i. for the in-line cylinder orientation: the drag force coefficient for both the upstream and downstream cylinder decreases with decreasing inter-cylinder spacing being more pronounced for downstream cylinder; the inertia force coefficient exhibits a "peak" at about S=1.75 and a "dip" at about S=1.25 within the closely spaced region (S<2~3) in the case of the downstream cylinder whereas very little influence is observed for the upstream cylinder in this region.

ii. for the transverse cylinder orientation: the drag force coefficient appears to decrease with increasing KC number and is near invariant against inter-cylinder spacing; the inertia force coefficient appears to increase markedly in the closely spaced region with decreasing inter-cylinder spacing and is almost invariant against KC number.

The results have extended the observations made in earlier work, to include a larger number of closely-spaced ordinates of inter-cylinder spacing for these two orientations.

REFERENCES

Haritos, N. & D. Smith 1994. Hydrodynamic force characteristics of closely spaced cylinder groups, *Proc. Australasian Structural Engineering Conference*, Sydney, September, 361-366.

Haritos, N. & D. Smith 1995. The effect of spacing transverse to the wave direction on the Morison force coefficients in two cylinder groups, *Proc. OMAE-95*, The Hague, June, 1A: 163-170.

He, Y. Q. 1993. *Hydrodynamic Interference and Interactions of Multiple Circular Cylinders*, Ph.D. Thesis, Dept. of Civil & Environ. Eng., Univ. of Melb.

Morison, J. R., M. P. O'Brien, R. W. Johnson & S. A Schaaf 1950. The force exerted by surface waves on piles, *Petroleum Trans.*, American Inst. of Mining and Metal. Eng., 189:149-154.

Rienecker, M. M. & J. D. Fenton 1981. A Fourier approximation method for steady water waves, *J. Fluid Mechanics*, 104:119-137.

Sarpkaya T. & M. Isaacson 1981. *Mechanics of Wave Forces on Structures*, Van Nostrand Reinhold.

Smith, D. & N. Haritos 1995. Vertical surface-piercing cylinders and the influence of grouping on the Morison force coefficients, *Proc. 12th Austn. Coastal and Offshore Engineering Conf.*, June, Melbourne, 317-322.

Smith, D. & N. Haritos 1996. The effect of in-line spacing of two cylinder groups on the Morison force coefficients, *Proc. OMAE-96*, Florence, June, 1A: 345-352.

Smith, D. 1997. *Hydrodynamic force characteristics of closely spaced cylinder groups*, M.Eng.Sci. thesis in prep., Dept. of Civil & Environ. Eng., Univ. of Melb.

Environmental loading, testing and response (fire)

The Mechanics of Structures and Materials, Grzebieta, Al-Mahaidi & Wilson (eds)
© *1997 Balkema, Rotterdam, ISBN 90 5410 900 9*

Numerical simulation of slender concrete walls in fire

D.A.Crozier & J.G.Sanjayan
Department of Civil Engineering, Monash University, Clayton Campus, Melbourne, Vic., Australia

ABSTRACT: A numerical model to simulate the buckling behaviour of slender reinforced concrete walls in fire is described. Both the thermal and structural responses are considered. Various parameters of the concrete constitutive model (tensile stress effects, ultimate strain, descending branch of the compressive stress-related strain curve and thermal strain) are examined and their significance in influencing the behaviour of concrete walls in fire discussed. A brief outline of planned experimental testing is provided. The numerically simulated results demonstrate the significance of slenderness effects, reinforcement location and thickness of the wall on the load-bearing capacity of slender concrete walls in fire.

1 INTRODUCTION

The increasing use of load-bearing reinforced concrete walls, particularly tilt-up and precast walls, is allowing Structural Engineers to design efficient and economical structures for buildings. Precast and tilt-up concrete walls serve a dual purpose in many modern structures. They provide both load-bearing support for the structure and a fire-separating function between compartments (ie. subjected to fire on one side only). Load-bearing concrete walls must therefore, satisfy all three requirements for fire safety (AS1530.4 1990a):

- structural adequacy - the ability of a concrete wall to maintain its load-carrying function during a fire,
- insulation - the ability of a concrete wall to provide thermal insulation between the fire compartment and an adjacent compartment by ensuring the average temperature rise on the unexposed face of the wall is less than 140°C, and
- integrity - the ability of a concrete wall to prevent transmission of hot gases and flames through cracks, fissures and the like.

The Concrete Structure code (AS3600 1994) sets out the requirements for assessing the fire resistance of concrete walls. The requirements for structural adequacy are based partly on a computer-based mathematical model (O'Meagher & Bennetts 1990) and partly on 'experience' (the limiting figure of 3% of crushing strength was apparently chosen to suit construction practices of the time). Neither the mathematical model, nor the current design guidelines, have been substantiated with experimental evidence.

This paper is the first in a series reporting the results of a long-term research programme to provide the necessary theoretical and experimental evidence to develop more appropriate design guidelines for assessing the effects of fire on structures incorporating slender reinforced concrete walls. The need for this research is crucial, particularly given the trend that requires Structural Engineers to provide a more detailed assessment of the fire safety of buildings or individual structural elements. In this paper, a mathematicall model, for numerically simulating the effects of fire on the structural behaviour of slender reinforced concrete walls, is proposed. The significance of various parameters in the model is examined. The effect of slenderness and reinforcement location and the interaction between thermal and structural response is also demonstrated. A brief outline of planned experimental testing is given.

2 THERMAL RESPONSE

Thermal response modelling is concerned with predicting the variation, with time, of the temperature of the concrete wall at varying depths when one side of the wall is subjected to a variable temperature environment (ie. a fire). In this case, the "fire" follows the requirements of the Standard Fire Test Conditions (AS1530.4 1990a).

Calculation of the temperature distribution in the wall is undertaken using the computer program TASEF-3 (Wickstrom 1988). In the analysis, the entire surface of one face of the wall is exposed to the fire environment. Variation of temperature is assumed to vary only with the thickness of the wall. The presence of reinforcement is ignored in the thermal analysis, however the temperature of the reinforcement used in the structural analysis is assumed as the temperature of the concrete at the centre-line of the reinforcement. Figure 1 shows the distribution of temperatures in a 150 mm thick concrete wall for varying durations of fire exposure.

Thermal response modelling forms the basis for assessing the insulation requirements of fire exposed concrete walls. Table 5.7.2 (AS3600 1994) specifies the minimum wall thickness required to meet the requirements of fire resistance for insulation. This table demonstrates that the load capacity of a concrete wall is improved by providing a wall thickness greater than that required to satisfy the fire resistance period for insulation.

Differences in thermal properties, as exhibited between the standard concrete material properties assumed by TASEF-3 and those reported in literature (Lie 1982), result in calculated differences of the thermal response of concrete walls exposed to fire. Moreover, the thermal properties of concretes composed of carbonaceous aggregate are more favourable than concretes composed of siliceous aggregates, from a heat transmission perspective (Lie 1982).

3 STRUCTURAL RESPONSE

Based on the thermal and structural response models being uncoupled, the time-temperature-depth data, generated from the thermal response model, is used to determine the temperature dependent mechanical properties of the materials.

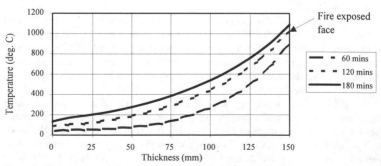

Figure 1. Temperature variation with time in a 150 mm thick concrete wall

An incremental solution procedure to assess the structural response of concrete walls in fire is adopted. For each time step during fire exposure, equilibrium is sought between the external load and internal resistance and deflection of the wall. Calculation of the equilibrium state of the wall, within a single time increment, is similar to that described elsewhere (El-Metwally 1990), except that a more elaborate finite difference scheme is employed in the current model to calculate wall deflections. The incremental procedure is repeated until equilibrium of the wall cannot be maintained, thus indicating structural collapse of the wall. The time at which this occurs is termed the 'Fire Resistance Period' for structural adequacy.

3.1 Constitutive equations and mechanical properties of concrete and reinforcement in fire

The total strain in the concrete ($\varepsilon_{tot,c}$) is expressed as:

$$\varepsilon_{tot,c} = \varepsilon_{th,c} + \varepsilon_{sr,c} \tag{1}$$

where $\varepsilon_{th,c}$ represents the unrestrained thermal strain and $\varepsilon_{sr,c}$ is the stress-related-strain of the concrete. The stresses in the concrete are calculated using stress-strain relations modified so that the maxima of the curves is shifted to higher strains for high temperatures to include creep effects. This form of constitutive modelling of concrete implicitly includes the transient creep strain component often considered separately by others (Anderberg & Thelanderson 1977, O'Meagher & Bennetts 1988).

In the numerical model, the variation, with temperature, of the mechanical properties of concrete (ultimate strength, ultimate strain, elastic modulus, thermal expansion), reflect experimental data (Anderberg & Thelanderson 1977). Figure 2 illustrates the concrete's compressive stress-strain relations. The equations which describe these curves are given elsewhere (Lie & Lin 1985). Tensile stresses are calculated using a previously proposed model (Massicotte et. al. 1990), with appropriate modification for the elevated temperature mechanical properties of concrete.

The total strain in the steel ($\varepsilon_{tot,s}$) is expressed similar to (1), with subscript c replaced with subscript s, to denote reference to the steel reinforcement. Calculation of the stresses in the reinforcement, and modelling of the variation, with temperature, of the mechanical properties of the reinforcement, is in accordance with the Steel Structures code (AS4100 1990b).

4 PARAMETRIC STUDY

4.1 Results of Parametric Study

Two specific wall configurations were considered in the current investigation. The test walls were 1000 mm wide, Grade 32 concrete, containing 0.3% total reinforcement by area (nominal room temperature yield strength of 450 MPa) and subjected to a compressive load acting at a distance of half the wall thickness towards the cool face. Table 1 summarises the numerically simulated load capacities generated by the proposed mathematical model.

The three main parameters of the constitutive relations for concrete investigated were the magnitudes of the ultimate strain at elevated temperatures, the slope of the descending branch of the compressive stress-strain curve and the thermal strain.

Numerical simulations indicate that ultimate strain at elevated temperatures and the slope of the descending branch of the compressive stress-strain curve have an insignificant effect on the structural response of concrete walls in fire. Therefore, detailed numerical results are not reported here, suffice to say, that for each parameter, the following modifications were considered:

- ultimate strain at elevated temperature - a movement to higher strains at higher temperatures, to reflect more severe thermal creep behaviour
- slope of the descending branch of the compressive stress-strain curve - increasing the slope, to reflect a higher rate of concrete unloading following attainment of the ultimate stress

Conversely, numerical simulations indicate that reduced estimation of the thermal strain (Lie & Lin 1985) significantly affects the structural response of concrete walls in fire. Figure 3 illustrates the thermal strain models investigated. Table 2 summarises the load capacities of a 150 mm wall thick using the reduced estimates of thermal strain. In these numerical simulations, it is assumed concrete in tension carries no stress.

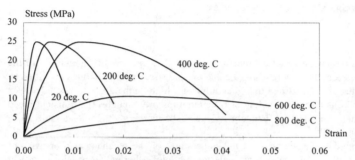

Figure 2. Elevated temperature stress-strain curves for concrete in compression (f'$_c$ at 20°C = 25 MPa)

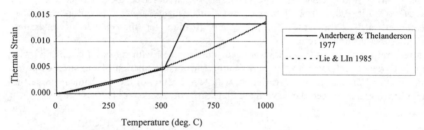

Figure 3. Alternative concrete thermal strain models investigated

Table 1. Load Capacities of 150 mm thick wall

H / t = 20	Central Reo.	Edge Reo.	H / t = 50	Central Reo.	Edge Reo.
FRP	P$_{NCT}$: P$_{WCT}$	P$_{NCT}$: P$_{WCT}$	FRP	P$_{NCT}$: P$_{WCT}$	P$_{NCT}$: P$_{WCT}$
Room Temp	224.0 : 277.0	270.0 : 302.0	Room Temp	52.0 : 192.0	75.0 : 202.0
60	117.0 : 122.5	78.0 : 84.0	60	21.0 : 22.0	13.8 : 14.8
90	109.0 : 115.0	62.0 : 67.0	90	19.5 : 20.5	11.0 : 11.9
120	95.0 : 101.0	50.0 : 56.0	120	17.0 : 18.0	8.7 : 9.5
100 mm thick wall					
H / t = 20	Central Reo.	Edge Reo.	H / t = 50	Central Reo.	Edge Reo.
FRP	P$_{NCT}$: P$_{WCT}$	P$_{NCT}$: P$_{WCT}$	FRP	P$_{NCT}$: P$_{WCT}$	P$_{NCT}$: P$_{WCT}$
Room Temp	149.0 : 188.0	183.0 : 202.0	Room Temp	35.0 : 130.0	51.0 : 136.0
60	66.0 : 71.0	37.0 : 43.0	60	11.7 : 12.5	6.6 : 7.2
90	52.0 : 57.0	26.0 : 31.0	90	9.4 : 10.1	4.8 : 5.5

NOTE: H = Wall height   t = Wall thickness   P = Applied Load (kN)
WCT/NCT = With/without concrete tension effects
FRP = Fire Resistance Period (minutes)
Edge reinforced walls have a centre-line cover of 20% of the wall thickness
100 mm thick wall limited to 85 minutes FRP for insulation

Table 2. Load capacity of 150 mm concrete walls using an alternative thermal strain model (Lie and Lin 1985)

H / t = 20	Central Reo.	Edge Reo.		H / t = 50	Central Reo.	Edge Reo.
FRP	P (kN)	P (kN)		FRP	P (kN)	P (kN)
60	155.0	94.0		60	30.0	17.0
90	133.0	69.0		90	25.0	12.4
120	111.0	52.0		120	20.4	9.2

4.2 Conclusions of Parametric Study

Analysing the results presented in Tables 1 and 2, and further examining the strain and stress profiles of the walls at the time of failure, the following conclusions are drawn:

- Accurate prediction of the thermal strain component is critical. Underestimating thermal strain leads to higher stress-related strains, for the same total strain. This is particularly evident close to the heated face, where the thermal strain is the dominant contributor to the total strain. Consequently, higher concrete stresses are predicted, leading to higher predictions of the load capacity for the wall.
- Modification of either the concrete's ultimate strain at elevated temperature, or the slope of the descending branch of the compressive stress-strain curve has an insignificant effect on the load capacity of concrete walls in fire. In fire situations, compressive stress-related strains are much lower than the ultimate strain. For very stocky walls (height-to-thickness ratio less than 20 - not considered in this study) however, these parameters may be significant and require further investigation.
- Including concrete tensile stresses slightly improves the load capacity of concrete walls in fire. Tensile strains in fire exposed concrete walls are usually beyond the point where tensile stresses result. Of the cases studied, typical increases of between 5% and 15% were obtained. Higher increases are obtained in fire exposed edge reinforced walls, while room temperature edge reinforced walls and fire exposed centrally reinforced walls show a lower increase. A similar trend is observed for the higher slenderness ratio, despite the fact that the room temperature response is significantly influenced by concrete tensile stress effects.
- Location of the reinforcement, with respect to the direction of fire exposure, is critical. Locating the reinforcement closer to the fire source dramatically reduces the load capacity of the concrete wall, despite an increase in the room temperature load capacity. In the cases studied, placing the total amount of reinforcement in two equal layers results in load capacities between 40 and 60% of corresponding centrally reinforced walls. This reduction was found to be independent of the wall's slenderness ratio and fire resistance period.
- The application of a particular fire resistance period (ie. 60, 90 or 120 minutes) has only a limited effect on the load capacity of fire exposed concrete walls. Typically, a reduction of less than 10% of the room temperature load capacity results when the fire resistance period changes from 60 minutes to 120 minutes.

5 FUTURE EXPERIMENTAL STUDIES

The behaviour of concrete walls in a fire environment involves many complex processes. A series of full-scale experimental fire tests will provide experimental data to validate the proposed mathematical model and provide a basis for evaluating current design guidelines for the design of fire exposed concrete walls. Moreover, confidence can only be given to the mathematical model if it is experimentally verified.

The experimental programme involves fire testing eight reinforced concrete walls, under Standard Fire Test Conditions (AS1530.4 1990a), using the fire testing facility at BHP's Melbourne Research Laboratories. A purpose-built experimental rig will test walls 3.6 m long by 1.2 m wide. The main parameters of interest are (i) height-to-thickness ratio : 24, 36 and 48,

(ii) wall thickness : 150 mm, 100 mm and 75 mm, and (iii) concrete strength : 40 MPa (normal strength) and 100 MPa (high strength). All wall thicknesses will be tested with a centrally located single layer of 0.25% total reinforcement. The 150 mm thick wall will be additionally tested with 0.25% total reinforcement, placed in two equal layers at the edge of the panel (30 mm centre-line cover to the reinforcement).

In addition to the wall testing programme, the high temperature properties of high strength concrete (up to 100 MPa compressive strength) is to be experimentally studied. A recent report (Phan 1996) concluded that current fire design recommendations for fire exposed concretes are only applicable to normal strength concrete and not to high strength concrete. Of particular importance is the need to experimentally quantify the conditions leading to explosive spalling of high strength concrete, particularly when subjected to the rapid heating process of fire.

6 SUMMARY

This paper has described the first stage of a long-term research programme investigating the behaviour of slender reinforced concrete walls in fire. A mathematical model to numerically simulate the structural behaviour of concrete walls in fire was proposed and various parameters influencing the structural response were studied.

The parameters identified in this study to be of particular importance in the structural response are the thermal strain, the height-to-thickness ratio and the reinforcement location. In a fire situation, placement of the reinforcement at the edge of the panel results in a load capacity lower than that of a corresponding centrally reinforced wall. Tensile stresses in the concrete have an effect under fire exposure conditions and should be included in the mathematical model. Selection of a particular fire resistance period has only a limited effect on the load capacity of concrete walls in fire. However, the presence of a fire significantly reduces the room temperature load capacity.

Experimental verification of the mathematical model was outlined. Further details of this long-term experimental and theoretical research programme will be reported in later publications.

7 REFERENCES

Anderberg, Y. & Thelanderson, S. 1977, A constitutive law for concrete at transient high temperature conditions, in *Douglas McHenry symposium volume, America Concrete Institute SP-55*:187-205, Detroit: ACI.
AS1530.4 1990a , *Methods for fire tests on building materials, components and structures, Part 4 : Fire-resistance tests of elements of building construction*, Sydney, Standards Association of Australia.
AS4100 1990b, *Steel Structures*, Sydney: Standards Association of Australia.
AS3600 1994, *Concrete Structures*, Sydney: Standards Association of Australia.
El-Metwally, S.E., Ashour, A.F. & Chen, W.F. 1990, Instability Analysis of Eccentrically Loaded Concrete Walls, *Journal of Structural Engineering*, 116(10):2862-2881, ASCE.
Lie, T.T. & Lin, T.D. 1985, Fire performance of reinforced concrete columns, in *Fire safety : Science and Engineering, ASTM STP 882*:76-205, Philadelphia: American Society for Testing and Materials.
Massicotte, B., Elwi, A.E. & MacGregor, J.G. 1990, Tension Stiffening Model for Planar Reinforced Concrete, *Journal of Structural Engineering, ASCE*, 116(11):3039-3058..
O'Meagher, A.J. & Bennetts, I.D. 1988, Modelling of concrete walls in fire, *Fire Safety Journal*, 17:315-335.
Phan, L.T. 1996, *Fire performance of high-strength concrete : A report of the state-of-the-art, NISTIR-5934*, Gaithersburg, Building and Fire Research Laboratory, National Institute of Standards and Technology,
Structural Fire Protection, Lie, T.T., ed. 1982, *Manuals and Reports on Engineering Practice No. 78*, ASCE, New York, N.Y.
Wickstrom, U. 1988, *TASEF-3, A Computer Program for Temperature Analysis of Structures Exposed to Fire*, Swedish National Testing Institute.

The Mechanics of Structures and Materials, Grzebieta, Al-Mahaidi & Wilson (eds)
© 1997 Balkema, Rotterdam, ISBN 90 5410 900 9

Time-dependent degradation of structures during fire – Analysis and design issues

J. Mohammadi & J. Zuo
Civil and Architectural Engineering Department, Illinois Institute of Technology, Chicago, Ill., USA

ABSTRACT: This paper describes the behavior of structures subject to high temperatures during a fire. Floor systems made up of reinforced concrete slabs with or without intermediate beams were investigated under application of a loading that consisted of dead load and fire exposure. The effect of high temperature on material strength and behavior; creep and cracking in concrete; formation of localized failures; and structural degradation resulting from loss of stiffness were investigated using a nonlinear finite element analysis. The analysis simulated the loading process and followed the step-by-step structural degradation of floor systems until a predetermined collapse criterion was reached. The results revealed a dramatic loss of structural stiffness after a temperature increase of about 400-500 °C. The paper also describes the significance of fire loads in structural analysis and design. Issues pertinent to design include: (1) prevention; and (2) design for safe performance in a fire. These issues as related to a performance-based design code are reviewed and discussed.

1 INTRODUCTION

Generally, any investigation into fire effects on buildings focuses primarily on casualties and property loss resulting from hazards associated with the burning capability of fire. Although potential casualties and losses constitute a major consideration in fire hazard investigations, an equally important issue is the effect of high temperatures, resulting from fire, on the performance of structures. A persistent fire may ultimately cause collapse of a structure. Understandably, the dramatic impact of fire-related casualties and losses on the society overshadows the structural collapse capability of fire. However, it is noted that the potential for structural collapse may impose a threat to the lives of fire fighters (whose duties require them to remain at or around the building battling the blaze) and may also increase the possibility for fire spread (flash-over) to adjacent buildings. Numerous studies have addressed the performance of materials and structural systems at elevated temperatures resulting from fire. Only a few studies, however, have addressed any relevant analysis and design issues for "fire loading" in a structure.

According to Zuo and Mohammadi (1996), a major effect of fire on structural systems is gradual degradation of the structural integrity due to: (1) changes in material properties at elevated temperatures; (2) changes in system geometry due to localized failures; and, (3) partial or total loss of supports that occurs during the fire exposure period. In terms of serviceability, a floor system can be assumed failed upon exposure to a severe fire. However in order to investigate its overall integrity, a floor system's performance throughout the loading process and up to the stage of collapse needs to be studied. As the fire exposure persists, the system undergoes changes leading to a substantial loss of stiffness and an increase in strain and deformation. Investigating these parameters offers a better understanding of how a structural sub-system (a floor system or wall) can survive a potential fire and how a sub-system should be designed such that the risk of its collapse can be reduced to an acceptable level. A rigorous investigation and analysis effort is required for understanding the response of structural subsystems (floor systems, walls, etc.) and to identify those parameters that best describe structural degradation during fire. Such an effort involves a combined heat transfer and stress analysis of the structure and may be quite complex. As a minimum, appropriate models for: (1) temperature distribution in a sub-system; (2) heat transfer from the point of fire initiation to other areas; and, (3) the material behavior, as the temperature rises, will be required for the analysis.

In a study conducted by the authors, several floor systems made up of reinforced concrete slabs were studied. The loading consisted of dead load and fire exposure to the floor systems. A nonlinear finite element analysis (using the program ABAQUS) was employed in each case to model: (1) the effect of high temperature on material strength and behavior; (2) the creep and cracking process in concrete; (3) formation of localized failures; and (4) structural degradation resulting from loss of stiffness. In each structure analyzed, the overall deformation and strains were investigated throughout the loading process. Furthermore, changes in effective cross section depth, partial loss of supports and changes in the geometry of the system resulting from severe localized cracking were also studied. Structural properties such as slab thickness, reinforcement ratio, intermediate beam spacing, etc. were changed in several simulations to develop a baseline for understanding the design attributes of fire loading.

The analysis for each system was conducted in stages with incremental increases considered for temperature, based on available temperature-time data typical of those reported for floor systems and walls. At the end of each stage, elements experiencing failure were removed. A re-configuration of the finite element mesh then followed; and the temperature increase was resumed. Collapse was assumed when system instability was detected.

2 FIRE EFFECT ON STRUCTURAL MATERIALS

The fire resistance of a structural member is defined as the ability to withstand exposure to fire without loss of load bearing capacity; and/or the capability to act as a barrier against spread of fire. Most codes express the requirement for fire resistance design as the length of time that a structural member is able to withstand exposure to "standard fire" without losing its bearing capacity or its fire isolating capability. This length of time is a measure of the performance of the member against fire and is termed the "fire resistance." Fire resistance rating of structural members has been determined by the results of the American Society for Testing and Materials (ASTM) standard tests as reported by ASTM (1989), Branson and Christiason (1970) and Lin and Abrams (1983). Generally, it is assumed that during a fire, the dead load and live load remain constant while the material strengths are reduced. A summary of specific effects of fire on structural materials and systems, as outlined by Zuo and Mohammadi (1996), are provided below:

2.1 Concrete

In concrete, major fire effects are on the coefficient of thermal expansion, transient and stress-related creep, compressive strength, and modulus of elasticity. At higher temperatures, the ability of the material to creep increases substantially especially if there is a long exposure period.

2.2 Steel

In steel, the effect of fire and high temperature is primarily on the stiffness and strength of the material. The modulus of elasticity reduces by small amount at lower temperatures. However, the reduction becomes dramatic at higher temperatures as does the yield strength.

2.3 Structural systems

Behavior of structural systems during fire depends on the material properties and also on such factors as the geometry of the system, dead and live load present at the time of fire, design of structural members (rebars, slab thickness, steel girders) used in the system, boundary conditions, the location where fire is initiated, general layout and area of openings in walls, and whether or not fire is confined to an area. In a typical floor system a combination of changes in the material properties will result in formation of cracks in the slab, formation of localized failures (loss of supports, reduction in effective depth) and large deformations. In addition to a change in material behavior, that affects the structural response, a floor system is also subject to a low-intensity pressure resulting from physical characteristics of fire, burning materials and heat.

3 EFFECT OF FIRE ON STRUCTURAL DEGRADATION

The amount of heat generated during a fire and the level of temperature rise in a building depends on many factors. Temperature may rise to such high levels as 650 ^{O}C and even higher if an ample supply of combustible materials and air to fuel the fire exists. In addition, the location of the fire in the building and its confinement to one area may have a dramatic effect on temperature rise and the behavior of structural members. In terms of effect on structural systems, the following characteristics related to the fire and structure are important: (1) the temperature rise; (2) the temperature difference between two surfaces of the system; (3) duration of fire; and (4) structural configuration. The surface pressure applied on floor systems and walls due to the characteristics of fire and heat is relatively small. Zuo and Mohammadi (1996) conducted a study on typical fires using the program FASF (developed by the US National Institute of Standard and Technology) and revealed that the pressure is only about 0.20 - 0.25 KN/m^2. Thus the major contributing factor to structural behavior is change in the material properties that occur during the temperature rise. In floor systems made up of reinforced concrete, these changes constitute a nonlinear behavior. Thus, a nonlinear finite element analysis (program ABAQUS) was used to consider: (1) changes in modulus of elasticity; (2) increase in the creep of concrete; (3) changes in the mechanical properties of the rebars; and (4) geometrical changes due to loss of supports and reduction in the effective cross section because of cracking.

The results for a typical 4.85x4.85-meter slab designed for an office building indicated that at about 400-500 ^{O}C a dramatic reduction in the system stiffness occurs. The change in stiffness occurs steadily up to about this temperature level. At higher temperatures, when the integrity of the structure is severely altered, and at the onset of collapse, the floor system may experience a noticeable level of vibration. This vibration has been expressed by fire fighters who have experienced a collapse during their fire fighting activities, (Zuo and Mohammadi 1996). The simulation studies indicated that a noticeable vibration can occur only when the integrity of the system has been severely affected and when the rate of change in the deformation has been increased dramatically. Accordingly, a behavior similar to that of structural dynamic response can be experienced by the structure prior to collapse. However, due to the limitation of the software used, the floor vibration could not be accurately simulated. During the loading, noticeable changes in strain and deformation occur especially at higher temperatures. These changes are not necessarily accompanied by floor vibrations. An investigation into trends in changes occurring to strains and deformations is of importance in providing any indication that a collapse may be forthcoming.

The simulation analysis was conducted on a variety of floor systems including: (1) two-way reinforced concrete slabs (4.85x4.85-meter dimensions as described above); (2) slab/beam monolithic system; (3) steel girder/reinforced concrete slab system; and (4) two-story/two-bay frame with slab systems. As an example, Fig. 1 shows a reinforced concrete beam-slab system analyzed. The main girders are each made up of a 15x32-cm rectangular section beam. The secondary beams form T-beam construction. The cross section of these beams are slightly tapered with approximate sizes of 8x25 cm each. The steel reinforcement in these beams is composed of two 12-mm bars. The spacing between beams is 1.2-1.5 meters. The slab is 7-8 cm in thickness. The slab reinforcement is composed of a wire mesh system. The panel shown in Fig. 1 is 4.25x5.5 meters. The finite element model consisted of shell and beam elements. As the temperature advanced, localized failures were reached at various locations and at the supports where the slab is attached to the beams.

Those elements experiencing failure were removed and a re-configuration of the finite element mesh followed. This process continued in a series of step-by-step incremental simulation analyses until a structural instability was detected. At this stage, collapse was assumed. At each temperature increment, and following the formation of localized failures, the deflection, system

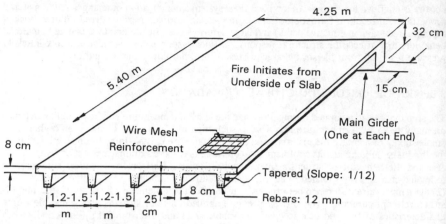

Fig.1 A joist/slab floor system analyzed

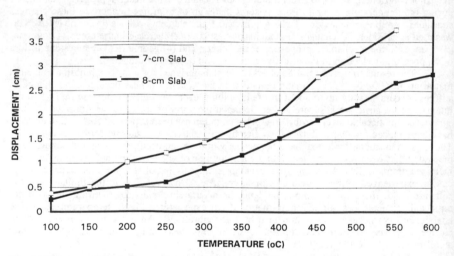

Fig. 2 Variation of displacement versus temperature (beam spacing: 1.2 m)

stiffness (as reflected in the fundamental natural frequency of vibration of the floor system) and strains at the cross section of the slab were obtained.

Figures 2 and 3 show the variation of displacement in the system of Fig. 1 for several design parameters (slab thickness and beam spacing). Similar results were also developed for other systems analyzed. A complete discussion on the final results is provided in Zuo and Mohammadi (1996). A brief discussion on the pertinent structural design issues is presented in the next section.

4 STRUCTURAL ANALYSIS AND DESIGN ISSUES -- DISCUSSION

In conventional design codes (prescribed-based design), fire loading is rarely considered as a determining factor in the required strength of structural members. In most parts, fire is addressed

288

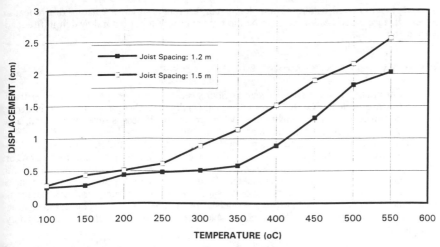

Fig. 3 Variation of displacement versus temperature (slab thickness: 7 cm)

through protection coating, and endurance rating of materials exposed to a prolonged fire. As it is evident in the results presented in this paper, fire can become an important loading factor in a structure. One argument about fire is the amount of damage that a potential fire causes to building contents. As a result, the structural design practice treats fire more as a non-structural issue in design. The study by Zuo and Mohammadi (1996) specifically addressed the potential destructiveness of fire in terms of structural collapse. Every year in the USA alone some 150 fire fighters become victims of structural collapse during a fire. Thus it seems reasonable to, at least, address the fire load issue in a fail-safe design criterion and from the view point of safety of fire fighting personnel. In this regard, the pertinent issue focuses on two possible alternatives. These are: (1) proper prevention; and (2) design to perform to a minimum acceptable level. Prevention covers a very broad base and depends on the level of protection desired for a building. Zuo and Mohammadi (1996) proposed the use of an intelligent warning system for use in buildings as a reliable means to enhance the safety of fire fighting personnel. A conceptual design was developed describing the requirements for implementing the system in a real prototype application. Since all sensors and data recording equipment must be protected from excessive heat, the conceptual design described the use of remote data transformation and isolation of the sensors and transmitter devices using commercially-available compounds that delay heat transfer and provide for adequate protection. Once fully operational, the system is expected to monitor changes occurring to the structural stability of the building and predict an upcoming collapse through an advanced remote signaling system. The system in concept should be able to warn fire fighters of an upcoming collapse and provide them with early advice to evacuate the building before it collapses. The development of such a system may take a major research effort with substantial initial and installation cost.

A decision to design a structure to perform to an acceptable minimum level may offer a more viable approach. Specifically, such a design can be implemented through a "performance-based" design code rather than through the conventional "prescribed-based" code. Performance-based design criteria are becoming popular as an alternative design approach. Design for fire effect through these criteria would allow a designer to decide on the level of performance which is considered feasible based on both the cost and the required safety. Furthermore, the designer may impose specific performance criterion to be implemented in design. For example, a performance criterion may require not only a safe passage for occupants during a fire, but also impose a requirement on fail-safe performance to prevent sudden collapse. This is a level of protection included in design for the purpose of preventing fire fighters' loss of lives.

Barnet (1995) presents a discussion on the performance-based design issues pertaining to fire protection and building design. Development of design requirements specific to a structure

289

is a concept that fits well within the complexity and 'prevention-oriented" nature of fire loads. The performance-based design seems to be the right approach for structural design in fire load environments.

5 SUMMARY AND CONCLUSIONS

The behavior of structural subsystems such as floor systems at high temperatures arising from major fires is investigated. The study concludes that: (1) dramatic changes to structural integrity of a floor system occur at high temperatures, when structural stiffness is severely reduced; and (2) changes in the structural properties may result in collapse.

The potential for structural collapse during fire brings about the issue of design. The decision to provide an adequate level of protection may be taken through either prevention, or through performance-based design criterion. Economic issues may prohibit the use of an advanced intelligent protection/warning system. Furthermore, the fabrication of such a system may require a comprehensive research and development effort. Accordingly, for most buildings with a typical occupancy category (i.e., residential or commercial), it seems reasonable to believe that "design to perform" offers a more viable alternative for enhancing structural safety and reliability.

6 REFERENCES

ASTM, 1989. *Standard test method for fire tests of building construction and materials*, ASTM E119-89, American Society for Testing and Materials, Philadelphia, Pennsylvania.

Barnet, J.R., 1995. *Structures subjected to accidental fire*, in *Proceedings of the 1995 Structural Lecture Series*, American Society of Civil Engineers, Illinois Section, 203 North Wabash, Chicago, Illinois.

Branson, D.E. and Christiason, M.L., 1970. *Time dependent properties related to design strength and elastic properties, creep and shrinkage*, in *Designing for effects of creep, shrinkage, temperature in concrete structures*, ACI Publication SP-27, American Concrete Institute, Detroit, Michigan.

Lin, T.D. and Abrams, M.S.1983. *Simulation of realistic thermal restraint during fire test of floor and roofs*, in *Fire safety of concrete structures*, ACI Publication SP-80, American Concrete Institute, Detroit, Michigan, 1983.

Zuo, J. and Mohammadi, J.,1996. *Time-dependent degradation of structural systems during fire*, Structural Engineering Series, Report # IIT-CAE-96-01, Department of Civil and Architectural Engineering, Illinois Institute of Technology, Chicago, Illinois, USA.

The Mechanics of Structures and Materials, Grzebieta, Al-Mahaidi & Wilson (eds)
© *1997 Balkema, Rotterdam, ISBN 90 5410 900 9*

Influence of coefficient of thermal expansion on steel frame strength

N. L. Patterson & M. B. Wong

Department of Civil Engineering, Monash University, Melbourne, Vic., Australia

ABSTRACT: The purpose of this paper is to investigate the magnitude of the possible error that could be introduced by assuming the coefficient of thermal expansion for steel is constant for all temperatures. In this paper the basic equation governing the determination of the elastic buckling load factor as a thermal problem is presented. The effects of elevated temperature on the behaviour of steel elements are modelled according to the requirements stipulated in AS4100. A comparison is made of the various limiting temperatures for structures subject to fire using different approximations for the coefficient of thermal expansion recommended by the ECCS and BISRA. Conclusions are drawn as to the acceptability of the assumption of using a constant coefficient of thermal expansion with regard to elastic critical load analysis.

1 INTRODUCTION

The structural performance of a building varies with the variation in its physical and mechanical properties under elevated temperatures. It is a major fire-safety requirement that adequate structural fire resistance is provided for buildings so that the stability and integrity of the structure can be maintained for a minimum period of time. If a fire occurred in a bare steel structure, members may be directly exposed to flames and heat, causing a loss of strength to the structure. In many cases the limiting temperature of a structure can be increased by encasing the bare steel beams with fire insulation materials. This is the preferred method but can lead to a significant increase in construction costs. For the purpose of this study all steel members are assumed to be unshielded from the heat effect.

Much work has been done on the properties of steel under elevated temperatures but the majority of the work has focused on the stress-strain behaviour of the materials used in the structure and the analysis of members in isolation, not as a system. A mathematical technique proposed to measure the performance of a structural system under elevated temperatures is to calculate the limiting temperature at which a structure will buckle elastically. These days computer packages are the most common design tool for structural analysis at elevated temperatures. The majority of these packages calculate the elastic response of the structure and require the user to input the specific thermal and mechanical properties of the steel members in the form of a table. This can be tedious where a large range of temperatures is required. The process is complicated when designers are given a range of data from different research groups to choose values from for a single coefficient in relation to elevated temperature behaviour.

When steel is subjected to a change in temperature, eg. fire, the steel element will undergo geometrical changes. These geometrical changes are due to changes in the mechanical and thermal properties at the increased temperature. When the temperature change is a uniform temperature increase throughout the section, the degree of the expansion will be uniform due to the axial forces induced by the heat. Aesthetically, the expansion of the structure may be unnoticeable but it may expand just enough to cause either itself or an object in contact with it to lose its design function.

The Australian Steel Code AS4100 (SAA, 1990) assumes implicitly that the change in geometry is linear with increasing temperature, ie. that the coefficient of thermal expansion is constant for all temperatures. In fact the value of the coefficient of thermal expansion changes with change in temperature of the member. Designers and authors have adopted many forms of the value of the coefficient of thermal expansion which result in different values of the parameter for the same temperature. This obviously leads to inconsistencies in structural design techniques and this paper will highlight the significance, or insignificance of these deviations in design standards.

2 MECHANICAL PROPERTIES OF STEEL AT ELEVATED TEMPERATURES

It is not only the loads on the structure which increase in a fire inducing structural failure, it is also the loss of strength and stiffness of the frame as a system which results in its ultimate collapse. When steel is subjected to an increase in temperature it becomes structurally weaker. This loss of strength of steel in excessively high temperatures can be reduced by protecting the member with some form of fire insulation. However, where bare steel members exist, as in the present study, relationships between strength and temperature exist by which the behaviour of the steel can be modelled. The behaviour of the bare steel can be approximated by varying mainly three fundamental coefficients with respect to the value of the temperature. These coefficients are the yield strength, the modulus of elasticity and the coefficient of thermal expansion. For the purpose of this paper the values adopted are those recommended by AS4100 (SAA, 1990) Steel Structures Code. This study is focused on the elastic response of the structure and thus yield strength is not required in the analysis process.

2.1 Young's modulus

The value of Young's modulus decreases with temperature due to the change in its σ-ε relationship at elevated temperatures. The values adopted by AS4100 (SAA, 1990) for the modulus of elasticity under elevated temperature conditions will be used in the present study.

2.2 Coefficient of thermal expansion

The coefficient of thermal expansion for steel is the variable which describes the rate at which it will expand when subject to a temperature change. Many tests have been done to obtain values for the coefficient of thermal expansion at different temperatures. Obviously the chemical composition of steel differs from section to section and slight variations will be seen in the various mechanical and thermal properties. In the case of thermal expansion, a wide variety of relationships have been identified to describe it as a function of temperature.

Uddin & Culver (1975) reported that in general the thermal expansion coefficient increases with an increase in temperature up until around 650°C and then decreases to zero at almost 820°C before increasing again to a high value. This form of behaviour in the steel is due to changes in its phase diagram at high temperatures.

ECCS (1983) has produced a linear equation to approximate the behaviour of steel at elevated temperatures. BISRA (1953) has come up with a series of constant values for temperature ranges. It is mainly from these two approximations that others have developed their own relationships. Purkiss(1988) commented on data produced by a RILEM report that the thermal expansion of steel was virtually linear and was independent of steel type. AS4100 (SAA, 1990) has assumed a value which is constant for all temperatures and is lower than any value given by either ECCS or BISRA test data. It is the accuracy of this assumption and its effects on the buckling strength of steel frames which will be studied in this paper. Figure 1 illustrates the various values of the coefficient for the different temperatures as recommended by different bodies.

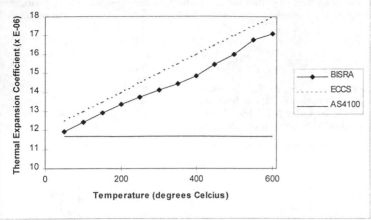

Figure 1 Variation of Coefficient of Thermal Expansion with Temperature

3 BUCKLING ANALYSIS

If a frame is subjected to increasing temperatures for a sufficient time, the frame may no longer be able to support its design loading. It has been shown (Wong & Patterson 1996b) that a temperature gradient across the cross-section of a member has little effect on the elastic critical load of a steel frame. In the case where the temperature is uniform throughout the frame axial forces will be induced in the members which will cause the frame to lose its strength. These axial forces must be included in the buckling analysis as increasing axial forces will lead to frame instability. An overall temperature will be assigned to the frame to aid comparisons in frame strength with changes in the coefficient of thermal expansion.

The method adopted for analysis is the method proposed by Wong & Patterson (1996a). This method involves the solution of an eigenvalue problem in which the contribution to instability by the internal forces due to the fire and external loads are calculated separately. The eigenvalue obtained in the analysis is the multiple by which all static loads must be multiplied for instability to occur in the structure under the prescribed temperature conditions. Generally as the temperature of the structure increases the strength of the structure decreases. This decrease in strength also relates to a decrease in the elastic buckling load factor.

The general equation for the solution of the eigenvalue can be derived as:

$$\left[\left(K' + K_G''\right)^{-1} K_G' + \beta I\right]\underline{D} = 0 \tag{1}$$

where K' is the elastic stiffness matrix,

K_G' is the geometric stiffness matrix using room temperature axial forces,

K_G'' is the geometric stiffness matrix using elevated temperature axial forces,

$\beta = 1/\lambda$ where λ is the buckling load factor,

I is the unit matrix,

D is the displacement vector.

This method is applicable to slender frame structures which are likely to fail by buckling before yielding takes place. With this method a buckling load factor of unity for the frame will correspond to its limiting temperature. Moreover, if the maximum temperature in a fire is known, by comparing the buckling load and the design load of the frame, it can be seen whether a structure will maintain its integrity during a fire. BHP(1990) found in a simulation

293

of an office fire involving bare steel that the maximum temperature reached in the bare steel was 600°C. For practical purposes, temperatures in excess of 600°C will not be considered in this study.

4 ILLUSTRATIVE EXAMPLES

Four examples have been selected to illustrate the effect of the coefficient of thermal expansion on the elastic buckling load factor. It has been assumed that fire and wind do not often occur simultaneously. This assumption holds true because most buildings are in a protected area (ie shielded against wind by other buildings). These examples are illustrated in Figure 2.

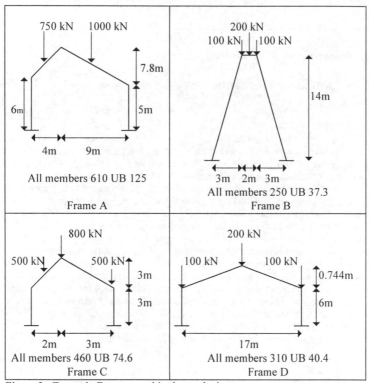

Figure 2. Example Frames used in the analysis

The frames depicted in Figure 2 were subjected to temperatures from 20°C up to 600°C in steps of 100°C. The elastic critical load factors were calculated at each of these temperatures and the results are shown in Table 1 below.

From the results of Table 1, it can be seen that there is not really a significant change to the elastic buckling load factor of steel frames under elevated temperatures due to the slight variations in the coefficient of thermal expansion. The maximum difference is 0.142% at 600°C. Generally it can be seen that the higher the value of the coefficient of thermal expansion the higher the value of the critical load multiplier. Therefore AS4100 (SAA,1990) will always yield results on the conservative side but the discrepancies will be negligible. It is also demonstrated that at 600°C the critical load factor is at least half of the room temperature critical load. This highlights the significant effect of heat on steel members and is due to the decrease in the value of Young's modulus to approximately half its room temperature value at about 500°C.

294

Table 1. Results of elastic critical buckling load factor analysis.

TEMPERATURE	FRAME	λ_E AS4100	% increase in λ_E ECCS	BISRA
20°C	A	36.428	0	0
	B	8.804	0	0
	C	38.393	0	0
	D	11.153	0	0
100°C	A	35.676	0	0
	B	8.621	0	0
	C	37.594	0	0
	D	10.925	0	0
200°C	A	34.306	0	0
	B	8.288	0	0
	C	36.145	0	0
	D	10.507	0	0
300°C	A	32.244	0.022	0.013
	B	7.789	0	0
	C	33.965	0.006	0.006
	D	9.878	0.030	0.010
400°C	A	29.253	0.034	0.031
	B	7.065	0	0
	C	30.809	0.007	0.007
	D	8.963	0.055	0.055
500°C	A	24.907	0.052	0.052
	B	6.014	0.017	0.017
	C	26.226	0.012	0.012
	D	7.633	0.092	0.092
600°C	A	18.426	0.109	0.0651
	B	4.448	0.023	0.023
	C	19.373	0.016	0.016
	D	5.646	0.142	0.124

5 CONCLUSIONS

Bare steel structures are most prone to structural failures when subjected to excessive elevated temperature conditions due to their generally low fire resistance. Nonetheless, where a bare steel structure has been designed and built to comply with fire standards it is rare that it will collapse under elevated temperature conditions, although depending on the severity of the fire it may need extensive repair.

It has been demonstrated that in the elastic critical buckling analysis of frames under elevated temperatures that a small variation in the coefficient of thermal expansion will only lead to a small difference in the strength of the frames. Therefore the assumption by AS4100 (SAA, 1990) Australian Steel Structures Code of a constant value of coefficient of thermal expansion results in a design which is conservative.

6 REFERENCES

British Iron and Steel Research Association (1953). *Physical Constants of Steels at Elevated temperatures*, BISRA, London.
BHP Melbourne Research Laboratories (1990). Fire in mixed occupancy buildings, *Report No. MRL/PS69/89/004* August 1989.
ECCS (1983). *European recommendation for the Fire Safety of Steel Structures ECCS Technical Committee 3 - Fire Safety of Steel Structures*, Elsevier, Amsterdam.

Purkiss, J. A (1988). Developments in the Fire Design of Structural Steelwork. *Journal Construct. Steel Research* 11, 149-173

Standards Australia (1990), AS4100 - Steel Structures, Section 12

Uddin, T & Culver, C. G (1975). Effects of elevated temperature on steel members. *Journal of Structural Division, Proc. ASCE*, 101, 1531-1549

Wong, M. B & Patterson, N. L (1996a). Buckling strength of steel frames under elevated temperatures. Third Asian-Pacific Conference on Computational Mechanics, Seoul, Korea, 16-18 September 1996:473-478

Wong, M. B & Patterson, N. L (1996b). Unit Load Factor Method for Limiting Temperature Analysis of Steel Frames with Elastic Buckling Failure Mode. *Fire Safety Journal 27*, 113-122.

The Mechanics of Structures and Materials, Grzebieta, Al-Mahaidi & Wilson (eds)
© 1997 Balkema, Rotterdam, ISBN 90 5410 900 9

Time effects on the behaviour of steel structures under elevated temperatures

M. B. Wong
Department of Civil Engineering, Monash University, Vic., Australia

J. I. Ghojel
Department of Mechanical Engineering, Monash University, Vic., Australia

ABSTRACT: This paper points out the importance of accurately predicting the temperature of steel members within a structure subjected to fire in relation to their structural performance. A new numerical model simulating the heat transfer mechanism between fire and steel members is presented. When compared with experimental data by others, this model provides realistic results for predicting steel temperatures. This paper also emphasises the use of real fire curves, rather than standard fire curve, in design. The difference of results in using both standard and real fire curves will be shown through examples.

1 INTRODUCTION

For design of structures subjected to fire, the primary objective is to assess the fire resistance time (period of structural adequacy) within which the structures will withstand the effects of high temperature without failure. The assessment of the period of structural adequacy for any structure can be divided into two parts : (1) the behaviour of structure under increasing temperature until failure; (2) the variation of the temperature of the structural members with time under certain fire scenarios. This paper focuses on the second part : how the temperature of steel members varies with time under different fire curves. As most structures are required to meet the minimum fire-resistance level according to certain construction classifications, the actual time history that a structural member acquires its temperature is crucial in determining the member size and the fire protection requirements which will affect the overall cost of construction. In the following, a model for the heat transfer mechanism between fire and steel members is described. Examples are given to demonstrate how the period of structural adequacy will be affected when the structure is subjected to both standard fire and real fire situations.

2 HEAT TRANSFER MODEL

The model assumes that the fire is centrally located in an insulated black enclosure. The rise in temperature in the steel member due to the fire is based on the following energy balance equation :

$$q_r + q_c = \frac{c_p V \rho}{A_r} \frac{dT}{dt} = \frac{c_p \rho}{P/A_s} \frac{dT}{dt} \tag{1}$$

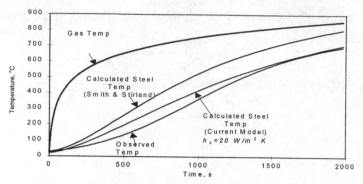

Figure 1(a) $h_c = 20W/m^2K$

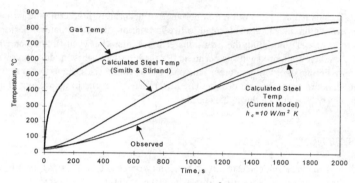

Figure 1(b) $h_c = 10W/m^2K$

Figure 1. Temperature prediction in standard fire.

where

q_r - radiative heat transfer per unit area from the fire in W/m^2,

q_c - convective heat transfer per unit area from the fire in W/m^2,

c_p, ρ - specific heat (J/kg K) and density (kg/m^3) of steel respectively,

V, A_r - volume (m^3) and exposed surface area (m^2) of the steel element respectively,

P, A_s - perimeter (m) and area of cross-section (m^2) of the steel element respectively,

dT/dt - rate of change of temperature of the steel element.

Equation (1) assumes that all the heat flux from the fire is absorbed by the element causing the rise of its average temperature. Temperature variations within the element and heat losses to the environment across the enclosure walls are neglected.

The radiative heat transfer component q_r accounts for the radiation properties (emissivity and absorptivity) of some of the combustion products, namely carbon dioxide, water vapour and

nitrogen. The convective heat transfer component q_c assumes a constant value for the coefficient of convection heat transfer. Details for radiative and convective heat transfer mechanisms can be found in textbooks (eg. Mills 1992).

298

2.1 Temperature profile in standard fire

Figure 1(a) shows the observed and two computed steel temperature profiles for an unprotected steel column exposed to a standard fire from four sides using data presented by Smith and Stirland (1983). The column is a British 305x305x198 kg/m with a P/A$_s$ ratio of 74 m^{-1}.

The calculated steel temperature profile obtained by Smith and Stirland (1983) was based on a simple model which assumes a gas (non-participating medium) perfectly transparent to radiation heat transfer. In this simple model, a constant convective heat transfer coefficient (h$_c$) equal to 25 W/m^{2}K was used. On the contrary, the model proposed in this paper which assumes the presence of participating gas medium and a value of h$_c$=20 W/m^{2}K gives a better correlation than the model given by Smith and Stirland (1983). However, the predicted temperature profile lies above the observed profile over most of the test duration. A trial value of h$_c$=10 W/m^{2}K gives an even better correlation with the observed steel temperatures over the entire test duration as shown in Figure 1(b).

The current proposed model is also used to predict the temperature rise of a 530UB82 and 360UB44.7 steel beams (P/A$_s$ ratio of 176 m^{-1} and 236 m^{-1} respectively). The results in both cases, as shown in Figure 2, are in good agreement with that obtained by an equation provided in AS4100 (1990). The latter is based on experimental data using a regression analysis technique and is applicable up to 750°C. The current model adopted a value of 10 W/m^{2}K for h$_c$.

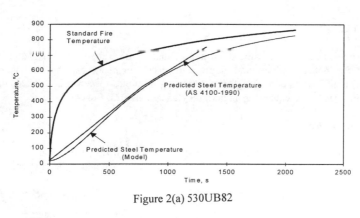

Figure 2(a) 530UB82

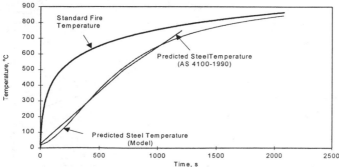

Figure 2(b) 360UB44.7

Figure 2. Temperature prediction in standard fire.

2.2 *Temperature profile in real fire*

While a standard fire follows an invariable fire curve simulating an extreme fire condition, the real fire curve simulates the actual fire which occurs in the structure. Although the current steel structures design code (AS4100 - 1990) requires the design to comply with the results obtained from a standard fire, it is more realistic and, in many cases, more economical to use the real fire curve to predict the temperature that the members of the structure may be subject to. Application of the above model to a number of documented cases of simulated real fires has indicated that h_c varies with time and is a function of fire load. However, from recent studies by the authors, reasonable results can be obtained for steel elements exposed to simulated real fires using h_c=20 W/m^{2}K. The reason for the need of a lower h_c than a standard fire value could be due to the lower degree of turbulence generated in standard fire tests where the air in the enclosure is essentially heated by burners supplied with fuel and air externally. For a real fire in an enclosure, both the reactants and oxidant are provided internally, with the flame being maintained by the constant inflow of air and outflow of combustion products inside the enclosure, causing turbulent mixing and flow of convection currents over the walls of the enclosure.

Figure 3 shows the results of temperature prediction for two unprotected steel beams (530UB82 and 360UB44.7) under real fire conditions. It is observed that the peak temperature of the steel members is always close to but lower than that of the gas temperature.

3 EXAMPLES

Design of simple beams can be carried out in accordance with AS4100 (1990). Detailed design procedures for both protected and unprotected steel members can be found in AISC publications (eg. Thomas et al 1992). The following examples demonstrate the calculation of the period of structural adequacy using both the standard and real fire curves.

3.1 *Simply supported beam*

The simply supported beam shown in Figure 4 is subject to a fire. The beam, of Grade 250, 530UB82 section, is fire-unprotected and the temperature is assumed to be uniform along the beam.

The design load, w, is 1.1G+0.4Q=21kN/m. Hence, the design moment is M*=212.6kNm.

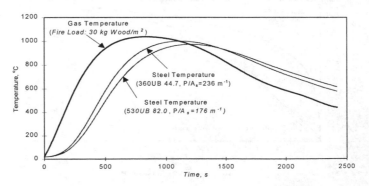

Figure 3. Steel temperature prediction in real fire.

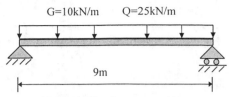

G=10kN/m   Q=25kN/m

9m

Figure 4. Simply supported beam.

Assuming that the beam is fully braced, the bending moment capacity at room temperature ϕM_s=463kNm. The strength limit state requirement for the beam at temperature T is

$$463\frac{f_y(T)}{f_y(20)} = 212.6 \tag{2}$$

where, according to AS4100 (1990), the yield stress ratio is

$$\frac{f_y(T)}{f_y(20)} = \frac{905-T}{690} \qquad \text{for } T > 215^{\circ}C.$$

The solution to Equation (2) gives T=588°C. From Figure 2(a) for a standard fire, the period of structural adequacy for the beam to reach this temperature is about 17 minutes. However, if the beam is subject to a real fire as that shown in Figure 3, the period of structural adequacy is about 10 minutes. It must be emphasised that the period of structural adequacy in a real fire varies according to, among other factors, the severity of the fire whereas it is a constant in a standard fire.

3.2 Fixed-end beam

The same beam in the previous example is analysed under the same fire conditions, but the ends of the beam are now fixed. The constraints at the ends of the beam induce an axial load to the beam when the temperature is increased, causing a possible failure due to bending and axial load interaction.

The design moment M* for the beam is found to be 141.8 kNm and the axial load, N*(T), in the beam due to temperature T is given as

$$N^*(T) = E(T)A\alpha(T-20) \tag{3}$$

where the coefficient of thermal expansion, $\alpha = 12\times10^{-6} / {}^{\circ}C$, Young's modulus, $E(20) = 2\times10^{8}$ kN/m^2, and ccross-sectional area, $A = 10400$ mm^2. The equation for the Young's modulus ratio, E(T)/E(20), can be found in AS4100 (1990). The effective length of the beam $L_e = 0.7\times9$ = 6.3m and therefore the member capacity under axial compression, ϕN_{cx}, at room temperature can be found to be 2060 kN.

Assuming that $T < 215^{\circ}C$ so that $f_y(T) = f_y(20)$ according to AS4100,

$$M^* \le \phi M_i(T) = \phi M_s(1 - \frac{N^*(T)}{\phi N_{cx}}) \tag{4}$$

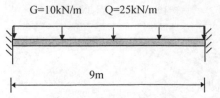

Figure 5. Fixed-end beam

Solving equation (4) for T gives T = 80°C. The corresponding period of structural adequacy for both the standard fire and the real fire can be found in Figure 2 and Figure 3 respectively.

It should be noted in this example that the beam being fixed at its ends has a detrimental effect on the beam's capability to resist the loads during a fire, which is in contrast to the fact that the beam can resist higher loads at room temperature when its ends are fixed.

4 CONCLUSION

A model for the calculation of the temperature rise in steel members under both standard fire and real fire is proposed. This model takes into account the heat transfer role of the gases produced as a result of combustion in a fire. It enables the temperature in the steel members to be predicted very accurately, resulting in an accurate assessment of the period of structural adequacy which is important for design of steel members in fire. This model can predict the temperature rise in steel subject to both standard and real fire conditions, provided that the fire curves are given.

REFERENCES

Mills, A.F. 1992. *Heat Transfer*, Irwin Inc.
Smith, C.I. & Stirland, C. 1983. Analytical methods and the design of steel framed buildings. *International Seminar on three Decades of Structural Fire Safety*, Fire Research Station: 155-200, Herts, UK.
Standards Australia 1990. AS4100 - 1990, Steel Structures, *SA*.
Thomas, I.R., Bennetts, I.D. & Proe, D.J. 1992. Design of steel structures for fire resistance in accordance with AS4100. *Steel Construction*, 26(3): 2-12, AISC.

Environmental loading, testing and response (seismic)

The Mechanics of Structures and Materials, Grzebieta, Al-Mahaidi & Wilson (eds)
© *1997 Balkema, Rotterdam, ISBN 90 5410 900 9*

Gap effects on the seismic ductility of brick infilled concrete frames

M.C.Griffith & R.Alaia
Department of Civil and Environmental Engineering, The University of Adelaide, S.A., Australia

ABSTRACT: This paper presents the results of an experimental study of reinforced concrete frames with brick infill wall panels subject to cyclic, horizontal loading. The concrete frame was detailed in accordance with the minimum seismic detailing requirements of the Australian Standard for Concrete Structures for an intermediate moment resisting frame. Four frames were tested. One had no infill and served as a reference for the three infilled frames. Three different construction gap sizes (15*mm*, 10*mm* and 5*mm*) between the concrete frame and brick infill were studied. The strength and ductility of the frames were evaluated from the experimental data and compared to the design values given in the Australian Loading Code - Part 4: Earthquake Loads. In these tests the brick infill increased the strength of the bare frame by 40% and improved the ductility of the bare frame, however, the system ductility was significantly less than the design value.

1 INTRODUCTION

One of the most common forms of construction for small to medium size buildings is concrete frame with clay brick infill. In order to design such construction, or assess the strength of existing construction, it is important to have an accurate estimate of the degree of interaction which can be expected to occur between the infill wall panel and the surrounding concrete frame during the design basis earthquake (DBE). Work by Page et al (1985), Stafford Smith (1962), Moghaddam and Dowling (1987), and Dawson and Ward (1972) indicate a wide range of interaction is possible with the frame geometries and material properties, as well as gap size all playing major parts.

Serviceability design considerations require that a gap be provided between the frame and infill so that cracking does not occur due to concrete creep and shrinkage, brick growth and thermal movement. For seismic design the in-plane strength and stiffness of the infill wall is usually neglected and some form of wall tie is used to maintain the wall's stability under seismically induced face loading. However, while the construction gap may be adequate to handle the slow, relative movement for which it was designed, it may not always be adequate to prevent interaction between the frame and brick infill during the large displacements expected to occur in the DBE.

In cases where no gap is provided, it is commonly assumed that the brick panel will fail, due to its lower strength, at an early stage of the DBE and the reinforced concrete frame will then be free to respond as if the wall was not in position. The complete, or partial, collapse of numerous buildings of this type of construction in Kobe, Japan (Pham and Griffith, 1995) and Northridge, California (EERI, 1995) suggests that this assumption may be unconservative, particularly for partial height infill walls.

Consequently, the Steel Reinforcement Institute of Australia and Adelaide University embarked on a two year research programme designed to investigate the seismic resistance of Australian designed concrete frame buildings with brick infill wall panels. The main focus was on the seismic response of the frame-infill wall system as a function of the size of gap between the infill wall and the surrounding frame members.

2 RESEARCH PLAN AND TEST CONFIGURATION

In order to study the effect of gap size on the seismic behaviour of concrete frames with brick infill, 4 frames were tested covering a range of 3 different gap sizes. The first test frame did not contain brick infill and served as a reference point for the three frames with infill panels. The three test frames containing brick infill were fabricated to have gaps between the columns and the brickwork of 15mm, 10mm, and 5mm, respectively.

The approximate frame size was 1.8m tall by 5m long. The tests were conducted in the Chapman Laboratory at the University of Adelaide. In this testing facility, the frames were anchored to the strong floor and loaded horizontally, with a series of progressively increasing quasi-static cyclic loads until failure was reached. This was achieved using a 1000kN hydraulic jack as shown below in figure 1.

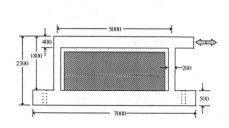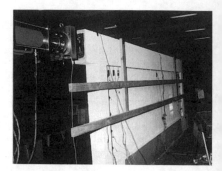

Figure 1: Frame dimensions and test frame prior to testing.

The beam and column sections were detailed in accordance with Appendix of AS 3600 (SA, 1988) which requires decreased spacing for lateral reinforcement adjacent to all beam-column connections. The reinforcement details, percentage steel reinforcement and material properties are shown in figure 2 and table 1, respectively. Further details of the design and fabrication of the test specimens has been reported previously (Alaia, Griffith and Freeman, 1997).

Table 1: Member details

Member	Dimensions (mm)	Longitudinal Reinforcement	% Steel
Column	200 × 200	8 Y16	4.0
Beam	400 × 200	8 Y16	2.0
Foundation	500 × 200	4 Y20	1.3
Property	Value	Property	Value
fsy	430MPa	Es	200,00MPa
f'c	32MPa	Ec	28,500MPa

Figure 2a: Column

Figure 2b: Beam

Figure 2: Reinforcing layout.

A total of 54 channels of data was collected for the test frames with brick infill using a data acquisition PC running the Visual Designer software. This package enabled triggered snapshot readings to be collected at the desired load increments. Readings were taken at 10kN load increments. The load history for test frame 3 is shown in figure 3. Three cycles of load were applied at each load amplitude until the specimen was no longer test worthy. The results of these tests are presented in the following section.

306

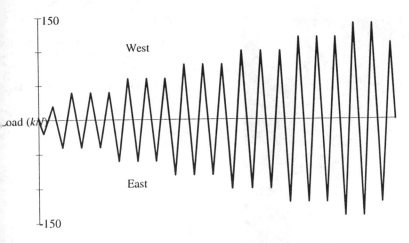

Figure 3: Load sequence graph for tests 3

EXPERIMENTAL RESULTS

.1 *Force-Displacement Data*

The force-displacement data for each frame test specimen is shown in figure 4. As can be seen, the three infilled frames exhibited reasonably stable post-yield behaviour although the hysteresis loops were pinched during the small amplitude loading cycles.. Substantial damping was only observed once the wall interacted with the frame.

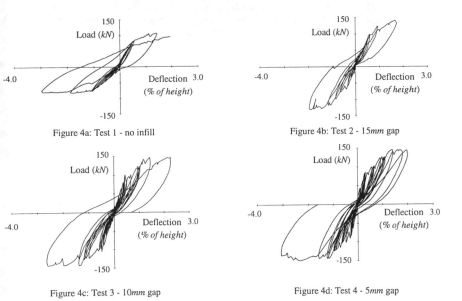

Figure 4a: Test 1 - no infill

Figure 4b: Test 2 - 15*mm* gap

Figure 4c: Test 3 - 10*mm* gap

Figure 4d: Test 4 - 5*mm* gap

Figure 4: Load-deflection graphs

307

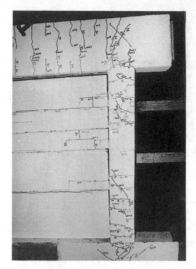

Figure 5: Horizontal planes in brickwork

3.2 *Failure Modes*

The test frames all failed with the columns forming plastic hinges at both top and bottom in classical column reverse curvature profile (see figure 5) The infill walls typically failed in shear However, the shear cracks did no propagate diagonally through the wall as expected. Instead, the wall failed in horizontal planes at approximately every 2nd or 3rd course of brickwork as shown in figure 5.

This was due to the wall's length to height ratio of 3.4 and the absence of vertical compression due to the gap between the wall and the beam. Hence the brickwork fractured into horizontal blocks and behaved as separate units With the transfer of load from the column to the wall, each unit slid along its bed until it came to bear against the opposite column. This caused the wall to exhibit the step appearance shown in figure 5 and became more apparent with reduced a gap size.

3.3 *Seismic Ductility*

Envelopes of the load-deflection plots for each test frame are shown in figure 6. It can be seen from the plot of data for test frame 1 (no infill) that the yield strength of the bare frame was about 95*kN* and that the frame/infill wall systems exhibited yield at similar load levels although at noticeably different displacements. It is also easy to identify the difference in strength between the bare frame and the frame/infill wall systems, with the frame/infill wall systems having ultimate strengths almost 50% greater than the bare frame.

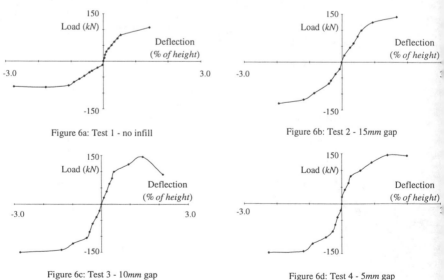

Figure 6a: Test 1 - no infill

Figure 6b: Test 2 - 15*mm* gap

Figure 6c: Test 3 - 10*mm* gap

Figure 6d: Test 4 - 5*mm* gap

Figure 6: Load-deflection envelopes

In order to estimate the ductility of the frame-wall system, the lateral displacement observed for which the frame could sustain at least 3 cycles of load without suffering more than a 30% drop in stiffness was divided by the lateral displacement recorded at onset of yield in the frame. This can best be illustrated by the load-deflection maxima plot for each cycle of test frame 3, shown in figure 7. Here, three values of displacement are plotted for each value of load. For a given force, the smallest value of displacement (Δ_{1MAX}) corresponds to the maximum displacement measured during the 1st loading cycle (F_{1MAX}) at that particular amplitude. The intermediate value of displacement (Δ_{2MAX}) corresponds to the maximum displacement measured during the 2nd loading cycle and similarly, the largest value of displacement corresponds to the value recorded during the 3rd loading cycle. In accordance with Californian practice, it was assumed that the system stiffness (F_{1MAX}/Δ_{3MAX}) during the 3rd loading cycle had to be greater than 70% of the initial cycle's stiffness (F_{1MAX}/Δ_{1MAX}) in order to be deemed acceptable as an indication of seismic ductility.

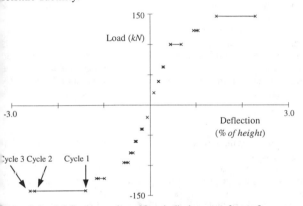

Figure 7: Load-deflection maxima of 3 cycle displacements for test 3

Using this criteria, the seismic ductility for test frames 3 and 4 was estimated to be 2.1 and 2.5, respectively. Test frames 1 and 2 were not tested with 3 loading cycles at each load amplitude. However, the static ductility for test frames 1 and 2 were estimated to be approximately 2. It would be expected that the seismic ductility would be substantially less.

The ductility values indicated from these tests is somewhat less than the value of 2.8 implied by the results of experiments at Melbourne University on beam-column joints (Corvetti et al, 1993). There are several possible reasons for the difference. First, the value of 2.8 was not estimated by applying a stiffness degradation limit. Furthermore, the absence of axial precompression in the columns during the frame tests reduced the strength and stiffness of the columns.

The Australian Loading Standard: Part 4 - Earthquake Loads (SA, 1993) specifies a structural response factor R_f of 6 for an intermediate moment resisting frame of concrete. The value of ductility implied by this factor is approximately 4 (R_f = 1.5 for unreinforced masonry construction). Clearly, there is a large discrepancy between the design value and that implied by the experimental results with the design value being unconservative by a factor of approximately 2. Nevertheless, the test results suggest that the frame/infill wall system ductility improves with a decreasing gap size.

4 IMPLICATIONS FOR DESIGN

It is widely accepted that unreinforced brick infill panels substantially increase the stiffness of a concrete frame building. On the other hand, it is also assumed that brick infill panels will fail (due to their low strength) very early on during the DBE, leaving the frame to resist the earthquake forces on its own. However, this assumption has been put under scrutiny and found to be conservative.

309

Based on this research, it appears a full-height brick infill wall can increase the strength and ductility of the bare concrete frame. Therefore, there may be potential cost savings if some account can be made for the strength and energy dissipating capacity of the infill walls.

On the other hand, if the infill wall is to be ignored, it is important to design the gap size to allow for the realistic storey drift which would occur in the DBE. Usually the gap size is fabricated to a standard detail which takes into account only concrete creep and shrinkage, brick growth and thermal effects. If a sufficient gap size is not provided, interaction with the infill wall will be inevitable and the earthquake response load paths will be redirected, perhaps with unfavourable consequences.

The results of this research also suggests that the structural response factor (R_f) for an intermediate moment resisting frame of concrete is unconservative. Part 4 of the Australian Loading Code: Earthquake Loads recommends a value for R_f of 6 whereas this research indicates that a value of 3 to 4 may be more appropriate. Research on beam-column joints at Melbourne University (Corvetti et al, 1993) also suggests that the code value may be unconservative, although not by as much as this work indicates.

Finally, it is recognised that for less ductile forms of construction, strength and stiffness properties play an important part in determining whether a building successfully withstands seismic events. The experiments reported on in this paper clearly show that the infill walls add substantially to both the strength and the stiffness of the concrete frame. This should be accounted for in the seismic design of such buildings unless the construction gap is sufficiently large to prevent frame/infill wall interaction.

5 ACKNOWLEDGMENTS

This paper presents the results of work which was sponsored by the Steel Reinforcement Institute of Australia (SA Division) and the Australian Research Council. In particular, the authors are grateful to Steve Freeman of Smorgan ARC who provided valuable technical advice and guidance to the project. Additional support provided by CSR, Abey Australia and Ramset is also gratefully acknowledged.

The findings and conclusions expressed are those of the authors and not necessarily those of the sponsors.

6 REFERENCES

Alaia, R., Griffith, M.C. and Freeman, S. 1997. An investigation into the seismic behaviour of reinforced concrete frames with brick infill panels. *Proceedings of Concrete 97, 18th Biennial Concrete Institute of Australia Conference*. Adelaide, Australia:433-439.

Corvetti, J., Goldsworth, H. and Mendis, P.A. 1993. Assessment of reinforced concrete exterior beam-column joints as specified in the new draft earthquake standard. *Proceedings, 13th Australasian Conference on the Mechanics of Structures and Materials, Wollongong*. Vol. 1: 225-232.

Dawson R. V. and Ward M. A. 1972. Dynamic response of framed structures with infilled walls. *Proceedings 5th World Conference on Earthquake Engineering*, Rome, Italy: 1507-1516.

Earthquake Engineering Research Institute. 1995. *The Hyogo-Ken Nanbu Earthquake - Preliminary Reconnaissance Report*. EERI report 95-04, Oakland, California.

Moghaddam, H. A. and Dowling, P. J. 1987. The state-of-the-art infilled frames. ESEE Imperial College of Science and Technology, Res. Rep. 87-2.

Page, A. W., Kleeman, P. W. and Dhanasekar, M. 1985. An In-Plane Finite Element Model for Brick Masonry. *New Analysis Techniques for Structural Masonry*, ASCE: 1-18.

Pham, L. and Griffith, M. 1995. *Report on the January 17, 1995 great Hyogo-Ken Nanbu (Kobe) earthquake*. CSIRO report DBCE 95/175, Melbourne, Australia.

Stafford Smith, B. 1962. Lateral stiffness of infilled frames. *Journal of the Structural Division, Proceedings of the American Society of Civil Engineers*. Vol 88, No ST6:183-199.

Standards Australia. 1988. *Australian Standard 3600 - Concrete Structures*. Sydney, Australia.

Standards Australia. 1993. *Australian Standard 1170.4 - Minimum design loads on structures - Part 4: Earthquake loads*. Sydney, Australia.

The Mechanics of Structures and Materials, Grzebieta, Al-Mahaidi & Wilson (eds)
© *1997 Balkema, Rotterdam, ISBN 90 5410 900 9*

Seismic effects on wide band beam construction in Australia: Review

K.Abdouka & H.Goldsworthy
Department of Civil and Environmental Engineering, The University of Melbourne, Vic., Australia

ABSTRACT: The Australian continent is within a tectonic plate and hence in a region of low seismicity in global terms. However there have been several moderate events of sufficient magnitude to justify a closer inspection of the detailing requirements specified in the Australian Standards for reinforced concrete structures. The aim of the research outlined in this paper and in the companion paper is to consider the probable behaviour of frames with wide band beams if subjected to moderate level seismic events. It will be a theoretical and experimental study of this type of frame. The focus of this paper is on the literature review of previous research and a description of the proposed experimental work. Some researchers in the USA and Japan have recently explored the possibility of using wide shallow beams in frames in regions of high seismicity. Key differences between tests conducted by those researchers and those to be performed as part of this project will be outlined here. This paper has a companion paper (Seismic Effects on Band Beam Construction in Australia - A preliminary analysis).

1 LITERATURE REVIEW

1.1 *Introduction*

The wide band beam type of construction is common in Australia mainly because of its smaller floor-to-ceiling heights and easier formwork construction. However, the acceptability of this type of construction in regions of low to moderate seismicity such as Australia needs to be verified. In regions of high seismicity current code provisions do not encourage the use of this type of construction for ductile earthquake resistant design. For example the ACI-ASCE Committee 352 (Monolithic Connections in Reinforced Concrete Framed Structures) does not recommend beams wider than columns. ACI 318-89 limits the beam width to " column width plus 1.5 beam depth". This limit is believed to be an arbitrary one (Gentry and Wight 1994). The New Zealand Standard for the Design of Concrete Structures NZS3101-1982 recommends an upper limit on the beam width, ie. that it should not exceed the width of the column plus a distance on each side of that column equal to one-fourth the overall depth of the column in the relevant direction, but not more than twice the width of the column. The Australian Standard for Concrete structures AS3600-1989 has been revised in AS3600-1994 due to the introduction of the Australian Standard for Earthquake Loading AS1170.4-1993. However the revisions of AS3600 do not specifically cover wide band beams .

1.2 *Relevant research on wide beam-column joints*

In their experimental investigation of the seismic behaviour of exterior wide beam-to-column connections, Gentry and Wight (1994) studied the effects of the following: (a) Joint shear stress level. (b) Fraction of longitudinal beam reinforcement anchored in the column core. (c) Beam width to column width ratio. (d) Column flexural strength to beam flexural strength. Four 3/4 scale reinforced concrete subassemblages were tested in a fixture with pin connections at column mid-height and beam mid-span. The pinned ends were at points of contraflexure assuming lateral loads only and a portal frame action. The specimens were loaded in a displacement controlled manner using a servo-controlled actuator. The loading process was quasi static and the displacement pattern was increased in cyclic increments of 0.5% to a maximum of 5% total interstorey drift. The most important finding of the above investigation was that the behaviour of an exterior wide beam-column connection depends mainly on the stiffness and strength of the transverse beam and its ability to transfer forces (by torsion) from the wide beam's longitudinal bars into the column core. The longitudinal bars that do not pass through the column core become ineffective in resisting forces when the transverse beam fails by torsional cracking (Figure 1). This observation was also reported by other researchers (Durrani and Zerbe 1987, French and Boroojerdi 1989). Gentry and Wight found that Hsu's formula $T_{cr} = 2\sqrt{fc'}\, x^2 y$ (Hsu 1968) gave a very good prediction of the torsional cracking strength of the transverse beam. It was concluded that wide beam-column connections can be used in zones of high seismic activity provided that the width of the beam is limited to column width plus twice the column depth. This limitation is indicative of the fact that the effective beam width is more closely related to the depth of the column than it is to the depth of the wide beam. It is more in line with the NZS 3101 approach, although allowing a larger width, rather than the ACI 318-89 approach. Another recommendation by the same investigators was that the ACI 318-89 seismic provision for the stirrup maximum spacing in the plastic hinge zone may be increased from d/4 to 3d/8 as long as the calculated shear stress does not exceed $2\sqrt{fc'}$ (in psi) . In addition the requirement for a stirrup leg around every other bar can be waived and instead a minimum of four stirrup legs may be provided for beams wider than 24 inches (approximately 610 mm). The observation that the slip of longitudinal column bars through the joint was not pronounced led to the recommendation that the lower limit of the ratio of beam depth to column bar diameter to be reduced from 20 to 16. The limit of 20 is currently required by ACI-ASCE Committee 352.

Another experimental research program on wide beam-column connections was conducted at the Kajima Institute of Construction Technology, Japan by Hatomoto, Bessho and Matsuzaki (1991). In their research program the investigators conducted two series of tests. The first series involved testing six one-sixth models to study the effect of the ratio of beam width to column width, with this ratio varying from 1 to 4. In the second series , four one-half scale subassemblages were tested to determine the maximum allowable amount of beam reinforcement placed outside the joint core. The main findings of the Kajima test series were:

(1) The width of the beam should not exceed twice the width of the column to ensure that all longitudinal beam bars were to yield at 2% lateral interstorey drift. This agrees with the upper limit on beam width in NZS 3101 and was determined experimentally from monitoring the strain distribution in the longitudinal beam bars for different drift ratios.

(2) The maximum amount of longitudinal beam reinforcement not passing through the column should be restricted in order to limit the resulting torsional stress in the transverse beam adjacent to the joint core. The ultimate torsional shear strength of $24\sqrt{fc'}$ as suggested by Kanoh and Yoshizaki (1979) was found to be a good criteria for limiting the torque at the side faces of the joint core.

(3) The stiffness of wide beam-to-column subassemblages is relatively low compared with that of ordinary frames. Therefore, large horizontal drifts under lateral seismic loads may be an important consideration in design .

Popov, Cohen , Koso-Thomas and Kasai (1992) tested one one-half scale wide beam-column specimen. They reported that the presence of large well reinforced transverse beams is essential to ensure the good performance of the wide beam frames under large seismic induced drifts.

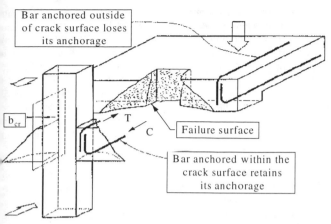

Figure 1. Failure surface of exterior wide beam-column joint. (after Durrani and Zerbe 1987)

1.3 *Relevant research on conventional beam-column-slab connections*

Numerous research projects have been conducted in the last 30 years on the seismic behaviour of conventional beam-column-slab connections, ie. where the width of the beam did not exceed the column width. Different aspects of the seismic behaviour were studied, the most important ones being the strength and stiffness degradation of the connection, the energy dissipation characteristics and the limits of interstorey lateral drifts that these connections can sustain. Early research considered subassemblages consisting of beams and columns only ie. the slab was not included.

Meinheit and Jirsa (1981) conducted a series of tests on beam-column joints to measure the basic shear strength characteristics of reinforced concrete beam-column connections and to examine methods of improving this shear strength. The major parameters studied were (1) Percentage of column reinforcement. (2) Nominal axial stress on the column. (3) Size and spacing of transverse (hoop) reinforcement in the connection core. (4) The effect of unloaded transverse beams. (5) The aspect ratio of the beam-column connection. The conclusions of the study were that the cross sectional area of the connection was the major factor in the strength of the connection and to a lesser degree the transverse reinforcement in the joint, and that the presence of transverse beams improved the shear capacity of the joint. It was observed that the column longitudinal percentage of reinforcement was not as important as the column's transverse reinforcement on the shear capacity. In addition the connection geometry had no effect on the shear strength of the joint provided that the shear area of the joint remained constant. It was found that the column axial load level did not influence the ultimate shear strength but with higher axial loads the shear at first cracking was increased. The former contradicts with the finding of other researchers (Park and Dai, 1988 , Cheung et al. ,1991) who have reported a strong relationship between ultimate shear strength and axial load level.

Ehsani and Wight (1985) investigated the effects of transverse beams and slabs on the behaviour of exterior reinforced concrete beam-column connections. The parameters varied in the tests were the ratio of the flexural capacity of the column to the flexural capacity of the beam, the percentage of transverse reinforcement and the shear stress within the joint. It was found that joint behaviour improved considerably if unloaded transverse beams were present. In fact the joint shear capacity did not improve with increase in joint reinforcement if transverse beams were present.

Leon (1989) tested four one-half scale interior beam-column joints without slabs. The objectives of this experimental study were to determine the interaction between shear strength and bond deterioration and to quantify the effects of column size on the hysteretic behaviour and shear performance of the joint. In addition the investigators sought to verify the current ACI 318 and ACI-ASCE 352 provisions. One important recommendation of the study was that the ratio of the sum of column moment capacities to beam moment capacities for design purposes should be taken as being between 1.6 and 1.8. This is greater than 1.4 which is required by ACI-ASCE 352 and than 1.2 which is required by ACI 318-83. It was also reported that some bond slip in the beam longitudinal bars may result in the closure of the flexural cracks at the beam-joint interface and hence enhance the concrete contribution in carrying the shear force in the core through the diagonal strut mechanism. This reduces the need for joint hoop reinforcement that is required to sustain the panel truss mechanism. The last observation was also reported by Cheung et al. (1991) and by Park and Dai (1988).

Paultre et al. (1989) conducted tests on reinforced concrete beam-slab-column subassemblages to investigate the seismic performance of frame structures designed in accordance with the Canadian Code for the design of Concrete Structures CSA standard CAN-A23.3 M84. Five full scale specimens representing external beam-column-slab subassemblages of a six storey moment resisting frame were tested. It was observed that by ignoring the contribution of the slab in tension an overestimate of the ratio of column flexural strength to beam flexural strength is obtained. This may lead to the undesirable weak column- strong beam mode of failure. The investigators reported also that the effectiveness of the slab reinforcement depends on the torsional capacity of the transverse beam.

Pantazapoulou et al. (1988) proposed a simple analytical model capable of predicting the contribution of the slab to the strength and stiffness of the beam. The factors that influenced the amount of slab participation were beam depth, steel material properties, maximum available slab-column width and maximum reinforcement strain.

Zerbe and Durrani (1990) studied the effect of slab continuity in two-bay concrete frames and concluded that the presence of the slab in multiple connection subassemblages increased lateral-load resistance by up to 40% and the loss of stiffness in these connections was smaller and more gradual than the loss in the continuous specimen without the slab.

1.4 Relevant research on flat plate construction

Pan and Moehle(1989) analysed available data from previous research and from their own experimental work to develop an understanding of the major parameters that influence the lateral displacement capacity and ductility of reinforced concrete flat-plates. The following main conclusions were made: (1) The magnitude of shear carried by the slab due to gravity load is a major factor affecting lateral displacement and ductility. (2) The magnitude of gravity loads and lateral interstorey drifts should be controlled to ensure the integrity of the slab-column connection under seismic loads. The investigators found that for an interstorey drift less than 1.5% the flat plate connection will perform adequately if the shear stress in the critical section does not exceed $1.5 \sqrt{fc'}$ (in psi).

Robertson and Durrani (1991 and 1992) conducted tests on two-bay flat plate subassemblages to assess the gravity load effect on the seismic behaviour of both exterior and interior joints. For interior joints they concluded that the slab gravity load significantly reduces the stiffness of the connection and its ability to transfer unbalanced moment due to lateral loading .

Durrani et al. (1995) investigated the seismic resistance of non-ductile slab-column connections in existing flat-slab buildings. The main conclusions of the research were: 1) The mode of failure of non-ductile flat-plate interior connections is flexural under service gravity loads. The connections fail in punching shear when the gravity load increases. 2) Cracking of the slab due to gravity load reduced the stiffness of the flat plate subassemblage under lateral loading leading to approximately 50% of the initial stiffness being lost during the first 1% drift and an additional 25% at 2% drift. 3) Exterior connections lost their stiffness faster than the interior connections. The exterior connection lost 70-75% of the stiffness in the negative

314

moment direction at 1% lateral drift. In the positive moment direction, the stiffness became negligible after the slab reached its cracking strength. As the exterior connection lost its stiffness, gravity shear was redistributed to the adjacent interior connection thus protecting the exterior connection from punching.

2 A BRIEF DESCRIPTION OF THE PROPOSED EXPERIMENTAL WORK:

A four storey six bay reinforced concrete frame with wide band beams has been analysed and designed in accordance with the AS3600-1994 and to loading codes AS1170.1, AS1170.3 and AS1170.4. Under earthquake lateral loads only, points of contraflexure occur in the columns at mid-height and at beam midspan. This observation allows testing of subassemblages of the whole frame between the points of contraflexure. The proposed experimental program consists of testing three external and three internal subassemblages representing the beams at the first floor level and the columns above and below that level. The test specimens will be one-half scale and will each consist of a lower column, an upper column and a wide beam. The ends of the beam and the columns will be pinned representing the points of contraflexure in the real structure. The effects of the gravity loads will be modelled using a combination of lead blocks on the beam and axial compressive loads on the columns. The axial column load will be provided by jacking prestressing cables against the ends of the columns. The lateral load simulation will be provided by a hydraulic actuator to the pin end of the upper column and will be in the form of displacement controlled quasi-static gradually increasing cycles of lateral displacement in a similar manner to the tests performed by Gentry and Wight (1994). The test specimens will be extensively instrumented with strain gauges to measure the strains in the reinforcement bars, extensometers , LVDTs, load cells and dial gauges The data acquisition system includes an Orion Data Logger, a Houston Instruments X-Y plotter and a 486-IBM compatible PC. The main objective of the experimental program is to determine the failure mechanism of the wide band beam-column connections due to the lateral loading.

The main difference between the proposed experimental program and that conducted in the USA and Japan is that the proposed specimens will be built according to non-ductile detailing requirements, in which case it is not compulsory to design to capacity design principles or to ensure that a strong column-weak beam hierarchy exists. The tests in the USA by Gentry and Wight (1994) and in Japan by Hatamoto et al. (1991) were directed towards assessing the behaviour of wide band beam frames in regions of high seismicity. Some of the differences can be summarised as follows: (1) The ratio of the width of the wide beam to column width in the proposed program will be as large as 6 compared with a maximum of 2.34 in the USA tests and 3.57 in the Japanese tests. (2) Gravity load effects will be modelled and applied to the test specimens. This effect was ignored in the USA and Japan tests and the specimens were tested only for lateral loading. (3) No shear reinforcement is required in the proposed test. This a major difference from the USA and Japan tests where beams were well reinforced for shear at the anticipated location of plastic hinge formation.

REFERENCES

American Concrete Institute 1976. *ACI-ASCE joint committee 352, Recommendations for Design of Beam-Column Joints in Monolithic Reinforced Concrete Structures.* Detroit:ACI
Cheung,P.C., Paulay,T.& Park, R. 1991. *Some possible revisions to the seismic provisions of the New Zealand Concrete Design Code for moment resiting frames.* Pacific Conference on Earthquake Engineering .New Zealand 20-23 Nov 1991: 79-90.
Durrani ,A.J. ,Du, Y. & Luo,Y.H. 1995. *Seismic resistance of non-ductile slab-column connections in existing flat-slab buildings.* ACI Structural journal Vol. 92 No. 4 :479-487
Durrani,A.J. & Zerbe,H. 1987. *Seismic Resistance of R/C exterior connections with floor slabs.*

Journal of Structural Engineering ASCE Vol.113 No.8: 1850-1864.

Ehsani, M.R. & Wight, J.K. 1985. *Effect of transverse beams and slab on behaviour of reinforced concrete beam-to-column connections.* ACI Journal March-April :188-195.

French,C. & Boroojerdi, A. 1989. *Contribution of R/C floor slabs in resisting lateral loads.* Journal of Structural Engineering Vol.115 No.1: 1-18.

Gentry,R.T. & Wight, J.K. 1994. Wide beam-column connections under earthquake-type loading. *Earthquake Spectra* Vol 10 No. 4: 675-703.

Hatamoto,H.,Bessho,S. & Matsuzaki, Y. 1991. *Reinforced concrete wide beam-to-column subassemblages subjected to lateral loads.* Design of beam-column joints for seismic resistance. American Concrete Institute's special publication SP-123. Michigan: ACI :291-316

Hsu, T. 1986. *Torsion of structural concrete-plain concrete rectangular sections.* Torsion of Structural Concrete. American Concrete Institute's special publication SP-18. Michigan :ACI:103-238.

Kanoh,Y. & Yoshizaki,S 1979. *Strength of slab-column connections transferring shear and moment.* ACI Journal, March: 461-478.

Leon, R. 1989. *Shear strength and hysteretic behaviour of interior beam-column joints.* ACI Structural Journal Vol. 87 No. 1 January-February : 2261-2275

Meinheit,D. & Jirsa, J. 1981. Shear strength of R/C beam-column connections. Journal of the Structural Division ASCE Vol 107 No. ST11: 2227-2245.

Pan, A. & Austin ,J.P. 1989. *Lateral displacement ductility of reinforced concrete flat plates.* ACI Structural Journal Vol. 86 No. 3: 250-258.

Pantazopolou,S.J., Moehle,J.P. & Shahrooz, B.M. 1988. *Simple analytical model for T-beam in flexure.* Journal of Structural Engineering ASCE Vol. 114 No. 7 : 1507-1523.

Park,R. & Dai,R. 1988. *A comparison of the behaviour of reinforced concrete beam-column joints designed for ductility and limited ductility.* Bulletin of the New Zealand National Society of Earthquake Engineering Vol.21 No. 4 : 255-278

Paultre,P., Castele,D., Rattary,S. & Mitchell,D. 1989 .*Seismic response of reinforced concrete subassemblages-a Canadian code perspective.* Canadian Journal of Civil Engineers Vol.16: 627-649 .

Popov,E.,Cohen,J.,Koso-Thomas, K. & Kasai, K. 1992. *Behaviour of interior narrow and wide beams.* ACI Structural Journal Vol.89 No.6 : 607-616

Robertson, I.N. & Durrani, A.J. 1992. *Gravity load effect on seismic behaviour of interior slab-column connections.* ACI Structural Journal Vol.89 No.1: 37-45

Robertson, J.N. & Durrani, A.J. 1991. *Gravity load effect on seismic behaviour of exterior slab-column connections.* ACI Structural Journal Vol. 88 No. 3 : 255-267

Standard Association of Australia 1993. *Minimum Design Loads on Structures part 4 Earthquake Loads AS1170.4.* Sydney:SAA

Standard Association of Australia 1994. *Concrete Structures Code AS3600* . Sydney: SAA

Standard association of New Zealand 1982. *The New Zealand Standard for the design of Concrete Structures NZS 3101.* Wellington:NZS.

Zerbe,H.E. & Durrani, A.J. 1990. *Seismic response of connections in two-bay reinforced concrete frame subassemblies with a floor slab.* ACI Structural Journal Vol.87 No.4 :406-415.

The Mechanics of Structures and Materials, Grzebieta, Al-Mahaidi & Wilson (eds)
© *1997 Balkema, Rotterdam, ISBN 90 5410 900 9*

Ductility of steel connections in concentrically braced frames under seismic forces

D.G.Wallace, H.M.Goldsworthy, D.S.Mansell & J.L.Wilson
Department of Civil and Environmental Engineering, The University of Melbourne, Vic., Australia

ABSTRACT: Steel Concentrically Braced Frames (CBFs) designed for seismic forces in accordance with current Australian practice are assumed to possess a level of ductility and overstrength sufficient to justify a response reduction factor (R) of five. This assumption implies that some earthquake energy will be dissipated through inelastic behaviour, specifically the buckling of the bracing members. However, the Australian steel code AS4100 does not require that the connections in the majority of CBFs be designed for the actual strength of the bracing members. Instead, the bracing member design load is used. A summary is presented of work currently under way to test the actual strength and ductility capacity of connections in a CBF designed for conditions typical of Melbourne. The results of a preliminary Finite Element Analysis (FEA) predict that the strength of the selected critical connection is less than the theoretical buckling capacity of the brace member. The series of practical investigations outlined is expected to confirm these FEA results.

1. INTRODUCTION

Until recently, design requirements for the structural resistance of seismic forces in Australia were essentially confined to large buildings located in Adelaide. This changed in 1993 with the introduction of the standard AS1170.4 Earthquake Loads. This new code specifies requirements for all types of buildings and is based predominantly on extensive Californian experience.

Current Australian construction practices and materials differ from those used in North America. In particular, field welding is used widely in North America whereas field bolting is preferred in Australia. The type of ground motion encountered in Australia also differs from that observed in California. Australia lies entirely within the Indo-Australian Plate so most earthquakes are due to localised intra-plate activity. These differences require investigation to verify the extrapolation of Californian experience to Australian conditions.

Adapted directly from the North American earthquake codes with little Australian experimental basis, the response reduction factor (R) is intended to account for the anticipated overstrength and ductility present in various forms of construction. It is assumed that the magnitude of the seismic forces attracted to a structure will be limited to a factor of R less than those that would occur if the structure were to remain fully elastic. Several factors, including serviceability deflection limits, contribute to overstrength in Concentrically Braced Frames (CBFs) whilst the ductility is assumed to be provided by the yielding of braces at overload. For this energy dissipation mechanism to develop, the brace connections must be at least as strong as the bracing member. Otherwise it is more likely that the connections will fail first. The Australian steel design code AS4100 only requires that the connections in CBFs be designed for the actual capacity of the bracing member in structures conforming to Earthquake Design Categories D&E. In most Australian population centres the only buildings to qualify for these categories are Type III structures (AS1170.4: 'buildings that are essential to post-earthquake

317

recovery') or Type II structures (AS1170.4: 'buildings that are designed to contain large numbers of people') on poor soil such as soft clays. Therefore, the connections in CBFs which fit the lower design categories (A to C) only need to be designed for the bracing member design load.

This study involves the testing of bracing members and connections from a design category B structure. The purpose of the study is to determine the level of ductility present in the selected CBF in order to assess the adequacy of current design requirements for connections.

2. CONCENTRICALLY BRACED FRAMES

By definition, a frame is said to be concentrically braced when the lines of action of the brace, beam and column intersect at a single point at each connection. The bracing may be designed to resist either tension only or both tension and compression forces. CBFs are used throughout the world in low- to high-rise buildings. Their inherent stiffness reduces lateral deflections, thereby minimising damage to non-structural elements and injury to occupants during significant seismic events.

CBFs with braces designed for tension only have been demonstrated to possess poor resistance to cyclic loading. This is due to the permanent stretching that occurs when the braces are overstressed which results in free sway of the frame on subsequent load reversals. Even when the braces are designed for both tension and compression, CBFs provide limited ductility under cyclic ultimate loads due to the decrease in compression capacity of buckled braces under repeated load reversals. This results in pinched hysteresis loops and therefore reduced energy absorption (Kahn & Hanson 1976; Popov 1980, pp. 1457-1458; Popov 1982, p. 148).

The seismic response of CBFs has been investigated and reported by many researchers. These investigations have typically involved full- or half-scale cyclic load tests (Maison & Popov 1980; Roeder 1989; Yamanouchi et al. 1989) or non-linear computer analyses using available computer packages (Goel & Hanson 1974; Anderson 1975; Gillies & Shepherd 1981). The researchers of these studies concluded that the buckling of the braces greatly reduces the ductility of CBFs and that the data obtained from tests on individual braces can be used directly in analytical studies to determine the overall behaviour of CBFs. This is because the braces contribute the major part of the total lateral stiffness to the CBF even when the beam-column connections are designed to develop the full moment capacity of the beam (Popov 1980, p. 1458).

3. SELECTION OF THE BRACING MEMBER AND CONNECTION TO BE TESTED

The sample building outlined by van der Kreek (1995) was selected as a suitable structure. This six-storey, rectangular office building is designed for weather and ground conditions typical of Melbourne and utilises steel frames for lateral load resistance. Moment Resisting Frames are located at the outer walls parallel to the longer building axis with CBFs at the lift core parallel to the shorter axis. AS1170.4 does not require that the building be designed for seismic forces on the assumption that it is ductile so wind loads constitute the design load for the bracing members.

Standard BHP 300PLUS grade sections were selected for the frame. UC sections were used for all beam and column members with 350MPa grade Square Hollow Section (SHS) used for the braces. The critical connection was denoted by the highest ratio of actual brace capacity to brace design load and was located in the first storey. The connection was then designed for a load of 591kN (factored for ultimate limit state) while the brace had a factored axial capacity (ϕN_c) of 826kN. The resulting connection is shown in Figure 1 below. The SHS is slotted along its centre-line on to the gusset plate, with 6mm welds at each of the four intersecting edges.

4. FINITE ELEMENT ANALYSIS

It is the nature of connections between steel structural elements that stress concentrations are induced in the connected parts. These zones of high stress can have a significant influence on the behaviour of a connection at ultimate load. In order to investigate these effects and establish an

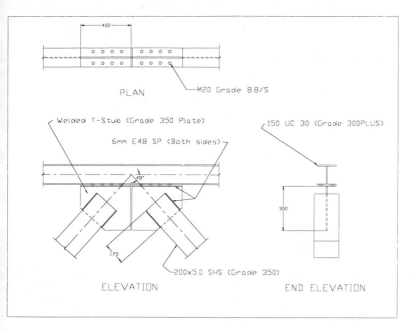

Figure 1. Brace connection geometry

estimate for the ultimate capacity of the particular connection to be tested for this study, a finite element analysis was conducted. The analysis package ANSYS 5.1 was used for this purpose.

4.1 Model Development

The SHS and gusset portions of the connection were modelled using 4-node plastic shell elements. Two longitudinal planes of symmetry were utilised so that only one quarter of the connection was modelled. The weld was simplified by a series of non-linear spring elements. At eleven points along the weld, three springs were used to connect the SHS to the gusset in each of the three translational degrees of freedom (DOFs). The three rotational DOFs were unrestrained.

A series of six Finite Element Models (FEMs) of increasing complexity were devised to ensure that valid results would be obtained. Initial FEMs assumed similar conditions to those adopted by Girard et al. (1995) to allow some verification of results. These models assumed a perfect connection between the brace and gusset plate (ie. the weld was ignored). The results agreed closely with those of Girard et al. (1995).

It was found that the coarseness of the mesh in the vicinity of the stress concentrations did not provide satisfactory resolution of the stress state. Hence, more refined FEMs were produced with a finer mesh around the weld. An illustration of the model is included as Figure 2 below.

The stress-strain curves required for the brace and gusset plate were developed using the Ramberg-Osgood function as described by Jennings (1964) and an assumed material yield stress of 350MPa. This function provides a smooth transition from elastic to plastic material behaviour which solved the convergence problems previously encountered when an elastic-plastic material model was used. The load-deflection curve for the springs in the transverse and longitudinal weld directions were determined using data and equations presented in papers by Miazga & Kennedy (1989) and Lesik & Kennedy (1990). These equations account for the effects of the direction of loading on the strength and ductility of welds, and are based on data collected for E48 welds connected to 300W grade plate.

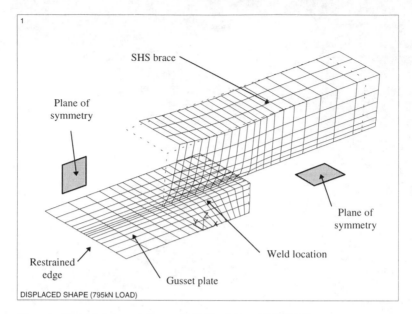

Figure 2. Model 5 deformed shape at 795kN load (from ANSYS 5.1)

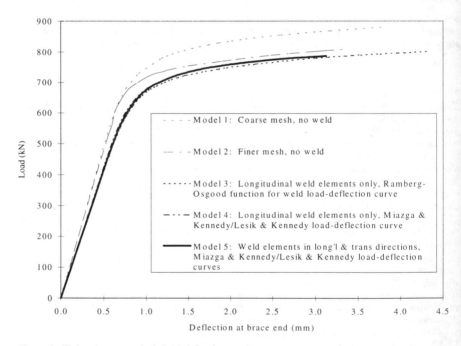

Figure 3. Finite element analysis load-deflection curves

4.2 Analysis Results

The force-deflection curves obtained from the non-linear analyses are shown in Figure 3 below. Generally, with increasing model complexity, the ultimate load decreased. This is due to the increased resolution of the stress state around the weld which reduced the influence of the localised failure zone on the rest of the connection, thereby reducing the ultimate load. The yield load for each FEM was approximately 600kN (slightly higher for the earlier models) and the ultimate load was around 800kN. The failure mechanism predicted by each model was a ductile tear along the SHS wall adjacent to the weld. The tear initiated at the end of the brace in the region subject to the highest stress concentration.

Throughout each analysis, it was observed that the gusset remained essentially elastic whereas significant yielding occurred in the SHS wall adjacent to the weld. At ultimate load, the stress distribution in the brace wall along the length of the weld was nearly uniform. The influence of the weld on the overall behaviour of the connection can be seen by comparing models 2&3 with the latter models. Although the ultimate strength was not greatly effected, the initial elastic stiffness is lower for the FEMs containing the weld.

5. PRACTICAL INVESTIGATION

A series of experiments of increasing complexity are to be performed. Initially, a simple static tension test of a specimen dimensionally similar to the FEM will allow verification of the FEM. The actual strength of the joint will be determined and compared with the nominal strength and that predicted by the FEM.

The next test will involve cyclic loading of the connection. This will highlight any effects induced by compression such as plate buckling. It will also exhibit the energy dissipation characteristics of this type of connection. The bolted connection between the gusset and the column or beam members will be included in these tests to identify the effect of bolt slippage or failure on the overall behaviour.

Once the behaviour of the joint has been studied and quantified, a scaled bracing member with connection will be tested under cyclic load. From this, the ultimate load, energy dissipation characteristics and failure mechanism for the system will be established.

6. CONCLUSION

The failure mode for a brace to gusset plate connection from a CBF designed for typical Melbourne conditions has been studied. The results of an initial FEA predict that failure will be initiated in the wall of the SHS brace adjacent to the weld at the end of the SHS. Even though the FEM predicts failure in the brace wall it is likely that the actual failure will occur in the weld. This is because the FEM cannot account for defects and residual stresses in the weld which reduce the weld capacity. The capacities determined in accordance with AS4100 also predict that the weld is weaker than the SHS wall. The ultimate load of approximately 800kN predicted by the FEM is less than the theoretical brace buckling capacity of around 900kN which indicates that the connection does not have sufficient strength.

At this stage, it is anticipated that the research work will highlight the need for an update of current Australian design practices relating to CBFs for low- to high-rise buildings. It is likely that the final recommendation will be that clauses in section 13 of AS4100 pertaining to design loads for CBF connections should be changed so that connections in all CBFs are designed for the full brace capacity.

REFERENCES

Anderson, J.C. 1975. Seismic behavior of K-braced framing systems. *Journal of the Structural Division, ASCE.* 101(ST10): 2147-2159.

Gillies, A.G. & R. Shepherd 1981. Post-elastic dynamics of three-dimensional frames. *Journal of the Structural Division, ASCE.* 107(ST8): 1485-1501.

Girard, C., A. Picard & M. Fafard 1995. Finite element modelling of the shear lag effects in an HSS welded to a gusset plate. *Canadian Journal of Civil Engineering.* 22(4): 651-659.

Goel, S.C. & R.D. Hanson 1974. Seismic behavior of multistory braced steel frames. *Journal of the Structural Division, ASCE.* 100(ST1): 79-95.

Jennings, P.C. 1964. Periodic response of a general yielding structure. *Journal of the Engineering Mechanics Division, ASCE.* EM2: 131-166.

Kahn, L.F. & R.D. Hanson 1976. Inelastic cycles of axially loaded members. *Journal of the Structural Division, ASCE.* 102(ST5): 947-959.

Lesik, D.F. & D.J.L. Kennedy 1990. Ultimate strength of fillet welded connections loaded in plane. *Canadian Journal of Civil Engineering.* 17: 55-67.

Maison, B.F. & E. P. Popov 1980. Cyclic response prediction for braced steel frames. *Journal of the Structural Division, ASCE.* 106(ST7): 1401-1416.

Miazga, G.S. & D.J.L. Kennedy 1989. Behaviour of fillet welds as a function of the angle of loading. *Canadian Journal of Civil Engineering.* 16: 583-599.

Popov, E.P. 1980. Seismic behavior of structural subassemblages. *Journal of the Structural Division, ASCE.* 106(ST7): 1451-1474.

Popov, E.P. & R.G. Black 1981. Steel struts under severe cyclic loadings. *Journal of the Structural Division, ASCE.* 107(ST9): 1857-1881.

Popov, E.P. 1982. Seismic steel framing systems for tall buildings. *AISC Engineering Journal.* 19(3): 141-149.

Roeder, C.W. 1989. Seismic behavior of concentrically braced frame. *Journal of Structural Engineering, ASCE.* 115(8): 1837-1856.

Standards Association of Australia 1990. *AS4100 Steel Structures.* Sydney: Standards Association of Australia.

Standards Association of Australia 1993. *AS1170.4 SAA Loading Code: Part 4 - Earthquake Loads.* Sydney: Standards Association of Australia.

van der Kreek, N. 1995. Steel lateral load resisting systems for buildings of up to 8 levels. *Journal of the Australian Institute of Steel Construction.* 29(2).

Yamanouchi, H., M. Midorikawa, I. Nishiyama & M. Watabe 1989. Seismic behavior of full-scale concentrically braced steel building structure. *Journal of Structural Engineering, ASCE.* 115(8): 1917-1948.

The Mechanics of Structures and Materials, Grzebieta, Al-Mahaidi & Wilson (eds)
© 1997 Balkema, Rotterdam, ISBN 90 5410 900 9

Interaction between brick veneer walls and domestic framed structures when subjected to earthquakes

E. F. Gad & C. F. Duffield
Department of Civil and Environmental Engineering, The University of Melbourne, Vic., Australia

ABSTRACT: During an earthquake event, domestic brick veneer structures experience significant interaction between the supporting frame and veneer walls. This interaction leads to load transfer between the brick veneer walls and frame through the brick ties. This paper investigates the interaction between out-of-plane veneer walls and frame. The paper reports the results of simulated earthquake tests on a full scale brick veneer one-room-house. An analytical model of the test house was developed, and verified against the experimental results, and subsequently used to predict the interaction between the frame and the veneer. A parametric study on the analytical model is also presented. Based on the experimental and analytical results it is concluded that brick veneer steel-framed domestic structures can survive vigorous ground excitation without damage. However, there are some deficiencies in the current method for evaluating the performance of brick ties when connected to steel studs.

1 INTRODUCTION

Framed structures are the most common type of housing construction in Australia. Such structures typically have brick veneer exterior wall cladding, plasterboard interior lining and roof tiles or steel roof cladding. Residential framed structures form a significant part of the construction industry. Indeed, houses make up almost 60% of the country's net wealth.

Brick veneer walls are normally attached to the frame via metal ties. A damp proof course is laid at the first course of the veneer to prevent the penetration of moisture. Therefore, veneer walls rely on the brick ties for their out-of-plane stability. Frames (whether steel or timber) are substantially more flexible than the rigid veneer cladding. The different mass and stiffness properties of the frame and the veneer walls result in some forces being transferred between them in the event of ground excitation. These forces are transferred by the brick ties.

In the Australian earthquake standard (AS1170.4-1993), brick veneer, steel or timber framed, structures are considered ductile and do not require any specific earthquake design. However, structural detailing may be required depending on the location and soil type. In such cases, the principal requirement for the structural detailing is that all parts of the structure to be tied together so that the forces, from all the structural and non-structural components, generated by the earthquake are carried to the foundation. Therefore, the brick ties are required to maintain connectivity between the frame and the veneer. In the earthquake standard, brick veneer walls are considered non-structural components and specifically exempted from any design for earthquakes.

This paper investigates the performance of typical brick veneer construction, in particular, failure modes, tie forces and base shear under earthquake loading. This investigation is part of a project to assess the performance of brick veneer cold-formed-steel framed domestic structures when subjected to lateral loads and in particular to earthquake induced loads.

2 BACKGROUND

A number of researchers have studied the interaction between out-of-plane veneer walls and supporting frames under wind loading. These include Arumala (1991) and Page et al (1996). However, the response of out-of-plane veneer to wind loading is different to its response under earthquake excitation. Under wind loading, the out-of-plane veneer may resist a proportion of the applied face load, while under earthquake loading the veneer may impose extra inertia loading on the frame.

The interaction between the veneer and frames has been investigated in New Zealand because of the significant proportion of this type of construction (Allen, 1991 and Lapish, 1991). In the New Zealand loading code (NZS 4203-1992), the face load on the veneer is dependent on the zone, height, natural period of vibration of the veneer, ductility of the building and wall ties. Face loads may be as high as 3g depending on the ductility of the building and attached components. To accommodate this expected level of force, there are more stringent requirements for the performance of brick ties used in New Zealand compared to Australia. This approach encourages designers to consider veneer as a designed building element having some degree of importance.

In assessing the performance of brick veneer cold-formed-steel framed structures it was required to assess the magnitude of loads imposed on the frame from the veneer via the brick ties and also determine the failure mode under earthquake excitation. Therefore, an experimental program was carried out to study the dynamic characteristics of both the frame and veneer and also the interaction between them.

3 EXPERIMENTAL PROGRAM

To study the interaction between the frame and veneer under earthquake loading a one-room-house "test house" was tested. It measured 2.3m x 2.4m x 2.4m high and was constructed from full scale components as shown in Figure 1. The test house simulates a section of a rectangular house with plan dimensions of 11m x 16m. A dead load corresponding to a house plan area of 11m x 2.4m was used on the test house. The mass of the roof tiles, insulation, ceiling cladding, battens, and trusses for that area was found to be 2350 kg. A concrete slab with the same weight was cast and supported on the East-West walls of the test house via steel C sections similar to those used for the bottom cord of typical roof trusses. The two walls in the North-South direction were non-load bearing and had standard 900 x 2100 mm door openings. The test house was built on a two degree of freedom shaking table at The University of Melbourne.

The test house was constructed by professional tradesmen according to detailing recommended by manufacturers. All construction details were standard except for the hold down details which were over designed to eliminate the potential failure mechanism through uplift or bending of the bottom plate. In order to get the maximum response from the out-of-plane brick walls, the veneer corners were not connected. These modelled walls are similar to wall sections which lie between door openings or expansion joints. The veneer walls were connected to the frame via clip-on steel ties every fifth course. These ties were bedded into the brickwork mortar and clipped onto the flanges of the c-section steel studs. A damp proof course was also provided. Full details of the testing program on the test house are described by Gad et al (1995).

The test house was subjected to El-Centro earthquake in the East-West direction. It survived the test with no apparent damage. The test house was then subjected to 200% El-Centro earthquake record. The 200% record is basically twice the acceleration amplitude of the original record. The test house also survived this record with some minor visible damage to the plasterboard interior cladding around the corners joints. The strap braces were also slack and close to failure. The brick veneer and ties were all intact. The test house was then subjected to approximately 300% El-Centro record. This caused the test house to fail. The strap braces failed which led to a sudden loading of the plasterboard which in turn incurred significant damage. This caused the test house to drift and shake vigorously. The out-of-plane walls and the brick ties could not accommodate this level of deformation and the walls snapped along the mortar at approximately mid-height and fell off. The in-plane walls slid along the damp proof course and brick ties lost connectivity with the frame. Table 1 presents the maximum accelerations of the shaking table, out-of-plane veneer, frame and relative displacement between the frame and veneer. The maximum table acceleration can be considered as the peak ground acceleration.

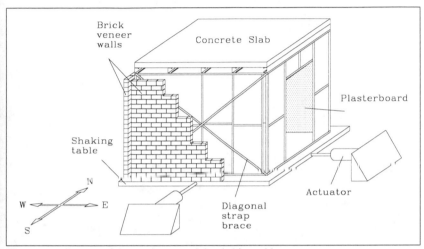

Figure 1. A diagram of the test house and the shaking table.

Table 1: Maximum response of the shaking table, frame and out-of-plane veneer.

	100% El-Centro	200% El-Centro	300% El-Centro
Maximum table acceleration	4.28 m/s^2	7.56 m/s^2	11.67 m/s^2
Maximum top of frame acceleration	6.51 m/s^2	10.37 m/s^2	failed
Maximum top of veneer acceleration	9.98 m/s^2	17.6 m/s^2	failed
Maximum relative displacement between frame & Veneer	5.43 mm	11.12 mm	failed

Although the top of veneer reached almost 2g and had a substantial movement relative to the frame, it did not cause premature failure of the test house. This is attributed to the fact that the veneer corners were not connected giving the brick walls the flexibility to ride the earthquake. It should be noted that a significant part of this relative movement was absorbed by flexing of flanges of the steel studs.

4 CONSTRUCTION OF THE ANALYTICAL MODEL

An analytical model was constructed to simulate the test house. The purpose of the model is to study the interaction between the out-of-plane veneer and frame. It is used to predict the response under various earthquakes and to conduct sensitivity analysis on parameters such as brick tie stiffness, veneer mass, frame stiffness and mass of the supported roof.

The model was constructed using Ruaumoko (Carr, 1995). Ruaumoko is designed to produce piece-wise time history response of a non-linear general two-dimensional framed structure to ground acceleration. In the test house model all the lateral bracing (plasterboard and strap braces) was simplified as one diagonal brace with the equivalent strength and stiffness, while the roof slab was simplified as two lumped masses as indicated in Figure 2. The diagonal bracing was modelled as a non-linear spring with degrading stiffness hysteresis. The out-of-plane veneer walls were modelled as beams with the equivalent properties and mass per unit length. The brick ties were simplified as linear springs based on tests conducted on stud-tie-veneer assemblies. The base of the veneer was modelled as a roller connection to simulate the damp proof course. The in-plane walls were not modelled because they had very little contribution to the response of the test house since the clip-on ties do not transfer shear forces.

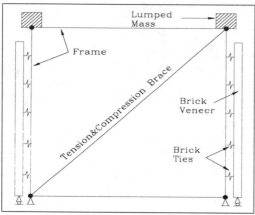

Figure 2. The equivalent model of the test house.

In testing the brick veneer ties, the method described by the Australian standard for wall ties (AS2699-1984) was used. In this method the whole stud-tie-veneer assembly is tested. The strength of the tie is assessed at 1.5mm axial deformation. This method is applicable to both timber and steel studs with no specific provisions for ties fixed to the web or the flange. It is believed that the assessment criteria proposed by the standard are not appropriate for veneer ties fixed to the flanges of steel studs because the flanges tend to bend, being much more flexible than the tie. Therefore, at 1.5mm deformation the stiffness of the assembly reflects the rotational stiffness of the stud flange. If the tie is close to the top/bottom plates or noggings, the rotation of the flange is greatly reduced. Hence, the strength at 1.5mm would greatly vary depending on the boundary conditions along the length of the stud section. Furthermore, the standard does not address the in-plane loading on tie assemblies. The brick ties must maintain connectivity between the frame and veneer walls in the direction of the attacking earthquake. Therefore, the standard should ensure that the ties do not disengage under shear displacement, otherwise, the in-plane veneer walls would become unstable in the out-of-plane direction.

5 VERIFICATION OF THE MODEL

The model was initially verified using modal analysis where the mode shapes and the natural frequencies matched the experimental results. The same earthquake records used for the experimental program were applied to the model. The experimental and analytical top of the frame accelerations are presented in Figures 3.

Although the analytical model is simple it did match the experimental results with a good degree of accuracy. The top of veneer acceleration and relative displacement between the veneer and frame were also compared and a very good match between the experimental and analytical results obtained.

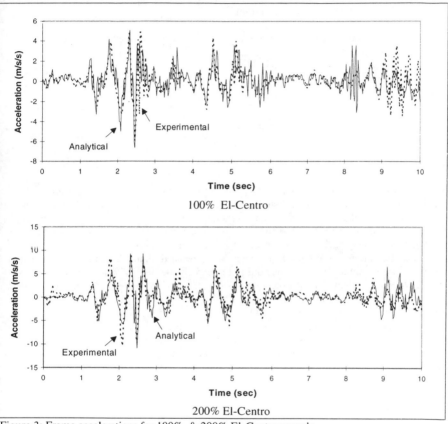

Figure 3: Frame accelerations for 100% & 200% El-Centro records.

6 SENSITIVITY ANALYSIS

Four earthquakes records were run through the model of the test house, with and without the brick veneer. For each run the base shear induced on the frame was obtained, Table 2. In all records the model with veneer had higher base shear than that without veneer. The percentage increase in base shear, due to the out-of-plane veneer, varied between 6% and 50%. This extra base shear was transferred from the veneer through the ties to the frame. The variation in the induced loads from the veneer is due to the fact that each earthquake record has a different spectral energy distribution. However, the Australian earthquake standard (AS1170.4-1993) has a response spectrum that is similar in shape to that produced by the El-Centro record (but lower in magnitude).

327

The clip-on tie is categorised as a light duty tie which is only required to have a minimum characteristic strength of 0.25 kN and 0.3 kN in tension and compression, respectively, according to the tie standard (AS2699-1984). Tie assembly tests revealed that ultimate capacities of 0.6 kN in tension and 2.0 kN in compression were possible. This additional capacity explains the survival of the out-of-plane veneer walls in the 200% El-Centro test in which the maximum tie force reached 0.52 kN. Therefore, for a brick veneer framed structure to survive such vigorous ground motion, brick ties have to be carefully selected to withstand the loads transferred between the veneer and frame.

Table 2. The base shear on the test house frame and maximum tie force.

Earthquake Record	Base shear (kN)		% Increase in Base Shear	Maximum Tie Force (kN)
	No veneer	With Veneer		
El-Centro - 1940 (USA)	12.2	15.3	25	0.30
El-Centro 200%	24.8	26.3	6	0.52
Parkfield - 1966 (USA)	15.2	21.1	39	0.38
Tenant Crk.- 1988 (Aust.)	3.2	4.8	50	0.09

7 CONCLUDING REMARKS

The current method presented in (AS2699-1984) for testing wall tie assemblies is not adequate for ties fixed to the flanges of steel studs. The results from this method would depend on the test setup, in particular, on boundary conditions along the length of the stud section being used. Furthermore, the standard does not specify any requirements for shear loading which could result in ties disengaging from the frame under in-plane movements.

Even though the top of out-of-plane veneer walls reached almost 2g and had a significant relative displacement to the frame, they did not collapse under the 200% El-Centro earthquake. One factor that contributed to the exceptional performance of the veneer is that the wall corners were not connected, giving the veneer walls the flexibility to ride the earthquake. However, further research is required to investigate the veneer response when wall corners are connected.

Generally, the test house performed very well not only resisting the loads induced by the roof mass but also the inertia forces from the out-of-plane veneer walls. The test house survived 100% and then 200% simulated El-Centro earthquakes before finally failing at approximately 300% El-Centro test.

ACKNOWLEDGMENTS

The contribution from the Australian Research Council (ARC), BHP Steel, Boral Plasterboard and Monier Bricks is gratefully acknowledged. Special mention and thanks for their support and contributions throughout the project to, Mr. G. Stark, Dr. L. Pham, Mr. A.D. Barton, Mr. J.L. Wilson, Prof. G.L. Hutchinson, A/Prof D.S. Mansell and Dr. N.T.K. Lam.

REFERENCES

Allen, D. 1991. Construction aspects of new masonry veneers. Proceedings of the Pacific Conference on Earthquake Engineering, New Zealand: 303-314.
Arumala, J O 1991. Mathematical modelling of brick veneer with steel stud backup wall systems, Journal of Structural Engineering, ASCE. 117(8): 2241-2257.
Carr, A. J. 1995. Ruaumoko. Dept. of Civil Engineering, Canterbury University. New Zealand

Gad, E. F., C. F. Duffield., G. Stark & L. Pham. 1995. Contribution of non-structural components to the dynamic performance of domestic steel framed structures. Pacific Conference on Earthquake Engineering. Melbourne. Australia. (3):177-186.

Lapish, E. B. 1991. A seismic designs of brick veneer and the New Zealand building codes. Proceedings of the Pacific Conference on Earthquake Engineering, New Zealand: 291-302.

Page, A. W., J. Kautto and P. W. Kleeman. 1996. A design procedure for cavity and veneer wall ties. Research Report No 137.5.1996. Dept. of Civil Engineering and Surveying. The University of Newcastle. Australia.

Standards Association of Australia. 1984. AS2699: Wall ties for masonry construction. Sydney

Standards Association of Australia. 1993. AS1170.4: Minium design load on structures, Part 4: Earthquake loads. Sydney.

Standards Association of New Zealand. 1992. Code of practice for general structural design loadings for buildings. NZS 4203. Wellington.

The Mechanics of Structures and Materials, Grzebieta, Al-Mahaidi & Wilson (eds)
© 1997 Balkema, Rotterdam, ISBN 90 5410 900 9

Modelling of plasterboard lined domestic steel frames when subjected to lateral loads

E. F. Gad, C. F. Duffield & D. S. Mansell
Department of Civil and Environmental Engineering, The University of Melbourne, Vic., Australia
G. Stark
Port Kembla Laboratories, BHP Research, BHP Steel, Australia

ABSTRACT: This paper presents the development of a detailed Finite Element (FE) model for plasterboard lined cold-formed-steel wall frames for residential construction. The model utilises non-linear element properties and three dimensional geometrical configurations. It is capable of simulating the influence of corner return walls as well as the contributions from the ceiling cornices. The analysis and results from the model correlate very well with experimental racking results. A sensitivity analysis was conducted on the model to study the influence of return walls and ceiling cornices. It is concluded that a wall with corner return walls, ceiling cornices and skirting boards has more than three times the lateral load capacity of an identical isolated wall panel.

1 INTRODUCTION

In Australia, typical domestic construction comprises a framed structure with plasterboard interior lining and brick-veneer exterior cladding, with terracotta or concrete roof tiles. Residential framed structures represents a large percentage of the total building construction. In fact, houses make up almost 60% of the country's net wealth compared with business capital, the second most important form of private wealth, of only 28%.

There is a continual and increasing pressure to effectively utilise existing building resources through material development and efficient structural design. Current design procedures for residential framed structures are mostly based on the strength of the bare frame. Little attention is given to the interaction between the various substructures and the so-called non-structural components. Plasterboard is considered a non-structural component and its contribution to the lateral strength and stiffness of domestic structures is largely ignored.

Plasterboard manufacturers allow strength contributions of 0.3 kN/m and 0.5 kN/m for timber walls lined on one side and both sides, respectively (Boral, 1992). In the domestic timber framing industry minimal allowance is commonly adopted for the bracing provided by the plasterboard wall lining. The capacity of diagonal bracing is increased by a factor of 1.4 to accommodate the bracing strength provided by the plasterboard (Groves, 1993). Although this approach is simple, it does not reflect the role of the plasterboard as its contribution should be independent of the performance of the diagonal bracing. These design recommendations are based on tests conducted on single isolated frames. No guidelines are present for walls with return corners and ceiling connection through cornices.

This paper presents a detailed Finite Element (FE) model for plasterboard lined cold-formed-steel domestic wall frames. The model is based on the commercially available general purpose finite element package (ANSYS, 1994). The model is not only for isolated wall panels but also for those walls with extra boundary conditions. Typical boundary conditions are return walls with set corner joints, skirting-boards and ceiling cornices. These extra boundary conditions dramatically affect the performance of these wall panels and change the failure mode from that of isolated panels. The FE model has been used to predict racking load-deflection behaviour as well as ultimate load carrying capacities for wall panels. It can accommodate any fixing pattern for the plasterboard and any wall geometrical configurations. Static, cyclic and dynamic loading can be applied.

The FE model is part of a project to assess the performance of brick veneer cold-formed steel-framed domestic structures when subjected to lateral loads and in particular to earthquake induced loads. A major part of the project was an experimental program on a 2.4m x 2.3m x 2.4m high one-room-house (Gad et al, 1995). Part of the experimental results is used to validate the analytical model presented in this paper.

2 BACKGROUND

Currently, there are three different methods for predicting the racking performance of lined framed walls. The first is based on empirical relations obtained from experimental data. These relations are limited to the material and configurations used for the test specimens. The second method uses simplified mathematical derivations. These mainly relate the performance of the connection between the frame and the lining material to the performance of the whole wall. Colins (1980) and McCutcheon (1985) developed these methods with some success. Assumptions are made in these derivations to make the mathematics manageable which limit the analysis to simple wall configurations. The third method is based on finite element modelling. This method was successfully used by Foschi (1977), Itani and Cheung (1984), Dolan and Foschi (1989) and others. These models were mostly for isolated wall panels. More recently, Kasal et al (1994) developed and verified a non-linear finite element model of a whole house using ANSYS. The researcher included the effects of openings and the non-linear properties of the connections. However, the corner connections were greatly simplified. Hence, this model would not be suitable for investigating the influence of the set corner joints. All the models were developed and verified for timber framed walls and mostly for plywood cladding. None of the models developed included other means of bracing (such as strap braces) in addition to the cladding material.

To truly quantify the contributions of plasterboard to racking performance of wall panels a model is required to accurately accommodate the effects of the set corner joints as well as the wall-ceiling connection via the cornice. Such a model would yield the modes of load transfer between the plasterboard and the framing members and the ultimate failure mode. The model could also be used for sensitivity analysis to identify critical parameters and provide design guidelines.

3 CONSTRUCTION OF THE FE MODEL

3.1 Typical load transfer between the frame and plasterboard

In single isolated lined frames, racking loads are primarily resisted by the shear strength of the cladding to frame connections (screws or nails). The failure mode of these walls is by tearing of plasterboard around the screws or nails. This generally occurs along the bottom connections or top and bottom connections. These mechanisms of load transfer and failure mode were recognised by many researchers including McCutcheon (1985) and Dowrick and Smith (1986). A section of a typical wall with two return walls is shown in Figure 1. When this wall is initially racked, the load is transferred from the frame to the plasterboard through the screws or

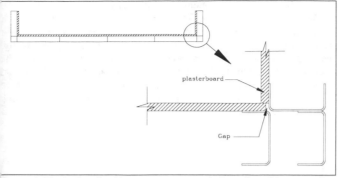

Figure 1. Typical wall with return corners and plasterboard layout at the corner.

nails in a similar fashion to a single frame. When the frame starts to move relative to the plasterboard, the gap closes between the plasterboard and the flange of the end stud of the return wall. This leads to a direct bearing of the plasterboard on the flange and hence another mode of load transfer between the frame and the plasterboard. When the racking displacement increases further, the bearing plasterboard edges start to crush. The crushing of the plasterboard edges propagate as the racking displacement increases. This becomes another failure mode.

3.2 Elements used for the models

At the early stage of modelling an isolated lined steel wall frame was modelled. The frame had tab-in-slot connections and was lined with standard 10mm plasterboard. The plasterboard was screwed to the frame according to the recommendations from the plasterboard industry. The developed model was verified against experimental results from a number of walls of identical construction details. The FE model accurately predicted the load-deflection curve and the ultimate failure load. The failure modes from the FE model and the experimental walls were quite similar.

Studs, top and bottom plates were modelled as beam elements while spar elements were used for the noggings. The connections between the studs and the top and bottom plates were modelled as pinned connections to represent the tab-in-slot connections.

Plasterboard was modelled as shell elements with a thickness. The screws connecting the plasterboard with the frame were modelled as non-linear spring elements. Each screw was actually modelled by four springs, two in the horizontal direction (one for the positive direction and one for the negative) and similarly two in the vertical direction. These spring elements had different load-deflection characteristics depending on the location of the screw being modelled. For example, the screws tearing through the plasterboard edges would be different to those in the middle of the board and yet different to those in the recessed part of the board. Hence, a number of shear tests were conducted on the plasterboard-stud screw connections to get the appropriate load-deflection curves. It should be noted that the performance of plasterboard screw connections are sensitive to construction quality. For example, over-driven screws or those driven not at a right angle to the plasterboard would have substantially lower shear resistance. Therefore, many of these connections had to be tested to eliminate unrepresentative results.

In modelling the return walls, the gap between the plasterboard and the flange was included as shown in Figure 1. To include the crushing of the plasterboard along the edges, the crushing capacity of the plasterboard had to be determined. Hence, a number of compression tests were conducted on plasterboard to determine its capacity when loaded along its edge. Based on these tests the load-deflection of a small segment in compression was obtained and then modelled as a non-linear spring. Hence, a series of spring elements with gaps were attached to the plasterboard edges and the flanges of end studs of the return walls.

The cornice performs three functions when combined with return walls. First, it prevents the out-of-plane buckling of the plasterboard when it is bearing against the return walls. Secondly, it assists in preventing the rotation of the plasterboard relative to the frame because it ties the wall plasterboard to that of the ceiling, Thirdly, it transfers a proportion of the racking load from the ceiling lining directly into the wall plasterboard (ie., not through the frame). These effects were considered in the model. The effect of the skirting-boards is to prevent out-of-plane buckling at the bottom. Hence, out-of-plane buckling was not considered.

3.3 Capabilities and limitations

The features of the model are:
- three dimensional effects can be included and any geometrical configurations can be accepted;
- although it was only used for plasterboard and steel frames it can accommodate any material properties, and non-linearities of connections;
- static, cyclic and dynamic loading can be performed.

This model has also been combined with another FE model for cross strap braces developed by Barton (1997). The combined models were used to confirm the load sharing between the plasterboard and the cross strap braces (Gad et al, 1995).

Plasterboard to frame nail or screw connections exhibit stiffness degradation and slip development under cyclic loading. ANSYS does not support elements with stiffness degradation; however, the developed models accommodated slip development. The models developed did not include door or window openings, but that can be accommodated in a similar fashion to that described by Kasal et al (1994). Glue connections between plasterboard and the framing members were not considered because of the uncertainty of the performance of the glue over the life span of the structure.

4 VERIFICATION OF THE MODEL

To verify the model with return walls the analytical results were compared with those obtained experimentally. The experimental results are based on tests conducted on a 2.4m x 2.3m x2.4m high one-room-house. The details of the experiment and the configuration of the test house are described by Gad et al (1995). The lateral load resisting elements in this test house were two plasterboard-lined (on one side) walls with return wall. The test house was also fitted with ceiling plasterboard, ceiling cornice and skirting-boards. Because of the test house symmetry each wall assembly resisted half of the lateral loads. The out-of-plane return walls were minimal, therefore their contribution by bending was negligible.

A finite element model was constructed using the above mentioned elements to model half of the test house. Half of the test house is basically a 2.4m long by 2.4m height cold formed steel framed wall with 10mm thick plasterboard lining on one side. At each end of the wall there is a return wall with set corner joints and along the top and bottom of the wall there is a ceiling cornice and skirting-board, respectively. The modelled wall had the same plan as that presented in Figure 1 but only the end studs of the return walls were modelled. The plasterboard was attached to the frame with the standard screw fixing method as recommended by the Australian plasterboard literature, in which screws are fixed at 200mm centres along the end studs, 400mm centres along the intermediate studs and 600mm centers along the top and bottom plates. The framing connections were tab-in-slot.

The analytical and experimental load-deflection curves are shown in Figure 2. The analytical model predicted the ultimate load capacity with a very good degree of accuracy. The frame

Figure 2. Comparison between the experimental and analytical results.

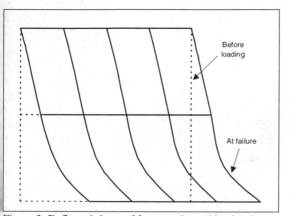

Figure 3. Deflected shape of frame under racking load.

deflected shape is shown in Figure 3 and matches what was observed in the experiment. It should be noted that in the experiment, the top of the frame was fixed while the bottom was racked. This produces the same net shear deformation as fixing the bottom and racking the top.

5 SENSITIVITY ANALYSIS

In order to illustrate and quantify the influence of boundary conditions on the performance of lined frames a sensitivity analysis was conducted on three wall configurations. These walls are identified as walls (a), (b) and (c) as illustrated in Figure 4. These three walls demonstrate the influence of ceiling to wall connection and return walls. The plasterboard was fixed in the same way for all the three walls as described previously. The framing members and connections were also identical.

In all these models the racking load was applied to the frame not the plasterboard. If the load is applied to the plaster directly then the load distribution may be different. For wall (b) the wall-plasterboard was prevented from moving vertically relative to the frame. This was the modelled contribution of the ceiling-wall connection through the cornice. The resulting load-deflection curves from the three walls are presented in Figure 5.

335

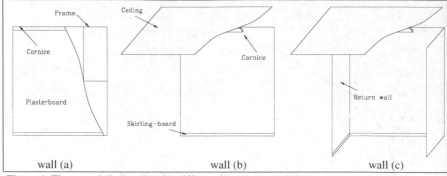

| wall (a) | wall (b) | wall (c) |

Figure 4. Three modelled walls with different boundary conditions.

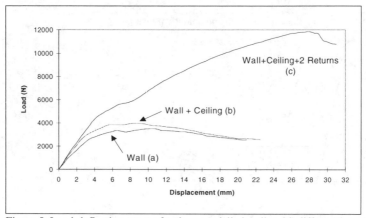

Figure 5. Load-deflection curves for three modelled walls with different boundary conditions.

From Figure 5 it is clear that return walls enhance the performance dramatically. The maximum load carrying capacity of walls (a), (b) and (c) are 3.5 kN, 4.0 kN and 11.9 kN respectively. By not allowing the plasterboard to rotate relative to the frame, wall (b), the load carrying capacity increased by approximately 15% while introducing the return walls increased the capacity by more than three times. This substantial increase is attributed to the bearing of the plasterboard edges on the return walls which forms another path of load transfer from the frame to the plasterboard. The ultimate failure mode for walls (a) and (b) was due to screws tearing into the plasterboard while wall (c) was a combination of that mode and crushing of the plasterboard at the edges.

Additional modelling has been undertaken to study the relationship between the wall length and the ultimate load carrying capacity. The flexibility of hold down details was also investigated as well as the variations in the strength and stiffness properties of the screws connecting the plasterboard to the frame. The results and conclusions from this work are beyond the scope of this paper and will be reported elsewhere.

6 CONCLUDING REMARKS

A finite element model has been described to model plasterboard-lined domestic cold-formed-steel wall frames under racking loads. The model is three-dimensional and capable of

336

accommodating boundary condition effects such as return walls, ceiling lining, ceiling cornices, and skirting-boards. The model was successfully verified against experimental results. The analytical and experimental load-deflection curves matched with a very good degree of accuracy as well as the deflected shapes and mode of failure. Based on a sensitivity analysis, contributions of return walls and ceiling cornices to lateral stability, were identified. It was found by including the return walls and ceiling cornice the load carrying capacity increases by more than three times. These boundary conditions enhance the racking resistance of plasterboard lined frames dramatically.

ACKNOWLEDGMENTS

The authors gratefully acknowledge the contribution from the Australian Research Council (ARC). The input from Prof. G.L. Hutchinson, J.W. Wilson, Dr. N.T.K Lam and Mr A. D. Barton of the University of Melbourne and Dr. L. Pham of the CSIRO is also gratefully acknowledged. The authors also like to extend their appreciation to BHP House Framing and Boral Australian Gypsum for the supply of materials and in particular Mr B. Poynter for his valuable input.

REFERENCES

ANSYS, 1994. Engineering Analysis System Manuals, Version 5.1, Swanson Analysis System Inc., Houston, Pennsylvania, USA.

Barton, A D. Performance of light gauge steel domestic frames when subjected to earthquake loads. A PhD thesis to be submitted to The university of Melbourne.

Boral Australian Gypsum 1992. Structural wall bracing. Technical Bulletin 3.10.

Colins, M J. 1980. Estimation of the diaphragm deflection due to nail slip. Unpublished Forest Products Laboratory Report No. FP/TE 112, New Zealand Forest Service, Forest Research Institute, Rotorua, New Zealand.

Dolan, J. D. & R. O. Foschi 1989. A structural Analysis Model For Timber Shear Walls. Structural Design, Analysis and Testing Proceedings of the sessions at Structures Congress, San Francisco, CA, USA: 143-152.

Dowrick, D. J. & P. C. Smith 1986. Timber sheathed walls for wind and earthquake resistance. Bulletin of The New Zealand National Society for Earthquake Engineering. 19(2): 123-134.

Foschi, R. O. 1977. Analysis of wood diaphragms and trusses, Part 1: diaphragms. Can J. Civil Engrg. 4(3):345-352.

Gad, E. F., C. F. Duffield., G. Stark & L. Pham. 1995. Contribution of non-structural components to the dynamic performance of domestic steel framed structures. Pacific Conference on Earthquake Engineering. Melbourne. Australia. (3):177-186.

Groves, B. J 1993. Understanding the timber framing code, A guide to AS1684-1992 The National Timber Framing Code. Standards Australia.

Itani, R. W. & K. C. Cheung. 1984. Non-linear analysis of sheathed wood diaphragms. Journal of Structural Engineering. 110(9):2137-2147.

Kasal, B., R. J. Leichti, & R. Y. Itani. Nonlinear finite element model of complete light frame wood structures. Journal of Structural Engineering, ASCE. 120(1):100-119.

McCutcheon, W. J. 1985. Racking Deformations in Wood Shear Walls. Journal of Structural Engineering, ASCE. 111(2): 257-269.

The Mechanics of Structures and Materials, Grzebieta, Al-Mahaidi & Wilson (eds)
© *1997 Balkema, Rotterdam, ISBN 90 5410 900 9*

The use of ductility for strength reduction in the seismic design of structures

N.T.K.Lam, J.L.Wilson & G.L.Hutchinson
Department of Civil and Environmental Engineering, The University of Melbourne, Vic., Australia

ABSTRACT: The relationship between ductility and strength reduction in the seismic response of structures is considered in this paper. Following a state-of-the-art review, the paper summarises the findings from a series of recent analytical studies. Recommendations for estimating the approximate strength reduction factors for systems with different ductility capacities are made for short period, medium period and long period structures. The effects of earthquake ground motion frequency content on the elastic response of structures is also included in the paper.

1 INTRODUCTION

Conventional aseismic design procedures specified in most contemporary earthquake loading standards and codes, including the U.S.A.'s Uniform Building Code (UBC-1991) use elastic analyses to estimate earthquake induced load effects on buildings. The earthquake loads are modified to account for the building's inelastic behaviour using the Structural Response Factor, or the Design Strength Reduction Factor (R), which is defined as the ratio :

$$R = \frac{S_{elastic}}{S_{design}} \qquad (1)$$

where $S_{elastic}$ = Estimated Elastic Strength Demand; S_{design} = Design Strength.

The UBC broadly classifies buildings according to their structural form and recommends an R value for design purposes (usually in terms of working stress). The R values have been developed from Californian research and experience and broadly account for the following characteristics of the structural system : energy absorbing capability, expected overstrength, likely degree of redundancy and performance in past earthquakes. A similar set of R values (modified where necessary for ultimate limit state design) have been adopted in earthquake loading standards worldwide including areas where the nature of the seismic hazard and the construction practices are very different to California.

Analytical methods have also been used to justify the "R" Factor. The strength reduction is considered to be partly attributed to the system's ductility and partly to the increase in strength as a result of strain hardening and plastic hinge formation. Thus, R is often expressed as :

$$R = R_\mu R_{os} \qquad (2)$$

where R_μ is the Ductility Reduction Factor and R_{os} is the Overstrength Factor (refer Figure 1).

$$R_\mu = \frac{S_{elastic}}{S_{mech}} \qquad (3)$$

where S_{mech} is the Mechanism Strength of the structural system obtained from a monotonic push-over analysis.
Similarly, R_{os} is defined as :

$$R_{os} = \frac{S_{mech}}{S_{design}} \qquad (4)$$

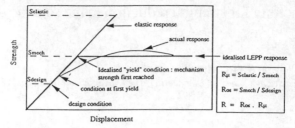

Figure 1 Definitions for the Design Strength Reduction Factor, Ductility Factor and Overstrength Factor

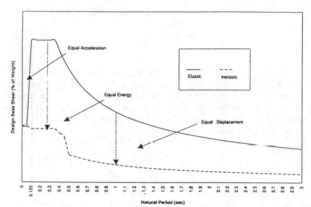

Figure 2 Nemark-Hall   Inelastic Response Spectrum
[after Newmark & Hall (1982)]

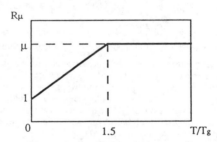

Figure 3 R_μ - T/T_g Relationship in Priestley's "Displacement Based Seismic Assessment Procedure"
[after Priestley (1995)]

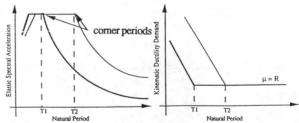

Figure 4 "Corner Periods" in Elastic Response Spectrum and Ductility Demand Spectrum
[after Lam, Wilson & Hutchinson (1995)]

340

Newmark and Hall (1982) related the kinematic ductility demand μ to R_μ by the following expressions :

$$R_\mu = \mu \qquad \text{(for T>0.5secs)} \qquad (5a)$$

$$R_\mu = \sqrt{2\mu-1} \qquad \text{(for 0.1<T<0.5secs)} \qquad (5b)$$

$$R_\mu = 1 \qquad \text{(for T<0.03)} \qquad (5c)$$

which are based on the "Equal Displacement" (Eq5a), "Equal Energy" (Eq5b) and "Equal Acceleration" (Eq5c) propositions respectively.

The Inelastic Design Response Spectrum (IDRS) can be approximately obtained by dividing the Linear Elastic Design Response Spectrum(LEDRS) by the product of R_{os} and R_μ for each of the period ranges recommended in Equation 5 (refer Figure 2).

This Newmark-Hall procedure has been the most well known and widely used analytical method in the aseismic design of ductile structural systems. However, the phenomena embraced in the procedure was only justified on the basis of dynamic response analyses carried out using a series of single displacement pulses. Only very limited recorded earthquake ground motions were available at the time to verify the results. In a follow-up study by Mahin & Bertero (1981) using five Californian accelerogram records, significant scatter in the results was observed which challenged the accuracy of the R_μ - μ relationships assumed in the Newmark-Hall procedure.

Simple SDOF systems have been used to model the buildings in these studies. Higher mode effects are therefore neglected and the results are only applicable to buildings in which the mode shape does not change significantly with yielding. Further, linear elasto-perfectly plastic (LEPP) hysteretic behaviour has been assumed as it has been found in previous studies that, provided the building's mechanism strength does not deteriorate under cyclic loading, stiffness degradation and the shape of the hysteretic loop does not significantly affect the inelastic response.

2 RECENT DEVELOPMENTS

Analytical studies (for example, Elghadamsi and Mohraz(1987)) using a much greater ensemble of earthquake records have been used to further derive inelastic design spectra. A recent study by Miranda (1993), employing mainly North American records from the West Coast has demonstrated the significant effect the ground motion frequency content has upon the ductility demand. These effects could have contributed to the large scatter previously observed by Mahin & Bertero (1981).

The R_μ factor defined in the "Displacement-Based Seismic Assessment Procedure" developed by Priestley (1995) for reinforced concrete buildings, is given by the expression (refer Figure 3):

$$R_\mu = 1 + (\mu-1)\frac{T}{1.5Tg} < \mu \qquad (6)$$

where T=natural period of SDOF model; Tg= site natural period (corresponding to the peak spectral response in the elastic response spectrum)

Eq(6) assumes that the "Equal Displacement" proposition ($R_\mu=\mu$) applies when $T > 1.5T_g$ and "Equal Acceleration" proposition ($R_\mu=1$) applies when T approaches zero. Within these two limits it is suggested that the value of R_μ can be interpolated.

In a recent study by the authors (1995,1996), analyses have been carried out on SDOF models using 180 random phase-angle artificially generated accelerograms and 81 recorded strong motion accelerograms collected from different regions worldwide. Importantly, the "Corner Period" of the ductility demand spectrum is consistent with the "Corner Period" of the corresponding elastic response spectrum suggesting that the ductility demand and the ground motion frequency content are strongly correlated (refer Figure 4).

The earthquake duration and phase angle effects on the ductility demand has also been studied in addition to the frequency content effects previously described. Ductility demand and damage levels associated with the elasto-perfectly plastic models were shown to be insensitive to the excitation duration. This result is not consistent with the observed damage levels in the field where duration is clearly important, suggesting that the effects of strength degradation should be included. There was also little evidence to suggest any systematic influence of the phase-angles on ductility demand, implying that there is no inherent difference between earthquakes of different source mechanisms (i.e. interplate or intraplate) in regard to the building's ductility demand provided that their elastic response spectrum (frequency contents) are similar.

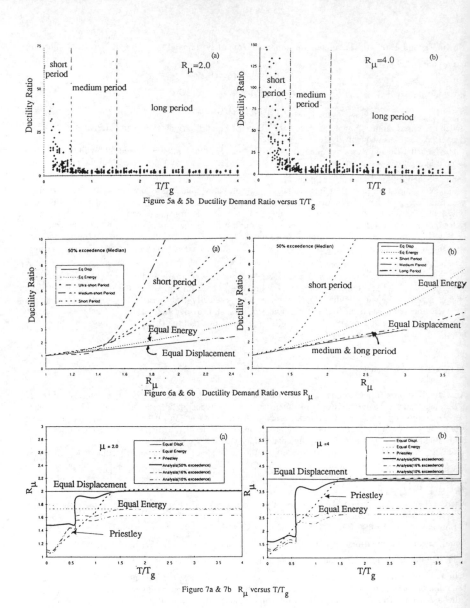

Figure 5a & 5b Ductility Demand Ratio versus T/T$_g$

Figure 6a & 6b Ductility Demand Ratio versus R$_\mu$

Figure 7a & 7b R$_\mu$ versus T/T$_g$

These recent studies have provided insight into the influence of various factors on ductility demand, but a thorough probabilistic treatment of the subject has been hindered by both the general lack of representative recorded data and the complications caused by the large number of variables involved. Analyses using artificially generated accelerograms are generally consistent with the real records although some discrepancies have been observed when the frequency content has not been accurately represented [Lam,Wilson & Hutchinson(1996)].

3 NEW DEVELOPMENTS

3.1 Objectives of Study

The aim of this study is to develop a reliable generic R_μ - μ relationship for elasto-perfectly plastic SDOF systems. This is part of a long term research programme carried out by the authors to rationalise seismic design provisions for codes of practice in intraplate regions. Recent studies examining the effects of frequency content, duration and phase-angles on ductility demand has greatly simplified a once seemingly complicated phenomenon. Further, a larger strong motion database has recently become available to facilitate accurate statistical treatment of the data.

In this study the distribution and scatter of the data has been analysed by counting the number of occurences in which particular ductility levels have been exceeded. This statistical study has enabled the estimation of the median value and a range of exceedence percentages to be carried out. Previous studies have typically assumed the results to be normally distributed and conveniently represented by calculating the mean and the standard deviation. Such assumptions, though commonly used, could be misleading should the distribution be asymmetrical or characterised by a cluster of outlying data points.

3.2 Modelling, Analyses and Sorting of Data

The modelling, analyses and sorting of data carried out in this study is summarised as follows :
(i) Sixty LEPP-SDOF models were created for analyses with natural periods, T , ranging from 0.1sec to 1.6 seconds. (ii) The yield strength of each model was adjusted to a fraction of the elastic strength demand predetermined by an elastic time-history analysis carried out for each earthquake ground motion. The fractions corresponded to R_μ factors of 1.5, 2.0, 2.5, 3.0, 3.5 and 4.0 . (iii) The ensemble of 81 recorded accelerogams, identical to that used in a recent investigation by the authors (1996), was used in the inelastic time-history analyses. The accelerograms covered a wide frequency content range. (iv) A total of 4860 inelastic time-history analyses were carried out. (v) The kinematic ductility demand ratio, μ , which is defined as the maximum displacement of the SDOF model divided by the "yield" displacement was evaluated for each analysis. The ductility demand ratio was then plotted against the ratio of the structure period to the site period, T/T_g, for a range of R_μ values. (vi) For the purpose of statistical analyses, results obtained for each SDOF model have been grouped into: the short period (T/T_g<0.6s), medium period (0.6s<T/T_g<1.5s) and long period (T/T_g>1.5s) categories. Results in the short period category have been further sub-divided into the ultra-short and the medium-short sub-categories at T/T_g =0.3s. Typical ductility ratios for R=1.5 and R=4.0 are plotted in Figure 5. (vii) Results for each category have been analysed statistically to identify the ductility levels corresponding to various exceedence levels.

3.3 Results and Discussion

The ductility demand imposed on the models depended on whether the T/T_g ratio was above or below 0.6 (refer Figure 5a & 5b). Above this limit, the ductility demand showed little scatter and was generally independent of T/T_g . Below this limit, the ductility demand was in contrast much higher and associated with a much larger scatter. The above observations, based on six plots each representing different R_μ values, highlights the importance of T/T_g as a controlling parameter for predicting the ductility demand. Importantly, the observed stepped change in the ductility demand was inconsistent with the steady transitional change for the short period range implied by the existing models (for example, equation(6) and Figure 3). The median (50% exceedence) ductility demand ratios observed from the analyses are plotted against R_μ in Figure 6a & 6b. Observations from the figures can be summarised as follows :-

(i) Structures in the medium period and long period categories ($T/T_g \geq 0.6$) :
The ductility demand can be reasonably predicted using the "Equal Displacement" propositions whereby μ increases directly in proportion to R_μ.
(ii) Structures in the short period categories ($T/T_g < 0.6$) :-
   (a) $R_\mu < 1.5$: the ductility demand can be predicted by the "Equal Displacement" or, more conservatively, by the "Equal Energy" proposition.

343

(b) $R_\mu > 1.5$: the ductility demand cannot be predicted using any of the conventional methods. The ductility demand appears to increase exponentially with R_μ.

Recommendations based on the 50% exceedence are as follows :

$R_\mu = \mu$ $(T>0.6T_g)$ (7a)

$R_\mu = 1.5$ $(T \leq 0.6T_g)$ (7b)

These recommendations are illustrated in Figures (7a) and (7b) for $R_\mu = 2.0$ and $R_\mu = 4.0$. The "Equal Displacement", "Equal Energy" and Priestley models are also illustrated for comparison together with the 10%, 16% and 50% exceedence curves. The recommendations are similar but considered more refined and less conservative than the Priestley model in the short period to medium period categories ($T/T_g \leq 1.5$).

4 CONCLUSIONS

(i) The ductility demand for medium period to short period structures ($T/T_g \leq 0.6$) can be reliably predicted using the "Equal Displacement" propositions.

(ii) The ductility demand for short period structures ($T/T_g < 0.6$) is very dependent on the design "R" factor.

(iii) $R_\mu = 1.5$ is recommended for short period structures to avoid extreme ductility demands.

(iv) The effects of the frequency content of earthquake ground motions has upon the inelastic ductility demand have been taken into account by the Design Strength Reduction Factor (R) recommendations.

5 ACKNOWLEDGEMENTS

The work reported in this paper forms part of the research undertaken in the Australian Research Council funded project titled "Earthquake ground motions and structural ductility factors for Australian condition".

6 REFERENCES

International Conference of Building Officials, 1991.*Uniform Building Code, chapter 23 part 3:Earthquake Design* .

Newmark,N.M. and Hall,W.J., 1982. *Earthquake Spectra and Design,* Engineering Monograph, Earthquake Engineering Research Institute, Berkeley, California.

Mahin,S.A. and Bertero,V.V., 1981. An evaluation of inelastic seismic design spectra, *Journal of Structural Division,* American Society of Civil Engineers, 107(ST9), 1777-1795.

Miranda, E.,1993. Evaluation of Site-Dependent Inelastic Seismic Design Spectra, *Journal of Structural Engineering,* American Society of Civil Engineers, 119(5), 1319-1338.

Elghadamsi,F.E. and Mohraz,B., 1987. Inelastic Earthquake Spectra, *Earthquake Engineering and Structural Dynamics,* 15, 91-104 .

Priestley, M.J.N. 1995. Displacement-Based Seismic Assessment of Existing Reinforced Concrete Buildings, *Proceedings of the PCEE5,* The University of Melbourne, Victoria, Australia, 225-244.

Lam,N.T.K., Wilson, J.L. and Hutchinson, G.L., 1996. Building Ductility Demand : Interplate versus Intraplate Earthquakes, *Earthquake Engineering and Structural Dynamics,* 25, 965-985.

Lam,N.T.K., Wilson,J.L. and Hutchinson,G.L. 1995. The dependence of ductility demand on the frequency characteristics of earthquake ground motion, *Proceedings of the 14th Australasian Conference on the Mechanics of Structures and Materials,* University of Tasmania, Hobart, 290-295.

The Mechanics of Structures and Materials, Grzebieta, Al-Mahaidi & Wilson (eds)
© *1997 Balkema, Rotterdam, ISBN 90 5410 900 9*

Introduction to a new procedure to construct site response spectrum

N.T.K.Lam, J.L.Wilson & G.L.Hutchinson
Department of Civil and Environmental Engineering, The University of Melbourne, Vic., Australia

ABSTRACT: A theoretically based procedure is presented in this paper to construct site response spectra based on the soil profile and the assumed bedrock response spectrum. The procedure consists of two stages : the prediction of the site natural period followed by the prediction of the peak spectral acceleration together with the construction of the response spectrum. The effect of the "Period Shift" due to non-linear behaviour of the soil, and the relationship between shear strain and damping has been taken into consideration.

1 INTRODUCTION

The spectral acceleration content of seismic shear waves measured at the ground surface could be significantly amplified by transmission through soft soils. Thus, site response is an important consideration in the aseismic design of structures. Different techniques have been developed to analyse site responses such as the computer program SHAKE developed by Schnabel et al. (1972) at the University of California, Berkeley. SHAKE is one of the many programs written to compute site responses based on 1-dimensional (1D) non-linear wave theory. The adoption of such a rigourous procedure requires detailed soil data, ground motion data and expertise. The procedure is therefore costly and cannot be justified for most projects.

Most earthquake loading standards do not require such a site response analysis but instead assign a site factor for each of the broadly defined site classifications. The Uniform Building Code (UBC) by the International Conference of Building Officials (1991) in the United States specifies site factors of between 1 and 2, depending on the site classification, to amplify spectral accelerations for natural periods greater than the "Corner Period" (natural period at which the curved section and flat section of the spectrum intercepts). The spectral accelerations specified for natural periods less than the "Corner Period" however remain unchanged. Many earthquake loading standards throughout the world, including the current Australian Earthquake Loading Standards (1993), adopt a similar approach for specifying the site response spectrum.

The site amplification provisions in the National Earthquake Hazard Reduction Program (NEHRP) in the U.S.A., as reported by Martin & Dobry (1994), are a major advancement in the modelling of site response. The amplification factors specified for different site classifications vary with the intensity and frequency content of the bedrock motion.

Both the UBC and the NEHRP models directly scale the reference bedrock response spectrum to obtain the site response spectrum. Therefore, the peak effective ground acceleration (PEGA) measured on the bedrock and on the soil has been assumed to be directly correlated. (PEGA is defined here as the peak response spectrum acceleration divided by 2.5). However, the two PEGA values have been found in a recent study by the authors (1996a) to be only loosely correlated. Thus, it is uncertain if the empirically based UBC or the NEHRP method can be applied to seismic conditions which are different to the origin of the data used in deriving the methods.

An alternative and theoretically based procedure developed by the authors (1996a &b) to construct site response spectrum is considered to be more general, and in particular, appropriate for areas where there is a paucity of local earthquake data. The procedure is divided into two stages :

First Stage - The site natural period is first determined by analyses of borehole records or by seismological field measurements. A period shift adjustment for non-linear effects is then made resulting in the "Corner Period" of the site response spectrum. Second Stage - To construct the site response spectrum, the peak spectral acceleration at the "Corner Period" is determined by modelling the dynamic behaviour of the soil profile as a tall building undergoing a beam-sway mechanism.

The procedure will be described in further details in the following sections.

2 SITE NATURAL PERIOD

2.1 Initial site period

A simple procedure to estimate the initial site natural period consists of the following steps:
(i) Estimation of the initial dynamic shear modulus for each layer of the soil profile - For cohesionless soil, the initial dynamic shear modulus G_o can be determined using the well known Imai-Tonouchi relationship :

$$G_o = 14N^{0.68} \tag{1}$$

where N is the Standard Penetration blow count.
Results obtained from Eq(1) and from other empirical studies have been found by the authors (1997) to be consistent. Standard penetration test values, where not given, may be estimated from the soil descriptions provided in the borehole record. For normally consolidated cohesive soils, Eq(1) has also been found to give consistent results for a plasticity index (PI) less than 40%. For PI≥40%, other more complicated models are available.
(ii) Calculation of the shear wave velocity V_i for each layer of the soil profile - The shear wave velocity for each layer of the soil profile can be calculated using the expression:

$$V_i = \sqrt{\frac{G_o}{\rho}} \tag{2}$$

(iii) Calculation of the weighed average shear wave velocity for the soil profile - The weighed averaged shear wave velocity V_s for the soil profile can be calculated using the expression :

$$V_s = \frac{\sum_i V_i H_i}{H} \tag{3}$$

(iv) Calculation of the initial site natural period T_i -
The initial site natural period T_i (for low amplitude vibration) can be calculated using the expression :

$$T_i = \frac{4H}{V_s} \tag{4}$$

It has been found, however, that the above procedure is not accurate for the following conditions :
(i) The presence of a substantial layer of very soft soil (N<10).
(ii) A substantial soft soil layer underlying, or sandwiched between, much stiffer soil layers.

Seismological measurements based on background "noise" excitations may alternatively be used to directly obtain the site natural period and the average shear wave velocity of the soil. Sample field measurements may also be used to calibrate the analyses of existing borehole records.

346

2.2 *Period shift adjustment for non-linear effects*

The period shift adjustment for non-linear effects is dependent upon the level of excitation experienced by the soil profile. In general, as the soil strains increase the dynamic shear modulus reduces and the soil damping increases. The spectral acceleration of the bedrock response spectrum Sa_i (at the initial site natural period T_i) has been found to be a good indicator of the extent of cyclic straining in the soil. The value of Sa_i , once established, can be used in the following relationship to predict the period shift ratio T_g / T_i :

$$\frac{T_g}{T_i} = \lambda + \frac{Sa_i}{10} \tag{5}$$

where T_g , the "Corner Period", has taken into account non-linear effects.

The parameter λ depends on the soil properties but appears not to be sensitive to the soil thickness. It has been found from a limited number of analyses that λ equals 1.05 for clay and 1.30 for sand.

3 SITE AMPLIFICATION AND RESPONSE SPECTRUM

The theoretically based model, as illustrated in Figure 1, is based on modelling the dynamic response of the soil profile as an imaginary tall building undergoing a beam-sway mechanism as described by Priestley (1995). The model assumes that there is no cross-coupling between the systems representing the soil and the structure which is consistent with the principles of free-field site response analyses. The peak spectral acceleration Sa_{max} can be obtained by the following relationships:

$$Sa_{max} = 2.5a_{max} \tag{6}$$

$$a_{max} = 1.18\beta Sa_i \tag{7}$$

where a_{max} = max. ground acceleration of site
   2.5 = Conventional spectral amplification factor
   1.18 = participation factor based on the parabolic displacement (refer Fig.1)
   β = correction factor for damping other than 5%
   Sa_i = bedrock spectral acc. at site natural period for 5% damping.

   A correction factor for damping other than 5% can be established using the following expression which is based on an extensive study by the authors (1997) :

$$\beta = \sqrt[\alpha]{\frac{\zeta\%}{5\%}} \tag{8}$$

   The exponent α in Eq(8) should assume a value of 2 (ie. a square root function) if the natural period of the bedrock and the site are comparable. Otherwise, α =4 should be assumed. For most near-field shallow earthquakes, up to Magnitude 6, the natural period of the bedrock excitation is normally well below 0.5sec. Under these conditions, α =2 is used for sites with natural period lower than 0.5sec and α =4 for sites with a larger natural period. The percentage of damping in the soil $\zeta\%$ may be estimated from the shear strain γ of the soil averaged over part of the soil profile as shown in Figure 2. The assumed uniform shear strain γ may be obtained from the following expressions :

$$\gamma = \frac{\Delta}{0.61H} \tag{9a}$$

where

$$\Delta = \frac{\beta Sa_i}{\left(2\pi/T_g\right)^2} \tag{9b}$$

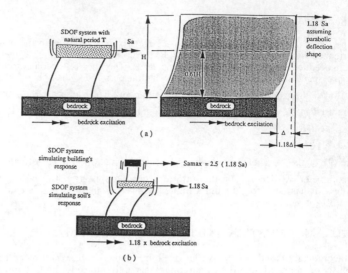

(a)

(b)

Figure 1   Principle of the theoretically based model

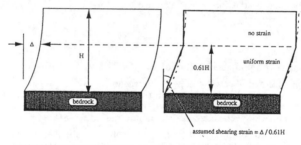

(a) Variable Strain assuming parabolic deflection   (b) Assumed Uniform Strain

Figure 2   The equivalent uniform shear strain

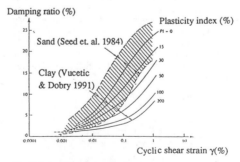

Figure 3   Relationship between cyclic shear strain and damping
(after Booth (1994))

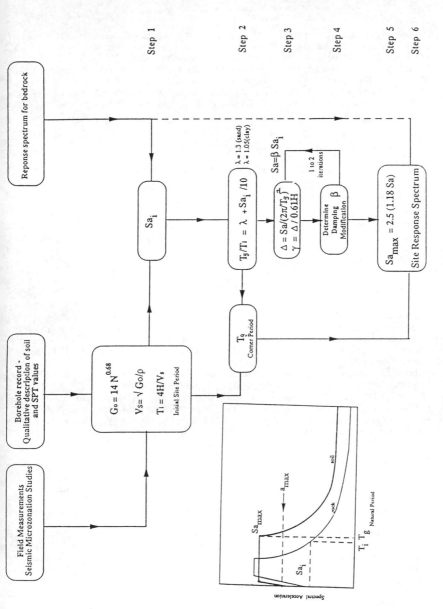

Figure 4 Flow chart representation of the proposed procedure

349

The relationship between γ and $\zeta\%$ has been the subject of considerable research interest. The most commonly used relationships have been presented by Seed & Idriss (1984) for sand and Vucetic & Dobry (1991) for clay (refer Figure 3). An initial value of $\zeta = 5\%$ (ie. $\beta=1$) is recommended for substitution into Eq(9b) in the first iteration. In the second iteration, the value of $\zeta\%$ so derived from Figure 3 can be used to estimate β. The value of $\zeta\%$ converges rapidly after only two iterations in the experience of the authors.

To facilitate calculation of Sa_{max}, Equations 6, 7 and 8 can be combined into the following expression :

$$Sa_{max} = 2.5(1.18)Sa_i \sqrt[\alpha]{\frac{\zeta\%}{5\%}} \frac{T_i}{T_g} \tag{10}$$

where the ratio T_i/T_g takes into account the small decrease in the value of Sa_i as a result of the period shift. (This ratio assumes that the bedrock response spectral acceleration is inversely proportional to the natural period). Finally, given T_g and Sa_{max}, the soil response spectrum can be constructed by making the following two assumptions :
(i) The response spectrum to the left of the "Corner Period" is flat.
(ii) The response spectrum to the right of the "Corner Period" is similar in shape to the bedrock response spectrum.
The procedure described above is summarised in Figure 4.

4 CONCLUSIONS

The site natural period is estimated using simplified analysis of the soil profile or by seismological field measurements. The "Corner Period" is then obtained by the application of a period shift to account for non-linear effects. The peak response spectral acceleration is calculated by modelling the soil profile as an imaginary tall building undergoing a beam-sway mechanism modified for the following two effects :
(i) Period shift adjustment for non-linear behaviour
(ii) Amplification adjustment for actual soil damping not equal to 5%
The final response spectrum is then constructed based on the "Corner Period", the peak response spectral acceleration of the soil site and the bedrock response spectrum shape.
This procedure is considered a more rational method than that usually specified in the UBC and that proposed by the NEHRP.

5 REFERENCES

Booth, E.D. 1994.Concrete Structures in Earthquake Regions, Longman Scientific and Technical Publisher
International Conference of Building Officials 1991. Uniform Building Codes, Ch23 Part 3: Earthquake Design.
Lam, N.T.K., Wilson, J.L. and Hutchinson,G.L., 1997. Analytical Studies on Intraplate Site Responses, Research Report : RR/STRUCT/04/97, Dept.of Civil & Environ. Engineering, The University of Melbourne.
Martin, G.R. and Dobry, R.,1994. Earthquake Site Response and Seismic Code ProvisionsNCEER Bulletin, 8(4),1-6.
Priestley, M.J.N. 1995. Displacement -Based Seismic Assessment of Existing Reinforced Concrete Building, Proceedings of the Pacific Conference on Earthquake Engineering, The University of Melbourne, Australia, 225-244.
Schnabel, P.B., Lysmer, J. and Seed, H.B.,1972. A computer program for earthquake response analysis of horizontally layered sites, EERC report : EERC72-12, University of California at Berkeley, U.S.A..
Seed, H.B.,Wong, R.T.,Idriss,I.M. and Tokimatsu,K. 1984. Moduli and Damping Factors for Dynamic Analysis of Cohesionless Soils, EERC report : UCB/EERC-84/14, University of California at Berkeley.
Standards Australia, 1993. Minimum Design Loads on Structures, Part 4, Earthquake Loads, AS1170.4
Vucetic, M. and Dobry, R., 1991. Effect of Soil Plasticity on cyclic response, Journal of Geotechnical Engineering, American Society of Civil Engineers, vol.117(1), pp. 89-109.

The Mechanics of Structures and Materials, Grzebieta, Al-Mahaidi & Wilson (eds)
© *1997 Balkema, Rotterdam, ISBN 90 5410 900 9*

Seismic effects on band beam construction in Australia: A preliminary analysis

J.S.Stehle, P.Mendis, H.Goldsworthy & J.L.Wilson
Department of Civil and Environmental Engineering, The University of Melbourne, Vic., Australia

ABSTRACT: This paper covers the preliminary analytical work for which the aim is to predict the probable failure mode in the joint region between reinforced concrete band beams and columns, and for the structure as a whole. Analytical results including linear static and nonlinear dynamic frame analyses are presented. The energy dissipation mechanism of a structure is investigated for a range of effective beam widths at the joint, with the implications for detailing requirements discussed briefly.

1. INTRODUCTION

This paper is presented as a companion to the paper entitled "Seismic effects on wide band beam construction in Australia: Review" (Abdouka & Goldsworthy 1997). Reinforced concrete band beam construction has been identified in the companion paper as a common construction method in Australia. Also, in the companion paper, limits on beam widths and methods for assessing joint strength are discussed. This paper covers the preliminary analytical work in which the main aim is to predict the probable failure mode in the joint region of band beams and columns, and for the structure as a whole. The experimental program described in the companion paper will be used to calibrate the analytical results presented here.

2. FAILURE MODES - A CASE STUDY

2.1 *Frame design according to Australian codes*

An interior frame of a 3 story, 2 bay building (see Figure 1) is designed in accordance with the earthquake provisions of AS3600 (SAA 1994) and AS1170.4 (SAA 1993) together with the usual load cases due to gravity and wind. The frame is designed as an ordinary moment resisting frame (OMRF), for which no extra detailing for seismic effects is required. Gravity loads are found to dominate the design. The resulting design, which is deliberately made simplistic and uniform is shown in Figure 2.

An intermediate moment resisting frame (IMRF) may be used, particularly if earthquake loads govern. This would require additional detailing considerations as specified in Appendix A of AS3600 (SAA 1994) to achieve better ductility without significantly increasing the strength of the structure. One of these requirements is found to congest the reinforcement in the beam.

Ligatures are required at a spacing of 0.25d for a distance of 2D at each beam end, which for a 400 mm deep beam would mean a ligature spacing of less than 100 mm. Recent research (Gentry & Wight 1992) has suggested that this ligature spacing may be increased up to 0.5d for wide beams in regions of high seismicity provided that experiments confirm their findings. This is one of the main aims of the research currently being undertaken at the University of Melbourne.

The code specified static earthquake loads on ordinary moment resisting band beam frames, are found to affect the design (where gravity loads do not dominate) in the following areas generally: shear and flexure in columns, and combined punching shear and torsion in exterior beam-column joints (assuming the exterior span is designed to take the full moment transfer to the exterior column). Hence, these are the most likely deficiencies in existing structures. The latter failure mechanism was investigated for wide beam - column joints by Gentry and Wight (Gentry and Wight 1992). Their studies, however, were limited to narrower beams, with strong columns and no significant gravity loads.

A significant aim of this research project is to assess the ductility available in band beam-column joints and whether it is sufficient for the ductility levels assumed in AS3600 (SAA 1994). Another project which is currently being undertaken at the University of Melbourne will assess the available ductility in columns with lapped splices near the base of the column. The results of the combined projects will enable a comprehensive understanding of the total structure with meaningful changes to the code then being recommended.

2.2. Strength of the joint in terms of an effective beam width

A range of beam width limits have been detailed in the companion paper. These widths are summarised in Table 1 and have been evaluated for the case study. The band beam framing system commonly used in Australia would not satisfy the limits given in Table 1. These limits are recommended in regions of high seismicity to ensure that the longitudinal reinforcement within the full beam width is effective.

The joint region is defined in this paper as the region through which moment is transferred between the band beam and the column or between two beams for the case of continuity around an interior column. For the case of a joint between a band beam and a column the moment transfer occurs through torsion on the side of the columns and direct moment transfer into the face of the column. Methods for assessing the strength of a joint are presented in the companion paper and in a more explanatory research report (Stehle et al 1996). Some of these strengths are evaluated in Table 2 in terms of an effective beam width. The effective beam width provides a width of band beam which would have the same moment capacity as the joint assuming the

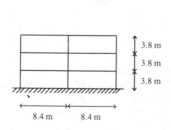

8.4 m   8.4 m

Figure 1: Case study frame dimensions.
Bay width = 8.4 m in out of plane direction.

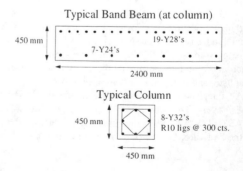

Typical Band Beam (at column)

Typical Column

Figure 2: Case study members

352

Table 1: Beam width limits

Source	Model	Equation	Beam width limit for case study (mm)
ACI-318 1989	Code limit for beam width	$b_w \leq b_c + 1.5d_b$	1125
NZS-3101 1982	Code limit for beam width	$b_w \leq b_c + 0.5h_c \leq 2b_c$	675
Gentry and Wight 1992	Suggested limit for beam width	$b_w \leq b_c + 2h_c$	1350

* Symbol definition: b_w = beam width

b_c = column width

d_b = beam depth

h_c = column width (dimension parallel to beam span)

Table 2: Effective beam widths

Source	Model	Equation	Effective beam width for case study (mm)	
			Interior Joint	Exterior Joint
Gentry and Wight 1992, using equation by Hsu 1968	Cracking torsional strength of a transverse beam on both sides of the joint using Hsu's equation.	$T_{cr} = 0.34f_c^{0.5}x^2y$-exterior $T_{cr} = 0.68f_c^{0.5}x^2y$-interior	1261	968
Warner, Rangan and Hall 1989	Ultimate punching shear strength of flat slab-column joints under combined shear, torsion and moment.	$T_{uc} = 2.2f_c^{0.5}d^2a$	2400*	2000*
Kanoh and Yoshizaki 1979	Ultimate torsional strength on both sides of an interior flat slab-column joint.	$T_{uc} = 2f_c^{0.5}d^2(c_1 + 0.66d)$	2400*	2400*

* Note: A linear interaction between torsional and shear strength assumed.

** Symbol definition: c_1 = column width (dimension parallel to beam span)

x = minimum of column width and beam depth

y = maximum of column width and beam depth

d = beam effective depth

a = depth of critical perimeter

same steel percentage. A direct moment transfer is assumed to occur over the width of the column face and a distance of d/4 on both sides. The value obtained using Hsu's formula represents the lowest possible bound. The full band beam width represents the maximum possible upper bound if slab effects are ignored.

2.3. Time history analysis

Time-history analyses of the selected structure were performed using the computer program "DRAIN2D". The joints were modelled by using semi-rigid connection elements as shown in Figures 3.a and 3.b. These elements were given moment capacities based on an effective beam width (see section 2.2) at the joint, which was varied so that its effect could be studied. The stiffness of these elements was also based on the effective beam width over a length equal to the depth of the column (see equation 1). These assumptions represent a first approximation of the joint behaviour which will be clarified with experimentation. It should be noted that the time-history analyses do not take into account any brittle modes of failure such as punching shear, shear in the beams, columns or joint core. Hence, the ductility demands are based solely on bending effects. Plastic hinging in the beams (which usually occurs over a distance 'd' from the beam end and over the full beam width) is distinguished from plastic hinging at the joint (which is assumed to occur over the effective beam width of the joint on the sides and face of the column).

$$k_j = \frac{(EI)_j}{h_c} \qquad\qquad (1)$$

where, k_j = Moment - rotation stiffness of joint

$(EI)_j$ = Flexural stiffness of an effective beam width at the joint

h_c = Column depth

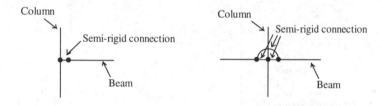

Figure 3.a: External connection   Figure 3.b: Internal connection

The frame was analysed using several acceleration records. Firstly, a number of artificially generated accelograms compatible with the response spectra specified in AS1170.4 with a peak ground acceleration coefficient of 0.1g were used. For these records, no inelastic behavior was observed which may be accounted for by a number of factors: (1) The overstrength built into the code via capacity reduction factors (2) Overstrength provided by the design for gravity loads (3) The natural period of the frame is much longer than the code determined value (0.9 seconds compared to 0.25 seconds from the code formula). However, the static earthquake forces which increase for shorter periods are only allowed to be reduced by 20% due to a refined determination of the natural period. Hence, the structure will be over-designed for earthquake forces. A soft soil site would be more likely to induce seismic forces at a level consistent with the original design value.

Using the North-South component of the 18th of May, 1940 El-Centro earthquake with a peak ground acceleration of 0.3g, significant excursions into the inelastic range were observed. The likelihood of an El-Centro type earthquake in Australia is low, however it is useful for evaluating the likely weaknesses present in the structure.

The variation in ductility demands for various joint strengths and stiffnesses in terms of effective beam width are presented in Figure 4. The analysis assumes that the structure would dissipate energy in a flexural mode and assumes that premature failure through brittle punching shear is prevented. The results show how the likely energy dissipation mechanism changes; from a column sidesway to a mixed mechanism and to a beam sidesway mechanism as the joint strength and stiffness decreases (see Figure 5). This highlights the importance of identifying the critical collapse mechanism to ensure appropriate detailing is provided. If Kanoh and Yoshizaki's suggested method for calculating the effective beam width (see Table 2) is used, it appears that a column sidesway mechanism is most likely. However, if the effective beam width is closer to the ACI code restrictions on beam width, a mixed joint-column sidesway mechanism is more likely. From these results a pure beam or joint sidesway mechanism appears to be unlikely. In the event of a brittle punching shear failure the preliminary results would need to be modified. For example, a combined punching shear and torsion failure in the exterior joints (these joints are identified in this study as requiring the most ductility), would lead to higher moments and shear forces on interior joints due to the redistribution of forces. This may result in a punching shear failure at the first interior joint and at successive interior joints until the collapse of the whole floor.

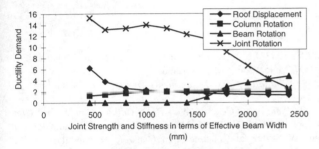

Figure 4: Trends in ductility demand for varying joint strength and stiffness

● = plastic hinge in beam or column   ✗ = plastic hinge in joint

c = critical location

| Effective joint width = 2400mm | Effective joint width = 1200mm | Effective joint width = 450mm |
| Column Sidesway Mechanism | Mixed Joint-Column Sidesway Mech. | Joint Sidesway Mech. |

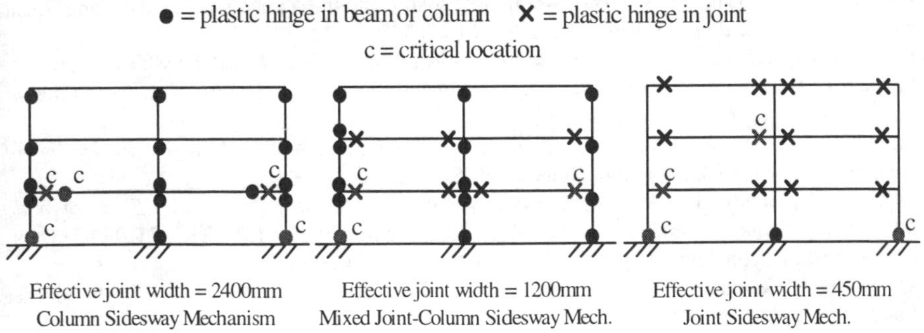

Figure 5: Plastic Hinge Locations

Plastic hinges in beams (away from the joints) are unlikely to require much special detailing for ductility, however a minimum quantity of beam ligatures may be desirable. According to AS3600 (SAA 1994), for an OMRF, no ligatures are required in the beam if shear stresses are small whereas for an IMRF, ligatures at 0.25d spacing are required near the beam ends. Plastic hinges in columns are likely to require special detailing of the type specified in AS3600 (SAA 1994) for an IMRF, but not for an OMRF. A further deficiency in AS3600 (SAA 1994) is the failure to specify a maximum spacing between column longitudinal reinforcement bars. This has an influence on concrete confinement. Plastic hinges in the joint regions outside the joint core is an important focus of the research currently being undertaken. The ductility capacity of this region requires quantification through experimental work.

3. CONCLUSIONS

(1) Preliminary results show that due to the flexibility and overstrength of reinforced concrete band beam frames, there may be very little or no damage sustained for the type of earthquake expected in Australia. A sensitivity analysis is required to confirm this.

(2) For a level of earthquake which did produce significant damage the following areas were identified as the most vulnerable, i.e. subject to significant ductility demands: the columns and the joint region. This outcome requires an experimental evaluation of the ductility capacity in these regions. The strength hierarchy within the frames and the importance of brittle failure modes also needs to be investigated. These brittle failure modes were not presented in the preliminary analysis here.

(3) The above issues are addressed in the theoretical and experimental program on band beams, currently underway at the University of Melbourne.

4. REFERENCES

Abdouka K., & Goldsworthy H. 1997. Seismic Effects on Wide Band Beam Construction in Australia: Review. *15th ACMSM*. Melbourne: to be published.

American Concrete Institute 1989. *Building Code Requirements for Reinforced Concrete (ACI-89)*. Detroit: ACI

Gentry, R.T. & Wight, J.K. 1992. *Reinforced Concrete Wide Beam-Column Connections Under Earthquake-Type Loading*. Report No. UMCEE 92-12, Michigan: The University of Michigan

Hsu, T. T. C. 1968. Torsion of Structural Concrete-Plain Concrete Rectangular Sections. *Torsion of Structural Concrete*. Detroit: ACI SP-18 :203-238.

Kanoh, Y. and Yoshizaki, S. 1979. Strength of Slab-Column Connections Transferring Shear and Moment., *ACI Journal*, Vol. 76, No. 2: 461-478.

Standards Association of Australia 1994. *Concrete Structures Code AS3600*. Sydney: SAA

Standards Association of Australia 1993. *Minimum Design Loads on Structures, Part 4: Earthquake Loads AS 1170.4*. Sydney: SAA

Standards Association of New Zealand 1982, *The New Zealand Standard for the Design of Concrete Structures NZS 3101*. Wellington: NZS

Stehle, J. S., Mendis, P., Goldsworthy, H., Wilson, J. L. 1996, *A Preliminary Study of Wide Band Beams Subject to Earthquake Loading*, Research Report No. RR/STRUCT/03/97, Dept. of Civil and Environmental Engineering, The University of Melbourne.

Warner, R. F., Rangan, B. V., and Hall, A. S. 1989. *Reinforced Concrete, 3rd ed*. Melbourne: Longman Cheshire Pty. Ltd.

5. ACKNOWLEDGMENTS

The authors gratefully acknowledge the financial assistance received from the Australian Research Council under large grant no. 896018878.

The Mechanics of Structures and Materials, Grzebieta, Al-Mahaidi & Wilson (eds)
© 1997 Balkema, Rotterdam, ISBN 90 5410 900 9

Earthquake response of coupled structural walls

M. H. Pradono, P. J. Moss & A. J. Carr
Department of Civil Engineering, University of Canterbury, Christchurch, New Zealand

ABSTRACT: In many tall buildings coupled structural walls provide the required stiffness and strength to resist lateral loading that arises from gravity, wind, and earthquake effects. In this paper, the inelastic response of such walls under earthquake loading is investigated and equivalent pushover analyses carried out to determine a measure of the dynamic amplification factors for these structures. The investigation looks at the use of different hysteretic models to describe the behaviour of both the inelastic wall members and the coupling beams. The walls are modelled either by beam-column elements or by wall elements developed to more accurately account for the actual distribution of reinforcing steel. Hysteresis loops measured in previous experimental work were used as the basis for the hysteresis models used to describe the behaviour of the coupling beams. The effect of different earthquakes on the magnitude of the dynamic amplification is also investigated.

1 INTRODUCTION

There is a large amount of data available about the elastic and inelastic behaviour of structural walls under cyclic loading. Some of this experimental research was carried out to investigate the various factors that influence the behaviour of structural walls. The results were then used to suggest ways to improve the strength and ductility of structural walls. However, it also became necessary to have information about structural wall behaviour in a building as a whole, hence there was a need to correlate the behaviour observed in small scale and full-scale component tests with the behaviour of structural wall components in a small scale or full-scale building acted on by seismic forces.

One way of correlating the behaviour of an isolated wall with that of the wall system is to carry out laboratory tests. These tests could be static reversing load tests, pseudodynamic tests, or shaking table tests. Another method is to simulate the structural behaviour using a numerical model in a nonlinear dynamic analysis. Two types of modelling are required for a reinforced concrete structure: (a) that for the distribution of stiffness along a member, i.e. a "member model", and (b) that for the force deformation relationship under stress reversals, i.e. "hysteresis models".

2 STRUCTURAL MODEL

The structure modelled is a symmetrical nine storey building designed according to NZS 4203 (1992) and NZS 3101 (1995) and as outlined in Paulay and Priestley (1992). Lateral forces are resisted in one direction by coupled walls positioned at each end of the building. In the other direction reinforced concrete frames of six 7.35 m long bays provide the necessary seismic resistance. The walls were each 5 m wide and 380 mm thick for the bottom two stories and 350 mm for the top stories. The 1 m long coupling beams were 1.5 m deep above

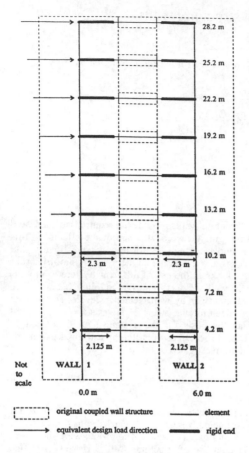

Figure 1 Details of the structural model analysed

the ground floor and 0.8 m deep elsewhere and 350 mm thick. The coupled walls are assumed to be fully fixed at the ground level by a full length, deep foundation wall. Details of the layout of the coupled shear walls are given in Figure 1.

The walls were modelled using either a beam-column element or a special wall element that modelled the reinforcing steel as well as the concrete. The beam-column element was used with its hysteretic behaviour modelled by either an elasto-plastic or Clough-degrading model. The wall element was one derived by Taylor (Carr 1996) and was used with either the Kent-Park hysteretic model based on the normal stress-strain relation for concrete or with an "elasto-plastic" model where the concrete stress-strain relation had an elasto-plastic backbone. While this latter case is not necessarily realistic, it was used to enable a direct comparison to be made between the results from using the wall element with those from the beam-column element. The coupling beams were diagonally reinforced and were modelled by a Giberson one-component beam element using one of three different hysteresis models, i.e. elasto-plastic model, Clough's degrading model, or the Ramberg-Osgood model. It was assumed the coupling beams would have similar hysteretic behaviour under cyclic loading to those tested previously (Paulay and Binney 1974) which could be best represented by a Ramberg-Osgood function. Clough's degrading model was a reasonable approximation to the experimental hysteresis loops.

Details of the walls are given in Figure 1, along with an outline of the equivalent frame model using beam-column elements. As can be seen, the beam-column elements were placed along the centrelines of the walls and the coupling beams with the beams being given rigid end blocks in order to model the coupling beams more appropriately. Details of the analyses carried out are given in Table 1.

3 PUSHOVER ANALYSES

These were carried out to determine the lateral strength of the coupled walls when subjected to the equivalent design load set out in NZS 4203 (1992). For this, the magnitude of the lateral force distribution on the walls was increased monotonically from zero until the maximum strength of the walls was reached; this was expressed in terms of the base shear carried by the walls. The total base shear used for the design was 2043 kN. Care needed to be taken in the analyses to ensure that the load was applied sufficiently slowly that the inertia effects did not generate an oscillatory response; in other words, the duration of loading should

Table 1

Analysis number	Coupling beam model	Wall model
A	Ramberg-Osgood	Taylor wall elements: Kent-Park concrete model
B	Clough degrading	as for A
C	Elasto-plastic	Taylor wall elements: elasto-plastic concrete behaviour
D	Clough degrading	Beam-column elements: Clough degrading
E	Ramberg-Osgood	Beam-column elements: elasto-plastic
F	Clough degrading	as for E
G	Elasto-plastic	as for E
H	Elasto-plastic	as for A
I	Elasto-plastic	as for D

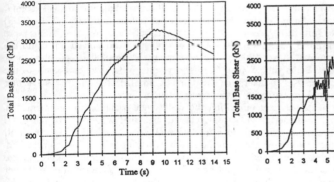

Figure 2   Total base shear as determined from a pushover analysis using the beam-column model

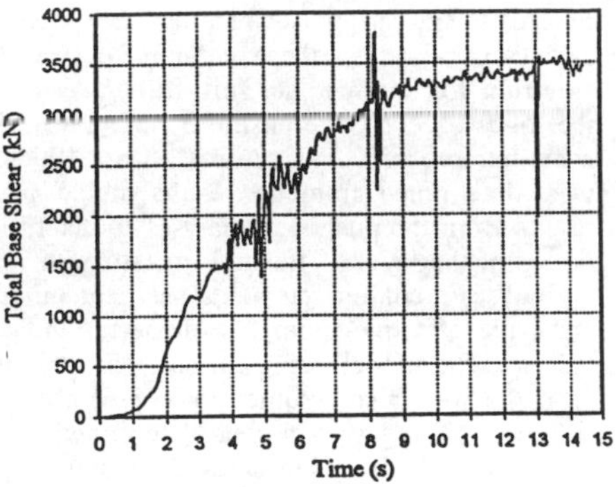

Figure 3   Total base shear as determined from a pushover analysis using a wall element model

be long compared with the fundamental period of the structure.

Figure 2 shows the results of a pushover analysis using the beam-column model from which it can be seen that the strength is such that the coupled walls are able to resist a maximum base shear of 3300 kN. The results for the pushover analysis using the Taylor wall elements are shown in Figure 3 where it can be seen that the maximum base shear that can be resisted is approximately 3350 kN (using Clough's degrading hysteresis model for the coupling beams). If the Ramberg-Osgood model is used, the maximum base shear is greater on account of the Ramberg-Osgood hysteresis model allowing the coupling beams to develop higher bending moments than the assumed yield values; this in turn meant that the walls were subjected to larger axial forces.

Figure 4 shows the bending moments developed at the base of the two coupled walls, while Figure 5 shows the actual shift in the neutral axes at the base of the two walls.

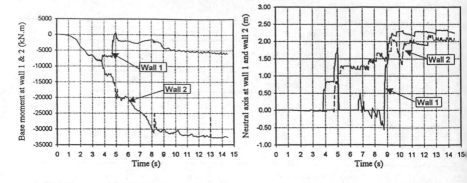

Figure 4   The base moments for the two walls as determined from a pushover analysis

Figure 5   The neutral axis shift for the two walls as determined from a pushover analysis

4 INELASTIC DYNAMIC ANALYSIS

Four different earthquake records were used: El Centro 1940 (corrected), Parkfield, Pacoima, and Taft. Each of these earthquake records was scaled to match the NZS 4203 (1992) spectrum for intermediate soil sites. The models used for the walls and coupling beams were the ones used for the pushover analyses. The inelastic analyses were carried out using the computer program "Ruaumoko" (Carr 1996). The maximum values of the total base shear were determined from the results of the analyses. The results were then compared to the results from the pushover analyses to determine the dynamic amplification of base shear that occurs at the level of flexural overstrength of the structural members.

The beam-column elements used for modelling the walls were assumed to be precracked and have effective moments of inertia and cross-sectional area that were 50% of the gross values. The models using these elements had a fundamental period of 0.81 secs. The wall elements were also assumed to be precracked to the extent of having 50% of their uncracked stiffness. The program thus determined how many of the outer filaments making up the element should be given cracked section properties at the start of the analysis in order to get as near as possible to the required level of cracking. In this case, it was found that the fundamental natural period of the wall element models depended on the sequence of the analysis. If the modal analysis was carried out before the static analysis, the fundamental period was found to be 0.785 secs, whereas if the modal analysis was carried out after the static analysis, the fundamental period was 0.699 secs. This latter period was lower because the forces applied in the static analysis caused the cracks to close up under the action of the compressive forces, hence the walls appeared stiffer.

Under the action of the seismic forces generated by an earthquake, the cracks in the wall elements could open and close depending on the direction of the forces, thus changing the apparent level of cracking and allowing a shift in the neutral axis position in the walls. In the case of the beam-column elements, there was no possible change in the stiffness values or the neutral axis.

5 DISCUSSION

The models analysed can be categorized according to the element type used to represent the walls: wall models (A, B, H and C) which use the Taylor wall element, or beam-column

360

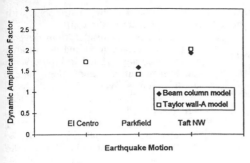

Figure 6 Dynamic amplification of Base Shear at flexural overstrength for different earthquakes using an elasto-plastic hysteresis for the coupling beam

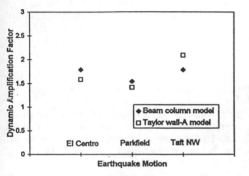

Figure 7 Dynamic amplification of Base Shear at flexural overstrength for different earthquakes using Clough degrading hysteresis for the coupling beam

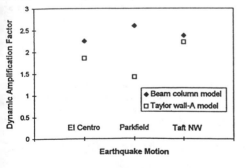

Figure 8 Dynamic amplification of Base Shear at flexural overstrength for different earthquakes using Clough degrading hysteresis for the coupling beam

element models (E, D, I, F, G) which use ordinary beam-column elements to respresent the wall members.

The results of using an elasto-plastic coupling beam in conjunction with elasto-plastic beam-column or wall elements are shown in Figure 6 where the dynamic amplification factor is the ratio of the base shear from a dynamic analysis to that determined from an equivalent static analysis. It can be seen that both types of elements used to model the walls give reasonably similar values for the dynamic amplification of the base shear, though the value varies depending on the earthquake record used; the wall elements giving slightly lower values than the beam-column elements for the El Centro and Parkfield earthquakes but a higher value when the Taft record was used.

In the case where the coupling beam is modelled using hysteresis behaviour represented by the Clough degrading model, the results for the dynamic amplification of the base shear are shown in Figure 7. In this case, there is some difference between the two models used with the wall element model showing a greater variation with the three earthquakes used than is the case with the beam-column model. The range of values is greater than for the case of elasto-plastic behaviour and varies with the earthquake record used in the same way.

When the Ramberg-Osgood hysteretic model is used for the coupling beam (Figure 8), there is considerable variation in the predicted dynamic amplification of the base shear with the wall element model giving lower values than the beam-column model. For the latter model, the dynamic amplification value is greater than two. The Ramberg-Osgood hysteresis model tends to produce larger amplification factors since this model allows the coupling beam moments to increase with increasing deformation; this means that the beam shear increases which in turn means that the axial force in the wall on the compressive side increases (with a corresponding reduction on the tension side). As the axial force in the wall is still below that corresponding to the balance point on the axial force-bending moment interaction curve for the wall section, this increase in axial force will correspond with an increase in the allowable bending moment and hence in the shear in the wall.

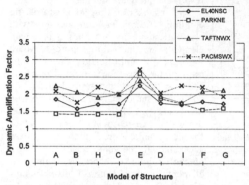

Figure 9 Dynamic amplification of Base Shear at flexural overstrength for different earthquakes and different models

A summary of results from several earthquake records used and the various models used is presented in Figure 9. It can be seen that, in general, the dynamic amplification is greatest for the Pacoima record and lowest for the Parkfield earthquake. There is some variation between the amplification factors predicted using the wall elements (A, B, H, C) and those determined using the beam-column elements. The variations result from the combination of earthquake characteristics (such as the many large peaks of similar magnitude in the Taft record as against the two large pulses close together that dominate the Parkfield earthquake) and the energy dissipation of the different hysteresis models.

6 SUMMARY AND CONCLUSIONS

- The dynamic magnification factor is influenced by the behaviour of the structure and the characteristics of the earthquake.
- If elasto-plastic behaviour is assumed for the coupling beam (and wall or column elements), this gives similar dynamic amplification regardless of whether the walls are modelled by column elements or wall elements. The wall element shows a greater variation with the earthquake than does the column element with the difference between the two values being dependent on the particular earthquake.
- When the coupling beams are assumed to behave in the manner of Clough's degrading model, the dynamic amplification will vary with different earthquakes. It is not possible to say which of the two models for the walls will predict the greater amplification.
- When the coupling beams are modelled by Ramberg-Osgood type behaviour, the dynamic amplification will vary with the earthquake but the column model will predict higher amplifications than does the more appropriate wall element. The magnitude of the difference in predictions between the two models appears to be a function of the earthquake.

REFERENCES

Carr, AJ, 1996. *Ruaumoko*, Computer Program Library, Department of Civil Engineering, University of Canterbury, Christchurch.

Paulay, T and Binney, JR, 1974. *Diagonally reinforced coupling beams of shear walls, Shear in Reinforced Concrete*, ACI Special Publication SP42-26, American Concrete Institute, Detroit, Vol. I, pp 579-598.

Paulay, T and Priestley, MJN, 1992. *Seismic design of concrete and masonry buildings*, Wiley-Interscience, New York.

Standards New Zealand (SNZ), 1992. *General structural design and loadings for buildings*, NZS 4203, Standards New Zealand, Wellington.

Standards New Zealand (SNZ), 1995. *The design of concrete structures*, NZS 3101, Standards New Zealand, Wellington.

The Mechanics of Structures and Materials, Grzebieta, Al-Mahaidi & Wilson (eds)
© 1997 Balkema, Rotterdam, ISBN 90 5410 900 9

Base isolation for segmental building structures

P.H.Charng, A.J.Carr & P.J.Moss
Department of Civil Engineering, University of Canterbury, Christchurch, New Zealand

ABSTRACT: The technique of base isolation has undergone considerable development over the last two decades in an attempt to mitigate the effects of earthquakes on buildings and their contents and has proved to be an effective solution for seismic design. The successful seismic isolation of a building depends on the installation of mechanisms which decouple the structure from potentially damaging earthquake-induced ground motions. Therefore, it is very important to have an adequate understanding of the influence of each parameter in the isolation system and the superstructure on the seismic performance of the isolated buildings.

1 INTRODUCTION AND BACKGROUND

From his research on base isolated structures, Andriono (1990, 1991a,b) developed design methods for base isolated buildings using lead-rubber isolation devices between the base of the structure and the rigid foundation. The present research seeks extend the design method to other isolation devices and to consider the effects of foundation compliance. The research also seeks to evaluate the effect of using a segmental building isolation system (Wei 1995) where the isolation devices are placed at various levels in the building in order to reduce the displacements imposed on each of the devices.

In order to do this, the main dynamic characteristics of shear buildings supported on non-linear seismic base isolation systems have been investigated. For the purpose of controlling the base displacement and resisting the wind load, most of the seismic base isolation systems are designed to have non-linear hysteretic characteristics. Since the successful seismic isolation of a building depends strongly on the selection of an appropriate base isolation system, it is necessary to choose the appropriate analytical model for use in the dynamic analysis of base-isolated buildings. Therefore, two kinds of models, bi-linear and elasto-plastic hysteretic models are considered in this research. The research develops models that are analysed using the computer program *Ruaumoko* (Carr 1995). Analyses have been carried out to compare the responses of base isolated buildings with those of fixed base structures and also considers the recently suggested segmental structures where isolation devices are located at various heights in the structure as well as at the base. The effects of foundation compliance effects on the response of the segmental structure is also considered. There is still a considerable amount of work required to identify the degree of foundation flexibility that will make significant changes in the response of the segmented structure.

2 STRUCTURAL MODEL

The prototype structures used in this research have the elevations shown in Figures 1-4. They have a 9.2 m double span and were designed as plane frames in accordance with NZS4203

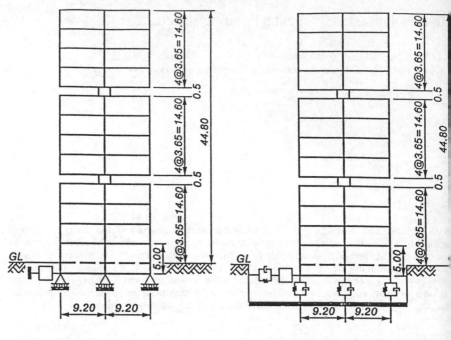

Figure 1 Segmental model on a rigid base

Figure 2 Segmental model on a compliant foundation

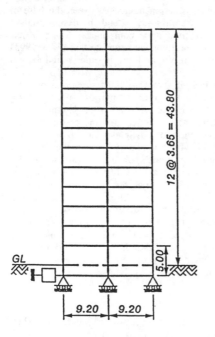

Figure 3 Base isolated model

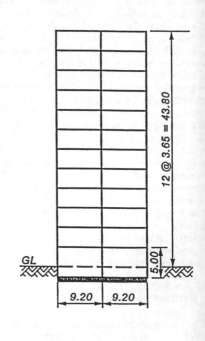

Figure 4 Fixed base model

(1992) and NZS3101 (1995) using the capacity design method. The interstorey height is 5.0 m for the ground floor and 3.65 m for the upper floors. As the structures were moment-resisting frames, shear deformation of the members was not taken into account. Based on the capacity design approach, plastic hinges were restricted to the beam ends and the ground floor column bases. The plastic hinge lengths at each end of the beams were assumed to be 0.5 times the member depth.

One of the problems in the use of conventional base-isolation in buildings is that under some excitations, the displacements required in the isolator are large and may not be readily available in many of the current devices. One suggested solution by Wei (1995) is to distribute the devices through the height of the building, the devices at each level providing part of the required displacement and limiting the dislocation of services, etc, at each isolation level. There is the problem, however, that the lower devices in the building may protect the upper structure so that the isolation devices there do not operate in their inelastic region and hence would not provide the expected displacement. The original work by Wei (1995) indicated a considerable reduction in acceleration in the upper part of the structure but as the analyses performed were only linearly elastic, further research is required to address the real inelastic responses.

In the case of the segmental structure, the building is divided into three segments that are interconnected by conventional isolation systems, such as lead-rubber bearings. The buildings are assumed to have either a rigid base or allow foundation compliance using a Voigt-Kelvin model as shown in Figures 1 and 2. The base-isolated and fixed base buildings are also shown in Figures 3 and 4 respectively. The ratio of the yield force of the isolator to the weight of the structure was taken to be 3% to 5% in the bi-linear and elasto-plastic hysteretic models. This isolator yield force is determined by the level of the wind loading used in design with a margin to prevent yield under wind-storm conditions.

The Voigt-Kelvin model (Lee 1968, Veletsos and Verbic 1973) was represented by a parallel combination of a single linear spring and a single dashpot with the requirement that the principle of equal strain between the two elements must be satisfied. In this research, the mechanical model used consists of a series of two Voigt-Kelvin models. One Voigt-Kelvin model is associated with the behaviour of the fictitious upper layer and the second one represents the behaviour of the lower soil layer. To be able to transform to the equivalent soil-mechanical model, the force equilibrium between two Voigt-Kelvin models in series must be satisfied and the first derivative of the displacement of both models obtained. Then, the total displacement of the system is the summation of each element. Finally, the equivalent stiffness and damping coefficients of the system can then be computed given the stiffness and damping coefficients of each Voigt-Kelvin model.

3 BASE ISOLATION SYSTEMS CONSIDERED

Many different forms of practical base-isolation systems have been developed to provide seismic protection for buildings, including laminated elastomeric rubber bearings, lead-rubber bearings, yielding steel devices, friction devices (PTFE sliding bearings) and lead extrusion devices. All of these systems can be categorized as displacement amplitude dependent devices. Their hysteretic behaviour is a function of the deformation imposed on the system. There is also recent interest in devices to increase the viscous damping in a structure. The yielding devices can generally be represented by a bilinear hysteresis model with the bilinear factor varying from 0% of the initial stiffness in the steel devices to approximately 15% in the lead-rubber bearings. Therefore, bi-linear and elasto-plastic hysteretic models will be considered to represent these various bearings; the bilinear hysteretic model for lead rubber bearings, and the elasto-plastichysteretic model for yielding steel devices, friction devices (PTFE sliding bearings), and lead extrusion bearings.

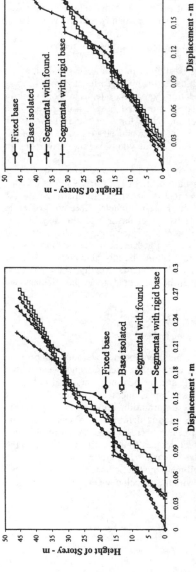

Figure 5 Comparison of displacement with storey height for fixed base, base-isolated and segmental buildings for the elasto-plastic hysteresis model

Figure 6 Comparison of displacement with storey height for fixed base, base-isolated and segmental buildings for the bilinear hysteresis model

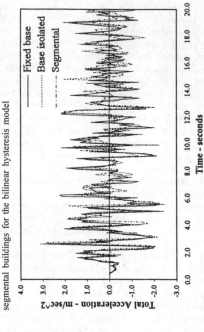

Figure 7 Comparison of displacement against time for fixed base, base-isolated and segmental buildings for the elasto-plastic hysteresis model

Figure 8 Comparison of total acceleration against time for fixed base, base-isolated and segmental buildings for the elasto-plastic hysteresis model

4 INELASTIC DYNAMIC ANALYSIS

The dynamic characteristics and seismic response of this building were determined using the N-S component of the 1940 El Centro earthquake ground motion as an earthquake input. However, at a later stage, other excitations with varying magnitudes of excitation and with differing spectral characteristics will be used in order to show that any conclusions reached are not just peculiar to the El Centro accelerogram. Since the members were assumed to be cracked, the effective member moments of inertia were calculated as 50% and 75% of the gross uncracked values for the beams and columns respectively. The weight of the structure was based on dead load plus 1/3 live load and concentrated at each floor of the structure. The inelastic analyses were carried out using the computer program *Ruaumoko* (Carr 1995). A Rayleigh damping model, where the damping is proportional to the mass and stiffness matrices of the structure, was used with 5% of critical damping being assumed to occur in the first and tenth modes for these twelve storey structures.

5 RESULTS

(a). Both the base-isolated and segmental building demonstrate the ability to decouple the building from the harmful horizontal earthquake ground motion.

(b). While keeping the ratio of yielding force of isolator to weight of structure low, segmentation of the super-structure reduces the base displacement response, as shown in Figures 5 and 6 for elasto-plastic and bi-linear hysteretic models respectively, compared with the values for fixed base and base-isolated buildings.

(c). Compared with a fixed base structure, base-isolated and segmental structures have much reduced interstorey drifts and, as shown in Figures 5 and 6, actually have much smaller displacements as well. This is true for both elasto-plastic and bi-linear devices.

(d). As shown in Figure 7, the dynamic characteristics do not change dramatically in that the overall natural period appears reasonably constant, though the fixed base structure shows a greater level of displacement response before twelve seconds. A similar response is shown in Figure 9. After twelve seconds, the segmental building shows the greatest response, though this is less than occurred earlier for the fixed base building. (Since the response for the segmental building with a fixed base and that with foundation compliance were similar, only one line is shown in Figures 7-10).

(e). It appears that the maximum total acceleration (ground acceleration + relative structural acceleration) is not greatly reduced by base isolation or segmental isolation as shown in Figures 8 and 10.

6 SUMMARY AND CONCLUSIONS

The benefits of implementing a base isolation system were demonstrated by the contrasting performance of base-isolated and non-isolated multistorey structures in the above results. With the inclusion of the base-isolation system, the inertia forces and the interstorey drifts can be significantly reduced. As a result, the superstructure can be designed to behave elastically during earthquakes. Also, base-isolation devices with low effective stiffness may cause significant lengthening of the fundamental period of the structures.

Further studies will be undertaken using equivalent added damping placed at each interior column of the superstructure to reduce the structure displacements. The seismic behaviour of the various types of building structures with flexible and stiff superstructures will also be investigated using a variety of earthquakes such as El Centro 1940 (N-S), Taft 1952 (N69W), Parkfield 1966 (N65E), and Mexico 1985 (SCT site S00E).

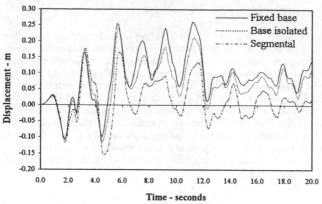

Figure 9 Comparison of displacement against time for fixed base, base-isolated and segmental buildings for the bilinear hysteresis model

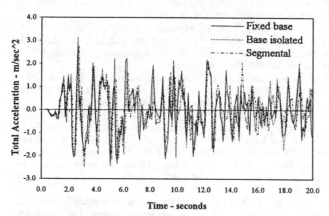

Figure 10 Comparison of total acceleration against time for fixed base, base-isolated and segmental buildings for the bilinear hysteresis model

REFERENCES

Andriono T. 1990. *Seismic Resistant Design of Base Isolated Multistorey Structures*, PhD Thesis, University of Canterbury, Christchurch, New Zealand.

Andriono T., and Carr A.J. 1991a. Reduction and Distribution of Lateral Seismic Inertia Forces on Base Isolated Multistorey Structures, *Bull. of the NZ National Society for Earthquake Engineering*, 24(3):225-237.

Andriono T. and Carr A.J. 1991b. A Simplified Earthquake Resistant Design Method for Base Isolated MultiStorey Buildings, *Bull. of the NZ National Society for Earthquake Engineering*, 24(3):238-250.

Cui, W. 1995. *Seismic Response of Non-Linear Base-Isolated Shear Buildings and Their Secondary Systems*, PhD Thesis, Nanyang Technological University, Singapore.

Carr, A.J. 1995. *Ruaumoko*, Computer Program Library, Dept. of Civil Engineering, University of Canterbury.

Lee, I.K. 1968. *Soil Mechanics*, Sydney:Butterworths, 504-515.

Standards New Zealand (SNZ), 1992. *General structural design and loadings for buildings*, NZS 4203, Standards New Zealand, Wellington.

Standards New Zealand (SNZ), 1995. *The design of concrete structures*, NZS 3101, Standards New Zealand, Wellington.

Veletsos, A.S and Verbic,B. 1973. Vibration of Viscoelastic Foundations, *Earthq. Engg. Struct. Dynam.*, 2:87-102.

The Mechanics of Structures and Materials, Grzebieta, Al-Mahaidi & Wilson (eds)
© 1997 Balkema, Rotterdam, ISBN 90 5410 900 9

Seismic performance of limited ductile frames including softening

S. de Silva
Connell Wagner Ltd, Melbourne, Vic., Australia

P.A. Mendis
Department of Civil and Environmental Engineering, University of Melbourne, Vic., Australia

ABSTRACT: A large portion of existing building stock in Australia has not been designed to seismic loading. Even most of the structures built now are still typically designed only for gravity loads or with irrational response modification factors leading to limited ductility. It is very important to assess the seismic performance of the limited ductile structures. Unlike steel structures which strain harden, reinforced concrete structures characteristically soften beyond the elastic, plastic phases. Therefore the softening phase has to be considered in the evaluation of seismic performance these structures. It is the intention of this paper to provide a method for assessing the ductility of a reinforced concrete frame by evaluating the structure softening slopes.

1 INTRODUCTION

The introduction of seismic resistant design to design standards of the low seismic intra-plate regions like Australia are discouraged due to the cost associated with improved seismic resistance and the misconception that the area concerned is seismically inactive. It is worth noting that prior to the release of the current Australian earthquake standard, AS 1170.4 (1993), a major part of Australia was considered as seismically inactive. Consequently a large portion of existing building stock in Australia has not been designed for seismic loading. Even most of the structures built now are still typically designed only for gravity loads or with irrational response modification factors leading to limited ductility. The current seismic design philosophy warrants a trade off between strength and ductility. Implicitly most design standards adopt a dual design criteria by ensuring the linear response for minor to moderate events (damageability limit state) and non-linear response for design credible earthquakes (ultimate limit state) through response modification factors. The direct application of response modification factors recommended for high seismic regions in low seismic conditions in Australia is not rational. The applicability of the response modification factors specified in AS1170.4 is questionable as they are based on Californian requirments. Therefore it is important to assess the sesimic performance of limited ductile existing and new structures using a rational procedure.

The seismic performance can be assessed through a forced-based process or a displacement-based procedure. The force-based seismic assessment procedure is based on determining the available static lateral load strength and ductility of the critical post-elastic mechanism of deformation of the structure. The displacement-based procedure, which is more rational, is based on the evaluation of the performance of a structure at different displacement levels. These procedures are discussed in detail in another publication. In both these procedures, it is essential to consider the full-range behaviour of a structure. Unlike steel structures, which

369

strain harden, reinforced concrete structures go through three phases elastic, plastic and softening.

2. CONCEPTS OF SEISMIC DESIGN AND SOFTENING

Current seismic design philosophy permits to compromise between elastic strength and ductility. The displacement ductility is defined as the maximum displacement to yield displacement ratio of the structural system. In order to meet the additional displacement ductility demands resulting from the permitted elastic strength reductions, the rotation capacities of the critical hinges should be adequate. Therefore, the strength degrading characteristics (or softening) must be considered when evaluating the appropriate level of lateral load reductions and the additional curvature ductility demands enforced as a consequence. Two major aspects of softening is of interest when investigating the impact of softening on current seismic elastic strength reduction criterion.

Firstly, the validity of the plastic analysis in the design of limited ductile reinforced concrete members is questionable. Plastic design method is based on the assumption of elastic perfectly plastic idealisation of the moment curvature characteristics. For example, the extensive ductility of steel and the subsequent strain hardening provide the realism of this basic assumption and the factor of safety respectively for steel structural systems. However, the limited rotation capacity and subsequent softening of limited ductile reinforced concrete sections prevent the realism of this basic assumption and reduces the safety margin extensively. Secondly, the realism of a favourably rational yielding mechanism, effective moment distribution and reserve over-strength of a limited ductile structural systems is in doubt. Obviously, these conditions are considered favourable under seismic response and explicitly constitute in a competent seismic design. Conventionally plastic analysis method assumes the structure can sustain its load carrying capacity until a yielding mechanism is formed via effective moment distribution and possess reserve over-strength as a consequence.

The above factors combined together lead to an upper bound approach to collapse limit state of limited ductile structural systems, if assessed using elastic perfectly plastic idealisation, ignoring the effects of softening.

3. SOFTENING CHARACTERISTICS OF LIMITED DUCTILE MEMBERS

Concrete sections characteristically soften in post-elastic behaviour (Darvall ,1983, Mendis 1986, Sanjayan 1988). Softening is defined as the reduction in bending moment capacity with increasing curvatures. This is more apparent when dealing with severe loading conditions such as earthquakes where reinforced concrete members are expected to perform non-linearly and likely to perform in its full range.

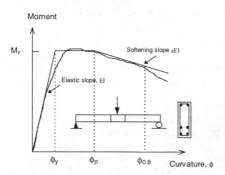

Figure 1. Moment-curvature diagram of a typical limited ductile reinforced concrete section

As given in Figure 1, the softening slope, a, is the ratio between the gradient of the softening branch and the gradient of the elastic branch of the M-ϕ curve. The rotation capacity, θ_p, is given in Eq. 1, where l_p is the hinge length and ϕ_y and ϕ_u are the curvature at yield and the end of the plastic plateau respectively as defined in Figure 5.1.

$$\theta_p = l_p (\phi_p - \phi_y) \tag{1}$$

The experimental observations based on limited ductile sections suggest that the full range behaviour of a typical reinforced concrete hinge subjected to severe loading conditions is better characterised by elastic-softening (moment reduction) rather than elastic-perfectly plastic idealisation (Mendis, 1986). Many researchers have observed the post elastic strength degrading characteristics of reinforced concrete possessing the said deficiencies when subjected to severe loading conditions (Sozen 1974, Mendis 1986, Sanjayan 1988, Corvetti 1993). This is more apparent in the low to medium ductile members, over-reinforced sections, sections with higher ratio of compression steel, very high strength concrete compression members, axially loaded flexural members and beam column joints. In addition load reversals can adversely contribute to the rate of strength degradation. The inappropriate proportioning of members (strong beams and weak columns combination), insufficient transverse steel and inadequate joint detailing that aggravate bond failure are some of the constitute deficiencies which contribute to softening. Arguably, the said deficiencies are inherent within a large population of the building stock located in the low seismic intra plate regions such as Australia. It has been demonstrated (de Silva 1997) that the most critical range of structures under intraplate earthquakes, are the low rise structural systems with high frequencies where quite often frames provide the lateral stability. Therefore, the current study is focused on the framed structures.

Mendis (1986) presented one of most comprehensive theoretical and experimental studies on the softening of reinforced concrete members. The work was aimed at investigating the softening parameters such as rotation capacity, hinge length and the softening slopes of critical regions of limited ductile structural members. Sanjayan (1988) investigated the effect of softening on the dynamic response of the structures. Corvetti et al. (1993) has clearly demonstrated the strength degrading characteristics of the external beam column joints designed to AS 1170.4 (1993) and AS 3600(1988) especially with the limited ductile detailing subjected to reversal loading. Two other projects recently completed at the University of Melbourne are also considered worth noting. Pendyala et al. (1996) and Kovacic (1995) investigated the full-range behaviour of high-strength concrete beams and columns respectively.

Findings of the above studies among several others, pauses a concern over the elastic perfectly plastic idealisation of limited ductile reinforced concrete members. Therefore, it is essential to investigate the response parameters of limited ductile systems using elastic-softening in order to realistically assess the seismic load carrying capacity.

The ductility of a elastic-palstic-softening section depends on:

- The difference between the area of compression steel, A_{sc} and the area of tension steel, A_{st}. On average, higher the difference, the steeper the softening slope and lesser the rotation capacity.
- Confinement pressure of the concrete core governed by stirrup spacing. The ability to develop higher confinement pressures within the core reduces the softening slope of a reinforced concrete section and increase the rotation capacity.
- Axial stresses due to external loads. Members with higher axial loads such as columns at lower floors, exhibit steeper softening slopes and lesser rotation capacities.
- Shear span. Shorter shear spans tends to generate steeper softening slopes and lesser rotation capacities due to higher shear forces acting on the section.

Mendis (1986) proposed empirical formulae to estimate the softening slope, rotation capacity and hinge lengths. The softening slope and rotation capacity can be also found by deriving theoretical moment curvature relationships (Mendis 1986). Spread-sheets are available at the University of Melbourne to find the confinement pressures and theoretical moment-curvature curves of reinforced concrete sections. These curves can be used to derive typical softening parameters for seismic design in the absence of more precise data.

4 STRUCTURE-SOFTENING

Closed form solutions to estimate overall structure softening, B, due to local member softening, a, for simple structures (eg. cantilever) can be derived with relative ease (de Silva 1997). However, the relationship between local member softening, a, and the overall structure softening, B, becomes too complicated as the redundancy of the structure increases. Not only it is tedious to ascertain but also cannot be generalised due to its case dependency. However reasonable estimates can be made on a case by case basis using numerical methods. It is evident that on average the structure softening slope is greater than the member softening slopes and depend on the hinge length ratio (Sanjayan, 1988).

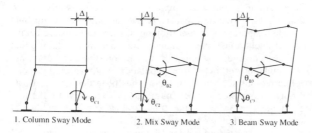

1. Column Sway Mode   2. Mix Sway Mode   3. Beam Sway Mode

Figure 2 Probable sway modes of frames subjected to seismic loads

The overall structure softening depends on the opted or realised yielding mechanism. Figure 2 illustrates the typical yielding mechanisms and sway modes of frames under seismic loading. The yielding mechanisms are arranged in the increasing order of preference where the beam side sway mode is the most preferred yielding mechanism in seismic design and the column side sway mode is attempted to be avoided in competent seismic design practice. The effect of yield mechanism on overall structure softening slope, B, depends on two contributing factors. Firstly, the softening slopes of axially loaded members (i.e. columns) are greater than that of the beams and therefore the presence of more column hinges tends to soften the structure more rapidly than that of the yielding occuring at the beams. Secondly, the curvature ductility demands at column hinges in the column sway mode shown in Figure 2, are much greater than that of the beam sway mechanism for a given displacement, Δ, which result in higher overall structure softening.

The effect of yielding mechanisms on the overall structure softening parameter, B is investigated and presented here using a two storey frame. The two storey single bay frame as given in Figure 3 is quoted in a number of studies (Mendis, 1986). In this study, the computer program "SOAPS" (Secon-Order Analysis with Plastic Softening, Mendis and Darvall, 1985) is modified to assess the post softening behaviour in order to obtain the full range response and the structure softening slopes. In order to obtain the required sway modes and hinge locations, the yield moments of beams and columns framing into a joint were slightly adjusted. The beam softening is selected as -1% whereas the column softening is increased to -3%, as the columns tends to exhibit more softening under axial loads than the beams. The lateral loads are progressively increased while the vertical loads are kept constant until the softening of the overall structure reaches 80% of the ultimate strength, simulating a push-over test condition. The structure softening slope ratio (B) is calculated by evaluating the ratio between the average softening slope (ultimate strength to 80% of ultimate strength) and the initial slope until yield.

The load-displacement relationships and the sequence of hinge formation are shown in Figures 4 and 5. The beam sway mode has the lowest ultimate strength (C_u=14.586kN) and the shallowest overall structure softening slope, (B= 6.2%). The column sway mechanism exhibit

372

the highest ultimate strength (C_u=19.2 kN) and the steepest softening slope (B= 24.8%) showing the least ductility. The B value for the mixed mode (6.9%) is between the B values for other two modes. As shown by this case study, the ductility of a frame beyond the plastic phase can be evaluated by deriving the structure softening slope.

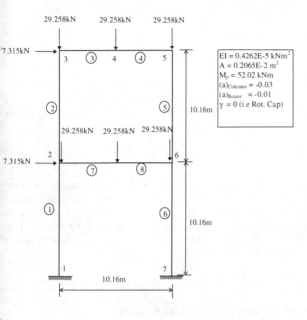

Figure 3. Two story frame layout, section properties and loads

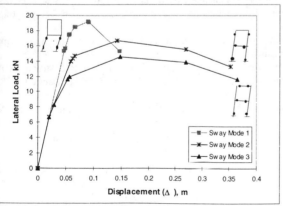

Figure 4. Load-displacement diagram

It is apparent that the overall strength degrading due to flexural softening is further enhanced by the presence of P-Δ effects and some allowance should be made to account for the combine effects especially if generalised response parameters are opted in design. Nevertheless, it is

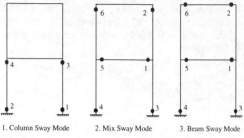

1. Column Sway Mode   2. Mix Sway Mode   3. Beam Sway Mode

Figure 5. Sequence of yielding

envisaged that similar behaviour with relatively less softening slopes may have resulted if higher redundancies combined with elastic-plastic-softening members were tried out. The actual load displacement curve of such a structure would be enveloped by the elastic-perfectly plastic case as the upper bound and the elastic-softening case as the lower bound.

5 CONCLUSIONS

The effects of flexural softening on the stability and the load carrying capacity of frame structures as discussed in the foregoing sections can be treated as the dominating factor causing the instability under seismic performance of softening or strength degrading limited ductile structural systems. However, the current seismic design procedures appears to have ignored the effects of flexural softening. The consideration of softening is essential in a force-based or displacement-based assessment of limited ductile systems.

REFERENCES

AS3600 1988. Australian concrete structures code. Standards Australia.

AS1170.4 1993. The design of earthquake resistant buildings. Standards Australia.

Corvetti, J. 1993. Assessment of reinforced concrete exterior beam-column joints as specified in the new draft earthquake standard, proceedings of 13th ACMSM conference, Wollongong, pp. 225-231.

Darvall, P.LeP. 1983 Some aspects of softening in flexural members, Civil Engineering Research Report, No. 3/1983, Monash University.

De Silva, K.S.P., Mendis, P. & Wilson, J., 1994. Design and detailing of R/C structures in accordance with the new australian earthquake standard, Civil Engineering Transactions, The Institution of Engineers, Australia, August.

De Silva, K.S.P. 1997. Response of softening reinforced concrete frames under intraplate earthquakes, Ph.D. Thesis, The University of Melbourne.

Kovacic, D. 1995. "Design of high-strength concrete columns", M.Eng. Thesis, The University of Melbourne.

Mendis, P.A., 1986. Softening of reinforced concrete structures, Ph.D. Thesis, Monash University.

Pendyala, R, Mendis, P. & Patnaikuni, I., 1996. Full-range behaviour of high-strength concrete flexural members, ACI Structures Journal, Vol. 93, No. 1, pp 30-35.

Sanjayan, G., 1988. Dynamic response of reinforced concrete structures with softening behaviour, Ph.D. Thesis, Monash University.

Sozen, M.A., 1974. Hysteresis in structural elements, Applied Mechanics in Earthquake Engineering, ASME, ASD8.

Finite element methods, optimisation and stability

The Mechanics of Structures and Materials, Grzebieta, Al-Mahaidi & Wilson (eds)
© *1997 Balkema, Rotterdam, ISBN 90 5410 900 9*

Computer modelling for the design of a brittle tensile specimen

M. P. Rajakaruna

School of Engineering, University of South Australia, Adelaide, S.A., Australia

ABSTRACT: Shape has a significant effect on the location of fracture in a brittle tensile specimen. Finite element analysis accurately predicts the location of fracture and enables the design of tensile test specimens that fracture within the central zone of the specimen.

INTRODUCTION

The static tensile test is well known as a simple method of determining the yield strength, tensile strength, modulus of elasticity and ductility of metals. Tension tests occupy a large share of attention in materials testing and a great deal of confidence is placed on results from these tests. The wealth of information obtained from a tension test is used not only to ensure compliance with specifications and in the development of new products but also in settlement of disputes in the event of failure. Tensile tests should therefore be carefully planned and designed to accurately predict the behaviour of the material in service.

Tensile tests are carried out on round and rectangular test pieces made in accordance with Australian Standard AS1391. Although a rectangular shape is favoured for sheet and plate stock, a circular cross section obtained by machining is commonly used for specimens when sufficient material is available. The most important feature of a tensile specimen is the slender uniform central portion known as the gauge length; the area of the reduced section relative to the gripped ends is usually determined by custom. The ends may be made plain, shouldered or threaded to suit the available gripping devices (Davis et al. 1982). AS 1391 recommends plain ends with circular fillets but does not provide details of gripped ends. The American Society of Testing Materials (ASTM) recommends plain ends with a gradual taper in the reduced section (ASTM A47M-90) or shouldered ends (ASTM A48-94a) for cast iron specimens and provides detailed dimensions of all features.

The presence of a fillet in a tensile specimen results in areas of localised high stress known as stress concentrations. Circular fillets are usually specified for simplicity in drafting and machining rather than for minimising stress concentrations. Stress concentrations can be reduced by increasing the radius of a circular fillet or decreasing the change in cross section immediately prior to the gauge length by providing a shoulder (ASTM A48-94a), which requires extra machining. Parabolic, elliptical or variable radius curves are more effective in reducing stress concentrations at fillets than arcs of fixed radius (Peterson 1974).

Tensile specimens made of mild steel according to AS 1391 always fracture within the central zone of the test piece. However, similar specimens made of cast iron frequently fracture near the fillets due to the low ductility of the material (Jayatilaka 1979). Fracture outside the gauge length is unacceptable and the usual practice is to ignore such results and repeat the test

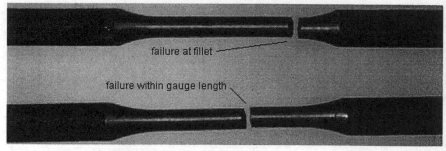

failure at fillet

failure within gauge length

Figure 1. Brittle fracture of cast iron

until failure is achieved in the central zone (AS1391-1991). Further investigations are needed to examine the likely causes for failures outside the central zone in brittle materials so that the wasteful practice of repeated testing is minimised or even eliminated (Figure 1).

Stress concentration factors for solid circular shafts under tension available in the literature (Lipson & Juvinall 1963) usually fall outside the geometric proportions commonly adopted for tensile specimens. Finite element analysis is a useful method to obtain detailed information on the distribution of stress in a specimen during a tensile test and can be used to identify probable locations of fracture.

This paper describes how finite element analysis was used to illustrate the incidence of peak tensile stresses at the fillets of a specimen made according to AS 1391 and offers an explanation for the large number of brittle fractures around this region. A slight taper in the gauge length and a substantial increase in transition radius moved the stress concentration away from the fillets towards the centre of the specimen so that fractures are more likely to occur within the gauge length as required by the Australian Standard.

MATERIALS AND METHODS

Finite element modelling

The state of stress arising in a specimen under an axial tensile load is complex at the change in cross section. The maximum normal stress theory asserts that failure of a material occurs when the maximum normal stress at a point reaches a critical value regardless of the other stresses. Experimental evidence indicates that the normal stress criterion applies to brittle materials in all ranges of stresses providing a tensile principal stress exists (Popov 1990).

A tensile specimen of circular cross section is an axisymmetric solid object with a geometry and material properties independent of the circumferential co-ordinates. Loads and supports can also be considered axisymmetric without significant loss of accuracy. The problem is therefore mathematically two-dimensional and a simplified analysis with axisymmetric finite elements was adopted for solution.

MSC/NASTRAN for Windows Version 2.0 was used in this investigation to obtain a detailed distribution of maximum normal stress in a tensile specimen with plain ends and circular fillets to identify the areas of stress concentrations. A finite element model was generated by considering only one quadrant of the specimen taking account of symmetry. The geometry was meshed with axisymmetric triangular elements with mid-side nodes; the only type supported by the software (Figure 2).

Gripping forces were applied as radial line loads over the circumference of the specimen with a radial line load of q units of force per unit of circumferential length acting at a radius r

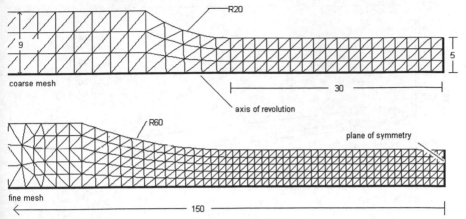

Figure 2. Part of finite element mesh; axisymmetric elements with mid side nodes

contributing a force of $2\pi r q$ to the load vector. Gripping forces were resolved into axial and radial components and distributed uniformly among the nodes. The axial component of the force was determined to impose a nominal stress of 100 MPa at the reduced section of the specimen. This value was arbitrarily selected to determine the stress concentration factors easily. The radial component of the force although statically equivalent to zero, could not be discarded from the load vector since it produces deformation and stress (Cook 1995). The radial component was calculated assuming a coefficient of friction of 0.2 between the metal surfaces of the specimen and holding grips.

Finite element modelling provided information on changes in magnitude and location of the peak tensile stress in the specimen when fillet radius and the taper in the reduced section were varied.

Materials

Tensile specimens were made of a close grained, predominantly pearlitic grey cast iron equivalent to BS 1452-1990, Grade 250.

Finite element modelling reduced the need for extensive laboratory testing of components otherwise needed for an investigation of this nature. Testing was limited to selected configurations based on the results of finite element analysis.

The tensile specimens had plain ends to suit the gripping devices available. The 10mm diameter reduced section had parallel or tapered sides 60mm in length to accommodate a 50mm gauge. Transition radii ranged from 20 to 75mm. The length of each specimen was 300mm.

These specimens were machined to tolerances specified in AS 1391 on a numerically controlled lathe and tested in a Mohr and Federhaff hydraulic testing machine until fracture. The load at failure and the location of fracture were observed.

RESULTS

The stress distribution at the transition played an important role in determining the probable location of fracture. Finite element analysis revealed a stress concentration at the root of the fillets in excess of the stress within the gauge length in a parallel-sided specimen. This region is quite substantial as shown in Figure 5 and fracture more likely to occur in this region as

379

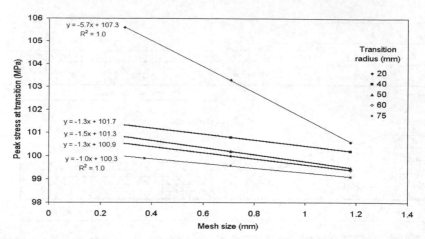

Figure 3. Variation of peak stress at fillets with mesh refinement; 10mm diameter specimen with 0.08mm taper

Table 1. Tensile test results on grey cast iron specimens

Fillet radius (mm)	Taper (mm)	Average stress at fracture (MPa)	Number of specimens	Location of fracture
20	-	Not observed	7	At fillet
		303	2	Within gauge length
30	-	294	1	At fillet
		307	1	Within gauge length
40	-	307	2	At fillet
50	-	298	1	At fillet
		316	1	Within gauge length
50	0.02	297	2	At fillet
50	0.08	301	4	At fillet
		323	1	Within gauge length
75	0.08	275	2	At fillet
		332	3	Within gauge length

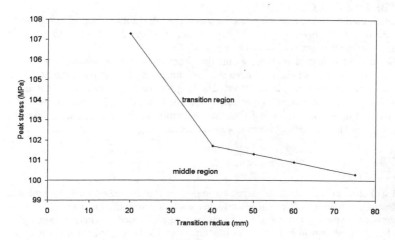

Figure 4. Expected peak stress at infinite mesh refinement; 10mm diameter specimens with a 0.08mm taper

380

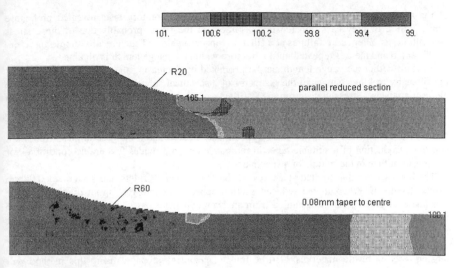

101.   100.6   100.2   99.8   99.4   99.

R20

parallel reduced section

R60

0.08mm taper to centre

Figure 5. MSC/NASTRAN output - contours of axisymmetric major principal stress

evident by laboratory results (Table 1). Tensile stress at the transition reduced when the fillet radius increased. However a large radius alone was not sufficient to minimise the stress concentration at the transition; a slight taper in the gauge length was needed in addition to move the stress concentration towards the middle of the specimen.

The results from a finite element analysis should converge towards the solution of the corresponding mathematical model that describes the problem, as the mesh is repeatedly refined, provided there are no errors in modelling. The results of three analyses done with moderate mesh refinement at each step for specimens with 20 and 75mm radii fillets showed a linear relationship between the peak stress and mesh size (expressed as square root of the approximate area of an axisymmetric element). The value of peak stress expected at infinite mesh refinement was obtained as the intercept of a straight line through these three points (Cook 1995). Expected values for peak stress for fillet radii between 20 and 75mm were obtained by linear extrapolation of computed results from two analyses (Figure 3).

Maximum tensile stress in the transition region reduced as the radius of the fillets increased. Transition radius and taper in the reduced section were increased until finite element analysis results indicated the peak stress in the fillets to be below the peak stress in the middle of the specimen (Figure 3). However the expected peak stress at zero mesh size was still marginally above the peak stress in the middle region as shown in Figure 4.

Laboratory tests on a limited number of specimens recorded a lower failure stress at a fillet compared to a failure within the gauge length (Table 1). Stress concentrations in the region of fillets are probably responsible for these premature failures and such results should therefore be discarded. Of the shapes studied in this investigation, the specimen with a 75mm radius fillet and a 0.08mm taper over a length of 30mm provided the best proportion of acceptable results. Since the calculated peak stresses in the fillets and in the middle of this specimen are almost the same, there is an equal probability of a fracture initiating in either of these regions.

DISCUSSION

The results from finite element analysis displayed a concentration of tensile stresses at the root of the fillets in specimens with parallel sides. Limited laboratory tests revealed premature brittle failure of grey cast iron tensile specimens at the fillets probably due to these stress concentrations. Premature failures at a fillet are unacceptable and such results should therefore be discarded and the test repeated until a fracture within the gauge length is obtained.

Fracture within the gauge length can be promoted by decreasing the stress concentration at the fillets by making changes to the geometry of a specimen supported by results from a finite element analysis. Stress concentrations at fillets can be decreased by increasing the transition radius and by introducing a suitable taper in the gauge length towards the middle of the specimen. Parallel-sided specimens with fillet radii close to the minimum specified values of AS 1391 are unsuitable for brittle tensile specimens. A substantial increase to these fillet radii and the introduction of a suitable taper in the gauge length provides favourable conditions for failures to initiate in the middle of a specimen.

The results form this limited study reveal that 75mm radius fillets and a taper of 0.08mm over a length of 30mm in the reduced section appears suitable for 10mm diameter tensile specimens of grey cast iron machined from an 18mm diameter bar. This taper is well outside the variation in diameter over the gauge length allowed by AS 1391 but is of a similar magnitude to that specified by ASTM A47M for machined cast iron specimens.

Finite element analysis provides a reliable method for predicting the location of failure in tension tests on brittle materials. Brittle tensile specimens prepared using this method will increase the likelihood of obtaining acceptable results in agreement with AS 1391.

REFERENCES

American Society for Testing and Materials, Standard Specification for Ferritic Malleable Iron Castings, ASTM A 47M – 90, 1990

American Society for Testing and Materials, Standard Specification for Grey Iron Castings, ASTM A 48 - 94a, 1994

British Standards Institution, Specification for Flake Graphite Cast Iron, BS1452, 1990

Cook R D, Finite Element Modelling for Stress Analysis, John Wiley & Sons Inc, 1995

Davis E D, Troxell E T, Hauck G F W, The Testing of Engineering Materials, 4th edition McGraw Hill Inc, 1982

Jayatilaka A de S, Fracture of Engineering Brittle Materials, Applied Science Publishers Ltd, London, 1979.

Lipson C & Juvinall R C, Handbook of Stress and Strength, Design and Material Applications, The Macmillan Company, 1963

MSC/NASTRAN for Windows, Installation and Application Manual, Version 2.0

Peterson R E, Stress Concentration Factors, John Wiley & Sons, 1974

Popov E P, Engineering Mechanics of Solids, Prentice Hall, 1990

Standards Australia, Methods for Tensile Testing of Metals, AS1391, 1991

The Mechanics of Structures and Materials, Grzebieta, Al-Mahaidi & Wilson (eds)
© *1997 Balkema, Rotterdam, ISBN 90 5410 900 9*

Finite element formulations for the analysis of open and closed cross sections

K.T. Kavanagh
Department of Civil Engineering, University of Western Australia, Nedlands, W.A., Australia

ABSTRACT: Two finite element models are compared for the analysis of beam cross sections in which the cross section is assumed to be rigid. The first model, a Bernoulli-Euler beam model with added shear flexibility, is shown to give good results for bending of the cross-section, but poor results for torsion of the cross section. A completely compatible cubic beam element is shown to provide improved results for the case of torsion. Differences between the results of the computer analyses and the classic Vlasov theory of torsion are highlighted.

1 INTRODUCTION

The analysis of thin-walled sections in bending and torsion has been the subject of considerable research [1 - 5]. The classical theories of Timoshenko, Bleich and Vlasov begin with the assumption of Bernoulli-Euler beam theory for individual elements that comprise a thin-walled cross-section. The assumption that plane sections remain plane and normal to the deformed axis of each segment requires that shear stresses (due to bending) be zero in order to completely satisfy equilibrium. The concept of shear flow in a cross section is obtained from Bernoulli-Euler theory by enforcing axial equilibrium along the length of the member under non-uniform moment. The concept of shear flow is useful as an engineering approximation, however, the presence of shear stresses is fundamentally incompatible with the Bernoulli-Euler assumption of cross-section normality.

The concept of shear flow can be subsequently used to calculate the location of the shear centre of any symmetric or unsymmetric member. Under pure torsion, the use of Bernoulli-Euler theory and the concept of shear flow leads to the conclusion that the shear centre is the centre of twist for any section. The simplest proof of this conclusion requires the use of Maxwell's reciprocal theorem, which in turn, assumes shear stresses derived from plane sections remaining plane and normal. When shear distortions are introduced, the normality condition is relaxed, and the centre of twist can be shown to move away from the shear centre.

A simple experiment can be used to demonstrate that the shear centre is not the true centre of twist. Consider the case of an open channel which is simply supported at both ends. If a pure torque is applied to the centre of the beam, Fig. 1b shows that the shear centre assumption requires the beam to rotate about a point outside of the channel. The mid-span of the beam, therefore, must move down. If the beam moves downward, a vertical shear is created in the web, and the vertical bending equilibrium is violated. To balance shear, the beam must move up to its original position, and the shear centre must also move up. In Fig. 1c, the final position of the beam requires that the section rotate about the web, not about the shear centre. This

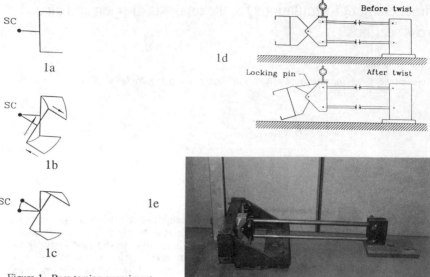

Figure 1. Pure torsion experiment.

ambiguity about the centre of twist is borne out by finite element representations of the torsion problem. The above experiment can be performed in the laboratory, with the result being most pronounced for short beam lengths.

Fig. 1d and Fig. 1e illustrate a laboratory experiment to demonstrate the shear centre movement under pure torque. A simply supported channel section is constrained to move vertically, and to twist about its calculated shear centre (pin position). A torque is applied at mid-span until a pre-set twist angle is achieved. A locking pin is then inserted to fix the angle of twist, and the torque is removed. Vertical motion of the shear centre is then measured by a dial guage. For the above experiment, a 100 mm deep channel on a 400 mm span produced an upward movement of 0.258 mm at the pin.

2 FINITE ELEMENT MODELLING

2.1 *A beam model*

The development of beam elements with shear flexibility dates back to the early stiffness matrix programs of the 1950's and 1960's. The intent of adding shear flexibility was to model shear walls as beam elements, where the contribution of shear often dominates the contribution of bending. A simple expedient to including the shear flexibility is to add the contributions of shear and bending from a statically determinate beam, Fig. 2, so that a modified stiffness arises from the addition of flexibilities. This modified stiffness form is a feature of most computer programs which are on the commercial market today.

From Fig. 2, it is apparent that the shear deformation is linear with length. Any point (A) on the extremity of the beam deforms longitudinally as a combination of bending rotation and shear deformation. The bending contribution to motion at (A) is quadratic with length, while the shear component is linear with length. thus, the addition of flexibilities to form an equivalent end rotation does not uniquely define the compatibility of adjacent elements along the element boundary (see Fig. 2b). It would normally be assumed that this lack of compatibility becomes inconsequential as the length of the element is decreased.

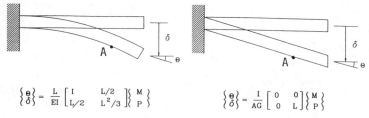

$$\begin{Bmatrix} \theta \\ \delta \end{Bmatrix} = \frac{L}{EI} \begin{bmatrix} I & L/2 \\ L/2 & L^2/3 \end{bmatrix} \begin{Bmatrix} M \\ P \end{Bmatrix}$$

$$\begin{Bmatrix} \theta \\ \delta \end{Bmatrix} = \frac{I}{AG} \begin{bmatrix} 0 & 0 \\ 0 & L \end{bmatrix} \begin{Bmatrix} M \\ P \end{Bmatrix}$$

Figure 2. Beam element with shear distortion.

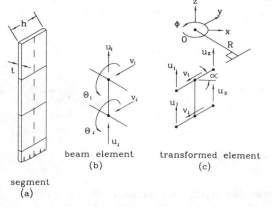

beam element
(b)

transformed element
(c)

segment
(a)

Figure 3. Cross-sectional element.

By changing variables in the beam model of Fig. 2, $(u_1 + u_2)/2 = u$, $(u_1 - u_2)/h = \theta$, the beam element can be used to represent a flat plate element with end joint compatibility at the element edges. This element can be assembled to form the geometry of any open or closed cross section. Fig. 3 shows an assembly of elements constrained in the end planes to move with an assumed origin of coordinates.

To test the element in bending and torsion, the channel section shown in Fig. 4 can be modelled as an assembly of three longitudinal flat plate elements. The application of three unit loads (X, Y, θ) to the end of a cantilever enables the computer to determine the location of the centre of stiffness (the shear centre) at the free end. The distance of the centre of stiffness from the web of the channel is given below for different length-to-depth ratios:

Table 1.

L/D	e
1	2.50
2	3.60
3	4.10
10	4.20

385

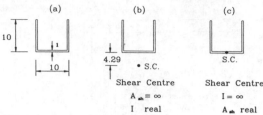

Figure 4. Open channel centres of twist.

For a preset length-to-depth ratio of 3, computed beam flexibilities for the channel can be obtained using three separate assumptions: (1) the shear area equal to infinity (the Bernoulli-Euler assumption); (2) the moment of inertia equal to infinity (the shear wall assumption), and (3) $I_{xx} = h^3t/12$, $A_{sh} = A = ht$ in each flat plate element. The uncoupled flexibilities at the top of the cantilever produce:

$$
\begin{array}{cccc}
& \begin{array}{c}(1)\\ \text{Bernoulli-Euler}\end{array} & \begin{array}{c}(2)\\ \text{Shear Wall}\end{array} & \begin{array}{c}(3)\\ \text{Actual}\end{array}\\[4pt]
\begin{array}{c}P_x:\\ P_y:\\ \theta\,:\end{array} &
\begin{bmatrix} 15.6 & 0 & 0 \\ 0 & 27 & 0 \\ 0 & 0 & 1.51 \end{bmatrix} &
\begin{bmatrix} 6.0 & 0 & 0 \\ 0 & 3.0 & 0 \\ 0 & 0 & 0.12 \end{bmatrix} &
\begin{bmatrix} 23.6 & 0 & 0 \\ 0 & 30 & 0 \\ 0 & 0 & 1.63 \end{bmatrix}
\end{array}
\tag{1}
$$

It can be seen that the flexibilities add in both the symmetric direction (Y) and in the rotation (θ), so that the actual beam agrees with the assumed summation of bending and shear deformation. In the unsymmetric (X) direction, however, the flexibilities do not sum to give the total response. The discrepancy in direction (X) arises from different centres of twist in the three cases. In the Bernoulli assumption, the centre of twist lies on the shear centre, Fig. 4b. In a shear wall, the centre of twist lies on the web, Fig. 4c. In the actual beam, the shear centre varies between the shear centre and web as a function of beam length.

When the beam element formulation is applied to a closed section, a more serious problem can be shown to arise. Complete compatibility of displacements between element boundaries requires that the cross section deform in pure shear. A box section is shown in Fig. 5. If the shear area of each plate element is driven to infinity, the element should become infinitely stiff. The lack of compatibility shown in Fig. 5b, however, allows the element to bend, and leads to the stiffness becoming a function of the element length. The lack of compatibility is shown schematically in Fig. 5b. This element, therefore, is not suitable for the analysis of closed or hybrid sections.

2.2 A compatible cubic beam element

Fig. 6 illustrates a fully compatible cubic beam element. Each of the displacements is described by a cubic Lagrange interpolation function in the length direction, and by a constant or linear variation across the element width. Internal energy terms are shown on Fig. 6, and the integrals can be carried out in closed form. The element has been tested on closed sections, and the well-known twisting resistance, $J = 4A^2/\Sigma\, l/t$, can be confirmed from the computed flexibilities. For open sections, the cubic beam model agrees with the simple beam model which was presented previously. A more serious test of the finite element model can be constructed from the cross section

386

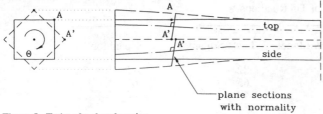

Figure 5. Twist of a closed section.

Shape Functions

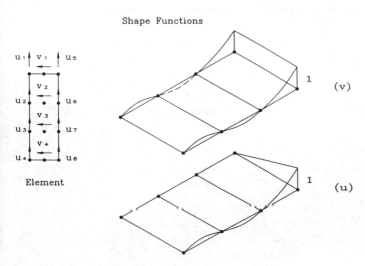

Figure 6. Cubic element.

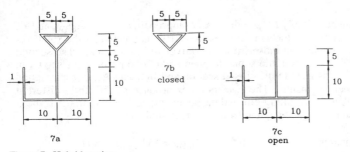

Figure 7. Hybrid section.

shown in Fig. 7. In this case, the cross section is a hybrid open/closed section, with contributions from both warping torsion and St. Venant Torsion. As shown in Ref 6, the computer model is at variance with the classical theory of torsion for thin walled beams.

An important characteristic of the cross section in Fig. 7 is the relative importance of St. Venant torsion (the closed triangle), and the warping torsion (the open flanges). Two methods can be used to calculate the relevant section properties of the hybrid

387

Table 2. Comparison of Tip Rotations

Height	Computer Model	Least Squares	Separation
H	1.05	.82	.81
4H/	.76	.65	.61
3H/5	.49	.49	.42
2H/5	.25	.32	.24
H/5	.07	.09	.06

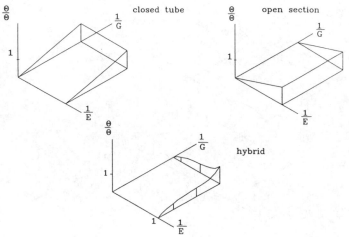

Figure 8. Parameterised behaviour of a hybrid section.

section. The first method simply fits the three section properties, (I_{ww}, A_{shw}, and J), by a least squares criterion which minimises the difference between the rotations from a full computer model and those from an equivalent beam model. The second method divides the cross section into separate open and closed portions, as shown in Fig. 7c, in which J is assigned to the closed section and I_{ww} and A_{shw} are assigned to the remaining open segments. The two methods produce significantly different section properties, but the resulting equivalent beam rotations are similar. In both cases, the comparison with the complete finite element model is poor.

Method 1: Least Squares $I_{ww} = 4 \times 10^6$ $A_{shw} = 152$ $J = 50.6$

$$(2)$$

Method 2: Separation $I_{ww} = 10^4$ $A_{shw} = 2000$ $J = 103.6$

In practice, engineers frequently assume that the stiffness of a hybrid section is controlled by St. Venant torsion in the closed region. This assumption can lead to gross errors in calculating the section behaviour. To demonstrate the differences between the behaviour of open, closed and hybrid sections. a parameterisation of the computer results has been constructed in Fig. 8. Since the cross section responds differently in shear and in bending, the sensitivity of the model can be examined by plotting the variation of the tip rotation against the two material constants, E and G.

Starting with $\alpha E = E$ and $\beta G = G$, and progressing to $\alpha E = \beta G = $ infinity, a surface representing the tip rotation can be obtained. Fig. 8 shows the surfaces produced for an open section, for a closed section, and for the above hybrid section. Note that the hybrid section does not respond in a similar manner to either the open or closed sections, and the hybrid cannot be formed by the addition of open and closed segments.

3 CONCLUSIONS

The representation of open and closed cross sections can be modelled by either a simple beam finite element or by a completely compatible cubic beam finite element. The former model is shown to be adequate for the analysis of open sections only. The latter is shown to give good results for both closed sections and open sections. In the cases of open and hybrid sections, the computer results differ from classical torsion theory for open sections. This difference is highlighted by the need to include a third section property, A_{sh}, and by the need to recognise that the centre of twist is not constant along the length of the member.

4 REFERENCES

1. Gere, J.M. & Timoshenko, S.P. 1961. *Theory of Elastic Stability*, McGraw-Hill.
2. Vlasov, V.Z. 1961. *Thin Walled Elastic Beams*, National Science Foundation.
3. Murray, N.W. 1984. *Introduction to the Theory of Thin Walled Structures*, Oxford Engineering Science Series.
4   Galambus, T.V. 1968. *Structural Members und Flumes*, Prentice Hall.
5. Megson, T.H.G. 1974. *Linear Analysis of Thin Walled Elastic Structures*, John Wiley.
6. Kavanagh, K.T. 1977. Equivalent Cross-Sectional Properties of Core Walls and Beams via the Finite Element Method, *15th Australasian Conference on the Mechanics of Structures and Materials*, Melbourne.

The Mechanics of Structures and Materials, Grzebieta, Al-Mahaidi & Wilson (eds)
© 1997 Balkema, Rotterdam, ISBN 90 5410 900 9

Equivalent properties of core walls and beams via the finite element method

K.T. Kavanagh

Department of Civil Engineering, University of Western Australia, Nedlands, W.A., Australia

ABSTRACT: Core walls and thin-walled beams generally comprise a series of flat segments which are joined to form a load resisting cross-sectional profile. The cross-sections can be open or closed, can be hybrids combining open and closed sections, or can be perforated along the length. A majority of cross sections are unsymmetric, and in general, the shear centre of the section does not coincide with the centroid. The determination of cross-sectional properties of any cross section necessarily includes: a) the moments of inertia (I_{xx}, I_{yy}, I_{xy}), b) the shear areas (A_{shx}, A_{shy}), c) the warping and torsion constants (I_{ww}, J), and d) the shear centre location (e_x, e_y). A simple finite element approach is presented for the determination of these properties from a series of unit load analyses.

1 INTRODUCTION

Fig. 1 illustrates a typical core wall built up from interconnecting flat segments with regular openings and a combination of open sections and closed box sections. Isolated segments of the core are constrained in their movements by a rigid floor diaphragm. The assumption of a rigid diaphragm enforces a condition of no cross-section distortion on both the individual elements of the core and in the total resisting system. Analysis of the core wall for wind or earthquake loading requires the computation of several properties of the assembled system: 1) the principal moments of inertia, 2) the effective shear areas 3) the center of twist, and 4) the torsional rigidity. All properties can be determined from a stiffness matrix formulation, by back-fitting the displacements obtained from a known loading condition.

2 EQUATIONS OF A SIMPLE CANTILEVER

Since the core wall or segment cross section is generally unsymmetric; an arbitrary origin is selected with a global X and Y axis which is different from the principal axes of the assembly. Equivalent moments of inertia, I_{mx} and I_{my} are defined by:

$$I_{mx} = (I_{xx}I_{yy} - I_{xy}^2)/I_{yy} \qquad (a)$$

$$\qquad\qquad\qquad\qquad\qquad\qquad\qquad (1)$$

$$I_{my} = (I_{xx}I_{yy} - I_{xy}^2)/I_{xx} \qquad (b)$$

and the deflections in direction X due to loads in direction Y, and direction Y due to loads in direction X, are produced by fictitious forces Q_x and Q_y:

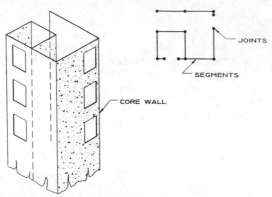

Figure 1. Typical Core Wall Cross Section.

$$Q_x = I_{xy}/I_{xx} P_y \qquad \text{(a)}$$

$$Q_y = I_{xy}/I_{yy} P_x. \qquad \text{(b)} \qquad\qquad (2)$$

The deflections at the top of a cantilevered core define the equivalent properties:

$$\text{Unit Load X:} \quad \delta_x = P_x L^3 / 3EI_{my} + P_x L / A_{shx}G \qquad \text{(a)}$$

$$\text{Unit Load Y:} \quad \delta_y = P_y L^3 / 3EI_{mx} + P_y L / A_{shy}G \qquad \text{(b)} \qquad (3)$$

Deflection ratios:

$$R_1 = \delta_x / \delta_y \,(\text{unit load Y}) = I_{xy} / I_{yy} \qquad \text{(a)}$$

$$R_2 = \delta_x / \delta_y \,(\text{unit load X}) = I_{xy} / I_{xx} \qquad \text{(b)} \qquad (4)$$

can be used to derive a matrix solution for the principal moments of inertia:

$$\begin{Bmatrix} I_{xx} \\ I_{yy} \end{Bmatrix} \frac{1}{(1 - R_1^2 R_2^2)} \begin{bmatrix} 1 & R_2^2 \\ R_1^2 & 1 \end{bmatrix} \begin{Bmatrix} I_{mx} \\ I_{my} \end{Bmatrix} \qquad (5)$$

with

$$I_{xy} = R_1 \, I_{yy}. \qquad (6)$$

The computation of bending properties, therefore, begins with the determination of I_{mx}, A_{shx}, I_{my}, and A_{shy}, and leads to values for I_{xx}, I_{yy}, and I_{xy}. For torsion, a similar sequence yields:

$\phi = T L^3 / 3EI_{ww}$ (No St. Venant Torsion) (a)

(7)

$\phi = T L / GJ$ (No Warping Torsion) (b)

3 MODELLING OF THE CORE WALL SYSTEM

The modelling of the core wall uses standard linear beam elements to represent either complete flat segments, or sections of flat segments. The rotational and longitudinal degrees of freedom are replaced by two longitudinal freedoms at the extremities of the element. Fig. 2 shows the beam, and its replacement, which defines the transformations:

$$u = (u_1 + u_2) / 2 \text{ and } \theta = (u_1 - u_2) / h \qquad (8)$$

The transverse degree of freedom is constrained to move with the X, Y, ϕ freedoms of the rigid diaphragm. In the diaphragm coordinates, the beam contains 5 degrees of freedom at each end. The transformation results from Fig. 2:

$$\begin{Bmatrix} \theta \\ u \\ v \end{Bmatrix} = \begin{bmatrix} 0 & 0 & 0 & 1/h & -1/h \\ 0 & 0 & 0 & 1/2 & 1/2 \\ c & s & R & 0 & 0 \end{bmatrix} \begin{Bmatrix} U \\ V \\ \phi \\ u_1 \\ u_2 \end{Bmatrix} \qquad \begin{array}{l} c = \cos(\alpha) \\ s = \sin(\alpha) \end{array} \qquad (9)$$

segment   beam element   transformed element
(a)   (b)   (c)

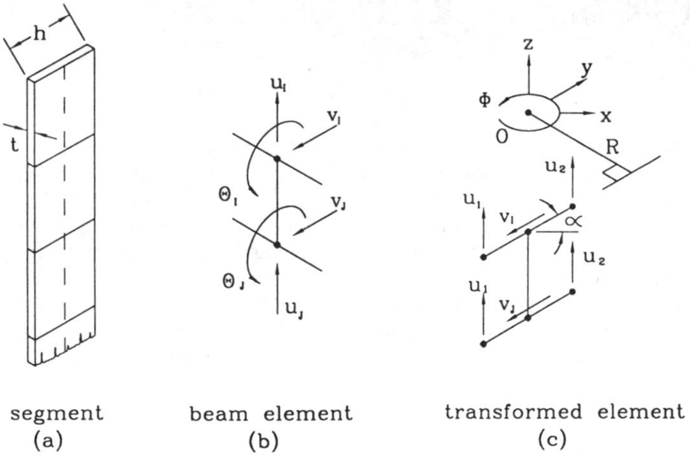

Figure 2. Finite element definition.

The resultant core wall model contains N + 3 degrees of freedom per level, where N is the number of joints defined over the wall system. A complete cantilever can be assembled from any number of vertical subdivisions of the model. Perforations can be modelled by omitting beam segments at regular intervals along the length.

St Venant torsion can be added to each beam element by including a linear rotational stiffness about the longitudinal axis. Each beam segment in the core wall, therefore, has the relevant properties: $I = h^3t/12$, $A = ht$, $A_{sh} = A$, $J = ht^3/3$.

Since the shear area, A_{sh} affects deflection in bending, the coupling between the beam deflection and the global rotation, ϕ, requires that the shear area also affects the torsion. It will be shown that there is an ambiguity in defining the role of J in traditional torsion theory.

4 SEQUENCING CALCULATIONS FOR THE DETERMINATION OF SECTION PROPERTIES

In Section 2, equations were presented for calculating section properties from the output of a computer program with unit loads placed at the end of a cantilever. The sequence of computer calculations requires:

(1) An initial computer run to determine the 3 x 3 stiffness matrix at the top of the cantilever using the arbitrary origin (O).

$$[F] = \begin{bmatrix} \delta 11 & \delta 12 & \delta 13 \\ \delta 21 & \delta 22 & \delta 23 \\ \delta 31 & \delta 32 & \delta 33 \end{bmatrix} \qquad [K] = [F]^{-1} \tag{10}$$

(2) The computation of shear centre offsets is then followed by a re-adjustment of the coordinates to locate the origin at the shear centre:

$$e_x = -K_{23}/K_{22} \qquad \text{(a)}$$

$$e_y = K_{13}/K_{11} \qquad \text{(b)} \tag{11}$$

(3) The recalculation of tip displacements with the new origin of coordinates:
a) having the element shear areas set to infinity (a Bernoulli-Euler beam model to determine I_{mx}, I_{my}, and I_{ww} from Eq. 1 and Eq. 7a), b) having the element moments of inertia set to infinity (a pure shear element to determine A_{shx} and A_{shy} from Eq. 3, or c) having the actual element values of I and A_{sh} (to determine A_{shx} and A_{shy} from Eqs. 1 and 3).

Lateral deflections are compared in the strong axis of the channel shown in Fig. 3 (the x-direction). At short lengths, (H/D =1), neither method (b) nor method (c) produces shear areas which agree with the computed results from a standard finite element beam analysis.

Table 1.

Height	2-D Model Finite Element Model	Equivalent Beam with Shear-Area by Method b	Equivalent Beam with Shear-Area by Method c
H	3.00	2.58	3.25
4H/5	2.33	2.01	2.54
3H/5	1.68	1.45	1.85
2H/5	1.06	0.92	1.19
H/5	0.49	0.43	0.57

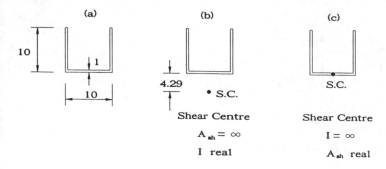

Figure 3. Centres of twist on an open channel.

At longer lengths (H/D=3 and longer), method (c) provides the best approximation. The difference between results from calculation (b) and calculation (c) is not computational, but rather is a function of the shear centre location. The determination of I_{XX} assumes no shear distortion, causing the centre of twist, Fig. 3b, to lie outside of the channel. The determination of A_{sh} assumes no bending, and the centre of twist, Fig. 3c, lies on the web. Thus, both the center of twist and the value of A_{sh} vary along the length of the cantilever whenever the shear centre and centroid do not coincide.

The above comparison for torsion illustrates another problem. The theoretical value of I_{ww} is obtained from the assumption of infinite shear area. Thus, it represents the stiffest possible contribution of bending of the flat plate elements. If J is assumed to be zero, the addition of shear distortion in the individual segments leads to a softening of the total system. Therefore, a stiffness form of

$$K_t = 1 / (L^3 / 3EI_{ww} + L / GA_{shw}) \qquad (12)$$

becomes the torsional equivalent of Eq. 3. The general form of torsion then becomes:

$$K_t = 1 / (L^3 / 3EI_{ww} + L / GA_{shw}) + GJ. \qquad (13)$$

Using the first form, the theoretical value of I_{ww} is calculated with the assumption of infinite shear area, and A_{shw} is computed from the assumption of infinite warping stiffness. These two assumptions produce different shear centres (or centres of twist). Since pure torsion does not associate with position, the comparison between the beam and computer values is excellent for all lengths.

The analysis of closed, or hybrid open-closed, sections cannot be modelled in torsion using the current finite element model (see Kavanagh 1997). The beam element in Fig. 2 is not completely compatible in longitudinal displacement along the element edges between levels, so that any attempt to calculate I_{ww} from Eq. 7a results in I_{ww} becoming a function of beam length. A fully compatible cubic element is required for any closed section (see also Kavanagh 1997).

Using a fully compatible cubic beam finite element, the determination of section properties becomes feasible, but not via Eqs. 3 or 7. In this instance, the differential equation solution is given by hyperbolic sines and cosines. An alternative to Eq. 3 and 7 is to fit the properties, I_{ww}, A_{shw}, and J, by least-squares to the computer solution at various heights. For the cross section shown in Fig 4, the computer calculates $I_{ww} = 4 \times 10^6$, $A_{shw} = 152$, and J = 50.6 by comparing rotations at x = H and x = 0.6H. The 'best fit' properties produce the comparitive rotations shown below:

395

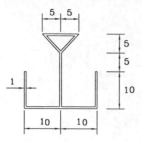

Figure 4. Hybrid open/closed cross section.

Table 2.

Height	Computer Model	Equivalent Beam Model
H	1.05	0.82
4H/5	.76	.65
3H/5	.49	.49
2H/5	.25	.32
H/5	.07	.09

From the comparison, it is apparent that the agreement between the two models is poor. The lack of agreement implies that the differential equation:

$$EI_{ww} \, \phi''' - GJ \, \phi' = T \tag{14}$$

is incorrect for hybrid sections. The previous comparison of deflections for the open cross section, implies that the theory is also incorrect for open sections, although the consequences are not so dramatic.

5 CONCLUSIONS

A core wall or beam cross-section has been modelled using Timoshenko beam elements, with shear distorion included. The model has been shown to produce deflections which can subsequently be used to back-calculate the theoretical section properties. By including shear distortion in the element, the model couples both shear and bending and shear and torsion. There becomes a need to define a new section property, the warping shear area, A_{shw}, in order to model the section behaviour. The addition of shear distortion is shown to influence the centre of twist, so that the shear centre location becomes a variable at short length-to-depth ratios.

   Hybrid sections require the development of a new element which is completely compatible in the longitudinal direction at the element faces. This element can correctly model both open and closed sections. Determination of the section properties is approached from a 'best-fit' solution for rotations along the length of the section. Comparisons between the 'best-fit' solution and traditional warping torsion theory are shown to be poor.

REFERENCES

Kavanagh, K.T. 1997. A Comparison of Finite Element Formulations for the Analysis of Open and Closed Cross Sections, *15th Australasian Conference on the Mechanics of Structures and Materials, Melbourne.*

Murray, N.W. 1984. *Introduction to the Theory of Thin Walled Structures,* Oxford Engineering Series.

Vlasov, V.Z. 1961. *Thin Walled Elastic Beams,* National Science Foundation.

Zetlin, L. & Winter, G. 1955. Unsymmetrical Bending of Beams With and Without Lateral Bracing, *Proceedings, ASCE Journal of the Structural Division, Vol* 81.

The Mechanics of Structures and Materials, Grzebieta, Al-Mahaidi & Wilson (eds)
© 1997 Balkema, Rotterdam, ISBN 90 5410 900 9

Identification of yield limits as a mathematical program with equilibrium constraints

H.Jiang & D.Ralph
Department of Mathematics, University of Melbourne, Parkville, Vic., Australia
F.Tin-Loi
School of Civil Engineering, University of New South Wales, Sydney, N.S.W., Australia

ABSTRACT: This paper deals with a special class of identification problems in structural engineering formulated as a Mathematical Program with Equilibrium Constraints or MPEC. More precisely, the equilibrium system is characterized, in our particular case, by linear complementarity conditions. We first consider the analysis problem, cast as a holonomic (path–independent) elastoplastic problem, before dealing with the identification or inverse problem. We then briefly introduce the topical and fertile area of MPECs before concentrating on a recent and promising algorithm for the solution of our identification problem. Its application is then illustrated by means of a numerical example.

1 INTRODUCTION

In this paper, we re-visit a problem considered briefly in the early 80s by Giulio Maier and his co-workers (Maier 1981; Maier *et al.* 1982). It concerns a particular class of structural identification problems in which, in its simplest form, it is required to identify some or all material yield limits presuming the availability of some measured information on the displacement response to a known loading condition. Our primary motivation for studying such problems is that the methodology appears to be a promising approach for identifying key fracture parameters for quasibrittle materials — a subject of current, intense research interest, as witnessed by the number of papers dedicated to this problem at the last FRAMCOS conference (Wittmann 1995). However, for the sake of simplicity, we consider only the formally simpler, albeit conceptually almost identical, problem of yield limit identification of discretized bar-like structures (e.g. trusses and frames).

In the next section, we review the forward or analysis problem for a class of discretized models exhibiting piecewise linearized holonomic plasticity. In essence, the displacement response is sought for a structure of known material properties. As is well-known (see references in Maier and Munro 1982; Maier and Lloyd Smith 1986), the governing relations lead to a particular mathematical programming problem known as a Linear Complementarity Problem (LCP) which involves, as a key mathematical structure, the orthogonality of two nonnegative vectors. The inverse problem is considered next. Briefly, we are now given the displacement response and we wish to obtain some important material properties, in particular yield limits in the present case. It is shown that this identification problem can be formulated as a special optimization problem involving the minimization of some quadratic error function subject to complementarity and, in some cases, additional linear constraints. The general class of problems under which this problem falls is known as Mathematical Programs with Equilibrium Constraints, MPECs for short (Luo *et al.*

1996). The study of MPECs is an important and new field in mathematical programming and is particularly challenging as far as the development of efficient numerical algorithms is concerned since MPECs cannot simply be treated and solved as nonlinear mathematical programs. We outline a recently developed algorithm (Ralph 1996, Luo *et al.* 1996), based on a straightforward extension of the classical sequential quadratic programming method, which appears to be a promising approach in view of its simplicity and desirable convergence properties. Application of this algorithm is illustrated using a hypothetical structural identification problem concerning an elastoplastic beam on elastoplastic foundation (Maier *et al.* 1982).

2 THE ANALYSIS PROBLEM

Consider a suitably space-discretized structural system. Under a holonomy assumption, we formulate the single step analysis problem simply by collecting and manipulating the relations describing the three key ingredients of the structural behaviour: statics, kinematics and constitutive laws. We further assume a small deformation theory and adopt suitably piecewise linearized yield surfaces. The governing relations (Maier 1970) for the whole structure can be written as

$$F = C^T Q, \tag{1}$$
$$q = Cu, \tag{2}$$
$$q = e + p, \tag{3}$$
$$Q = Se, \tag{4}$$
$$p = N\lambda, \tag{5}$$
$$\phi = N^T Q - H\lambda - r \le 0, \quad \lambda \ge 0, \quad \phi^T \lambda = 0. \tag{6}$$

As is typical, vector and matrix quantities represent the unassembled contributions of corresponding elemental entities, as concatenated vectors and block diagonal matrices, respectively. For a structure with d degrees of freedom, l member generalized quantities and y yield functions, equilibrium between the nodal loads $F \in \Re^{d \times 1}$ and the natural generalized stresses $Q \in \Re^{l \times 1}$ is expressed by (1) through the compatibility matrix $C \in \Re^{l \times d}$. Equation (2) represents linear compatibility of strains $q \in \Re^{l \times 1}$ with the nodal displacements $u \in \Re^{d \times 1}$. Relations (3)–(6) embody the holonomic constitutive laws: additivity of elastic $e \in \Re^{l \times 1}$ and plastic $p \in \Re^{l \times 1}$ strains in (3); linear elasticity in (4), where $S \in \Re^{l \times l}$ is an elastic matrix of unassembled element stiffnesses; plastic strains p in (5) defined by an associated flow rule and expressed as functions of the plastic multipliers $\lambda \in \Re^{y \times 1}$ through the constant matrix of outward normals $N \in \Re^{l \times y}$ to the yield surface; a piecewise linear yield function $\phi \in \Re^{y \times 1}$ in (6) which accommodates, through $H \in \Re^{y \times y}$, a class of hardening models with known yield limits $r \in \Re^{y \times 1}$; and finally, a complementarity relationship in (6) between the sign-constrained total quantities ϕ and λ.

Simple manipulations of relation set (1)–(6) lead to the following LCP:

$$\phi = N^T Q^e - A\lambda - r \le 0, \quad \lambda \ge 0, \quad \phi^T \lambda = 0 \tag{7}$$

where both A and Q^e (the elastic response to F) are known from the given data. After solution of the LCP for λ, the displacements u can be easily calculated.

3 THE IDENTIFICATION PROBLEM

The inverse problem, originally investigated by Maier (1981), can be briefly described as follows. It is assumed that some displacements $u^m \in \Re^{k \times 1}$ are known (measured) deterministic quantities, whereas the vector of yield limits r is unknown, except possibly for an *a priori* grouping due to knowledge that certain structural members are identical. For simplicity, assume that only one test to measure displacements is carried out and only one load level is applied; extension to the case of multiple tests and several load levels is straightforward. Further, if we denote by $u^c \in \Re^{k \times 1}$ the displacement values corresponding to u^m that would be obtained from the structural model for the same loading, then a natural measure of the discrepancy (or error ω) between the measured and the theoretical displacements is provided by the square of the Euclidean norm of the difference between u^m and u^c, or

$$\omega = (u^m - u^c)^T (u^m - u^c). \tag{8}$$

The identification problem obviously requires the *global* minimum of ω subject to (7), any prior knowledge that certain members are identical, and any known bounds on the yield limits to be identified. This can be formally stated as the optimization problem

$$
\begin{aligned}
&\min && \omega = \lambda^T M \lambda + b^T \lambda + c \\
&\text{subject to} && \phi = N^T Q^e - A\lambda - BR \leq 0, \quad \lambda \geq 0, \quad \phi^T \lambda = 0 \\
& && R^L \leq R \leq R^U
\end{aligned}
\tag{9}
$$

where ω, a convex function, has been expressed, through some straightforward manipulations, as a function of λ (M, b and c are known quantities); the yield limits r have been grouped through a Boolean matrix B so that R represents the unknown and required parameters (i.e. $r = BR$); and any lower and upper bounds on R are specified as in (9).

Optimization problem (9) is a special case of an MPEC (Luo *et al.* 1996) for which the so-called "leader" variables R do not appear in the objective function, and the equilibrium system takes the form of a complementarity condition. We introduce such problems in the next section and describe a promising algorithm for solving (9), or at least for finding a local optimal point for it.

4 MPEC: INTRODUCTION AND PROPOSED SOLUTION ALGORITHM

Given any $f : \Re^{n_1 + n_2} \to \Re$, $G_1 \in \Re^{n_3 \times n_1}$, $G_2 \in \Re^{n_3 \times n_2}$, $a_1 \in \Re^{n_3}$, $N_1 \in \Re^{n_2 \times n_1}$, $N_2 \in \Re^{n_2 \times n_2}$, $a_2 \in \Re^{n_2}$, a relatively simple form of the MPEC is the mathematical program with linear complementarity constraints:

$$
\begin{aligned}
&\min && f(x, z) \\
&\text{subject to} && G_1 x + G_2 z + a_1 \leq 0 \\
& && N_1 x + N_2 z + a_2 \geq 0, z \geq 0, z^T (N_1 x + N_2 z + a_2) = 0.
\end{aligned}
\tag{10}
$$

It is easy to see that (9) is of form (10), and that both problems have been written as nonlinear programs, NLPs, that is optimization problems involving nonlinear objective functions and nonlinear constraints.

On one hand, due to the presence of the complementarity constraints — i.e. $\phi^T \lambda = 0$ in (9) and $z^T (N_1 x + N_2 z + a_2) = 0$ in (10) — traditional constraint qualifications such as the Mangasarian-Fromovitz constraint qualification are never satisfied, and hence traditional numerical methods for solving nonlinear programming may be expected to have some difficulties when applied to MPECs. This is also confirmed by numerical experiments.

On the other hand, we wish to take advantage of the vast body of experience already obtained in solving nonlinear programs. In particular, we will adapt the method of sequential quadratic programming (SQP), which is amongst the best and most popular of numerical methods for NLP, to MPEC. The method of piecewise sequential quadratic programming or PSQP (Ralph 1996, Luo *et al.* 1996) for (10) uses the SQP approach.

PSQP is based on a *decomposition* technique. Simply speaking, given a feasible point (x, z) of the MPEC (10), the feasible region of (10) is decomposed into several linearly constrained regions, or polyhedra, each containing (x, z). A nonlinear program with the original objective function f and constraints corresponding to any one of these linearly constrained regions is called a "branch" of the MPEC at the current point (x, z). We choose an arbitrary branch and then apply one step of SQP to that NLP to obtain a new iterate (x, z). An example will be given later.

PSQP can be summarised as follows.

Step 0 Find a feasible starting point for the MPEC (10).

Step 1 Given the current point (x, z), decompose the MPEC into several NLP problems or branches. Choose any NLP branch that has not been examined in this iteration, and approximate it by a quadratic program: Replace the objective function with a quadratic approximation of f. Find a solution (x', z') of the quadratic program.

If (x, z) does not solve (within some tolerance) the quadratic program, then go to Step 2.

Else if there exists an unexplored branch of the MPEC at (x, z), then repeat this step.

Else PSQP terminates (at a local minimizer of the MPEC).

Step 2 Perform a line search based on the objective function f to obtain a suitable step size $t > 0$ such that $f((1 - t)(x, z) + t(x', z')) < f(x, z)$.

Step 3 Update the iteration point, $(x, z) = (1 - t)(x, z) + t(x', z')$, and go to Step 1.

PSQP has fast local convergence properties under some reasonable conditions (Ralph 1996, Luo *et al.* 1996). If the objective function f of (10) is quadratic, then (10) is called a QPEC, and the algorithm terminates in finitely many iterations.

Let us see how PSQP works for the following simple example which is a QPEC:

$$\begin{aligned} \min \quad & f(x, z) = (x + 1)^2 + (z - 3)^2 \\ \text{subject to} \quad & z - x \geq 0, z \geq 0, z(z - x) = 0. \end{aligned}$$

It is clear that the feasible set of this problem can be "globally" decomposed into 2 polyhedra $\{(x, z) : z - x \geq 0, z = 0\} \cup \{(x, z) : z - x = 0, z \geq 0\}$. Note this decomposition is simply the union of two rays, $\{(x, z) : x \leq 0\} \cup \{(x, x) : x \geq 0\}$.

Each iteration will identify one or more of these polyhedra that contain the current iterate, to form branches. Also note that each branch of a QPEC at a feasible point is a quadratic program (the problem of minimizing a quadratic function subject to linear constraints) so there is no need for approximation of the objective function as in Step 1.

Suppose PSQP starts from an initial point $(x^0, z^0) = (x^0, 0)$ with $x^0 < 0$. The MPEC only has one branch at this point, which is the quadratic program:

$$\begin{aligned} \min \quad & f(x, z) \\ \text{subject to} \quad & z - x \geq 0, z = 0 \end{aligned} \tag{11}$$

402

whose solution is $(x', z') = (-1, 0)$. Thus with a step size of $t = 1$, see Step 2 above, PSQP produces the next point $(-1, 0)$ which is a local solution of the MPEC with objective value $f(-1, 0) = 9$. Similarly, if the starting point of PSQP is (x^0, z^0) with $x^0 = z^0 > 0$, then there is only one branch:

$$
\begin{aligned}
\min \quad & f(x, z) \\
\text{subject to} \quad & z - x = 0, z \geq 0
\end{aligned}
\tag{12}
$$

whose solution is $(1, 1)$, another local solution of the QPEC to which PSQP converges in one iteration. The final objective value is $f(1, 1) = 8$ which is the global optimal value of the QPEC.

However if $(x^0, z^0) = (0, 0)$ is chosen as the starting point, then there are two branches of QP problems because each of the polyhedra mentioned above contains this point. By decomposition, the two quadratic programming branches are exactly (11) and (12). Depending on which branch is chosen, PSQP converges in one iteration to either $(-1, 0)$ or $(1, 1)$, thus to either a local or a global minimizer.

ILLUSTRATIVE EXAMPLE

This example, used by Maier et al. (1982), concerns the elastoplastic beam on tensionless elastoplastic springs loaded as shown in Fig. 1. The springs are spaced at 32 cm and material properties are as indicated. Our aim was to identify $R \in \Re^{4 \times 1}$ assuming that all springs (and beams) are identical. We first analysed the structure, using the R values shown in Fig. 1. and some assumed diagonal H matrix, to obtain its displacement response u. Then, we identified R using u, or a slightly perturbed u to simulate inexact measurements, as an input.

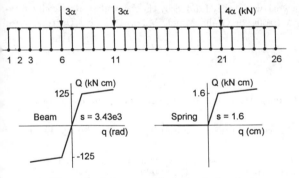

Figure 1. Elastoplastic beam on elastoplastic foundation.

Table 1. Computational Results.

Measured u	Comments	$R(1)$	$R(2)$	$R(3)$	$R(4)$	Iterations
All	$H = 0, E$	124.96	121.44	1.60	0	17
All	$H = 0, I$	124.66	120.30	1.61	0	17
17,19,21,23,25	$H = 0, E$	124.96	121.44	1.60	0	23
All	$H > 0, E$	124.91	125.06	1.60	0	28
All	$H > 0, I$	125.14	122.91	1.60	0	28
17,19,21,23,25	$H > 0, E$	124.91	125.06	1.60	0	29

403

Table 1 displays some of the results obtained; they all match closely expected values. Note that $H = 0$ means perfect plasticity, $H > 0$ indicates a hardening set to $S/20$ except for spring tension-hardening which was assumed to be zero; E indicates exact u; I means inexact u perturbed by multiplying the exact values with a uniform distribution with a range $[0.9\ 1.1]$; for $H = 0$ the load parameter $\alpha = 2.13$ and for $H > 0$ parameter $\alpha = 4.26$. In all cases, we imposed the bounds $0 \le R \le 1000$, and initiated a PSQP MATLAB code with all R components set to 500. Theoretically, it should be possible in all cases to identify R completely, except its second component when $H = 0$.

6 CONCLUDING REMARKS

We briefly consider an inverse problem in elastoplasticity, first studied by Maier (1981), which is believed to be a representative model for the important and challenging task of identifying quasibrittle fracture parameters. This particular parameter identification problem can be cast as a special type of optimization problem belonging to the broad class of so-called Mathematical Programs with Equilibrium Constraints or MPECs. A simple and promising algorithm based on a decomposition principle and on the classical sequential quadratic programming approach is proposed for solving this difficult optimization problem. A numerical example is given to illustrate suitability of the approach. In spite of encouraging results obtained to date, improvements are still needed. Future work will be directed at enhancing the robustness of the method for the identification problems described, at solving actual (formally more complex) fracture identification problems, and at developing reliable schemes for accommodating imperfect measurements. Finding efficiently a global minimum to such MPECS will, of course, remain a challenging research goal.

ACKNOWLEDGMENTS: This work was partly funded by the Australian Research Council. We would also like to thank Professor Giulio Maier, Milan Technical University for motivating this work.

REFERENCES

Luo, Z.Q, Pang, J.S. and Ralph, D. 1996. *Mathematical Programs with Equilibrium Constraints*, Cambridge University Press.

Maier, G. 1970. A matrix theory of piecewise linear elastoplasticity with interacting yield planes, *Meccanica*, 5, 54–66.

Maier, G. 1981. Inverse problem in engineering plasticity: a quadratic programming approach, *Accademia Nazionale dei Lincei*, Serie VIII, vol LXX, 203–209.

Maier, G., Giannessi, F. and Nappi, A. 1982. Indirect identification of yield limits by mathematical programming, *Engineering Structures*, 4, 86–98.

Maier, G. and Lloyd Smith, D. 1986. Update to "Mathematical programming applications to engineering plastic analysis", *Applied Mechanics Update, ASME*, 377–383.

Maier, G. and Munro, J. 1982. Mathematical programming applications to engineering plastic analysis, *Applied Mechanics Reviews*, 35, 1631–1643

Ralph, D. 1996. A piecewise sequential quadratic programming method for mathematical programs with linear complementarity constraints, *Proceedings of the seventh conference on Computational Techniques and Applications (CTAC95)*, R.L. May and A.K. Easton, editors, Scientific Press, Singapore, 663-668.

Wittmann, F.H. (ed) 1995. *Fracture Mechanics of Concrete Structures*, AEDIFICATIO Publishers.

The Mechanics of Structures and Materials, Grzebieta, Al-Mahaidi & Wilson (eds)
© *1997 Balkema, Rotterdam, ISBN 90 5410 900 9*

Finite element modelling of plastics fittings

J.P. Lu & L.S. Burn
CSIRO Building, Construction and Engineering, Melbourne, Vic., Australia

ABSTRACT: A finite element analysis has been performed on plastic tees to minimise and/or distribute more evenly the high stress in the stress concentration regions, by redesigning the tee in the area of high stress. Many geometries (17 cases) have been analysed by adding local reinforcement to the region of high stress, as well as by varying parameters such as external/internal area fillet radii, internal reinforcement thickness, internal reinforcement length, and so on. Each case included two profiles, i.e. tees only, and tees with pipes cemented into all three outlets of the fittings. The effect of each parameter and the local reinforcement is also discussed. It is found that the finite element method can be used as an effective approach to the design of plastic fittings, and the linear analysis is sufficient to effectively determine the stress concentration areas which are likely sites for failure initiation, when tested for fatigue under cyclic pressure.

1 INTRODUCTION

Under aggressive operating conditions such as those that occur in pumping operations, plastic fittings can fail by a process of crack initiation and propagation. These failures initiate in areas where flaws exist, such as in the weld lines, or in areas of high stress due to the design of the fitting. In the case of tees, these fittings usually fail at the internal intersection, due to a change in geometry which gives rise to high stress concentration at the discontinuity. By redesigning the tee in this area of high stress, lower stress levels and more uniform distributions should be achievable and consequently longer lifetimes should occur under cyclic stress operating conditions.

The levels of stress in the wall can be reduced by the development of experimental prototypes. However, experimental prototype testing is expensive and often impractical or impossible for complicated configurations and loading conditions. It is therefore better to use a reliable numerical technique as an alternative to experimental prototyping. Finite element analysis has been found to be an effective approach to reduce design and manufacturing costs, due to its ability to model complicated geometry and boundary conditions.

In this research, the ANSYS (Swanson Analysis System, Inc. 1992) finite element package is employed to minimise stress concentration by adding local reinforcement and by changing parameters such as fillet radii (internal-r or external-R), fitting and pipe wall thickness (t and t_2), internal reinforcement thickness (t_1) and internal reinforcement length (L). For simplicity of presentation, an equivalent thickness t_e (defined as the sum of the pipe wall thickness t_2 and fitting wall thickness t) is introduced for the future discussion. Fittings with a range of parameters, as mentioned above, were assessed with different loading and constraint conditions, to simulate conditions experienced by the fitting under test. Because of the expense of retooling, only easily adjustable parameters such as fillet radii and reinforcement thicknesses were examined, as shown in Figure 1.

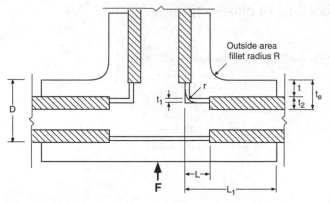

Figure 1. Symbols used in the model.

2   BRIEF INTRODUCTION TO THE ANSYS PACKAGE

ANSYS is a general-purpose, large-scale finite element computer package for engineering analysis. It includes many powerful capabilities such as preprocessing; solid modelling; structural, thermal, magnetic, fluid flow and coupled field analyses; postprocessing; premier graphics capabilities; and design optimisation. Not only can it model both material and geometric non-linearity, but it also provides a technique to estimate the amount of solution error due to mesh discretisation. It has been designed according to functional requirements for the solution of an extremely wide range of large and complex problems in engineering analysis, with high accuracy and computational efficiency, which are met by developing advanced mathematical models of the physical phenomena and incorporating their computational algorithms into the program.

3   DESCRIPTION OF THE FITTINGS AND MATERIAL PROPERTIES

The fittings assessed in these analyses were DN 100 mm tees. The outside diameter ($D = 142$ mm) and the thickness ($t = 14$ mm) of the fitting were kept constant in the analyses. Two profiles, as shown in Figure 2, were assessed: tees only (profile 1); and tees with L_1 long pipes cemented into all three outlets of the fittings (profile 2). L_1 was defined as: $L_1 = 180$ mm, when $L = 0$ or 74 mm (the L_1 chosen was the minimum required to minimise the end effects according to Saint–Venant's Principle); $L_1 = 189-L$ when L is any other values (to meet the end of the reinforcement L, which also served as a locating lug). The linear material properties used in the analyses were: Young's modulus $E = 1850$ MPa, Poisson's ratio $v = 0.35$ and density $\rho = 1050$ kg/m^3.

4   FINITE ELEMENT MODELLING

Because of symmetry, only one-quarter of the tee needs to be modelled. To ensure that accurate stresses in the tee junction can be calculated and the fillet area can be modelled in sufficient detail, 3-D 4-node tetrahedron and 8-node solid model brick elements were used in the mesh generation to ensure efficient model definition. To allow accurate stress definition, greater mesh refinement is needed near the intersection of the tee, while coarser meshes are used away from the intersection. In order to achieve this, the fitting has been made from many individually meshed blocks. Nodes and elements are automatically generated within the block. Two or three layers of elements through the wall thickness were used near the tee junction to increase accuracy. ANSYS's error-estimation technique was employed to ensure that in the region of high stress the amount of solution error due to mesh discretisation was small.

The symmetric boundary conditions applied were zero out-of-plane translations and in-plane rotations on all nodes in the symmetry planes. The boundary conditions at the ends of the

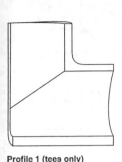
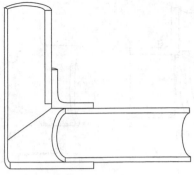

Profile 1 (tees only)   **Profile 2 (tees with pipes solvent-cemented into the tee)**

Figure 2. Two profiles considered in each case.

longitudinal and vertical pipes in the model were fully built-in (zero translations and rotations). A uniform pressure of 3.9 MPa was applied normal to the internal surface of the model. Analyses were carried out with and without node forces applied at the bottom of the tee, to simulate the thrust block force.

5   RESULTS AND DISCUSSION

5.1 *General*

A total of 17 cases were analysed by varying the parameters shown in Figure 1 and also by adding local reinforcement to the region of high stress (Figure 3). Each case includes the two profiles illustrated in Figure 2. As the fitting is under combined triaxial stress, Von Mises maximum distortion energy criterion is used to express the equivalent maximum stress. The details of the cases studied and the associated maximum stresses are given in Table 1. Also included in this table is a stress ratio defined as $SR = \sigma_m/\sigma_n$, where σ_m denotes the highest Von Mises stress in the fitting and σ_n is the nominal stress in the pipe, calculated by $\sigma_n = P(D - t)/2t$, where P is the internal pressure, D is the outside diameter and t is the thickness of the pipe.

A typical Von Mises contour display showing how the equivalent stresses vary over the model is given in Figure 4.

The initial results on unmodified tees (case 1 as detailed in Table 1) show that a very high stress concentration occurs in the internal intersection of the tee for both profile 1 and profile 2. This is the same location where the failure initiation point occurred when tested for fatigue under cyclic pressure. The maximum stress levels reduced significantly when pipes were cemented into the outlets of the tees (profile 2), compared with profile 1 (tees only).

The analysis results of all the other cases also show the same concentration location (except for cases with local reinforcement added). In addition, relatively high stress levels also exist at the following areas for profile 2:

• centre of external surface;
• bottom of the internal surface, particularly for tees under both internal pressure and node forces applied at the bottom of the tee simulating the thrust block force; and
• internal corners of pipes where long pipes meet the locating lug (end of the internal rein-forcement) close to the junction and bottom of tees.

5.2 *Effect of area fillet radius (external-R and internal-r)*

For the internal area fillet, introducing an internal area fillet radius $r = 5$ mm does not reduce the maximum equivalent stress for profile 1, although it does lead to more even stress distribution (cases 9 and 11). However, this internal fillet radius does reduce the maximum stress for profile 2. No apparent reason has been found as to why it affects the two profiles differently.

407

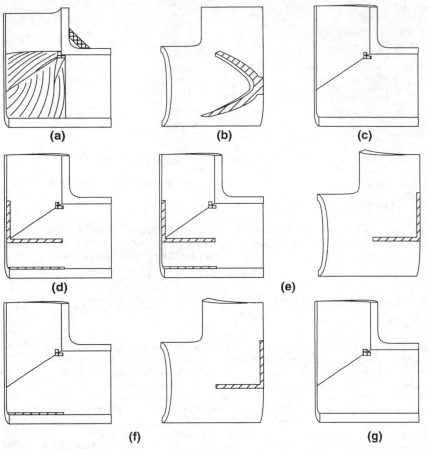

Figure 3. Location and geometry of local reinforcement (category g is applicable to profile 1 only; all the other categories are applicable to both profiles).

The increase of external fillet radius from $R = 5$ to 10 mm doesn't significantly affect the level of stress concentration experienced in the fittings, as detailed in Table 1 (cases 2 and 3). This is also the case for $L = 21$ mm and $t_1 = 2$ (cases 8 and 9) when the external area fillet radius has been increased from 0 to 5 mm. It is seen that increasing the external fillet radius for this fitting design does not minimise the stress levels. This is because, with larger fillet radius, the geometric discontinuity becomes more severe, even though the local stiffness is increased. The introduction of both internal and external fillet radii simultaneously (cases 8 and 9) gives similar results.

However, with a uniform internal reinforcement ($L = 9$ mm and $t_1 = 2$ mm) (cases 1 and 4), increasing an external fillet radius from 5 to 10 mm leads to an unexpected increase of maximum stresses, yet with much the same pattern of distribution.

Because these two parameters (external and internal fillet radii R and r) give little benefit, their influence on the design process will not be further considered.

5.3 Effect of the internal reinforcement thickness (t_1)

For a uniform internal reinforcement, only three thicknesses ($t_1 = 0$, 2 and 5 mm) are considered in the analyses. If all the other parameters are kept constant, increasing the internal reinforcement thickness from 0 to 2 mm results in a significant increase of the stress levels at the internal wall

Table 1. Analysis results of maximum stress at fitting internal wall for all cases studied.

Case no.	Parameter				Reinf. loc.	Profile 1 (fitting)				Profile 2 (fitting + pipe)			
						With F		Without F		With F		Without F	
	L	t_1	R	r		σ_m	SR	σ_m	SR	σ_m	SR	σ_m	SR
1*	9	2	5	0	–	79.5	4.5	72.8	4.08	40.7	2.28	36.4	2.04
2†	0	0	5	0	–	69.2	3.88	64.4	3.61	40.7	2.28	36.0	2.02
3	0	0	10	0	–	67.7	3.80	62.7	3.52	41.4	2.32	36.8	2.06
4	9	2	10	0	–	82.2	4.61	74.9	4.2	43.1	2.42	38.1	2.14
5	9	5	5	0	–	104.0	5.83	96.6	5.42	66.3	3.72	56.6	3.17
6	9	2	5	0	a	80.5	4.52	72.2	4.05	39.8	2.23	35.5	1.99
7	9	2	5	0	b	78.8	4.42	69.0	3.87	39.9	2.24	35.7	2.00
8	21	2	5	5	–	–	–	59.5	3.34	–	–	49.5	2.78
9	21	2	0	0	–	66.7	3.74	60.7	3.4	58.5	3.28	52.4	2.94
10	21	2	5	0	–	67.6	3.79	62.6	3.51	59.1	3.31	53.5	3.00
11	21	2	0	5	–	–	–	62.1	3.48	–	–	47.0	2.64
12	26	2	5	0	–	66.7	3.74	59.8	3.35	69.7	3.91	62.5	3.51
13	0	0	5	0	c	72.0	4.04	67.0	3.76	42.3	2.37	37.7	2.11
14	0	0	5	0	d	71.8	4.03	66.8	3.75	41.9	2.35	37.3	2.09
15	0	0	5	0	e	71.3	4.00	66.2	3.71	40.4	2.27	35.9	2.01
16	0	0	5	0	f	71.6	4.02	66.5	3.73	40.8	2.29	36.3	2.04
17	0	0	5	0	g	73.0	4.09	66.0	3.70	–	–	–	–

* Existing fitting design
† Without any reinforcement
$SR = \sigma_m/\sigma_n$
F = thrust block force
a, b, c, d, e, f and g denote seven categories of location and geometry of local reinforcement defined as (see Figure 3):
    category a – crossed ribs at centre of external surface (Figure 3a);
    category b – external rib at elbow region (Figure 3b);
    category c – internal ribs close to internal junction corner (Figure 3c);
    category d – internal ribs close to internal junction corner (same as c), at centre and bottom of internal surface (Figure 3d);
    category e – both internal ribs at the same locations as indicated in category d and external ribs at centre of external surface (Figure 3e);
    category f – internal ribs close to internal junction corner and at bottom of internal surface, and external ribs at centre of external surface (Figure 3f); and
    category g – internal ribs at internal junction corner (Figure 3g).

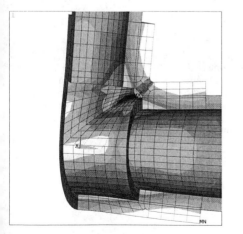

```
ANSYS 5.2
JAN 21 1997
14:46:56
     .825672
    5.255
    9.684
   14.112
   18.541
   22.97
   27.399
   31.828
   36.257
   40.686
```

Figure 4. Von Mises contour displays for case 1 (with F).

(Table 1) for profile 1 but not for profile 2 (cases 1 and 2). Increasing the thickness from 2 to 5 mm (cases 1 and 5) results in extremely high stress concentrations at the internal intersection of the tee, though with less concentration locations. With a larger reinforcement thickness, the geometric discontinuity becomes more severe, even though the fitting stiffness is increased. This indicates that the general rule of larger thicknesses furnishing lower stresses for straight pipes is not applicable to fittings of complex geometry. Although the sensitivity of maximum stresses to various internal reinforcement thicknesses is not regular, it is seen that, among the three thicknesses considered here, the optimum thickness with which to achieve the minimum stress levels is 0.

5.4 Effect of the internal reinforcement length (L)

For profile 1, generally, the sensitivity of maximum stresses to various internal reinforcement length is not regular. For profile 2, if the pipe length within the socket is kept constant, the then longer the reinforcement, the lower the maximum stress. However, the opposite is true if the pipe length situated within the socket is decreased to meet the end of the reinforcement (which also served as a locating lug that is the seat for the bonding of pipes) due to the reduction of total stiffness of the fitting.

5.5 Effect of the local reinforcement

Seven categories of the location and geometry of local reinforcement are defined in Figure 3 (also see note in Table 1 for a detailed description).

For fittings without internal reinforcement, a very high stress concentration occurs at the internal intersection close to the internal junction corner (case 2). To reduce this concentration, internal ribs were added to the inside of the fitting, as shown in Figure 3c. It is found that the high stressed area shifts slightly towards the centre of the fitting, however, the concentration is even higher (case 13). The extra two internal ribs added to the centre and bottom of the internal surface (case 14, Figure 3d) reduce the stress levels at the centre of the external surface and the bottom of the internal surface. Further, the addition of external ribs to the centre of the external surface (case 15, Figure 3e) not only reduces the concentration in this area, but also shifts another high concentration away from the internal junction. Very little difference between the results of categories e and f (cases 15 and 16) demonstrates that the effect of internal ribs at the centre of internal surface is insignificant.

Local reinforcement category g (case 15) is added to the tees only (profile 1). It is found that, although the high stressed area shifts slightly towards the centre of the fitting, the concentration is even higher.

6 CONCLUSION

A total of 17 cases have been analysed by adding local reinforcement to the region of high stress as well as by varying parameters such as external/internal area fillet radii, internal reinforcement thickness, internal reinforcement length, and so on. The effect of fillet radii, internal reinforcement, internal reinforcement length and the local reinforcement has also been discussed in great detail.

It is found that the finite element method can be used as an effective approach to the design of plastic fittings, and the linear analysis is sufficient to effectively determine the stress concentration areas which are likely sites for failure initiation, when tested for fatigue under cyclic pressure.

ACKNOWLEDGMENTS

The authors are grateful to the financial support of Euratech Pty Ltd in carrying out this work.

REFERENCES

Swanson Analysis System, Inc. 1992. *ANSYS User's Manual Revision 5.0*, Houston, PA: Swanson Analysis System, Inc.

The Mechanics of Structures and Materials, Grzebieta, Al-Mahaidi & Wilson (eds)
© 1997 Balkema, Rotterdam, ISBN 90 5410 900 9

Optimum design of discrete structures with stress, stiffness and stability constraints

D. Manickarajah & Y. M. Xie
Department of Civil and Building Engineering, Victoria University of Technology, Melbourne, Vic., Australia

G. P. Steven
University of Sydney, N.S.W., Australia

ABSTRACT: This paper presents a simple evolutionary method for the optimum design of structures with stress, stiffness and stability constraints. The evolutionary structural optimization (ESO) method is based on the concept of slowly removing the inefficient material and/or gradually shifting the material from the strongest part of the structure to the weakest part until the structure evolves towards the desired optimum. The iterative method presented here involves two phases. In first step, the design variables are scaled uniformly to satisfy the most critical constraint. In the second step, a sensitivity number is calculated for each element depending on its influence on the strength, stiffness and buckling load of the structure. Based on these element sensitivities, material is shifted from the strongest to the weakest part of the structure. These two steps are repeated in cycles until the desired optimum design is obtained. Illustrative examples are given to show the applicability of the method to the optimum design of frames and trusses with a large number of design variables.

1 INTRODUCTION

Although there has been considerable amount of work carried out on the optimum design of frame structures, most of these studies fail to treat the stability constraint in parallel with other common constraints such as strength and displacement limits. Optimum design of frame structures including stability constraint along with stress and displacement constraints have been reported by Lin and Liu (1988), Barson (1993) and Pezeshk and Hjelmstad (1991) using optimality criteria methods and by Karihaloo and Kanagasundaram (1993) using non-linear mathematical programming method.

Recently, a simple method for shape and layout optimization, called Evolutionary Structural Optimization (ESO), has been proposed by Xie and Steven, which is based on the concept of gradually removing redundant elements to achieve optimal design. Some examples of ESO method for problems with stress or frequency or stiffness constraints can be found in Xie & Steven (1993, 1996) and Chu et al. (1996). Manickarajah et al. (1995) developed an evolutionary method for the optimum design of structures against buckling. This paper presents a new development of ESO method for the cross-sectional optimization of structures considering stress, stiffness and stability constraints simultaneously.

2 OPTIMIZATION PROBLEM

Consider a 2-dimensional linear-elastic structures consisting of truss or beam elements. The design variables are the cross-sectional areas of elements and the objective is to minimize the total structural weight. The layout of the structure is not changed. It is assumed that the cross-sectional area at any section x, $A(x)$ is related to its flexural stiffness, $I(x)$ by $I(x)=cA(x)^{n}$ in which c and n are constants determined by the cross-sectional shape. For example, $n = 1$ represents a rectangular section of constant depth, $n = 2$ represents a geometrically similar cross-section, say circular or square, and $n = 3$ represents a rectangular section of constant width.

2.1 Stiffness constraint

In the finite element method, the static behaviour of a discrete structure is given by

$$[K]\{d\} = \{P\} \tag{1}$$

where $[K]$ is the global stiffness matrix, $\{d\}$ and $\{P\}$ are the global nodal displacement and nodal load vectors respectively. The inverse measure of the overall stiffness of a structure is known as the mean compliance, C, and is defined as

$$C = \tfrac{1}{2}\{P\}^T\{d\} \tag{2}$$

The overall stiffness of a structure is maximized by minimizing its mean compliance. Hence the stiffness constraint is given in the form $C \le C_{all}$, where C_{all} is the prescribed limit for C.

2.2 Stability constraint

The linear buckling behaviour of a structure is governed by the following eigenvalue problem:

$$\left([K] - \lambda_j[K_g]\right)\{u_j\} = \{0\} \tag{3}$$

where $[K_g]$ is the global geometric stiffness matrix, λ_j is the jth eigenvalue and $\{u_j\}$ is the corresponding eigenvector. The eigenvalues from (3) are those which scale the applied loading to give the buckling load. The critical buckling load factor, λ_{cr} is the first eigenvalue λ_1 at which the instability occurs. Thus the stability constraint of a structure is given in the form $\lambda_{cr} \ge FS$, where FS is the factor of safety against buckling.

2.3 Stress constraint

The stress level at a point can be measured by some means of average of the normal and shear components of stress. For this purpose, the von Mises stress, σ_{vm} has been frequently used for isotropic materials. The von Mises stress at any point of the structure should not exceed the allowable yield stress σ_{all}. For plane structures, the von Mises stress is reduced to $\sigma_{vm} = (\sigma^2 + 3\tau^2)^{1/2}$ where σ and τ are normal and shear stresses respectively. The stress levels are examined at certain preselected points per section at both ends of each element and the critical stress of each element is found. For example, von Mises stress at a point on a rectangular cross-section of a beam element of a planar structure as shown in Figure 1 is given by

$$\sigma_{vm} = \sqrt{\left(\frac{12My}{bd^3} + \frac{P}{bd}\right)^2 + \frac{108S^2}{b^2d^6}\left(\frac{d^2}{4} - y^2\right)^2} \tag{4}$$

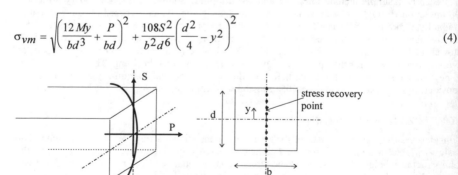

Figure 1. Rectangular cross-section of a beam element

The stress recovery points considered are sufficient to identify the critical stress in the section under combined normal and shear stress conditions. In addition to these constraints, the cross-sectional areas of elements are bound by maximum area, A_{max} and minimum area, A_{min}.

3 UNIFORM SCALING AND CRITICAL SCALING FACTORS

During the optimization process, it is convenient to obtain a feasible design after each iteration by scaling the design uniformly (all the design variables are scaled by a single factor) in order to satisfy the specified constraints. This helps keep track the reduction in the weight of the structure after each iteration and identifies the most active constraints. The uniform scaling factor is calculated from the condition, that the scaled design should be critical to the most active constraint. The structural responses of frames primarily depend on the moment of inertia and for trusses these depend on the cross-sectional area. For the structures with linear size-stiffness relationship (trusses and frames with beams of $n = 1$), uniform scaling factor of constraints are given as follows: stress scaling factor $S_s = \sigma_{vm}^{max}/\sigma_{all}$; stability scaling factor $S_b = FS/\lambda_{cr}$; and stiffness scaling factor $S_c = C/C_{all}$ where σ_{vm}^{max} is the maximum von Mises stress in the whole structure.

For the beams with other values of n, $S_b \approx (FS/\lambda_{cr})^{1/n}$; $S_c \approx (C/C_{all})^{1/n}$ and it may be necessary to make more than one analysis to determine S_b or S_c. In general the structural response is not very sensitive to changes in the ratio of the area to the moment of inertia. Consequently one or two iterations are generally found to be sufficient. Similarly for these sections, von Mises stress may not be linearly related to the design variable. Hence additional iterations are needed to resolve S_s.

4 THE SENSITIVITY NUMBER FOR ELEMENT RESIZING

Sensitivity analysis plays a central role in structural optimization, since virtually all the optimization methods require the computation of the derivatives of structural responses quantities with respect to design variables. In ESO, for shape and layout optimization (which involves element removal), the contribution of each element to the structural responses such as stress, displacement, frequency, buckling load etc. is assessed. For cross-sectional optimization, the effects on these structural responses due to the local modification of each element need to be estimated.

Consider a small change in the cross-sectional area of a particular element, i. For a statically determinate structure, the cross-sectional changes do not affect the element resultant forces. In the following analysis it is assumed that the internal forces acting in the structure be statically determinate. For a statically indeterminate structure, if the cross-sectional modifications at each iteration are small, they do not cause significant changes in the element forces. For frames and trusses this assumption is reasonably accurate.

4.1 Change in mean compliance, ΔC

From (3), the change in mean compliance due to the change in element i, ΔC_i is given by

$$\Delta C_i = -\tfrac{1}{2}\{d\}^T[\Delta K]\{d\} = -\tfrac{1}{2}\{d_i\}^T[\Delta k_i]\{d_i\} \tag{5}$$

where the change in the global stiffness matrix, $[\Delta K]$ is equal to the change in the element stiffness matrix, $[\Delta k_i]$ and $\{d_i\}$ is the displacement vector associated with the element, i. Details of this analysis can be found in Chu et al. (1996).

4.2 Change in buckling load factor, $\Delta\lambda_{cr}$

Multiplying (4) by the transpose of the eigenvector $\{u_j\}$ produces the following Rayleigh quotient for the eigenvalue:

$$\lambda_j = \frac{\{u_j\}^T [K]\{u_j\}}{\{u_j\}^T [K_g]\{u_j\}}$$ (6)

Neglecting the change in stress matrix, $[\Delta K_g]$ and normalising the eigenvector with respect to $[K_g]$, from (6) the change in the first eigenvalue due to the change in the element i, $\Delta\lambda_{1i}$ is derived. For single modal structures,

$$\Delta\lambda_{1i} = \{u_{1i}\}^T [\Delta k_i]\{u_{1i}\}$$ (7)

where $\{u_{1i}\}$ is the fundamental eigenvector associated with the element, i. For multimodal structures, the effect on the first eigenvalue due to all participating eigenvectors need to be considered. Details of multimodal analysis can be found in Manickarajah et al. (1995).

4.3 Change in stress

Since we assume a small change in the cross-sectional area of an element does not cause significant changes in the element forces, new stress in that element can be directly calculated from the element forces and new cross-sectional dimensions using equations such as (4).

4.4 Sensitivity number

Now let us describe how these element sensitivities with respect to each constraint can be used in the optimization algorithm.

Optimization with stiffness constraint: The aim is to minimize the mean compliance C. Defining $\alpha_{ic} = \{d_i\}^T [\Delta k_i]\{d_i\}$, from (5), the cross-sectional area of elements with highest α_{ic} values has to be increased to minimize C. If the structural weight is to be kept constant, the elements with highest α_{ic} values need to be strengthened and the elements with lowest α_{ic} values may be weakened. In the case of an increase in the cross-sectional area A of an element i,

$$[\Delta k_i] = [\Delta k_i]^+ = [k_i(A + \Delta A) - k_i(A)]$$ (9)

and in the case of a reduction in the cross-sectional area

$$\Delta k_i] = [\Delta k_i]^- = [k_i(A - \Delta A) - k_i(A)]$$ (10)

Hence the following two sensitivity numbers need to be calculated for each element.

$$\alpha_{ic}^+ = \{d_i\}^T [\Delta k_i]^+ \{d_i\} \qquad ; \qquad \alpha_{ic}^- = \{d_i\}^T [\Delta k_i]^- \{d_i\}$$ (11a,b)

When elements are of different lengths, the sensitivities of elements depend also on their lengths. When comparing two elements with same α_{ic}^+, increasing the cross-sectional area of shorter element will result in a lighter design. Similarly for elements with same α_{ic}^-, cross-sectional area of longer element need to decreased. Consequently, the element sensitivities for stiffness constraint are redefined below, where l_i is the length of the element i.

$$\alpha_{ic}^+ = \{d_i\}^T [\Delta k_i]^+ \{d_i\} / l_i \qquad ; \qquad \alpha_{ic}^- = \{d_i\}^T [\Delta k_i]^- \{d_i\} / l_i$$ (12a,b)

Optimization with stability constraint: The objective is to raise the fundamental buckling load factor λ_1. Based on the similar argument as above and from (7), the following element sensitivities are defined for single mode structures.

$$\alpha_{ib}^+ = \{u_{i1}\}^T [\Delta k_i]^+ \{u_{i1}\} / l_i \qquad ; \qquad \alpha_{ib}^- = \{u_{i1}\}^T [\Delta k_i]^- \{u_{i1}\} / l_i$$ (13a,b)

To raise the buckling load factor it will be most effective to increase the cross-sectional areas of elements with highest α_{ib}^+ and reduce those with the lowest α_{ib}^- values.

Optimization with stress constraint: In practice, the strength criterion is satisfied by using the fully stressed design (FSD) concept which is one of the early optimality criteria. If the stress distribution of a structure is to be brought to uniform, highly stressed elements need to be strengthened and lowly stressed elements need to be weakened. The following element sensitivity numbers are defined for stress constraint.

$$\alpha_{is}^{+} = \sigma_{ivm}^{+} / l_i = \sigma_{ivm}(A + \Delta A) / l_i \quad ; \quad \alpha_{is}^{-} = \sigma_{ivm}^{-} / l_i = \sigma_{ivm}(A - \Delta A) / l_i \quad (14a,b)$$

where σ_{ivm}^{+} is the maximum stress in the element i, when the area is increased by ΔA. Thus to bring the stress distribution uniform, increase the cross-sectional areas of elements with highest α_{is}^{+} and reduce those with the lowest α_{ic}^{-} values.

Optimization with all the constraints considered simultaneously: Ideally cross-sectional area of the element whose all α_{ic}^{+}, α_{ib}^{+} and α_{is}^{+} are highest should be increased. However this situation generally does not exist. To overcome this difficulty the element sensitivity is evaluated by the sum of its relative sensitiveness with regard to each constraint. Further it is necessary to treat each constraint separately depending on how active it is in the current design. To measure the influence of each constraint, the critical scaling factors defined in section 3 are used as weighting parameters. Taking all these into account, finally for each element the following two new sensitivity numbers are defined.

$$\alpha_i^{+} = S_b \frac{\alpha_{ib}^{+}}{\alpha_{b,av}^{+}} + S_c \frac{\alpha_{ic}^{+}}{\alpha_{c,av}^{+}} + S_s \frac{\alpha_{is}^{+}}{\alpha_{s,av}^{+}} \quad ; \quad \alpha_i^{-} = S_b \frac{\alpha_{ib}^{-}}{\alpha_{b,av}^{-}} + S_c \frac{\alpha_{ic}^{-}}{\alpha_{c,av}^{-}} + S_s \frac{\alpha_{is}^{-}}{\alpha_{s,av}^{-}} \quad (15a,b)$$

where $\alpha_{b,av}^{+}$ is the average of the α_{ib}^{+} values of all elements and other average values are similarly defined. The design variable of the cross-section could be one or several from width or depth of a section, web or flange of an I section, radius of circular section etc. This design variable is allowed to vary in small steps in a prescribed manner. Hence the change in the element stiffness matrix, $[\Delta k_i]$ can be easily calculated. All the other information required for the calculation of sensitivity numbers is readily available from the finite element solution.

5 OPTIMIZATION PROCEDURE

An iterative procedure is set up for uniform scaling and resizing the elements so that the weight of the structure is systematically reduced and the material is gradually shifted from the strongest to the weakest part of the structure. The procedure is given as follows:

Step 1: Select an initial design and discretize the structure using a finite number of elements.
Step 2: Perform FSD two or three times.
Step 3: Solve static and buckling analyses and determine the critical scale factors S_b, S_c and S_s.
Step 4: Scale the design variables uniformly using the most critical scaling factor, i.e. maximum of S_b, S_c and S_s. Impose the sizing constraint, i.e. if the area of an element, $A > A_{max}$, Let $A=A_{max}$ and if $A < A_{min}$, Let $A=A_{min}$.
Step 5: Solve the static and buckling analyses and calculate the sensitivity numbers α_i^{+} and α_i^{-} for each element.
Step 6: Increase the cross-sectional area of elements which have the highest α_i^{+} values and decrease the cross-sectional area of the same number of elements which have the lowest α_i^{-} values.
Step 7: Repeat Steps 3 to 6 until the weight of the structure cannot be reduced any further.

6 EXAMPLES

6.1 50-bar truss tower

A minimum weight design for a 50-bar planar truss tower as shown in Figure 2 is sought using the proposed method. Similar structure has been analysed previously by Lin & Liu (1993) and Khot et al. (1976) with different dimensions and loadings. However in both examples, the final optimum design was governed by only buckling and minimum size constraints. Here the loadings and dimensions of the structure are intentionally chosen so that the structure is critical to all the constraints.

The constraints taken into account are: $\sigma_{vm} \leq \sigma_{all} = 150$ MPa; $C \leq C_{all} = 10$ kNm; and $\lambda_{cr} \geq FS = 2.0$. Young's modulus $E = 200$ GPa. The initial cross-sectional areas of all members are uniform and equal to 15.625 cm^2. This initial area is chosen so that the uniform design just satisfies the most critical constraint. For this example it is the buckling constraint and for the initial design $\lambda_{cr} = 2.0$ ($S_b = 1$). The cross-sectional areas are allowed to vary to a maximum 20 cm^2 and to a minimum 1 cm^2. The area step used for resizing is equal to 0.5 cm^2 and 20% of the elements are subjected to resizing at each iteration. The optimum design is obtained after 11 iterations. The optimization history of scaling factors and the current design to initial design weight ratio, w/w_o is given in Figure 3(a). Initially FSD is carried out to bring the variables closer to the optimum design. The optimum design is governed by all the constraints and at optimum, $S_b = S_S = S_c = 1$. Optimum designs are also obtained considering each constraints separately. Optimization histories for these designs are shown in Figure 3(b,c&d). The weight of the optimum design with all the constraints is reduced to 46.9% of the initial design.

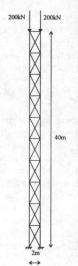

Figure 2. 50-bar truss

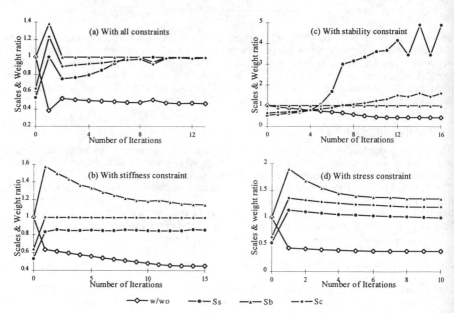

Figure 3. Iteration histories of the 50-bar truss

6.2 5-Storey frame

A 2-bay, 5-storey frame, as shown in Figure 4 is analysed. The following values are taken for the analysis: $\sigma_{all} = 400$ MPa; $C_{all} = 40$ kNm; $FS = 2.5$; and $E = 200$ GPa. All the members are of rectangular cross-section with constant depth $d = 120$mm ($n = 1$). Initial uniform breadth, $b = 141$ mm for all members and b is allowed to vary to maximum 200 mm and to minimum 40 mm. This initial uniform design rightly satisfies the most critical, the stress constraint ($S_s = 1$). The step value used for resizing is equal to 5 mm and 20% of the elements are subjected to resizing at each iteration. Since the beams are subjected to uniformly distributed loads, each beam is divided into 10 sections for the analysis and the stress constraints are evaluated at each section. However optimum designs are obtained by treating each beam as a segment, thus avoiding stiffness jump within the beam. The optimum design is obtained after 13 iterations. The iteration history is given in Figure 5. At optimum $S_b = S_S = 1$ and $S_c = 0.92$. The weight of the optimum design with all the constraints is reduced to 67.9% of the initial design.

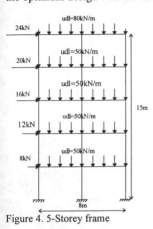

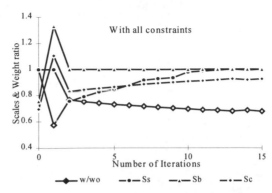

Figure 4. 5-Storey frame   Figure 5. Iteration history of the 5-storey frame

7 CONCLUSIONS

A new approach to optimizing the structures with stress, stiffness and stability constraints is illustrated with examples. The optimum designs obtained for the 50-bar truss and the 5-storey frame result in significant volume reductions. The proposed method is simple and the optimum designs are easily achieved without resorting to any complex mathematics. This method is suitable for designing practical structures with a large number of design variables and it can be readily implemented in any of the existing finite element codes.

REFERENCES

Barsan, G.M. 1994. Optimal design of planar frames based on structural performance criteria. *Computers & Structures*. 53(6):1395-1400.
Chu, D.N., Xie, Y.M., Hira, A. & Steven, G.P. 1996. Evolutionary structural optimization for problems with stiffness constraints. *Finite Elements in Analysis and Design*. 21:239-251.
Karihaloo, B.L. & Kanagasundaram, S. 1993. Optimum design of plane structural frames by non-linear programming. In G.I.N. Rozvany (ed.), *Optimization of Large Structural Systems*, Vol. *II*: 897-926. Netherlands: Kluwer Academic Publishers.
Khot, N.S., Venkayya, V.B. & Berke, L. 1976. Optimum structural design with stability constraints. *Int. J. Numer. Meth. Engng*. 10:1097-1114.
Lin, C.C. & Liu, I.W. 1989. Optimal design based on optimality criterion for frame structures including buckling constraint. *Computers & Structures*. 31(4):535-544.
Manickarajah, D., Xie, Y.M. & Steven, G.P. 1995. A simple method for the optimization of columns, frames and plates against buckling. In S. Kitipornchai, G.J. Hancock & M.A. Bradford (eds.), *Structural Stability and Design*: 175-180. Rotterdam: Balkema.
Pezeshk, S. & Hjelmstad, K.D. 1991. Optimal design of planar frames based on stability criterion. *J. Struct. Engng, ASCE*. 117(3):896-913.
Xie, Y.M & Steven, G.P. 1993. A simple evolutionary procedure for structural optimization. *Computers & Structures*. 49(5):885-896.
Xie, Y.M & Steven, G.P. 1996. Evolutionary structural optimization for dynamic problems. *Computers & Structures*. 58(6):1067-1073.

The Mechanics of Structures and Materials, Grzebieta, Al-Mahaidi & Wilson (eds)
© 1997 Balkema, Rotterdam, ISBN 90 5410 900 9

The application of new sets of static beam functions to the vibration analysis of rectangular plates with varying complexity

Y. K. Cheung
Department of Civil and Structural Engineering, The University of Hong Kong, People's Republic of China

Zhou Ding
School of Mechanical Engineering, Nanjing University of Science and Technology, People's Republic of China

ABSTRACT: Several sets of static beam functions have been established by the authors, which are the static solutions of beams under arbitrary static loads. The beam is a strip taken from the rectangular plate to be considered, and the static load acting on the beam is expanded into a series which may be a series of point loads or a series of sinusoid loads or a series of polynomial loads. These sets of static beam functions have been successfully applied to the vibration analysis of rectangular plates with varying complexity by the Rayleigh-Ritz method and finite strip method. Some numerical results are given. By comparing the results with those available from literature it can be shown that the present method is extremely effective for solving the aforementioned problems and usually a few static beam functions will be sufficient to give acceptable results.

1 INTRODUCTION

The rectangular plate is one of the most widely used structural elements in engineering. In practical application, some line supports may be placed at intermediate locations in order to reduce the magnitude of dynamic and static stresses and displacements of the structure.

Much of the work on the vibration of uniform rectangular plates with classical boundary conditions was reported before the early 1970's (Leissa 1973). However the vibration problem of rectangular plates with varying complexity, such as variable thickness, intermediate elastic or rigid line supports and elastic boundary constraints etc., has received rather less attention. The receptance method was used to analyse the two opposite edges simply supported plates with multi-spans in one direction (Azimi et al. 1984). The modified beam vibration functions (Zhou 1994a) and a set of orthogonal polynomials (Kim & Dickinson 1987) were used to analyse the plates with intermediate line supports in one or two directions by the Rayleigh-Ritz method. Such plates were also investigated by the finite strip method (Wu & Cheung 1974). Beam characteristic functions were used to study stiffened skew plates on non-periodic elastic supports in one direction (Bhandari et al. 1979). The differential quadrature method was used to analyse plates simply supported at two opposite edges with linearly varying thickness in one direction (Bert & Malik 1996). The superposition method was used to analyse plates with symmetrically distributed uniform elastic edge supports (Gorman 1989). Recently, the authors developed several sets of static beam functions which have been successfully applied to the free vibration of uniform rectangular plates with classical and elastic boundary conditions (Zhou 1994b, 1995, 1996), uniform rectangular plates with intermediate rigid or elastic line supports (Zhou & Cheung 1996, Cheung & Zhou 1996), non-uniform beams (Zhou & Cheung 1997) etc. by the Rayleigh-Ritz method. The sets of static beam functions may also be applied directly as the strip functions in the finite strip method (Cheung & Kong 1995).

2 THE EIGENFREQUENCY EQUATION

Consider a tapered rectangular plate which lies in the x-y plane and is bounded by edges

$x = \alpha a, a$, $y = \beta b, b$ where α and β are referred to as truncation factors of the plate in x and y directions respectively. The truncated plate is part of a sharp ended plate. There are S_x intermediate elastic supports in the x direction, whose coordinates are $x_s(s=1,2,...,S_x)$ and whose translational and rotational stiffnesses are k_{Ts}^x and k_{Rs}^x $(s=1,2,...,S_x)$ respectively. Similarly, there are S_y intermediate elastic supports in the y direction, whose coordinates are $y_s(s=1,2,...,S_y)$ and whose translational and rotational stiffnesses are k_{Ts}^y and k_{Rs}^y $(s=1,2,...,S_y)$ respectively. k_{T0}^x, k_{R0}^x and k_{Ta}^x, k_{Ra}^x are the stiffnesses of translational and rotational constraints at the edges $x = \alpha a$ and $x = a$ respectively and k_{T0}^y, k_{R0}^y and k_{Tb}^y, k_{Rb}^y are those at the edges $y = \beta b$ and $y = b$ respectively. It is assumed that the thickness of the plate may be described by a power function as follows

$$h(x,y) = h_0 (x / a)^r (y / b)^t \tag{1}$$

where r and t are referred to as taper factors of the plate in x and y directions respectively. The flexural rigidity of the plate is $D(x,y) = D_0(x / a)^{3r}(y / b)^{3t}$ where $D_0 = Eh_0^3 / 12(1 - v^2)$. It is further assumed that the variables in modal shape function $W(x,y)$ of the plate are separable and may be expressed in terms of a series as follows

$$W(x,y) = \sum_{m=m_0}^{\infty} \sum_{n=n_0}^{\infty} A_{mn} \varphi_m (x / a) \psi_n (y / b) \tag{2}$$

where m_0 and n_0 are the beginning orders of the modal series of the plate in x and y directions respectively. Letting $x_0 = \alpha a$, $k_{Ts}^x = k_{Ta}^x$, $k_{Rs}^x = k_{Ra}^x$, $x_s=a$ for $s = S_x + 1$ and $y_0 = \beta b$, $k_{Ts}^y = k_{Tb}^y$, $k_{Rs}^y = k_{Rb}^y$, $y_s=b$ for $s = S_y + 1$ and introducing non-dimensional parameters $\xi = x / a$, $\xi_s = x_s / a$, $K_{Ts}^x = k_{Ts}^x a^3 / D_0$, $K_{Rs}^x = k_{Rs}^x a / D_0$, $s = 0,1,2,...,S_x + 1$ and $\eta = y / b$, $\eta_s = y_s / b$, $K_{Ts}^y = k_{Ts}^y b^3 / D_0$, $K_{Rs}^y = k_{Rs}^y b / D_0$, $s = 0,1,2,...,S_y+1$ and minimizing the total energy of the plate about the coefficients A_{mn} leads to the eigenvalue equation

$$\sum_{m=m_0}^{\infty} \sum_{n=n_0}^{\infty} [C_{mnij} - \lambda^2 \overline{E}_{mi} \overline{F}_{nj}] A_{mn} = 0, \quad i = m_0 + 1, m_0 + 2,...,\infty, j = n_0 + 1, n_0 + 2,...,\infty \tag{3a}$$

where

$$C_{mnij} = E_{mi}^{(2,2)} F_{nj}^{(0,0)} + 2\gamma^2 (1 - v) E_{mi}^{(1,1)} F_{nj}^{(1,1)} + \gamma^4 E_{mi}^{(0,0)} F_{nj}^{(2,2)} + v\gamma^2 (E_{mi}^{(0,2)} F_{nj}^{(2,0)} + E_{mi}^{(2,0)} F_{nj}^{(0,2)})$$

$$+ \sum_{s=0}^{S_x+1} K_{Ts}^x B_{mi}^s F_{nj}^{(0,0)} + \sum_{s=0}^{S_x+1} K_{Rs}^x D_{mi}^s F_{nj}^{(1,1)} + \gamma^4 (\sum_{s=0}^{S_y+1} K_{Ts}^y G_{nj}^s E_{mi}^{(0,0)} + \sum_{s=0}^{S_y+1} K_{Rs}^y I_{nj}^s E_{mi}^{(1,1)}),$$

$$\lambda^2 = \rho h_0 \omega^2 a^4 / D_0 = \Omega^2 / (1 - \alpha)^4, \quad E_{mi}^{(p,q)} = \int_{\alpha}^{1} \xi^{3r} (d^p \varphi_m / d\xi^p)(d^q \varphi_i / d\xi^q) d\xi,$$

$$F_{nj}^{(p,q)} = \int_{\beta}^{1} \eta^{3t} (d^p \psi_n / d\eta^p)(d^q \psi_j / d\eta^q) d\eta, \quad p,q=0,1,2, \quad \overline{E}_{mi} = \int_{\alpha}^{1} \xi^r \varphi_m \varphi_i d\xi,$$

$$\overline{F}_{nj} = \int_{\beta}^{1} \eta^t \psi_n \psi_j d\eta, \quad B_{mi}^s = \varphi_m(\xi_s) \varphi_i(\xi_s), \quad D_{mi}^s = (d\varphi_m / d\xi \cdot d\varphi_i / d\xi)_{\xi=\xi_s},$$

$$G_{nj}^s = \psi_n(\eta_s) \psi_j(\eta_s), \quad I_{nj}^s = (d\psi_n / d\eta \cdot d\psi_j / d\eta)_{\eta=\eta_s}, \quad \gamma = a / b \tag{3b}$$

3 THE SETS OF STATIC BEAM FUNCTIONS

A unit width of the strip is taken out as a beam with the same variation of the depth as the rectangular plates in one or the other direction. Without losing generality, only the strip in the x

direction is considered here. The static deflection z of the non-uniform beam with intermediate point supports under a static load must satisfy the governing differential equation

$$\frac{d^2}{d\xi^2}(\xi^{3r}\frac{d^2 z}{d\xi^2}) = \sum_{s=1}^{S_x} P_s \delta(\xi-\xi_s) + \sum_{s=1}^{S_x} T_s \delta'(\xi-\xi_s) + Q(\xi) \tag{4}$$

where P_s and T_s ($s=1,2,...,S_x$) are ,respectively, the non-dimensional reaction forces and reaction moments of the sth intermediate point supports on the beam. $\delta(\xi-\xi_s)$ and $\delta'(\xi-\xi_s)$ are, respectively, the Dirac delta functions and the doublet functions. Correspondingly, the intermediate support conditions and the boundary conditions of the beam are respectively

$$P_s = -K_{Ts}^x z, \qquad T_s = K_{Rs}^x \frac{dz}{d\xi}, \qquad \text{at} \quad \xi = \xi_s \, (s=1,2,..., S_x) \tag{5a}$$

$$K_{R0}^x \frac{dz}{d\xi} = \xi^{3r}\frac{d^2 z}{d\xi^2}, \qquad -K_{T0}^x z = \frac{d}{d\xi}(\xi^{3r}\frac{d^2 z}{d\xi^2}), \qquad \text{at} \quad \xi = \alpha \tag{5b}$$

$$-K_{Ra}^x \frac{dz}{d\xi} = \xi^{3r}\frac{d^2 z}{d\xi^2}, \qquad K_{Ta}^x z = \frac{d}{d\xi}(\xi^{3r}\frac{d^2 z}{d\xi^2}), \qquad \text{at} \quad \xi = 1 \tag{5c}$$

An arbitrary static load $Q(\xi)$ may be expanded into a series as follows

$$Q(\xi) = \sum_{m=m_0}^{M} Q_m F_m(\xi) \tag{6}$$

where Q_m can be uniquely determined if $Q(\xi)$ is given m_0, M and $F_m(\zeta)$ are decided by the type of the series. For example, if $Q(\xi)$ is expanded into a Taylor polynomial series, then $m_0=0$, $M = \infty$ and $F_m(\xi) = \xi^m$. If $Q(\xi)$ is expanded into a Fourier sine series, then $m_0=1$, $M = \infty$ and $F_m(\xi) = \sin(m\pi\xi)$, and if $Q(\xi)$ is expanded approximately into a discrete series of N points then $m_0=1$, $M=N$ and $F_m(\xi) = \delta(\xi-\xi'_m)$ in which ξ'_m are the coordinates of the mth point loads.

The solution of equation (4) may be written in a unified form of

$$z(\xi) = \sum_{m=m_0}^{M} Q_m z_m(\xi), \qquad z_m(\xi) = \bar{z}_m(\xi) + \sum_{s=1}^{S_x} P_s^m \tilde{z}_p^s(\xi) + \sum_{s=1}^{S_x} T_s^m \tilde{z}_t^s(\xi) + \tilde{z}_q^m(\xi) \tag{7a,b}$$

where $\bar{z}_m(\xi)$ are the homogeneous solutions of equation (4), $\tilde{z}_p^s(\xi)$ and $\tilde{z}_t^s(\xi)$ are the special solutions corresponding to the reaction forces and reaction moments of the sth point supports respectively, $\tilde{z}_q^m(\xi)$ are the special solutions corresponding to the mth series loads. Substituting equations (7) into equation (4), one has

$$\bar{z}_m(\xi) = b_0^m + b_1^m\xi + b_2^m\xi^{-3r+2} + b_3^m\xi^{-3r+3}, \quad \text{for } r \neq 1/3, 2/3, 1,$$

$$\bar{z}_m(\xi) = b_0^m + b_1^m\xi + b_2^m\xi(\ln\xi - 1) + b_3^m\xi^2, \quad \text{for } r=1/3,$$

$$\bar{z}_m(\xi) = b_0^m + b_1^m\xi + b_2^m \ln\xi + b_3^m\xi(\ln\xi - 1), \quad \text{for } r=2/3,$$

$$\bar{z}_m(\xi) = b_0^m + b_1^m\xi + b_2^m/\xi + b_3^m \ln\xi, \qquad \text{for } r=1 \tag{8}$$

where b_s^m ($s = 0,1,2,3$) are the unknown constants, and

$$\tilde{z}_p^s(\xi) = \frac{1}{2-3r}(\xi^{3-3r}/(3-3r) - \xi_s\xi^{2-3r}/(1-3r) + \xi_s^{2-3r}\xi/(1-3r)$$

$$-\xi_s^{3-3r}/(3-3r))U(\xi-\xi_s), \qquad \text{for } r \neq 1/3, 2/3, 1,$$

421

$$\tilde{z}_p^s(\xi) = ((\xi^2 - \xi_s^2)/2 - \xi_s\xi \ln(\xi/\xi_s))U(\xi - \xi_s), \quad \text{for } r=1/3,$$

$$\tilde{z}_p^s(\xi) = ((\xi + \xi_s)\ln(\xi/\xi_s) - 2(\xi - \xi_s))U(\xi - \xi_s), \quad \text{for } r=2/3,$$

$$\tilde{z}_p^s(\xi) = (\ln(\xi_s/\xi) - \xi_s/2\xi + \xi/2\xi_s)U(\xi - \xi_s), \quad \text{for } r=1 \qquad (9)$$

where $U(\xi - \xi_s)$ are the Heaviside functions, and

$$\tilde{z}_t^s(\xi) = -\frac{1}{1-3r}(\frac{1}{2-3r}\xi^{2-3r} - \xi_s^{1-3r}\xi + \frac{1-3r}{2-3r}\xi_s^{2-3r})U(\xi - \xi_s), \quad \text{for } r \neq 1/3, 2/3,$$

$$\tilde{z}_t^s(\xi) = -(\xi \ln(\xi/\xi_s) + \xi - \xi_s)U(\xi - \xi_s), \quad \text{for } r=1/3,$$

$$\tilde{z}_t^s(\xi) = (\ln(\xi/\xi_s) - \xi/\xi_s + 1)U(\xi - \xi_s), \quad \text{for } r=2/3 \qquad (10)$$

If the static load is expanded into a Taylor polynomial series, one has

$$\tilde{z}_q^m(\xi) = \xi^{m+4-3r}/(m+3-3r)(m+4-3r)(m+1)(m+2), \quad \text{for } m \neq 3r-3, \ 3r-4,$$

$$\tilde{z}_q^m(\xi) = \xi(\ln\xi - 1)/(m+1)(m+2), \quad \text{for } m=3r-3,$$

$$\tilde{z}_q^m(\xi) = -\ln(\xi)/(m+1)(m+2), \quad \text{for } m=3r-4 \qquad (11)$$

If the static load is expanded into a Fourier sine series, one has

$$\tilde{z}_q^m(\xi) = \sin(m\pi\xi)/(m\pi)^4 \qquad (12a)$$

for the uniform beam and

$$\tilde{z}_q^m(\xi) = -\frac{1}{(m\pi)^2} \iint \frac{\sin(m\pi\xi)}{\xi^{3r}} d\xi \qquad (12b)$$

for the tapered beam. And if the static load is expressed approximately into a point load series, the special solution $\tilde{z}_q^m(\xi)$ are the same as those in equations (9) but with ξ_s being replaced by ξ_m' ($\xi_m' \neq \xi_s$, $m=1,2,...,N$, $s=1,2,...,S_x$).

In equations (7), there are $4+2S_x$ unknown constants involving $b_s^m (s = 0,1,2,3)$, and P_s^m, $T_s^m (s = 1,2,...,S_x)$, which can be uniquely decided by equations (5) for a beam without rigid body motions. A modified approach has already been described for a beam with rigid body motions (Zhou & Cheung 1997).

4 THE APPLICATIONS OF THE SETS OF STATIC BEAM FUNCTIONS

In order to illustrate the accuracy, convergency and usefulness of the approach described above, some numerical results are reported and compared with those available from literature. In the numerical computation, double precision is used for the static beam functions of both Fourier sine series loads and point series loads and quadruple precision is used for those of Taylor polynomial series loads. Table 1 gives the first five eigenfrequency parameters of uniform square plates clamped at all edges by using the three sets of static beam functions respectively and they are then compared with those from the literature (Leissa 1973). Six terms of static beam functions are used in each direction. Good accuracy has been observed for all the three sets of static beam functions.

Table 2 gives the first five eigenfrequency parameters of a square plate with symmetrically distributed elastic stiffnesses on all edges. K_T and K_R represent the non-dimensional translational and rotational stiffnesses respectively. The side length of the square plate is $2a$ and $v = 0.3$. Seven terms of the static beam functions developed from the Fourier sine series loads are used in each direction. Comparing the results with those available from the reference (Gorman 1989), good agreements have been observed.

Table 1. The first five eigenfrequency parameters of square plates clamped at all edges

Type of static loads	Ω_1	Ω_2	Ω_3	Ω_4	Ω_5
sine loads	35.99	73.41	73.41	108.27	131.64
polynomial loads	36.00	73.43	73.43	108.27	131.83
point loads	36.00	73.47	73.47	108.36	131.98
Leissa 1973	35.99	73.41	73.41	108.27	131.83

Table 2. The first five eigenfrequency parameters of a square plate with symmetrically distributed elastic stiffnesses on all edges, the values in ( ) are from the reference (Gorman 1989)

K_T	K_R	Ω_1	Ω_2	Ω_3	Ω_4	Ω_5
200	50	5.370	8.189	8.189	10.70	12.92
		(5.368)	(8.185)	(8.185)	(10.69)	(12.90)
400	50	6.349	10.38	10.38	13.59	15.35
		(6.347)	(10.37)	(10.37)	(13.57)	(15.32)
600	50	6.833	11.69	11.69	15.46	17.23
		(6.830)	(11.58)	(11.58)	(15.43)	(17.19)

Table 3 gives the first five eigenfrequency parameters of a one-way rectangular plate with four equal spans in the x direction. The aspect ratio of the plate equals 4 and the plate is simply supported at two opposite edges $y=0$ and $y=b$. Eight static beam functions in the x direction developed from the Fourier sine series loads and four of those in the y direction are used. The results have been compared with those from the reference (Azimi et al. 1984) and good agreement has been observed.

Table 3. The first five eigenfrequency parameters of a one-way rectangular plate with four equal spans, the values in ( ) are from the reference (Azimi et al. 1984)

Type of edges	Ω_1	Ω_2	Ω_3	Ω_4	Ω_5
S-S-S-S	19.74	20.81	23.65	27.12	49.35
	(19.74)	(20.81)	(23.65)	(27.12)	(49.35)
C-C-S-S	20.81	23.65	27.13	28.96	(49.99)
	(20.81)	(23.65)	(27.12)	(28.95)	(49.98)
C-S-S-S	20.01	22.06	25.41	28.44	49.51
	(20.01)	(22.06)	(25.41)	(28.44)	(49.51)

Table 4 gives the first five eigenfrequency parameters of a simply supported square plate with linearly variable thickness in one direction for different truncation factor α. Five terms of the static beam functions developed from the Taylor polynomial series loads are used in each direction. The results are compared with those from the reference (Bert & Malik 1996). Good agreement has also been observed.

Table 4. The first five eigenfrequency parameters of the simply supported square plate with linearly variable thickness, the values in ( ) are from the reference (Bert & Malik 1996)

α	Ω_1/α	Ω_2/α	Ω_3/α	Ω_4/α	Ω_5/α
1/5	55.54	128.9	134.8	219.0	232.8
1/3	38.22	91.76	94.06	151.8	173.2
1/2	29.21	71.76	72.54	116.5	139.8
5/7	23.61	58.78	58.94	94.38	116.7
	(23.61)	(58.77)	(58.93)	(94.38)	(116.7)

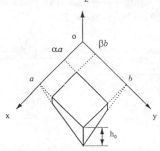

Figure 1 A rectangular plate with variable thickness in two directions

5. CONCLUSIONS

Three sets of admissible functions which represent the static solutions of a strip taken from the rectangular plate with varying complexity in one or the other direction under an arbitrary static load expanded into a Fourier sine series or a Taylor polynomial series or a point series respectively, have been developed in a unified manner. The sets of static beam functions can vary appropriately with the complexity of the plate. The basic concept to form the sets of static beam functions is very simple and clear and requires no complicated mathematical knowledge. The sets of beam functions may be directly applied to the vibration analysis of rectangular plates with varying complexity in Rayleigh-Ritz method or finite strip method. The approach may be further expanded to develop other new strip functions such as computed static beam functions under a point load series (Cheung & Kong 1995) etc. The numerical results show that the first few eigenfrequencies can be obtained with good accuracy by using only a few of the static beam functions.

REFERENCES

Azimi, S., J. F. Hamilton and W. Soedel 1984. The receptance method applied to the free vibration of continuous rectangular plates. *J. Sound Vibr.* 93: 9-29.
Bert, C. W. & M. Malik 1996. Free vibration analysis of tapered rectangular plates by differential quadrature method: a semi-analytical approach. *J. Sound Vibr.* 190: 41-63.
Bhandari, N. C., B. L. Juneja & K. K. Pujara 1979. Free vibration and transient forced response of integrally stiffened skew plates on irregularly spaced elastic supports. *J. Sound Vibr.* 64: 475-495.
Cheung, Y. K. & J. Kong 1995. The application of a new finite strip to the free vibration of rectangular plates of varying complexity. *J. Sound Vibr.* 181: 341-353.
Cheung, Y. K. & Zhou Ding 1996. Free vibrations of elastically line supported rectangular plates. *Int. J. Mech. Sci.* (submitted).
Gorman, D. J. 1989. A comprehensive study of the free vibration of rectangular plates resting on symmetrically-distributed uniform elastic edge supports. *ASME J. Appl. Mech.* 56: 893-899.
Kim, C. S. & S. M. Dickinson 1987. The flexural vibration of line supported rectangular plate systems. *J. Sound Vibr.* 114: 129-142
Leissa, A. W. 1973. The vibration of rectangular plates. *J. Sound Vibr.* 31: 257-293.
Wu, C. I & Y. K. Cheung 1974. Frequency analysis of rectangular plates. *Earthq. Engg. Struct. Dynam.* 3: 3-14.
Zhou Ding 1994a. Eigenfrequencies of line supported rectangular plates. *Int. J. Solids Struct.* 31: 347-358.
Zhou Ding 1994b. The application of a type of new admissible function to the vibration of rectangular plates. *Comput. Struct.* 52: 199-204.
Zhou Ding 1995. Natural frequencies of elastically restrained rectangular plates using a set of static beam functions in the Rayleigh-Ritz method. *Comput. Struct.* 57: 731-735.
Zhou Ding 1996. Natural frequencies of rectangular plates using a set of static beam functions in Rayleigh-Ritz method. *J. Sound Vibr.* 189: 81-87.
Zhou Ding & Y. K. Cheung 1996. Free vibration of line supported rectangular plates using a set of new beam functions in Rayleigh-Ritz method. *ASCE J. Engg. Mech.*(submitted).
Zhou Ding & Y. K. Cheung 1997. The free vibration of a type of tapered beams.(submitted for publication).

424

The Mechanics of Structures and Materials, Grzebieta, Al-Mahaidi & Wilson (eds)
© *1997 Balkema, Rotterdam, ISBN 90 5410 900 9*

Three dimensional FE analysis of high strength concrete columns

S.J.Foster, J.Liu & M.M.Attard
School of Civil Engineering, The University of New South Wales, Sydney, N.S.W., Australia

ABSTRACT: In the continuing research on large and high strength concrete structures there is a need for improved finite element models. Such models should be capable of analysing the full range of non-linear behaviour observed in the laboratory, including confining effects provided by steel reinforcement. Developed in this paper is a finite element model to investigate the response of concrete columns with concrete strengths upto 70 MPa. A numerical example is presented for an axially loaded column with varying eccentricities with the results compared with various simplified stress blocks. Analyses were also undertaken for columns with reduced cover strengths and compared with proposed strength reduction formulae. The results compare well with stress block parameters adopted by the Canadian and New Zealand codes of practice.

1. INTRODUCTION

Concrete structures which rely on confinement from steel reinforcement for ductility, can not be modelled in two-dimensions. Such structures include columns, prestressed concrete anchorage regions, column-beam and column-slab connections, etc. For studies on the ductility of concrete columns, it is important to investigate the strain softening behaviour of confined and unconfined normal and high strength concrete. The application of the finite element (FE) method to confined concrete is, however, in the infancy stage due to the complicated constitutive relationships. A successful FE analysis of concrete structures strongly depends on obtaining workable and reliable material laws.

The increase in ductility of normal strength concrete (NSC) columns afforded by well detailed lateral confinement reinforcement is well documented (Sheikh and Uzumeri, 1980). Questions however have been asked as to whether or not similar detailing is suitable for high strength concrete (HSC) columns, and if not, what amount of confinement reinforcement is necessary to obtain a similar level of ductility in HSC columns as is currently available in NSC columns. The literature is also contradictory as to whether or not increasing the strength of tie reinforcement leads to an increase in ductility of HSC columns.

In this paper a three dimensional finite element model, based on the microplane formulation of Carol et al. (1992), has been developed to investigate the behaviour of eccentrically loaded concrete columns.

2. A BRIEF REVIEW OF THE MICROPLANE MODEL

The explicit microplane model formulation is fully discussed in Carol et al. (1992) and is not discussed further here. The incorporation of the material laws into the 3D model are outlined in Liu et al. (1997) and are only briefly discussed below.

The empirical stress strain laws on which the micro strain-stress relationships are founded are given by

Volumetric law

$$\sigma_V = E_V^o \, \varepsilon_V \left[\left(1 + |\varepsilon_V|/a\right)^{-P} + \left(|\varepsilon_V|/b\right)^q \right] \quad \text{... for compression}$$

$$\sigma_V = E_V^o \, \varepsilon_V \, e^{-\left(|\varepsilon_V|/a_1\right)^{p_1}} \quad \text{... for tension} \tag{1}$$

Deviatoric law

$$\sigma_D = E_D^o \, \varepsilon_D \, e^{-\left(|\varepsilon_D|/a_2\right)^{p_2}} \quad \text{... for compression}$$

$$\sigma_D = E_D^o \, \varepsilon_D \, e^{-\left(|\varepsilon_D|/a_1\right)^{p_1}} \quad \text{...for tension} \tag{2}$$

Tangential law

$$\sigma_{Tr} = \tau \, \varepsilon_{Tr}/\nu$$

$$\tau = E_{Tr}^o \, \nu \, e^{-\left(\nu/a_3\right)^{p_3}} \tag{3}$$

where σ is the stress, ε the strain, E_V^o, E_D^o and E_{Tr}^o are the initial volumetric, deviatoric and tangential modulus of elasticity, respectively, ν is the Euclidean length of the transverse strain vector $\left(\nu = \sqrt{\varepsilon_{Tr}\varepsilon_{Tr}}\right)$ and the subscripts V, D and Tr refer to the volumetric, deviatoric and tangential directions, respectively. The confinement parameter a_3, can be defined as

$$a_3 = a_3^o + k_a \, \sigma_1/\sigma_3 \tag{4}$$

where σ_1 and σ_3 are the major and minor principal stresses, respectively. In developing Eq. (3) it is assumed that the tangential stress vector on each microplane remains parallel to the corresponding tangential strain vector.

The parameters a, b, p, q, a_1, p_1, a_2, p_2, a_3^o, p_3 and k_a are positive empirical material constants required for calibration of the constitutive model.

Applying the volumetric law the parameters a, b, p and q can be independently identified. Bazant and Prat (1988) gave these parameters as $a = 0.005$, $b = 0.225$, $p = 0.25$ and, $q = 2.25$ and are nearly the same for all concrete strengths. Further analysis of the test data by Bazant and Prat also showed that a_1, p_1 and p_2 can be fixed as $a_1 = 5 \times 10^{-5}$, $p_1 = 0.5$ and $p_2 = 1.5$ and that p_3 can be fixed at $p_3 = 1.5$ for conventional strength concrete. However, in order to obtain the steep descending curves necessary to model high strength concrete, the authors found it is necessary to take p_3 as variable. The ratio between the deviatoric modulus to the initial volumetric modulus $\left(E_D^o/E_V^o\right)$ is taken as 0.5 for all analyses.

In studies to date, the remaining material parameters required for Eqs. (1) to (4) have been chosen to best match selected experimental data. Some of the parameters, however, vary greatly with the type of specimen and loading conditions (Liu et al., 1997). For concrete compressive strengths from 20 MPa to 70 MPa, Liu et al. (1997) proposed relationships for the microplane parameters a_2, p_3, a_3^o and k_a as functions of the ratio of the major and the minor principal stresses $\left(\sigma_1/\sigma_3\right)$, the initial elastic modulus $\left(E_o\right)$ in MPa and the in-situ concrete compressive strength $\left(f_c\right)$ in MPa. The confinement parameters proposed are given by

$$a_3^o = -7.5 \times 10^{-7}\left(E_o/f_c\right) + 0.0023845$$

$$k_a = f_c/3000 \tag{5}$$

426

The strength parameter (a_2) and descending branch parameter (p_3) are defined using the cubic spline

$$a_2, p_3 = s_0 + s_1(\sigma_1/\sigma_3) + s_2(\sigma_1/\sigma_3)^2 + s_3(\sigma_1/\sigma_3)^3 \qquad (6)$$

where s_0, s_1, s_2 and s_3 are spline coefficients (given in Liu et al. 1997).

Figure 1(a) shows results of the microplane model generated with parameters given in Liu et al. compared to the uniaxial test data of Dahl (1992). Figure 1(b) shows the stress versus strain results for the numerical model compared to the confined experimental data of Attard and Setunge (1994) for various confining pressures. In both figures a good correlation is shown between the finite element results and the test data.

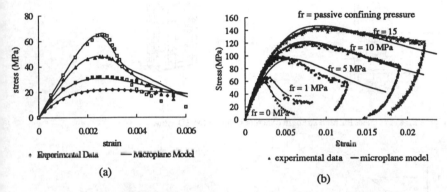

Figure 1. Numerical stress versus strain curves versus a) uniaxial cylinder tests of Dahl (1992);
b) experimental data of Attard and Setunge (1994).

3. NUMERICAL EXAMPLE

A 300 mm × 200 mm rectangular column was analysed for axial compression and bending with varying load eccentricities. With symmetries about two axes only one quarter of the column was meshed as shown in Figure 2. Analyses were undertaken with the cover concrete of equal strength to the core concrete, and with a cover strength of half the core uniaxial strength. The reasoning behind this is that experimental observations have shown that for some HSC columns failure has occurred at lower loads than should be theoretically so using traditional formulae for concentrically loaded columns. A possible explanation, put forward by Paultre et al. (1996), is that the cover concrete buckles prior the core concrete reaching its capacity. In tests by Foster and Attard (1997), on columns loaded under low eccentricities, splitting cracks along the line of the cover required by the buckling theory were not observed. In fact concrete outside of the hinge zones did not appear to be significantly damaged. Thus the authors have considerable doubts as to the validity of the buckling theory. One other possible explanation is that the cover concrete is of lower strength than the core concrete possibly due to thermal variations across the section in the first day after pouring, shrinkage in the concrete cover shell or other effects. If the concrete in the cover region is of lower compressive strength (but of comparable stiffness) to the concrete core zone, then at the peak strain the strength of the cover concrete is further reduced as shown in Figure 3.

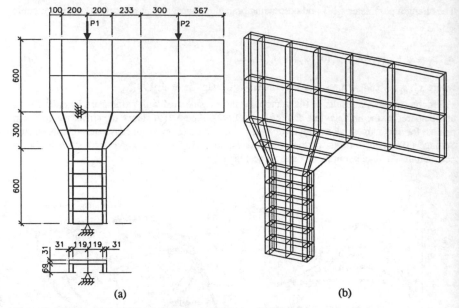

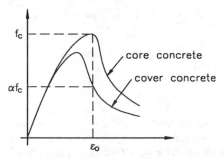

Figure 2. (a) FE mesh for high strength concrete column; (b) 3D view.

Figure 3. Comparison of core and cover stress-strain curves assuming a reduction cover strength.

For the FE model, the in-situ concrete strength was taken as $f_c = 70$ MPa. With an in-situ strength to cylinder strength ratio of 0.85 a 70 MPa in-place strength is equivalent to a cylinder strength of $f'_c = 82.4$ MPa. The columns contained 1.3% of longitudinal reinforcement (400 mm^2 top and bottom) and 1.1% (by volume) of lateral steel. The concrete was modelled using 20 node isoparametric solid elements with the element stiffness matrix formulated using numerical integration on a $2 \times 2 \times 2$ gaussian quadrature. The concrete material parameters used were $f_c = 70$ MPa $E_o = 36$ GPa and $\mu = 0.18$. For the reduced cover strength analyses, the cover concrete material properties were $f_c = 35$ MPa $E_o = 27$ GPa and $\mu = 0.18$ with the core concrete properties as before. The microplane parameters a, b, p, q, p_1 and p_2 are as discussed above and a_2, a_3 and p_3 were obtained from Eqs. (4), (5) and (6).

The longitudinal and tie reinforcements were modelled using three node truss elements. An elasto-plastic stress versus strain relationship was used with perfect bond being taken between

428

the steel and the concrete. The yield stress and initial elastic modulus of the longitudinal reinforcing bars were taken as $f_{sy} = 430$ MPa and $E_o = 190$ GPa, respectively, and for the tie reinforcement $f_{sy} = 400$ MPa and $E_o = 220$ GPa. The columns were loaded as shown in Figure 2, with the ratio of P_1/P_2 adjusted accordingly.

Peak loads are shown on Figure 4 and compared with interactive diagrams computed using the rectangular stress block of AS3600 (1994), a triangular stress block and the stress block models in CSA94 (1994) and NZS3101 (1995) design codes (which have been modified for the design of HSC structures). As is expected the current Australian code, not having been modified for HSC, over-predicts the capacity of columns failing in primary compression (similarly for ACI-318, 1995), while the triangular stress block is reasonably conservative. The numerical results compare reasonably well with both the Canadian and New Zealand codes of practice, particularly in the primary failure compression region.

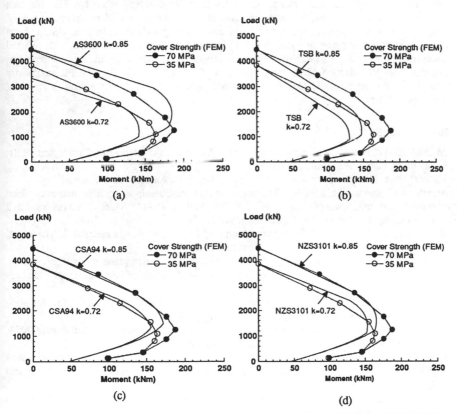

Figure 4 - Comparison between FE results and (a) AS3600; (b) triangular; (c) CSA94 and (d) NZS3101 stress blocks.

To take into account the lower peak loads observed in some HSC columns failing in compression, Collins et al. (1993) proposed that the ratio of in-situ concrete strength to cylinder strength be taken as

$$k = 0.6 + 10/f'_c \leq 0.85 \tag{7}$$

Eq. (7) was calibrated against a range of experimental data with concrete strengths from 30 MPa to 130 MPa. The interaction curves produced using Eq. (7) compares well with the reduced cover strength model, particularly for the Canadian code (as shown in Figure 4c).

4. CONCLUSIONS

A 3 dimensional FE model has been developed which takes into account confinement effects imposed either directly due to externally imposed loads, or indirectly due to reinforcing steel. The constitutive relationships used are based on the explicit microplane model developed by Carol et al. (1992) with material parameters developed by the authors. A numerical example is presented for a column loaded under various eccentricities. The FE data are compared with data produced using various simplified stress blocks. Good results were observed when the FE data was compared with the stress block parameters used in the Canadian and New Zealand codes of practice. It was also hypothesised that the reduction in strength observed in some HSC column tests is due to lower cover strengths. FE analyses were undertaken for columns with reduced cover strength and compared with the interaction diagrams produced using the in-situ strength to cylinder strength reduction parameter proposed by Collins et al. (1993). The FE results are consistent with the proposed Collins et al. model which was calibrated against a range of experimental data.

REFERENCES

ACI318-95 1995. *Building code requirements for structural concrete.* Farmington Hills: American Concrete Institute.

AS3600 1994. *Concrete structures code.* Sydney: Standards Association of Australia.

Attard, M.M. & Setunge, S., 1994. The stress - strain relationship of confined and unconfined normal and high strength concretes. *UNICIV Report R-341*, Sydney: School of Civil Engineering, The University of New South Wales.

Bazant, Z. P. & Prat, P. C. 1988. Microplane model for brittle - plastic material (I. Theory, II. Verification).*J. Engrg. Mech., ASCE.* 114(10): 1672-1702.

Carol, I., Prat, P. C. & Bazant, Z. P. 1992. New explicit microplane model for concrete: theoretical aspects and numerical implementation. *Int. J. Solids and Structures,* 29(9): 1173-1191.

Collins, M.P., MacGregor, J.G. and Mitchell D. 1993. Structural design considerations for high-strength concrete.*Concrete International, ACI,* May: 27-34.

CSA94 1994. CSA Technical Committee. *Design of concrete structures for buildings.* CAN3-A23.3-M94. Rexdale: Canadian Standards Association.

Dahl, K. B. 1992. Uniaxial stress-strain curves for normal and high strength concrete. *Research Report Series No. 282.* Copenhagen: Department of Structural Engineering, Technical University of Denmark.

Foster, S. J., and Attard, M. M. 1997. Experimental tests on eccentrically loaded high strength concrete columns.*ACI Structures Journal* (accepted for publication).

Liu, J., Foster S. J. & Attard, M. M. 1997. Axi-symmetric finite element analysis of confined concrete structures. *Second China-Australia Symposium on Computational Mechanics.*

NZS3101 1995. *Design of Concrete Structures.* Wellington: Standards New Zealand.

Paultre, P., Khayat, K.H., Langlois, A., and Cusson, D. 1996. Structural performance of some special concretes. *Fourth Int. Symposium on Utilization of High-Strength/High-Performance Concrete,* Paris: 787-796.

Sheik, S.A., and Uzumeri, S.M. 1980. Strength and ductility of tied concrete columns, *Journal of Str. Div., ASCE,* 106(5): 1079-1102.

The Mechanics of Structures and Materials, Grzebieta, Al-Mahaidi & Wilson (eds)
© 1997 Balkema, Rotterdam, ISBN 90 5410 900 9

Nonlinear discrete crack model of concrete arch dams

V. Lotfi
Civil Engineering Department, Amirkabir University, Tehran, Iran

ABSTRACT: There has been extensive research to examine the effect of contraction joints opening in the nonlinear response of arch dams. However, a complete nonlinear dynamic study is required to focus on possible failure mechanisms formed by certain potential separated blocks. In this paper the dynamic stability of these blocks is investigated for a typical concrete arch dam.

1 INTRODUCTION

Dynamic analysis of thin arch dams in high seismic zones is a complex task. Since contraction joints usually open and high arch stresses are released. Meanwhile, cantilever stresses in the downstream face increase. These stresses are often much higher than the tensile strength of lift surfaces, causing horizontal cracks which grow and form separated blocks. The stability of these separated blocks for the remaining part of excitation poses a major question for designers.

In this paper, this will be investigated for a typical thin Shahid Rajaee concrete arch dam. This is a dam with a height of 130 m and crest length of 420 m. The dam is being constructed in north of Iran, in the seismically active foothills of Alborz mountain near the city of Sari. A special finite element program called "MAP-73" (Lotfi 1994) and a corresponding Pre- and post-processing program "MAP-P" are utilized. The program is suitable for modeling the behavior of three dimensional solids as well as contraction joints openings or cracking at any potentially weak planes by isoparametric interface elements similar to the works of O'Connor (1985), and Fenves et al. (1993).

2 MODELS AND ASSUMPTIONS

Two cases are studied. Case A, a linear model used mainly for comparative purposes. Case B, a nonlinear discrete crack model in which interface elements (Lotfi 1994) with initial limited tensile strength are utilized at some predefined surfaces where cracks deemed to occur based on the linear analysis results. The foundation is also taken as rigid to keep the computational time realistic. It is clear that this could influence the boundary stresses in case of linear analysis drastically. However, it is less important for the nonlinear case due to joint openings at the boundaries. It should also be noted that rigid foundation assumption has less impact on stresses in the vicinity of the spillway which the major failure mode is deemed to exist.

In the nonlinear case, the total tangent stiffness matrix is obtained by assemblage of different elemental tangent stiffness matrices which can be defined for isoparametric interface crack elements and stiffness of different solid elements. The nonlinearities are limited to interface elements. It should be noted that the solid elements can be taken as usual linearly elastic isoparametric elements with stiffnesses calculated once at the beginning, assembled

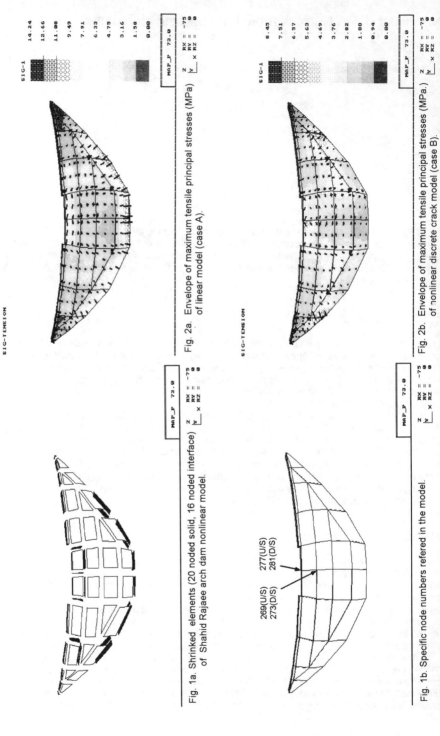

SIG-TENSION

SIG-1	
	14.24
	12.66
	11.08
	9.49
	7.91
	6.33
	4.75
	3.16
	1.58
	0.00

MAP_P 73.0

RX = -75
RY = 0
RZ = 0

Fig. 2a. Envelope of maximum tensile principal stresses (MPa) of linear model (case A).

MAP_P 73.0

RX = -75
RY = 0
RZ = 0

Fig. 1a. Shrinked elements (20 noded solid, 16 noded interface) of Shahid Rajaee arch dam nonlinear model.

SIG-TENSION

SIG-1	
	8.45
	7.51
	6.57
	5.63
	4.69
	3.76
	2.82
	1.88
	0.94
	0.00

MAP_P 73.0

RX = -75
RY = 0
RZ = 0

Fig. 2b. Envelope of maximum tensile principal stresses (MPa.) of nonlinear discrete crack model (case B).

269(U/S)   277(U/S)
273(D/S)   281(D/S)

MAP_P 73.0

RX = -75
RY = 0
RZ = 0

Fig. 1b. Specific node numbers refered in the model.

432

and stored for recursive application throughout the analysis. While the interface elements stiffnesses need to be updated at each iteration.

In the analysis carried out, the Rayleigh damping matrix is applied and the corresponding coefficients are determined such that the equivalent damping for frequencies close to the first and sixth modes of vibration would be 12% of the critical damping.

There are also several precise procedures to consider the effect of hydrodynamic pressures induced by dam-reservoir interaction in linear dynamic analysis. For instance, finite element methods based on compressible or incompressible water theories in frequency or time domain may be used. Although some of these are also appropriate for nonlinear dynamic analysis, In this work a conservative and computationally efficient modified Westergaard method (Clough et al. 1985) is employed. In this approach, consistent added mass matrices are introduced which can easily be combined with the mass of the dam body.

3 IDEALIZATION AND BASIC PARAMETERS

3.1 Finite element mesh

As mentioned two cases are considered, linear model (case A), and discrete crack model (case B). In both cases, the finite element mesh consists of 660 nodes and 76 isoparametric 20 noded elements. Meanwhile in case B, there are 26 isoparametric 16 noded interface elements to model cracks and opening of joints (Fig. 1). Noteworthy, for cases A and B the same number of nodes is used. However , the nodes corresponding to interface elements of case B are constrained in the linear case(A), to guarantee that every two adjacent nodes with the same coordinates have similar displacements.

3.2 Basic parameters

The concrete is assumed to have the following basic properties:

Elastic modulus = 30.0 GPa.
Poisson's ratio = 0.18
Unit weight = 24.0 kN./m^3

The interface elements, utilized for the nonlinear case are applied with the following parameters:

Elastic modulus (Ec) = 30.0 GPa.
k1=k2 (tangential stiffness) = 0.2 * Ec
k3 (normal stiffness) = 50. * Ec
σ^* (tensile stress limit for contraction joints and boundaries) = 1.5 MPa.
σ^* (tensile stress limit for horizontal lift joints) = 3.0 MPa.

The water is taken as incompressible, inviscid fluid, with weight density of 10. kN/m^3 and the water level to be at elevation 485.0 m.a.s.l (h=122.0 m)

3.3 Loading

It should be mentioned that static loads (weight, hydrostatic pressures) are visualized as being incrementally increasing in time until they reach their full magnitude. Therefore the same time step of 0.01 second which is chosen in dynamic analysis, is considered as time increment of static loads application. It is noted that time for static analysis is just a convenient tool for applying the load incrementally, but it is obvious that inertia and damping effects are disregarded in the process. In this respect, the dead load is applied in one increment and hydrostatic pressures thereafter in nine increments at negative range of time. At time zero, the actual nonlinear dynamic analysis begins with the static displacements and stresses being applied as initial conditions.

433

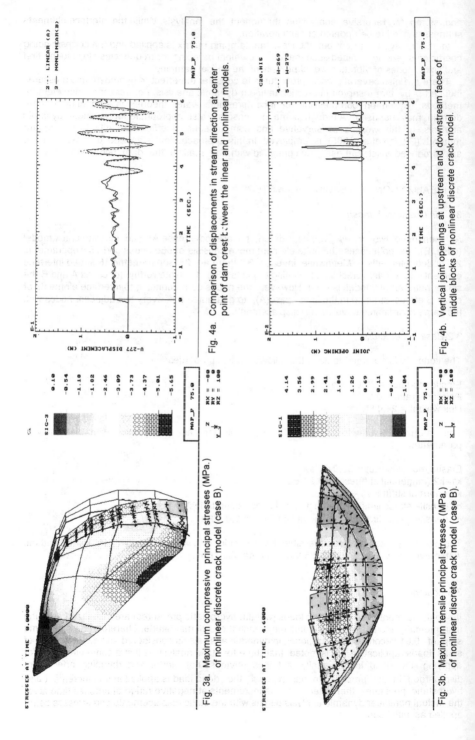

STRESSES AT TIME 0.0000

SIG-3

0.10
-0.54
-1.18
-1.82
-2.46
-3.09
-3.73
-4.37
-5.01
-5.65

MAP_P 75.0

RX = -90
RY = 0
RZ = 100

Fig. 3a. Maximum compressive principal stresses (MPa.) of nonlinear discrete crack model (case B).

STRESSES AT TIME 4.6000

SIG-1

4.14
3.56
2.99
2.41
1.84
1.26
0.69
0.11
-0.46
-1.04

MAP_P 75.0

RX = -90
RY = 0
RZ = 100

Fig. 3b. Maximum tensile principal stresses (MPa.) of nonlinear discrete crack model (case B).

Fig. 4a. Comparison of displacements in stream direction at center point of dam crest between the linear and nonlinear models.

2 --- LINEAR (A)
3 —— NONLINEAR(B)

MAP_P 75.0

Fig. 4b. Vertical joint openings at upstream and downstream faces of middle blocks of nonlinear discrete crack model.

C30.HIS

4 --- N-269
5 —— N-273

MAP_P 75.0

Table 1. Maximum tensile and compressive stresses.

Case	Location	Maximum principal stress σ_1 (MPa.)		Minimum principal stress σ_3 (MPa.)	
		upstream	downstream	upstream	downstream
A (linear)	Spillway	7.9	7.9	-16.7	-14.8
	Base and abutments	14.2	3.1	-9.2	-13.0
B (Disc. crack)	Spillway	5.6	6.0	-15.0	-17.6
	Base and abutments	7.5	2.0	-7.0	-13.5

The dynamic excitations include the three components of Friuli-Tolmezzo earthquake records normalized based on the frequency content for MDE condition with a peak ground acceleration of 0.42g. It needs to be mentioned that even though a time duration of 20 seconds was applied on the initial trial cases, it was noticed that the response declines drastically after 6.0 seconds. For this reason and due to long execution times, the time duration was limited to 6.0 seconds in the main analyses carried out.

4 RESULTS

The two models are analyzed, and the results of envelope of maximum principal stresses throughout the time are obtained and they are summarized in Table 1 below. Meanwhile, the tensile stress envelopes are displayed in Fig. 2.

It is observed that there are very high tensile stresses (Fig. 2a) occurring at the base of the dam for the linear model, which is due basically to rigid foundation model deficiency, and high tensile stresses in the area of spillway which in reality are released with opening of contraction joints. The maximum compressive stress occur in the upstream face nearby spillway region with a magnitude of -16.7 MPa..

For the discrete crack model, high tensile stresses at the base of the dam and high tensile arch stresses In the spillway region are released by opening of the interface elements (Fig. 2b), although there are still tensile stresses of lower magnitude in inclined directions. Furthermore, the compressive stresses increase slightly in the downstream face (σ_3= -17.6 MPa.) while decreasing on the upstream face in comparison with the linear case. A snapshot of state of stresses are also shown at two intermediate time steps (Fig. 3).

As for displacements, three nodes are monitored at dam crest. These are located approximately at left quarter point, center, and right quarter point. The maximum displacements through time for each component of these nodes are summarized on Table 2 below.

Furthermore, the center point displacement component in stream direction are compared between the linear case and the nonlinear model (Fig. 4a). It is apparent that for the nonlinear model, displacements are close to the linear case.

The joint opening in horizontal direction for middle separated blocks under the spillway (Fig. 1) is displayed in Fig. 4b. The maximum joint opening is 1.8 cm occurring in the horizontal direction under the spillway on upstream face.

Table 2. Maximum displacements at dam crest.

Case	Component	Displacements of Dam Crest (cm)		
		Left 1/4 point	Center point	Right 1/4 point
A (linear)	U(cross-canyon)	+3.0	-1.6	-4.6
	V(stream)	+4.2	+11.0	+8.0
	W(vertical)	-0.4	-1.4	+0.6
B (Disc. crack)	U(cross-canyon)	+4.0	-2.4	-5.3
	V(stream)	+6.4	+12.0	+9.4
	W(vertical)	+0.8	+2.1	+1.3

5 CONCLUSIONS

Two cases are considered, linear case (A), and nonlinear discrete crack model (B). Overall, the main conclusions obtained can be listed as follows:

- There is basically the same state of stresses and deformation at the end of self-weight analysis for both cases.
- At the end of static analysis, high tensile stresses are observed in the linear case at the upstream face of the base of the dam. This is reduced to nil for the discrete crack model. At the same time, compressive stresses increase in the downstream face on the base and abutments for discrete crack case.
- In the spillway region, tensile stresses decrease from 7.9 MPa. in the case of linear model to a value of 6.0 MPa. for discrete crack model. Meanwhile, the compressive stresses are increased from -16.7 MPa. in linear case to -17.6 MPa. for discrete crack model.
- In the nonlinear model, the maximum joint opening is 1.8 cm, displacements are moderately increased in the central portion and the difference is more pronounced for the quarter points. However, the dam has remained stable throughout the analysis.

REFERENCES

Clough, R.W., Chang, K.T., Chen, H.Q. and Ghanaat, Y., 1985, "Dynamic interaction effects in arch dams", Report No. UCB/EERC-85/11 Univ. of Calif., Berkeley, Calif
Fenves, G. L., Mojtahedi, S., 1993,"Earthquake response of an arch dam with contraction joint opening", Dam Engineering, Vol. 4, No. 2.
Lotfi, V., 1994, "MAP-73, Mahab Ghodss and Tehran Polythechnique University Software", Mahab Ghodss Consulting Engineers, Tehran, Iran.
O' Connor, J.P.F.,1985, "The modeling of cracks, potential crack surfaces and construction joints in arch dams by curved surface interface elements", Proc. of the 15th Int. Conference on Large Dams, Lausanne, Switzerland.

The Mechanics of Structures and Materials, Grzebieta, Al-Mahaidi & Wilson (eds)
© *1997 Balkema, Rotterdam, ISBN 90 5410 900 9*

Stochastic linearization coefficients of the Bouc-Wen oscillator

G.C. Foliente
CSIRO Building, Construction and Engineering, Highett, Vic., Australia

ABSTRACT: The Bouc-Wen oscillator has been popular in non-linear random vibration analysis because of its versatility and mathematical tractability but its stochastic linearization coefficients that have been presented in the literature have different forms and may cause some confusion. This paper presents a general form of the coefficients that is valid for the full range of possible values of the correlation coefficient between response velocity and hysteretic restoring force, and of the hysteresis shape parameter that controls the degree of non-linearity. The complete set of linearization coefficients are used in a random vibration analysis of a single-degree-of-freedom hysteretic, degrading and pinching system. Results under white noise excitation compare favourably with those from Monte Carlo simulation.

1 INTRODUCTION

The Bouc-Wen oscillator and its extensions have been widely used in deterministic and stochastic dynamic analyses of a wide variety of hysteretic systems, including reinforced concrete, steel and timber buildings, and base isolation systems [see recent reviews (Wen 1989; Foliente et al. 1996)]. Its importance in non-linear dynamic analysis, damage modelling and reliability studies continues to increase. Extensions to include strength and stiffness degradation (Baber and Wen 1981) and general pinching effects (Foliente et al. 1996) and to reduce local violation of Drucker's stability postulate of plasticity when the system undergoes intermediate loading-reloading without change of the load sign (Casciati 1987; Wong et al. 1994) have made it even more versatile and applicable to a wide variety of structural systems and materials.

This paper compares various published forms of the stochastic linearization coefficients of the Bouc-Wen oscillator, and proposes coefficients that are valid for the full range of possible values of the correlation coefficient between response velocity and hysteretic restoring force, and of the hysteresis shape parameter that controls the degree of non-linearity.

2 THE BOUC-WEN OSCILLATOR

The Bouc-Wen oscillator has three parallel elements: a linear viscous damper, a linear spring and a hysteretic element (Figure 1a). The equation of motion of a single-degree-of freedom (SDOF) Bouc-Wen oscillator, in standard form, is

$$\ddot{u} + 2\,\xi_o\omega_o\,\dot{u} + \alpha\omega_0^2\,u + (1 - \alpha)\,\omega_0^2\,z = f(t) \tag{1}$$

where u is the relative displacement of the mass with respect to the ground (dots designate derivatives with respect to time t), ξ_o is the system damping ratio, ω_o is the linear natural frequency, α is the ratio of post yielding to pre-yielding stiffness (or rigidity ratio), z is the hysteretic displacement and $f(t)$ is the mass-normalised forcing function. The constitutive law is given by the first-order non-linear differential equation:

$$\dot{z} = \dot{u} - \beta |\dot{u}| |z|^{n-1} z + \gamma \dot{u} |z|^n \tag{2}$$

where β, γ and n are the hysteresis shape parameters (if $n = \infty$, the elasto-plastic case is obtained in the z-u plane). Different values of β and γ give various softening and hardening systems while n controls the degree of non-linearity or smoothness of the hysteresis curve (Figure 1) (Baber and Wen 1981; Wong et al. 1994). Additional parameters to introduce stiffness degradation, strength degradation and pinching can be included (Appendix I).

3 STOCHASTIC EQUIVALENT LINEARIZATION

The popularity of the Bouc-Wen oscillator is due to its versatility and mathematical tractability in random vibration analysis. The differential constitutive relation allows stochastic equivalent linearization without recourse to the Krylov-Bogoliubov approximation based on a direct linearization scheme (Baber and Wen 1981).

The equation of motion, Eq. (1), is linear so only Eq. (2) needs to be linearized into:

$$\dot{z} = c_e \dot{u} + k_e z \tag{3}$$

where c_e and k_e are time-dependent linearization coefficients, obtained by minimizing the expected value of the square of the error process resulting from linearization. Following a direct linearization scheme (Atalik and Utku 1976),

$$c_e = 1 - \beta F_1 - \gamma F_2 \tag{4}$$

$$k_e = - \beta F_3 - \gamma F_4 \tag{5}$$

where F_i's are linearization coefficients (presented in the next section).

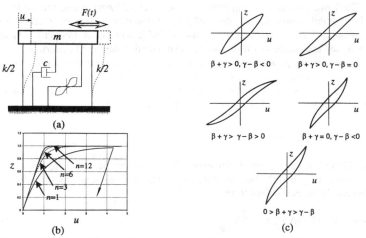

(a)

(b)

(c)

Figure 1. The Bouc-Wen oscillator: (a) single-degree-of freedom (SDOF) mechanical model, (b) and (c) possible hysteresis shapes using the differential Bouc-Wen restoring force.

The linearized system of equations can then be analysed using normal tools and techniques in structural dynamics (e.g., modal analysis). Thus, it is important that proper coefficients that minimise the error due to linearization be obtained.

In random vibration analysis, the linearized system of equations (using state-space formulation) leads to the equation for the zero time lag covariance response, if the excitation is nonstationary or for a transient solution of the system under stationary excitation (e.g., solution for degrading hysteretic systems) (Baber and Wen 1981). The desired response statistics (root mean square or RMS displacement, σ_u, RMS velocity, $\sigma_{\dot{u}}$, RMS restoring force, σ_z, and mean energy dissipation, μ_e) are obtained from the covariance matrix after numerical integration. If the excitation is stationary, the same set of equations leads to the well known Liapunov matrix equation, solutions for which are well established. Details about stochastic equivalent linearization of the Bouc-Wen oscillator and its extensions are given elsewhere (Baber and Wen 1981; Wen 1989; Foliente et al. 1996). With the degrading, pinching case (Appendix I), nonstationary response statistics of a SDOF system subjected to white noise excitations have been computed using equivalent linearization and Monte Carlo simulation. Figures 2a, b and c, reproduced from Foliente et al. (1996), show that linearization results compare favourably with those from simulation at all excitation levels.

Figure 2d shows that although the correlation coefficient, $\rho_{\dot{u}z}$ (or ρ for simplicity), of response velocity $(\dot{u})$ and hysteretic restoring force (z) obtained by simulation assumes positive values most of the time, it can also assume negative values. Thus, linearization coefficients should be applicable for the full range of possible values of ρ, that is $-1 \leq \rho \leq 1$.

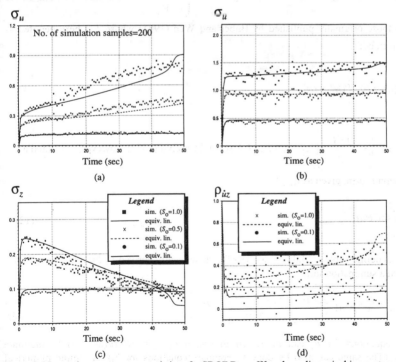

Figure 2. Nonstationary response statistics of a SDOF Bouc-Wen degrading, pinching system under white noise (S_o = power spectral density): (a) RMS displacement, (b) RMS velocity, (c) mean energy dissipation [from Foliente et al. (1996)], (d) correlation coefficient of velocity and hysteretic restoring force.

439

4 LINEARIZATION COEFFICIENTS

The linearization results in Figure 2 were obtained using the following coefficients in Eqs. (4) and (5) (Foliente et al. 1996):

$$F_1 = \frac{1}{\pi}(2)^{n/2}\sigma_z^n\Gamma\left(\frac{n+2}{2}\right)I_{sn} \tag{6}$$

$$F_2 = \frac{1}{\sqrt{\pi}}(2)^{n/2}\sigma_z^n\Gamma\left(\frac{n+1}{2}\right) \tag{7}$$

$$F_3 = \frac{n}{\pi}(2)^{n/2}\sigma_{\dot{u}}\sigma_z^{n-1}\Gamma\left(\frac{n+2}{2}\right)\left[\frac{2}{n}\left(1-\rho^2\right)^{(n+1)/2} + \rho I_{sn}\right] \tag{8}$$

$$F_4 = \frac{n}{\sqrt{\pi}}(2)^{n/2}\rho\sigma_{\dot{u}}\sigma_z^{n-1}\Gamma\left(\frac{n+1}{2}\right) \tag{9}$$

where

$$I_{sn} = 2\,\text{sgn}\,(\rho)\int_{\theta_0}^{\pi/2}\sin^n\theta d\theta \tag{10}$$

in which sgn($\cdot$) is the signum function and $\theta_o = \tan^{-1}\left(\dfrac{\sqrt{1-\rho^2}}{\rho}\right)$. The forms of F_1 and F_3 are different to those originally published by Baber and Wen (1981) whose I_{sn}, here denoted as $I_{sn\text{-}B}$, was given by:

$$I_{sn\text{-}B} = \int_o^{\theta_1}\sin^n\theta d\theta - \int_{\theta_1}^{\pi}\sin^n\theta d\theta \tag{11}$$

where $\theta_1 = \tan^{-1}\left(-\dfrac{\sqrt{1-\rho^2}}{\rho}\right)$. Chang et al. (1986) and Casciati (1987), however, presented a slightly different form, given as $I_{sn\text{-}C}$:

$$I_{sn-C} = 2\int_{\theta_o}^{\pi/2}\sin^n\theta d\theta \tag{12}$$

where θ_0 is as defined above. Maldonado (1992) independently derived F_1 for odd values of n and proposed a summation form:

$$F_1 = \frac{1}{\sqrt{\pi}}\,(2)^{n/2}\sigma^n\rho^n n!\sum_{j=0}^{\left(\frac{n-1}{2}\right)}\frac{[(n-2j-1)/2]!}{j!(n-2j)!}\left[\frac{1-\rho^2}{4\rho^2}\right] \tag{13}$$

The above equations for F_1 can be recast to obtain $F_{1,\text{norm}} = (F_1\pi)/(2^{n/2}\sigma_z^n)$ to facilitate comparison. (The same can also be done to compare equations for F_3). A simple numerical exercise shows that values obtained using the above equations, given the same input, do not agree (Table 1). F_1 with Eq. (10) agrees with Eqs. (11) and (13) when n is odd and $\rho < 0$, and

440

with Eq. (12) when $\rho > 0$ (for all values of n). None of the equations agree when n is even and $\rho < 0$. It is shown in the detailed derivation (Foliente 199_) that the equations for F_1 and F_3 given in Eqs. (6), (8) and (10) are valid for the full range of possible values of ρ (from -1 to 1). Thus, the form of F_1 with Eq. (12) is valid only when $\rho > 0$ (for all values of n), and the form originally derived by Baber, with Eq. (11), is valid only when $\rho < 0$ and n is odd. The case where n is even and $\rho < 0$ had not been properly covered.

Table 1. Comparison of proposed linearization coefficients

Form of F_1	Input Parameters		$F_{1,norm}$
	n	ρ	
Eqs. (6) and (10)	1 (odd)	-0.5	-0.8862
[from Foliente et al. 1996]		0.5	0.8862
	2 (even)	-0.5	-2.1849
		0.5	0.9566
Eqs. (6) and (11)	1	-0.5	-0.8862
[from Baber and Wen		0.5	-0.8862
1981]	2	-0.5	-0.9566
		0.5	-2.1849
Eqs. (6) and (12)	1	-0.5	0.8862
[from Chang et al. 1986;		0.5	0.8862
Casciati 1987]	2	-0.5	2.1849
		0.5	0.9566
Eq. (13)	1	-0.5	-0.8862
[from Maldonado		0.5	0.8862
1992]	2	-0.5	NA
		0.5	NA

SUMMARY AND CONCLUSION

The most common forms of stochastic linearization coefficients of the Bouc-Wen oscillator are not valid for the full range of possible values of the correlation coefficient ρ between response velocity and hysteretic restoring force, and of the hysteresis shape parameter n that controls the degree of non-linearity. None of the common forms covered the case where n is even and ρ is negative. The increasing use and importance of the Bouc-Wen oscillator in many engineering applications require the correct forms of linearization coefficients for dynamic analysis and random vibration analysis. A general form of the coefficients valid for the full range of possible values of ρ and n has been presented in this paper. Nonstationary response statistics using equivalent linearization compare favourably with those from Monte Carlo simulation. This finding should increase confidence in the use of the Bouc-Wen oscillator and its extensions in the inelastic analyses of many structures under severe dynamic loads, such as extreme wind, wave action and earthquake ground motion, using an equivalent linearized system.

REFERENCES

Atalik, T.S. and S. Utku 1976. Stochastic linearization of multidegree of freedom nonlinear systems. *Earthq. Engg. Struct. Dynam.* 4:411-420
Baber, T.T. and Y.-K. Wen 1981. Random vibration of hysteretic degrading systems. *J. Engg. Mech. Div. ASCE* 107(EM6):1069-1089.
Casciati, F. 1987. Approximate methods in non-linear stochastic dynamics. Chapter 5 and Appendix A *in* Schueller, G.I. and M. Shinozuka (eds.) 1987. *Stochastic methods in structural dynamics.* Dordrecht: Martinus Nijhoff Publishers.

Chang, T.-P., T. Mochio and E. Samaras 1986. Seismic response analysis of nonlinear structures. *Prob. Engg. Mech.* 1(3):157-166.

Foliente, G.C., M.P. Singh and M.N. Noori 1996. Equivalent linearization of generally pinching hysteretic, degrading system. *Earthq. Engg. Struct. Dynam.* 25:611-629.

Foliente, G.C. 199_. On the derivation of linearization coefficients of the Bouc-Wen oscillator and its extensions. To be submitted, *Intl. J. Nonlinear Mech.*

Maldonado, G.O. 1992. Stochastic and seismic design response of linear and nonlinear structures. PhD Dissertation, Virginia Polytechnic Institute and State University, Blacksburg, VA.

Wen, Y.-K. 1989. Methods of random vibration for inelastic structures. *Applied Mech. Review ASME.* 42(2):39-52.

Wong, C.W., Y.Q. Ni and J.M. Ko 1994. Steady state oscillation of hysteretic differential model. II: performance analysis. *J. Engg. Mech. ASCE.* 120(11):2299-2325.

APPENDIX I. Restoring Force Model With Degradation and Pinching Functions

To introduce strength and stiffness degradation and generalised pinching, Eq. (2) is replaced by:

$$\dot{z} = h(z)\left\{\frac{\dot{u} - v\left(\beta|\dot{u}\|z|^{n-1}z + \gamma\dot{u}|z|^{n}\right)}{\eta}\right\} \tag{14}$$

where $h(z)$ is the pinching function and v and η are strength and degradation parameters, respectively, modelled by (Baber and Wen (1981):

$$v(\varepsilon) = 1.0 + \delta_v\varepsilon \tag{15}$$

$$\eta(\varepsilon) = 1.0 + \delta_\eta\varepsilon \tag{16}$$

where δ_v and δ_η control the rate of degradation and ε is the hysteretic energy dissipation:

$$\varepsilon = \int_{t_o}^{t_f} z\dot{u}\, dt \tag{17}$$

Generalised pinching is modelled by (Foliente et al. 1996)

$$h(z) = 1.0 - \zeta_1 \exp[-(z\,\mathrm{sgn}\,(\dot{u}) - qz_u)^2/\zeta_2^2] \tag{18}$$

where ζ_1 controls the severity of pinching, ζ_2 controls pinching spread, z_u is the ultimate value of z:

$$z_u = \left[\frac{1}{v(\beta+\gamma)}\right]^{1/n} \tag{19}$$

and q is a constant that sets a fraction of z_u as the pinching level. Both ζ_1 and ζ_2 are functions of ε:

$$\zeta_1(\varepsilon) = \zeta_{1_o}\left[1.0 - \exp(-\rho\varepsilon)\right] \tag{20}$$

$$\zeta_2(\varepsilon) = \left(\psi_o + \delta_\psi\varepsilon\right)(\lambda + \zeta_1) \tag{21}$$

where p controls the rate of initial drop in slope, ζ_{1o} is the measure of total slip, ψ_o is the parameter that contributes to the amount of pinching, δ_ψ is the constant specified for the desired rate of pinching spread, and λ is the small parameter that controls the rate of change of ζ_2 as ζ_1 changes.

If $h(z)$ is set to unity, the degrading Bouc-Wen model (Baber and Wen 1981) is obtained. If in addition to this, δ_v and δ_η are set to zero, the original Bouc-Wen model [i.e., Eq (2)] is obtained.

The Mechanics of Structures and Materials, Grzebieta, Al-Mahaidi & Wilson (eds)
© 1997 Balkema, Rotterdam, ISBN 90 5410 900 9

Stability analysis of finite anisotropic panels with elliptical cutouts and cracks

Zhao Qi
Shanghai Institute of Applied Mathematics and Mechanics, People's Republic of China

ABSTRACT: An approximate analysis for stability of anisotropic finite panels with centrally located elliptical cutouts is presented. The analysis is composed of two parts: a plane stress analysis and a stability analysis. The plane stress distribution is determined using Lekhnitskii's complex variable equations of plane elastostatics combined with a Laurent series approximation constructed by the conformal mapping function, and a boundary collocation method. Its solutions satisfy the conditions along the interior boundary and at a discrete number of points along the exterior panel ones. The stability analysis is conducted using the different equations which result from the Hamilton's principle and the classical plate theory. The relation of vibration frequency, load parameter and stability of panels is investigated by solving the fundamental equations using separating variables technique, so as to obtain the critical loads. Finally, comparisons with documented experimental results and finite element analysis are done. Results of a parameter study are presented.

1. INTRODUCTION

The flat panels with cutouts are found in several aircraft structural components such as wing ribs and spars. Because of their light weight and tailor-ability, composite materials are becoming increasingly popular for the manufacturing of these panels. Therefore, the research on stability of finite anisotropic panels with cutouts is important in future aircraft design.

The majority of investigations on the stability of composite panels with cutouts have been conducted using finite element methods. For example, Lin & Kou(1989) analyzed rectangular composite laminates with circular cutouts using a nine-node Lagrangian finite element technique, etc. However, under satisfying the accurate of solutions and performing parameter or design studies, they are costly methods. So classical analysis methods have also been developed to efficiently analyze the buckling behavior of composite panels with cutouts. For example, Nemeth (1983) approached the problem of compressive buckling of composite panels with square or circular cutouts by approximating the in-plane and out-of-plane displacements with truncated kinematically admissible series. Owen(1990) used an analysis based on Lekhnitskii's complex variable equations, boundary collocation and a Laurent series approximation to solve for the prebuckling stresses and the principle of minimum potential energy to determine the buckling loads of simple supported, shear-loaded panels with circular cutouts. Jones and Klang(1992) used this method to study the combined shear and compression loading of rectangular panels with circular cutouts. Britt(1988) expanded this method to include elliptical cutout conditions. However, the forementioned methods have a common drawback, i.e., the prebuckling stresses can't satisfy accurately the boundary condition of the cutouts, especially the prebuckling stresses obtained using Owen's method only satisfy the conditions at a discrete number of points along the interior and exterior boundaries of the panels. It will result in larger calculation errors, and even wrong conclusions if the cutout size is larger.

The objective of the present paper is to improve the Owen's method and to analyze the stability of anisotropic finite panels with elliptical cutouts and cracks. The analysis is composed of two parts: a plane stress analysis and a stability analysis. The plane stress distribution is determined using Lekhnitskii's complex variable equations of plane elastostatics combined with a Laurent series approximation constructed by the conformal mapping function, and a boundary collocation method. Its solutions satisfy the conditions along the interior boundary and at a discrete number of points along the exterior panel ones. This procedure can be applied for analysis of plane stresses of arbitrary shaped plates. The stability analysis is conducted using the different equations which result from the Hamilton's principle and the classical plate theory. The relationship of vibration frequency, load parameter and stability of panels is investigated by solving the fundamental equations using the technique of separating variables, so as to obtain the critical loads. Finally, comparisons with documented experimental results and finite element analysis are done. Results of a parameter study are presented.

2. ANALYSIS

The objective of this analysis is to determine the critical load of the dynamic stability of a finite anisotropic panel with an elliptical cutout. The rectangular anisotropic panel of same thickness, symmetry and homogeneity, has an elliptical cutout at its center and is subjected to in-plane biaxial- and shear-loads, P and T . The edge of the cutout is traction free. The panel is of length L and width W. The elliptical cutout has a major axis 2a and a minor axis 2b, of which the major axis is inclined at some angle φ to x' axis. This panel is assumed to be in a state of little transverse vibration, neglecting the effectiveness of volume forces and damping characteristics of the panel. The analysis of this model consists of two parts: a plane stress analysis and a stability analysis.

2.1 *Plane stress analysis*

This panel is assumed to be in a state of plane stress and controlled by Lekhnitskii's beharmonic equation as follows,

$$\frac{\partial^4 F}{\partial y^4} - 2\eta_{xy \cdot x}\frac{\partial^4 F}{\partial y^3 \partial x} - \left(2\gamma_{xy} - \frac{E_x}{G_{xy}}\right)\frac{\partial^4 F}{\partial x^2 \partial y^2} - 2\eta_{xy \cdot y}\frac{E_x}{E_y}\frac{\partial^4 F}{\partial y \partial x^3} + \frac{E_x}{E_y}\frac{\partial^4 F}{\partial x^4} = 0 \qquad (1)$$

where F is an Airy stress function，$E_x, E_y, G_{xy}, \gamma_{xy}, \eta_{xy \cdot x}, \eta_{xy \cdot y}$ are average material properties under the coordinate xoy，determined by the properties under the coordinate x'oy'.

This biharmonic equation can be simplified using the transformation

$$Z_k = x + \mu_k y \qquad (k=1,2) \qquad (2)$$

where μ_k and its complex conjugates are the roots of characteristic equation

$$\mu^4 - 2\eta_{xy,x}\mu^3 - \left(2\gamma_{xy} - \frac{E_x}{G_{xy}}\right)\mu^2 - 2\eta_{xy,y}\frac{E_x}{E_y}\mu + \frac{E_x}{E_y} = 0 \qquad (3)$$

The solution for the stress function F is in the form

$$F = f_1(Z_1) + f_2(Z_2) + \overline{f_1(Z_1)} + \overline{f_2(Z_2)} \qquad (4)$$

Along the panel boundary, applied traction in the x direction, X_n, and in the y direction, Y_n, are related to the stress function by

$$2\operatorname{Re}\left[\phi_1(Z_1) + \phi_2(Z_2)\right]_{Z_0}^Z = \pm\left(-\int_0^s Y_n ds\right), \qquad 2\operatorname{Re}\left[\mu_1\phi_1(Z_1) + \mu_2\phi_2(Z_2)\right]_{Z_0}^Z = \pm\int_0^s X_n ds \qquad (5)$$

Applied displacements, u and v, are related to the stress function by

$$2\operatorname{Re}\left[p_1\phi_1(Z_1) + p_2\phi_2(Z_2)\right] = u, \qquad 2\operatorname{Re}\left[q_1\phi_1(Z_1) + q_2\phi_2(Z_2)\right] = v \qquad (6)$$

where p_k and q_k are defined as

444

$$P_k = \frac{1}{E_x}\mu_k^2 - \frac{\eta_{xy,x}}{E_x}\mu_k - \frac{\upsilon_{xy}}{E_x}, \qquad q_k = \frac{\upsilon_{xy}}{E_x}\mu_k - \frac{\eta_{xy,y}}{E_y} - \frac{1}{E_y}\frac{1}{\mu_k} \qquad (k=1,2) \tag{7}$$

$\phi_k(Z_k)$ in Eq. （5） and （6） is defined as

$$\phi_k(Z_k) = \frac{\partial f_k}{\partial Z_k} \qquad (k=1,2) \tag{8}$$

S denotes the boundary arc-length， and begins at a point Z_0 and ends at Z。 the sign of positive is corresponding to the exterior boundary of panel.

If the value of $\phi_k(Z_k)$ is known for every point within the panel boundary， the stress field in the panel can be determined. In the present analysis， $\phi_k(Z_k)$ will be determined by the traction boundary condition （5）. First， it is represented by a truncated Laurent series

$$\phi_k(Z_k) = \sum_{-N}^{N} B_{kn}\sigma_k^n \tag{9}$$

where σ_k is conformal mapping function from elliptical to unit circular area，

$$\sigma_k = \frac{1}{a - i\mu_k b}\left(Z_k + \sqrt{Z_k^2 - a^2 - \mu_k^2 b^2}\right) \tag{10}$$

$B_{kn} = B'_{kn} + iB''_{kn}$， is unknown constant coefficients of 8N， decided by means of satisfying the conditions of panel's interior boundary exactly and exterior boundary at a discrete number of points. On panel's interior boundary $(x = a\cos\theta, y = b\sin\theta)$, the traction is free, i.e. $X_n=Y_n=0$, and the complex variable σ_k becomes unity， i.e. $\sigma_k = e^{i\theta}$. Satisfying the traction conditions of panel's interior boundary leads to following relationships

$$B'_{1n} + B'_{2n} + B'_{1-n} + B'_{2-n} = 0$$
$$B''_{1n} + B''_{2n} - B''_{1-n} - B''_{2-n} = 0$$
$$\alpha_1 B'_{1n} - \beta_1 B''_{1n} + \alpha_2 B'_{2n} - \beta_2 B''_{2n} + \alpha_1 B'_{1-n} - \beta_1 B''_{1-n} + \alpha_2 B'_{2-n} - \beta_2 B''_{2-n} = 0 \tag{11}$$
$$\alpha_1 B''_{1n} + \beta_1 B'_{1n} + \alpha_2 B''_{2n} + \beta_2 B'_{2n} - \alpha_1 B''_{1-n} - \beta_1 B'_{1-n} - \alpha_2 B''_{2-n} - \beta_2 B'_{2-n} = 0$$

where $\mu_k = \alpha_k + i\beta_k$， above equations may decide unknowns of 4N.

Because of not satisfying exactly the conditions of panel's interior boundary， 2N points of same distance are collected along panel's exterior boundary， i.e. $Z_i(x_i, y_i)(i=1,2,\cdots,2N)$, at which the traction conditions （5） are satisfied to yield 4N equations， which together with Eq. （11） lead to the following system of 8N algebraic equations that can be solved for the unknown Laurent series constants

$$[C_{mkn}]\{B_{kn}\} = \{F_m\} \tag{12}$$

After B_{kn} is determined， the in-plane stresses in the panel can be calculated using following stress equations,

$$\sigma_x = \frac{\partial^2 F}{\partial y^2}, \sigma_y = \frac{\partial^2 F}{\partial x^2}, \sigma_{xy} = -\frac{\partial^2 F}{\partial x \partial y} \tag{13}$$

The stress values of panel under coordinate x'oy' can be obtained.

2.2 Kinetic stability analysis

The stability analysis of transverse vibration of the panel is conducted using the Hamilton theorem and the classical plate theory. The relation of the dynamic stability of the panels, load parameter λ and vibration frequency ω is investigated using solving the fundamental equations by the technique of separating variables, so as to obtain the critical loads l_{crit}. In the coordinate system x'oy',

445

for convenience the superscript '' is omitted in following formulation, so the Hamilton formulation of the panel can be written as

$$\delta \int_{t_1}^{t_2} (U - T - W_e)\, dt = 0 \tag{14}$$

where δ is the symbol of first variation. U, T and W_e denote respectively strain energy, the kinetic energy of the panel and the exterior force work, and take the following form

$$U = \frac{1}{2}\int_S [D_{11}(\frac{\partial^2 w}{\partial x^2})^2 + 2D_{12}\frac{\partial^2 w}{\partial x^2}\frac{\partial^2 w}{\partial y^2} + D_{22}(\frac{\partial^2 w}{\partial y^2})^2 +$$
$$4D_{66}(\frac{\partial^2 w}{\partial x \partial y})^2 + 4D_{16}\frac{\partial^2 w}{\partial x^2}\frac{\partial^2 w}{\partial x \partial y} + 4D_{26}\frac{\partial^2 w}{\partial y^2}\frac{\partial^2 w}{\partial x \partial y}]ds \tag{15}$$

$$T = \frac{1}{2}\int_S \rho h (\frac{\partial w}{\partial t})^2 ds \tag{16}$$

$$W_e = \frac{1}{2}\int_S [N_x(\frac{\partial w}{\partial x})^2 + N_y(\frac{\partial w}{\partial y})^2 + 2N_{xy}(\frac{\partial w}{\partial x}\frac{\partial w}{\partial y})]ds \tag{17}$$

where D_{ij} is bending stiffness constants. w is out-of-plane displacement. r is panel density. H is panel thick. In-plane interior forces is $N_x = \lambda \sigma_x$, $N_y = \lambda \sigma_y$, $N_{xy} = \lambda \sigma_{xy}$, where $\sigma_x, \sigma_y, \sigma_{xy}$ have been determined in the plane stress analysis.

Assume $w = p(t)\phi(x, y)$ (18)

Substitute Eq.(18) into the first variation of Eq.(14), then yields,

$$\int_{t_1}^{t_2} \frac{\partial^2 p(t)}{\partial t^2} \cdot p(t)\, dt \cdot \delta T(\phi) + \int_{t_1}^{t_2} p^2(t)\, dt \cdot [\delta U(\phi) - \lambda \delta W_e(\phi)] = 0 \tag{19}$$

Let $$\frac{\delta U(\phi) - \lambda\, \delta W_e(\phi)}{\delta T(\phi)} = -\frac{\int_{t_1}^{t_2} \frac{\partial^2 p(t)}{\partial t^2} \cdot p(t)\, dt}{\int_{t_1}^{t_2} p^2(t)\, dt} = \omega \tag{20}$$

from Eq.(20) obtain the two relations,

$$\int_{t_1}^{t_2} (\frac{\partial^2 p(t)}{\partial t^2} + \omega p(t)) \cdot p(t)\, dt = 0 \tag{21}$$

$$\delta U(\phi) - \lambda\, \delta W_i(\phi) - \omega\, \delta T(\phi) = 0 \tag{22}$$

Firstly, from Eq.(21) and $p(t) \neq 0$ yields,

$$\frac{\partial^2 p(t)}{\partial t^2} + \omega\, p(t) = 0 \tag{23}$$

Eq.(23) has the solution as follows,

$$p(t) = B \sin(\sqrt{\omega}\, t) + C \cos(\sqrt{\omega}\, t) \tag{24}$$

where B and C are integration constants. From Eq.(24) know, If $\omega \in R$ and $\omega > 0$, $w = p(t) \cdot \phi(x, y)$ keeps finite value, this panel is of stability.

Secondly, the relation of load parameter and vibration frequency can be established from Eq.(22). The vibration shape w can be expressed as a truncated kinematically admissible series and take the following form,

$$\phi(x, y) = \sum_{i=1}^{N}\sum_{j=1}^{N} \phi_{ij}\, f_i(x)\, g_j(y) \tag{25}$$

For a simply supported panel, the vibration shape w is chosen to be

446

$$\phi(x,y) = \sum_{i=1}^{N}\sum_{j=1}^{N}\phi_{ij}\sin\frac{i\pi}{L}(x+\frac{L}{2})\cos\frac{j\pi}{W}(y+\frac{W}{2}) \tag{26}$$

and for a clamped panel the vibration shape w is chosen to be

$$\phi(x,y) = \sum_{i=1}^{N}\sum_{j=1}^{N}\phi_{ij}[\cos\frac{(i-1)\pi}{L}(x+\frac{L}{2}) - \cos\frac{(i+1)\pi}{L}(x+\frac{L}{2})]\times$$

$$[\cos\frac{(j-1)\pi}{W}(y+\frac{W}{2}) - \cos\frac{(j+1)\pi}{W}(y+\frac{W}{2})] \tag{27}$$

Substitute Eq.(25) into Eq.(22), obtain the following matrix equation,

$$([U_{mn}] - \lambda[W_{emn}] - \omega[T_{mn}])\{\phi_n\} = 0 \tag{28}$$

Eq.(28) constitutes an eigenvalue problem that can be solved to determine the relations of l and w. In the case of w=0, the smallest eigenvalue l_{min} is the critical load of the panel l_{crit} and the corresponding eigenvector ϕ_{crit} gives the critical mode. The eigenvalue problem, Eq.(28), was solved using the Choleskey decomposition and the Jacobi technique.

3. RESULTS AND DISCUSSION

Because of Limited space, the effects of the shape, size, orientation of the elliptical cutout and combined load on the stability of composite panels $[(\pm 30^0)_6]_s$ are only investigated in paper. The lamina properties of AS4/3502 graphite-epoxy material is, $E_1 = 18.5 \times 10^6$ psi, $E_2 = 1.6 \times 10^6$ psi, $G = 0.832 \times 10^6$ psi, $v_{12} = 0.35$. For convenient, the critical load will be described by the non-dimentioned critical load coefficient $K = N^{cr}W^2/(\pi^2\sqrt{D_{11}D_{22}})$, where w is the panel width, N^{cr} is the critical load.

A comparison between the present analysis, finite element method(FEM) and the experimental results(1988,M.P.Nemeth) is shown in table 1. The results are for a square, compression-loaded panel with a circular cutout. The compression load is applied by uniform edge displacements. The loaded edges of the panel are clamped, and the unloaded edges of the panel are simply supported. For all case, the finite element results agree well with the results from the present analysis. The experimental and present analysis' results agree well only in the case of smaller cutout size. When the cutout size becomes larger, the experimental results have great difference with the results of the present analysis and the FEM, and have the trend of moving up. The reason needs to be studied further.

Table 1.Results from Experiment, FEM & Present

d/w	d	Critical load, kN		
	(cm)	Experiment	FEM	Analysis
0.105	2.54	40.510	42.040	41.912
0.316	7.62	37.668	37.783	37.850
0.600	14.48	38.686	25.947	26.013
0.660	15.88	38.922	5.204	5.951

Table 2. Relations of φ, b & k for compression load

b/a	0^0	30^0	60^0	90^0
0.0	12.166	12.514	12.979	13.098
0.2	11.502	11.836	12.300	12.427
0.4	10.883	11.153	11.542	11.657
0.6	10.356	10.527	10.786	10.873
0.8	9.949	10.021	10.139	10.186
1.0	9.663	9.663	9.663	9.663

The relations of the orientation angle φ, minar axis b of the elliptical cutout and the critical load coefficient k for compression load are shown in Table 2. The results are for a square, compression-loaded panel with a elliptical cutout (major axis a=1.5). The loaded edges of the panel are clamped, and the unloaded edges of the panel are simply supported. In the case of b=0, the elliptical cutout degenerates into a crack. The coefficient k increases while b becomes smaller or φ becomes larger.

The relations of the orientation φ ,minar axis b of the elliptical cutout and the critical load coefficient k for shear load are shown in Table 3. Compared with Table 2., the difference is, in the case of minar axis b being smaller, the coefficient k firstly increases and then decreases as the cutout orientation angle φ becomes larger ; In the case of minar axis b being larger, the coefficient k firstly decreases and then increases as the angle φ becomes larger.

The relations of the orientation φ ,minar axis b of the elliptical cutout and the critical load coefficient k for combined compression & shear load($N_x^0 = N_{xy}^0$) are shown in Table 4.

Table 3. Relations of φ ,b & k for shear load

b/a	0^0	30^0	60^0	90^0
0.0	19.481	17.752	17.773	19.245
0.2	18.005	16.543	16.542	17.793
0.4	16.558	15.663	15.660	16.350
0.6	15.249	15.159	15.136	15.071
0.8	14.160	15.018	14.988	14.054
1.0	13.313	13.313	13.313	13.313

Table 4. Relations of φ ,b & k for combined compression & shear load($N_x^0 = N_{xy}^0$)

b/a	0^0	30^0	60^0	90^0
0.0	9.970	9.170	9.586	10.539
0.2	9.319	8.571	8.963	9.860
0.4	8.706	8.082	8.393	9.131
0.6	8.168	7.727	7.921	8.435
0.8	7.733	7.512	7.590	7.848
1.0	7.404	7.404	7.404	7.404

4. CONCLUDING REMARKS

The stability analysis of rectangular anisotropic panels with elliptical cutouts or cracks objected to shear and/or axial compression loading has been presented. The analysis consists of two parts, a plane stress analysis and a stability analysis. The stress distribution of the panel determined in a plane stress analysis can satisfy accurately the conditions along cutout boundaries and Lekhnitskii's complex variable equations of plane elastostatics. Thus this procedure can be applied for the plane stress analysis and the stability analysis of finite anisotropic panels with cracks of line-shaped. Because of limited space, this paper only investigates the effects of the shape ,size and orientation of the elliptical cutout and combined loads on stability of the composite panel $[(\pm30^0)_6]_s$, and focuses on the effect extent of circular cutouts, elliptical cutouts and cracks on the stability of the panels. Results show that, the shape ,size and orientation of the elliptical cutout and combined loads affect deeply the panel stability. The critical load for the elliptical cutout always is less than the one for the circular cutout with the diameter of major axis of the elliptical cutout, and greater than the one for the crack from the degenerated elliptical cutout. The critical load for combined loads is much lower than the one for the single load. The critical load for the line-shaped crack is of the smallest value at the angle $\varphi = 0^0$ in the case of compression loading, and at the angle $\varphi = 45^0$ in the case of combined compression and shear loading($N_x^0 = N_{xy}^0$).

REFERENCES

Britt, V. O.1988. Shear and Compression Buckling Analysis for Anisotropic Panels with Elliptical Cutouts, *AIAA Journal*, Vol.32, No.11, pp.330-336.

Jones,K.M.,and Klang,E.C.1992. Buckling Analysis of Fully Anisotropic Plates Containing Cutouts and Elastically Restrained Edges, *AIAA Paper 90-2279*.

Lin,C., and kuo, C.1989. Buckling of Laminated Plates with Holes, *Journal of Composite Materials*, Vol.23, June, pp.536-553.

Nemeth, M. P.1983. Buckling Behavior of Orthotropic Composite Plates with Centrally Located Cutouts, Ph. D Dissertation, Virginia Polytechnic Inst. and State Univ., Blacksburg, VA.

Nemeth,M.P.1988. Buckling Behavior of Compression-Loaded Symmetrically Laminated Angle-Ply Plates with Holes, *AIAA Journal*, Vol. 26, No.3, pp. 330-336.

Owen,V. L.1990. Shear Buckling of Anisotropic Plates with Centrally Located Circular Cutouts, M.S. Thesis, North Carolina State Univ., Raleigh, NC.

The Mechanics of Structures and Materials, Grzebieta, Al-Mahaidi & Wilson (eds)
© *1997 Balkema, Rotterdam, ISBN 90 5410 900 9*

Nonlinear FE analysis of RC wall panels with openings

R. Al-Mahaidi
Department of Civil Engineering, Monash University, Clayton, Vic., Australia

K. Nicholson
McConnell Dowell Constructors, Melbourne, Vic., Australia

ABSTRACT: This paper presents the case for utilising nonlinear finite elements to determine the strength, stiffness and cracking patterns of slender reinforced concrete wall panels containing openings. The panels are analysed for one-way and two-way actions. Advanced constitutive models for material behaviour coupled with a geometric nonlinearity model available in the software DIANA are used for this purpose. The FE predictions of behaviour and strength are assessed on the basis of experimental results.

1 INTRODUCTION

Door and window openings are often present in load bearing concrete wall panels. When such walls are slender, the walls become susceptible to buckling, which is normally not a design consideration in conventional concrete members. Further, the presence of openings results in excessive cracking in regions around the openings, which makes the failure characteristics of the panels unpredictable.

At present, the ultimate capacity of slender concrete walls is mostly based on empirical formulae derived from tests carried out on solid panels with standard boundary conditions. Extension of such equations to panels with openings and non-classical boundary conditions can lead to faulty and unsafe designs. There have been relatively few studies conducted on the effect that openings have on the ultimate strength. The most noteworthy work in this regard is the investigation of Saheb and Desayi (1990b) in which the test results of twelve concrete wall panels with openings were presented. The panels tested had different configurations of openings and span in one and two directions. They were loaded in plane, at an eccentricity of thickness/6.

This paper presents the results of FE analysis using the software DIANA to predict the stiffness and ultimate strength of eight slender concrete wall panels selected from the experimental investigations of Saheb and Desayi (1990b). Material as well as geometric sources of nonlinearity were taken into consideration. The capability of nonlinear FE models to predict the strength and behaviour of such panels will be demonstrated. More detailed features of FE modelling of concrete structures can be found in Al-Mahaidi (1996) and Kotsovos and Pavlovic (1995).

2 DESCRIPTION OF WALL PANELS

Four panels from the test series of Saheb and Desayi (1990b) with different opening configurations were selected for the analysis. These are representative of panels containing windows and/or doors. Each panel was analysed for one-way and two-way action. Figure 1 shows the full details of panel WWO1 and outlines of the other three. Panels named WWO1a to WWO6a indicate testing under one-way action, while panels named WWO1b to WWO6b indicate testing under two-way action. In order to ensure that the influence of the type and location of the openings was the only variable in the testing of the panels, the panel characteristics were kept as similar as possible. All the test panels had the same dimensions, 600 mm high, 900 mm long and 50 mm thick. This meant the aspect ratio and slenderness ratio were equal for all panels. The percentage of horizontal and vertical reinforcement was kept the same for all panels. Window openings were 240 mm x 240 mm and door openings were 210 mm x 420 mm.

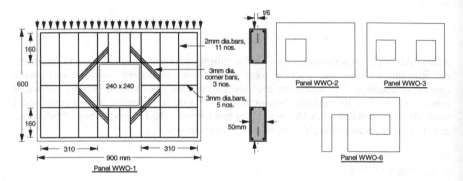

Figure 1: Wall Panels with Openings of Saheb and Desayi (1990b)

3 NONLINEAR FE ANALYSIS

3.1 Modelling of concrete and reinforcement

The primary features of concrete structures to be addressed are the composite nature of reinforced concrete and, in some cases, the interaction between the reinforcement and concrete (bond slip). A proper representation of these characteristics requires the employment of two types of elements; plane or solid elements to represent concrete and bar elements to represent the reinforcement. These are standard elements that can be found in many non-linear FE packages.

In this paper, the type of element used to model the concrete panels was a quadratic eight-node, 40 d.o.f. curved shell element. Each element has 28 integration points, 2x2 in the plane of the element and 7 layers divided evenly across the thickness. The layering across the thickness allows for the modelling of reinforcement placed at an eccentricity from the centroidal plane of the panel and for the gradual crack penetration into the thickness of the panel. A representative FE mesh for panel WWO6 is shown in Figure 2.

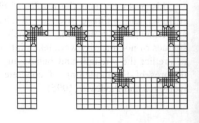

Figure 2: FE mesh for wall panel WWO6

450

A *smeared* cracking method is used to model cracking of concrete. It is based on the concept of displacement continuity across a crack but achieves a stress discontinuity state consistent with the existence of a crack. It assumes that a crack is smeared or distributed over a finite bandwidth. When the principal tensile strain in an element reaches a limiting value, the stiffness of the element is reduced suddenly or gradually to zero in the direction of the principal tensile strain. In DIANA (1996), the unloading or the "tension-softening" part of the stress-strain relation may be brittle, linear, bi-linear or non-linear. To account for the ability of cracks to transfer shear stresses by aggregate interlock, a reduced material shear stiffness, known as shear retention, is used. DIANA supports full, constant and variable shear retention relations.

This approach of crack modelling has been widely used because it enjoys the advantage of permitting automatic crack propagation with a relatively small effort. Also it offers a complete generality in regard to the cracking directions independent of the mesh configuration and the type of elements used. With this approach, it is possible to model multiple fixed cracks in any one element, an important feature in cases of cyclic loading.

Two methods of reinforcement modelling were used. For the grid like reinforcement in the panel, a thin continuous plate, with equivalent steel areas in each direction was used. This saves time, as the location of each particular bar does not have to be specified individually. The reinforcing bars at the corners of the openings were entered individually as discrete bar elements.

3.2 Loading and boundary conditions

In the experiments, the panels were loaded via a stiff steel distribution beam, which is equivalent to applying uniform displacement. For this reason it was decided to apply a uniform vertical displacement incrementally across the top of the panel, rather than a uniformly distributed load. In the experiments, the load was applied at an eccentricity of 8.33-mm (thickness/6). In order to model this in the FE model, stiff beam elements of length 8.33 mm in the lateral direction were fixed to the nodes at the top of the panels. Displacements were then applied to the end of these beams. The displacement steps were applied in 0.025-mm increments initially followed by smaller increments of 0.0125 mm up to failure. As a result of the displacements applied to the eccentric beams, the FE program produced corresponding forces that are formed at these nodes. These were summed at each load step to form the total load corresponding to a given applied displacement.

The panels were modelled in the x-y plane, with z being the direction of lateral deflection. The loaded edge was fixed from moving in the z direction. The bottom edge was supported with rollers, to allow for movement along the length of the panel, with one corner node hinged. For a symmetrical panel, only one-half of the panel was modelled with appropriate restraints applied at its lines of symmetry. For analysis under one way action, the side edges were left free. For two-way action the side edges were restrained from moving in the z direction.

4 RESULTS AND DISCUSSION

The results obtained from the FE analysis included failure loads, lateral deflections, cracking patterns and reinforcement stresses. Where available, the predicted results were compared with experimental ones. This section presents the results and appropriate comparisons, and then discusses the relevance and accuracy of the results.

4.1   Failure loads

A load versus deflection curve was plotted for each wall, which enabled the failure load to be clearly identified. Only the final failure load was reported experimentally. The FE load versus deflection curves are plotted in Figure 3 against the experimental failure loads. Table 1 lists the predicted failure loads and compares them with the experimental ones.

All of the load-deflection curves have a similar shape for each of the walls. The initial part of the graph is linear, which indicates that at this stage the walls are in their linear elastic region. Most of the curves begin to unload after reaching a long plateau near the ultimate failure load. This gives a good indication that the FE predicted failure occurred because of structural failure rather than due to numerical instability.

It can be seen form Table 1 that in all cases the predicted failure loads were larger than those found experimentally. It appears that the most accurate results were obtained for wall WWO3 (two symmetrically located windows). The result which was least accurately predicted was that of WWO2 (unsymmetrically located window). It was expected that the ultimate load of panel WWO2 would have been lower than the ultimate load of panel WWO1 (symmetrically located window). This did occur experimentally, but was not the case in the FE predictions.

In order to see the effect of one way and two-way action, the ultimate failure loads were expressed as a ratio. For each of the four opening configurations analysed, the ratio divided

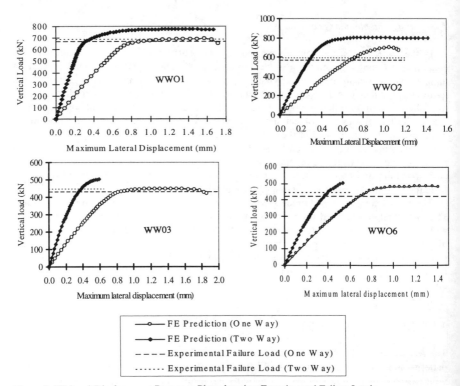

Figure 3: FE Load-Displacement Response Plotted against Experimental Failure Loads

the two-way failure load by the one-way failure load. Table 2 lists these ratios for both the predicted and the experimental results. In the table, P_a represents the ultimate failure load for one-way action, while P_b represents the ultimate failure load for two-way action. This clearly shows that the predicted and experimental failure loads increased slightly when the sides of the panel were supported against lateral movement (two-way action). Experimental results indicate that the difference in failure load between one and two-way actions for walls with openings is not nearly as much as the difference found in panels without openings tested by Saheb and Desayi (1989, 1990a). It was concluded that, under two-way action, the presence of openings considerably reduced the increases in cracking and ultimate load.

Table 1: Comparison of FE and Experimental Failure Loads

Panel	FE Failure Load (kN)	Exp Failure Load (kN)	(%) Diff
One-way action			
WWO1a	694.1	672.6	3.1
WWO2a	700.8	568.9	18.7
WWO3a	448.9	433.5	3.4
WWO6a	486.8	423.5	13.0
Two way action			
WWO1b	776.6	6925	10.8
WWO2b	803.5	592.8	26.2
WWO3b	503.4	448.4	10.9
WWO6b	504.5	448.4	11.1

Table 2: Comparison of FE and Experimental Failure Load Ratios

Panel	P_b/P_a FE	P_b/P_a Exp	% Diff.
WWO1	1.119	1.030	8
WWO2	1.147	1.042	9.2
WWO3	1.122	1.035	7.8
WWO6	1.036	1.059	-2.2

Saheb and Desayi (1990b) indicated that the lateral deflections were found to vary linearly with the applied load up to approximately 60 - 70 % of the ultimate load for most specimens. In the FE analysis the load deflection curve remained linear for 50 - 70 % of the ultimate load for the panels analysed.

4.2 Cracking patterns

The cracking patterns of the eight walls analysed can be found in Nicholson and Al-Mahaidi (1996). Presented in Figure 4 are the predicted cracking patterns for WWO3 .

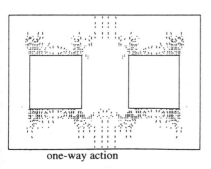

one-way action   two-way action

Figure 4: Predicted cracking patterns for panel WWO3

As would be expected, one-way action produced primarily vertical cracking, whereas two-way action produced vertical as well as diagonal cracking. Further, two-way action produced cracks near the top left and right corners of the panel. These are due to twisting moments generated by the corner restraints.

5 CONCLUSIONS

The following points summarise the conclusions drawn from this study:

- Refined finite elements coupled with advanced constitutive models for material behaviour and the geometric nonlinearity modelling available in the FE analysis package Diana have resulted in good prediction of strength and stiffness of RC slender wall panels with openings.
- In three of the panels analysed, the predicted failure loads exceeded by 3.1 to 13 % the experimental failure loads. For panel WWO2, the margin was higher, 18.7% for one-way action and 26.2% for two-way action.
- The presence of openings considerably reduces the load capacity enhancement evident in solid panels under two-way action. This was observed experimentally and predicted by the FE models.
- The smeared cracking approach was found to be adequate in tracing the crack propagation in the wall panels.
- Displacement controlled increments were found to give more reliable predictions than load controlled increments. A Newton-Raphson nonlinear solver with variable stiffness followed by increments with constant stiffness predicted the ultimate strength and stiffness with reasonable accuracy. Newton-Raphson increments with variable stiffness alone resulted in premature failure due to numerical instability.

6 REFERENCES

Al-Mahaidi, R.S. 1996, *Non-Linear Finite Element Analysis: A Research and Verification Tool for Concrete Structures* Monash Industry Geomechanics and Structures Symposium, Melbourne, pp.27-32.

DIANA Finite Element Analysis User's Manual, Release 6.1, 1996, TNO Building and Construction Research, Amsterdam.

Kotsovos, M.D. and Pavlovic, M.N. 1995, *Structural Concrete - Finite Element Analysis for Limit State Design,* London, Thomas Telford.

Nicholson, K., Al-Mahaidi, R. 1996, *Nonlinear FE Analysis of Reinforced Concrete Wall Panels with Openings,* Department of Civil Engineering, Monash University, Melbourne, July 1996.

Saheb, M. and Desayi, P. 1989, *Ultimate Strength of RC Wall Panels in One-Way In-Plane Action,* Journal of Structural Engineering, Vol 115, No. 10,.pp. 2617-2630.

Saheb, M. and Desayi, P. 1990a, *Ultimate Strength of RC Wall Panels in Two-Way In-Plane Action,* Journal of Structural Engineering, Vol 116, No. 5, May 1990, pp. 1384-1402

Saheb, M. and Desayi, P.1990b, *Ultimate Strength of RC Wall Panels with Openings,* Journal of Structural Engineering, Vol 116, No. 6, June 1990, pp. 1565-1578

The Mechanics of Structures and Materials, Grzebieta, Al-Mahaidi & Wilson (eds)
© *1997 Balkema, Rotterdam, ISBN 90 5410 900 9*

Nonlinear finite element analysis of slender HSC walls

A. Raviskanthan, R. Al-Mahaidi & J.G. Sanjayan
Department of Civil Engineering, Monash University, Clayton, Vic., Australia

ABSTRACT: This paper presents the case for utilising the nonlinear finite element method in the prediction of strength and load-displacement characteristics of slender high strength concrete walls. It reports the results of FE analysis of walls having different amount of reinforcement and acted upon by in-plane forces with variable eccentricity using the FE software DIANA. Material sources of non-linearity, such as cracking, plasticity of concrete in compression, yielding of steel as well as geometric nonlinerarities are taken into consideration. The FE predictions of stiffness and ultimate strength are assessed on the basis of experimental results.

1 INTRODUCTION

Reinforced concrete walls are widely used in construction as structural elements. Their capacities are mostly estimated from empirical formulae based on a limited number of tests. Recent advances in concrete technology have made it possible to construct slender walls. Present knowledge concerning the stability of such walls is limited. Raviskanthan et. al. (1995) showed that present codes do not provide design guidelines, which sufficiently cover the increased slenderness ratios and higher strengths of concrete used in construction. Fragemoni (1994) proposed a modified equation for the design of HSC walls, for concrete strengths up to 80 MPa. This however, does not account for the effect of increased slenderness. Thus a better understanding of wall behaviour and more sophisticated design guide lines is needed for satisfactory design of HSC walls.

Non linear finite element analysis provides a method by which structures could be analysed to progressive failure. The paper outlines an effort by the authors to model the test results of six wall panels, so that a model for a parametric study using nonlinear FEM techniques could be obtained.

2 EXPERIMENTAL STUDY

Maheswaran (1996) carried out a series of tests on walls loaded axially, supported on all four sides. The testing procedure was different to conventional tests as the loading was carried out using a series of jacks, which provided constant loading, as opposed to constant displacements obtained in conventional tests. The Walls were tested in a horizontal position and loaded eccentrically. The I-sections providing the support conditions were designed to simulate a simply supported pinned edge boundary condition. Table 1 gives details of walls used in experiments, where d_x and d_y are the reinforcement depths and A_{sx} and A_{sy} are the area of reinforcements respectively in X and Y directions.

455

Table 1 - Details of experimental study of Maheswaran (1996)

Wall	Length (mm)	Width (mm)	Thickness (mm)	Ecc. e (mm)	d_y (mm)	A_{sy} (mm²)	d_x (mm)	A_{sx} (mm²)	f'_c (MPa)
3	2000	1500	50	8	29	635	21	846	90.5
4	2000	1500	50	25	29	635	21	846	96.0
5	2000	1500	50	8	25	635	17	846	50.5
6	2000	1500	50	8	25	1270	17	846	65.0
7	2000	1500	50	25	25	1270	17	846	82.5
8	2000	1500	50	8	25	1270	17	846	83.0

3 NONLINEAR FINITE ELEMENT MODELLING

Six wall panels were chosen for finite element modelling. Only a quarter of the wall divided into 100 elements was modelled to reduce processing time, while allowing a finer mesh. Refer to Figure 1 for details. Modelling was carried out using DIANA, a finite element package developed by TNO Building and Construction Research, Delft, Netherlands

3.1 FE mesh and boundary conditions

An eight-node layered curved quadrilateral shell element (CQ40S) was used instead of a four-node element (Q20SH), to get a more refined solution. These elements allowed a geometric nonlinear analysis (total and updated Lagrange) to incorporate physical non-linearities such as plasticity and cracking among many others. These elements also allowed the inclusion of reinforcement in them as embedded elements. Reinforcement was specified as a mesh in two directions for simplicity, even though it could have been specified as discrete bars. The presence of jacks was incorporated into the wall model by modifying the boundary conditions at the top of the wall. Factors such as boundary conditions, tensile strength, Young's modulus, and cracking model were found to have an effect on the FE predictions.

The boundary conditions played a major role in modelling. It was not unusual to see an increase or decrease of 50% of the final load by varying the boundary conditions. The left edge and bottom edges were part of the centre lines and therefore had the symmetry about them. The top edge was restrained against movement where the jacks were present. Figure 1 shows the boundary conditions used for the model.

The analysis also included two other sets of boundary conditions, which attempted to simulate the influence of the loading frames by introducing beam elements at the right edge or to restrain the X rotation of the edges where loading frames were connected, to introduce a very large stiffness. Refer to Figure 2 for the effect of all three sets of boundary conditions on

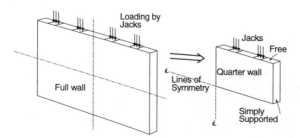

Figure 1 Loading and boundary conditions of walls

456

the model. The boundary conditions shown in Figure 1 were used for the subsequent analysis, as they suited the problem better.

3.2 Material nonlinearity

A number of researchers have studied high strength concrete and proposed equations to predict its modulus of Elasticity (E_c) of High Strength Concrete (HSC). As the behaviour of concrete is not linear it is not easy to determine the Young's modulus (or modulus of Elasticity), directly. Carrasquillo (1981) proposed the relationship:

$$E_c = [3320 (f'_c)^{0.5} + 6900] [\rho / 2320]^{1.5} \qquad (1)$$

where f'_c = Uniaxial compressive strength of concrete, and ρ = density of concrete. The validity of equation 1 was found to be $21 < f'_c < 81$ MPa for normal weight concretes and $21 < f'_c < 62$ MPa for light weight concrete. This was later adopted by the American Concrete Institute (ACI). The Australian Standard (AS 3600) uses the ACI equation.

Aitcin et.al. (1987) suggested that the Carrasquillo equation's validity be extended to include concretes with compressive strengths, up to 100 MPa. In this study the author has considered Carrasquillo's findings based on Aitcin's extension of validity of the relationship for E_c. The Norwegian Standard NS 3473 gives,

$$E_c = 9500 (f_{cm})^{0.3} \qquad (2)$$

where f_{cm} = concrete cube compressive strength in MPa. The best equations to predict the Young's modulus are those put forward by Carrasquillo et.al (1981) and the Norwegian Design Code of concrete structures. In this investigation, the equations proposed by Carrasquillo, and Norwegian standard NS 3473 were considered.

Carrasquillo et.al. (1981) showed that Poisson's ratio for light weight aggregate high strength concrete having uniaxial compressive strengths up to 73 MPa to be around 0.20 regardless of compressive strengths. Lloyd and Rangan (1993) concluded that Poisson's ratio could be taken as 0.20 for high strength concrete. The authors found that the wall model was not sensitive to Poisson's ratio and a value of 0.20 was adopted in line with recommendations from other researchers.

Tensile strength was found to be a sensitive parameter with the wall models. This could be attributed to the fact that the wall reinforcement was placed in the centre, which made it behave more like a plain concrete wall causing the tensile strength of the concrete to become an important strength factor. Tensile strength is generally expressed as a measure of direct tensile strength or splitting tensile strength or beam tensile strength (modulus of rupture).

Very little data is available on the direct tensile strength of high strength concrete. Direct tensile strength is also reported to vary from 3 MPa for grade C60 concrete to 5 MPa for grade C110 concrete by Moksnes and Jakobsen (1985). Norwegian standard NS 3473, gives the direct tensile strength (f_{kt}) as

$$f_{kt} = 0.3 [f_{ck}]^{0.5} \qquad (3)$$

where f_{ck} = cube strength of concrete.

Carrasquillo (1981) proposed that the tensile splitting strength as,

$$f_{sp} = 0.59 (f'_c)^{0.5} \qquad (4)$$

Equation 4 was later adopted by ACI. Lloyd and Rangan (1993) report that for low strength concrete the splitting tensile strength is approximately 10% - 11% of the compressive strength. The Australian code AS 3600, gives one of the most conservative of all expressions, which gives

$$f_{sp} = 0.4 \, (f'_{c})^{0.5} \tag{5}$$

In this analysis the authors considered the values recommended by AS 3600, Norwegian standard NS 3473, and Carrasquillo (later adopted by ACI).

Even though many material models are available the models proposed by von Mises for steel and Drucker-Prager for concrete were used in this analysis. While the single parameter von Mises criterion for steel needs only the yield stress, the multi parameter Drucker-Prager model for concrete needs cohesion, angle of friction and angle of dilatancy.

Three different tension-softening models were considered in this analysis, as shown in Figure 3. The linear tension-softening model has the tensile stress reducing to zero after the initiation of the crack. The second model has a non-linear relationship with the tensile stress reducing exponentially after the initiation of the crack. The third model used a relationship proposed by Hillerborg. After comparing the effects of the tension models in Figure 3, the authors selected linear tension softening, as it agreed closer to experimental results.

The constitutive strength parameters of concrete and steel had a significant effect on the wall models. In particular, the tensile strength of concrete had a significant influence on the walls as illustrated by Figure 4. Three different sets of constitutive strength parameters were used in these tests. Figure 5 shows the behaviour of wall 6, with parameters from Carrasquillo combined with ACI and AS 3600 recommendations for tensile strengths and parameters from NS 3473. It can be seen that NS 3473 gives the closest agreement of the three to experimental behaviour.

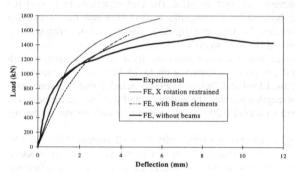

Figure 2 - Effect of boundary conditions, Wall 6

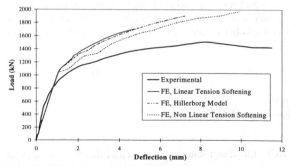

Figure 3 - Effect of tension softening models

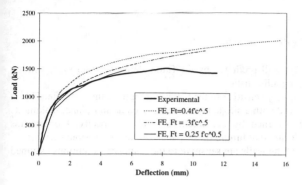

Figure 4 - Effect of tensile strength , Wall 6

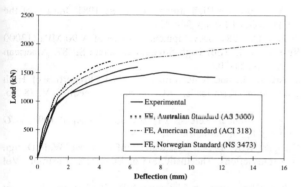

Figure 5 - Effect of code recommendations, wall 6

4 COMPARISON OF FE ANALYSIS WITH EXPERIMENTAL RESULTS

Table 2 gives a comparison of results for all six walls in this study. Most of the failure loads predicted by non-linear analysis are close to experimental values, except wall 8. The higher difference could be attributed to experimental errors, such as boundary conditions. It is concluded that the non-linear analysis could be used to obtain a good prediction of the failure loads.

Table 2 - Comparison of experimental and predicted values of failure loads

Wall	Exp. (kN)	FE (kN)	% diff.
3	1431.0	1363.5	-4.7%
4	560.0	489.6	-12.6%
5	870.0	864.0	-0.7%
6	1510.0	1597.5	5.8%
7	647.3	720.0	11.2%
8	1532.6	1957.5	27.7%

5 CONCLUSIONS

The following conclusions can be drawn from this study:

a)   The behaviour of the walls is well predicted by the finite element model incorporating geometric as well as material nonlinearities.
b)   The finite element results are very sensitive to the tensile strength of the concrete. This is because the walls are reinforced with a single layer mesh placed at mid-plane, making its ultimate moment capacity governed by the cracking moment capacity. Further, this sensitivity leads to the conclusion that failure of the panels are due to torsional effects.
c)   The behaviour and strength of the walls are sensitive to the boundary conditions assumed in the finite element models.

6 REFERENCES

ACI Committee 363, State of the Art Report on High strength Concrete 1981. Journal of the American Concrete Institute, Detroit, 81 (4), pp. 364-411.

Aitcin, P. C., Laplante, P. and Bedard, C., 1987 "Development and Use of A 90 MPa (13000 psi) Field Concrete." In Russell, H. G. (Ed) High Strength Concrete SP 87. American Concrete Institute, Detroit, 1987, pp. 51-70.

Carrasquillo, Ramon L.; Nilson, Arthur H.; and Slate, Floyd O., 1981 "Properties of High Strength Concrete Subjected to Short-Term loads," ACI Journal, Proceedings V. 78 No. 3, May-June 1981, pp. 171-178.

DIANA Finite Element Analysis User's Manual, Release 5.1, Revision A - April 1993, *TNO* Building and Construction Research, Amsterdam.

Fragomeni, S., Mendis, P.A., Grayson W.R. 1994 " Assessment of Concrete Wall Design Methods", Civil Engineering transactions, The Institution of Engineers, Australia, Vol 36, No. 1, Jan. 1994.

Lloyd, N.A. and Rangan, B.V., 1993 "High Strength Concrete: A Review" Research Report 1/93, School; of Civil Engineering, Curtin University of Technology, Perth 1993.

Maheswaran, S., 1996 "Buckling Behaviour of Two-way walls," M.Eng.Sc. Thesis, Monash University, Melbourne, 1996.

Moksnes, J. and Jakobsen, B. 1985 "High Strength Concrete Development and Potential for Platform Design." 17th Annual offshore Technology Conference, 1985, pp. 485-495.

Norwegian Standard, "Design of Concrete Structures - NS 3473"

Raviskanthan, A., Sanjayan, J. and Al-Mahaidi, R.S., 1995 "Reinforced Concrete Wall Designs Compared with Experimental Results", Fourteenth Australian Conference on the Mechanics of Structures and Materials, Hobart, Australia December 1995.

Standards Association of Australia, 1994 "Australian Standard 3600, Concrete Structures", North Sydney, 1994.

The Mechanics of Structures and Materials, Grzebieta, Al-Mahaidi & Wilson (eds)
© *1997 Balkema, Rotterdam, ISBN 90 5410 900 9*

Effect of foam-filler on the elastic post-buckling strength of a thin plate with in-plane compression

L. Sironic, N.W. Murray & R.H. Grzebieta
Department of Civil Engineering, Monash University, Melbourne, Vic., Australia

ABSTRACT: This paper presents a method for studying the effectiveness of foam support on the elastic post-buckling behaviour of an axially loaded, simply supported plate. This is seen as a first step to gain some understanding of the effect of foam on member strength when used to fill rectangular sections. The effects of varying the width to plate thickness (b/t), filler stiffness and plate imperfections are investigated. Results show there is generally little increase in strength by using low density (approx. 0 to 200kg/m^3) fillers.

INTRODUCTION

It has been suggested that foam-filled thin-walled closed sections would be efficient members capable of dissipating high levels of impact energy. Recently some vehicle manufacturers have used light-weight epoxies to fill roof pillars in car subframes to improve performance in roll-over situations. However, it was shown that for thicker profiles void filling may be inefficient (Sironic et. al. 1996). Laboratory studies in the elastic and post-elastic ranges using polyurethane foam of different densities have been carried out by Thornton (1980), Lampinen & Jeryan (1982), Reid et. al.(1986), Reddy & Wall (1988). The general conclusions from these studies were that (a) the filler did increase the failure load, (b) the elastic buckling and plastic mechanism patterns were influenced, (c) the greater the density of foam the greater the failure load, and (d) the strength of a foam filled tube was not simply the sum of the strengths of the tube and the foam. Reid et al.(1986) appears to be the only reference where a theoretical analysis has been attempted to calculate the elastic critical buckling load using light weight (flexible) fillers, but no similar attempt has been made of the post-buckling strength.

Hollow tube sections and isolated plates generally possess a considerable amount of post-buckling strength, and hence it is this capacity and not the critical elastic buckling capacity, which gives a measure of the strength of the section. This paper follows on from earlier work by the Authors (Sironic et. al. 1997) where the critical elastic buckling load for a rectangular tube with an elastic foam filler for both axial compressive and bending load cases was derived. Using the same approach, viz., a plate on an elastic foundation model, the large-deflection governing equations for an isolated plate with a lightweight elastic foam filler under axial compressive loading are derived. The equations are solved using Galerkin's method (Murray 1984).

THEORETICAL ANALYSIS

Marguerre's governing differential equations for an isolated plate (Murray 1984), subjected to an in-plane uniform compressive force, N_y (N/unit width), allowing for a fully adhesive elastic filler support are (Figure 1);

$$DV^4w - N_y\frac{\partial^2(w_o + w)}{\partial y^2} + k_f w = 0 \tag{1a}$$

$$\nabla^4\phi + Et[\frac{\partial^2 w_o}{\partial y^2}\frac{\partial^2 w}{\partial x^2} - 2\frac{\partial^2 w_o}{\partial x\partial y}\frac{\partial^2 w}{\partial x\partial y} + \frac{\partial^2 w_o}{\partial x^2}\frac{\partial^2 w}{\partial y^2} + \frac{\partial^2 w}{\partial y^2}\frac{\partial^2 w}{\partial x^2} - (\frac{\partial^2 w}{\partial y\partial x})^2] = 0 \tag{1b}$$

where w = the lateral deflection, w_o = initial imperfection, t = the plate thickness, E = steel Young's modulus, υ = Poisson's ratio, k_f= filler stiffness (N/mm^3), ϕ = a stress function, and

$$D = \frac{Et^3}{12(1-\upsilon^2)} \quad \text{and} \quad \nabla^4() = \frac{\partial^4()}{\partial x^4} + 2\frac{\partial^4()}{\partial x^2\partial y^2} + \frac{\partial^4()}{\partial y_4} \tag{1c}$$

Equations (1a) and (1b) are derived by considering equilibrium and compatibility respectively. The introduction of filler into the system only affects plate equilibrium (equation (1a)) with the introduction of an additional term, $k_f w$, which represents the out-of-plane restraint (force per unit area) afforded to the plate by the filler. These equations are solved simultaneously for the stress function, ϕ, and the deflection, w, subject to the given boundary conditions. However, except for very simple cases they have no known closed form solution. Various numerical techniques are used to solve these equations, such as the finite difference method, Galerkin's method, numerical iterative methods and the perturbation technique. Similar to previous investigators (Murray (1984), Abdel-Sayed (1969) & Dayawansa (1987)), an approximate solution of the equilibrium equation is found using Galerkin's 'minimum energy' method.

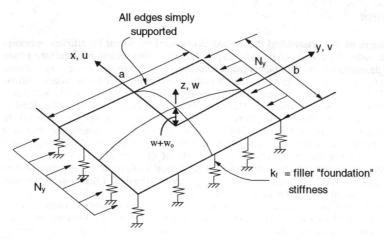

Figure 1: Simply-supported rectangular plate with initial imperfection, w_o, axial loading, N_y, supported laterally by filler of stiffness, k_f (N/mm^3).

The plate model analysed here is shown in Figure 1, where the filler is assumed to be fully adhesive and acts over the whole plate in the $\pm z$ direction. Here we consider a length of plate equal to a single half-wavelength buckle, a, being representative of an infinitely long rectangular section under compressive loading. The buckled form and initial imperfection assumed for this analysis are, respectively,

$$w = W\cos(\frac{\pi x}{b})\cos(\frac{\pi y}{a}) \quad \text{and} \quad w_o = W_o\cos(\frac{\pi x}{b})\cos(\frac{\pi y}{a}) \tag{2}$$

where b = the plate width and the critical buckle half wavelength, a, as a function of filler stiffness (Sironic et. al (1997)) is

$$a = b\left(1+\left(\frac{b}{\pi}\right)^4 \frac{k_f}{D}\right)^{-0.25} \tag{3}$$

Plate longitudinal edges are assumed to be free to move in the plane of the plate, as connecting plates of a hollow rectangular section provide no restraint against in-plane movement. In-plane restraint from the filler compressing is assumed to be negligible, therefore normal membrane stresses on these edges are zero i.e. $\sigma_{x(Memb)} = 0$ at $x = \pm b/2$.

There is, however, some question as to whether or when the waveform assumed in this paper, given by equation (2), occurs in practice. Much research (Wright (1993)) has recently been concentrated on the effect of filling hollow sections with concrete, where it is assumed that the concrete filler constrains the plate to buckle only away from the concrete, separating from the filler. Preliminary experimental work by the authors, however, indicates that a buckle mode similar to that assumed in equation (2) occurs with low density polyurethane fillers (0-200kg/m³) changing only with more rigid fillers to a mode similar to that assumed by Wright (1993). Therefore the following work relates to the use of light weight fillers only.

A solution of equations (1a) and (1b), with the assumed buckled form given by equations (2), results in the same expressions for membrane stresses similar to the case of no filler support (Abdel-Sayed (1969). The effect of the filler appears only in the term for the average load applied per unit length, n_a, increasing its value (and the value of $\sigma_{y(Memb)}$) by $k_f Wab/4$, as follows,

$$n_a = \frac{t}{b}\int_{-b/2}^{+b/2} \sigma_{y(Memb)} dx$$

$$= \frac{4}{\lambda^2 ba(W+W_o)}\left[DW[\beta^2+\lambda^2]^2\frac{ab}{4} + \frac{Etba}{64}(W+W_o)(W^2+2WW_o)[\beta^4+\lambda^4]\right.$$

$$- \frac{Et\lambda\pi}{32}(W+W_o)(W^2+2WW_o)\left\{\frac{\beta^2}{2\lambda}[(M-N)\sinh\lambda b + N\lambda b\cosh\lambda b]\right.$$

$$+ \frac{\beta^2\lambda M\sinh\lambda b}{2[\beta^2+\lambda^2]} + \frac{\beta^2\lambda^3 N\sinh\lambda b}{[\beta^2+\lambda^2]^2} + \frac{\beta^2\lambda^2 Nb\cosh\lambda b}{2[\beta^2+\lambda^2]} \tag{4a}$$

$$\left.\left.- \frac{\beta^2\lambda N[\lambda^2-\beta^2]\sinh\lambda b}{2[\beta^2+\lambda^2]^2}\right\} + \frac{k_f Wab}{4}\right]$$

where $\beta = \dfrac{\pi}{b}$; $\quad \lambda = \dfrac{\pi}{a}$; $\quad \alpha = \dfrac{\pi b}{a}$ $\tag{4b}$

$$N = -\left(\frac{a}{b}\right)^2\frac{\sinh\alpha}{\alpha+\cosh\alpha\sinh\alpha} \qquad M = \left(\frac{a}{b}\right)^2\frac{\alpha\cosh\alpha+\sinh\alpha}{\alpha+\cosh\alpha\sinh\alpha} \tag{4c}$$

The total stresses experienced by the plate, are a function of membrane and bending stresses (Murray (1984)). As soon as a plate experiences out-of-plane deflections, bending stresses occur based on the well-known relationships between moment and curvature (Timoshenko (1961)). These bending stresses, as expected, reach their maximum compression and tension values at the outer fibres of the plate thickness, making these outer fibres the location of first yield.

If a plate is perfectly flat ($w=w_o=0$ i.e. pre-buckling, no imperfections), the only stress experienced by the plate is that of uniform compression in the direction of loading, equal to n_a/t. If a plate has an imperfection, membrane and bending stresses develop as soon as load is applied.

463

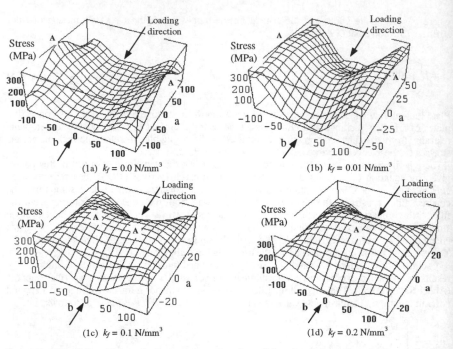

Figure 2. Von Mises Failure Surfaces at the top surface of the plate (where first yield occurs), for a 250mm wide by 1mm thick simply-supported plate loaded in compression, with $W_o/t=0.0$, and (1a) $k_f = 0$ N/mm^3, (1b) $k_f = 0.01$ N/mm^3, (1c) $k_f = 0.1$N/mm^3. and (1d) $k_f = 0.2$N/mm^3. First yield occurs at Pts A.

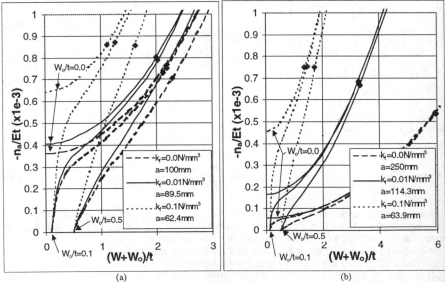

Figure 3. Graphs of average stress applied (made dimensionless by division by Et) versus central deflection for simply-supported plates, (a) 100x1mm (b) 250x1mm with various initial imperfections and filler stiffnesses. Points ($\blacklozenge$) represent initiation of von Mises yield.

464

Because of the highly non-uniform nature of these stresses, it is difficult to locate the point of maximum total stress by inspection. The von Mises failure criterion is used to find the location of the point of first yield and the value of n_a at which yielding commences (here defined as the failure load), whereby;

$$\sigma_x^2 - \sigma_x\sigma_y + \sigma_y^2 + 3\tau_{xy}^2 = \sigma_{yield}^2 \tag{6}$$

where σ_x, σ_y, and τ_{xy} = the total stresses as a sum of membrane and bending stresses in the plate resulting from uniaxial compression loading.

DISCUSSION

Figure (2) represent typical 3-D plots of the von Mises failure surfaces in the top surface (where first yield occurs) of a 250mm wide by 1mm thick simply-supported plate experiencing uniaxial compression loading in the y direction, with $k_f = 0, 0.01, 0.1$ and $0.2N/mm^3$. No similar such diagrams of stress are known to the Authors. The material yield stress, Young's modulus and Poisson's ratio were assumed to be 380MPa, 200,000MPa and 0.3 respectively.

Figures (3a & b) plot post-buckling curves for two b/t ratios, 100mm wide by 1mm thick and 250mm wide by 1mm thick plates, with varying W_o/t and k_f. Table 1 further summarises information for a 250mm wide by 1mm thick plate regarding the location of the point at which first yielding occurs and the ratio W/t and the average applied stress when yielding commences for various values of W_o/t and k_f.

In the case of a perfect plate ($W_o=0$), the effect of the filler, as noted in earlier work by the Authors (Sironic et. al. 1997), is to increase the critical buckling load, and the more slender the plate and the greater the filler stiffness the greater the level of improvement found. This is clearly seen in Figure (3) by comparing the values of the curves for $W_o/t=0$ at $(W+W_o)/t=0$. However the noted increase in critical buckling stress due to the filler does not carry through at such a level to the post-buckling failure stress. The occurrence of von Mises yield failure are noted on the post-buckling curves by points (♦), and are also tabulated in column 3 of Table 1. For the 250x1mm plate with $k_f = 0.1N/mm^3$, and $W_o/t = 0$ there is greater than 700% increase in the critical buckling stress but only a 40% increase in post-buckling failure stress. For the corresponding 100x1mm plate the percentages increases are approximately 70% and 13% respectively.

The effect of the filler on the shape of the failure surface is shown in Figure (2) and column 4 of Table 1. The location of the points of first yield move from the mid-point of the longitudinal edges back towards the centre of the plate as the filler stiffness increases. Murray (1986) showed that for a single plate (with no filler support) the location of first yield, occurring at the mid-point of the longitudinal edges or at the centre of the plate, dictates the type and geometry of the local plate mechanism which occurs, and therefore the failure stress. He found that for b/t less than about 80 a so-called roof mechanism developed and for higher ratios a flip-disc mechanism appears. Figure 2 and Table 1 show that for a 250x1mm plate the location of first yield moves from the edge (flip-disc mechanism) with no filler, to the plate centre (roof mechanism) with $k_f = 0.2N/mm^3$ (for $W_o/t=$ 0), thereby implying a change in the local plastic mechanism with filler of sufficient stiffness.

In contrast to increased filler stiffness, initial imperfections show the opposite trend and move the failure locations back towards the edges (refer to Table 1). Post-buckling failure stresses are shown to be fairly insensitive to initial imperfections, by comparing the proximity of the points of first yield (♦), plotted on Figure 3 for a given filler stiffness, for imperfections, W_o/t, ranging from 0 to 0.5.. Further to this, Table 1 shows a reduction in failure stress of the order of 1-2% for 250x1mm plates with $W_o/t=0.5$, compared with the perfect condition, for the range of filler stiffnesses tabulated.

465

Table 1: 250x1mm plate under uniaxial compression loading, with varying filler stiffness, k_f, and varying imperfection, W_o/t. (E)=Midpoint of longitudinal edges. (C)= Centre of the plate.

	Imperfection/ thickness W_o/t	Applied average failure stress (N/mm^2) n_a/t	Location of von Mises Failure (mm) (x & y coord)	Deflection/ thickness W/t
k_f = 0.0 N/mm^3 a=250mm	0.0	107.7	125, 0.0 (E)	5.9408
	0.1	107.6	125, 0.0 (E)	5.843196
	0.5	107.0	125, 0.0 (E)	5.469567
k_f = 0.01N/mm^3 a=114.3mm	0.0	137.1	125, 0.0 (E)	3.26188
	0.1	136.3	125, 0.0 (E)	3.168234
	0.5	133.5	125, 0.0 (E)	2.8233
k_f = 0.05N/mm^3 a=77.1mm	0.0	145.6	82.3, 0.0	1.944618
	0.1	145.9	85.6, 0.0	1.891435
	0.5	145.3	94.6, 0.0	1.678399
k_f = 0.1N/mm^3 a=63.9mm	0.0	150.4	67.4, 0.0	1.392055
	0.1	151.0	78.8, 0.0	1.373988
	0.5	149.7	88.9, 0.0	1.241414
k_f = 0.2N/mm^3 a=54.62mm	0.0	159.9	0.0, 0.0 (C)	0.890887
	0.1	160.6	41.5, 0.0	0.955215
	0.5	157.3	83.0, 0.0	0.932134

CONCLUSION

The effect of filler on the post-buckling curve is a function of the same factors affecting critical buckling of a single plate, viz., b/t, and k_f, as well as newly introduced initial plate imperfections, W_o/t. The presence of a filler improves the plate post-buckling strength by providing lateral restraint to the plate, as is the case for critical elastic buckling. The more slender the plate and the greater the filler stiffness, the greater the level of improvement found. However, failure loads were found to be largely insensitive to initial imperfections. It is felt that these trends can qualitatively be applied to rectangular hollow filled sections, to gauge the benefit of a light weight filler on their post-buckling failure strength. Further theoretical investigation and experimental testing are required to verify and confirm these theoretical findings.

REFERENCES

Abdel-Sayed, G. (1969). *Effective Width of Thin Plates in Compression. J. Struct. Div. ASCE. Vol. 95. Oct. p. 2183-2203.*
Dayawansa, P. H. (1987). *Elastic Post-buckling Behaviour of Isolated and Stiffened Plates in the Large Deflection Range. PhD thesis, Monash University.*
Lampinen B. H. & Jeryan R. A.(1982). *Effectiveness of Polyurethane Foam in Energy Absorbing Structures. SAE paper 820494.*
Murray, N.W. (1984). *Introduction to the Theory of Thin-Walled Structures, Oxford Engineering Science Series, Claredon Press, U.K.*
Murray, N.W. (1986). *Recent Research into the Behaviour of Thin-walled Steel Structures, Steel Structures: Recent Research Advances and their Applications to Design, ed. M.N. Pavlovic, Elsevier Applied Science Publishers.*
Reddy T.Y. & Wall R.J.(1988). *Axial Compression of Foam-filled Thin-walled Circular Tubes, Int. J. Impact Engng Vol. 7, No. 2, pp. 151-166.*
Reid S.R., Reddy T.Y. & Gray M.D.(1986). *Static and Dynamic Axial Crushing of Foam-Filled Sheet Metal tubes, Int. J. Mech. Sci. Vol. 28, No. 5, pp. 295-322.*
Sironic, E. & Grzebieta, R.H. (1996). *Should Car Pillars be Epoxy-Filled for Increased Roll-over Strength?, 15th ESV Conference, Melbourne, Australia.*
Sironic, E., Murray, N. W. & Grzebieta, R. H. (1997). *Elastic Stability of Foam-Filled Thin-Walled Rectangular Sections under Static Loading, Accepted for presentation at the 5th International Colloquium on Stability and Ductility of Steel Structures, July 1997.*
Timoshenko S. P., & Gere J. M., (1961), *Theory of Elastic Stability, McGraw-Hill, New York, N.Y.*
Thornton P.H.,(1980), *Energy Absorption of Foam Filled Structures, SAE technical paper series 800081, Congress & Exposition Cobo Hall, Detroit, Feb. 25-29.*
Wright, H. D (1993). *Buckling of plates in contact with a rigid medium., The structural Engineer, Vol. 71 No. 12, pp 209-215*

General structures and design

The Mechanics of Structures and Materials, Grzebieta, Al-Mahaidi & Wilson (eds)
© *1997 Balkema, Rotterdam, ISBN 90 5410 900 9*

Large displacement nonlinear analysis of space trusses

S.H.Xia & F.Tin-Loi
School of Civil Engineering, University of New South Wales, Sydney, N.S.W., Australia

ABSTRACT: This paper presents a method for the nonlinear analysis of large-scale space trusses in the presence of both geometric and material nonlinearities. The incremental formulation is derived from a discrete model of the structure using the fundamental relations of equilibrium, compatibility and constitutive laws. This formulation involves complementarity (orthogonality of two sign-constrained vectors) as a key mathematical structure. For ease of solution, the exact nonholonomic (path-dependent) constitutive rate problem is transformed into its finite incremental stepwise holonomic (path-independent) counterpart by adopting a backward difference approximation scheme. The resulting governing equations lead to a nonlinear complementarity problem (NCP) which exhibits symmetry as a result of the introduction of so called "nonlinear residuals". The proposed solution algorithm for the NCP is based on an iterative predictor-corrector computational scheme. An example is given to illustrate application of the approach to a fairly large size and realistic space truss problem in the presence of softening.

1 INTRODUCTION

The analysis of large space trusses in the presence of both geometric and material nonlinearities (in particular softening) is an important and difficult problem in structural mechanics. The use of limit state design principles (Supple & Collins 1981) for the design of such structures has made it necessary to carry out this type of analysis. A number of approaches (Gioncu 1995), each with its own advantages of use and accuracy, have been proposed for elastoplastic analysis of space trusses under large displacements.

In this paper, we consider the formulation and numerical solution of the space truss analysis problem in the presence of both geometric and material nonlinearities. We first present the governing equations for the exact rate problem in which irreversibility (nonholonomy hypothesis) is adopted. In view of the difficulties involved in solving the rate nonholonomic (path-dependent) problem directly, an approximate and well-known stepwise holonomic (path-independent) representation is used (DeDonato & Maier 1973) together with the artifice (DeFreitas & Lloyd Smith 1984-85) of collecting terms that destroy the symmetry of the key operators in so-called "nonlinear residuals". This finite incremental holonomic form of formulation, similar to the one DeFreitas & Ribeiro (1992) proposed, is particularly amenable for solution by an iterative predictor-corrector computational scheme that will be briefly described; DeFreitas & Ribeiro (1992) used a perturbation technique. In view of paucity of space, detailed derivations are omitted, with emphasis being placed on conceptual rather than formal aspects. The present work is computation-oriented in that theoretical considerations such as extremum characterizations, stability, existence and uniqueness of solutions are not dealt with.

A word regarding notation is in order. We do not use any special convention to distinguish between scalars, vectors and matrices, and between functional dependence and multiplication; these should be clear from the context. Vectors are assumed to be column vectors. Transpose is indicated by the superscript T, the inverse of a matrix by the superscript -1 and a superimposed dot represents a time-like derivative.

2 NONHOLONOMIC FORMULATION

Consider a space truss discretized as an aggregate of finite elements. We assume constitutive laws which reflect directly the behaviour of individual components at the element level (Corradi 1978), rather than a stress-strain relation at the material level.

Following well-known notation and description (Maier 1970) in terms of natural (unaffected by rigid body motions) generalized quantities, the general nonholonomic rate problem governing the response of the elastoplastic structure under large displacements can be compactly described as the following set of relations. Vector and matrix quantities collect the contributions of each element, namely of concatenated vectors and block diagonal matrices, respectively.

$$q = q(u) \tag{1}$$

$$F = C^T Q, \quad C = \frac{\partial q}{\partial u} \tag{2}$$

$$q = e + p \tag{3}$$

$$Q = Se \tag{4}$$

$$\dot{p} = N\dot{\lambda}, \quad N = \frac{\partial \Phi^T}{\partial Q} \tag{5}$$

$$\Phi = \Phi(Q, \lambda) = N^T Q - H\lambda - R \tag{6}$$

$$\Phi \leq 0, \quad \dot{\lambda} \geq 0, \quad \Phi^T \dot{\lambda} = 0 \tag{7}$$

Briefly, Eq. (1) represents compatibility involving a highly nonlinear dependence of strains q on the nodal displacements u. Equilibrium between the nodal load vector F and the generalized stresses Q is expressed by Eq. (2) through the compatibility matrix C.

Relations (3)-(7) establish the nonholonomic constitutive laws expressed in rate form, under the assumption of the flow theory of plasticity. In particular, total strains q are given as the sum of elastic e and plastic p components in Eq. (3). Elasticity is described in a Lagrangian form by Eq. (4), where S is a stiffness (possibly nonlinear) matrix of unassembled elements. The plastic strain rates $\dot{p}$ are defined in Eq. (5) by an associated flow rule and expressed as functions of the plastic multiplier rates $\dot{\lambda}$ through the matrix of unit outward normal vectors N to the yield surface. The generally nonlinear yield function Φ is specified in Eq. (6a) and specialized to Maier's remarkable piecewise linear representation (Maier 1970) in Eq. (6b); H is a hardening/softening matrix, R is a vector of yield limits and N is now constant.

Finally, a complementary (orthogonality) relationship between the sign-constrained vectors Φ and $\dot{\lambda}$ establishes the nonholonomic nature of plasticity in which unloading is fully allowed for. The mechanical interpretation of this condition is that: (i) if $\Phi < 0$ (no yielding) then $\dot{\lambda} = 0$ (no plastic flow) or (ii) if $\Phi = 0$ (active yield mode) then either $\dot{\lambda} > 0$ (plastic flow) or $\dot{\lambda} = 0$ (elastic unloading with $\dot{\Phi} < 0$ or neutral state with $\dot{\Phi} = 0$). It must be noted that, in view of the sign constraints on the orthogonal vectors, the complementarity condition holds componentwise.

3 STEPWISE HOLONOMIC FORMULATION

Since the nonholonomic problem (1)-(7) is difficult to solve directly, the entire structural response evolution is best approximated as a sequence of finite incremental problems, each concerning a configuration change $\Delta\Sigma$ caused by a finite load increment ΔF, from a previously known state $\hat{\Sigma}$ to a final unknown state $\Sigma = \hat{\Sigma} + \Delta\Sigma$.

The nonholonomic constitutive laws can then be simply transformed through an implicit backward difference integration scheme into a stepwise holonomic format (DeDonato & Maier 1973). The final incremental formulation then becomes:

$$\Delta q = q(\hat{u} + \Delta u) - q(\hat{u}) \tag{8}$$

$$F = \hat{F} + \Delta F = C^T(\hat{Q} + \Delta Q) \tag{9}$$

$$\Delta q = \Delta e + \Delta p \tag{10}$$

$$\Delta Q = \hat{S}^* \Delta e + \Delta R_s \tag{11}$$

$$\Delta p = N \Delta \lambda \tag{12}$$

$$\Phi = (\hat{\Phi} + \Delta\Phi) = N^T(\hat{Q} + \Delta Q) - H(\hat{\lambda} + \Delta\lambda) - R \tag{13}$$

$$\Phi \leq 0, \quad \Delta\lambda \geq 0, \quad \Phi^T \Delta\lambda = 0 \tag{14}$$

This set of relations solves the stepwise holonomic problem. It is in fact a special mathematical programming problem known as a nonlinear complementarity problem (NCP). Direct solution of this NCP, involving relations (8)-(14), is not immediately obvious in view of its complex nonlinear dependence on some variables (such as u and Q) of the formulation.

At this stage, a few points are worthy of note: (i) as is expected, any unloading can only be accommodated at the beginning of each step; (ii) $\hat{S}^*$ represents the symmetric incremental form of the stiffness matrix with ΔR_s collecting various nonlinear functions; and (iii) matrix C, of course, is not constant and needs to be calculated at the end of each step.

4 NUMERICAL ALGORITHM FOR STEPWISE HOLONOMIC PROBLEM

The stepwise holonomic problem, as expressed by Eqs. (8)-(14), is not easily amenable to numerical computation. Also, there does not appear to be a simple way of writing this problem in the form of a standard NCP for direct solution by any of the many algorithms that exist at present (Ferris & Pang 1995). We therefore propose an extension of the classical initial stress algorithm (Franchi & Genna 1984) for the numerical solution process.

Without giving the details of the straightforward manipulations involved, we can rewrite relations (8)-(14) as a so-called "mixed" nonlinear complementarity problem to facilitate the application of a simple iterative scheme. The final governing form for this problem in variables $(\Delta u, \Delta\lambda)$ is given as

$$\begin{bmatrix} \hat{K}_{uu} & \hat{K}_{u\lambda} \\ \hat{K}_{u\lambda}^T & \hat{K}_{\lambda\lambda} \end{bmatrix} \begin{Bmatrix} \Delta u \\ \Delta\lambda \end{Bmatrix} = \begin{Bmatrix} 0 \\ -\Phi \end{Bmatrix} + \begin{Bmatrix} \Delta F + \Delta R_1 \\ \hat{\Phi} + \Delta R_2 \end{Bmatrix} \tag{15}$$

$$\Phi \leq 0, \quad \Delta\lambda \geq 0, \quad \Phi^T \Delta\lambda = 0 \tag{16}$$

where ΔR_1 and ΔR_2 collect all nonlinear residuals. It is noted that (i) the symmetry of $\hat{K}$ has been preserved by the forcible introduction of nonlinear residuals; (ii) submatrix $\hat{K}_{uu}$ can be singular in the presence of instability, and (iii) the linear case can be easily recovered from Eqs. (15)-(16) by setting all residuals to zero.

A natural method of solving Eqs. (15)-(16), and one that we have used, is through an iterative scheme involving a sequence of prediction of Δu and correction of $\Delta \lambda$. In particular, we first use Eq. (15a) to estimate Δu for known (initially assumed) $\Delta \lambda$ and ΔR_1, and follow this by solving Eq. (16) for $\Delta \lambda$ which is in turn used to refine our estimate of Δu, etc.

The algorithmic steps of this predictor-corrector scheme can be briefly described as follows; superscript i denotes an iteration number.

Step 1 (Initialization)

- $i = 1$, $\Delta \lambda^{i-1} = 0$, $\Delta R_1^{i-1} = 0$, set convergence tolerance ε (e.g. 10^{-4}).
- Evaluate $\hat{K}_{uu}, \hat{K}_{u\lambda}$ and $\hat{K}_{\lambda\lambda}$.

Step 2 (Prediction)

- $\Delta u^i = \hat{K}_{uu}^{-1}(\Delta F - \hat{K}_{u\lambda}\Delta\lambda^{i-1} + \Delta R_1^{i-1})$.
- Calculate Δq^i, ΔQ^i, ..., ΔR_2^i.
- If norm of out-of-balance load $\left\| \hat{F} + \Delta F - (C^{i-1})^T Q^{i-1} \right\| \le \varepsilon \|\Delta F\|$ then stop.

Step 3 (Correction)

- Solve the linear complementarity problem (LCP):

$$\hat{K}_{u\lambda}^T \Delta u^i + \hat{K}_{\lambda\lambda}\Delta\lambda^i = -\Phi^i + \hat{\Phi} + \Delta R_2^{i-1}$$
$$\Phi^i \le 0, \quad \Delta\lambda^i \ge 0, \quad \Phi^{iT}\Delta\lambda^i = 0$$

- Calculate ΔR_1^i.
- $i = i+1$, go to Step 2.

Several points are worth noting.

(i) We have embedded an arc-length orthogonality procedure (Forde & Stiemer 1987) within the algorithm so that we can traverse critical points and trace unstable equilibrium paths. An alternative would be to use, as the path-following parameter, a quantity (e.g. some work index) known to increase monotonically during the process.

(ii) The mathematical programming problem in the corrector phase consists of small-size, uncoupled LCPs with positive semi-definite matrices and is therefore easy to solve; we recommend use of the standard Lemke's algorithm (Cottle *et al.* 1992).

(iii) While explicit evaluation of closed-form expressions for the residuals can be achieved relatively easily, it is more convenient to indirectly calculate them (Tin-Loi & Misa 1996).

(iv) The predictor steps can be speeded up by the use of a tangent stiffness but may not ensure guarantee of convergence. In our implementation we start with a tangent predictor but revert to an initial stiffness if convergence is not achieved after a preset number of iterations.

(v) If required, critical events such as unloading and activation of hinges can be captured exactly by iterating on the load step sizes.

472

5 EXAMPLE

The hypothetical example presented here is that of a space truss roof shown in Figure 1(a). This double-layer parallel grid is 50 m by 60 m in plan size and 3.5355 m high. It is restrained vertically at each top node along the perimeter, and in all directions at the four corner supports to avoid rigid body motion. All truss members are made up of the same tubular cross-section with 0.132 m inner diameter and 0.004 m wall thickness, with a material assumed to obey a bilinear constitutive law with elastic modulus of 206 kN/mm^2 and yield limit of 235 N/mm^2 in tension and compression. We assumed equal axial capacities in tension and compression for all cases so that the influence of softening alone can be compared.

The structural model consists of 960 members and 263 nodes. It was loaded by nodal vertical loads applied to the top nodes to simulate a uniformly distributed loading of 2.0μ kN/m^2. Three different softening slopes in compression were considered, viz. 0.0, 13.75 kN/mm^2 and 36.20 kN/mm^2, the numbers being arbitrarily chosen to represent different degrees of softening.

The results obtained are plotted in Figure 1(b) as graphs of load parameter μ versus the vertical deflection at the centre of the truss; solid diamonds, crosses and solid circles refer to results corresponding, respectively, to softening slopes of 0.0 (perfectly elastoplastic), 13.75 and 36.20 kN/mm^2. As expected, an increase in softening leads to a reduction in maximum load carrying capacity. For the perfectly elastoplastic case, the effect of geometry appears large enough to stabilize the structure even with the development of plasticity. For the 13.75 kN/mm^2 softening case, however, symmetric (primary) and asymmetric (bifurcation) paths can be followed. An increase in softening to 36.20 kN/mm^2 leads to a severe brittle type behaviour, possibly involving snap-back.

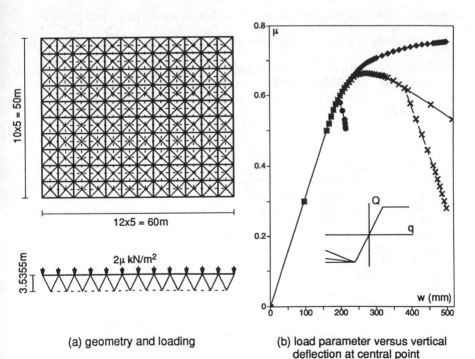

(a) geometry and loading

(b) load parameter versus vertical deflection at central point

Figure 1. Double-layer space truss example.

473

6 CONCLUSIONS

A general methodology for the elastoplastic analysis of space trusses within the large displacement regime has been presented. Starting from the space discretized nonholonomic problem in rates, we transform the elastoplastic analysis into its stepwise holonomic counterpart by an approximate backward difference scheme. The problem can then be rearranged in a form suitable for numerical computation. In particular, we propose a variant of the classical predictor-corrector algorithm to solve the resulting mathematical programming problem known as a mixed nonlinear complementarity problem. The methodology has been implemented for large-scale space trusses, and a numerical example is presented to illustrate application of the method. The numerical procedure has been found to be generally robust and efficient. However, work still needs to be carried out on devising procedures capable of detecting efficiently equilibrium branch points and on the subsequent path-following process.

ACKNOWLEDGMENTS

This work was partly supported by the Australian Research Council. We would also like to thank Professor Michael Ferris, Computer Sciences Department, University of Wisconsin (Madison) for helpful discussions regarding the solution of large-scale mixed complementarity problems.

REFERENCES

Corradi, L. 1978. On compatible finite element methods for elastic plastic analysis. *Meccanica* 13: 133-150.

Cottle, R.W., Pang, J.S. & Stone, R.E. 1992. *The Linear Complementarity Problem.* Academic Press.

DeDonato, O. & Maier, G. 1973. Finite element elastoplastic analysis by quadratic programming: the multistage method. *Proceedings, 2nd International Conference on Structural Mechanics in Reactor Technology (SMiRT).* Berlin, Vol. V, Paper M2/8, 1-12.

DeFreitas, J.A.T. & Lloyd Smith, D. 1984-85. Elastoplastic analysis of planar structures for large displacements. *Journal of Structural Mechanics* 12: 419-445.

DeFreitas, J.A.T. & Ribeiro, A.C.B.S. 1992. Large displacement elastoplastic analysis of space trusses. *Computers & Structures* 44: 1007-1016.

Ferris, M.C. & Pang, J.S. 1995. Engineering and economic applications of complementarity problems, Computer Sciences Department. University of Wisconsin, Madison, *Technical Report 95-07* (to appear in SIAM Review).

Forde, B.W.R. & Stiemer, S.F. 1987. Improved arc length orthogonality methods for nonlinear finite element analysis. *Computers & Structures* 27: 625-630.

Franchi, A. & Genna, F. 1984. Minimum principles and initial stress method in elastic-plastic analysis. *Engineering Structures* 6: 65-69.

Gioncu, V. 1995. Buckling of reticulated shells: state-of-the-art. *International Journal of Space Structures* 10: 1-46.

Maier, G. 1970. A matrix theory of piecewise linear elastoplasticity with interacting yield planes. *Meccanica* 5: 54-66.

Supple, W.J. & Collins, I. 1981. Limit state analysis of double-layer grids, in: *Analysis, Design and Construction of Double-layer Grids.* ed. Z.S. Makowski, Elsevier Applied Science Publishers, London, 93-117.

Tin-Loi, F. & Misa, J. 1996. Large displacement elastoplastic analysis of semirigid steel frames. *International Journal of Numerical Methods in Engineering* 39: 741-762.

The Mechanics of Structures and Materials, Grzebieta, Al-Mahaidi & Wilson (eds)
© *1997 Balkema, Rotterdam, ISBN 90 5410 900 9*

Methodology for assessment of serviceability of aged transmission lines

E. Martin & A. Basu
Department of Mechanical Engineering, University of Wollongong, N.S.W., Australia

G. Brennan
Integral Energy, Wollongong, N.S.W., Australia

ABSTRACT: The degradation of transmission line conductors is attributed to annealing, fatigue, corrosion, creep and in the case of an aluminium conductor steel reinforced (ACSR) construction, stress redistribution in the aluminium and steel wires. A methodology for monitoring and testing for the degradation mechanisms has been devised. Simple life expectancy models give confidence in the long term serviceability of conductors. The methodology and life expectancy models developed are based on examining and testing conductor samples that are representative of various stages of each of the degradation mechanisms. The conductor samples were removed from in service transmission lines that had over twenty years service.

1. INTRODUCTION

The transmission of power in New South Wales, Australia, is achieved by over 70,000 km of steel reinforced aluminium conductors of varying size. The bulk of these conductors were manufactured and erected between 1955 and 1970, hence making some of these conductors over 30 years old. Given time, conductors experience a variety of in service conditions that may have varied from emergency operating conditions creating elevated temperatures to long exposures of low velocity winds inducing aeolian vibration. In addition, pollution levels may vary from light to very heavy.

The objective of these studies was to ascertain the level of degradation of the conductors, the continued serviceability of the conductors and develop a methodology to test conductors in the future. The project resulted in the development of a simple conductor life expectancy model, based upon actual conductor samples, which were removed from service for testing. These samples had been subjected to pollution environments ranging from light to heavy. Testing and examinations for degradation of the removed conductor samples cover metallurgical examinations, wire tests and full scale conductor tests. The degradation mechanisms include corrosion, fatigue, creep and annealing.

2 CONDUCTOR TENSION AND SAG

The basic theory of transmission line conductor tension and sag has been well established over many years. Three methods of determining conductor, tension and sag have been examined. Each method requires an increasing understanding and knowledge of conductor behaviour. The continuous catenary method (Barrien 1975) is seen as being a useful aid in mountainous terrain where changes in reduced levels of conductor attachment points may be considerable.

The problem of determining tension and sag design for transmission line and distribution conductors is generally restricted to determining the conductor mechanical loading conditions within the bounds of the following constraints;

(1) Everyday Tension (EDT) determines a suitable level of conductor stress to limit the likelihood of long term conductor fatigue caused by aeolian vibration.

(2) The Maximum Working Tension (MWT) usually occurs at the maximum design wind and an everyday ambient temperature. In short spans the maximum working tension can be at low ambient temperatures. For elevations greater than 1200 m a combination of maximum design wind and maximum design ice mass will produce a maximum working tension. The maximum working tension ensures adequate safety factors for all line equipment, structures and foundations.

(3) The Maximum Operating Temperature (MOT) ensures that at a maximum operating temperature, the required minimum ground clearance is maintained and the loss of ultimate tensile strength of the conductor due to annealing is minimised.

(4) Stringing Constraint involves any allowances in sag for the inelastic stretch in the conductor before final termination or prestressing of the conductor to stabilise the inelastic stretch prior to final termination.

(5) The Maintenance Constraint allows for the compliance of statutory safety factors during maintenance operations.

From the above mentioned constraints, it is easy to conclude that the fundamental requirements for the mechanical design of transmission lines may be restricted to the need to determine conductor vertical sag, conductor tension, long term conductor creep behaviour and the loss of ultimate tensile strength due to annealing.

3 MECHANICAL PROPERTIES OF AGED CONDUCTORS

In examining the performance of conductors, two categories of testing were available: non destructive testing and destructive testing.

Non destructive testing will in most cases disclose the occurrence of broken wires and the presence of corrosion products in the case of the aluminium wires of an ACSR. There are several methods of non destructive testing that provide valuable information on the short term performance and the need to repair damaged conductors. Non destructive testing does not provide an indication of the long term and component performance of the conductor. Destructive testing of samples however, does provide a true measure of the deterioration of the components and the composite of the conductor. An assessment of the test results can determine the long term performance of the aged conductor.

Three categories of test were used to give an overall indication of the level of degradation of the conductor. They are: metallography examination, mechanical tests and chemical tests examination. Hence in determining the level of degradation (Brennan 1989), samples of aged transmission lines were collected and subjected to each of the above tests.

3.1 Transmission line conductor samples

As part of the transmission line conductor sag and tension determination, two samples of transmission line having similar conductors were selected giving representations of differing pollutant environments. The criteria for the various pollution environments was adopted from IEC Recommendation 815, (IEC, 1986). Characteristics of the sites chosen for conductor removal include relatively flat terrain, free from rocky outcrops that may have damaged the conductor, areas of good access and also included a section of conductor from a suspension point.

3.2 Metallographic Examination

Macro-examination consisted of an inspection and visual examination of the various layers that constitute the conductor and recording the observations. Observations revealed in cases, surface pitting, corrosion, foreign particles, closing die depressions, surface abrasion, black and white powders, weeping tars and greases, wire fretting, wire fatigue cracks, broken wires and in steel cores, a loss of zinc coating and in some cases rust. Foreign particles typically sulphates and nitrates were found on and within the voids of the various layers.

Where sites of fretting fatigue were observed under macro-examination, these were further examined using scanning electron microscopy to detect evidence of crack formation. Longitudinal sections of aluminium and steel wires were examined to quantify the depth of pit corrosion and the condition of the galvanising coating. In the conductors examined, no aluminium wire fatigue crack formation was present, nor was there evidence of pitting or loss of the zinc coating of the steel wires. In some samples, the outer layer aluminium pitting corrosion was present. In general the condition of a conductor was consistent with the age and the level of the pollution environment.

3.3 Mechanical tests

Mechanical tests of conductors fall into two categories, the component or wire tests and the composite or conductor tests.

Wire tests to the determine the tensile strength and conductivity are an elementary means of determining the level of degradation in a conductor. Mechanical Properties of a wire are assessed by determining the breaking load, ultimate tensile strength and elongation. To eliminate variations in the mechanical properties of the wire caused by variations in cross sectional area and to assess the extent of annealing, the ultimate tensile strength of wire is determined. Annealing of the wires can cause appreciable loss of tensile strength. All conductor samples used for the mechanical property tests were taken away from the suspension points. Generally samples complied with the minimum breaking load, ultimate tensile strength and elongation.

The mechanical behaviour of a conductor differs considerably from that of a wire because of the stranding factors of lay length, lay angle and diameter of lay. Individual wire tests on destranded conductors provide valuable information on the mechanical performance of the components of the conductor, however this information requires translation in order to obtain an overall conductor performance. In the case of aged transmission line conductors, actual conductor tests rather than a reliance on theoretically determined conductor data is required in order to gain a useful insight to the continued serviceability of conductors. Furthermore, wire tests do not provide information on the permanent elongation of aged transmission line conductors.

The basic data of the mechanical properties of a conductor are obtained from a stress strain test in which a prepared conductor sample is subjected to an increasing axial load to 30% of the calculated breaking load, after which the load is held constant for 30 minutes. The load is released gradually until approximately 2% of the calculated breaking load is reached. The load is then reapplied gradually until 50% of the calculated breaking load is obtained and again held constant for 30 minutes. Again it is relived until 2% of the calculated breaking load is reached and reapplied a third time to reach 70% of the calculated breaking load and held. After one hour it is gradually released to zero. The final modulus of the conductor (Nigol and Barrett) is determined from the slope of the regression line fitted to the unloading curve from the final hold point at 70% of the calculated breaking load to the slope transition point. The breaking loads for the tested conductor samples were found to be in good agreement with that of the value quoted in AS C75-1963, despite premature termination failures. Premature termination failures are defined as failures that occur outside the gauge test length.

3.4 Chemical tests

The chemical tests that are possible on aged transmission conductors, fall into three areas, material composition, grease drop point and grease mass. Material properties of a conductor are a function of the composition and structure of the material. If the properties are suspected of changing with time then either or both the metallurgical structure and the chemical composition have changed. Chemical composition of a conductor cannot change with time however, so to eliminate chemical composition as a factor in determining a property change it is necessary to carry out tests to establish a bench mark for the material.

The principal reason of including a grease or tar in a conductor is to provide corrosion resistance to the wires that constitute the conductor. The grease/tar plays an important role in preventing zinc aluminium interaction. The grease/tar drop point is determined in accordance with IP Standards (ASTM D566). It was found on some conductor samples that the grease drop points were less than the maximum design operating temperature, which would indicate a loss of grease if the conductor temperature was greater than the grease drop point for a sufficient time to allow migration of the grease from the core to the surface.

The effects of grease/tar flow results in the creation of voids where water and corrosive products will accumulate hence enable corrosion. Also electrolytic corrosion will occur as the

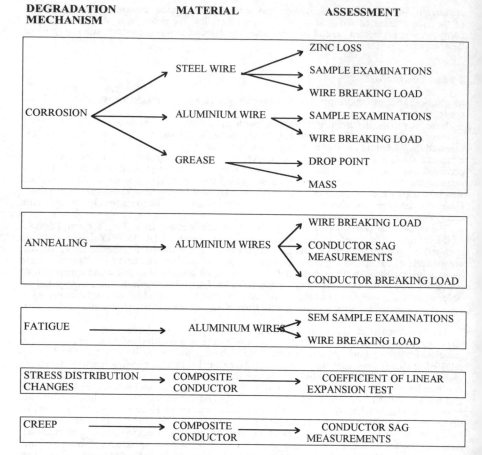

DEGRADATION MECHANISM	MATERIAL	ASSESSMENT

Figure 1. Summary of degradation mechanisms and assessment procedures for aged conductors

protective interface between the aluminium and zinc materials has been diminished and finally radio interference voltage caused by electrostatic discharge from the droplets.

3.5 Methodology for testing aged conductors

Conductor degradation with time can be attributed to one or all of five mechanisms, corrosion, fatigue, annealing, electrical and mechanical. Also in an ACSR construction, stress distribution changes can occur due to creep of the aluminium wires over the steel wire. A methodology for conductor sampling, testing and assessing the level of degradation is discussed as follows.

The level of corrosion in a conductor is a function of the pollution level of the environment and the topography of the route of the transmission line. Climate factors can also influence the extent of the corrosion in a conductor. Results suggest that negligible corrosion will occur in pollution environments ranging from light to heavy. Conductor life expectancy is predicted by a simple philosophy. If negligible degradation has occurred in the grease and zinc coating for a particular service history then provided the environment remains the same, one would expect a similar future exposure period without significant additional degradation to occur from the time of initial sampling.

Two design philosophies are used to account for material fatigue, damage tolerant and damage free. Damage free determines a critical value of stress in the material, which if not exceeded, inhibits crack initiation.

Annealing is the heating of a material generally followed by a cooling period. During this process the material experiences a change in the microstructure and for aluminium this includes a reduction in tensile strength and an increase in conductivity. Life expectancy is measured by determining the future date when statutory safety factors cannot be achieved, indicated by a linear degradation model.

High localised temperatures in conductors due to lightening strikes, conductor clashing and insulator flashovers may result in the conductor materials being vaporised, causing pitting and deposit of arc products. The effect of this degradation is the reduction of cross sectional area, localised annealing, wire breaks and in the worst case, conductor failure. Damaged sections may require total replacement.

Degradation due to a mechanical source resulting in the failure of the conductor is generally attributed to flying objects coming in contact with the conductor. The conductor is likely to sustain considerable mechanical damage, necking damage may have occurred in the affected section of the conductor. The effective cross sectional area reduction will reduce the mechanical integrity of the conductor, warranting replacement of the affected sections of the conductor.

4. SUMMARY AND CONCLUSIONS

A summary of the degradation mechanisms and assessment for aged conductors is shown (Figure 1) Routine annual inspections provide assurance of the continued serviceability and integrity of the conductor. Such inspections are useful because of the complex nature of a conductor, deterioration rates may in some circumstances be indeterminate and unpredictable. Creep and fatigue mechanisms of conductor degradation need to be treated as design parameters rather than allowing with time a level of acceptable degradation to occur. However it is necessary to examine conductor samples and line spans as a means of comparison.

When choosing the site for the subsequent removal of transmission line conductor sample, it must be decided that the location is a representative of the exposures to the extremes of pollution and frequent low velocity winds. Topography and climatic conditions will also have a major influence on the degradation of conductor and must be included in the sampling criteria.

5. REFERENCES

Barrien, J. Precise sags and tensions in multiple span transmission lines. *Elec. Eng. Trans. IE Aust, Vol EE II No. 1 1975:6-11*

Brennan, G. 1989 *Methodology for assessment of serviceability of aged transmission line conductors.* University of Wollongong M.E. Thesis.

Nigol, O. & Barrett, J.S. Characteristics of ACSR conductors at high temperatures and stresses *IEEE Trans on Pas, Vol PAS-100, No. 2, 1981:485-493*

IEC 60815 *Guide for the selection of insulators of polluted conditions.* IEC Recommendation (1986-05).

ASTM D566-64 *Dropping Point of Lubricating Grease.* IP standards for petroleum and its product part I Methods for analysis and testing section 1 IP Methods 1.

The Mechanics of Structures and Materials, Grzebieta, Al-Mahaidi & Wilson (eds)
© *1997 Balkema, Rotterdam, ISBN 90 5410 900 9*

Use of spreadsheets in plastic analysis and design

M. B. Wong & K. H. McKenry
Department of Civil Engineering, Monash University, Melbourne, Vic., Australia

ABSTRACT: The ability of spreadsheets to perform both elastic and plastic analyses of simple structures is demonstrated. The linear elastic solution is implemented as a "user-defined" function that takes an array input and produces an array output. The plastic analysis is an iterative process that detects the formation of successive plastic hinges, an elastic solution with an added hinge being required at each stage. A second development is to allow interaction of member axial load and plastic moment to be satisfied with a further iteration process. The resources of spreadsheets are demonstrated and analysis procedures optimised for spreadsheet operation are shown.

1 INTRODUCTION

The ability of a spreadsheet to store and manipulate numbers, formulae and text makes it a useful platform on for performing structural analysis, further enhanced with capacity for plotting and interaction with word-processed documents.

For structural calculations the ability to store and identify the input, the equations and the output makes checking operations efficient. The capability of an established procedure to recalculate a new set of answers for no more than a new input allows sensitivity and optimisation determinations. The response for trivial, symmetrical or other verifiable output allows ready validation of the calculation.

This paper will describe how a plastic collapse analysis involving several levels of iteration can be implemented using the "macro" features of the "EXCEL v5®" worksheet supplied by the Microsoft Corporation as part of its graphical user interface known as "Windows". Within these "macro" facilities, users may automate spreadsheet operations and provide their own functions. The spreadsheet capabilities are discussed at length in Person (1993), and Davies (1995). McKenry & Wong (1996) discuss their application to structural analysis.

2 THE STRATEGY FOR ANALYSIS OF PLASTIC COLLAPSE

To implement limit state design, a plastic analysis of the structure may be needed to assess its strength capacity. The plastic collapse load is to be calculated by an iterative procedure that starts with an initial elastic solution of the structure with applied loads of nominal magnitude. The location of the first plastic hinge to form as the loading magnitude is increased from zero is found with an associated load factor. The elastic analysis is repeated for the structure with a pin inserted at this plastic hinge location and subsequent plastic hinges and load factors are found. This is repeated until the structure becomes a mechanism, the sum of the component load factors from each iteration giving the collapse load.

Where a significant compressive axial load acts in a member, the plastic moment is reduced as given by an interaction equation such as that in the steel structures code AS4100 (1990). This effect can be accommodated by a further iteration process. An initial collapse analysis using the section plastic moments (ie no allowance for axial load) gives an estimate of the member axial loads that are used to find new plastic moments for the members. The analysis repeated using the new plastic moments until there is no significant change in the value of the collapse load.

3 THE ELASTIC ANALYSIS

The matrix method of analysis of rigid jointed frames as given in Mohr & Milner (1986) was used because of its flexibility in accepting frame geometry, loading and section properties. Here the frame is identified by nodes defining linear elements that are given specific Young's modulus (E), cross-sectional area (A) and second moment of area (I) with loading applied only at nodes. A flag for each node allows its joint to be either fixed or pinned.

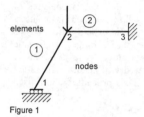

Figure 1

Figure 1 shows a simple frame of three nodes and two elements with Table 1 giving the corresponding spreadsheet input for its analysis. The identity number of an entity is set by its order in its sub-table.

The function "Pframe()" that implements the elastic frame analysis takes the numerical portion of Table 1 as the input argument. The output is set in the array of cells containing the function call as shown in Table 2, where rows of text are added for data identification.

Table 1.

	Nodes	Elements	Fixings	Loads	Sections
Numbers	3	2	5	1	1
	Young's modulus	Area	2nd m of area		
Sections	2.0e8	1.05e-2	4.77e-4		
	X coord	Y coord	Pin condition	<- "0" = rigid	"1" = pin
Coordinates	0.0	0.0	0		
	1.0	4.0	0		
	5.0	4.0	0		
	Section No	Left node of element	Right node of element		
Elements	1	1	2		
	1	2	3		
	Node No	Horz Load	Vert Load	B/Mmt	
Loads	2	0	-100.0	0	
	Node No	Fix No	<- "1" = horizontal; "2" = vertical; "3" = rotation		
Fixings	1	2			
	3	1			
	3	2			
	3	3			
	1	3			

Table 2.

X-coord	Y-coord	u	v	theta
0.0	0.0	-0.001537	0.0	0.0
1.0	4.0	0.0	-0.548447	-0.000187
5.0	4.0	0.0	0.0	0.0
1	1	-81.016	20.254	46.075
1	2	-81.016	-20.254	-7.434
2	2	0.0	-16.490	-37.434
2	3	0.0	16.490	28.527
element	node	axial	shear	b/mmt

For the plastic analysis the general elastic analysis function was modified to have separate input arrays for basic frame data and for pin condition, with the output condensed to only element number, axial load and bending moment.

4 PLASTIC ANALYSIS

Two plastic analyses will be described, firstly the case where the plastic moments are constant, then a modification of this method where plastic moment is determined by including the effect of the member axial load.

4.1 Plastic analysis with constant plastic moment

Tables 3 to 10 refer to the same spreadsheet and so cell references in any table can access cells throughout the spreadsheet. The array, named "data", describing the frame shown in Figure 1 comes from Table 1, with the constant plastic bending moment property being entered in column J of Table 3. Initially the fixed joint condition is set by the array named "p0" containing all zeros (cells B3 to B5, not shown). Table 3a shows the output of the initial elastic analysis from the array function "Pframe()", entered as shown in Table 3b.

Table 3a.

	J	K	L	M	N
1	Input plastic	Member	Joint	Axial Load	Moment M_o
2	515	1	1	-81.02	46.08
3	515	1	2	-81.02	-37.43
4	515	2	2	0.00	-37.43
5	515	2	3	0.00	28.53

Table 3b.

2	515	{=Pframe(data, p0)}	{=Pframe(data, p0)}	{=Pframe(data,p0)}	{=Pframe(data,p0)}

Table 4a.

	O	P	Q
1	Residual Plastic Moment Mp-Mi	Plastic Moment	Load Factor f = (Mp -Mi) /Mo
2	515.00	515.00	11.18
3	-515.00	-515.00	13.76
4	-515.00	-515.00	13.76
5	515.00	515.00	18.05

Table 4b.

2	=SIGN(N2)*J2	=IF(N2<0, -J2,J2)	=IF(N2<>0,O2/N2,10000)

For the first iteration values of "Residual Plastic Moment", "Plastic Moment" and "Load Factor" are shown in Table 4a, as calculated by the typical entry given in Table 4b. The moments reflect the sign of the elastic moment, and the "Load Factor" is set to a large value for zero elastic moment.

The determination of the Critical Load Factor component, and its location, is shown in Table 5a with a typical cell entries shown in Figure 5b.

Table 5a.

	R		S	T
1	Critical Load Factor	fcr	Location of next plastic hinge	Hinge reference
2	11.18		1	1
3			0	2
4			0	2
5			0	3

Table 5b.

2	=MIN(Q2:Q5)	=IF(Q2=R2,1,0)	=L2

In Table 5, column R shows a single entry, the Critical Load Factor, the minimum from column Q in Table 4. The location of this minimum is indicated by an entry of "1" based on a logic comparison operation; note the absolute reference "R2" for the Critical Load Factor. Evaluation of initial cumulative plastic moments and axial loads is shown in Table 6. The plastic hinge is located at node 1 in Figure 1.

Table 6a.

	U	V
1	Cumulative Moment Mi+1 =Mi + fcr*Mo	Cumulative axial load Ni+1
2	515	-905.55
3	-418.4176	-905.55
4	-418.4176	0.00
5	318.8627	0.00

Table 6b.

2	=N2*R2	=M8*R2

The pin condition array for the second iteration is shown in Table 7 (cellsB9 to B11), for the inserted plastic hinge.

Table 7a

	B
7	1
9	1
10	0
11	0

Table 7b.

	B
	=VLOOKUP(1,S2:T5,2,FALSE)
	=IF(B7=1,1,B3)
	=IF(B7=2,1,B4)
	=IF(B7=3,1,B5)

A further elastic analysis, similar to that shown in Table 3, is performed in columns J to N but rows 9 to 12 using the input array "data" and array B9 to B11 (Table 7)for the pin fixity. The Plasticity calculation for subsequent iterations is shown in Table 8, where computations involve values from the previous calculation. Where the current elastic moment is zero, the plastic moment takes its value from that of the previous iteration.

Table 8a.

	O	P	Q
1	Residual Plastic Moment Mp-Mi	Plastic Moment Mp	Load Factor f = (Mp - Mi)/Mo
9	0.00	515	10000.00
10	-96.58	-515	1.41
11	-96.58	-515	1.41
12	196.14	515	3.38

Table 8b.

9	=P9-U2	=IF(N9<0, -J2,IF(N9=0,P2,J2))	=IF(N9<>0,O9/N9,10000)

As shown in Table 9 the Cumulative Plastic Moments and Axial Loads are now truly cumulative, and the Critical Load Factor is found in cell R9 from cell Q10 in Table 8 in this case.

The process is repeated, assisted by the spreadsheet copying facility, until the elastic analysis indicates that a mechanism has formed.

Table 9a.

	U	V
1	Cumulative Moment Mi+1 =Mi + fcr.Mo	Cumulative plastic axial load Ni+1
9	515	-999.25
10	-515	-999.25
11	-515	0.00
12	400.75	0.00

Table 9b.

9	=U2+N9*R9	=V2+M9*R9

The collapse load factor is the sum of the critical load factors from all calculations. The resulting cumulative forces and moment in each element are given by the cumulative values of the last iteration. The collapse load factor for this analysis is 12.87.

4.2 *Plastic analysis by including the effect of element axial load*

In Australian Standard AS4100 - 1990, clause 8.4.3.4 gives the plastic moment capacity, reduced for axial force, as follows:

$$\Phi M_{PRX} = 1.18.\ \Phi M_{SX}\ (1 - N^* / \Phi N_S)\ <= \Phi M_{SX} \tag{1}$$

where ΦM_{SX} is the design section moment capacity,
    N^* is the design axial force, and
    ΦN_S is the design section capacity in compression.

This sequence builds on an existing spreadsheet solution for the plastic analysis given in section 4.1. In Table 10, columns C and D contain the plastic capacities for bending and compression. Column F is a copy of the cumulative axial load from Column V (see Table 9). Column G implements equation 1.

The values of column G containing the new reduced plastic moment capacity are copied into the input Plastic Moment (column J of Table 3) and the spreadsheet automatically performs a new plastic analysis. Repeated transfer from column G to column J results in new and old values converging, eventually satisfying equation 1. This process of data transfer and

485

a record of the intermediate Collapse Load Factors can be implemented by a procedure macro.

Table 10a.

	C	D	F	G
1	Pure Plastic Moment	Pure Squash load	Last cum. axial load	New plastic moment
2	515	-2600	-999.25	374.15
3	515	-2600	-999.25	374.15
4	515	-2600	0.00	515.00
5	515	-2600	0.00	515.00

Table 10b.

2	515	-2600	=V16	=IF(F2/D2>0.18,1.18*C2*(1-F2/D2),C2)

The final collapse load factor from the repeated analysis is 10.69.

4.3 Plastic analysis function

Because adjusting the spreadsheet for different sizes of problem needs significant formula entry and due to limited assistance from the spreadsheet copy facility for some of the relationships, it is effective to implement the logic of section 4.1 and 4.2 in a macro function of the same type as "Pframe()".

5 CONCLUSION

The versatility of the modern spreadsheet allows rapid solution of framed structures for both elastic and plastic conditions. Once the basic problem is solved various input parameters for load, section property and geometry may be altered to assess sensitivity or optimise the structure.

REFERENCES

Davies, S.R. 1995. *Spreadsheets in Structural Design* Harlow: Longman Scientific & Technical.

McKenry K. H. & M.B. Wong 1996. Structural Analysis assisted by Spreadsheets. *18th Conference on Science and Technological Development*. Hanoi: Hanoi University of Technology.

Microsoft Corporation 1993. *Microsoft Excel Version 5.0.* Redmond, WA

Mohr, G. A. & H.R. Milner H R 1986. *A microcomputer introduction to the Finite Element Method.* Melbourne: Pitman.

Person, Ron 1993. *Using Excel 5 for Windows, Special Edition.* Indianapolis: Que.

Standards Australia 1990. *AS4100-1990 Steel Structures.* Homebush: Standards Australia.

The Mechanics of Structures and Materials, Grzebieta, Al-Mahaidi & Wilson (eds)
© *1997 Balkema, Rotterdam, ISBN 90 5410 900 9*

A users survey on building structural design: Safety vs economy

C.Q.Li
Department of Civil Engineering, Monash University, Vic., Melbourne, Australia

ABSTRACT: This is a summary report on the users survey about building structural design: safety *vs* economy, which was carried out all over Australia. The survey had generated great interest among the users of Australian Building Codes and Standards with a very successful reply rate of 46%. Australia is now reaching the end of a generation of developing structural codes and standards characterised by the 'limit state' design format. It is time to start thinking about the next generation of structural design codes and standards, and laying the principles for their development in the same way that the principles for limit state format codes and standards were laid nearly twenty years ago. What are these principles? The results of the survey have shown that one of the principles could be economic optimisation in structural design. The following question is then: do we need research in this area ?

1 INTRODUCTION

It appears that Australia is now reaching the end of a generation of developing structural codes and standards characterised by the 'limit state' design format (Walker, 1994). It is time to start thinking about the next generation of structural design codes and standards, and laying the principles for their development in the same way that the principles for limit state format codes and standards were laid nearly twenty years ago. What are these principles? In other words, what do we really want in the next generation of structural codes and standards. While some aspects may be political, the technical aspects are more important as far as structural design criteria and formats are concerned. This paper only concerns with technical aspects of codes and standards development.

It may be recalled that the key element of the principles laid for the limit state design format is the introduction of probability theory so that the safety of the structures (components) to be designed can be quantitatively evaluated. By using structural reliability theory, a set of load and resistance factors can be derived for different load combinations and/or different limit states when a 'target' reliability level (called target safety index in the context of structural reliability) is selected (Melchers, 1987). There is no doubt that the limit state design format is a breakthrough from the 'old' permissible stress design format. It appears that in Australia, the limit state design format is well accepted in nearly all building structural design codes and standards.

As appreciated in the development of limit state codes, the application of structural reliability theory to code and standard work was not completely satisfactory. One limitation is that the 'target' safety index that is used to derive the load and resistance factors was determined based on the calibration of existing codes and standards so that the safety level of the newly developed codes and standards is consistent with that of the old one although consideration had been given to optimising cost and safety in a qualitative manner (including experience). It is not obvious that the target safety index selected in this way is the rational safety level that should be adopted in design codes and standards.

In fact, one of the most important advantages of reliability-based design methodology is the potential economical benefit, i.e., reliability techniques may be used in choosing the target safety index in structural design so that an optimal balance between the total cost of a structure (including initial construction cost, maintenance cost, the cost for structural failure, etc.) and the safety required for the structure can be achieved. Theoretically this is possible (Li, 1995). Although the concept of economic optimisation through design codification is not new, it has not been fully explored in the development of codes and standards. Whether it is desirable to adopt the concept in the development for the next generation of codes and standards is an interesting question It would be ideal that economic optimisation could be incorporated in design format, and designers are given a certain number of options (such as three) to select minimum requirements for structural safety according to their economic status.

Given this circumstance, the present paper would suggest that one of the principles for the development of next generation of design codes and standards could be economic optimisation in structural design. To substantiate the suggestion, a users survey on building structural design safety vs economy has been carried out. By presenting the results of the survey, the intention of the paper is to make more public awareness of the importance of the concept of economic optimisation and the recognition of its role in the development of next generation of structural design codes and standards. It is hoped the paper could serve as a base for wider discussion on whether we need research in the area of reliability-based economic optimisation.

2 SCOPE OF THE QUESTIONNAIRE

The target of the questionnaire is of course, the users of Australian building structural codes and standards. It is natural to think that the primary users are consulting structural engineers and therefore the questionnaire was sent to nearly all Australian consulting structural engineering companies. Five hundred and four of these companies (which is 76% of the total number of questionnaires) were surveyed. It should be pointed out that only one questionnaire was sent to each consulting company because it was intended that one reply would represent the opinion of the company unless there are different opinions. Also, only those who have more than three years experience in structural engineering were invited to participate.

The second major users of building codes and standards are engineers of City and Shire councils. For this purpose, all councils of Australian capital cities were selected for the survey. In all 77 councils were surveyed which constitutes 12% of the total number of questionnaires. Again it was intended one reply represents the view of one council.

To account for the individual consulting engineers, 65 copies of the questionnaire were directed to the Institution of Engineers, Australia through major regional divisions which was 10% of the total number of questionnaires.

Six Universities were selected for survey and four of them are believed to have research expertise in the area of structural reliability, i.e. Newcastle University, Monash University, Sydney University and Queensland University of Technology. As can be seen, universities only constitute 1% of the total number surveyed.

The final 15 questionnaires (2%) went to individuals of personal contacts, some of whom are builders or building product manufacturers or researchers not at universities.

As said in an attached letter with the questionnaire, only those codes and standards that are used for design of building structures are of concern in the survey. These are the Loading Code (AS 1170); the Concrete Code (AS 3600); the Masonry Code (AS 3700); the Steel Code (AS 4100); the Timber Code (AS 1720); the Cold-formed Steel Code (AS 1538) etc.

The most difficult, but nevertheless, the most important part of the survey is the compilation of questions. The scope of the questions is confined to the aspect of safety vs economy. The contents of the questionnaire were composed of three parts (see Appendix where four questions are not included due to the length limit of the paper). The first part (questions 1–6) was intended to ask very general, fundamental questions such as 'how do you feel about current codes?', 'is safety vs economy important?', 'do we need change or research?'. The result of Part 1 will provide a basic platform on which the following question can be answered, 'is it worthwhile to include safety vs economy as a key element of the principles for the development of the next generation of structural codes?'.

Part 2 of the questionnaire (questions 7–10) is a more detailed technical aspect as to how to implement or incorporate safety vs economy in codes and standards. Part 3 (questions 11–14) is basically asking the user's preference for description or expression of design format in codes and standards, such as partial factors and prescriptive codes, and other aspects, such as simplicity etc. The last two questions are not technical but for information only.

It needs to be pointed out that great care was taken to make the questionnaire as representative, as objective and as unbiased as possible, and at the same time as succinct as possible. Before the final version of the questionnaire was sent out, the draft of the questionnaire was widely discussed within the structural engineering group of CSIRO, including researchers who are key members of various code committees and who were among those that laid the principles for the development of limit state codes and standards. To some extent, the questionnaire came out of the collective thought.

It is also admitted that deficiencies existed in the survey. For example, knowing the background of responders, their expertise and seniority would certainly help to analyse the results of the survey. But this also involves tremendous work and it is most likely that the result could be too diverse to mean anything. However, one thing that could have brought better results was to identify the category of users, i.e. a designer or a council engineer etc, which would provide the knowledge of 'who like what'.

3 ANALYSIS OF RESULTS

The survey on building structural design safety vs economy had generated great responses and enthusiasm among the users. The response to the survey was overwhelming with a reply rate of 46%. Previous experience of this kind of survey suggests that this is a very high reply rate (usually 10–20%), which means the questionnaire has really caught the interest of the users and the issue of safety vs economy is a real issue in structural design. Although the identity of the responders is unknown, it is believed that the majority of the replies came from consulting practising engineers.

The overall statistical analysis of the results from the survey is shown in Figures of questions 1 to 14. Details are as follows:

Question 1.   The question is essentially asking if the current codes are conservative. It is not a surprise that 73% of users feel they are conservative. The figure should be interpreted in a way that whether there needs a change is not the issue at this question, otherwise, the answer could be different. It might be noted that 14% of the responders chose the 'other' and all of the 14% specified (more or less) 'reasonably conservative'. It is believed that this proportion is too large to ignore. Also it may be of interest to note that 4% of responders said 'unconservative'. Most of the 4% are council engineers, which is quite understandable since when they check designs they prefer more conservative codes, and as said in a letter from a council engineer 'our first consideration is always safety'.

Question 2.   It is the overwhelming opinion that safety criteria should be regulated. There should not be any doubt about it. But the intention of this question is to increase the designers' role in decision making regarding safety requirements because they are the people that are doing the work. As can be seen, 19% of responders selected 'Designers'. Again it is believed that most of the 19% are experienced practising engineers. Also only 1% of the responders said 'they don't know', so it is a clear question.

Question 3.   The accurate answer to this question is that safety is one of the most important technical aspects and although 72% of responders ticked 'yes' to this question, many wrote '+ serviceability' next to answers No. 1, 2 or 4.

Question 4.   It is clear that 79% of responders regarded it worthwhile to find the balance between safety and cost. The majority of those who responded negatively (15%) were not against the idea to find a balance and in fact, they desire the same as those who are positive

(79%), but as they pointed out in their replies, the fact is that, in reality, it is nearly impossible practically to implement it even though theoretically it is feasible. Therefore, if research is needed in this aspect the major effort should be on codification so that the outcome is useable.

Questions 5 and 6.　It is clear that the change of design codes to incorporate the philosophy of economically optimised design is not imperatively needed at present. This is consistent with the previous answers, and 34% of the responders do not want change because they believed that economic optimisation, i.e. safety vs economy in questionnaire, would further complicate the design process. But they did think relevant research in this direction should be carried out. The message from these two questions is that it is premature to change the codes and standards unless sufficient research suggests so. As can be seen, nearly 90% of all responders (including those who said 'no' to change) said 'yes' to research. All those said 'no' to research (7%) were because they said 'no' to change. Therefore, it can be concluded that relevant research in the area of economical optimisation in structural design should go ahead to meet the demands of the industry and the community.

Questions 7 to 10.　These questions are more technical details as to how to implement safety vs economy principles. It should be noted that some aspects have been included in codes and standards, but it is not uniform as a whole. The answers to these questions will help to work out details in code provisions, which can be further explored later on.

Questions 11 and 12.　These questions asked the preference for expressions of design formulae. It can be seen that the majority of code users have got used to the load and resistance design format (partial factors in the questionnaire). It is also clearly seen that quite a few people still prefer a single factor design format. It seems that most of this choice (single factor) is from users with little knowledge of limit state design principles and little practical experience.

It appears that practising engineers (74%) overwhelmingly prefer prescriptive design codes because it has been noted that most of those preferring performance-based codes (19%) were not consulting design engineers, but were council engineers, and engineers from building product manufacture companies.

Question 13.　The real story behind the bar chart is that when people ticked 'no' or 'other' they added 'as an option'. Therefore, it is reasonable to believe that if there is a method to optimise the safety and economy, a large majority of the users would prefer it to be included in the codes and standards either as a formal provision or as an option. Again, this is consistent with previous answers, that is, research needs to be done first, if it can come out with a feasible method, the codes and standards will adopt it, i.e change is then necessary.

Questions 14 to 16. These questions are not the major purpose of the survey. Question 14 shows that simplicity and classification are of concern to most responders. Questions 15 and 16 indicated that the survey has covered all categories of building structures and all structural materials, which need not be shown graphically.

4 OBSERVATIONS

Firstly, the intention of the questionnaire was not to address some of the problems with the current structural codes and standards. This had been misunderstood by some responders, particularly practising engineers. The major purpose of the survey was to get an idea of users feeling and concerns with respect to the issue of safety vs economy and hence to justify the research in this direction.

Secondly, confusion was caused due to the use of the term performance-based codes. It must be pointed out, first of all, that this survey did not in any way intend to generate the debate on what kind of code description is needed. The following comments may be helpful in clarifying the confusion.

A performance-based code is an alternative form of provision expression or description of structural design. In other words, a code can be written in a prescriptive form or a performance-based form or a combination of both. Performance of a structure (or component) can be measured qualitatively as well as quantitatively. In prescriptive codes, a design criterion has to be identified and then to be quantified and in the end to be expressed by a design formula. If the same provision is in a performance-based code, only a quantitative performance requirement will be given to a design criterion. Compared to prescriptive codes, this is relatively easier technically because a solution is not needed. Therefore, the issue of safety *vs* economy remains in the development of both codes.

There are advantages in performance-based codes, which is why it is a viable alternative to structural engineering as a profession. The most noticeable advantage, for example, is that it can breakdown the barrier between professions, or between countries where different methodologies and/or principles might be employed to derive the solution in a prescriptive form, which makes it nearly impossible to compare the design formula on the same basis. If a quantitative performance is to be compared in this case it is much easier to reach a consensus.

Thirdly, what balance is preferred? As pointed out in an attached letter by a prominent professor, 'many structural design engineers are primarily interested in minimising the time taken to produce a design rather than the total cost. This is to maximise their earnings. It is to be expected therefore, that an engineer with this focus would automatically prefer a design methodology of greater simplicity and highly conservative safety factors regardless of the cost implications for the community as a whole. There will therefore be, in general, a gulf between those practicing the profession and those non-practitioners like academics and researchers trying to improve the quality of the products of the profession.'

Finally, as noted, 'commercial influences are significant and often have detrimental influences on the design process'. But the concept of reliability-based economic optimisation will not complicate the design formula. In fact, in reliability-based economic optimisation, a set of factors are derived to be used in structural design, which is exactly the same form as current design format. Therefore, there is no 'extra' design cost for optimisation since this has been done and represented by the set of optimal load and resistance factors.

5 CONCLUSIONS

From the analysis of the results of the survey, the most important conclusions are:
- As far as structural safety is concerned, current Australian structural codes and standards are perceived to be conservative although they may be reasonably conservative.
- It is the responsibility of regulators to determine the safety levels of structures. It is desirable that such safety levels vary for different classes of structures so that designers have options to select the class of structures according to economic status.
- Safety and serviceability are the most important technical aspects in structural design.
- Research is needed in the area of economic optimisation for the development of next generation of structural design codes and standards.
- Practising engineers require the design format as simple as technically possible.

6 REFERENCES

Li, C.Q., 1995, Optimisation of Reliability-Based Structural Design, *Civil Engrg. Trans.*, **37**, (**4**), 303 - 308.
Melchers, R.E., 1987, *Structural Reliability Analysis and Prediction*, John Wiley and Sons, Chichester, U.K..
Walker, G.R., 1994, Structural Codes and Standards - The Next Generation, *Proc. Aust. Struct. Engrg. Conf.*, 21-23 Sept., Sydney, 1047 - 1052.

7 ACKNOWLEDGMENT

The survey was carried out while the author worked at CSIRO. He wishes to thank all those who generously supported his work, in particular, Dr R H Leicester, Dr J D Holmes, Dr L Pham, Dr G Foliente and Mr M Symes.

8 APPENDIX: SUMMARY OF THE RESULTS

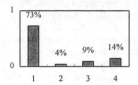

Q1. As far as safety is concerned, current Australian Codes and Standards for structural design are generally:
1 Conservative
2 Unconservative
3 Don't know
4 Other

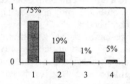

Q2. Whose responsibility should it be to determine the safety level of structures?
1 Regulators (Codes)
2 Designers
3 Don't know
4 Other

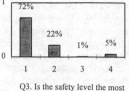

Q3. Is the safety level the most important technical aspect in structural design?
1 Yes
2 No
3 Don't know
4 Other

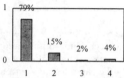

Q4. Is it worthwhile to find a balance between safety and cost, i.e., using an optimal safety index?
1 Yes
2 No
3 Don't know
4 Other

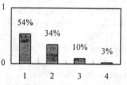

Q5. Is there a need for a change of Building Codes and Standards to provide guidance as far as safety vs cost is concerned?
1 Yes
2 No
3 Don't know
4 Other

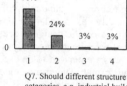

Q7. Should different structure categories, e.g. industrial buildings, office buildings, houses etc., have different safety levels (i.e. load and resistance factors)?
1 Yes
2 No
3 Don't know
4 Other

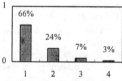

Q8. Should different design limit states, e.g. bending, shear, serviceability etc., have different safety levels?
1 Yes
2 No
3 Don't know
4 Other

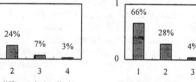

Q10. Should different structural materials, e.g. concrete, steel, timber etc., have different safety levels?
1 Yes
2 No
3 Don't know
4 Other

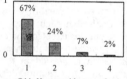

Q11. How would you prefer the safety level be incorporated in design codes and standards?
1 Partial factors
2 Single factor
3 Don't know
4 Other

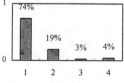

Q12. As a design engineer, what kind of codes and standards do you prefer in your work?
1 Prescriptive
2 Performance-based
3 Don't know
4 Other

Q13. If a methodology to optimise the safety and economy is developed, should it be adopted by codes and standards?
1 Yes
2 No
3 Don't know
4 Other

Q14. Are there any other aspects of structural design, in which changes are needed as far as codes and standards are concerned?
1 Yes
2 No
3 Don't know
4 Other

The Mechanics of Structures and Materials, Grzebieta, Al-Mahaidi & Wilson (eds)
© 1997 Balkema, Rotterdam, ISBN 90 5410 900 9

Structural testing of a 200 tonne mobile structure

B.W.Golley, D.J.Sharp & J.Sneddon
*School of Civil Engineering, University College, Australian Defence Force Academy, Canberra,
A.C.T., Australia*

B.B.Hughes
Facilities Branch, Defence Centre Perth, Freemantle, W.A., Australia

ABSTRACT: The Australian Defence Force operates a non-flying mock-up wide bodied aircraft, similar in appearance to a short-bodied Boeing 747. After a short period
of use, the mock-up suffered a major structural failure of the rear undercarriage assemblies. Extensive testing was conducted to determine tyre stiffness parameters required to
redesign the supporting structure. Following the completion of extensive modifications,
strain gauges were placed on large box sections, and forces transmitted from the bogies
to the superstructure were measured during a series of trials. In this paper, the strain
gauging system is described, and the method for obtaining forces, moments and torques
is discussed. Typical results obtained during towing around curved paths are presented.

1 INTRODUCTION

As part of its training facilities, the Australian Defence Force operates a non-flying mock-
up wide bodied aircraft. A functional requirement is that the mock-up must be able to be
towed to any position on the paved areas of the airfield where it is located. A photograph
of the 200 tonne structure, which is similar in size and appearance to a short-bodied
Boeing 747, is shown in Figure 1. The mock-up is supported on a tricycle undercarriage,
with a steerable front bogie.

After a short period of use, the mock-up suffered major structural failure of the rear
undercarriage assemblies. The probable causes of the failure were determined by the
authors, and have been described in some detail elsewhere (Golley et al, 1996), and are
summarised here. Photographs taken after the failure showed tyre marks on the pavement,
from which it was concluded that there was tow-in, or misalignment, of the undercarriage
of about 5° just prior to the failure. The tow-in produced large lateral forces, which were
sufficient to cause bending failure of 150 mm diameter steel stub columns connecting the
undercarriages to the superstructure. The large tow-in was attributable to two causes.
Firstly, large torques about vertical axes are generated between the undercarriages and the
superstructure when the mock-up moves on a curved path. Owing to the flexibility of the
superstructure, these torques caused elastic rotation to occur. Secondly, a tubular member
providing resistance to this rotation was partially detached at an important connection,
adding to the flexibility of the superstructure and hence increasing the rotation.

Prior to redesigning the supporting system it was necessary to obtain stiffness parameters for the tyres. Both rear bogies were supported by eight large earth compactor tyres,
and the front bogie was supported by four of the same tyres. The tyres were all mounted
in pairs separated by a 100 mm gap, with each pair on a common hub. Mounting the tyres
in this way was necessary because of the braking system used, but the system results in
larger torques than if the tyres were on separate hubs. Relevant stiffness parameters for
the tyres, which are 1320 mm diameter, and 400 mm wide, were not available from the

manufacturer. The parameters were determined in a series of laboratory tests conducted in which a wheel/tyre assembly was subjected to vertical forces of 61 kN and 92 kN and the contact patch was subjected to lateral, rotational and circumferential deflections. The tests, and test results have been described by Sneddon (1993,1994a,1994b).

Figure 1. The mock-up during reconstruction.

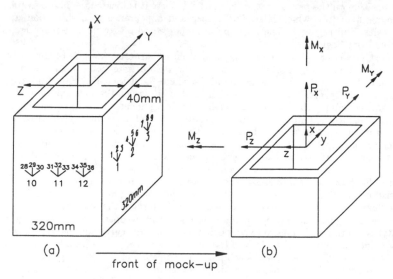

Figure 2. Rosette locations, action notation and coordinate system.

Following a detailed study of re-design options, it was decided to retain the existing type of undercarriage. However, the mock-up body would be supported on a shallow trailer, which would be both strong enough to resist large torques and lateral forces and stiff enough to prevent distortion of the cladding and fittings in the mock-up. The re-designed supporting system was completed in June, 1996.

It was decided to monitor the new system to ensure it would operate safely. Strain gauges were installed, and actions arising during towing operations were measured during a series of trials in July, 1996. As part of a more detailed study, two Global Positioning System (GPS) receivers were mounted on the mock-up, working in differential mode with a third fixed station near the test site. The GPS readings were accurate to within 30 mm, and readings will be used in future to further assess the relationships between the various actions and path curvatures and to assess other factors such as inertia effects. Only the strain gauging system, and the method used for determining actions from the strains are discussed in this paper.

2 ACTION DETERMINATION

The new support structure is a shallow box truss, which is attached to the front and rear bogies using square box sections with outer dimensions 320×320 mm, and with 40 mm wall thickness. These box sections provided an ideal location for determining the six stress resultants transmitted from the bogies to the superstructure. Twelve 45° strain gauge rosettes were glued to each box section, from which 12 longitudinal strains, 12 transverse strains and 12 shear strains could be determined. A typical configuration is shown in Figure 2(a) which also shows the rosette and gauge numbering. The rosettes were positioned at the centres of each face of the box section, and 100 mm on each side of the centre. The coordinate system and notation for forces, moments and torques (hereafter referred to collectively as actions) at the gauged section are shown in Figure 2(b). Moments follow the right hand rule. During the field tests, each gauge was read in 2.5 milliseconds, and hence the twelve rosettes were scanned in 90 milliseconds.

Actions were obtained using the procedure described below, and were calculated in real time and displayed during the towing operations to ensure design values were not exceeded. To determine actions from the strain measurements, an approximate theoretical relationship between strains and actions was obtained using the following assumptions:

(a) Normal strain and the actions P_x, M_y and M_z: The relationships were obtained using engineering beam theory, that is plane sections remain plane. Transverse stresses were assumed to be zero.

(b) Shear strain and the torsion M_x: St Venant's theory of torsion was used with Prandtl's stress function being determined by finite element analysis.

(c) Shear strain and the shear forces P_y and P_z: these relationships were determined using thin-walled beam theory.

Using these approximate theoretical relationships, the effects of normal and shear strains uncouple. Accordingly, P_x, M_y and M_z, were obtained from the twelve longitudinal strains and P_y, P_z and M_x were obtained from the twelve shear strains. The actions were determined using the least squares procedure, which is equivalent in this case to a weighted average approach. The theoretical relationships are given for longitudinal strain effects to show the form of solution. The matrix coefficients are given in newton and millimetre dimensions. In the formulas, ϵ_i is the longitudinal strain measured at rosette i, using the numbering system in Figure 2(a). Coefficients used to determine actions due to shear strains are not presented here. Using engineering beam theory, the relationship between longitudinal strains and actions is

$$\begin{bmatrix} \epsilon_1 \\ \epsilon_2 \\ \epsilon_3 \\ \epsilon_4 \\ \epsilon_5 \\ \epsilon_6 \\ \epsilon_7 \\ \epsilon_8 \\ \epsilon_9 \\ \epsilon_{10} \\ \epsilon_{11} \\ \epsilon_{12} \end{bmatrix} = 10^{-12} \begin{bmatrix} 111.6 & -1.339 & 0.837 \\ 111.6 & -1.339 & 0.000 \\ 111.6 & -1.339 & -0.837 \\ 111.6 & -0.837 & -1.339 \\ 111.6 & 0.000 & -1.339 \\ 111.6 & 0.837 & -1.339 \\ 111.6 & 1.339 & -0.837 \\ 111.6 & 1.339 & 0.000 \\ 111.6 & 1.339 & 0.837 \\ 111.6 & 0.837 & 1.339 \\ 111.6 & 0.000 & 1.339 \\ 111.6 & -1.339 & 1.339 \end{bmatrix} \begin{bmatrix} P_x \\ M_y \\ M_z \end{bmatrix} \tag{1a}$$

or

$$\mathbf{B}_1 = \mathbf{A}_1 \mathbf{X}_1 \tag{1b}$$

The actions $\mathbf{X}_1$ were then obtained from the measured normal strains $\mathbf{B}_1$ using the classical least squares equation, ie

$$\mathbf{X}_1 = (\mathbf{A}_1^T \mathbf{A}_1)^{-1} \mathbf{A}_1^T \mathbf{B}_1 = \mathbf{C}_1 \mathbf{B}_1 \tag{2}$$

Carrying out the matrix operations gives

$$\mathbf{C}_1 = 10^9 \begin{bmatrix} 0.7467 & -98.75 & 61.73 \\ 0.7467 & -98.75 & 0.00 \\ 0.7467 & -98.75 & -61.73 \\ 0.7467 & -61.73 & -98.75 \\ 0.7467 & 0.00 & -98.75 \\ 0.7467 & 61.73 & -98.75 \\ 0.7467 & 98.75 & -61.73 \\ 0.7467 & 98.75 & 0.00 \\ 0.7467 & 98.75 & 61.73 \\ 0.7467 & 61.73 & 98.75 \\ 0.7467 & 0.00 & 98.75 \\ 0.7467 & -61.73 & 98.75 \end{bmatrix}^T \tag{3}$$

These formulas are presented in full, as they show that the least squares procedure in this case is equivalent to determining the actions using weighted sums. Perhaps not surprisingly, the normal force P_x is obtained from the average of the longitudinal strains, as the terms of the first row of $\mathbf{C}_1$ are the same. The terms of the second row of $\mathbf{C}_1$ may be visualised as weights, and hence the least squares method gives identical results for M_y as would be obtained from a weighted sum. The "weights" are seen to vary linearly with distance from the Y axis of the beam, as do bending strains under that moment.

A three metre tow arm is attached to the front bogie below a slew ring to enable the mock-up to be steered. Towing was effected by a pin passing through a ring on the end of the tow arm, and hence there were only three significant stress resultants in the tow arm, which could be estimated by determining the axial force and a bending moment at a section. The tow arm is a box section, which was instrumented with six strain gauges measuring longitudinal strains only from which these actions were determined assuming simple beam theory applied.

3 FIELD TESTS

The mock-up was towed approximately three kilometres from its dedicated parking area to a large apron for testing. As the major forces on the undercarriage are transmitted during motion around curves, a series of tests were conducted in which the mock-up was towed around circular curves. To this end, circles of 50, 45, 40 and 35 m radius were marked on the apron, and these curves were followed by the towing vehicle, commencing at the 50 m radius. During each test at a particular radius, the towing vehicle completed two circuits before moving to the next smaller radius. The following factors are noted:

(a) Towing stopped after each half circuit to determine if any recovery occurred following stopping.

(b) The centreline of the mock-up at the rear bogies tracked several metres inside that being followed by the towing vehicle, for example when the towing vehicle was at 35 m radius the mock-up centreline was 7 m inside at 28 m radius.

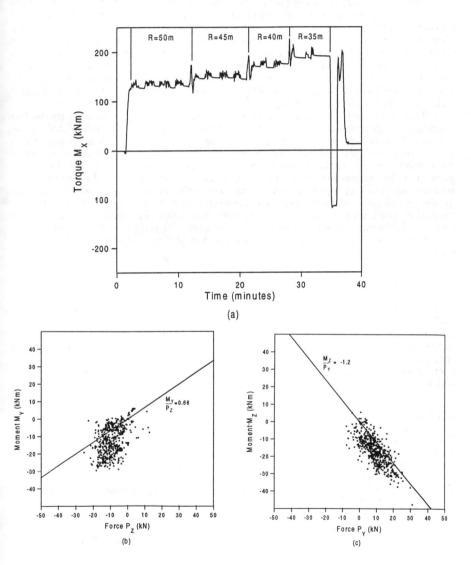

Figure 3. Typical test results.

497

Data logging commenced while the mock-up was travelling in a straight line, and ended generally when the mock-up was taken to a parking position, again in straight line mode. Actions from all bogies and the tow arm were evaluated in real time, and displayed on a monitor located on the mock-up to enable comparison with design values.

4 FIELD TEST RESULTS

Representative results of actions on a rear bogie during a typical test are presented in this section. The torsional moment, M_x, is shown plotted against time in Figure 3(a), a scatter plot of the moment M_y against P_z is shown in Figure 3(b), and a scatter plot of the moment M_z against P_y is shown in Figure 3(c). The torsional moment is seen to have increased as the radius of curvature decreased, as expected. The essentially horizontal portions of the graphs correspond to the stationary positions every half circle, and indicate slight relaxation of the torque while stopped. During motion, the torque fluctuated, as a consequence of tyre slipping. The final portion of the graph corresponds to parking, and the small residual torque is because the mock-up was not perfectly straight in the parking bay. It should be noted here that the maximum torque measured during the tests (304 kNm) was less than the design value (393 kNm), and did not occur during the controlled radius tests but during manoeuvring.

The scatter plots were drawn to check the analytical model. Assuming frictionless bearings and ignoring inertia effects, forces and moments in the bogies are related by the distances from the appropriate bearing to the gauged section, being 0.66 m for the actions in the first scatter plot and 1.2 m in the second. These "theoretical" lines are shown in Figures 4(b) and (c). In the first case, there is only slight agreement between the scatter points and the "theoretical" line, and it was concluded that two sets of bearings, which were subjected to very high forces and torques were stiff, and required careful monitoring. In the second case, although there is considerable scatter, the agreement is generally reasonable given a small zero offset. Although not shown here, measurements of axial forces P_x in the rear bogies were sufficiently accurate to detect the 0.8% crossfall on the test apron, which caused differences in axial force as the mock-up reversed direction on the apron.

5 CONCLUSIONS

Only brief details of the results of field tests have been shown in this paper. Detailed studies of the results of all tests showed that the actions transmitted through the bogies to the superstructure during turning at the minimum design radius were below design values suggested as a result of laboratory testing of a tyre. No design values were exceeded during the tests. Cross checking results through a range of procedures such as scatter plots showed that the theoretical model used in conjunction with the least squares method enabled undercarriage forces to be determined with sufficient accuracy.

REFERENCES

Golley, B.W., B.B. Hughes, P. Reynolds, J. Sneddon, J.E. Tick &D.M. Devenish 1996. Undercarriage repair and failure of a mock-up wide-bodied aircraft. In F. Escrig &C. A. Brebbia(eds), *Mobile and Rapidly Assembled Structures II, Seville, June 1996*: 293–304. Southampton: Computational Mechanics Publications.
Sneddon, J. 1993. *Mock-up Tyre Tests-Preliminary Report*. Sydney: Unisearch Ltd, University of NSW.
Sneddon, J. 1994a. *Mock-up Tyre Tests-Preliminary Report 2*. Sydney: Unisearch Ltd, University of NSW. Unisearch Ltd, University of NSW.
Sneddon, J. 1994b. *Mock-up Tyre Tests-Additional Cicumferential Test Results and Some Further Observations*. Sydney: Unisearch Ltd, University of NSW.

The Mechanics of Structures and Materials, Grzebieta, Al-Mahaidi & Wilson (eds)
© *1997 Balkema, Rotterdam, ISBN 90 5410 900 9*

Static analysis of framed structures founded on elastic media

R. Izadnegahdar & E. S. Melerski
Department of Civil and Mechanical Engineering, University of Tasmania, Hobart, Tas., Australia

ABSTRACT: A displacement-type numerical technique for the elastic solution to problem of the interaction between a bar structure and the supporting soil medium is outlined. The behaviour of soil is modelled by means of homogenous, isotropic, elastic half-space, whilst in the structural analysis of the superstructure the very well-known Direct Stiffness Method is used. The Boussinesq and Cerutti force-displacement formulae are utilised to develop the flexibility matrix for the elastic half-space, related to the interaction forces. The combination of the inverse of this matrix with the stiffness matrix of the superstructure, constitutes the stiffness matrix of the whole system. An example is included to illustrate the obtained results.

1 INTRODUCTION

In civil engineering practice, it is widely recognised that the response of a structure to the applied load is also very strongly dependent upon behaviour of the soil underneath. Due to its nature, the soil behaves in a very complex way in response to loads. A number of soil models have been proposed over the years of research on the matter that exhibit various levels of complexity and, of course, distinct levels of ability to describe the behaviour of the real medium. One of these, which is relatively simple and yet quite adequate for many situations, is the concept of homogenous, isotropic, elastic half-space. A rigorous theoretical solution to the case of a point load normal to, and acting at, the surface of such a medium was obtained by Boussinesq (Poulos & Davis 1974). Cerutti (Poulos & Davis 1974), on the other hand, obtained eqations for dispalcemnets within the elastic half-space caused by a horizontal concentrated force applied at the surface. In the present work, the Boussinesq and Cerutti solutions are taken as the basis for the development of the soil flexibility matrix related to forces of interaction between the foundation bed and a two-dimensional bar structure (such as a frame or a truss).

The inverse of this flexibility matrix constitutes the relevant stiffness matrix of the soil, which is appropriately combined with the stiffness matrix of the superstructure. The resulting stiffness matrix of the statical system considered is then used to formulate the stiffness equations governing the problem. The stiffness matrix of the superstructure used in this method is derived from the Direct Stiffness Method.

A computer programme has been developed that implements the proposed approach. The main feature of this programme is that it requires minimal data to produce the solution to the problem. A number of example problems have been solved to test the proposed method and the computer programme implementing it. Here, only one example is shown for illustration. However, the results of all these tests have attested to the adequacy of the approach.

2 GENERAL FORMULATION OF THE PROBLEM

As in any displacement-based method of analysis, the set of governing equations can be expressed in the following way:

$$[K] \{D\} = ([K_s] + [K_f]) \{D\} = \{P\} \tag{1}$$

where [K] is the stiffness matrix of the statical system considered and is composed of the stiffness matrix of the superstructure [K_s] and the stiffness matrix of the foundation bed [K_f]; {D} is the vector of nodal displacements of the superstructure; and {P} is the vector of equivalent nodal forces representing the applied loads.

Matrix [K_s] is the typical stiffness matrix of the relevant bar structure that is derived from the Direct Stiffness Method (Ghali & Nevill 1982). Since this method is very well known, no further discussion on the matter is included. The stiffness matrix [K_f] representing the soil must, of course, be related to the same degrees of freedom as those governing its counterpart [K_s]. Hence, in general, it may contain quite a large number of zero-elements.

Since in the present approach, the soil stiffness matrix [K_f] is obtained by inverting the flexibility matrix [F_f], further discussion will be focussed on the development of the latter.

3 DEVELOPMENT OF SOIL FLEXIBILITY MATRIX

As already mentioned, the concept of homogenous, elastic half-space with isotropic material properties is used to model the foundation bed (Poulus & Davis 1974). The Boussinesq force-displacement equations are utilised to develop flexibility relations for the vertical reactions, whilst the corresponding relationships associated with the horizontal reactions are obtained from the relevant Cerutti solution. The Boussinesq solution is also used to take account of support moments in the soil flexibility matrix. In this study, it is assumed that the points (areas) of contact between the superstructure and the foundation bed coincide with the surface of the half-space medium. Consequently, only the force-displacement relations for forces applied at, and displacements on, the surface are of interest here.

3.1 Vertical force - displacement relationships

Figure 1 shows a vertical force P_i applied at point i at the surface of an elastic half-space. In addition, it shows the profile of the resulting surface displacements with the vertical and horizontal displacements at some point B designated by Δ_j (v_B) and Δ_k (u_B). The Boussinesq equations governing the case and associated with force P_i and displacements v_B and u_B can be expressed as:

$$v_B = \Delta_j = \frac{(1 - v_f^2)}{\pi E_f R_{ji}} P_i = f_{ji} P_i \tag{2a}$$

$$u_B = \Delta_k = \frac{(1 + v_f)(1 - 2v_f)}{\pi E_f R_{ki}} P_i = f_{ki} P_i \tag{2b}$$

where R_{ji} and R_{ki} are the distances between point i and the point associated with the vertical and horizontal displacements (point B, in this case); and where E_f and v_f are the Young's modulus and the Poisson's ratio of the soil.

In equations (2a) and (2b), the quantities f_{ji} and f_{ki} represent the coefficients of flexibility relating force P_i to displacements Δ_j and Δ_k.

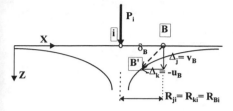

Figure 1. Profile of the surface displacements due to a vertical force and given by Equations 2.

A simple inspection of equations (2a) and (2b) reveals that they can directly be applied to evaluate coefficients of flexibility where the force application point and the point associated with the displacement are distinct points. It can also be noticed that these theoretical solutions feature singularities as the distances between the points of force application and the points of the calculated displacements (R_{ji} and R_{ki}) tend to zero. However, because of axial symmetry, the horizontal displacement under the force must be zero. Hence, the coefficient of flexibility given by equation (2b) and governing the horizontal displacement under a vertical point load should also be zero. To evaluate a coefficient of flexibility relating a vertical concentrated force (vertical reaction) to the vertical displacement under the force, a numerically equivalent uniform load distributed over an area can be used. For such a case, the application of equation (2a) to an infinitesimal load and integration over the load area leads to a finite value of the coefficient of flexibility. Since the approach involves a fair degree of discretion, a number of distinct types of load areas can be considered. A circular area has this advantage over other types of load areas that it leads to a closed-form coefficient of flexibility (Melerski 1990). Such an area has been used in this work, yielding the flexibility coefficient as follows:

$$f_{ii} = \frac{2(1 - v_f^2)}{\pi E_f r} \tag{3}$$

where r is the radius of the load area considered.

3.2 Horizontal force-displacement relationship

A case of a horizontal point force P_l applied to the surface of a half-space is shown in Fig. 2. Referring to Figure 2, the Cerutti solutions for the vertical and horizontal displacements of a surface point C (horizontal displacement coinciding with the direction of the force) can be expressed as follows:

$$v_C = \frac{(1 + v_f)(3 - 2v_f)}{4\pi E_f R_{ml}} P_l \tag{4a}$$

$$u_C = \frac{(1 + v_f)}{\pi E_f R_{nl}} P_l \tag{4b}$$

where the used symbols follow the convention introduced in section 3.1.

Equations (4a) and (4b) exhibit the same singularity shortcomings as those associated with the Boussinesq solution. Hence, a similar remedy has been applied to solve the problem. In this case, however, the vertical displacement at the point of application of the force as well as the related coefficient of flexibility can be assumed to be zero. As regards the coefficients of

501

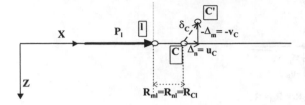

Figure 2. Profile of the surface displacements due to a horizontal force and given by Equations 4.

flexibility governing the horizontal displacements (in the direction of applied forces), these can also be evaluated with the aid of numerically equivalent, uniformly distributed forces. The approach involves treatment similar to that described in section 3.1.

4 SOFTWARE DEVELOPED

To apply the outlined method for the elastic solution to problem of the interaction between plane bar structures and a homogenous elastic half-space, a user friendly computer programme (Soil and Structure Interaction Analysis Package SASIAP) has been developed in Turbo Pascal. The data input and computation handling are facilitated by use of windows, pull-down menus and dialogue boxes. The programme requires minimal data and yields all structural response fields of importance to design. Moreover, the computation time requirements are very modest.

5 EXAMPLE

To numerically illustrate the proposed method of elastic analysis, a two-bay plane frame founded on a homogenous, isotropic half-space is considered (See Figure 3). For simplicity, member cross section and material properties are assumed to be uniform and of the following values:

 cross-sectional area, $A=0.0001$ m^2

 relevant second moment of cross-sectional area, $I=0.0001$ m^4

 Young's modulus, $E= 200$ GPa

The properties of the elastic half-space are defined by the Young's modulus $E_f = 20$ MPa and the Poisson's ratio $v_f =0.45$.

The variations of bending moments and nodal displacements returned by the computer programme described in section 4 are shown in Figures 4 and 5, respectively (values with no brackets). The bracketed values provided in these figures illustrate the solution to the case of the same frame but attached to an infinitely rigid foundation bed (using SPACE GASS package). Consequently, the differences between the bracketed values and the corresponding unbracketed numbers represent the effect of the interaction between the superstructure and the elastic half-space considered.

To verify the obtained solution for the interaction case of this example, the frame was again analysed under conditions of the loading applied and the obtained support displacements. The Space Gass software was used for this purpose and the returned structure response fields entirely coincided with the corresponding quantities from the interaction analysis. This, of course, verified only the adequacy of the structural analysis of the superstructure.

Subsequently, the reactions from the interaction analysis were applied to the half-space and appropriate Boussinesq and Cerutti force-displacement relations were employed to check the

502

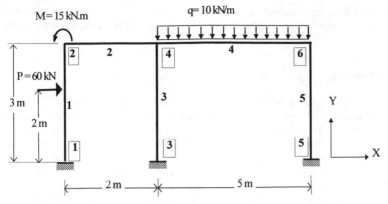

Figure 3. Structure and the applied loading.

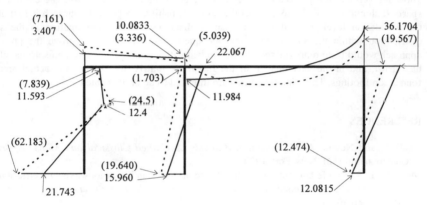

Figure 4. Bending moment variation (kN.m). Results from the SASIAP (solid line) and Space-Gass (broken line) analyses.

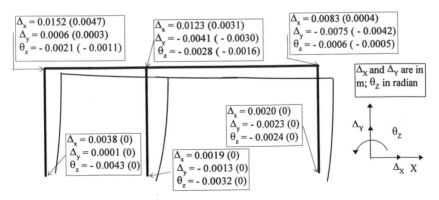

Figure 5. Displacement diagram (Results from the SASIAP and Space-Gass analysis).

support displacements obtained from the interaction analysis. Good agreement of the corresponding results testifies to adequacy of the developed stiffness matrix of the elastic half-space.

6 CONCLUSIONS

In this paper, a displacement-based technique for the statical solution of a linear elastic problem of interaction between a bar structure and an isotropic elastic half-space representing the supporting soil is outlined. The structural analysis of the superstructure involves the classical, direct stiffness method, whilst the Boussinesq and Cerutti force-displacement relations are utilised to develop the soil flexibility matrix, and hence to obtain the corresponding stiffness matrix. The application of the Boussinesq and Cerutti solutions infers that the superstructure-soil interaction takes place at the surface of the half-space. This, of course, is a simplifying assumption, which can also be considered as a shortcoming of the approach. However, this problem can easily be rectified by using the Mindlin solutions that allow for vertical and horizontal forces to be applied under the surface of the elastice half-space. It should be added, though, that the rationale behind the present approach is that above the foundation level the soil layer is usually disturbed and its participation in the stress distribution within the soil medium may be questioned. To further improve the presented approach would be to represent the foundation bed with a homogenous cross-anisotropic elastic half-space, and in the development of the flexibility matrix, to consider the actual areas of foundation footings. These improvements are intended to be implemented soon.

REFERENCES

Ghali A. and Neville A. M. *Structural Analysis; A Unified Classical and Matrix Approach* Chapman and Hall, New York 1989.

Melerski. E S. Simple computer analysis of circular rafts under various axisymmetric loading and elastic foundation conditions. *Proc. Institution of Civil Engineers, Part 2, September 1990,* 89, pp 407-431

Poulos H. G. and Davis E. H. *Elastic Solutions for Soil and Rock Mechanics* John Wiley & Sons, Inc. Sydney, 1974.

The Mechanics of Structures and Materials, Grzebieta, Al-Mahaidi & Wilson (eds)
© *1997 Balkema, Rotterdam, ISBN 90 5410 900 9*

Base settlement effects on fixed roof storage tank shells

M. Jonaidi & P. Ansourian
Department of Civil Engineering, The University of Sydney, N.S.W., Australia

ABSTRACT: The effects of peripheral differential settlement on cylindrical storage tanks with fixed roofs are evaluated for uniform or tapered shells; the imposed settlement is of harmonic form. Stresses are evaluated using membrane theory in the case of uniform thickness, compared with accurate finite element analysis, also used to evaluate the tapered thickness behaviour. The significance of stresses is assessed in special relation to the degree of tangential restraint afforded by a roof. It is shown that stresses in the shell are strongly influenced by the degree of restraint, in the practical range of harmonic number n.

1 INTRODUCTION

Foundations of large steel tanks for fluid storage tend to be shallow and consequently suffer differential settlement under load. But these tanks are ductile and can tolerate limited settlement without distress. Tanks containing fluids such as oil products or water are normally covered by floating or fixed roofs, and sometimes by a fixed and an internal floating roof. Concerns related to settlement include distortion of the tank shell which may affect the smooth operation of a floating roof; and overstressing of the shell. The total settlement of the periphery is normally measured during tank construction, at hydro-testing, and periodically in service. From these measurements, the three components are derived: uniform settlement, planar tilt, and differential settlement. The uniform component does not cause serious problems to the shell but may damage pipework and connections; planar tilt has an effect on stresses in the shell, due to the change in liquid height (Palmer 1992); the differential component is however the primary source of serious stress and distortion.

In order to establish a relationship between membrane stresses in the shell and vertical movement of the base, several shell theories ranging from inextensional theory to modified Donnell theory have been applied by researchers. Most previous research in this area has centred on open-top tanks, and are therefore of limited relevance to fixed-roof tanks. In this paper, membrane theory is used to relate the meridional stress resultant at the base to the settlement, in the case of uniform thickness; finite element analysis is then used to assess the membrane solution and define behaviour when the wall has tapered thickness.

2 SHELL BASE SETTLEMENT

Measurement of settlement is normally made effected with survey levels at a number of equally spaced stations around the base. API 653 (1991) specifies at least 8 points, spaced at most 9 m apart. Since the differential components are defined relative to the tilt plane, this plane must be established from the measured levels. Several methods are suggested in the literature, and some controversy exists. A graphical procedure is given by de Beer (1969), and a first harmonic Fourier analysis, initially suggested by Malik et al (1977) is recommended by Marr et al (1982). D'Orazio (1989) has also developed a practical method of evaluating settlement data and shell distortion. The best method would appear to be that which fits an optimum tilt plane, using computer methods. The differential settlement component therefore becomes the vertical distance between the data points and the cosine curve-of-best-fit representing the tilt plane.

3 SHELL BEHAVIOUR

3.1 Boundary conditions

The boundary conditions at the top and bottom of the shell need definition in terms of the structural connections. Radial restraint at the base can be achieved through the annular plate, the bottom plate and bolting. The bottom plate alone because of its slenderness is only effective in tension, but provides little restraint where it is in compression. Its restraining action in the absence of bolting to the foundation can be determined in a non-linear finite element analysis that includes the dual response of the plate. At the top, the roof structure including the plates can provide relatively stiff restraint against radial movement. On the other hand, circumferential restraint is strongly dependent on the interaction between shell and roof or annular plate; the magnitude of this restraint depends on the axial rigidity of the *effective cross sectional area* (A) at these junctions. Even in anchored tanks, full circumferential restraint may not be effective because of clearance in the anchor holes. It is generally recognised that full circumferential restraint is difficult to achieve in laboratory or prototype.

The *compression area* concept of BS 2654 (1989) can provide physical insight into this circumferential restraint (Fig. 2). Although the rigidity of the roof structure may have some effect, the major contribution comes from the action of this area.

3.2 Uniform wall thickness

The measured settlement values when plotted relative to the tilt plane have an irregular shape that can be expressed as a Fourier series: by considering the n-th component of that series, and referring to the coordinate system of Fig. 1, the differential settlement at the base becomes $u_{x=0} = u_n \cos(n\theta)$, the meridional stress resultant $N_{x=0} = N_n \cos(n\theta)$ and the tangential and radial displacements over the shell height respectively $v = v_n \sin(n\theta)$ and $w = w_n \cos(n\theta)$.

Membrane analysis is normally based on the neglect of circumferential direct stress, and therefore the only significant stress resultants on the shell are $N_{x\theta}$ and N_x. On this assumption, the shell is statically determinate and the equations are solved directly; this is followed by Kamyab and Palmer (1989) for 'open-top' tanks. In closed-top tanks the radial displacement at the top is assumed zero and the corresponding expression for N_n becomes:

$$N_n = \frac{Eu_n}{\dfrac{(2+v)r^2}{n^2 t h} + \dfrac{h}{3t}} \tag{1}$$

where t = wall thickness, E = elastic modulus, v = Poisson ratio; in dimensionless form, the maximum meridional stress at the base as a function of differential settlement u_n becomes:

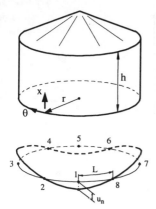

Fig. 1 Saddle settlement (n = 2)

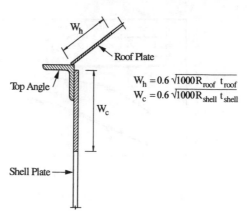

$W_h = 0.6 \sqrt{1000 R_{roof}} \, t_{roof}$

$W_c = 0.6 \sqrt{1000 R_{shell}} \, t_{shell}$

Fig. 2 'Compression' area (BS 2654 - 1989)

$$\frac{\sigma_x / E}{u_n / h} = \frac{1}{(2 + v)(\dfrac{r}{nh})^2 + \dfrac{1}{3}} \qquad (2)$$

A comparison of this solution with a highly accurate linear finite element solution (Abaqus 1995) is given in Fig. 3 for a slender shell (r/t = 1000) and two aspect ratios h/r = 0.5, 1.5. The error is less than 6% at harmonic numbers below 9, and is probably acceptable in practice, especially since Equ. 2 gives conservative values. At higher values of n, Equ. 2 becomes highly inaccurate particularly in taller tanks: the error at n = 12 for h/r = 1.5 exceeds 40%. At high n, the shell behaviour is strongly influenced by circumferential bending neglected in Equ. 2. In these cases, finite element analysis is warranted, or alternatively the more complex semi-bending theory incorporating the beam-on-elastic-foundation analogy (Greiner 1983). Equ. 2 does provide a simple means of relating meridional stress at the base to u_n in the practical range of n, for uniform wall thickness and restraint against radial and circumferential translation.

The field evaluation of settlement involves the taking of survey levels at a number of equally spaced stations around the base. The differential settlement is therefore related to L = πr/2n, where L is the arc length between stations (see Fig. 1). Assuming that three consecutive stations correspond to one half wave, u_n becomes the maximum distortion between these points. Equ. 2 can therefore be expressed as:

$$u_n = \{\frac{4L^2(2 + v)}{\pi^2 h} + \frac{h}{3}\}\frac{\sigma_x}{E}$$

$$\approx (\frac{L^2}{h} + \frac{h}{3})\frac{\sigma_x}{E} \qquad (3)$$

Replacing $\dfrac{4(2 + v)}{\pi^2}$ by unity in Equ. 3 involves an error of 7%. The equation provides a limitation on settlement u_n for a given limit on stress $\sigma_x = N_{nx}/t_{max}$. In a zone where u_n is *downward*, the yield stress F_y might be used as the basis of design. When u_n is *upward*, the limiting settlement must involve the compressive buckling behaviour of the imperfect shell. Marr et al (1982) have studied cases of large or unusual deformations showing that some tanks

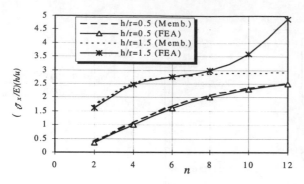

Fig. 3 FEA results and membrane solution for range of n (r/t=1000)

with fixed roofs can tolerate settlements greater than set by Equ. 3. This may be due to increased tank flexibility due to taper, and/or to the assumed full restraint condition not being achieved.

3.3 Tapered shells with elastic circumferential restraint at the top

The membrane theory may be applied to tanks having wall thickness tapering from the base to the top, but the solution is lengthy and complex. For these cases, the finite element method is ideal, as it easily accounts for the several thicknesses, the presence of a stiff roof and a range of boundary conditions. In this method, the modeling of the shell-roof junction can be effected by the use of 3 - dimensional beam elements B31 (Abaqus 5.5 (1995)). Such analyses using highly refined meshes have been made for tanks of wall slenderness r/t_{base} = 1,000, geometry h/r = 0.3 - 3.0, taper T_{ratio} = t_{max}/t_{min} = 1 and 3, and shell-roof junction stiffness ratios A_r = 10 - 60, where A_r is a non-dimensional parameter equal to A/t^2. Radial and tangential translational restraints were applied at the base, and harmonic meridional translations or forces in the range n = 2 - 8 were imposed to the base. When *translations* were imposed at n = 2, with a weak circumferential restraint (eg., A_r = 10), inaccuracies arose in the finite element solution, attributed to the very low stiffness of the resulting structural form; for those cases, harmonic *forces* are preferable. The elements used were 4-node shell elements S4R (Abaqus 5.5 (1995)) and the solution therefore represented the most accurate results of full shell bending theory. Typical comparisons with predictions of the membrane theory for a uniform wall thickness and fixed conditions at the top are shown in Fig. 4a,c,e for n = 2, 3 and 4 corresponding to 8, 12 and 16 measuring stations.

Good agreement exists between the finite element results and the membrane solution for full tangential restraint. It is also clear from Fig. 4 that the effect of A decreases at higher n. Also clear from Equ. 2 is that σ_x is independent of the (uniform) wall thickness t. Further analysis has shown that in the range 0.5 < h/r < 1.5, the curves of Fig. 4 can also be used in the range 700 < r/t < 1200 for 'normal' taper. It is concluded that the results cover the majority of practical tanks.

It should be noted that the present membrane theory neglects $N_{n\theta}$. Although this assumption is justified over the major part of the shell, it breaks down at the base and at the top. In the presence of full tangential restraint at the top, straining in that direction is prevented and therefore $N_{n\theta}$ is almost zero in the neighbouring shell; with elastic tangential restraint however, $N_{n\theta}$ is no longer zero for a small depth of shell at both ends.

In the case of *open-top* tanks, circumferential stresses are critical because the absence of radial restraint allows strong bending of the primary wind girder (Jonaidi and Ansourian

(1996)); this is a distinct action from that discussed here, but still related to the presence of the top ring. With small effective area A, a low level of circumferential restraint is realised, leading to high circumferential stress at the top, but lower meridional stress at the base.

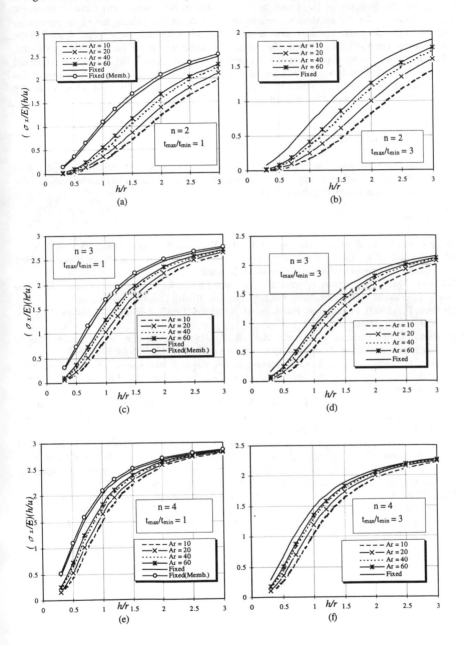

Fig. 4 Membrane solution and FEA results for range of h/r (r/t = 1000)

4 CONCLUSION

The membrane solution given in this paper and finite element analysis have provided new insight into the response of fixed-roof tanks subjected to differential settlement of harmonic form. Solutions for the maximum meridional stress σ_x at the bottom based on membrane analysis, full tangential restraint and a uniform wall thickness have been found to be conservative, because in a real tank, the restraint is never complete, and the wall is tapered. Solutions have also been provided for a tapered wall thickness. The circumferential/axial stiffness of the shell/roof junction has a considerable effect on the reduction of base meridional stress; when this junction is relatively flexible (low A), the circumferential stress at the top rises, but the vertical stress at the bottom falls.

ACKNOWLEDGMENT

The first author is supported by a Scholarship of the Ministry of Culture and Higher Education of Iran. The second author is supported by a Grant from the Australian Research Council. The authors gratefully acknowledge this support.

REFERENCES

Abaqus, (1995). *Abaqus Users Manual 5.5*, Habbit, Karlsson & Sorensen Inc., USA.

API Standard 653, (1991). *Tank Inspection, Repair, Alteration, and Reconstruction.* First edition, Washington: American Petroleum Institute.

BS 2654 (1989). *Manufacture of vertical steel welded non-refrigerated storage tanks with butt-welded shells for the petroleum industry.* London: British Standard Institution.

De Beer, E., (1969). Foundation Problems of Petroleum Tanks. *Annales de l'Institut Belge de Petrole.* 6: 25-40.

D'Orazio, T. B., Duncan, J. M., and Bell, R. A. (1989). Distortion of steel tanks due to settlement of their walls. *Journal of Geotechnical Eng.* ASCE 115(6): 871-890.

Greiner, R., (1983). Zur Ingenieurmassigen Berechnung und Konstruktion Zylindrischer Behalter aus Stahl Allgemeiner Belastung, *Stahlbauseminar der FH Biberach, Lindau, Neu-Ulm.*

Jonaidi, M. and Ansourian, P., (1996). Effects of Differential settlement on storage tank shells. SL Chan and JG Teng (eds), *ICASS'96 Intl. Conf. on advances in steel structures, Hong Kong, Dec. 1996:* 821-826. Pergamon.

Kamyab, H., and Palmer, S. C., (1989). Analysis of displacements and stresses in oil storage tanks caused by differential settlement. *Proc. Instn Mech. Engrs, Part C.* 203(C1): 60-70.

Malik, Z., Morton, J. and Ruiz, C. (1977). Ovalization of cylindrical tanks as a result of foundation settlement. *Journal of Strain Analysis.* 12(4): 339-348.

Marr, W. A., Ramos, J. A., and Lambe, T. W., (1982). Criteria for settlement of tanks, *Journal of The Geotechnical Engineering Division.* Proc. ASCE 18:GT8: 1017-1039.

Palmer, S. C., (1992). Structural effects of foundation tilt on storage tanks. *Proc. Instn Mech. Engrs, Part E.* 206:E2: 83-92.

Pavements as structures subject to repeated loadings

I. F.Collins & M. Boulbibane
School of Engineering, University of Auckland, New Zealand

ABSTRACT: The concepts and methods of shakedown theory, which describes the ultimate response of an elastic/plastic structure to cyclic loads, are used to analyse the response of idealised unbound pavements. The pavements are modeled as multilayered structures of Mohr-Coulomb material, characterised by their elastic moduli, cohesion and friction angle. The use of the upper and lower bound theorems of classical shakedown theory to estimate the critical shakedown load is described, and the former is employed to predict the relative importance of various failure mechanisms such as subsurface slip and rut formation.

1 INTRODUCTION

Design procedures for pavements are based on a model of the response of the pavement to traffic loading together with some 'failure criterion'. A summary of currently used design procedures may be found in Ullidtz (1987). The 'failure mode' depends on the type of pavement and on the environment. Here we are concerned principally with unbound pavements, where the thin top asphaltic layer acts only as a weatherproofing coating and does not contribute to the structural strength of the pavement. This is determined by the physical characteristics of the subgrade aggregates and basecourse soils. These are well described by rate-independent elastic/plastic models, so that creep effects which are of importance in asphaltic pavements can be ignored.

The eventual 'failure' of a pavement is determined by the degree of rutting, subsurface shearing and/or the formation of surface cracks, all of which are a result of the accumulation of plastic strains. In current design procedures, a representative plastic strain ε_p is calculated from an empirical formula of the form:

$$\varepsilon_p = A \, N^B \sigma^C \tag{1}$$

where N is the number of load applications, σ is the 'axial stress' and A, B, C are constants. No attempt is made to use the plastic strength parameters of the aggregate materials, such as cohesion and internal angle of friction. This, of course, is in marked contrast to other geotechnical design procedures, such as slope stability or foundation design. The main reason for this disparity is that pavements are subjected to *repeated* rather than monotonic loading processes. The branch of plasticity theory which deals with the irreversible response of structure to repeated loadings is "shakedown theory", and our principal concern here is to explore the possible application of this theory to the development of new pavement design procedures.

Sharp and Booker (1984) seem to be the first researchers to suggest the use of shakedown concepts to pavement analyses. The essential concept of shakedown theory is that of the limiting *shakedown load*. If the structure is subjected to a repeated loading cycle with the maximum load *below* this critical load, then the long term response of the pavement will be purely elastic, although the response may well be plastic for a finite number of initial load applications. Some irrecoverable strains will be accumulated, but the total plastic strain will be bounded. In this situation the structure is said to have 'shakedown'. On the other hand if the load level exceeds the critical shakedown limit, the response of the structure will be plastic for *all* load applications. In this regime two types of long term behaviour can occur:

(a) The plastic strains accumulate indefinitely, resulting in the continual development of ruts or subsurface shearing. This behaviour is referred to as *ratchetting*.

(b) The ultimate response of the structure cycles between two plastic states, so that there is no further build up of irreversible strains. This is referred to as *alternating plasticity* or *plastic shakedown* and will most likely result in fatigue failure.

It is hence suggested that this shakedown limit load is a rationally based parameter on which to base a design procedure.

2 CALCULATION OF SHAKEDOWN LIMIT

Estimates of the shakedown limit load can be found by application of the two extremum shakedown principles for elastic/perfectly plastic structures (Lubliner 1990). The lower bound theorem, due to Melan states that "if *any* equilibrium residual stress distribution can be found, which together with the stress field produced by the passage of a load does not exceed the yield condition at any time, then shakedown will occur." To implement this theorem the elastic stress field induced in the layered pavement must be calculated. This is readily achieved using a finite element procedure or a standard package such as BISAR or CIRCLY. The problem of finding the shakedown limit load can then be formulated as a linear programming problem. This approach was pioneered by Sharp and Booker (1984) and extended by Raad *et al* (1989) , Boulbibane (1995) and Boulbibane and Weichert (1997), who have also incorporated elastic anisotropy and non-associated flow rules. This approach works well for *two-dimensional* models where the loaded wheel is replaced by a cylinder of infinite extent in the direction perpendicular to the travel direction. However the size of the linear programming problem becomes prohibitively large when the more realistic three dimensional problem is attempted.

Collins and Cliffe (1987) employed the dual kinematic theorem, due to Koiter, to obtain upper estimates of the shakedown load, and showed that in the two-dimensional case the results were identical with those obtained by Sharp and Booker's lower bound approach; these estimates were hence, in fact, exact. More importantly however, it was demonstrated that upper bounds could be obtained with relative ease in the much more realistic three-dimensional case, where the load is assumed to be applied over one or more circular areas. This extended the findings of the parallel analysis of Ponter, Hearle and Johnson (1985), who calculated the shakedown loads on uniform metal surfaces. The upper bound theorem states that "incremental collapse will occur if any kinematically admissible plastic collapse mechanism can be found in which the rate of working of the elastic stress exceeds the rate of plastic energy dissipation." The application of this procedure hence requires us to postulate a failure mechanism or family of failure mechanisms. The "elastic" and "plastic work-rates" can then be calculated, again using BISAR or some similar elastic, stress analysis program, and the solution optimised. This optimisation problem is now a nonlinear one, but with very few variables and much more manageable than in the corresponding lower bound approach.

In the two dimensional model the only possible such mechanism is sliding over a plane or shearing in the direction of travel. Ponter *et al* (1985) and Collins and Cliffe (1987) showed that in fact the optimal solution involves block sliding over a plane and the only optimisation necessary is to find the critical depth of this plane which gives the lowest shakedown load estimate. However in three-dimensions many more types of deformation are possible. Collins, Wang and Saunders (1993 (a) and (b)) and Collins and Wang (1994) have calculated the shakedown load associated with slip in V-shaped channels in the direction of travel. Such failure would be recognised by the formulation of surface shear cracks on either side of the loaded area. However a much more common failure mechanism is rut-formation in which the load sinks into the pavement and material is displaced sideways. The purpose of this paper is to present the results of initial calculations based upon such mechanisms.

3 SHAKEDOWN DUE TO LATERAL MOVEMENT

The shakedown calculations are most conveniently represented in terms of a dimensionless load parameter defined by:

$$\lambda = P/A\, c_1 \tag{2}$$

where P is the applied load, A the cross-sectional area of the loaded area ($= \pi a^2$, if this is circular with radius a) and c_1 is the cohesion of the basecourse. The elastic stresses induced are proportional to λ. Once λ_c corresponding to the critical shakedown, the load, corresponding to a given load area and basecourse cohesion can be deduced from (2). Upper bounds to λ_c can then be obtained from the basic inequality:

$$\lambda_c \leq \frac{\int\limits_v \sigma_{ij}^p\, e_{ij}^p\, dV}{\int\limits_v \sigma_{ij}^e\, e_{ij}^p\, dV} \tag{3}$$

where e_{ij}^p and σ_{ij}^p, are the time-independent strain-rate and plastic stress fields associated with some kinematically admissible velocity field and σ_{ij}^e is the corresponding elastic stress field.

The form of the proposed family of simple block sliding mechanisms is shown in Figure 1. The plastic strain-rate e_{ij}^p is only non zero on the velocity discontinuities and the plastic work rate can be computed as in the well known limit analysis calculation for monotonic loading (e.g. Chen 1975). The elastic work-rate is readily calculated by integration along the velocity discontinuities, the relevant components of the elastic stress being computed using BISAR. On a given discontinuity the plastic work-rate per unit length, is simply $c[v_t]$, the cohesion times the jump in the tangential velocity components. The corresponding expression for the elastic work-rate is $c^e[v_t]$, where:

$$c^e = |\sigma_{nt}^e| - \sigma_{nn}^e \tan\phi \tag{4}$$

may be termed the "elastic cohesion", by comparison with the corresponding expression for the "plastic" cohesion c. σ_{nt}^e and σ_{nn}^e are the elastic shear and normal stress components on the velocity discontinuity, whilst ϕ is the angle of friction of the material.

In a multilayer system, where ϕ is different for each layer the orientation of the velocity discontinuities must also be discontinuous across the layer boundaries. A simple two layer model, corresponding to a single basecourse plus subgrade is illustrated in Figure 3, with the associated hodograph diagram. The continuity of velocity across the layer boundary leads to the simple relation:

$$\alpha_i - \phi_1 = \alpha_i' - \phi_2 \tag{5}$$

where ϕ_1, ϕ_2 are the friction angles in the two layers. This result enables α_i' to be computed for a given α_i.

Both mechanisms are determined by the single parameter α and the best value can be obtained by using a standard nonlinear optimisation routine to minimise the upper bound estimate of λ_c obtained from (3).

513

4 SOME REPRESENTATIVE RESULTS

In Figure 2, the dimensionless shakedown limit load is plotted against the friction angle and the family of curves obtained are for slip and rut formation. For any one case, the shakedown limit increases as the friction angle increases. The better of the two rutting modes is seen to occur at significantly lower loads than the earlier translational slip solutions.

Some aspects of two-layer pavements are examined in this section. It is assumed that material properties are homogeneous and isotropic in each layer.

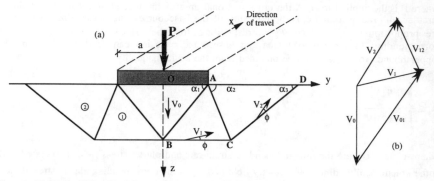

Figure 1. (a) Simple sliding block for one-layer system , (b) Hodograph.

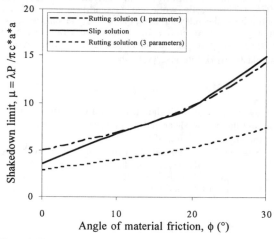

Figure 2. The comparison between slip solution and this study for one-layer system.

Table 1. The parameters for shakedown analysis

Layer	E (Mpa)	ν	c (kPa)	φ (°)	h (mm)
Basecourse	234	0.4	87	42	200
Subgrade	85	0.4	40	0	infinite

514

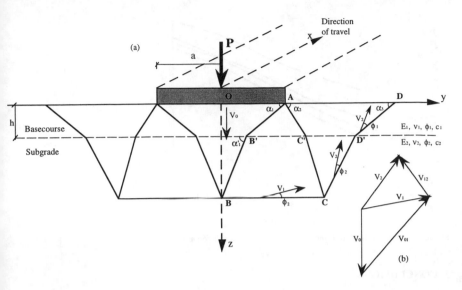

Figure 3. (a) Simple sliding block for two-layer system, (b) Hodograph.

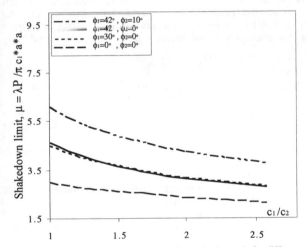

Figure 4. The dependence of shakedown load on cohesion ratio for different values of friction angles (two-layer half space under moving surface load).

As would be expected it is found that the presence of the weaker subgrade lowers the critical shakedown load. However, it is be noted that the change in the basecourse friction angle has little effect on this load provided the friction angle of the subgrade is kept constant. The effect of varying the depth of the basecourse is shown in Figure 5.

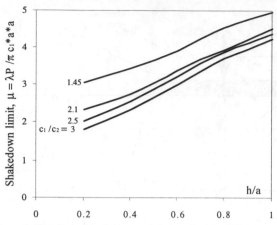

Figure 5. The effect of depth layer on shakedown load for different cohesion ratio.

5 CONCLUSIONS

This investigation is at an early stage, but the computations so far performed indicate that shakedown theory provides a rational framework within which to analyse different failure mechanisms. In all situations so far studied the shakedown load associated with rut formation is significantly lower than that produced by subsurface slip, indicates that the former is the more critical failure criterion. Ongoing research involves looking at further families of failure mechanisms, improving the minimization procedures and introducing more realistic constitutive models.

REFERENCES

Boulbibane, M. 1995. *Application de la théorie d'adaption aux milieux élastoplastiques non-standards: cas des géomateriaux.* Thèse de Doctorat, Université des Sciences et Technologies de Lille, France.
Boulbibane, M. and Weichert, D., 1997. Application of Shakedown theory to soils with non associated flow rules. *Mechanics Research Communications* (in press).
Chen, W.F. 1975. *Limit analysis and soil plasticity.* Elsevier,Amsterdam.
Collins, I.F. and Cliffe, P.F., 1987. Shakedown in frictional materials under moving surface loads. *Int. J. Num. Anal. Meth. Geomechanics* 11: 409-420.
Collins, I.F. and Wang, A.P., 1994. Shakedown theory and pavement design. *Proceedings of the 8th International Conference on Computer Methods abd Advances in Geomechanics, Morgantown, West Virginia, USA,* 1465-1470.
Collins, I.F., Wang, A.P. and Saunders, L.R., 1993. Shakedown theory and the design of unbound pavements. *Road and Transport Research,* 2: 29-38.
Collins, I.F., Wang, A.P. and Saunders, L.R., 1993. Shakedown in layered pavements under moving surface loads. *Int. J. Num. Anal. meth. Geomechanics* 17: 165-174.
Lubliner, J. 1990 *Plasticity theory,* Macmillan, New York.
Ponter, A.R., Hearle, A.D. and Johnson, K.L. 1985. Application of the kinematic shakedown theorem to rolling and sliding contact points. *J. Mech. Phys.Solids 33:339-362.*
Raad,L., Weichert,D. and Haidar,A. 1989 Shakedown and fatigue of pavements with granular bases. *Transportation Research Record* 1227:159-172.
Sharp, R.W. and Booker, J.R. 1984. Shakedown of pavements under moving surface loads. *American Society of Civil Engineers, J. of Transport Engineering* 110: 1-14.
Ullidtz, P. 1987. *Pavement Analysis,* Elsevier, Amsterdam.

Reliability and geomechanics (geo)

The Mechanics of Structures and Materials, Grzebieta, Al-Mahaidi & Wilson (eds)
© *1997 Balkema, Rotterdam, ISBN 90 5410 900 9*

Compressibility and crushability of calcareous soils

H.A. Joer, M. Ismail & M.F. Randolph
The University of Western Australia, Perth, W.A., Australia

ABSTRACT: The crushability of soil can have important consequences for engineering design. An obvious example is the low shaft capacity of driven piles in calcareous material, where compression and crushing of the soil leads to low normal effective stresses at the pile-soil interface. This paper describes the results of a study of the crushability of two calcareous soils, one offshore and one coastal, from the Western coast of Australia. Stress levels of up to 55 MPa were applied, in a high-pressure oedometer. Different crushability criteria were explored, and a modified criterion proposed based on the average grain size above 0.005 mm. This criterion leads to a clear relationship between the pressure applied and the degree of crushing, with a limited relative breakage of about 0.4 at high stress levels.

1. INTRODUCTION

Piles driven into calcareous soils are known to exhibit low friction capacity, attributed mainly to the compressibility and crushability of the soil, leading to low normal effective stress at the pile-soil interface. Extensive studies of calcareous sediments have been conducted at the University of Western Australia (UWA) on soils from off the North West coast Australia and also from an onshore site about 100 km North of Perth. A comparative study of the offshore and onshore soils is being carried out at UWA.

This paper presents the compression behaviour of two different calcareous soils (one offshore and one onshore). The tests were performed using a high pressure oedometer apparatus, with a maximum vertical pressure of 55 MPa. In order to quantify the amount of crushability, grading curves of the samples were determined, before and after the tests. Three different criteria for crushability were explored, based on suggestions by Datta et al (1979), Lee and Farhoomand (1967) and Hardin (1985). The criterion proposed by Datta et al (1979), which compares the percentage of particles finer than D_{10} before and after compression to different stress levels, was found to be sensitive to the level of fine particles contained in the soil, as noted previously by Morrison et al (1988). A modified Hardin criterion is proposed here as more useful for soils with a significant proportion of fine particles.

2. SAMPLE PREPARATION

The tests were carried out on samples of 102.5 mm diameter, with average initial height of 55 mm (loosest state). The soil was first dried in an oven at 50 °C for a minimum period of 48 hours. It was then weighed and poured into the mould and compacted (using an air hammer) to the desired density. The offshore material is from the vicinity of the North Rankin 'A' platform (NR) and the onshore material is from Ledge Point beach, 100 km North of Perth (LP). The grading curves of the two soils tested are shown in Figure 1. The curves show clearly the uniform grain size of the LP soil, with a D_{50} of about 0.32. The offshore material (NR) contains a larger range of particle size, with a D_{50} of about 0.1 mm. The carbonate content of the soils is about 94 % for the offshore material and about 85 % for the onshore material.

Figures 2 shows photographs obtained from microscopic analysis of the NR and LP soils respectively. It can be observed that, following the classification scheme of Fookes (1988), the NR soil comprises particles of various shapes identified as skeletal grains, detrital grains, coated grains, pellets and lumps. The onshore material comprises mainly coarse rounded particles which could be classified as pellets.

3. TEST RESULTS

3.1 Compression response

Figure 3 shows the axial stress versus the axial strain for the NR and LP soils. Three tests with different initial void ratios were undertaken for each soil. The tests were carried out with continuous (displacement controlled) loading, rather than in increments of stress. As may be seen from the curves, there was a gradual increase in stiffness with increasing load level (and decreasing void ratio). The tests were terminated at a maximum stress of about 55 MPa. Unload-reload loops were performed at various stress levels.

The compression responses are shown in Figure 4 for the two soils, in terms of void ratio versus axial stress. The three curves for each soil tend to merge as the stress level is increased, showing that the effect of the initial void ratio is gradually eliminated at high pressure. The relationship between the voids ratio e and the axial effective stress σ_v can be described in terms of the critical state parameters, λ and κ, for loading and unloading as following:

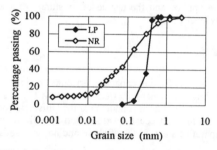

Figure 1 Original grading curves of the two soils tested

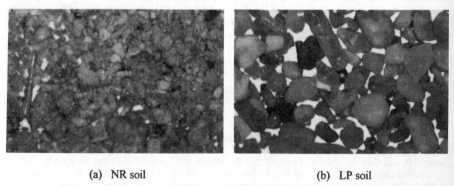

(a) NR soil   (b) LP soil

Figure 2 Photographs using an ordinary microscope of the two soils tested (magnification x 30).

520

$$e_c = e_1 - \lambda \ln \sigma_v' \quad \text{(loading)} \qquad (1)$$

$$e_s = e_2 - \kappa \ln \sigma_v' \quad \text{(unloading)} \qquad (2)$$

Values for λ of 0.27 and 0.23 and for κ of 0.0114 and 0.006 were obtained for the NR and LP soils respectively. Carter et al (1988) reported values of λ of 0.18 to 0.24 and 0.0065 for κ for core samples from the North West Shelf of Western Australia. The ratio λ/κ is about 24 and 38 for NR and LP soils respectively, compared with a value of 30 obtained for the natural material (Carter et al, 1988).

Equation (1) has a conceptual limitation at high pressures, where the voids ratio could become negative, although in the present tests the minimum void ratios recorded at the maximum stress of 55 MPa were 0.2 and 0.58 for the NR and LP soils respectively. The limitation of equation (1) has been discussed at length by Butterfield (1979) and Pestana and Whittle (1995), who have proposed relationships more consistent with the shape of the above curves, which show a decreasing gradient of voids ratio with increasing logarithm of stress towards the end of the curve. This is more pronounced for the NR soil, which contains a larger amount of fines, and less uniform shaped particles, than the LP soil. The Pestana-Whittle model has been found to provide a good fit to the measured compression responses, and this is addressed in a separate paper.

3.2 Breakage behaviour

Calcareous sediments are known for their fragile characteristics, which leads to breakage of particles when stressed. This has been reported by various researchers (Jewell and Andrews, 1988; Jewell and Khorshid, 1988). An investigation of particle crushing was carried out by loading the soil samples to various levels of stress, between 1.8 MPa and 55 MPa. After unloading, the samples were carefully removed from the mould and particle size analyses were carried out using the entire sample.

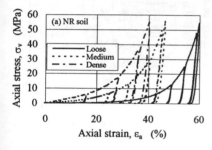
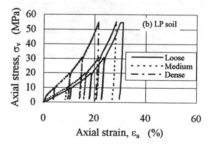

Figure 3   Stress-strain response of (a) NR and (b) LP soils.

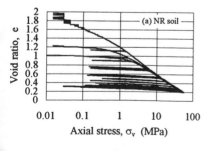
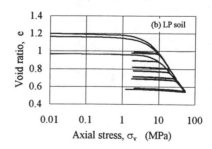

Figure 4   Void ratio versus axial stress for (a) NR and (b) LP soils.

Figure 5 shows the particle size distributions for the NR and LP soils for the original soil and for the soil after compression to different stress levels. It can be seen that the average grain size of the material reduces with increasing stress level for both soil types.

In order to quantify the degree of crushability, Datta et al. (1979) proposed a crushability coefficient, C_c, defined as:

$$C_c = \frac{(D_{10})_a}{(D_{10})_i} \qquad (3)$$

with $(D_{10})_a$ and $(D_{10})_i$ the percentage finer than D_{10} after and before compression respectively. As may be seen from Figure 5, $(D_{10})_a$ and $(D_{10})_i$ for the NR soil cannot easily be determined due to the large quantity of fine particles in this material. Lee and Farhoomand (1967), proposed a breakage factor, B, defined as:

$$B = \frac{(D_{15})_i}{(D_{15})_a} \qquad (4)$$

with $(D_{15})_a$ and $(D_{15})_i$ the percentage finer than D_{15} after and before compression respectively.

Note that both criteria are based on a factor expressing the change at a single point on the particle size distribution curve, as indicated in Figure 6. Figure 7 show the ratios of percentage finer than D_α, for α varying between 10 % and 85 %, versus the axial stress (σ_v) for both soils.

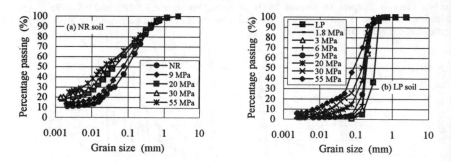

Figure 5   Particle size distributions before and after compression to different stress levels.

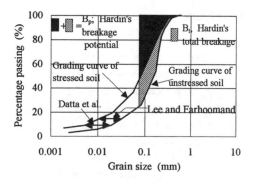

Figure 6   Illustration of various breakage criteria

As expected, the results show an increase in the ratio R_α with increasing σ_v for all values of α (noting that a lower limit for R_α of unity indicates that no crushing occurs). As for C_c, the breakage factor, B, could not be deduced for the NR soil due to the high fines content, once the stress level exceeded 10 MPa. However for the LP soil, a C_c value of about 24 and a B value of about 14 are obtained for the maximum vertical stress of 55 MPa.

Hardin (1985), proposed a criterion based on the average particle size distributions before and after compression, concentrating on particles larger than 0.075 mm. He proposed a relative breakage parameter, B_r, defined as:

$$B_r = \frac{B_t}{B_p} \qquad (5)$$

where B_p is the breakage potential and is defined as the area between the original grading curve of the soil and the size of particles of 0.075 mm. B_t is defined as the area between the original grading curve and the final grading curve. These two quantities are illustrated in Figure 6.

Unlike the criteria discussed earlier, Hardin's criterion shows that B_r reaches a threshold at large stress (see Figure 8). This indicates that a maximum density curve for the soils is approached (Lade et al, 1996). For all tests carried out on the NR soil, and all the tests on the LP soil for stresses larger than 9 MPa, it was observed that 'fusion' seemed to occur between particles after compression, with the samples appearing like 'cemented cakes' after extraction from the mould. However, when the 'cemented cake' was submerged in water it broke down slowly (after about 2 hours) into the constituent particles, allowing the particle size distribution to be assessed. The apparent 'fusion' therefore appears to be due to suction effects with very small pore spaces.

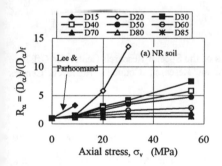 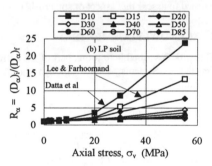

Figure 7 Ratios of D_α for both NR and LP soils, for different values of α.

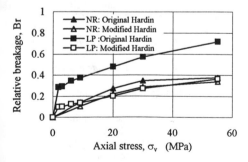

Figure 8 Relative breakage according to Hardin (1985) for NR and LP soils.

In order to take into account the large proportion of silt sized particles, a modified Hardin criterion is suggested here, with the breakage potential B_p redefined as the area between the grading curve of the soil and particles of size 0.005 mm (silt size according to ASTM standard). The results obtained from this modified criterion are shown in Figure 8. It may be seen that for the NR soil the original and modified criteria are practically identical, while the modification produces a large difference for the LP soil, mainly due to the lower initial amount of fines. Using the modified criterion, the relative breakage for both soil types shows a similar variation with stress level, and a similar limit of just under 0.4 at high stress levels.

4. CONCLUSIONS

One-dimensional compression tests at high levels of vertical stress (up to 55 MPa) were carried out on two different calcareous soils from the coast of Western Australia. The degree of crushability was investigated for the two soils, one of which had large quantities of fines in its original state while the other was narrowly graded sand. Three different approaches for measuring the crushability of the soils were investigated.

From this study, the crushability of the two soils was found to increase with increasing stress level. The crushability coefficient, C_c, proposed by Datta et al. (1979) and the breakage factor, B, proposed by Lee and Farhoomand (1967) could not easily be used for soils with large initial proportion of fines. The criterion of Hardin (1985), based on the average grain size over 0.075 mm, was more applicable, and indicated a limiting relative breakage at high stress levels. A modified Hardin criterion was proposed for these types of soil, based on the mean particle size over 0.005 mm (silt size). With this criterion, both soil types showed a limiting relative breakage at high stress levels of just under 0.4.

Crushability of soils may also be affected by the stress path the soil is subjected to, in particular the relative amount of shear and volumetric strain. Further investigations are being pursued to assess this aspect of soil crushability.

5. ACKNOWLEDGEMENTS

The work described here forms part of the activities of the Special Research Centre for Offshore Foundation Systems, funded through the Australian Research Council's Research Centres Program.

6. REFERENCES

Butterfield R. 1979. A natural compression law for soils (an advance on e - log p). *Geotechnique*, **29**(3), 469-480.

Carter J.P., Kaggwa W.S., Johnston I.W., Novello E.A., Fahey M. and Chapman G.A, 1988. Triaxial testing of North Rankin calcarenite. *Engng for Calcareous Sediments*, **2**, 515-530.

Datta M., Gulhati S.K. and Rao G.V., 1979. Crushing of calcareous sands during shearing. *11th Annual offshore Technology Conference*. Houston, Paper OTC 3525, 1459-1467.

Fookes P.G., 1988. The geologie of carbonate soils and rocks and their engineering characterisation and description. *Engng for Calcareous Sediments*, **2**, 787-806.

Hardin B.O., 1985. Crushing of soil particles. *J. Geotech. Eng. ASCE*, **111** (10), 1177-1192.

Jewell R.J. and Andrews D.C. (Eds) 1988. *Engineering for Calcareous Sediments*, Vol.1, Perth: Balkema.

Jewell R.J. and Khorshid M.S. (Eds) 1988. *Engineering for Calcareous Sediments*, Vol.2, Perth: Balkema.

Lee K.L., and Farhoomand I., 1967. Compressibility and crushing of granular soils in anisotropic triaxial compression. *Can. Geotech. J.*, **4**(1), 68-86.

Lade P.V., Yamamuro J.A., and Bopp P.A., 1996. Significance of particle crushing in granular materials. *J. Geotech. Eng. ASCE*, **122** (4), 309-316.

Morrison M.J., McIntyre P.D., Sauls D.P. and Oosthuizen M. 1988 Laboratory test results for carbonate soils from offshore Africa. *Engineering for Calcareous Sediments*, **1**, 109-118.

Reliability and geomechanics (rel)

The Mechanics of Structures and Materials, Grzebieta, Al-Mahaidi & Wilson (eds) © 19
97 Balkema, Rotterdam, ISBN 90 5410 900 9

Reliability of multi-storey reinforced concrete buildings during construction

Deepthi Epaarachchi & Mark G. Stewart
Department of Civil, Surveying and Environmental Engineering, The University of Newcastle, N.S.W., Australia

David V. Rosowsky
Department of Civil Engineering, Clemson University, S.C., USA

ABSTRACT: The paper develops a probabilistic model to estimate the probability of structural collapse during the construction of typical multi-storey reinforced concrete buildings. The influence of concreting workmanship (curing, compaction) on concrete compressive strength is included in the analysis. It was found that concreting workmanship is more detrimental to system risk than reducing the construction cycle by a few days.

1. INTRODUCTION

A significant proportion of reinforced concrete (RC) building structural failures occur during construction. The current use of rapid construction techniques places pressure on contractors to meet predetermined scheduling targets by reducing the time between placement of successive floors. The minimum number of levels of shoring needed (from two to five floors) and the time between placement of successive floors are specified by the Australian Formwork Code AS 3610-1990. However, it appears that the existing requirements were developed from a deterministic approach, with no consideration of the influence of the requirements on fatality and financial risks. This is not particularly surprising since very little is known about the performance of a structure during its construction because of the complexity of modelling time dependent processes; nearly all of the existing literature is concerned with the system risk of the finished structure. Consequently, the aim of the present paper is to estimate system risks for the construction of multi-storey concrete buildings. At a later stage the effect of construction errors and error control measures on system risk will be investigated.

A Probabilistic Risk Assessment (PRA) model is developed to estimate the system risk (i.e., probability of structural collapse) during the construction of typical multi-storey RC buildings. Collapse will occur when the slab capacity (in flexure or shear) is exceeded. The model proposed herein, although preliminary in nature, provides a more detailed approach than that used by El-Shahhat, et.al. (1993). For instance, El-Shahhat, et.al. (1993) assumed uncorrelated shear and flexural capacities (and modes of failure), what appears to be an overly conservative construction live load model [Stewart, 1996], and relatively simple probabilistic capacity models. The present PRA model simulates construction scheduling (e.g., timing of concrete placement and removal of shoring) and includes the influence of dead and construction live loads, material properties and dimensional variations in the computation of system risk. It is assumed that existing design code specifications are complied with. This risk-based approach can be used to select optimal construction procedures. The PRA model is used to assess the influence that the number of levels of shoring/reshoring, quality of concrete workmanship (concrete compressive strength) and scheduling (3, 5 or 10 days) have on system risks. The structural configuration considered herein is a four storey RC office building with flat plate construction.

2.   STRUCTURAL CONFIGURATION

The structural configuration considered herein is a four storey RC office building with flat plate construction (i.e., flat slab without drop panels). Three realistic shoring configurations are considered:
(i)   two floors shored;
(ii) two floors shored and one floor reshored (i.e., backpropped) with no pre-compression (see Figure 1); and
(ii) three floors shored.
It is assumed that floors are shored with braced steel "V-Shore Frames" (1.2m spacing). It was found by El-Shahhat, et.al. (1993) that changing the design shore spacing from 1.0m to 2.0m has a negligible influence on system risk; hence only one shore spacing is considered herein. The slabs are designed to Australian code specifications [AS3600-1994] considering two strength limit states:
(i)   "Service Loading": $1.25G+1.5Q = 10.5kPa$ and
(ii) "Construction Loading": $\max[1.25G+1.5Q, 2.0*(1.25G+1.5Q_C)] = 15.0kPa$.
where G, Q and Q_c is the design dead load, office floor live load (3kPa) and construction live load (1kPa) respectively [AS1170.1-1989; AS3610-1995]. The "service loading" limit state is normally non-conservative because the construction process often requires that some structural members support construction loads that are greater than their design service loads. Thus, design loads are max(Construction Loading, Service Loading)=10.5kPa. Design slab depth is 200mm for 6m column spacing and slabs are assumed to be fully restrained at supports.

Note it has been shown by Epaarachchi, et.al. (1996) that if slabs are designed to the non-conservative "service loading" limit state then the system risk is significantly (up to four times) greater than that obtained for the "construction loading" limit state and indicates that ignoring structural effects that occur during the construction process can cause a dramatic loss of structural safety.

3.   PROBABILISTIC RISK ASSESSMENT MODEL

3.1   Structural Analysis

Grundy and Kabaila (1963) developed one of the earliest structural analysis model for estimating the actions in shore-slab structural systems. A number of other models have been developed since this "simplified" model. A model with significant improvements over the "simplified" model is the "refined approach" [Liu, et.al., 1986]. The model includes the effect of time-dependent concrete stiffness, and assumes linear elastic behaviour, infinitely rigid foundations, shore and reshore stiffness, and pinned connections between shores and slabs. It was found that a three dimensional structural system can be accurately modelled by a two dimensional model, and that the creep and shrinkage have little effect on structural actions [Liu and Chen, 1987]. Other models have been developed since the "refined" model (e.g., "improved analysis method" - El-Shahhat and Chen, 1992); however, the limited gain in accuracy is offset by the significant increase in model complexity. Consequently, the "refined approach" is incorporated into the PRA model developed herein.

3.2   Concrete Workmanship

Parameters used in the PRA model include the compressive and tensile strengths of concrete cylinder specimens. Some data is available that relates the effects of curing and compaction to the compressive strength of cylinder specimens. The concrete compressive strength of a cylinder specimen is thus expressed as

$$f'_c = k_{cp}k_{cr}f'_{cyl} \qquad (1)$$

where k_{cp} is the compaction coefficient; k_{cr} is the curing coefficient; and f'_{cyl} is the concrete compressive strength of a standard test cylinder. The statistical parameters for k_{cp} and k_{cr} will be influenced by the workmanship of the construction workers. It is assumed that compaction and curing are subject to poor, fair or good levels of workmanship. A fair level of performance refers to a task outcome that satisfies minimum Australian Concrete Structures

Code [AS3600-1994] requirements. Probabilistic models for the variables given in Eqn. (1) are described elsewhere [Stewart, 1995].

The variables k_{cp}, k_{cr}, and f'_{cyl} are assumed independent and normally distributed, so f'_c is normally distributed also. The statistical parameters for the concrete compressive strength of cylinder specimens (f'_c) are shown in Table 1, when all tasks are performed to poor, fair or good levels of performance. Differences between these distributions will significantly influence axial and punching shear capacities.

Concrete strengths and elastic modulus are time dependent variables; namely,

$$f'_c(t) = \frac{t}{\lambda + \omega t} f'_c(28)$$ (2)

where $\lambda=4.0$ and $\omega=0.85$ for moist cured normal (type I) Portland cement [ACI 209, 1978], and concrete tensile strength (f_{ct}) and elastic modulus (E_c) are dependent variables.

Element dimensions, material properties, prediction models and loads are treated as random variables, see Table 2. Slab dead load is a dependent variable and variability of shore capacity is negligible. It is assumed that the construction live load for floors supporting shores/reshores is 25% of the construction live load for the uppermost floor [AS3610-1995].

3.3 Structural Reliability

Collapse will occur when the slab capacity (in flexure or shear) is exceeded. The structure will be re-analysed whenever failure of one or more shores/reshores occurs so that the influence of the resulting load redistribution is considered. Thus, system risk is defined as the probability of exceeding (i) flexural slab capacity at midspan; (ii) flexural slab capacity at column support

Table 1. Statistical Parameters for f'_c

Worker Performance	Minimum Curing Times			
	3 days		7 days	
	Mean	COV	Mean	COV
Poor	0.623F'$_c$	0.16	0.623F'$_c$	0.16
Fair	0.862F'$_c$	0.16	1.027F'$_c$	0.15
Good	1.18F'$_c$	0.14	1.18F'$_c$	0.14

Table 2. Statistical Parameters of Random Variables.

Parameter	Mean	COV	Distribution	Source
Dead Load of Shores	0.11G	0.05	Normal	El-Shahhat et.al. (1993)
Construction Live Load	0.30kPa	σ=0.32	Weibull	Karshenas & Ayoub (1994)
Concrete Density	24kN/m^3	0.03	Normal	Ellingwood, et.al. (1980)
Slab Depth	D_{nom}+0.80mm	σ=11.9mm	Normal	Mirza & MacGregor (1979a)
Top Cover	C_{nom}+3.55mm	σ=12.7mm	Normal	Morgan et.al. (1982)
Bottom Cover	C_{nom}-0.22mm	σ=2.5mm	Normal	Morgan et.al. (1982)
A_{st}-0.91A_{stnom}	0.1A_{stnom}	0.04	Lognormal	Mirza & MacGregor (1979a)
f_{cyl}	1.18F'$_c$	0.14	Normal	Pham (1985)
$f_{ct}(t)$	0.69$\sqrt{f'_c(t)}$	0.20	Normal	Mirza, et.al. (1979)
$E'_c(t)$	4400$f'_c(t)^{0.516}$	0.15	Normal	Attard & Stewart (1996)
f_{sy}	465 MPa	0.098	Beta	Mirza & MacGregor (1979b)
Model Error (Flexure)	1.01	0.046	Normal	Ellingwood, et.al. (1980)
Model Error (Shear)	1.36	0.19	Normal	Ellingwood, et.al. (1983)

Step	Operation	Status of Structure and Concrete Age (in days)	Poor			Fair			Good		
			V	M	V&M	V	M	V&M	V	M	V&M
1	Place Level 1 Concrete	 0	–	–	–	–	–	–	–	–	–
2	Place Level 2 Concrete	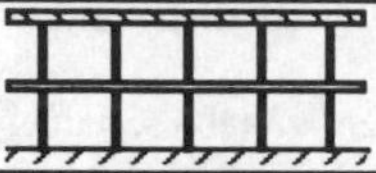 0	–	–	–	–	–	–	–	–	–
		7	0	0.1	0	0	0	0	0	0	0
3	Remove Shores under Level 1	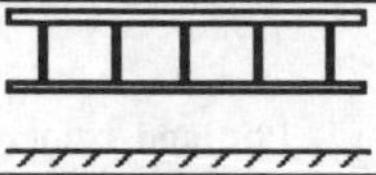 7	0.2	0.1	0	0.3	0	0	0	0	0
		14	6.7	1.0	0.3	6.0	0	0	10.0	0	0
4	Place Reshores Under Level 1	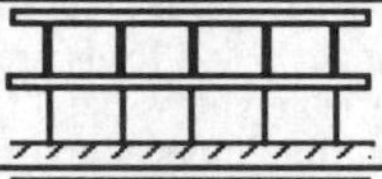 7	0	0	0	0	0	0	0	0	0
		14	0	0	0	0	0	0	0	0	0
5	Place Level 3 Concrete	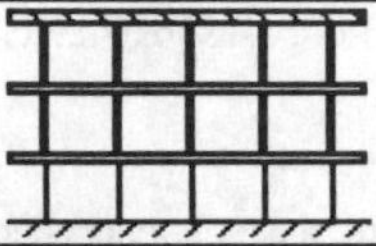 0	–	–	–	–	–	–	–	–	–
		7	1.8	0.5	0.1	1.4	0	0	4.0	0	0
		14	3.9	0.3	0	3.5	0	0	7.0	0	0
6	Remove Reshores Under Level 1	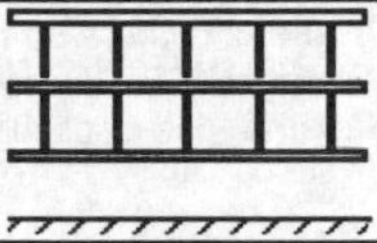 7	0	0	0	0	0	0	0	0	0
		14	0.3	0	0	0.5	0	0	1.0	0	0
		21	9.1	0.4	0	10.9	0	0	9.0	0	0
7	Remove Shores Under Level 2	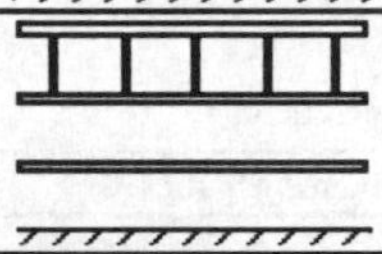 7	0	0.2	0	0	0	0	0	0	0
		14	20.5	1.9	0.8	24.9	0	0	15.0	0	0
		21	0	0	0	0	0	0	0	0	0
8	Reshore Below Level 2	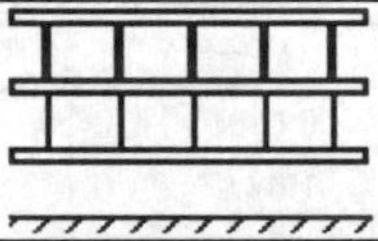 7	0	0	0	0	0	0	0	0	0
		14	0	0	0	0	0	0	0	0	0
		21	0	0	0	0	0	0	0	0	0
9	Place Level 4 Concrete	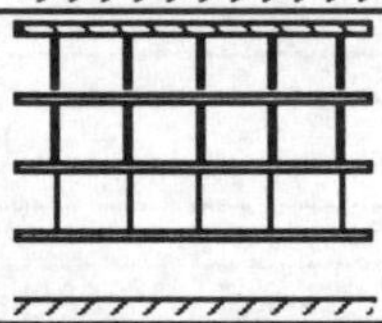 0	–	–	–	–	–	–	–	–	–
		7	0.3	0.2	0	0.8	0	0	0	0	0
		14	45.8	1.1	0.4	47.6	0.3	0	52.0	0	0
		21	0.1	0.4	0	0	0	0	0	0	0
10	Remove Reshores Under Level 2	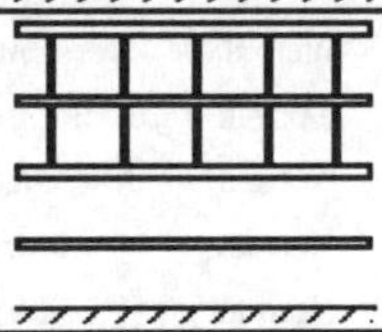 7	0	0	0	0	0	0	0	0	0
		14	0	0	0	0	0	0	0	0	0
		21	3.2	0.1	0	3.8	0	0	2.0	0	0
		28	0	0	0	0	0	0	0	0	0
		Total (%)	89.1	6.5	1.6	99.7	0.3	0	100.0	0	0

Note: ▌= shore; ▏= reshore; V=punching shear failure; M=flexural failure; V&M=combined failure

Figure 1. Seven Day Construction Cycle (2 shores, 1 reshore) of a Multi-Storey Concrete Building and Proportion of Failures, for Poor, Fair and Good Levels of Concreting Performance.

Table 3. System Risks

Construction Cycle	2 shores			2 shores / 1 reshore			3 shores		
	Poor	Fair	Good	Poor	Fair	Good	Poor	Fair	Good
5 days	0.0491	0.0109	0.0027	0.0231	0.0051	0.0013	0.0101	0.0023	0.0008
7 days	0.0341	0.0082	0.0019	0.01622	0.0037	0.0010	0.0076	0.0017	0.0006
10 days	0.0258	0.0056	0.0015	0.01227	0.0029	0.0008	0.0061	0.0014	0.0005

or (iii) punching shear capacity of slab adjacent to the column. System risk is calculated from Monte-Carlo computer simulation analysis.

4.   RESULTS

System risks obtained from Monte-Carlo simulation analysis (100,000 runs) of the PRA model are given in Table 3, for (i) specified concrete compressive strength (F'_c) of 32MPa; (ii) construction cycles of 5, 7 and 10 days; (iii) concrete workmanship where all tasks are performed to poor, fair or good levels of performance; (iv) three shoring configurations (2 shores, 2 shores / 1 reshore, and 3 shores), and (v) specified curing time is three days. The relative proportion of failures attributed to each failure mode (punching shear, flexural or both), construction step and floor level is shown in Figure 1.

System risks obtained by El-Shahhat, et.al. (1993) for a similar structure (2 shores / 1 reshore, but with timber shores) appear to be unrealistically high (0.05-0.12); this is most likely due to the very high construction live loads used in that study. The system risks estimated herein are significantly less than these figures and so are probably more realistic. However, the system risks calculated in the present paper are still "notional" and so are used herein for comparative purposes only.

It is observed from Figure 1 that the dominant failure mode is punching shear, particularly at construction steps 7 (Level 2) and 9 (Level 2). Punching shear is the failure mode for approximately 90% of failures for other shoring configurations also. This suggests that concrete compressive strength and slab depth are important variables.

The results show that a shoring configuration comprising of 2 shores produces system risks approximately twice than that observed for a 2 shore / 1 reshore configuration. A similar increase in system risks is observed between the 2 shore / 1 reshore and 3 shore configurations. This is expected, increasing the number of floors that are reshored (or even better - fully shored) increases the age of the slab when it experiences its greatest load. The age of the slab at first or maximum loading is important since early-age affects considerably influence concrete compressive strength, and similarly punching shear capacity. This explains also why poor concreting performance (which significantly affects f'_c, see Table 1) is more detrimental to system risk than shortening the construction cycle by 2 days. For example, system risk may be reduced by several orders of magnitude if concreting workmanship is improved from poor to good. However, increasing the construction cycle from 7 to 10 days reduces system risk by no more than only 30%. The important influence of concrete compressive strength on system risk shown herein is in broad agreement with El-Shahhat, et.al. (1993).

5.   FURTHER WORK

Recent reviews of construction failures shows that a significant proportion of failures can be attributed to construction errors [e.g., Eldukair and Ayyub, 1991]. Failure may result from construction errors that cause: (i) high extraordinary loads (e.g., material stacking); (ii) reduction in structural resistance of shores, beams or slabs (e.g., reinforcement placement errors), (iii) incorrect shore installation or (iv) premature removal of shoring. The PRA model will be extended at a later date to include the effects of these and other human errors, other shoring and structural configurations, and a risk-cost-benefit analysis to optimise the construction process.

531

6. CONCLUSION

A probabilistic model has been developed to estimate the probability of failure during the construction of multi-storey concrete buildings. Results suggest that poor concreting workmanship (poor curing and compaction that reduces concrete compressive strength) is more detrimental to system risk than shortening the construction cycle. System risks for different shoring configurations were estimated also.

REFERENCES

ACI 209 (1978), *Prediction of Creep, Shrinkage and Temperature Effects: 2*, ACI Committee 209, Subcommittee II, Draft Report, Detroit.

Attard, M.M. and Stewart, M.G. (1996), A Two Parameter Stress Block for High Strength Concrete, *UNICIV Report No. R-357*, School of Civil Engineering, The University of New South Wales, Australia.

Eldukair, Z.A. and Ayyub, B.M. (1991), Analysis of Recent U.S. Structural and Construction Failures, *Journal of Performance of Constructed Facilities*, ASCE, 5(1), 57-73.

Ellingwood, B., Galambos, T.V., MacGregor, J.G. and Cornell, C.A. (1980), Development of a Probability Based Load Criterion for American National Standard A58, NBS Special Publication 577, Washington D.C.

El-Shahhat, A.M. and Chen, W.F. (1992), Improved Analysis of Shore-Slab Interaction, *ACI Structural Journal*, 89(5), 528-537.

El-Shahhat, A.M., Rosowsky, D.V. and Chen, W-F (1993), Construction Safety of Multistory Concrete Buildings, *ACI Structural Journal*, 90(4), 335-341.

Epaarachchi, D., Stewart, M.G. and Rosowsky, D.V. (1996), System Risk for Multi-storey Reinforced Concrete Building Construction, *ASCE Specialty Conference on Probabilistic Mechanics and Structural Reliability*, Worcester, Massachusetts, 230-233.

Grundy, P. and Kabaila, A. (1963), Construction Loads on Slabs with Shored Formwork in Multistory Buildings, *ACI Journal*, 60(12), 1729-1738.

Karshenas, S. and Ayoub, H. (1994), Analysis of Concrete Construction Live Loads on Newly Poured Slabs, *Journal of Structural Engineering*, ASCE, 120(5), 1525-1542 (see also pp. 1543-1562).

Liu, Xi-L. and Chen, W-F (1987). Effect of Creep on Load Distribution in Multistory Reinforced Concrete Buildings during Construction, *ACI Structural Journal*, 84, 192-200.

Liu, Xi-L., Chen, W-F and Bowman, M.D. (1986). Shore-Slab Interaction in Concrete Buildings, *Journal of Structural Engineering*, ASCE, 112(2), 227-244.

MacGregor, J.G, Mirza, S.A. and Ellingwood, B. (1983), Statistical Analysis of Resistance of Reinforced and Prestressed Concrete Members, *ACI Journal*, 80, 167-176.

Mirza, S.A. and MacGregor, J.G. (1979a), Variations in Dimensions of Reinforced Concrete Members, *Journal of the Structural Division, ASCE*, 105(ST4), 751-766.

Mirza, S.A. and MacGregor, J.G. (1979b), Variability of Mechanical Properties of Reinforcing Bars, *Journal of the Structural Division, ASCE*, 105(ST5), 921-937.

Mirza, S.A., Hatzinikolas, M. and MacGregor, J.G. (1979), Statistical Descriptions of Strength of Concrete, *Journal of the Structural Division*, ASCE, 105(ST6), 1021-1037.

Morgan, P.R., Ng, T.E., Smith, N.H.M. and Base, G.D. (1982), How Accurately Can Reinforcing Steel Be Placed? Field Measurement Tolerances Compared to Codes, *Concrete International*, 4(10), 54-65.

Pham, L. (1985), Reliability Analyses of Reinforced Concrete and Composite Column Sections Under Concentric Loads, *Civil Engineering Transactions*, IEAust, CE27(1), 68-72.

Stewart, M.G. (1995), Workmanship and its Influence on Probabilistic Models of Concrete Compressive Strength, *ACI Materials Journal*, 92(4), 361-372.

Stewart, M.G. (1996), Construction Error, Proof Loading and the Reliability of Service Proven Structures, Research Report No. 127.1.1996, Department of Civil Engineering and Surveying, The University of Newcastle, Australia.

The Mechanics of Structures and Materials, Grzebieta, Al-Mahaidi & Wilson (eds)
© 1997 Balkema, Rotterdam, ISBN 90 5410 900 9

Sea-spray corrosion and time-dependent structural reliability of concrete bridges

Mark G. Stewart
Department of Civil, Surveying and Environmental Engineering, The University of Newcastle, N.S.W., Australia

David V. Rosowsky
Department of Civil Engineering, Clemson University, S.C., USA

ABSTRACT: A structural deterioration reliability model is developed to calculate probabilities of structural failure (flexure) for a typical reinforced concrete bridge. Corrosion is initiated from atmospheric exposure in a marine environment. Monte-Carlo simulation analysis is used to calculate probabilities of failure for annual increments over the lifetime of the structure. It was found that atmospheric marine exposure causes significant long-term deterioration and reduction in structural safety.

1 INTRODUCTION

Corrosion may significantly influence the long-term performance of reinforced concrete (RC) structures, particularly in aggressive environments. Of most concern among concrete structures is the deterioration of reinforced concrete bridges, and an aggressive chloride environment often exists for bridges sited in a marine environment, say within 1-2km from a coast. Bridge Management Systems may be used for the optimal allocation of resources for bridge design, construction and maintenance; this may include, amongst other things, design specifications, construction quality, inspection and maintenance (repair) strategies and bridge (or load) ratings. A useful assessment tool is probabilistic analysis since it provides a rational criterion for the comparison of the effectiveness of decisions taken under uncertainty. For example, an updated reliability of a bridge (after an inspection) may be compared with some minimum acceptable reliability; from this comparison the relative "safety" of the bridge can be ascertained, a load rating assigned or a "residual life" estimated. Consequently, the present paper develops a preliminary probabilistic framework for time-dependent life-cycle (reliability) analysis.

The application of probabilistic methods to the evaluation of deterioration of RC bridges is not new [e.g., Thoft-Christensen, 1992]. However, existing methods are relatively limited because they ignore loss of bond, focus on effects of de-icing salts only, ignore cracking, etc. The present paper will develop an improved probabilistic framework for estimating the time-dependent reliability for RC bridges subject to chloride-induced corrosion. This will involve the development of probabilistic models for behaviour-based corrosion initiation and propagation phenomena. Initiation is deemed to occur if the chloride concentration exceeds a critical threshold value or if flexural crack widths are sufficiently large to allow the direct ingress of chlorides. The analysis includes the random variability of chloride diffusion, critical threshold chloride concentration, crack width, corrosion rates, material properties, element dimensions, reinforcement placement, environmental conditions and loads. It is assumed that corrosion will lead to a reduction in the cross-sectional area of the reinforcing steel. Loss of bond is not considered. For illustrative purposes, a reliability analysis is performed on a three span RC slab bridge. Monte-Carlo simulation analysis will help quantify the extent to which time-dependent actions such as corrosion affect long-term durability and reliability of RC bridge decks.

2. TIME TO INITIATION

2.1 Chloride Diffusion

In a coastal (atmospheric marine) zone, salt spray (water-borne chloride ions carried by the wind) will accumulate on the concrete surface and so it is likely that the surface chloride content will increase with time in service. The chloride content [C(x,t)] at a distance x from the concrete surface and at time t is given empirically by Fick's second law:

$$C(x,t) = 2W \left\{ \sqrt{\frac{t}{\pi D}} \; \exp\left(-\frac{x^2}{4Dt} \right) - \frac{x}{2D}\left[1 - \text{erf}\left(\frac{x}{2\sqrt{tD}} \right) \right] \right\}$$ (1)

where W is the diffusion flux on the concrete surface (kg/cm^{3}s), D is the apparent diffusion coefficient of chloride and t is the length of exposure [e.g., Uji, et.al., 1990]. A constant W implies that the chloride content in the environment is constant and that the chloride content on the surface accumulates with the square root of the time in service. The chloride concentrations used herein are acid-soluble chlorides.

Note that W_o denotes the diffusion flux for structures sited adjacent (or very close) to the coastline. The diffusion flux (W) reduces as the distance from the ocean increases; for example, the diffusion flux for structures sited at 250m and 1km from a coastline are $W \approx 0.33W_o$ and $W \approx 0.10W_o$ respectively. The chloride concentration must reach a critical threshold chloride concentration (Cr) to cause dissolution of the protective passive film around the reinforcement, thereby initiating corrosion of reinforcement.

Statistical parameters for these random variables (D, W_o, Cr) are shown in Table 1.

2.2 Flexural Cracking

Cracks exist in concrete due to construction, shrinkage, thermal and loading effects. For the present, flexural cracking is considered since it is assumed that the diffusion process described previously includes (though not explicitly) the influence of microcracking. In general, it can be stated that crack width may affect the initiation of corrosion, but does not affect the subsequent propagation process [Beeby, 1983]. The maximum crack width (w_{max}) at any point is given by the BS 8110 crack control equation. However, this and other equations show significantly increased crack widths as concrete cover increases. Yet experimental results suggest that crack width is markedly less sensitive to concrete cover [Makhlouf and Malhas, 1996]. This suggests that existing crack width equations are of limited accuracy.

The maximum surface crack width may be expressed as w_{max} (t)=[1+k(t)] w_{max} where k(t) is a factor incorporating the time-dependent growth of crack width and w_{max} is the maximum surface crack width due to a sustained load. A negative exponential relationship for k(t) is used since it is assumed that the crack width will double after three years [e.g., Nawy, 1992]. It has been shown that corrosion initiation is not significantly affected by surface crack widths less than about 0.3-0.6mm [e.g., Gergely, 1981]. Statistical parameters for the crack width random variables are shown in Table 1.

3. PROPAGATION

It is assumed herein that the deterioration process caused by corrosion of reinforcement will lead to a reduction in the bar diameter of the reinforcing steel D(t); thus

$$D(t) = \begin{cases} D_i & t \leq T_i \\ D_i - 2\lambda(t - T_i) & T_i < t \leq T_i + (D_i / 2\lambda) \\ 0 & t > T_i + (D_i / 2\lambda) \end{cases}$$ (2)

Table 1. Statistical Parameters for Corrosion Models [Stewart and Rosowsky, 1997].

Parameter	Mean	COV	Distribution
D	$2.0E-8$ cm^2/s	0.75	Lognormal
W_o	$7.5E-15$ kg/cm^{3}s	0.6	Lognormal
Cr	0.9kg/m^3	0.19	Uniform [0.6-1.2]
i_{corr}	1.0	0.2	Normal
R	3.0	0.33	Normal
w_{max}	1.0	0.4	Normal
Critical crack width	0.45mm	0.19	Uniform [0.3-0.6]

where D_i is the initial bar diameter, T_i is the time to initiation (see Section 2) and λ is the corrosion rate (at a surface) in mm/yr. The corrosion rate is measured from field/experimental studies as the current density i_{corr} (normally expressed in μA/cm^2), where $\lambda=0.0116Ri_{corr}$ (mm/yr) and R is a factor that includes the effect of highly localised pitting normally associated with chloride contamination. In this case, $R=P_{max}/P_{av}$ where P_{max} is the maximum penetration of corrosion and P_{av} is the penetration calculated solely from i_{corr} values since measured values of i_{corr} are average values for the overall reinforcing bar surface. Statistical parameters for i_{corr} and R are given in Table 1.

4. TIME-DEPENDENT RELIABILITY ANALYSIS

A corrosion-induced deterioration process will reduce the structural resistance of one or more structural elements. Hence the structural resistance is time-dependent, and is denoted herein as $R(t)$. Further, structural loads may occur randomly in time and/or in intensity. If it is assumed that n independent load events S_j occur within the time interval (0, t_L) at deterministic times t_j, $j=1,2,....,n$, then the probability that the structure fails anytime during this time interval is

$$p_f(t_L) = 1 - Pr\left[R(t_1) > S_1 \cap R(t_2) > S_2 \cap .... \cap R(t_n) > S_n\right] \qquad (3)$$

in which $t_1<t_2<....<t_-$ [e.g., Kameda and Koike, 1975]. The analysis in this study is further complicated by the fact that both the load and resistance are time-dependent quantities, as well as the presence of highly nonlinear limit state functions, non-normal random variables, and time-varying load and/or resistance quantities. Thus, closed-form solutions are not tractable and Monte-Carlo simulation will be used herein to evaluate Eqn. (3).

5. ILLUSTRATIVE EXAMPLE - THREE SPAN RC SLAB BRIDGE

5.1 Structural Configuration and Computational Procedure

The bridge considered in this study is a three-span continuous RC slab bridge. The end and mid spans are 9.75m and 12.2m respectively, and the bridge width is 11m (2 lanes). The deck slab is 412mm thick, including a 19mm integrated wearing course. Reinforcing steel is Grade 40 (276MPa), specified concrete strength is 34.5MPa, and top and bottom covers are 46mm and 32mm respectively.

The statistical parameters for dimensions, materials and dead and truck live loads appropriate for the three-span continuous RC concrete bridge are given elsewhere [Stewart and Rosowsky, 1997]. It is assumed that the slab deck and soffit are exposed to atmospheric marine chlorides and that slab depth, all material properties and all corrosion parameters are constant across the entire bridge.

The analysis uses statistically independent annual maximum truck loads; hence time-dependent failure probabilities can be calculated from Eqn. (3) considering annual maximum loads and the cracking/damage they cause. Failure probabilities are calculated for

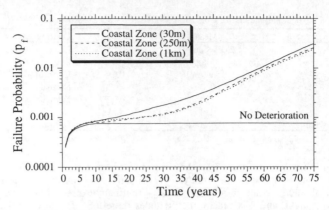

Figure 1. Time-Dependent Failure Probabilities

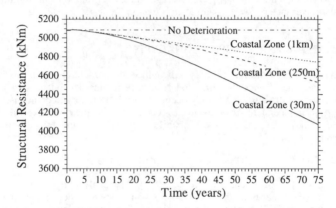

Figure 2. Mean Structural Resistance over Support

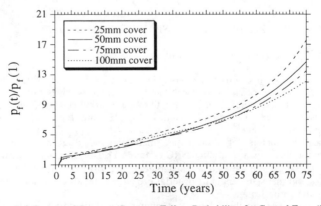

Figure 3. Influence of Concrete Cover on Failure Probability, for Coastal Zone (30m)

75 successive annual time increments. Flexural actions are computed for: (i) middle of central span "midspan" and (ii) over the supports. Failure is deemed to occur if the flexural actions exceed the structural resistance at either of these cross-sections. Axle spacings and distribution of axle loads are based upon those of the standard HS-20 truck and these were used to calculate peak flexural actions at these cross-sections.

5.2 Results

Probabilities of failure as obtained from Monte-Carlo simulation analysis are presented in Figure 1 for deterioration of reinforcing steel caused by atmospheric marine exposure (coastal zones). For comparative purposes, failure probabilities for the ideal case of no deterioration (i.e., time invariant resistance) are shown also. Mean structural resistances (time-dependent), in this case over the supports, are shown in Figure 2.
It is observed that atmospheric marine exposure causes a large loss of structural safety for the bridge considered herein. Diffusion and flexural cracking induced initiation for the cross-section over the supports. Flexural cracking is the dominant cause of initiation only for top reinforcement over the support. This is due mainly to the increased cover for the top reinforcement, and the characteristic of the cracking model to increase crack width as cover increases. For other slab surfaces, penetration of chlorides by diffusion is the dominant cause of corrosion initiation.
Reliability analyses were conducted to assess the influence of cover on failure probabilities. In order to isolate cover as the main variable, only the cross-section over the supports was considered. For example, Figure 3 shows comparisons of $p_f(t)/p_f(1)$, where $p_f(1)$ is the probability of failure prior to any deterioration, for an element exposed to atmospheric marine salts (30m from coast). It appears that reliabilities are not significantly affected by cover. However, note that whilst increasing cover further impedes the diffusion of chlorides, existing crack width models suggest increasing cover also leads to wider surface cracks and thus increased crack-induced initiation.

6. DISCUSSION

The parameter values selected herein are obtained from limited empirical evidence. Further, the parameter values are generally obtained from field and laboratory tests conducted in the US, Japan, Ireland, the UK, Canada, Germany, etc.- so there is likely to be some inconsistencies in materials tested, experimental techniques, and so forth. Note also that existing studies are limited mainly to concrete that is unstressed and uncracked. There is growing evidence to suggest that in the uncracked tensile zone damage at the grain/paste interface results in the increased diffusion of chloride and other aggressive ions. Further, bridges receive intermittent exposure to chlorides (rain, direction of wind, etc.), yet most chloride diffusion models assume continuous exposure to chlorides.
The present analysis (i) ignores system effects and load redistribution in the calculation of flexural capacity and actions; (ii) assumes all reinforcement corrodes uniformly across the entire bridge deck (11m); (iii) assumes that corrosion will occur only at the critical cross-sections, and (iv) assumes the absence of inspections and repairs (hence "failure" will occur whereas in practice signs of structural distress would indicate a need for repair) - this will necessarily lead to the overestimation of failure probabilities. Further, the values assigned to the parameters and the selection of material, behavioural and load models used herein are subject to considerable uncertainty. Hence, the failure probabilities calculated herein are "notional" and so should be used for comparative purposes only.
Finally, note that the "notional" failure probabilities calculated herein refer to failure probabilities for "generic" bridges that reflect a variety of construction practices, environmental conditions, etc. A reliability analysis may be conducted also for any specific bridge. In such cases the coefficient of variation for some parameters (e.g., D, W_0, C_0, i_{corr}) would most likely be lower than the values used herein (i.e., increased certainty about their variability), although the mean values may change also. In general, this will result in a

calculated failure probability lower than that obtained in the present paper.

More details of the probabilistic models, method, results (including a sensitivity analysis) and discussion (and their limitations) are described elsewhere [Stewart and Rosowsky, 1997]. The influence of the application of de-icing salts and exposure to atmospheric CO_2 were investigated also by Stewart and Rosowsky (1997).

7. FURTHER WORK

The models of corrosion initiation and propagation developed and used herein should be considered preliminary only. Further work is clearly needed in refining these models and their associated parameters, and in the inclusion of other factors affecting corrosion initiation and propagation. It is expected that the procedure and techniques proposed herein for time-dependent reliability analysis eventually will have important practical implications for load ratings and the optimisation of design, construction, inspection and repair strategies. It is clear that much scope exists for further work, and as such, a programme of research is now in progress [e.g., Val, et.al., 1997].

8. CONCLUSION

Probabilistic models have been proposed to represent the structural deterioration of RC bridge decks. Corrosion initiation and propagation reduces the cross-sectional area of the steel reinforcing bars, leading to a loss in structural strength. The influence of chloride diffusion and flexural cracking have been incorporated into the proposed probabilistic models. A reliability analysis of a three span RC slab bridge found that atmospheric marine exposure causes significant long-term deterioration.

REFERENCES

Beeby, A.W. (1983), Cracking, Cover and Corrosion of Reinforcement, *Concrete International,* 5(2): 35-40.

Gergely, P. (1981), Role of Cover and Bar Spacing in Reinforced Concrete, *Significant Developments in Engineering Practice and Research,* ACI SP-72, American Concrete Institute, Detroit, 133-147.

Kameda, H. and Koike, T. (1975), Reliability Analysis of Deteriorating Structures, *Reliability Approach in Structural Engineering,* Maruzen Co., Tokyo, 61-76.

Makhlouf, H.M. and Malhas, F.A. (1996), The Effect of Thick Concrete Cover on the Maximum Flexural Crack Width under Service Load, *ACI Structural Journal,* 93(3): 257-265.

Nawy, E.G. (1992), Macro-Cracking and Crack Control in Concrete Structures - a State of the Art, *Designing Concrete Structures for Serviceability and Safety,* E.G. Nawy and A. Scanlon (Eds.), ACI SP-133, American Concrete Institute, Detroit, 1-32.

Stewart, M.G. and Rosowsky, D.V. (1997), Corrosion, Deterioration and Time-Dependent Reliability of Reinforced Concrete Bridge Decks, *Structural Engineering Research Report, STR-97-01,* Dept. of Civil Engineering, Clemson University, SC.

Thoft-Christensen, P. (1992), A Reliability Based Expert System for Bridge Maintenance, *Tekno Vision Conference on Bridge and Road Maintenance Management Systems,* Copenhagen, Denmark, May 25-26.

Uji, K., Matsuoka, Y. and Maruya, T. (1990), Formulation of an Equation for Surface Chloride Content of Concrete Due to Permeation of Chloride, *Corrosion of Reinforcement in Concrete,* C.L. Page, K.W.J. Treadway and P.B. Bamforth (Eds.), Elsevier Science, Barking, U.K., 258-267.

Val, D.V., Stewart, M.G. and Melchers, R.E. (1997), Effect of Reinforcement Corrosion on Reliability of Highway Bridges, Journal of Engineering Review (in press).

Steel design and welding

The Mechanics of Structures and Materials, Grzebieta, Al-Mahaidi & Wilson (eds)
© *1997 Balkema, Rotterdam, ISBN 90 5410 900 9*

The effect of hole size and position in circular plates

J. Vogwell
School of Mechanical Engineering, University of Bath, UK

J. M. Minguez
Facultad De Ciencias, Universidad Del Pais Vasco, Bilbao, Spain

ABSTRACT : A study has been carried out to investigate the effect of hole size and position on the bending strength and stiffness of circular plates subjected to pressure loading. This has involved developing theory for validating basic finite element analyses for similar annular plates and then extending models for dealing with eccentric hole arrangements.

This paper summarises equations which have been derived specifically for circular plates (or diaphragms) containing a central hole when firmly held around the outer perimeter and supporting a uniformly distributed (pressure) load. It presents the resulting hoop and radial stress distributions which are obtained for a range of different hole sizes and compares these with those obtained using finite element analysis.

Although the results compare very well for most geometries considered (especially at the no hole / small hole transition), there are other geometries where results do not compare favourably. As this characteristic has been found with other component shapes studied, the paper questions whether well established analytical data is generally as precise as previously assumed before the advent of finite element solutions. In addition, the desirability for initially validating the results from simple finite element models with reliable results is advocated.

NOMENCLATURE

				suffix	
e	Eccentricity of hole position	(mm)			
E	Elastic modulus of plate material	(GN/m^2)	h	hoop	
R	Radius	(mm)	i	inner	
t	Thickness of plate	(mm)	o	outer	
w	Distributed (pressure) load	(MN/m^2)	r	radial	
Q	Shear force per unit length	(kN/m)			
s	Stress	(MN/m^2)		accent	
n	Poisson's ratio		^	maximum value	

1 INTRODUCTION

Circular plates with centrally positioned holes occur widely in engineering applications. A common example is the rod end wall of a hydraulic actuator but they also occur in pressure vessels, as diaphragm discs or as the platen on a press. However, holes of non-circular shape and also those eccentrically positioned (perhaps not located at the centre for reasons of access or observation convenience) also occur and these factors greatly affect the resulting hoop and radial stress distributions and the deflection (thus the plate strength and stiffness). Conventional theory, and even that displayed in data form in reference texts such as Roark (1989), may appear at first to be comprehensive but can prove unsatisfactory as they are limited to symmetrical cases and do not deal, for example, with non-circular or eccentrically positioned holes.

To investigate the effect of changing hole size and position on a pressure loaded circular plate, a series of finite element analyses have been carried out although due to space restrictions only the former case is included. Comparison has been made with annular plate theory, as direct analysis can readily be applied, and this has identified some interesting shortcomings in the theory which had not been recognised previously.

Also, the anomaly in stress states which occurs in the respective theories between a plate having, and one not having a hole present, has been studied using finite element analysis; this has lead to furthering understanding of this transitional state discontinuity. Various hole shapes have been considered ranging from circular, square and even an elongated hybrid form and the radial and hoop stress distributions have been recorded.

2 ANNULAR PLATE THEORY

The fundamental theory for circular plates is presented in many structural mechanics texts such as Benham [1996] and Ryder [1981]. The general form of the differential equation which has been derived to determine expressions for slopes and deflections for specific cases is given by :-

$$\frac{d}{dR}\left[\frac{d}{RdR}\left(\frac{RdY}{dR}\right)\right] = -\frac{Q}{D} \qquad \text{where plate constant, } D = \frac{Et^3}{12(1-v^2)} \qquad (1)$$

Once a hole is introduced, however, developing specific equations for even a simple pressure loaded annular plate becomes relatively complex. Although very general equations are available in reference texts such as Timoshenko [1970] and Young's Roark [1989] for determining bending moments, stresses and deflections, obtaining solutions typically necessitates calculating many factors; this is found very cumbersome and has the disadvantage that the effect of individual parameters is not directly apparent. As this work has been concerned with establishing trends so specific theory has been developed.

2.1 Edge Supported Annular Plate

A case which is worthy of special attention is that of a circular plate (of outer radius R_o) containing a central hole (with inner radius R_i), subjected to a uniform pressure loading, w and rigidly held around the outer circumference as shown in Figure 1. This plate arrangement occurs in practise, for example, the rod end wall of a hydraulic actuator.

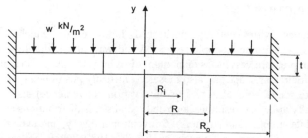

Figure 1   Annular plate in bending with edge firmly supported

The equations for the deflection, hoop and radial bending moment equations (and thus the respective stresses) are derived from applying appropriate boundary conditions to equation (1). There are three boundary conditions to be satisfied :

at the inner radius $R = R_i$ radial moment, $M_r = 0$

at the outer radius $R = R_o$ plate slope, $\theta = \dfrac{dy}{dR} = 0$ and deflection, $y = 0$

The salient equations which result from solving equation (1) for these conditions are :-

$$\text{deflection, } y = -\frac{w}{64D}\left[4R_o^2(R_o^2 - K)\ln\frac{R_o}{R} + (R_o^2 - R^2)(2K - R_o^2 - R^2)\right] \tag{2}$$

$$\text{where } K = \frac{(1-v)R_o^4 + (3+v)R_i^4}{(1-v)R_o^2 + (1+v)R_i^2} \qquad \text{a constant which is unique to this case.} \tag{3}$$

The resulting radial and hoop stresses are respectively :-

$$\sigma_r = \frac{6D}{t^2}\left(\frac{d^2y}{dR^2} + \frac{v}{R}\frac{dy}{dR}\right) = \frac{3w}{8t^2}\left[(1+v)K - (1-v)\frac{R_o^2}{R^2}(R_o^2 - K) - (3+v)R^2\right] \tag{4}$$

$$\sigma_h = \frac{6D}{t^2}\left(\frac{dy}{RdR} + \frac{vd^2y}{dR^2}\right) = \frac{3w}{8t^2}\left[(1+v)K + (1-v)\frac{R_o^2}{R^2}(R_o^2 - K) - (1+3v)R^2\right] \tag{5}$$

An expression for the radius at which the maximum radial stress, σ_r occurs may readily be obtained by differentiating equation (4), setting to zero and rearranging in terms of radius, R. The maximum and minimum hoop stresses occur at the hole edge and outer edges, respectively, as illustrated in Figure 2 for a particular plate arrangement.

2.2 Plate without a Hole

It is interesting to observe that when there is no hole present (that is when $R_i \rightarrow 0$) the constant, $K \rightarrow R_o^2$ which means that the radial and hoop stress equations (4) and (5) reduce to :

$$\sigma_r = \frac{3w}{8t^2}[(1+v)R_o^2 - (3+v)R^2] \tag{6}$$

$$\sigma_h = \frac{3w}{8t^2}[(1+v)R_o^2 - (1+3v)R^2] \tag{7}$$

Also the central deflection , $\hat{y} = -\dfrac{wR_o^4}{64D} = -\dfrac{3(1-v^2)wR_o^4}{16Et^3}$ (8)

Equations (6) , (7) and (8) are the standard equations which also result from independent derivation for a solid circular plate rigidly held around the outer edge.

2.3 Some Important Values

The greatest radial stress magnitude is at the rigidly held edge (where $R = R_o$)

$\hat{\sigma}_r = -\dfrac{3wR_o^2}{4t^2}$ This will be compressive at the bottom surface and tensile at the top.

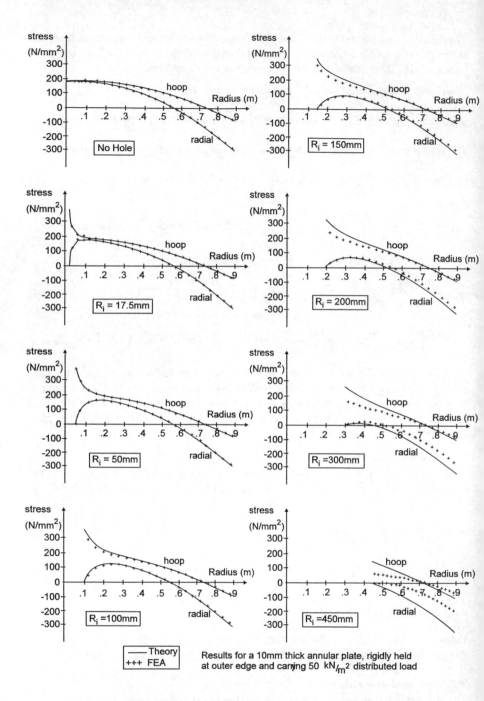

Figure 2 Radial and hoop stress distributions for plates with different hole sizes

Note that when $R = 0$ equation (6) yields a radial stress magnitude $\sigma_r = \dfrac{3(1+v)wR_o^2}{8t^2}$

for the centre of the plate. For an annular plate, even when having a very small hole in the centre, the theory assumes zero radial moment at the edge of the hole, at $R = R_i$. Consequently, there is no radial stress at this very small radius unlike the finite value which occurs if there were no hole. This anomaly identifies an interesting transition in the two theories which is well illustrated in both sets of results and this is discussed later.

The greatest hoop stress occurs at the centre of the plate (and has the same magnitude as the radial stress at the plate centre), that is $\hat{\sigma}_h = \dfrac{3(1+v)wR_o^2}{8t^2}$

No theory is given for plates with non-circular holes or for ones in which the hole in not centrally positioned. Annular plate theory was considered adequate for the purpose of validating the initial finite element results. Non-symmetrical cases are not covered in reference texts such as Young's edited Roark [1989] although Ollerton [1974] does present some data for stresses and deflections in circular plates with eccentric circular holes in his papers.

3 FINITE ELEMENT MODEL

The ABAQUS finite element package has been used and circular plates have been modelled using quadrilateral shell elements (type S4R5). The mesh used is shown in Figure 3 for a circular hole containing an eccentrically positioned circular hole.

Stress and deflection distributions have been obtained for plates which are rigidly constrained around the outer circumference and subjected to lateral pressure loading. The plate outer diameter, thickness and material properties have been taken to be constant to enable the effect of hole size and position to be studied. A range of different hole sizes and positions have been considered in order to compare accuracy and establish any deviations which occur with theoretically derived stresses.

Results have been based upon an aluminium plate, having an elastic modulus, $E = 70$ GN/m^2 and $v = 0.3$, an outer radius, $R_o = 900$ mm, a thickness, $t = 10$ mm and subjected to a uniformly distributed load per unit area, $w = 50$ kN/m^2.

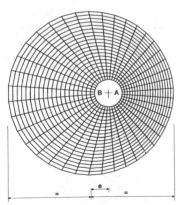

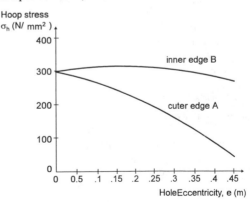

Figure 3   F.E. Model of plate with hole eccentrically positioned

Figure 4   Effect of hole position on maximum hoop stress

4 DISCUSSION OF RESULTS

The hoop and radial stress distributions for a range of plates with central holes are displayed in Figure 2. When summarised in such a way it is possible to observe a number of effects. For example, it would seem that the hoop and radial stresses at the rigidly supported edge are little affected by the size of the central hole according to the theory, although they are to a greater extent according to the finite element results.

Generally, the radial stress at the outer edge (regardless of hole size) has the greatest magnitude, with the exception of plates with central holes with diameters (or radii) up to about a third of the outer diameter (or radius) where the hoop stress at the inner hole diameter is slightly larger. This stress does not alter appreciably with hole size except at the transition between there being no hole to the inclusion of a very small hole. In contrast to other effects the difference between having or not having a hole is very pronounced and this is identified very well by the theory and finite element results.

For eccentrically positioned holes, the hole location is not found to make a significant difference in terms of maximum stress, as illustrated in Figure 4. This figure displays the extreme magnitudes of hoop stress in circular plates containing a 30 mm diameter hole at different eccentricities measured from the plate centre. These occur at the inner and outer edge of hole positions, taken radially from the plate centre, and it is the edge nearest the plate centre which is critical in terms of the greatest hoop stress magnitude. Indeed, the maximum stress magnitude remains virtually constant despite the position of the hole.

5 CONCLUSIONS

Very good agreement exists between theory and finite element stress results with annular plates in which the central hole diameter is smaller than about 20% of the plate outer diameter. For larger holes, the difference between the methods increases with hole size. It is believed that the derived equation becomes less accurate with increasing hole diameter because of the simplified constant load assumption made in the theory.

Both the derived theory and the finite element analysis methods accurately identified the stress anomaly between the no hole / small hole transition.

The hole position has not been found to be critical in terms of maximum hoop stress magnitude and thus is not a significant factor when determining the plate strength.

Because the finite element results are found to be accurate and as such packages are versatile, a revision of the format of published plate data is clearly possible and this is considered very desirable. An extension of the plate shapes considered is also now possible as it need not be limited just to those which can be analysed.

6 REFERENCES

Benham, P.P. Crawford, R.J. Armstrong, C.G. [1996] *Mechanics of Engineering Materials* - Chapter 16 Thin Plates and Shells, Longman Group Limited

Ollerton, E [1976a] *Bending Stresses in a Thin Circular Plate with Single Eccentric Circular Hole*, Inst. Mech. Eng. J. Strain Analysis, vol. 11, no. 4

Ollerton, E [1976b] *The Deflection of a Thin Circular Plate with an Eccentric Circular Hole* Inst. Mech. Eng. J. Strain Analysis, vol. 11, no. 2

Ryder,G [1981] *Strength of Materials* MacMillan Publishers London

Szilard, R [1974] *Theory and Analysis of Plates; Classical and Num. Methods*, Prentice-Hall

Timoshenko,S Woinowsky-Kreiger,S[1970] *Theory of Plates and Shells*, McGrawHill Int. NY

Young, W.C. [1989] *Roark's Formulas for stress and strain* McGraw Hill Int. New York

The Mechanics of Structures and Materials, Grzebieta, Al-Mahaidi & Wilson (eds)
© *1997 Balkema, Rotterdam, ISBN 90 5410 900 9*

Baseplates for structural steel hollow sections subject to axial tension

J.E. Mills, E.T. Barone & P.J. Graham
University of South Australia, Adelaide, S.A., Australia

ABSTRACT: Design recommendations for baseplates for structural steel hollow sections subject to axial tension were examined. Current methods are based on adaptations of formulae developed for baseplates of hot-rolled I sections using yield line methods. A testing programme was carried out to verify the existing theory on hot rolled I sections and then to examine baseplates for square hollow sections (SHS) with four bolts, in two different configurations. Both configurations used gave distinct and repeated yield line patterns. Analysis of results showed that the current design recommendations were non-conservative for the bolt configuration most commonly used in practice.

1. INTRODUCTION

Lightly loaded flexible baseplates for structural steel hollow sections are commonly found in many low-rise conventional building frames and pre-engineered metal structures such as industrial and farm sheds. In southern hemisphere countries such as Australia, wind loads will frequently govern the design of these types of structures due to their light weight, low roof slopes and the fact that snow loads are not relevant. These wind loads will commonly result in uplift loads on columns and hence their baseplates.

Current design guidance has been based on adaptations of formulae developed for baseplates of hot-rolled I sections, which in turn have been based on limited research and testing.

2. CURRENT THEORY AND DESIGN RECOMMENDATIONS

Existing methods of analysis and design of column base plates under axial tension rely either on classical plate bending solutions or yield line analysis. The yield line method is an energy method requiring the least upper bound of all possible failure mechanisms. Determination of the critical yield line pattern must be a trial and error process but can be assisted by laboratory testing or finite element analysis. Limited testing has been published in this area, particularly with respect to steel hollow sections. De Wolf (1990) relied on the work of Murray (1983) and the yield line analysis of Murray has been adopted by the AISC in Hogan and Thomas (1994). Melchers (1988) also carried out testing of larger I section baseplates. This analysis was based on a two bolt configuration for a hot-rolled I beam and has been extended by Hogan and Thomas to two and four bolt configurations for channels, rectangular, square and circular hollow sections.

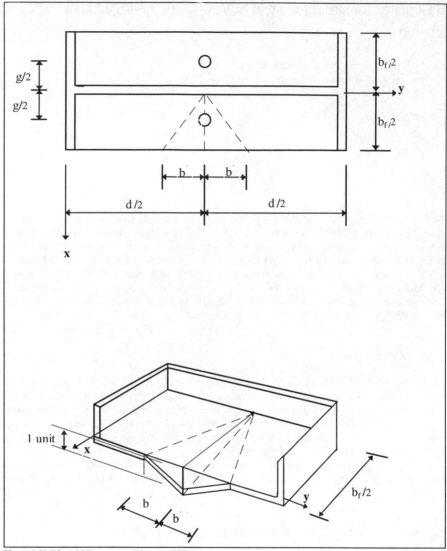

Figure 1:Uplift yield-line pattern (Murray, 1983)

The yield line pattern analysed by Murray is illustrated in Figure 1. From this Murray proposed that the baseplate thickness could be calculated as follows:

$$t = \sqrt{\frac{\sqrt{2}P_u g}{4b_f F_y}} \quad \text{for} \quad \sqrt{2}b_f \leq d \tag{1}$$

$$t = \sqrt{\frac{P_u g d}{F_y\left(d^2 + 2b_f^2\right)}} \text{ for } \sqrt{2}b_f > d \tag{2}$$

where P_u = the total uplift force on the baseplate

t = baseplate thickness

F_y = yield stress of baseplate component

These equations were modified for design by Hogan and Thomas to:

$$\phi N_s = \frac{\phi \times 4b_{fo} f_{yi} t_i^2}{\sqrt{2}s_g} \times \frac{n_b}{2} \text{ when } \sqrt{2}b_{fo} \leq d_c \tag{3}$$

$$\phi N_s = \frac{\phi \times f_{yi}\left(d_c^2 + 2b_{fo}^2\right)t_i^2}{s_g d_c} \times \frac{n_b}{2} \text{ when } \sqrt{2}b_{fo} > d_c \tag{4}$$

where ϕN_s = design strength of steel base plate due to axial tension in column

n_b = number of bolts

d_c, s_g, b_{fo} are defined in Fig 2 below

$t_i, f_{yi} = t, F_y$ defined above

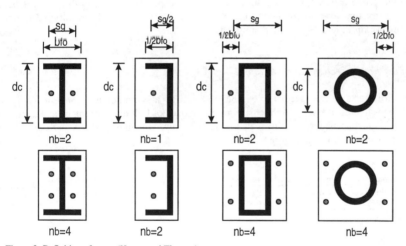

Figure 2: Definition of terms (Hogan and Thomas)

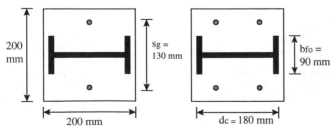

Figure 3: Bolt group arrangements used in testing

3. TESTING PROGRAMME

3.1 *Hot rolled I sections*

A series of eight tests were carried out to verify Murray's recommendations for I sections. The bolt group arrangements tested are shown in Figure 3. The I section used was a 180 UB 22, bolts were M20-8.8/S, the UB was welded all round to the baseplate with a 6mm fillet weld (bolts and weld were oversized to ensure failure occurred in the baseplate) and baseplate thicknesses of 6mm and 8mm were used. Testing was performed on an 1800 kN capacity Baldwin Universal Testing Machine. The base of the test frame was a 380 Parallel Flange Channel (PFC)with web horizontal. Holes were drilled through the web to match the bolt configuration of the test baseplates. The web of the PFC was stiffened with 10mm flat plate stiffeners to prevent local buckling of the bed during testing. The test frame was attached to the bottom platen of the Baldwin with four strongbacks, (one above and one below the channel section, on each side) using four high strength steel bolts. The upper end of the UB was clamped into the platen of the test rig by cutting off the flanges and trimming the web.

Baseplate deformations were monitored during loading by the use of dial gauges supplemented by strain gauging in some of the tests. Plots of load vs deformation were then used to determine the average yield loads for each test and these values were compared to the unfactored design recommended load N_s, in accordance with Hogan and Thomas, in Table 1. As indicated in the table, the design values were conservative by a minimum factor of 2.0.

The yield line patterns observed during testing corresponded closely with those predicted by Murray for the two bolt configuration. Yield line patterns for the four bolt configuration were more complex and difficult to distinguish.

3.2 *Square hollow sections-configuration 1*

In the initial tests carried out using baseplates for SHS a symmetrical arrangement of four bolts was used to determine yield line patterns and modes of failure. The bolt arrangement and observed yield line pattern is shown in Figure 4 below, with the test bed and column cap arrangements used shown in Figure 5.

Table 1: Comparison of experimental and recommended design values for hot rolled I section baseplate

Baseplate thickness and no. of bolts	No. of tests	F_y (Mpa)	Design yield load (kN)	Ave. test yield load (kN)	Ratio
6mm - 2 bolts	2	289	20	40	2.0
6mm - 4 bolts	2	289	40	95	2.4
8mm - 2 bolts	2	291	35	76	2.2
8mm - 4 bolts	2	291	69	136	2.0

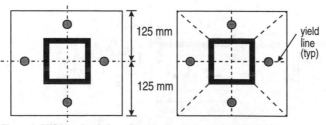

Figure 4: SHS baseplate bolt configuration 1 and observed yield line pattern

550

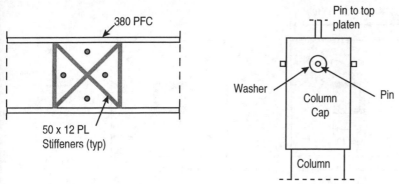

Figure 5: Configuration 1 column cap and test bed arrangement (viewed from underside) (Micke)

Table 2: Comparison of experimental and recommended design values for SHS baseplate, configuration 1

Baseplate thickness (mm)	Failure mode	F_y (Mpa)	0.8 x Ult. failure load (kN)	Design yield load (kN)	Ratio
5.9	weld failure	301	n.a.	66	
5.9	baseplate yield	301	241	66	3.7
5.9	baseplate yield	301	242	66	3.7
4.6	baseplate yield	361	142	48	3.0
4.6	baseplate yield	361	158	48	3.3
4.6	baseplate yield	361	145	48	3.0

Three tests were carried out for each of two baseplate thicknesses (5.9mm and 4.6mm). Baseplates were 250mm x 250mm, the column was 100 x 100 x 4 SHS, bolts were M20-8.8/S and the column was welded all round to the baseplate with an 8 mm fillet weld (bolts and weld oversized as before).

In this series of experiments records of ultimate failure loads and modes were made, but detailed load vs deformation readings of the baseplates were not, hence it is difficult to compare recommended design and observed yield loads. However, using a ratio of 0.8 for yield strength to tensile strength as an approximation a comparison has been given in Table 2. The comparison indicated that the design recommendations are very conservative (and possibly the definitions of geometric terms are inappropriate) for this bolt configuration.

3.3 Square hollow sections-configuration 2

A second series of tests was then carried out on baseplates for SHS using a symmetrical arrangement of four bolts more commonly found in industry practice, with bolts located at the corners of the baseplate. The bolt arrangement and observed yield line pattern is shown in Figure 6. The yield line pattern was consistent in all tests. The test bed and column cap arrangements used were similar to those of the previous tests but with a different configuration of stiffeners in the PFC bed to suit the revised bolt arrangement.

Eight tests were carried out, four on each of 6mm and 8mm baseplate thicknesses. Baseplates were 300mm x 300mm, the column was 100 x 100 x 4 SHS, bolts were M16-8.8/S and the column was welded all round to the baseplate with an 8 mm fillet weld (bolts and weld oversized as before). Baseplate deformations were monitored during loading and average yield loads determined in a similar manner to the hot rolled UB section tests. These values were compared to the unfactored design recommended load N_s in accordance with Hogan and Thomas in Table 3.

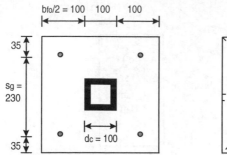
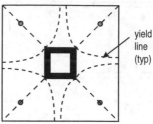

Figure 6: SHS baseplate bolt configuration 2 and observed yield line pattern

Table 3: Comparison of experimental and recommended design values for SHS baseplate - configuration 2

Baseplate thickness (mm)	No. of tests	F_y (Mpa)	Design yield load (kN)	Ave. test yield load (kN)	Ratio
6.03	2	326	93	47	0.5
6.13	2	310	91	54	0.6
7.92	2	304	150	84	0.6
8.03	2	301	152	80	0.5

As seen in Table 3, the design values were non-conservative, i.e. observed values were down to a half of the design recommended values.

4. SUMMARY AND CONCLUSIONS

The results of this work indicate that the current design guidance for baseplates subject to axial tension is conservative when applied to hot rolled I section baseplates but the theoretical basis for the recommendations, i.e. the assumed yield line pattern, is valid. However, the extension of this theory to baseplates for SHS (and consequently one would expect Rectangular and Circular Hollow Sections also) appears questionable. The recommendations can be either very conservative or unacceptably non-conservative depending on the bolt configuration used. Yield line patterns do not correspond with those assumed in the current recommendations.

Further testing and analysis of results is required to develop design recommendations that are valid for hollow section baseplates subjected to axial tension.

REFERENCES

De Wolf, J.T. 1990. Column base plates. *American Institute of Steel Construction, Design Guide Series No 1*

Hogan, T.J. & Thomas, I.R. (4th ed.) 1994. *Design of structural connections.* Sydney: Australian Institute of Steel Construction

Melchers, R.E. 1988 Modelling of column base behaviour. In R.Bjorhovde et al (eds) *Connections is Steel Structures*: New York: Elsevier Applied Science

Micke, S.L. 1995 *Validation of the existing design theory for baseplate connections for steel hollow sections subject to axial tension* Unpublished Honours Thesis University of South Australia

Murray, T.M. 1983. Design of lightly loaded steel column base plates. *Engineering Journal, American Institute of Steel Construction.*20-4:143-152.

The Mechanics of Structures and Materials, Grzebieta, Al-Mahaidi & Wilson (eds)
© *1997 Balkema, Rotterdam, ISBN 90 5410 900 9*

Corrosion of steel in seawater – A poorly formulated but important practical problem

Robert E. Melchers
Department of Civil, Surveying and Environmental Engineering, The University of Newcastle, N.S.W., Australia

ABSTRACT: Mild and low alloy steels are used extensively for off–shore structures and for ship construction as well as for sheet piling and harbour–side facilities. Despite more than 60 years of effort, and some quite extensive, long term experimental test programs, the precise understanding of the mechanics of the marine corrosion of these steels and the influences of the various factors which are now known to be important has been slow to develop and, in the context of mathematical formulation, have been neglected by material scientists/engineers. The various factors of importance in marine corrosion and the models which have been proposed to describe time–dependent material loss as a function of time are reviewed.

1 INTRODUCTION

In the reliability assessment of structures which deteriorate through metal corrosion, appropriate mathematical models are required to describe the (monotonic) time varying reduction and increasing uncertainty in the structural strength; see Figure 1 for the simplest reliability problem. The probability density function of the resistance flattens with time and the strength probability distribution has a reducing mean. Hence the reliability of the system reduces with time. Although loading models are well developed, models for strength deterioration have received little attention in the literature. Herein a model for material loss is considered – from this a time dependent strength model may be deduced. Although there is a wealth of material science research findings, surprisingly little quantitative information exists for the development of quantitative models of material loss (or pitting, etc.) due to corrosion.

Herein only the corrosion of mild and low alloy steels under marine conditions (such as at sea) is considered – these steels are of considerable economic importance and comprise the majority of steels used in off–shore platforms, ships, sub–sea pipelines and similar structures.

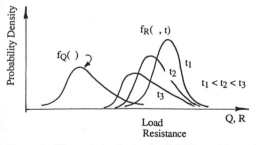

Figure 1. Elementary structural reliability problem with strength deterioration.

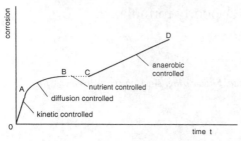

Figure 2. Proposed conceptual model for marine corrosion.

For simplicity attention will be restricted also to uniform' immersion corrosion; that is, the nominal loss of thickness of material as derived from weight loss data. Experimentally this is obtained from the exposure of small scale specimens of metal. Apart from structures which extend a considerable way above the splash zone, such results generally are directly translatable to full–scale structures.

The main factors which influence marine corrosion of steel have been identified (Schumacher, 1979); space precludes a detailed discussion. Biological factors, oxygen availability, water temperature, pH, pollutants, water velocity and suspended solids have all been suggested as potentially important. However, quantitative measures of their effect have not been identified.

2 PROPOSED MODELS

A probabilistic 'phenomenological' model, that is, one which makes no attempt to model the fine details of the actual corrosion processes, but which nevertheless attempts to take account of the broad principles of the corrosion processes, has been described previously (Melchers, 1995a). The model is shown in generic form in Figure 2. Each of the model components is described in more detail below. Because the precise effects of the many factors influencing corrosion are not well understood (IJsseling, 1989), probabilistic models are appropriate, for example (Melchers, 1995a):

$$c(t,\mathbf{E}) = fn(t,\mathbf{E}) + \varepsilon(t,\mathbf{E}) \tag{1}$$

where $c(t,\mathbf{E})$ is the weight loss of material, $fn(t,\mathbf{E})$ is a mean valued function, $\varepsilon(t,\mathbf{E})$ is a zero mean error function, t is time and $\mathbf{E}$ is a vector of environmental conditions.

3 COMPONENTS OF THE MODEL

3.1 *Initial corrosion (Kinetic)*

Immediately upon being placed in seawater, the surfaces of steel specimens are colonized by micro–organisms, mainly bacteria and diatoms; forming the primary biofilm. Subsequent microbial and larger biological fouling growth (macro–fouling), when sufficiently thick and dense, may assist in excluding oxygen completely from the metal surface. Initially, however, the corrosion products and the biological materials do not interfere signicantly with the corrosion process (the process is therefore "kinetically controlled") (IJsseling, 1989; Melchers, 1996).

Since the corrosion process is dependent on the availability of oxygen (Schumacher, 1979) the weightloss x (per unit area) of material due to corrosion can be expressed as:

$$x = \lambda.k_i.C_i^n \tag{2}$$

where C_i is the oxygen concentration at the corrosion interface, n is an exponent (the *order* of the reaction), k_i is a constant for the rate of corrosion and λ is a surface roughness coefficient to allow for the known effect of surface roughness on corrosion rate. Field data for general corrosion suggests that kinetic corrosion is a first order rate process (Schumacher, 1979), thus $n = 1$. It follows that the corrosion process under kinetic conditions is a linear process (see Figure 2)

3.2   Diffusion limited corrosion

The (increasing) thickness of the corrosion product (and the biomass) will eventually limit the rate of oxygen supply to the corroding surface. The corrosion process now becomes "diffusion controllled". Various factors influence this: including the possible removal of corrosion product (and/or biomass) due to impact, abrasion or erosion (such as due to water current) and the effects on biomass of UV radiation, pollutants and, possibly, drying–out (as in tidal and atmospheric conditions). The basic diffusion–controlled model becomes (e.g. Evans, 1960):

$$\frac{dO_2}{dt} = oxygen\ mass\ flow = \frac{k_1(C_L - C_i)}{R} \tag{3}$$

where O_2 represents the mass of oxygen transfer, C_L and C_i are the concentrations of dissolved oxygen in the seawater and at the corrosion interface respectively, R is the resistance to diffusion and k_1 is a constant. The mass transfer of oxygen is governed directly by the difference in the dissolved oxygen concentrations. If it is assumed that the seawater is fully oxygenated at a given temperature (see below) and the concentration at the corroding surface is close to zero (due to it being immediately consumed by the corrosion process), the difference is oxygen concentration across the corrosion product layer can be taken as a constant. Since the amount of corrosion is a linear function of the dissolved oxygen arriving at the corrosion interface, it follows that, with η = constant:

$$\frac{dO_2}{dt} = \eta.\frac{dx}{dt} \tag{4}$$

It would be expected that the corrosion products (and the biomass) together provide the resistance to oxygen transfer. For the special case of uniform density corrosion product with depth and with the curve passing through the origin (which means that the process is diffusion controlled from the very beginning – an impossibility as there is no corrosion product), the above expression reduces to:

$$x = k_2.t^{\frac{1}{2}} \tag{5}$$

With k_2 = constant, this will be recognized as the diffusion–controlled corrosion rate equation widely quoted in the *atmospheric* corrosion literature. However, rather little attention appears to have been given to the various assumptions which must be made in its derivation. The only discussions of this aspect and with a corresponding set of equations for the intial and oxygen diffusion part of the process are those given by Evans (1960), work which appears to have been ignored by subsequent workers when discussing theoretical bases for corrosion curves (i.e. $fn(t,\mathbf{E})$) (cf. Bernarie and Lipfert, 1986; Feliu, et al., 1993).

   The theoretical relationship (5) does not fit experimental data for *atmospheric* corrosion very well over the full range of observations, even though the same general principles hold as for immersion corrosion. The most common approach in the *atmospheric* corrosion literature has been to modify the exponent. Theoretically, it may be shown that non–uniform corrosion products with depth reduce the exponent to less than 0.5 (as has been suggested is the case also for *atmospheric* corrosion). However, removal of corrosion products (such as through velocity) does not increase the exponent to more than 0.5 (as suggested in the *atmospheric* corrosion literature (e.g. Bernard and Lipfert, 1986), but rather modifies the constant (Melchers, 1995b).

3.3 *Nutrient controlled conditions*

This aspect of the model has not yet been developed, even though it is known from the marine and micro–biology literature that the activation of the anaerobic bacteria (see below) requires an adequate food supply, thought to be made available through the aerobic bacteria (Melchers, 1995b).

3.4 *Anaerobic corrosion*

Besides the aerobic bacteria, anaerobic micro–organisms are always present in the primary biofilm. Of these the most important are the sulphate reducing bacteria (SRB), which control the rate of corrosion under low to zero oxygen conditions in the region of the corrosion interface, producing sulphides (and hence ferrous sulphide) (e.g. Hamilton, 1994). This part of the proposed model can be represented by

$$x(t) = \mu . D(t - t_t) + E \tag{6}$$

where μ describes the availability of moisture, D is a rate constant and E is a constant describing the depth of corrosion at time $t = t_t$ at which time anaerobic activity concludes or the nutrient requirements for the anaerobic bacteria are satisfied.

4   EFFECT OF WATER TEMPERATURE

As noted above, water temperature will influence the corrosion process. Initially, with a temperature increase, two effects operate: (i) the kinetic rate of the corrosion process increases and (ii) the dissolved oxygen content in the surrounding seawater decreases. Usually the first one is more important and higher temperatures lead to greater corrosion, as observed experimentally for short term exposure (LaQue, 1975). For longer exposures, there is the additional effect of biological activity which tends to increase with temperature under constant oxygen supply. In principle, this will increase biological growth which in turn tends to increase resistance to oxygen diffusion (if the reaction is diffusion controlled). This will tend to reduce the corrosion rate.

Despite these observations, the conventional wisdom for marine corrosion is to make no allowance at all for the effect of temperature. Perhaps this is because of the apparent difficulty in clearly differentiating in practice between the various processes which are involved. However, Melchers and Ahammed (1994), re–evaluating test data, argued that a first order temperature correction could be taken as:

$$d_2 = \frac{(t_2 - t_0)}{(t_1 - t_0)} d_1 \tag{7}$$

where d_2 is the depth of penetration due to weight loss at mean temperature t_2, d_1 is the corresponding depth at t_1 and t_0 is the mean temperature below which there is no corrosion throughout the year. Examination of the data suggested that $t_0 = -10^{\circ}C$ was a useful approximation. This was based on the common observation that corrosion does not occur below sea water freezing point (approximately $-2^{\circ}C$ at 3.5% salinity). [The reason for this has been discussed (Melchers, 1995b)]. The proposed correction (7) appears to be consistent both with more recent field data and with laboratory data.

Figure 3 shows corrosion observations for 5 years of exposure of mild steel specimens (to ASTM K01051) at 14 different sites at various locations around the world (Phull, Pikul and

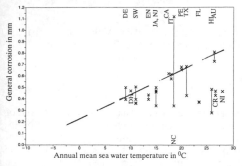

Figure 3. Five year corrosion as a function of annual mean water temperature.

Kain, 1996). (Three other data sources shown as DH (Den Helder), CR (Christobel) and NI (Naos Island) are shown also). Evidently, there is considerable scatter. It appears that where the specimens were exposed in harbour or channel conditions, the scatter in data points (including those for earlier sampling periods) was greater than elsewhere.

The reported results were accompanied by some information about environmental conditions, including temperature range, pH, salinity and dissolved oxygen content. Unfortunately this was insufficient to determine trends. Contact was made with those responsible for each test site and additional information was requested. This resulted in some useful extra information being obtained. Details will be reported elsewhere; suffice to note that the following observations and tentative subjective judgements were then made:

1. sites AU, CA, DH, DK, EN, IT, NJ, PE, SW corresponded closely to unpolluted fully aerated conditions,
2. sites CR, FL, JP, NC, NI, TX have less than fully aerated conditions,
3. the specimens at HI were horizontal rather than vertical (effect not clear),
4. there was a remarkable but unexplained (so far) variability in the results for NC.

Considering only the results for sites under (1) leaves a clear trend, shown with the broken line. It corresponds to equation (7). Moreover, this trend is consistent with recent laboratory results for small–scale specimens exposed to natural and artificial sea–water and areated to an experimentally–controlled level (Mercer and Lumbard, 1995) and with the earlier limited observations of Fink (see Schumacher, 1979).

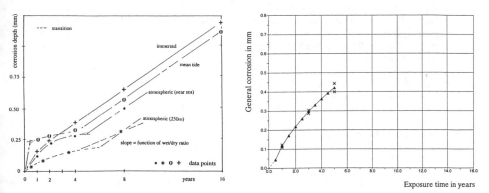

Figure 4. Typical application of data to proposed model.

5 MODELS

The generic model (1) has been applied to Panama Canal zone data (Melchers, 1995b), see Figure 4(a). It is also not inconsistent with the data reported in Phull, Pikul and Kain (1996). One example is given in Figure 4(b).

6 DISCUSSION

Some of the features in the model postulated herein apparently either have not been observed or have not been recorded. One possible reason is that not all the influences have been easily observable without a theoretical model to guide observation. Another is that longer term corrosion test programs could not afford to deal in detail with, say, the early part of the corrosion process, or had observations so far apart in time that only general trends could be discerned. Re-analysis of available data, using better information about environmental factors, is currently being undertaken. This is necessary to derive a satisfactory model for $fn(t,\mathbf{E})$. It would appear, however, that the present data base is insufficient to derive models for $\varepsilon(t,\mathbf{E})$ and that further experimental (field) work is required.

7 CONCLUSION

A phenomenological model for the probabilistic description of the corrosion of steel under given marine conditions, using weight loss as a function of time has been outlined. Mathematical descriptions for three of the various components of the proposed model are given. It is shown that a linear temperature correction previously proposed is consistent with trends underlying recently reported field and laboratory data.

REFERENCES

Benarie, M. and Lipfert, F.L., 1986, *A General Corrosion Function in Terms of Atmospheric Pollutant Concentrations and Rain pH*, Atmos. Environment, 20(10), 1947–1958.

Evans, U.R., 1960, *The Corrosion and Oxidation of Metals: Scientific Principles and Practical Application*, Arnold, London.

Feliu, S., Morcillo and Feliu, Jr. S., 1993, *The Prediction of Atmospheric Corrosion From Meteorological and Pollution Parameters: 1 – Annual Corrosion; and 2 – Long–Term Forecasts*, Corrosion Science, 34(3), 403–422.

Hamilton, A.W., 1994, *Bio–Corrosion: The Action of Sulphate Reducing Bacteria*, (In) Ratledge, C. (Ed.), Biochemistry of Microbial Degradation, Kluwer, pp. 555–570.

IJsseling, F.P., 1989, *General Guidelines For Corrosion Testing For Marine Applications*, British Corrosion Journal, 24(1), 55–78.

LaQue, F.N., 1976, *Marine Corrosion*, Wiley–Interscience, New York.

Melchers, R.E., 1996, *Modeling of Marine Corrosion of Steel Specimens*, (In) Corrosion Testing in Natural Waters, Second Volume, Kain, R.M. and Young, W.T. (Ed.), ASTM STP 1300, Philadelphia.

Melchers R.E., 1995a, *Probabilistic Modelling of Marine Corrosion of Steel Specimens*, Proceedings, ISOPE'95, The Hague, The Netherlands, 12–15 June.

Melchers, R.E., 1995b, *Marine Corrosion of Steel Specimens – Phenomenological Modelling*, Research Report No. 125.12.1995, DCE&S, The University of Newcastle.

Melchers, R.E. and Ahammed, M., 1994, *Nonlinear Modelling of Corrosion of Steel in Marine Environments*, Research Report 106.09.1994, DCE&S, The University of Newcastle.

Mercer, A.D. and Lumbard, E.A., 1995, *Corrosion of Mild Steel in Water*, British Corr. J., 30(1), 43–55.

Phull, B.S., Pikul, S.J. and Kain, R.M., 1996, *Seawater Corrosivity Around The World – Results From Five Years of Testing*, (In) Corrosion Testing in Natural Waters; Second Volume, Kain, R.M. and Young, W.T. (Ed.), ASTM STP 1300, Philadelphia.

Schumacher, M. (ed), 1979, *Seawater Corrosion Handbook*, Noyes Data Corporation, New Jersey.

The Mechanics of Structures and Materials, Grzebieta, Al-Mahaidi & Wilson (eds)
© *1997 Balkema, Rotterdam, ISBN 90 5410 900 9*

Tests to investigate the web slenderness limits for cold-formed RHS in bending

T.Wilkinson & G.J.Hancock
Centre for Advanced Structural Engineering, The Department of Civil Engineering, The University of Sydney, N.S.W., Australia

ABSTRACT: Steel beam cross sections are normally classified according to the effect of local buckling on the moment capacity of the section. This is achieved by defining slenderness limits for the flange and the web. This paper describes a series of bending tests on cold-formed RHS to examine these limits. The resulting experimental relationship between web slenderness and plastic rotation capacity is given. It is found that the web slenderness limits in steel design specifications for compact sections are not conservative for RHS when used in plastic design. There is clear interaction between the flange and web slenderness which should be accounted for in the limits. A tentative proposal for a new compact web slenderness limit for RHS is suggested.

1   INTRODUCTION

Steel design specifications give plate element slenderness (b/t) limits to assess the suitability of sections for plastic design, and the effect of local buckling on moment capacity. A *Slender* section experiences local buckling before the yield moment, M_y, is reached. *Compact* sections can maintain the plastic moment M_p for sufficiently large rotations suitable for plastic design. *Non-Compact* sections buckle inelastically between M_y and M_p.

Table 1 shows the b/t limits in AS 4100 (Standards Australia (1990)) for cold-formed RHS and hot-rolled I-sections bending about the major principal axis. The symbols for RHS are defined in Figure 1. An I-section has flange dimensions $b_f \times t_f$, and web dimensions $d \times t_w$.

Table 1: Plate element slenderness limits in AS 4100 for bending about the major principal axis.

Element	Slenderness definition		Compact		Non-Compact	
	RHS	I-section	RHS	I-section	RHS	I-Section
Web	$\lambda_w = \dfrac{d-2t}{t}\sqrt{\dfrac{f_y}{250}}$	$\dfrac{d-2t_f}{t_w}\sqrt{\dfrac{f_y}{250}}$	82	82	115	115
Flange	$\lambda_f = \dfrac{b-2t}{t}\sqrt{\dfrac{f_y}{250}}$	$\dfrac{b_f-t_w}{2t_f}\sqrt{\dfrac{f_y}{250}}$	30	9	40	16

The stress distribution and support conditions for the webs are the same (even though an RHS has two webs). This is reflected in the identical limits for RHS and I-sections webs given in all steel design codes. These limits are based on tests of I-sections (summarised by Sedlacek and Feldmann (1995)). Zhao and Hancock ((1991a) and (1992)) observed inelastic web local buckling in some RHS. This happened at low rotation values for specimens with web slenderness below the compact limit. This provided the justification for the current series of tests.

AS 4100 allows compact I-sections to be used in plastic design, but not compact RHS. Other design standards, such as AISC LRFD (AISC (1994)), permit plastic design in compact RHS. Hence the applicability of the compact limit is important. Indeed, the compact limit in AS 4100 for RHS flanges was based on use in plastic design should it be permitted for RHS in the future. It is intended to use the current research to support the use of cold-formed RHS for plastic design.

Steel design specifications have varying rotation capacity requirements for plastic design. The wide variety of loading patterns and structural frame shapes results in a large range of required plastic rotations. Hence it is appropriate to adopt some arbitrary value of rotation capacity which is satisfactory for most practical situations. Eurocode 3 (European Committee for Standardisation (1992)) Class 1 and AISC LRFD Compact limits are based on a rotation capacity of $R = 3$. However AISC LRFD states that greater rotation capacity may be required in seismic regions. A value of $R = 4$ was adopted in tests (Hasan and Hancock (1988), Zhao and Hancock (1991b)) to examine the RHS flange limits in AS 4100. The value of $R = 4$ is deemed suitable in this paper for use in plastic design.

2   TEST PROGRAM

2.1   Test specimens

A variety of cold-formed RHS was chosen for the plastic bending tests. The test specimens were manufactured by BHP Steel Structural and Pipeline Products. Two strength grades were selected, Grade C350L0 and C450L0, manufactured to AS 1163 (Standards Australia (1991)). The Grade C450 specimens were *DuraGal* sections, produced using a unique cold-forming and in-line galvanising process. The cross section of a typical RHS is shown in Figure 1.

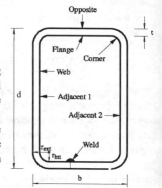

Figure 1. RHS Section

2.2   Tensile coupons

Three coupons were taken from the flats of each tube. One was cut from the face opposite the weld, and one from each of the sides adjacent to the weld. Figure 1 shows the position of these faces with respect to the weld. The coupons were prepared and tested in accordance with AS 1391 (Standards Australia (1991)) in a 250 kN capacity INSTRON Universal Testing Machine.

Since the steel was cold-formed, the yielding was gradual. Accordingly the yield stress (f_y) used is the 0.2% proof stress. The average of the yield stress from both of the adjacent faces is used in the determination of section capacities (plastic moment (M_p)) and slenderness values. The average of the elastic modulus (E) from both of the adjacent faces is used in stiffness calculations. The yield stress of the opposite face is on average 10% higher than that of the adjacent faces. This is a result of the cold-forming process.

2.3   Bending tests - procedure

The bending tests were performed in a 2000 kN capacity DARTEC testing machine, using a servo-controlled hydraulic ram as shown schematically in Figure 2. The four point bending arrangement provided a central region of uniform bending moment and zero shear force. Specimens were supported on half rounds resting on greased Teflon pads, simulating a set of simple supports.

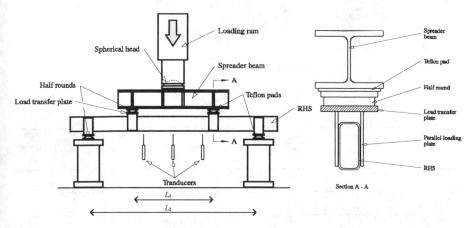

Figure 2. Loading mechanism for bending tests

The loading method adopted was one that has been used at The University of Sydney for many years (Hasan and Hancock (1988), Zhao and Hancock (1991b)). A greased Teflon pad was placed between the bottom of the spreader beam and the half round. This allowed for the load points to move due to the axial shortening of the beam caused by curvature. The half round bore upon a thick load transfer plate, which in turn transmitted the force by bearing to loading plates. These loading plates were welded to each web of the RHS and were parallel to the webs. Two other loading methods were used in selected tests. Details of these have been omitted but were found not to produce significantly different results. For RHS with depth $d \geq 100$ mm, the length between the loading points (L_1) was 800 mm, and the distance between the supports (L_2) was 1700 mm. For sections with $d \leq 75$ mm, L_1 was 500 mm, and L_2 was 1300 mm. Generally, two samples of each specimen size were tested.

2.4   Bending tests - results

All specimens except the two 150 × 50 × 5.0 C450 specimens and the 100 × 100 × 3.0 C450 sample experienced web local buckling. Local buckling produced a rapid shedding of load with increased deflection. In all cases, the buckle formed adjacent to one of the loading plates. The two 150 × 50 × 5.0 C450 RHS exhibited large deflections and an inelastic lateral (flexural-torsional) buckle formed at high curvatures. There was no sudden unloading associated with this lateral buckle. The square hollow section 100 × 100 × 3.0 C450 SHS, as expected, failed by flange buckling. No specimen failed due to insufficient material ductility.

561

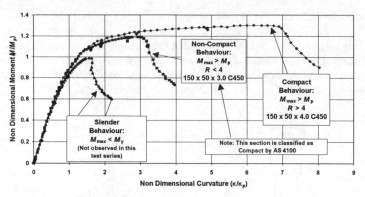

Figure 3. Typical moment curvature relationships

Table 2. Summary of results of bending tests

Section	Slenderness Web λ_w	Flange λ_f	$\dfrac{M_{max}}{M_p}$	Rotation capacity R	Section	Slenderness Web λ_w	Flange λ_f	$\dfrac{M_{max}}{M_p}$	Rotation capacity R
150 × 50 × 5.0 C450	38	11	1.23	>13	75 × 25 × 2.0 C450	49	15	1.11	5.7
	38	11	1.17	>8		49	15	1.13	n/a
150 × 50 × 4.0 C450	49	15	1.27	6.6	75 × 25 × 1.6 C450	61	19	1.03	2.4
	50	15	1.19	7.7		61	19	1.15	2.7
150 × 50 × 3.0 C450	65	20	1.15	2.7	75 × 25 × 1.6 C350	61	18	1.03	1.9
	65	20	1.16	2.3		60	18	1.00	2.6
	65	20	1.13	2.9					
150 × 50 × 2.5 C450	75	23	1.02	1.4	150 × 50 × 3.0 C350	59	18	1.21	4.1
	75	23	1.00	1.2		59	18	1.15	3.6
150 × 50 × 2.3 C450	86	27	0.98	0.0	100 × 50 × 2.0 C350	60	28	1.00	1.1
	85	27	1.01	0.0		60	29	1.00	1.3
	86	27	0.98	0.0					
100 × 50 × 2.0 C450	62	30	1.07	0.8	125 × 75 × 3.0 C350	52	30	1.03	1.5
	62	30	1.01	0.8		51	30	1.03	1.6
75 × 50 × 2.0 C450	47	31	1.04	1.7	100 × 100 × 3.0 C450	44	44	1.00	0.8
	47	30	1.02	1.9					

Three typical moment rotation curves are depicted in Figure 3. The results of all the plastic bending tests are presented in Table 2. It lists the ratio M_{max}/M_p and R. M_{max} is the maximum moment reached during the test. M_p is based on the measured dimensions and the average measured yield stress. The rotation capacity (R) is defined as $\kappa/\kappa_p - 1$, where κ/κ_p is the dimensionless curvature at which the moment falls below M_p. κ_p is the curvature at which the plastic moment is reached assuming an elastic rigidity (EI) and is given by $\kappa_p = M_p/EI$.

Figure 4 displays the measured rotation capacity versus web slenderness and indicates the current position of the web compact limit. Approximate iso-rotation curves for the bending tests, taking into account both the flange and web slenderness are shown in Figure 5. Since some tests were repeated the rotation capacity shown is the average capacity. The results of the tests by Hasan and Hancock (1988), and Zhao and Hancock (1991b) are also included in this figure. These were chosen as the same method of loading was used in these tests.

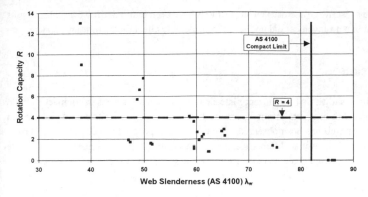

Figure 4. Rotation capacity - web slenderness relationship.

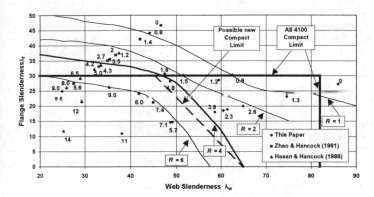

Figure 5. Approximate iso-rotation curves and possible new limit

3   DISCUSSION

There are a large number of sections which are currently classified as Compact which demonstrate insufficient rotation capacity. These results indicate that the current web slenderness limit in AS 4100 for Compact RHS is non-conservative for use in plastic design. Comparisons with other design standards also indicate a similar non-conservative limit.

The iso-rotation curves in Figure 5 give the best indication as to the behaviour of these specimens. There is a most definite interaction between the flange and the web. The sections with higher aspect ratio (d/b) have a relatively stiffer flange which provides restraint against web local buckling. A section with the same web but more slender flange has less restraint and consequently a lower rotation capacity.

Figure 5 indicates that the current flange slenderness limits are reasonably satisfactory for RHS sections with stocky webs. However there is a significant region for members with more slender webs in which the AS 4100 compact limit does not allow adequate rotation capacity for plastic design. A simple linear interaction formula may be appropriate. A *preliminary* proposal is also drawn in Figure 5.

563

Results of the tests using alternate loading methods have been omitted. However these results indicate that the loading arrangement did not have a considerable affect on the rotation capacity.

4   CONCLUSION

The bending tests have shown that the current plate slenderness limits for webs in compact RHS do not produce sections with adequate rotation capacity for plastic design. Considerable interaction between the flange and web has been observed. It may not be appropriate to continue the current design philosophy in which the web and flange slenderness limits are prescribed separately. A simple straight line interaction formula could be more useful for classification of sections in bending.

5   REFERENCES

AISC, (1994), *Metric Load and Resistance Factor Design Specification for Structural Steel Buildings*, American Institute of Steel Construction, Chicago, Il.

European Committee for Standardisation, (1992), *Design of Steel Structures: Part 1.1 - General Rules and Rules for Buildings*, DD ENV. 1993-1-1, Eurocode 3 Editorial Group.

Hasan S. W. and Hancock G. J., (1988), "Plastic Bending Tests of Cold-Formed Rectangular Hollow Sections", *Research Report*, No. R586, School of Civil and Mining Engineering, The University of Sydney, Sydney.

Sedlacek G. And Feldmann M., (1995), *The b/t ratios Controlling the Applicability of Analysis models in Eurocode 3, Part 1.1*, Aachen, Germany.

Standards Australia (1990), Australian Standard *AS 4100 Steel Structures*, Standards Australia, Sydney.

Standards Australia (1991), Australian Standard *AS 1163 Structural Steel Hollow Sections*, Standards Australia, Sydney.

Standards Australia (1991), Australian Standard *AS 1391 Methods for Tensile Testing of Metals*, Standards Australia, Sydney.

Zhao X. L. and Hancock G. J., (1991a), "T-Joints in Rectangular Hollow Sections Subject to Combined Actions", *Journal of Structural Engineering*, ASCE, Vol 117, No. 8, pp 2258-2277.

Zhao X.L. and Hancock G.J. (1991b), "Tests to Determine Plate Slenderness Limits for Cold-Formed Rectangular Hollow Sections of Grade C450", *Steel Construction*, Journal of Australian Institute of Steel Construction, 25 (4), Nov 1991, pp 2-16.

Zhao X. L. and Hancock G. J., (1992), "Square and Rectangular Hollow Sections Subject to Combined Actions", *Journal of Structural Engineering*, ASCE, Vol 118, No. 3, pp 648-668.

6   NOTATION

b	Width of RHS flange	L_2	Length between supports	t_f	Thickness of I-section flange
b_f	Width of I-section flange	M_{max}	Maximum bending moment	t_f	Thickness of I-section web
d	Depth of RHS or I-section web	M_p	Plastic bending moment	κ	Curvature
E	Young's modulus of elasticity	M_y	Yield bending moment	κ_p	Plastic curvature ($= M_p/EI$)
f_y	Yield stress	R	Rotation capacity	λ_f	Flange slenderness
I	Second moment of area	t	Thickness of RHS	λ_w	Web slenderness
L_1	Length between loading plates				

7   ACKNOWLEDGEMENTS

This research is funded by CIDECT. Tube specimens were provided by BHP Steel Structural and Pipeline Products. The experiments were carried out in the J.W. Roderick Materials and Structures Laboratory, Department of Civil Engineering, University of Sydney.

The Mechanics of Structures and Materials, Grzebieta, Al-Mahaidi & Wilson (eds)
© *1997 Balkema, Rotterdam, ISBN 90 5410 900 9*

Tests and design of longitudinal fillet welds in cold-formed C450 RHS members

Kwong-Ping Kiew, Xiao-Ling Zhao & Riadh Al-Mahaidi
Department of Civil Engineering, Monash University, Vic., Australia

ABSTRACT: The paper describes a series of tests on longitudinal fillet welds in cold-formed C450 (450MPa nominal yield) RHS members which have a thickness of less than 3 mm. The typical failure mode is base metal failure rather than weld failure. The effect of end return welds, RHS orientation and load conditions on weld strength is investigated. The experimental weld strength is compared with the prediction using existing Australian, American, Canadian, and European design standards. The test results are also compared with the proposed design rules for longitudinal fillet welds in RHS members derived from the previous research project on C350 (350MPa nominal yield) RHS members. A reliability analysis method is used for calibration

1 INTRODUCTION

Tests on longitudinal fillet welds in cold-formed C350 RHS members were performed by Zhao and Hancock (1995b). A design formula based on the parent (base) metal failure was proposed for cold-formed steel standards [Zhao and Hancock (1995b)]. Another design formula based on the weld metal failure was proposed for the Amendment No.3 to AS4100-1990 [Zhao, Hancock and Sully (1996)]. This paper will verify if those design rules are applicable for C450 RHS members. The C450 RHS sections have a nominal yield stress of 450MPa and are manufactured by Palmer Tube Mills (Aust) Pty Ltd, whereas the C350 RHS sections have a nominal yield stress of 350MPa and are manufactured by BHP Steel-Tubemakers of Australia. Only one load condition (see case 1 in Figure 1) was used in the tests of C350 RHS members. Three load cases are used in the current test program to investigate if different load transfer path will affect the strength of longitudinal fillet welds. Parameters varied in the tests also include RHS orientation and end return welds. The test results are compared with design rules based on parent metal failure and those based on weld metal failure. The reliability analysis method is used for calibration.

2 EXPERIMENTAL STUDY

2.1 Material Properties

Tensile properties were determined by using two post yield strain gauges attached on either side of the test coupon with cross reference to the Australian Standards AS1391[SAA (1991)]. Two corner coupons, two adjacent coupons and one opposite coupon were taken on each specimen to

565

consider the effect of the stress variation around the RHS sections. The nominal dimensions $(D \times B \times t)$ for the three RHS sections (G1,G2,G3) are $50 \times 50 \times 2.3$, $75 \times 50 \times 2.3$ and $50 \times 50 \times 2.8$, respectively. The mean of both measured tensile yield stress and tensile strength (492 MPa and 553 MPa) were found to be lower than those values (520 MPa and 568 MPa) provided by Palmer Tube Mills. A similar phenomenon was found for C350 sections [Zhao and Hancock (1995a)]. The welds had a mean measured tensile strength of 528 MPa with a coefficient of variation of 0.0597. The nominal tensile strength is 480 MPa in AS4100-1990.

2.2 Test Program

Twenty four different combinations of test specimens were carried out in this program. They are designated as EY & WY for RHS Orientation with end return weld and EN & WN for RHS Orientation without end return weld. Figure 1 gives the layout for both load combinations and RHS orientations. A specimen label is assigned to each specimen as shown in Table 1, where the last number in a label refers to the loading case. All the samples were tested to failure in a 500 kN capacity Baldwin Universal Machine. The ultimate loads are given in Table 1.

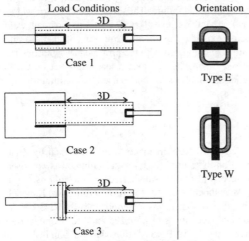

Figure 1: Schematic View of Load Combination and RHS Orientations

Table 1: Comparison of Test Results

Specimen Label	P_{max} (kN)	$\dfrac{P_{max}}{P_p}$	$\dfrac{P_{max}}{P_W}$
G1EY1	183.0	0.97	0.74
G1EY2	184.6	1.03	0.92
G1EY3	192.0	1.12	1.00
G1EN1	139.5	0.89	0.73
G1EN2	161.4	1.02	0.90
G1EN3	147.8	0.93	0.75
G2EY1	170.5	0.96	0.85
G2EY2	176.8	0.99	0.84
G2EY3	167.4	1.11	0.95
G2EN1	142.5	0.91	0.88
G2EN2	141.4	0.92	0.71
G2EN3	141.3	0.90	0.87
G2WY1	174.0	1.06	0.88
G2WY2	174.9	1.06	0.89
G2WY3	169.1	1.00	0.84
G2WN1	139.6	0.96	0.75
G2WN2	138.8	0.96	0.81
G2WN3	139.9	0.95	0.78
G3EY1	215.8	1.05	1.19
G3EY2	230.8	1.15	1.15
G3EY3	220.7	1.10	1.02
G3EN1	183.7	1.02	1.28
G3EN2	192.1	1.06	1.07
G3EN3	183.1	1.05	0.93
Mean	---	1.01	0.90
COV	---	0.07	0.17

2.3 Failure Mode

The dominant failure mode observed in the tests was tearing of the parent metal along the weld contour. For test types EN and WN, where no end return welds were applied, the tearing was usually found to initiate at the end of the weld adjacent to the end of the plate and then progressed along the contour of the weld. This sort of failure may be due to the higher stress concentration at the termination point. As for test type EY and WY, where end return welds were applied, the rupture is starting at the transition zone from the end of the weld to the end return weld. This failure profile is most likely due to the additional elements around the end return weld.

2.4 Effect of Load Conditions

One of the main concerns on the proposed formula [Zhao and Hancock (1995b)] is that the effect of load conditions was not taken as a variable in predicting the ultimate strength of fillet weld joints. This may cause a concern if the contribution of the applied loads to the reaction points is considerably large. From the experimental results shown in Table 2, it can be seen that load case 1 has the lowest load capacity and load case 2 has the highest . This may be due to the fact that a higher percentage of the applied load in load case 1 was propagated directly to the support end whilst the applied loads in load case 2 required a longer path to cross the member which results in lower contributions of the applied load components to the reactions. On the basis of the experimental evidence, it can be concluded that the strength of fillet welds is insignificantly dependent upon the load orientations. This is true provided that the domain between the application point to the restrained end is greater or equal to 3D as used in all tests, where D is the overall depth of RHS section.

2.5 Effect of End Return Weld

A study of the test results revealed that the ultimate strength of the specimens with a continuous end return weld is on average 9.6% higher than the one without end return weld. Similar results were reported by Zhao and Hancock (1995b) on grade 350 MPa RHS cold-formed sections and a mean comparison is presented in Table 2. The reason for the increase in weld strength is most likely due to the end return weld, which permitted the redistribution of the stresses around the critical zones. The detailed experimental and theoretical analysis of stress distribution around the welds can be found in Kiew, Al-Mahaidi and Zhao (1997). The advantage of end return welds has been adopted by various structural design codes such as Eurocode 3 Part 1.1 (1992) and AISC-LRFD (1993), but not yet by AS4100.

Table 2 : Comparison with Zhao and Hancock

Parameters	C450 RHS				C450 RHS		C350 RHS	
					Mean	Cov	Mean	Cov
	Mean	Cov	Effect of end return weld	$\dfrac{erw}{Werw} = 1.096$		0.05	$\dfrac{erw}{Werw} = 1.125$	0.05
Load conditions	$\dfrac{LC2}{LC3} = 1.029$	0.04						
	$\dfrac{LC2}{LC1} = 1.039$	0.05	RHS Orien.	$\dfrac{TypeE}{TypeW} = 1.005$		0.02	$\dfrac{TypeE}{TypeW} = 0.92$	0.04
	$\dfrac{LC3}{LC1} = 1.009$	0.03						

erw: end return weld (W: without)

2.6 Effect of RHS Orientation

Table 2 shows that the effect of RHS orientation is 1.005, which is higher than that of 0.92 observed for C350 RHS members. This may be due to the fact that the stress variation around the section ($\sigma_{utO}/\sigma_{utA} \approx 1.006$) for C450 RHS members is smaller than that for C350 members, and that the seam weld is applied at opposite face rather than adjacent face.

3 DESIGN RULES

3.1 Formula Based on Parent Metal Strength

A design formula based on parent metal strength AS/NZS 4600-1996 [SAA(1996)] was proposed by Zhao and Hancock (1995a) as

$$P_p = 0.75 F_u t L \tag{1}$$

where F_u, t and L are tensile strength of parent metal, wall thickness of RHS and weld length respectively. It has the same expression as that in AISI-1996 [AISI(1996)] and CSA-S136-M94 [CSA(1994)] for $L_w / t \geq 25$. This formula gives a good estimation on the static strength of the longitudinal fillet welds as shown in Table 1 where the average mean ratio of the observed to the predicted strength is 1.01. The corresponding coefficient of variation (COV) is 0.07. A capacity factor of 0.72 was proposed based on the tests with end return welds [Zhao and Hancock (1995a)]. This value is higher than that of 0.55 specified in AS/NZS 4600-1996 and AISI-1996, and that of 0.67 in CSA-S136-M94. A reliability analysis method will be used to calibrate the results obtained for C450 RHS members.

3.2 Formula Based on Weld Metal Strength

A design formula based on weld metal strength (X_u) was proposed by Zhao and Hancock (1995a) as

$$P_w = 0.6 X_u a L \tag{2}$$

where X_u and a are tensile strength of weld metal and weld throat thickness respectively. It has the same expression as in AS4100-1990, AISC-1993 and CSA-S16.1-M89 [CSA(1989)]. This formula overestimates the weld strength as shown in Table 2. The corresponding COV is 0.166. A capacity factor of 0.7 was proposed, based on tests including those without end return welds, for Amendment No.3 to AS4100 [Zhao, Hancock and Sully(1996)]. This value is lower than that of 0.8 specified in AS4100-1990 and 0.75 in AISC-1993, but slightly higher than that of 0.67 specified in CSA-S16.1-M89. A reliability analysis method will be used to calibrate the results for C450 RHS members.

4 RELIABILITY ANALYSIS

A reliability index described by Cornell (1969) and Galambos and Ravindra (1973) is used to assess the safety indices of the proposed formula. This method is called FOSM (First Order Second Method) mean value method, with both Resistance and Load assumed lognormal distributions. A detailed review of this method can be found in Zhao and Hancock (1993). The same statistical parameters of load are used in this paper. The statistical parameters of Resistance for P_p are given in Table 3. The values of P_m and V_p are determined from Table 1. The values of F_m, V_F and V_m are supplied by Palmer. The value of M_m was calculated as the product of the mean ratio of measured to nominal tensile strength supplied by Palmer, and the mean ratio of measured tensile strengths are given by Monash University to those given by Palmer. The statistical parameters of Resistance for P_w are given in Table 4. The values of P_m and V_p are determined from Table 1. The values of M_m, V_m, F_m and V_F are taken as those in Zhao and Hancock (1993).

The reliability index β versus $D_n / (D_n + L_n)$ curves are plotted in Figure 2 for design model based on parent metal strength as shown in Eq (1). The calibration point is chosen as L_n / D_n of 3, ie $D_n / (D_n + L_n)$ of 0.25 as explained in Galambos (1995). The target reliability index for cold-formed connections is 3.5 as specified in AISI-1990 [AISI(1990)]. It can be seen in Figure 2 that

Table 3 : Statistical Parameters of Resistance for Proposed Formula [Zhao and Hancock (1995b)]		
Variables	Statistical Parameters	Values
Design Model (0.75F$_u$tL)	P$_m$ (EY&WY)	1.050
	V$_P$ (EY&WY)	0.060
	P$_m$ (All Tests)	1.008
	V$_P$ (All Tests)	0.073
Tensile Strength (Parent Metal)	M$_m$	1.114
	V$_M$	0.057
Thickness (Parent Metal)	F$_m$	0.976
	V$_F$	0.043

Table 4 : Statistical Parameters of Resistance for AS4100-1990		
Variables	Statistical Parameters	Values
Design Model (0.6X$_u$aL)	P$_m$ (All Tests)	0.905
	V$_P$ (All Tests)	0.166
Tensile Strength (Weld Metal)	M$_m$	1.099
	V$_M$	0.0597
Thickness (Weld Metal)	F$_m$	1.470
	V$_F$	0.0670

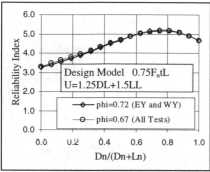

Figure 2 : Reliability Analysis Curves for Model 0.75F$_u$tL

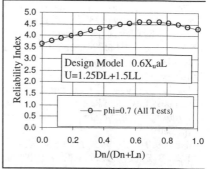

Figure 3 : Reliability Analysis Curves for Model 0.6X$_u$aL

the proposed design formula (0.75F$_u$tL) with a capacity factor of 0.72 applies to longitudinal fillet welds in C450 RHS members provided that the end return welds are applied. If the end return welds are not specified as a requirement, the capacity factor should reduce to 0.67. The reliability index β versus $D_n/(D_n + L_n)$ curves are plotted in Figure 3 for design model based on weld metal strength as shown in Eq (2). It can be seen in Figure 3 that the proposed design formula (0.6X$_u$aL) with a capacity factor of 0.7 applies to longitudinal fillet welds in C450 RHS members

5 CONCLUSION

The dominant failure observed in the tests was base metal failure rather than weld failure. The effect of load conditions and RHS orientation are found to be insignificant to the strength of fillet welds because of the interaction domain (3D) and a smaller stress variation around the section. End return welds are found to be on average of 9.6% higher than the one without end return weld and this advantage should be adopted in the design codes as proposed by Zhao and Hancock (1995a). The design formula proposed by Zhao and Hancock (1995a), based on parent metal strength is applicable for C450 RHS members. The Amendment No.3 to AS4100 is also applicable for C450 RHS members.

6 ACKNOWLEDGMENTS

The authors are grateful to the DEETYA for an OPRS, Monash University for a MGS and Faculty of Engineering for additional financial support. The authors are also grateful to Palmer Tube Mills (Aust) Pty Ltd for providing test specimens.

REFERENCES

AISC (1993), *Load and resistance factor design specification for structural steel buildings.* Am. Inst. of Steel Constr., Chicago. Ill.

AISI (1996), *Load and resistance factor design specification for cold-formed steel structural members.* Am. Iron and Steel Inst., Washington, D.C.

AISI (1990), *Commentary on the LRFD design specification for cold-formed steel structural members.* Report CF 90-2, August, Am. Iron and Steel Inst., Washington, D.C.

Cornell, C. A (1969). *A probability baaed structural code.* ACI J., 66(12), 974-985.

CSA - S16.1 - M89 (1989), *Steel structures for buildings (limit state design).* Can. Standards Assoc., Rexdale, Ont., Canada.

CSA - S136 - M94 (1994), *Cold-formed steel structural members.* Can. Standards Assoc. Rexdale, Ont., Canada.

Eurocode 3 Part 1.1; Design of steel structure: Part 1.1- *General rules and rules for buildings,* DD ENV, 1993-1. (1992). Eurocode 3 Editorial Group, London, U.K.

Galambos, T. V., and Ravindra, M. K (1973). *Tentative load and resistance factor design criteria for steel building.* Res. Rep. No. 18, Civ. and Envir. Engrg. Dept., Washington Univ., St. Louis, Mo.

Galambos, T. V., (1995). *Public Safety - is it compromised by new LRFD design standards?[a]* . Res. . J. Struct. Div., ASCE, Vol.121, No.1, pp 143 -144.

Kiew, K. P., Al-Mahaidi, R. and Zhao, X. L. (1997). *Theoretical analysis of the longitudinal welds in cold-formed RHS members using FEM..* Proc., 2nd Int. Conference on the Application of Numerical Methods in Engineering, (submitted for publication).

SAA (1990), *Steel structures,* Australian standard AS4100, Sydney.

SAA (1991), *Methods for tensile testing of material,* Australian standard AS1391, Sydney.

SAA (1996), *Cold-formed steel structures,* Australian standard AS/NZS 4600, Sydney.

Zhao, X. L., and Hancock, G. J. (1993). *Tests and design of butt welds and transverse fillet welds in thin cold-formed RHS members.* Res. Rep. R681, School of Civ. and Min. Engrg., Univ. of Sydney, Sydney, Australia.

Zhao, X. L., and Hancock, G. J. (1995a). *Butt welds and transverse fillet welds in thin cold-formed RHS members.* J. Struct. Div., ASCE, Vol.121, No.11, pp 1674-1682.

Zhao, X. L., and Hancock, G. J. (1995b). *longitudinal fillet welds in thin cold-formed RHS members.* J. Struct. Div., ASCE, Vol.121, No.11, pp 1683-1690.

Zhao, X. L., Hancock, G. J. and Sully, R. M. (1996). *Design of tubular members and connections using amendment number 3 to AS4100.* J. AISC, 30(4), pp 2-15.

The Mechanics of Structures and Materials, Grzebieta, Al-Mahaidi & Wilson (eds)
© *1997 Balkema, Rotterdam, ISBN 90 5410 900 9*

Steel membrane floors for bodies of large rear-dump mining trucks

P. H. Dayawansa, K. W. Poh, A. W. Dickerson, I. R. Thomas & H. Bartosiewicz
BHP Research, Melbourne Laboratories, Vic., Australia

ABSTRACT: This paper presents the development of an innovative curved steel membrane floor for the bodies of large rear-dump mining trucks. Three profiles of curved steel membrane floors were considered. The performance of the profiles were evaluated by comparing the dent resistance of the structures against rock impacts. A finite-element dynamic analysis program MSC/DYNA was used to model the impact behaviour of these membrane floors and on a typical conventional flat-plate type floor. The results of the analyses are presented in this paper. A prototype of the recommended structure based on this study has since been built and successfully implemented in a mine-site.

1 INTRODUCTION

The body of a large rear-dump mining truck generally accounts for 20 to 25% of the total empty vehicle mass and is the heaviest single item of the truck. One way to reduce the mass of the truck and hence increase its load carrying capacity is to reduce the mass of the body.

In a conventional body, the floor plate is welded to a system of beams located underneath the plate. In many applications the floor and to a lesser extent the sides of the body are lined with additional wear plates (generally 25 mm thick) to prevent wearing of the original plates due to sliding of material during the unloading process. The wear plates also reduce the impact damage to the floor structure. A conventional truck body may use approximately 10 tonnes of wear plates on the floor, and they are replaced several times during the life of the truck body.

The stresses in a conventional body under a range of static loading conditions were investigated by Kowalczyk (1989) using finite element analysis. Although it was not the main objective of this study, Kowalczyk (1989) suggested that a double curvature steel membrane type load entraining structure could form an efficient structural system for the body.

Impact loads acting on the floor during loading of large boulders is a main loading condition considered in the design of truck bodies. The spacing of the floor beams is generally determined by this condition of loading.

A new structural concept for truck bodies where the conventional floor is replaced with a thin steel membrane (e.g. 19 mm thick) was investigated. Several membrane profiles were analysed and the most suitable profile was used in the design.

It was envisaged that a membrane truck body floor would reduce the magnitudes of the impact loads, in comparison to conventional bodies, due to its reduced stiffness. It should also reduce fatigue cracking because of the significantly reduced number of welded joints. Because the steel membrane would be attached to the basic frame of the truck body only at its edges, it offers the prospects for rapid replacement and the elimination of the use of separate wear plates. In addition, a truck body with a membrane floor was also envisaged to have a reduced mass in comparison to a conventional truck body.

The structural concept which employs a membrane floor was found to be satisfactory in all other aspects, and it was necessary to undertake a detailed investigation to determine the performance of the steel membrane under impact loads.

The investigation was conducted using the finite element analysis program MSC/DYNA (1991). A conventional floor was also analysed for comparison purposes. The results of the analyses are presented in this paper. Details of the analyses can be found in the report by Poh, Dayawansa and Dickerson (1992).

2 IMPACT MODELLING

The magnitude and area extent of the impact forces on the floor of the truck body depends mainly on the following factors:
• the mass and shape of the individual rocks;
• the velocity of the rock at contact (proportional to the square root of the drop height);
• the material properties of the rocks;
• the stiffness of the body floor, the inertia and stiffness of the body support system;
• the angle of incidence between the rock trajectory and the floor of the body;
• the amount of fine material already on the floor of the truck body; and
• the point on the truck floor at which the impact occurs.

It was decided to develop a standardised impactor (a standard rock impacting at a constant speed) and to compare the effect of this impactor on various membrane type structures with the effect of the same impactor on the floor of a conventional truck body. It was decided that the standard impactor should represent the top end of rock mass and impact speed likely to occur in normal mining operations.

2.1 Assumptions for the Impact Analysis

The basic assumptions made with regards to the impact analyses are outlined below:
• The truck body was empty and the impactor hit the floor plates directly.
• The impactor hit the floor plate at right angle to its surface without prior spinning or rotation.
• The impactor hit the floor plate only once. After the impact, the impactor would rebound freely (without the effects of gravity) and lose contact with the floor plate.
• The properties of the impactor and the floor plate material were elastic-plastic.
• Strain-rate effects on the mechanical properties of the materials were ignored.
• Cracking or fracture of the materials is not considered.
• The effect of damping on initial impact forces is negligible.
• The centre of a floor plate panel was a critical position for impact.

2.2 Standard Impactor

The standard impactor developed for the purpose of the analysis is shown in Fig. 1. The 200 mm diameter hemi-spherical tip formed the contact surface. The total mass of the impactor was 3,000 kg.

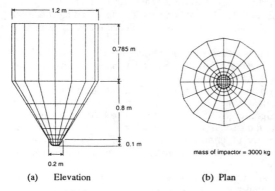

| 1.2 m |
| 0.785 m |
| 0.8 m |
| 0.1 m |
| 0.2 m |

mass of impactor = 3000 kg

(a)   Elevation   (b) Plan

Figure 1. Standard Impactor

The surface of the spherical tip was lined with a layer of 4-node *contact* elements. The shapes of these contact elements (and the brick elements they were attached to) were chosen to match the mesh over the contact point on the floor plate, which was also lined with another layer of similar contact elements.

The properties of the impactor material used in the analyses are as follows:

modulus of elasticity	= 25,000 MPa	yield stress	= 40 MPa
modulus of strain hardening	= 25 MPa	Poisson's ratio	= 0.2
density	= 2,400 kg/m^3		

A friction coefficient of 0.8 was assumed at the contact surface between the impactor and the floor plates.

At the start of each analysis, the impactor was positioned with its tip just above (1 mm) the surface of the floor plate. A downward initial velocity of 7.67 m/s was assigned for the entire impactor. This simulates the conditions of the impactor being dropped freely under gravity from a height of 3 m.

2.3 General Features of the Floor Plates in the Models

The finite-element models of the various floor plates described in subsequent sections were all created using 4-node quadrilateral shell elements, with three integration points across the thickness of each element.

The properties of the steel plate material used in the analyses are as follows:

modulus of elasticity	= 210,000 MPa	Poisson's ratio	= 0.3
modulus of strain hardening	= 210 MPa	density	= 7,850 kg/m^3

The plate thicknesses and the yield stresses varied between the different models.

3 IMPACT PERFORMANCE OF A CONVENTIONAL FLAT-PLATE TYPE BODY

A reference study on the impact performance of a conventional flat-plate type body was conducted in the three stages shown in Fig. 2.

In Stage 1, the impactor impacted onto a rectangular panel which was simply supported along all edges. In Stage 2, the area of the floor plate was extended to represent the central section of the floor of the truck body. The floor structure was rigidly supported at the ends of the longitudinal beams. In Stage 3, the section of the body floor used in Stage 2 was supported on an assembly of beams, masses and springs which approximated the vertical dynamic

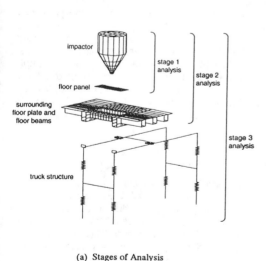

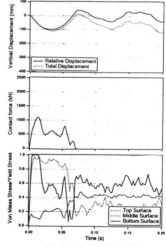

(a) Stages of Analysis

(b) Results for Stage 3(a)

Figure 2. Stages of Analysis and Results for Stage 3(a)

573

Table 1. Summary of Analysis Results for Impact onto Conventional Flat-Plate Type Floor

Analysis Stage	Max Relative Deflection (mm)	Time at Max Deflection (ms)	Max Contact Force (kN)	Time at Max Contact Force (ms)	Contact Time (ms)	Estimate of Max Permanent Deflection (mm)
1a	33	15	2072	16	24	22.5
1b	12	14	2267	13	21	2.4
2a	37	17	1767	16	28	18.5
2b	11	14	2051	15	22	1.4
3a	36	16	1707	14	27	17.3
3b	10	12	1894	15	22	0.7

characteristics of the remainder of the truck. The truck structure was created to represent a typical 172-tonne payload capacity truck.

3.1 Analysis and Results

The analysis was conducted to simulate the impact processes for the first 0.1 seconds which was sufficient to capture the full impact process. For each stage, analyses were conducted for plate thicknesses of 19 mm (case a) and 35 mm (case b). The 35 mm thickness represented a floor plate supplemented by typical wear plates. The displacements, stresses and the contact forces were output at regular intervals of one hundredth of the analysis time.

Fig 2(b) shows the displacements and stresses in the floor plates at the centre of impact, and the total forces at the contact surfaces for Stage 3(a) are shown against the impact time. The contact force exerted by the impactor onto each floor plate (and vice versa) during impact were obtained by summing the individual force output of the contact elements on the steel plates.

The Von Mises stresses (top, middle and bottom surface) at the centre of impact of each floor plate are also given in Fig. 2(b). The stresses are normalised against the yield stress of the floor plates. The results for the six analysis cases are summarised in Table 1.

3.2 Discussion of Results

With 19 mm of 1,000 MPa yield strength steel floor plate, the permanent deflection of the floor plate for a single impact ranges between approximately 22 mm (for the simply supported 500 mm wide section of plate) and approximately 17 mm when the complete truck structure is included in the model. This result fits with field observations that 19 mm thick floor plates are unsatisfactory for severe impact applications.

For the plate thickness of 35 mm and the yield strength of 1,050 MPa, the permanent deformations were approximately 2.4 mm for the simply supported plate and 0.7 mm for the complete truck structure. These floor plate deformations are consistent with mine-site experiences for trucks with similar floor plate thicknesses and severe impact conditions.

The maximum contact forces for the 35 mm thick plate range from 2,267 kN for the simply supported plate to 1,894 kN for the complete truck system. They follow the expected trend of reducing with reduced stiffness in the plate support system. For the 35 mm thick floor plate, the ratio of the peak contact force to the weight of the impactor varied from 77 for the simply supported plate to 64 for the complete truck system. The corresponding ratios for the 19 mm thick floor plate are 70 and 58 respectively.

4 IMPACT PERFORMANCE OF MEMBRANE TYPE BODIES

This study concentrated only on single curvature membrane shapes and membrane thicknesses of 19 mm. The material used for the membranes was "Bisalloy 500" grade of quenched and tempered plate which is widely used as a wear plating for truck bodies, particularly in the impact regions of the floor. It satisfies the weldability and formability requirements for a single curvature membrane, and is available in suitable sizes at reasonable cost. This material has a yield strength of 1,480 MPa and a tensile strength of 1,585 MPa. Consequently a material yield strength of 1,480 MPa was used for all of the membrane profiles investigated.

The three membrane profiles considered in this preliminary study are shown in Fig. 3.

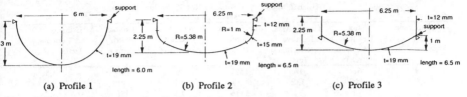

(a) Profile 1   (b) Profile 2   (c) Profile 3

Figure 3 Membrane Profiles Considered in the Analysis.

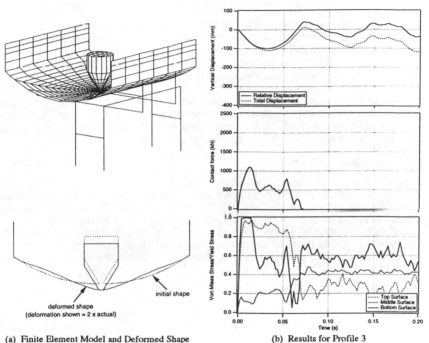

(a) Finite Element Model and Deformed Shape   (b) Results for Profile 3

Figure 4. Finite Element Model and Results for Profile 3

4.1 Analysis and Results

Profile 1—The semi-circular shape membrane was chosen as the simplest membrane shape with the smallest possible radius of curvature and therefore likely to be the strongest possible membrane shape within the width constraints of the size of a typical 172-tonne truck being considered. The floor area covered by the membrane is 6 m × 6m.

The results of this investigation indicated that the prospects for success using a single-curvature membrane of about 19 mm thickness were reasonable. The key to success appeared to be the lower stiffness and therefore lower impact forces on the membrane compared to conventional flat-plate type truck bodies.

Profile 2—This profile was developed as a practical membrane shape in comparison to Profile 1. The results showed that an open-ended membrane of this shape (case a) achieved the desired result of significantly reducing the peak contact force and almost eliminating yielding in the plate. However, the total deflection at the centre of the plate (about 340 mm) is impractically high. The same membrane was also analysed with one end closed (case b). The maximum deflection for this case was 74 mm. A maximum deflection in excess of 150 mm would generally be regarded as unacceptable.

575

TABLE 2. Summary of Analysis Results for Impact onto Membrane Type Floors

Membrane Profile	Max Relative Deflection (mm)	Time at Max Relative Deflection (ms)	Max Contact Force (kN)	Time at Max Contact Force (ms)	Contact Time (ms)	Estimate of Max Permanent Deflection (mm)
1	61	18	1597	20	40	4.5
2a	346	132	856	10	46	-
2b	74	22	1119	12	58	5.5
3	99	32	1103	12	72	4.0

Profile 3—This membrane profile resulted from consideration of how Profile 2 would handle distributed vertical and transverse loads. The 5.38 m radius was retained from Profile 2 and is the smallest radius which could be considered as possibly practical for this size of truck. The finite element model of the membrane and the cross-sectional shape of the membrane at maximum deflection under impact are shown in Fig. 4(a).

The results from the impact analysis of Profile 3 are shown in Fig. 4(b). The maximum deflection at the centre of impact was about 100 mm which is considered acceptable. The permanent deflection at the location of impact was about 4.0 mm.

As with membrane Profile 2b, the contact force was extended over a longer time and there are second and third peaks in the contact force-time relationship.

4.2 Discussion of Results

The results of the analysis are summarised in Table 2.

The results of the analysis showed that a simple single curvature steel membrane of about 19 mm thickness and a shape similar to that of Profile 3 may be able to perform satisfactorily under the impact loads which will be experienced in typical large mining trucks.

A preliminary design of a steel membrane type body for a truck having a nominal payload rating of 172 tonnes indicated that mass reductions of up to approximately 8 tonnes can be achieved by using the membrane type body concept described in this investigation.

5 PROTOTYPE STRUCTURE

A prototype steel membrane truck body floor similar to Profile 3 was built and installed on a truck with a rated payload of approximately 180 tonnes. At the time of writing, it had performed very satisfactorily for 6,500 operating hours in a very arduous conditions (many large, sharp and hard rocks). It is expected that it will continue to perform satisfactorily for many thousands of additional hours.

6 CONCLUSIONS

A simple single-curvature steel membrane with a thickness of about 19 mm has been shown to be able to provide acceptable resistance to impact damage during normal loading processes for large mining trucks.

7 REFERENCES

Kowalczyk, W. 1989. Truck Tray Mass Reduction. *BHP Melbourne Research Laboratories Report MRL/PM58/89/001*. Melbourne, Australia.

Poh, K.W., Dayawansa, P.H. and Dickerson, A.W. 1992. Steel Membrane Floors for Bodies of Large Rear-Dump Mining Trucks - A Preliminary Study. *BHP Melbourne Research Laboratories Report BHPR/ENG/R/92/089/KH43*. Melbourne, Australia.

MacNeal Schwendler Corporation 1991. *MSC/DYNA, Version 2A*. Los Angeles, California, USA.

The Mechanics of Structures and Materials, Grzebieta, Al-Mahaidi & Wilson (eds)
© *1997 Balkema, Rotterdam, ISBN 90 5410 900 9*

Testing of the web-side-plate steel connection

J.C.Adams, M.Patrick & P.H.Dayawansa
BHP Research, Melbourne Laboratories, Vic., Australia

ABSTRACT: The web-side-plate steel connection specified in the Australian Institute of Steel Construction (AISC) design manual on structural connections is commonly used in steel-frame buildings. When a concrete slab acts compositely with the steel beams and passes between their ends, the action of the web-side-plate connection is quite different to the bare steel state. A special test rig described in the paper has been constructed to simulate these states. Small plate specimens are used which avoids the need for a steel beam, thus considerably reducing the cost and time required to perform each test. Therefore, it has been possible to conduct a large number of tests and to examine the effects that cleat and beam web thickness, steel grade, bolt group geometry, bolt tension, etc. have on rotational behaviour and moment capacity. Some additional tests have been performed to investigate interference between the beam bottom and column flanges. Selected results from the test program are presented and briefly discussed.

1 INTRODUCTION

The web-side-plate steel connection specified in the Australian Institute of Steel Construction (AISC) design manual on structural connections (Hogan and Thomas 1994) consists of a holed cleat that is fillet welded to a supporting column, beam or embedded plate, and to which a beam is bolted. Owing to its simplicity and the small number of components involved, fabrication and erection are quick and its use is very economical. In composite steel-frame buildings, the floor steelwork is often assembled using this connection.

This connection in its bare steel state is assumed to be flexible and subject to reaction shear force, with relatively small bending moments developing in its components due to shear force eccentricity. Therefore, in multi-storey construction the steel frame must be braced against sway. Normally, no significant axial force is present in the connection and at ultimate load the beam end effectively rotates about the centre of the bolt group.

When a concrete slab acts compositely with the steel beams and passes between their ends in a composite steel-frame building, the action of the web-side-plate connection changes significantly. The height of the centre of rotation is affected by the relative stiffness of the steel connection and the slab, and in the extreme, rotation occurs about a position level with the longitudinal reinforcement in the slab. This causes the rotational stiffness to increase greatly, and under vertical loading potentially large tensile and compressive forces can develop in the reinforcement and cleat, respectively. Consequently, the moment capacity of the composite connection is considerably larger than that of the web-side-plate connection acting alone.

A test program is being conducted at BHP Research — Melbourne Laboratories to improve knowledge of the behaviour of the web-side-plate connection in both the bare steel and composite states.

2 AISC CONNECTION DETAILS

For brevity, the reader is referred to the AISC design manual for details of the connection. However, the following aspects are of particular importance to this discussion:

(1) The cleat is normally fabricated from 10 mm thick, Grade 250 BHP square edge flat bar with a nominal yield stress of 260 MPa (which is likely to increase in the future to 300 MPa). This is the maximum thickness allowable, and a new limitation introduced into the latest edition of the manual requires that the cleat thickness (t_i) does not exceed the beam web thickness (t_w) by more than one millimetre, i.e. $t_i \leq t_w + 1.0$ mm ≤ 10 mm. This limitation was apparently placed in recognition of the important contribution that deformation of the cleat makes towards rotation capacity. However, it could conceivably cause practical problems on site ensuring that this requirement is met and therefore should be reviewed.

(2) The top edge of the cleat is located a standard 65 mm below the top of the steel beam.

(3) The beams will normally be BHP hot-rolled universal beams (maximum 610 UB) or BHP welded-plate beams (700WB or larger) of Grade 300 MPa.

(4) The 20 mm diameter, high-strength structural bolts are normally snug-tightened, i.e. M20/8.8/S, with their thread in the shear plane.

(5) A 20 mm gap must be provided between the bottom flange of the steel beam and the column face, which is expected to typically allow the beam end to rotate 20 mrad before interference occurs.

3 DEVELOPMENT OF A SPECIAL TEST RIG

In Australia, the web-side-plate connection has previously been tested in the rigid support condition using a relatively complex procedure by lowering the free end of a long steel beam attached by the connection to a column (Pham and Mansell 1982, Patrick et al.1986). No tests have to date been performed on composite specimens incorporating the web-side-plate connection.

However, Xiao et al. (1994) and other overseas researchers have performed tests of this latter type, noting that in the UK it is referred to as the fin plate connection. Only small numbers of tests have been performed though, typically in a conventional cruciform arrangement involving twin cantilever steel beams connected to a continuous composite slab. Patrick et al. (1986) have explained some of the technical disadvantages of conducting tests of this type. The economic disadvantage is obvious, and largely explains why so few tests have been performed. Also, the compressive force in the steel part of the connection has not been measured and nor can it necessarily be determined accurately from the test results, even though this is a critical consideration during the modelling or design of the connection components.

Ren (1995) claims to have performed the first tests of their type on several other types of bare steel connections (e.g. double angle cleats) to investigate the effect of axial compressive force on moment-rotation behaviour. She has attempted to simulate the conditions experienced by a steel connection at a support in the hogging moment region of a continuous composite beam. She devised a test rig whereby a vertical force was applied at the end of a steel beam which could be rotated and fixed in one of either four positions, ranging from a cantilever (no axial force with beam horizontal) through to a column (no bending with beam vertical). Otherwise the load was applied at either 30 or 60 degrees to the longitudinal axis of the steel beam. A disadvantage with this system is that the ratio of axial force to shear force effectively remains constant during a test and is not affected by the behaviour of the connection.

3.1 Test rig, test specimens and test procedure

A special test rig has been constructed which uses small plate specimens to simulate the conditions experienced by a web-side-plate connection subjected to rotation in either the bare steel or composite states. Because the test specimens do not incorporate steel beams or a concrete slab, the cost and time associated with their construction and testing is considerably reduced. Therefore, it has been possible to conduct a large number of tests, and thus systematically examine the effect that a number of major variables have on behaviour.

A composite web-side-plate connection is shown schematically in Figure 1 alongside the test rig (shown on its side) to illustrate their relationship. In the test rig the reinforcement is represented by a fixed pin which may be removed during testing to simulate the bare steel state.

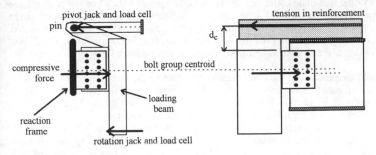

Figure 1 BHP Research's connection test rig and a composite connection

The test rig comprises the following major items (see Figure 2(a)): (1) a rigid loading beam supported from below at each of its ends by a load cell and a 1000 kN capacity hydraulic jack operated in position control, (2) a test specimen housed in a slot in the loading beam to restrain the bottom edge rotationally which is also machined flat for bearing, and bolted above to a rigid reaction frame, (3) a steel pin which engages one end of the loading beam when simulating the composite state, otherwise it is left out, (4) a frame at each end of the loading beam to provide lateral support, and (5) a roller box to prevent horizontal restraint forces from building up in the system. The jack at the end of the loading beam which can accommodate the pin will be referred to as the *pivot jack*, and the other as the *rotation jack*. Linear potentiometers are used to measure the vertical movement of the loading beam relative to the reaction frame, and also the lateral movement of the connection. All measurements are taken using computer-controlled equipment. The test rig is statically determinate in the vertical plane. Therefore, the resultant compressive force acting on the connection and its line of action can be determined at any stage in a test, whereby the bending moment acting on the connection at any particular vertical cross-section can also be calculated.

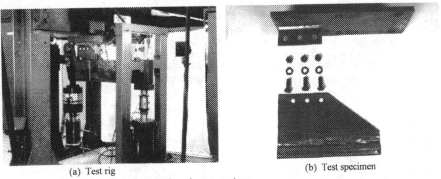

(a) Test rig   (b) Test specimen

Figure 2 BHP Research's connection test rig and a test specimen

The test specimens are intended to conform with the AISC design manual. Each specimen comprises a web plate, a set of bolts and a cleat welded to a reusable support plate (see Figure 2(b)). The web plate and cleat have been either match-drilled or punched separately depending upon the conditions being investigated. The web plate approximately simulates the web of a double-coped beam with the stiffening effect of beam flanges absent. The height that the plate extends above the loading beam measured to the top edge along the bolt holes is 210 mm, which

579

naturally can affect the load at which the connection plates buckle out-of-plane. A small experimental investigation was undertaken which established that for double-column bolt groups, and for the range of plate thicknesses to be studied, that with this arrangement at least 80 per cent of the squash load of the web plate could be reached in the composite tests when the compressive force is a maximum. As can be seen in Figure 2(b), the plate also extends beyond the bolt group to assist in this regard, but is bevelled so that it does not interfere with the top plate. The largest standardised connection that can currently be tested is a nine-row by two column (9×2) bolt group, which arises due to the limited width of the reaction frame measured in the direction of the loading beam.

The influence that the position of the cleat measured relative to the slab reinforcement has on behaviour can be investigated by varying the distance between the top of the cleat and the centre of the pin (dimension d_c in Figure 1). For example, for a three-row bolt group d_c can be varied from 140 mm to 555 mm, while there is less scope to move deeper connections.

Tests are performed by the rotation jack thrusting upwards with the pivot jack (if engaged) maintaining the position of the pin constant. When the pin is omitted, the steel connection must also support the force applied by the rotation jack. Motion of the rotation jack is interrupted each time computer readings are taken.

3.2 *Typical test results*

Some typical test results are shown in Figures 3(a) and 3(b) for a 6×2 connection in the bare steel and composite states, respectively. For comparison purposes, both results are shown in Figure 3(b), although the part of the bare steel curve at high rotations has been omitted. It can be seen that during the composite state both the secant stiffness and moment capacity were significantly increased, while rotation capacity was reduced. It is interesting to note that in the composite test, the compressive force on the cleat reached 988 kN and caused it to buckle, while in the bare steel test it only reached 167 kN (see also Section 3.3).

The moment-rotation behaviour of the web-side-plate connection in either the composite or bare steel states is characterised by four distinct phases (see Figure 3(b)), viz.: (1) friction - the initial response is essentially linear-elastic as friction at the faying surfaces prevents slip, (2) slip - the rotation increases rapidly as slip progresses and the bending moment may reduce slightly, (3) bearing - slip is arrested when the bolts bear into the plies and moment increases linearly, and (4) failure - the plates buckle or a component fails leading to a reduction in moment capacity, which may be a sudden event. Other researchers have found similar moment-rotation behaviour for this type of connection (e.g. Lipson 1968).

Single-column bolt group specimens failed by either shearing an extreme bolt, which occurs predominantly in the bare steel state, or by buckling. Composite specimens with double-column bolt groups always failed by buckling. Bare steel specimens with double-column bolt groups generally failed by progressively tearing the cleat along the net section of the inner bolt column starting from the edge of the cleat nearest the pivot jack and possibly extending through several bolt holes, except that the deepest nine-row connection buckled.

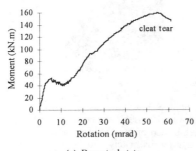

(a) Bare steel state

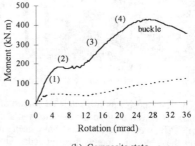

(b) Composite state

Figure 3 Moment-rotation curves (6×2 bolt group)

3.3 *Effects of compressive and shear forces acting on the cleat*

In practice, the connection in its bare steel state is predominantly subjected to shear force. However, as previously mentioned, in the test some resultant compressive force is always present and shear is omitted. Consideration of the effects that these factors have on the moment-rotation curves obtained from tests is given below.

Firstly, the compressive force that develops in the cleat is typically less than 10 per cent of its compressive capacity in the absence of bending. Tests on bare steel connections at different positions in the rig have confirmed that compressive force of this magnitude can be ignored.

Concerning shear force, a series of double-cantilever tests was performed varying the shear-to-moment ratio using the test set-up shown in Figure 4(a). The results are shown in Figure 4(b) where V is the peak shear force obtained in a test and V_c is the nominal shear capacity of the cleat calculated according to the AISC design manual. A practical upper-limit to the value of V/V_c is 0.75. Therefore, it can be seen from Figure 4(b) that shear force does not normally have a major effect on the shape of the moment-rotation curve, although using results obtained with shear force absent will to some extent overestimate the moment and rotation capacities of the bare steel connection. This conclusion has yet to be confirmed for the composite state. The differences between the curves in Figure 4(b) at lower moments and rotations can be attributed to variations in friction and bearing due to the bolts.

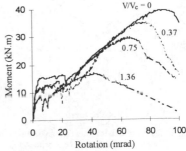

| (a) Test set-up | (b) Moment-rotation curves |

Figure 4 Effect of shear force on moment-rotation curve (3×2 bolt group, bare steel state)

4 PARAMETRIC INVESTIGATION

The variables studied in the tests completed to date are: (1) 3, 6, and 9 row single- and double-column bolt groups, (2) cleat thickness of 10 or 16 mm (Grade 250), (3) web plate thickness of 12 mm (Grade 250), 6.1, 8 or 10 mm (Grade 300), (4) bolt tightness, snug or friction with bare steel faying surfaces, (5) position of specimen along loading beam, and (6) interference.

Cleat thickness: Changing the cleat thickness from 10 mm to 16 mm on a 6×2 connection tested in the bare steel state did not affect moment capacity, while the rotation capacity reduced by about 20 per cent from 65 to 52 mrad. The web plate was 12 mm thick and no buckling was observed. Although the 16 mm cleat was thicker than allowed by the AISC design manual, the performance of the connection would be satisfactory in many situations.

The AISC design manual allows thin cleats and beams with thin webs to be used. However, the results in Figure 5 show the detrimental effect that web (or cleat) buckling can have on moment-rotation behaviour during the composite state and due consideration will need to be given to this matter.

Steel grade: Increasing the strength of the steel plates reduces rotation capacity by reducing the amount of deformation locally at the bolts. Steel ductility is critical when tearing occurs.

Bolt group geometry: Rotation capacity is reduced if the number of bolt rows increases or the steel connection is located further down the beam web, all other parameters remaining constant. However, the moment at first slip, moment capacity and secant stiffness all increase.

Bolt tensioning: Bolts have been either snug- or friction-tightened in accordance with standard recommendations (Firkins and Hogan 1984). The test results show that the moment at first slip increases significantly, whilst both the rotation and moment capacities are effectively unchanged. The increased restraint from friction tightening may be a useful effect to consider in design when beam deflection governs, provided it is economical.

Interference: In some tests, an additional load cell was connected to the reaction frame and allowed to come into contact with the loading beam after a predetermined amount of rotation, thus simulating interference between the beam bottom and column flanges. The results of two tests involving different rotational intervals before interference occurred are shown in Figure 6 with those for a test on an otherwise identical specimen but without interference. The bending moment acting on the connection increased rapidly when interference occurred, and similar peak values were reached for the two specimens. However, rotation capacity was much less affected by interference. The tests showed that the bolt group centroid was the centre of rotation. Therefore, it can be shown by simple calculation that the 20 mm gap required in the AISC design manual will normally allow the steel beam to rotate significantly more than 20 milli-radians in the bare steel state (as exhibited in Figure 6), and that the method of calculating the necessary gap to avoid interference should be amended to account for this fact.

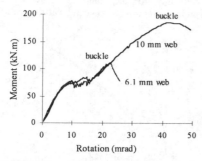

Figure 5 Effect of web thickness on moment-rotation curve (3×2 bolt group, composite state)

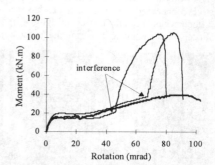

Figure 6 Effect of interference on moment-rotation curve (3×2 bolt group, bare steel state)

5 CONCLUSIONS

The development of a special rig for testing the web-side-plate connection in either the bare steel or composite states has been described. The capability of this unique piece of equipment has been demonstrated with test results obtained while investigating a large number of complex effects. It has been explained that very similar results can be obtained using small plate specimens rather than expensive, time-consuming full-scale testing. Therefore, it will be possible to obtain many more test results than has previously been possible. The information gained will be used to improve detailing rules and design methods applicable to the web-side-plate connection in both the bare steel or composite states.

6 REFERENCES

Firkins, A. & Hogan, T. J. 2nd Edition 1984. *Bolting of steel structures*. Sydney: Australian Institute of Steel Construction.
Hogan, T. J. & Thomas, I. R. 4th Edition 1994. *Design of structural connections*. Sydney: Australian Institute of Steel Construction.
Lipson, S. L. 1968. Single angle and single plate beam framing connections. *Canadian Structural Engineering Conference*. Toronto. 141-162.
Patrick, M., Thomas, I. R. & Bennetts I. D. 1986. Testing of the web side plate connection. *Australian Welding Research*. AWRA Report P6-1-87.
Pham, L. & Mansell D. S. 1982. Testing of standardised connections. *Australian Welding Research*. AWRA Report P6-22-81.
Ren, P. 1995. Numerical modelling and experimental analysis of steel beam-to-column connections allowing for the influence of reinforced-concrete slabs. *Swiss Federal Institute of Technology, Lausanne*. Thesis No 1369.
Xiao, Y., Choo, B. S. & Nethercot, D. A. 1994. Composite connections in steel and concrete. I. Experimental behaviour of composite beam-to-column connections. *J. Construct. Steel Research*. V31. 3-30.

The Mechanics of Structures and Materials, Grzebieta, Al-Mahaidi & Wilson (eds)
© *1997 Balkema, Rotterdam, ISBN 90 5410 900 9*

Deep collapse of square thin-walled cantilever-beams subjected to combined bending and torsion

G.J.White
A.C.S. Consulting Pty Ltd, Melbourne, Vic., Australia

R.H.Grzebieta
Monash University, Clayton, Vic., Australia

ABSTRACT: This paper describes a number of quasi-static combined bending and torsional load tests carried out on closed-hat section, thin-walled steel cantilever beams. How certain forms of localised collapse mechanisms develop in each of these beams for a particular load combination are described. These observations provide insight into how kinematically admissible collapse models may be constructed to assess the crashworthiness of a car roof during a roll-over crash.

1 INTRODUCTION

Engineers have recently fooussed their attention on the collapse mechanisms of thin-walled structures subjected to different loading configurations. Designers of bridges, aircraft and other thin-walled structures have analysed in detail how such mechanisms form in order to predict both the maximum strength of the structure [White et. al. (1993)] and its ductility during collapse [Murray (1984)]. The results of this research are particularly useful when analysing the collapse behaviour of energy absorbing structures such as passenger automobile bodies. The aim of such analyses is to increase survivability in crashes by improving the energy dissipation characteristics of structural components. The work carried out to date has concentrated on simple load cases, namely uniaxial bending [Kecman (1983) and Cimpoeru (1992)], biaxial bending [Brown et. al. (1983)], axial compression [Wierzbicki et. al. (1991)], and combined axial compression with torsion [Mahendran (1984)]. This paper deals with the special case of a cantilever beam loaded in torsion and bending and has a practical application in the design of automobile roof structures.

Figure 1. Automobile roof pillar under combined loading.

In the event of a rollover accident, the pillars of a car roof are subjected to a combination of loads including compression, bending and torsion. The pillars are usually heavily loaded because of the weight of the vehicle, and often exhibit gross deformation in a crash as indicated in Figure 1. To analyse the failure modes of such components, a number of experiments were carried out on thin-walled closed-hat section cantilever beams of different width to wall thickness ratios. The results of these tests are presented in this paper. Observations of the localised mechanism provide valuable information for constructing a kinematically admissible plastic collapse model for crashworthiness analyses.

2 COLLAPSE BEHAVIOUR

Consider the simple cantilever beam subjected to the combined bending and torsion load shown in Figure 2. The beam was manufactured as sketched in Figure 3. The typical load-deflection curve displays three distinct phases during loading and collapse of the section as indicated in Figure 4. The beam initially responds in a linear-elastic manner (region "A") both in bending and torsion. The material has neither yielded or buckled significantly. Both the bending and torsional angular deflections of the beam are related to the undeformed section properties, the applied load, the length of the beam C and the length of the torsion arm L. While the torsional moment T is constant along the length of the beam, the bending moment M increases linearly to a maximum at the built-in end. The largest stresses and curvature due to bending are therefore at the built-in end of the beam.

When the beam is deformed beyond the elastic limit of the material into region "B", local buckling is observed and eventually localised yielding occurs at or near the peak load. The point of first yield and the maximum load the section can withstand can be predicted using the failure criterion model proposed by White et. al. (1993). Once the peak load is reached, a plastic mechanism forms on the compression flange and the specimen enters the deep collapse region "C". This results in a rapid loss of residual moment capacity the beam can support.

In the case of pure bending, the collapse curve of the section can be predicted using a plastic hinge model and the work dissipation method proposed by Kecman (1983). This approach can also be used for beams subjected to combined bending and torsion. In this paper discussion is limited to a description of the experimental observations during collapse.

3 EXPERIMENT

Thin-walled cantilever beams were deformed into deep collapse to a bend angle of approximately 13 degrees. Several cantilever specimens with different ratios of bending cantilever to torsion

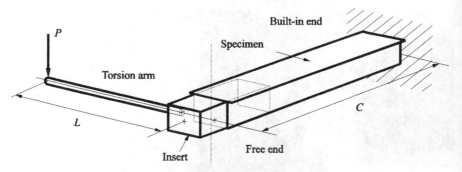

Figure 2: Cantilever beam subjected to combined loading.

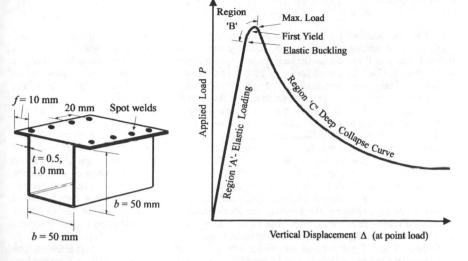

$f = 10$ mm

20 mm   Spot welds

$t = 0.5,$
1.0 mm

$b = 50$ mm

$b = 50$ mm

Region 'B'

Max. Load

First Yield

Elastic Buckling

Applied Load P

Region 'A'- Elastic Loading

Region 'C' Deep Collapse Curve

Vertical Displacement Δ (at point load)

Figure 3: Specimen manufacture

Figure 4: Typical load-deflection curve for the thin-walled beam loaded as shown in Figure 2.

arm length (C/L) were tested for two b/t ratios of 50 and 100. Details of the test equipment and specimens are explained in White et. al.(1993). A view of the loading rig and how the beam is subjected to a combined load is shown in Figure 5. Tests were performed under quasi-static conditions. The displacement rate of the Baldwin testing machine base plate was between 1.25 and 3.75 mm/min. Movement of the point of application of the load on the torsion arm was facilitated by lubricating the contacting surfaces. This meant that the failure mode of the section was unrestricted, allowing the beam to deflect from its original vertical plane of symmetry.

Figure 5. Specimen under combined load in test rig.

585

4   DISCUSSION

The following section is a qualitative description of the collapse process for the combined bending-torsion loading. As the applied load approaches the maximum strength of the beam, it first behaves in a linear-elastic manner (region "A" in Figure 4). Local buckling then becomes visible near the built-in end of the beam as a precursor to failure of the section (region "B" of Figure 4). Peak load corresponds to the visible formation of a plastic hinge line on the compression flange. This is to be expected as it is the most highly stressed part of the section.

Once the peak load is reached and the plastic mechanism forms, the mode of failure is predominantly in bending. All deformations of the section are subsequently confined to the mechanism formed near the built-in end. It is observed that the section rotates about a simple inclined hinge line that forms through the upper or tension flange, the angle of inclination of which appears to be determined by the C/L ratio. Thus the whole hinge mechanism distorts about this inclined hinge to accommodate twisting of the beam. Moreover, the angle of deformation associated with cantilever bending begins to increase rapidly while the observed deformation angle associated with twisting displays a reduced rate of increase. The beam also begins to unload elastically, reducing the bending curvature along the elastic portion of the cantilever outside the hinge mechanism region. Elastic unloading of torsional deflection is also observed, but to a much lesser degree than for the case of bending.

Figures 6 and 7 show failed specimens for b/t ratios of 50 and 100 respectively. The proportion of bending to torsion moment applied ranges from pure torsion ($C/L = 0$) on the left, to cantilever bending on the right ($C/L = \infty$). The distortion of the hinge mechanism with increasing torsional moment can be seen clearly in these views. Four distinct types of hinge mechanism form depending on the loading condition. The hinge mechanisms can be categorised as:

1: Bending only (Figures 6 and 7 right end specimens).
2: High bending-low torsion (Figure 6, specimens 3-6, and Figure 7, specimen 7).
3: Low bending-high torsion (Figure 6 specimen 2, and Figure 7 specimens 2-6).
4: Torsion only (Figures 6 and 7, left end specimen).

The transition from the mechanism type 1 to 2 occurs once a torsional moment is present. The transition from mechanism type 3 to 4 occurs once the bending moment is no longer present. The transition from mechanism type 2 to 3 is dependent upon the b/t ratio of the section and the ratio of bending moment to torsion moment (C/L). For high b/t (slender sections), the transition from 2 to 3 is at a higher C/L ratio. For example, specimens with $b/t=100$ exhibited a transition from mechanism type 2 to 3 at $C/L=4.40$, but the same transition occurred for $b/t=50$ specimens at approximately $C/L=1.98$. Further investigation is required to determine the reasons for triggering each particular type of hinge mechanism.

5   CONCLUSION

Observations were made of the localised plastic deep collapse failure mechanisms of more than thirty-five cantilever beams loaded in torsion and bending. The peak load was observed to correspond to the formation of a plastic hinge line in the compression flange of the beam. When the structure begins to collapse after reaching a peak load capacity, the predominant mode of failure is bending, characterised by the formation of a collapse mechanism at the built-in end of the beam.

Four distinct types of mechanism were observed, depending upon the section properties and the C/L ratio. The first was associated with pure bending ($C/L = \infty$), but with the reversed hinge form as first observed by Cimpoeru (1992). For small proportions of torsional load (high C/L), the second type of mechanism resembled a slightly distorted form of the mechanism observed in pure bending. As the proportion of bending moment to torsion moment decreases

1: 0   *2*: 0.72   *3*: 1.11   *4*: 1.49   *5*: 1.98   *6*: 4.32   *7*: ∞

Figure 6. Failed specimens with C/L from 0 (left) to ∞ (right) for b/t=50.

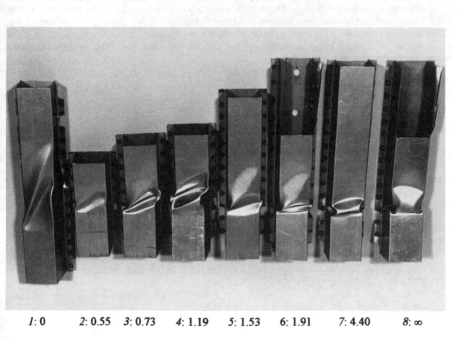

1: 0   *2*: 0.55   *3*: 0.73   *4*: 1.19   *5*: 1.53   *6*: 1.91   *7*: 4.40   *8*: ∞

Figure 7. Failed specimens with C/L from 0 (left) to ∞ (right) for b/t=100.

587

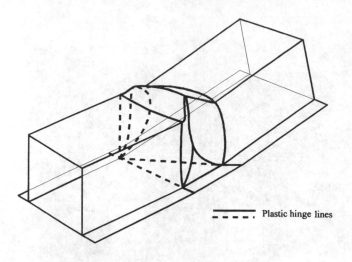

Plastic hinge lines

Figure 8. Proposed type 2 hinge mechanism under combined loading.

($C/L \to 0$) the degree of distortion progressively increases. Thus at low C/L values the mechanism begins to exhibit a third form resembling that observed in combined axial and torsion tests by Mahendran (1984). Finally, the predominance of shear stresses completely distorts the mechanism to the shape observed in pure torsion and shown in specimen 1 in Figures 6 and 7.

Based on the above observations two models were constructed to describe the formation of the plastic hinge mechanism for the combined bending-torsion load case. An example of one of the mechanisms is sketched in Figure 8. However, further investigation is required to determine the effect of the b/t ratio on hinge mechanism shape and consequently the effect of different mode shapes on energy absorption capacity.

6 REFERENCES

Brown, J.C. and G. H. Tidbury, G.H. 1983. An Investigation of the Collapse of Thin-Walled Rectangular Beams in Biaxial Bending, *Int. J. Mech. Sci.*, Vol. 25, No. 9-10: pp 733-746.

Cimpoeru, S., 1992. *The Modelling of the Collapse Behaviour During Rollover of Bus Frames Consisting of Square Thin-Walled Tubes.* Ph.D. Thesis, Monash University, Australia.

Kecman, D. 1983. Bending Collapse of Rectangular and Square Section Tubes. *Int. J. Mech. Sci.*, Vol. 25, No. 9-10: pp 623-636.

Mahendran, M. 1984. *Box Columns with Combined Axial Compressive and Torsional Loading.* Ph.D. Thesis, Monash University, Australia.

Murray, N.W. 1984. *Introduction to the Theory of Thin-Walled Structures*, Oxford Engineering Science Series. U.K: Clarendon Press.

White, G.J., Grzebieta, R.H. and Murray, N.W. 1993. Maximum Strength of Square Thin-Walled Sections Subjected to Combined Loading of Torsion and Bending. *Int. J. Impact Engng.*, Vol. 13, No. 2, pp 203-214.

Wierzbicki, T. and Huang, J. 1991. Initiation of Plastic Folding Mechanism in Crushed Box Columns. *Thin-Walled Structures* **13**: pp 115-143.

The Mechanics of Structures and Materials, Grzebieta, Al-Mahaidi & Wilson (eds)
© 1997 Balkema, Rotterdam, ISBN 90 5410 900 9

The effect of geometrical parameters, corrosion and residual stresses on the fatigue behaviour of welded structures

M.A.Wahab, N.T.Nguyen & M.Demir
Department of Mechanical Engineering, University of Adelaide, S.A., Australia

ABSTRACT: This paper addresses several practical problems concerning fatigue behaviour of welded structures in both air and corrosive environments. Initially, a numerical model has been developed using linear elastic fracture mechanics (LEFM), weight function technique and superposition principle to assess fatigue behaviour of butt-joints in air environment subjected to various weld geometries (including undercut) and residual stresses. This model has also been used for the assessment of corrosion fatigue of welded joints in the corrosive environment by modifying the Paris' crack growth law. It has been found that the fatigue life and fatigue strength of butt welded joints in air and corrosive environments can be improved by decreasing the tip radius of undercut at weld toes or eliminating the weld toe undercut completely.

NOMENCLATURE

r	- radius of weld toe of butt welded joint
r'	- tip radius of weld toe undercut
θ	- weld flank angle
ϕ	- plate edge preparation angle
t	- plate thickness
a	- crack length (edge crack); also half-length of minor elliptical surface crack
b	- half-width of cracked plate
c	- half-length of major elliptical surface crack
h	- half-length of cracked plate
ϕ_0	- parametric angle of the ellipse
a_i	- crack initiation length
a_f	- final crack length
N	- number of fatigue cycles
N_p	- fatigue crack propagation life
C, m	- material constants in Paris' equation
K_{IC}	- fracture toughness of the material
K_I	- stress intensity factor in mode-I loading
$K_{I.eff}$	- effective stress intensity factor in mode-I loading
ΔK	- range of stress intensity factor
ΔK_{eff}	- range of effective stress intensity factor
Y_a	- stress intensity geometry-configuration correction factor in axial loading mode for welds
$Y_{o,a}$	- stress intensity geometry-configuration correction factor in axial loading (flat plate)
$Y_{o,b}$	- stress intensity geometry-configuration correction factor in bending loading (flat plate)
$M_{k,r}$	- stress intensity magnification factor produced by weld profile geometry and residual stress
$M_{k,a}$	- stress intensity magnification factor produced by weld profile geometry in axial loading
$M_{k,b}$	- stress intensity magnification factor produced by weld profile geometry in bending loading

$M_{k,eff}$ - effective stress intensity magnification factor produced by weld profile geometry and residual stresses in combined loading mode (axial and bending)

S_A - axial nominal stress range

S_B - bending nominal stress range

S_y - yield strength of parent material

R_{ba} - ratio between bending and axial nominal stress range ($R_{ba} = S_B/S_A$)

S_r - maximum residual stress at weld toe surface

S   - fatigue strength of butt welded joint subjected to residual stresses and combined loading

1 INTRODUCTION

Most of the steel structures in engineering practice today are fabricated by welding. These welded structures are often subjected to dynamic service loads ranging from cyclic fluctuations to completely random loads. Welded structures such as pipelines, cranes, ships, ground vehicles, aircraft, bridges, offshore structures and pressure vessels are nearly always affected by fatigue loading. Fatigue behaviour of these welded structures is complicated by many factors intrinsic to the nature of welded joints. Normally crack-like defects such as slag inclusions, gas pores, lack of penetration at weld root or undercut at weld toes may be introduced in welded joints. Stress concentrations usually arise at locations of crack-like defects and at weld toes due to unpredictable variations in weld geometry profile. Locked-in residual stresses, which arise in the welded joints as a consequence of incompatible thermal strains during welding process also affect the fatigue behaviour of welds. In particular, tensile residual stress of yield magnitude may exist in as-welded structures and may cause detrimental effects to the fatigue behaviour of welded structures (Maddox 1991).

Furthermore, during the last two decades the number of the offshore welded structures have been increased significantly (Ouchi et al. 1990) and the safety assessment of such structures incorporating all the above mentioned aspects becomes more complicated. Particularly, the corrosion fatigue behaviour of these structures are subjected to some additional parameters such as cyclic frequency, shapes of cyclic wave forms and stress ratios which are reported to have little influence in air environment (Scott &Cottis 1990). Therefore, in this study, a simple but reliable model for the fatigue assessment of welded joints subjected to weld geometry (e.g. weld toe undercut) and residual stresses in both air and corrosive environments was developed and some interesting results have been found and are reported in this paper.

2 THEORETICAL ANALYSIS

2.1 *Concept of crack propagation using Linear Elastic Fracture Mechanics (LEFM)*

Using Fracture Mechanics, the rate of fatigue crack growth under cyclic loading can be expressed in terms of the range of stress intensity factor (ΔK) through Paris-Erdogan's power law (Paris & Erdogan 1963) as follows:

$$da \, / \, dN_p = C.(\Delta K)^m \tag{1}$$

Stress intensity factor in mode-I loading has the following form:

$$K_I = Y_a . S_A . \sqrt{\pi \, a} \tag{2}$$

Equation (2) can be rewritten in the following form to allow for the effect of weld geometry and residual stress in combined loading mode (effect of R_{ba}) as follows :

$$K_{I,eff} = Y_{o,a} . M_{k,eff} . S_A . \sqrt{\pi \, a} \tag{3}$$

590

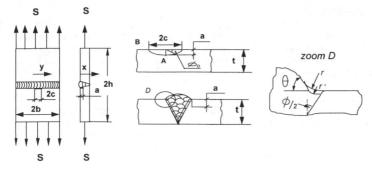

Figure 1. Surface crack model for transverse butt joint

where

$$M_{k,eff} = M_{k,a} + \frac{Y_{o,b}}{Y_{o,a}} \cdot M_{k,b} \cdot R_{ba} + M_{k,r} \cdot \frac{S_r}{S_A}$$

If the range of the stress intensity factor of a cracked body is known, the fatigue crack propagation life N_p can be calculated by integrating Eq. (1) between the initial crack length a_i and the final crack length at failure a_f. In this study, the range of stress intensity factor is replaced by the range of effective stress intensity factor (ΔK_{eff}) to allow for the effect of weld geometry and residual stresses in combined loading mode.

A semi-elliptical surface crack with initial aspect ratio $a/c = 0.2$ (Figure 1) is assumed to be located at a weld toe with a weld defect of 0.1 mm ($a_i = 0.1$ mm). Then the total fatigue life of the welded plate can be considered as number of cycles needed for this already initiated semi-elliptical surface crack to propagate through the thickness of the welded plate. The material constants of Paris' equation are m = 3, C= 3 x 10^{-13} mm/cycle in air and C = 2.3 x 10^{-12} mm/cycle in a marine environment at temperature up to 20°C as recommended by BS PD 6493 [3]. It would be more appropriate to experimentally determine the calue of C, for the marine environment condition, if possible. At this stage the present research is in progress to determine the value of C in corrosive environment and the best value that can be used in this simulation model is to take a value from BSI PD:6493. The The failure criterion is chosen for an instant when the effective range of stress intensity factor has exceeded the fracture toughness of the material ($\Delta K_{eff} > K_{IC}$) or the depth of the semi-elliptical surface crack in through thickness direction reaches a half of the plate thickness ($a_f = 0.5t$) (whichever occurs earlier). The value of ΔK_{eff} takes into consideration the combined effect of geometrical parameters, residual stresses and loading conditions. For a given stress ration R=0.0, $\Delta K_{eff, max} = K_{eff,max}$, and for that particular condition the failure criteria reduces to $K_{eff, max} > K_{IC}$. More details about numerical procedures used in this study can be found elsewhere (Nguyen & Wahab 1995).

2.2 *Undercut*

Undercut is considered as a defect and is described (Jubb 1981) as an irregular groove caused by the welding process, situated along the toe of a weld in the parent metal or in the weld metal already deposited during a previous run. Undercuts may be divided into three types such as: (1) wide and curve, (2) narrow and very narrow (crack-like) and (3) shallow and narrow (micro-flaw with depth up to about 0.25 mm) (Figure 2). Type 1 undercut is the most common form of undercut occurs during automatic welding of long single run fillet welds in the horizontal-vertical position with large heat input . This type of undercut can be found in manual metal arc welding due to the careless electrode control. As a result, a lack of sufficient weld metal deposit occurs at the weld toe. In this study, type 1 undercut is chosen for the modeling as it is the most common form of undercut in welded joints and therefore, types 2 and 3 will not be discussed further.

591

3 RESULTS AND DISCUSSION

Figure 3 shows the effect of tip radius of undercut at weld toe (*r'*) on the *S-N* curve of aligned joints in both air and sea water environments. It is obvious from this figure that the environmental effect play a significant role in the corrosion fatigue behaviour as shown below in S-N curves. The fatigue lives of the welded joints and flat plates have significantly been decreased in the sea water in comparison to that in air.

The fatigue limit of welded joint in air with an undercut is decreased by 58.3 % (from 120 to 50 MPa) compared to that of undercut-free joint and by up to 75 % (from 200 to 50 MPa) compared to that of flush-ground welded plate or parent material. In contrast, fatigue limit of undercut-free joint in air is decreased only by 40 % (from 200 to 120 MPa) compared with flush-ground welded plate or parent material. It means that the reduction of fatigue life and fatigue strength in comparison to flush-ground welded plate, caused by an introduction of weld toe undercut is as much as twice of that due to welded joint without undercut. This also means that the fatigue limit of undercut joint in air can be improved by up to 140 % (from 50 to 120 MPa) and 300 % (from 50 to 200 MPa) by eliminating the undercut at the weld toe and by flush-grinding the weld bead to the level of the parent plate respectively. Similar conclusions can be obtained for the corrosion fatigue curves.

Furthermore, it can be seen from Figure 3 that the fatigue curve of the joint with an undercut (r'=0.25 mm) in air falls in between the fatigue curves of the undercut-free and flat plate (or flash-ground joints) in sea water environment.

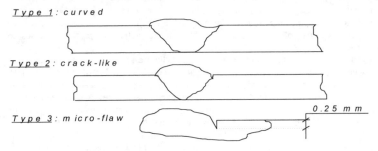

Fig. 2. A classification of undercut at the weld toe of butt welded joints

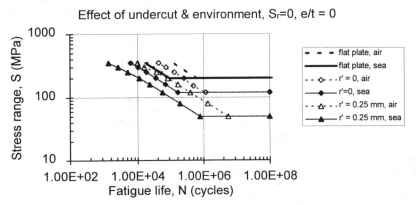

Figure 3. Effect of weld toe undercut (r') and the environment on S-N curve

In this study, various levels of compressive residual stress field induced by several post-weld surface treatments such as single or multiple hammer peening (S_r=-152 MPa and -172 MPa), glass or steel shot peened (S_r= -62 MPa and -110 MPa) and stress peened (loaded and peened) (S_r=-172 MPa) were considered from available experimental data (Bellow et al. 1986). Various levels of tensile residual stress in as-welded condition are assumed up to the level of yield strength of the parent material (S_r=300 MPa). In stress-relieved conditions residual stresses are assumed to vary from S_r=0 and S_r=14 MPa (Bellow et al.1986).

Figure 4. shows the effect of residual stresses in butt welded joints on the fatigue S-N curves in both air and the sea water environment. These plots were drawn using calculated data from theory. For the calculation of fatigue life, the levels of residual stresses are taken from the above mentioned experimental data. It is obvious from this figure that the fatigue lives of the welded joints in sea water environment are significantly reduced in comparison with those in the air under the same residual stress conditions.

Figure 4. also shows that regardless of the effect of environment, fatigue life was improved by compressive residual stresses while it was reduced by tensile residual stresses. This conclusion is consistent with the results reported by many other authors. Furthermore, under the influence of high tensile residual stresses (S_r = 300 MPa) the S-N curve plotted in log-log scale will no longer be linear as expected in the normal conditions.

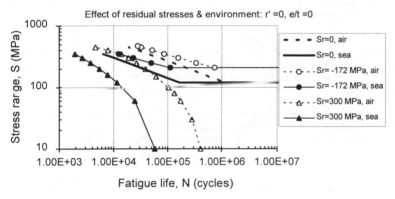

Figure 4. Effect of residual stresses and environment on fatigue of butt-joints

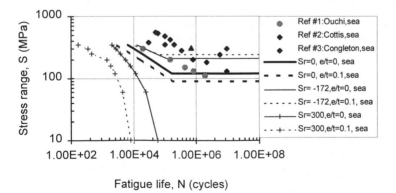

Fatigue life, N (cycles)

Figure 5. A comparison between calculated results and experimental data in sea water

Furthermore, Figure 4 also shows that the fatigue behaviour of the undercut-free as-welded joints with high tensile residual stresses conditions (S_r=300 MPa) tested in air is comparable with that of welded joints under surface treated or stress-relieved conditions tested in sea water. This means that there is a possibility that as-welded specimens with high tensile residual stresses of yield magnitude of the base metal may be used to simulate the behaviour of these joints under sea water environment. More experimental evidence is needed to prove this hypothesis.

Figure 5 shows the experimental results (in sea - water condition from literature : Reference: Ouchi, Cottis, and Congleton) and theoretical modelling developed in this study for various level of misalignment and surface treated residual stress conditions. The theory underpredicts fatigue life which may be due to uncertainties in the actual value of C, Paris's crack propagation constant and effect of other minor geometric and process parameters. Experiments are underway to confirm this findings.

4 CONCLUSIONS

This study gives the following important conclusions:

(1) Fatigue life and fatigue strength of butt welded joints in air & corrosive environment can be significantly improved by decreasing the tip radius of undercut at weld toes or eliminating the weld toe undercut completely.

(2) Fatigue life and fatigue strength of welded joints decreases significantly in corrosive environment in comparison with that in air. It is possible to use the welded specimens with an undercut, tested in air to simulate the corrosion fatigue behaviour of the undercut-free joints in the corrosive environments. More experimental evidence is needed to confirm this hypothesis.

(3) Regardless of the effect of environment, compressive residual stresses induced on the surface of welded joints by various surface treatment methods are beneficial; while post-weld tensile residual stresses are detrimental for the fatigue life and fatigue strength of butt welded joints.

(4) The fatigue behaviour of the undercut-free as-welded joints with high tensile residual stresses (S_r=300 MPa) tested in air is comparable with that of welded joints under surface treated or stress-relieved conditions tested in sea water.

(5)The fatigue model developed in this study can be used to predict the effect of weld geometry and residual stresses in both air and corrosive environment satisfactorily.

This approach is relatively simple and practical for the "Fitness-for-purpose" assessment of fatigue of welded joints and structures in both air and corrosive environment with satisfactory results.

REFERENCES

Jubb, J.E.M. 1981 Undercut or Toe Groove - the Cinderella Defect, *Metal Construction*, February, Vol. 13(2), 94-98.
Maddox, S. J. 1988 Revision of the Fatigue Clauses in BS PD 6493. *Proc. Inter. Conf. on Weld Failures* (ed. J. D. Harrison), 307-321.
Maddox, S. J. 1991 Fatigue Strength of Welded Structures, Cambridge University Press, Abington, Paris, P. C. and Erdogan, F., A Critical Analysis of Crack Propagation Laws, *Trans. ASME*, Vol. 85, Series D, 1963, pp. 528-534.
Nguyen, Ninh T. and Wahab, M. A. 1995. A theoretical study of the effect of weld geometry parameters on fatigue crack propagation life, *Engineering Fracture Mechanics*, Vol. 51, No.1, May 1995, pp. 1-18.

Paris, P. C. and Erdogan, F. 1963 A Critical Analysis of Crack Propagation Laws, *Trans. ASME*, Vol. 85, Series D, 1963, pp. 528-534.

Petershagen, H. 1990 The Influence of Undercut on The Fatigue Strength of Welds - A Literature Survey, *Welding in the World*, Vol. 28, No. 7/8, 1990.

Scott P. &   Cottis R.A. 1990   Environment Assisted Fatigue, *EGF Publication 7, Mechanical Engineering Publications Ltd.*, London, 1990.

Bellow, D.G., Wahab, M.A. and Faulkner, M.G. 1986 Residual Stresses and Fatigue of Surface Treated Welded Specimens, *Advance in Surface Treatments*, Pergamon Press, Volume 2, 1986, pp. 85-94.

Ouchi H., Soya I., Ebara and Yamada Y. 1990   Effects of Temperature and Dissolved Oxygen in Seawater on the Fatigue of Welded Steel Joints, Environment Assissted Fatigue, *EGF7 (edited by P.Scott), 1990, Mech. Eng. Publication,* London, pp. 17-30.

Cottis R.A, Markfield A and Haritopoulos, P. 1990 The Role of Corrosion in the Initiation and Growth Corrosion Fatigue Cracks, Environment Assisted Fatigue, EGF7, (Edited by P.Scott), Mechanical Engineering Publication.

Cogleton J and Wilks T.P, 1990, The corrosion Fatigue of a 13 percent Chromium Turbine Blade Steel in a condensing steam environment, Environment Assisted Fatigue, EGF7 (Edited by P.Scott), Mechanical Engineering Publication, London pp323 - 337.

The Mechanics of Structures and Materials, Grzebieta, Al-Mahaidi & Wilson (eds)
© *1997 Balkema, Rotterdam, ISBN 90 5410 900 9*

Finite element prediction of post-weld distortion using the linear elastic shrinkage volume method

A.W. Bachorski & M.A. Wahab
Department of Mechanical Engineering, The University of Adelaide, S.A., Australia
M.J. Painter
Commonwealth Scientific and Industrial Research Organisation (CSIRO), Division of Manufacturing Technology, Woodville North, S.A., Australia

ABSTRACT: Despite comprehensive research efforts in recent years, very little of the finite element modelling of welding processes has found an application in industry. In most cases the very long times required to obtain a solution continue to be the key hindrance. This fact has motivated the authors to investigate simplified approaches to the modelling of welding processes. Using the NISA™ finite element code, a simple method has been developed to predict the structural distortions which result from the transient non-linear localised heating phenomena inherent to arc welding processes. The model is based on the simplifying assumption that the volumetric changes occurring during fusion welding can be approximated by the linear thermal contraction of the designated shrinkage volume. This shrinkage volume then generates the internal forces which interact with the rigidity of a given structure to produce distortion. This approach enables the use of linear elastic steady-state finite element methods instead of non-linear transient formulations. The net results are significantly reduced solution times and more importantly the ability to predict distortions in large, highly complex welded structures. An experimental program, which looks at the distortion in a number of simple butt joint preparations was embarked upon. Initial results were in good qualitative agreement with the modelled predictions.

1 INTRODUCTION

Over the last decade Finite Element Analysis (FEA) has firmly established itself as an accepted design method across most engineering disciplines. The unique insights which numerical techniques such as FEA can afford make these methods very attractive to industry. Furthermore, modelling and simulation techniques are often a significantly cheaper and more flexible alternative to the rigorous experimental testing and prototype development inherent in product or process design.

Although computer modelling techniques, and Finite Element Analysis in particular, have become prevalent industrial tools, the use of some such modelling techniques in welding and welded fabrication has up to date been very limited. This means that welding and welding procedure qualification continue to be a largely trial and error process based on experience rather than scientific and mathematical principles. Not only is this expensive, it is unreliable because the engineer is rarely aware of the sensitivity to design parameters (Goldak 1992)

In contrast, FEA modelling of welding processes has almost a twenty year history in many research institutions (Smith 1991). The majority of the work in this area has set out to computationally simulate the complex non-linear physics which occur during arc welding and in so doing to gain an understanding of the influence of the many variables on weld quality and behaviour. Although this type of work has produced many valuable and unique insights into welding processes, the modelling techniques which have been developed to produce such results have often been too complex and labour intensive to be applied industrially. This fact has been recognised by numerous authors in the literature, and in one of the more recent, it is pointed out that although significant progress has been made in the finite element modelling of welding processes many of the techniques are still far short of being successfully used for the control of residual stress and distortion in actual structures (Masubuchi 1996).

Numerical models are generally used for the purposes of process simulation or process understanding. However these two requirements place differing demands on the modelling strategy. Whilst process simulation typically implies a quick, largely interactive and approximate approach, process understanding requires the model to closely duplicate the physics of the process (Painter et al 1994). Another author (Radaj 1992) makes a similar distinction but goes on to say that attempting to model the complex physics of arc welding to the maximum possible detail constitutes an 'unintelligent solution'. He contrasts this with the intelligent approach of only modelling the dominant parameters which influence weld behaviour in a given process.

The selection of a particular modelling approach is therefore critically dependant not only on the desired output accuracy but also on the required solution time and the anticipated model complexity.

One of the few groups in recent time (Tsai et al 1995) to have recognised the main obstacles facing the use of FEM for large scale welded fabrications has developed simplified approaches to tackling welding distortion problems. One such method has involved the use of linear spring elements (Figure 1) to model the thermal contraction of weld metal in a welded joint. Although some such models are crude in comparison to fully transient thermal elastic-plastic modelling regimes, they are capable of reasonably accurate predictions in a fraction of the time.

This paper presents one such simplified method for the prediction of welding distortion. Given the composition of the parent and weld metal, the parent plate thickness', the joint preparation and the parent plate constraints, the model is able to predict the post weld displacement of any point along the welded plates. The objective is to produce a simple and time efficient modelling strategy which is able to provide information for the welding engineer in the design stage. In this way designs can be optimised long before fabrication is due to commence.

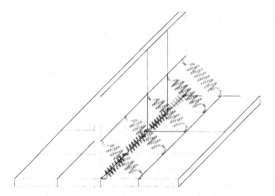

Figure 1. Spring element modelling approach (Tsai et al 1995)

598

2 METHOD

The linear-elastic shrinkage volume method is a steady-state finite element approach which assumes that the main driving force for distortion is linear thermal contraction of the weld metal as it cools from elevated to room temperature. This contraction is resisted by the surrounding parent metal resulting in the formation of internal forces. The parent metal distorts to accommodate these shrinkage forces until equilibrium is achieved.

Previous work in this area (Tsai *et al* 1995) had investigated the difference in the amount of distortion between a square-butt and a vee-butt joint preparation. The focus of this paper has been to verify the validity of the shrinkage volume approach by investigating the effect of different vee-angles on the magnitude of distortion in single-welded plain carbon steel butt welded joints.

The diagrams below illustrate the types of single vee joint preparations which were modelled (Figure 2a) together with the typical finite element mesh used (Figure 2b). Being a steady-state linear elastic method, the shrinkage volume approach is able to tolerate a reasonably coarse mesh without the incidence of significant error. However, it was found that the accuracy of the models was strongly dependant on the element type used in the analysis. Eight noded brick elements, which allow for a parabolic strain distribution along the element edge, were therefore used for all models.

Having created a model of the welded joint using the FEA software, elements representing the weld metal are assigned an elevated initial temperature value. For carbon steel weldments this temperature is assumed to be 800°C, representing the point at which the yield strength of steel drops-off to a very small value. Above this temperature the weld metal is perfectly plastic and so unable to carry any load. Linear thermal strains (-α.ΔT) are imposed as the weld metal elements cool from the assigned temperature and these elastically distort the weldment. In this way thermal strains produced in the temperature range of 800°-1500°C are not considered to have an effect on the overall distortion. It is significant to note that by assigning an initial nodal temperature to the weld metal elements, the heating-up thermal cycle of the welding process is ignored. This makes possible the adoption of a steady-state approach to the modelling of welding induced distortion. However, this assumption also causes the stress history during the heating-up thermal cycle to be ignored. This means that the magnitude of the predicted residual stress field in a given analysis will not be accurate.

An experimental program to verify the model results was embarked upon. Experimental repeatability was achieved by using a Motoman welding robot which was outfitted with a Gas Metal Arc Welding (GMAW) torch, in a configuration similar to that illustrated above (Figure 3).

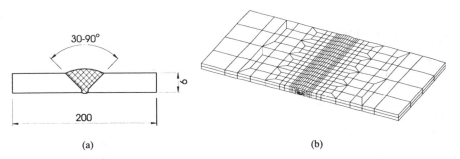

(a)	(b)

Figure 2. (a) Typical joint preparations modelled. (b) Finite element mesh for a 90° included angle single-vee butt joint

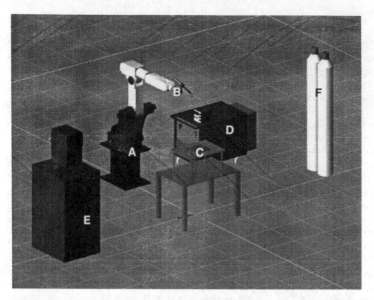

Figure 3. Experimental setup used to verify the welding distortion models. (a)Motoman welding robot (b) welding torch (c) work piece (d) power source (e) robot controller (f) shielding gas.

The welding was carried out using 1.2mm LW1(ER70S-4) welding wire under an argon rich gas shield. The typical current values were between 300 to 350 amps with a travel speeds ranging from 450 - 550 mm/min. It was decided that tacking the plates prior to welding would unduly influence the distortion data. Although tacks could be modelled, this was not seen to be the focus of the experiment and so tacking was ruled out. However the plates could not be welded if left unrestrained.

A procedure was therefore developed where by one of the plates was firmly clamped to the work bench while the other plate was pressed-up to small vertical pins. The function of the pins was to prevent transverse distortion of the plates during welding but to allow free vertical movement of the un-clamped plate.

3 RESULTS AND DISCUSSION

The graph below (Figure 4) presents a comparison of the modelled and experimentally measured angular distortions in 6 millimetre single-welded butt joints. Four different vee-butt preparations were investigated. The experimental and modelled results are in good agreement with one another for included angles of 90° (Figure 5), 75° and 60° but a significant discrepancy exists for the 30° vee preparation.   Both sets of data possess an upward trend in the magnitude of distortion for increasing included angles. The finite element models indicate an almost linear relationship between distortion and vee preparation angle, whilst the experimental results seem to indicate a limiting value of the included angle below which the magnitude of distortion remains constant.

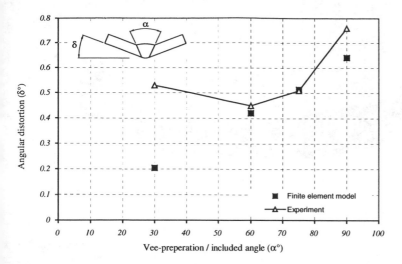

Figure 4. Graph comparing finite element predictions to experimental results, for 6mm thick plates.

For these preliminary studies the shrinkage or molten zone had been taken as the weld preparation geometry. However this assumption is less acceptable for small included angles where the welding process begins to significantly influence the shape of the molten zone. The underestimation of the deflection for the 30° weld preparation is therefore understandable.

Solution times for these models were generally about 150 CPU seconds which is relatively fast. Moreover, the model mesh size has not yet been optimised so further reductions may be possible.

These models are primarily aimed at predicting distortion. They rely on simulating the complex non-linear, elastic plastic deformation as a linear elastic one. Consequently the calculated stresses are expected to be above the yield strength of the material, and as such can not be considered as realistic residual stress levels. Nonetheless, the general pattern of stress, namely a narrow region of longitudinal tension along the weld metal surrounded by bands of compression, is predicted by the models. So whilst the magnitude of stress is not accurate, the pattern and sense of the residual stress field is correct and can be useful in identifying regions of stress concentration in more complex welded joints.

4 CONCLUSIONS

The shrinkage volume approach, which assumes that linear thermal contraction is the main driving force for distortion, is able to predict the magnitude of these distortions reasonably accurately and therefore has the potential to serve as a very useful tool in the prediction of distortions in large, complex welded structures

The magnitude of angular distortion in single-vee butt welded joints is influenced by the included angle of the vee preparation. As the included angle increases, so does the resultant angular distortion.

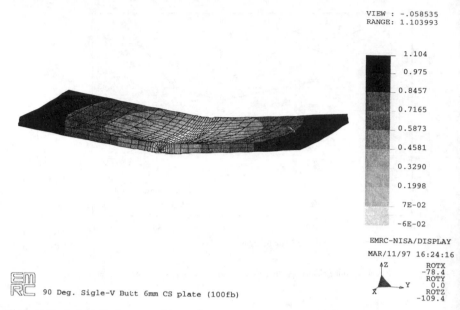

VIEW : -.058535
RANGE: 1.103993

1.104
0.975
0.8457
0.7165
0.5873
0.4581
0.3290
0.1998
7E-02
-6E-02

EMRC-NISA/DISPLAY
MAR/11/97 16:24:16

Z   ROTX
         -78.4
         ROTY
Y   0.0
X   ROTZ
        -109.4

90 Deg. Sigle-V Butt 6mm CS plate (100fb)

Figure 5. Vertical (z) displacement of butt welded plates with a vee preparation angle of 90°

ACKNOWLEDGMENTS

The financial support of the CRC for Materials Welding and Joining (95-25) is gratefully acknowledged. Our thanks also to Mr Ken Barton from the CSIRO workshop for his welding expertise and to Mr Rezak Kargas for his assistance with RobCad.

REFERENCES

Goldak,J., Gu,M., Paramjeet,K., & M.Bibby 1992. Computer simulation of welding processes. *International Trends in Welding Science and Technology. International Conference Proceedings*, ASM International.

Masubuchi,K. 1996. Prediction and Control of Residual Stress and Distortion in Welded Structures. *International Symposium on Theoretical Prediction in Joining and Welding, JWRI Osaka University:* 71-88.

Painter,M.J., Davies,M.H., Battersby,S., Jarvis,L., & M.A.Wahab 1994. *Numerical Modelling the Gas Metal Arc Welding Process: A Literature Review.* Report for CRC for Materials Welding & Joining.

Radaj,D. 1992. *The Heat Effects of Welding.* Springer-Verlag Berlin.

Smith,S.D. 1991. *A review of numerical modelling of fusion welding for the prediction of residual stress and distortion.* TWI Report 431/1991.

Tsai,C.L., Cheng,W.T., Lee,T. 1995. Modelling Strategy for Control of Welding-Induced Distortion. Modelling of Casting, Welding and Advanced Solidification Processes VII. The Minerals, Metals and Materials Society: 335-345.

The Mechanics of Structures and Materials, Grzebieta, Al-Mahaidi & Wilson (eds)
© *1997 Balkema, Rotterdam, ISBN 90 5410 900 9*

Computer simulation of the ultimate behaviour of purlin-roof sheeting systems

R.M.Lucas, F.G.A.Al-Bermani & S.Kitipornchai
The University of Queensland, Brisbane, Qld, Australia

ABSTRACT: Purlin-sheeting systems used for roofs and walls commonly take the form of cold-formed channel or zed section purlins, screw-connected to corrugated sheeting. This paper presents a nonlinear elasto-plastic finite element model, capable of simulating the ultimate behaviour of purlin-sheeting systems. The model incorporates both the sheeting and the purlin, and is able to account for cross-sectional distortion of the purlin, the flexural and membrane restraining effects of the sheeting, and failure of the purlin by local buckling or yielding. The validity of the model is shown by their good correlation with experimental results.

1. INTRODUCTION

Channel and zed cold-formed section members are widely used as purlins or girts. In Australia, purlins are connected to the corrugated roof sheeting by way of a screw through the crest of the corrugated sheeting and the purlin flange. The cross-sectional configurations of the zed and channel section purlins are such that they undergo both bending and twist from the beginning of loading. Due to the restraining action of the sheeting they tend to fail, not by overall flexural-torsional buckling, but by local plastic collapse, yielding or local buckling. Fig. 1 shows the general deflected shape of these purlins. The two purlins are shown under wind uplift loading which tends to be the dominant factor in Australian design.

The corrugated sheeting attached to the purlin provides two main restraining effects, shear stiffness k_{ry} and rotational stiffness k_{rs}.as shown in Fig. 2. The rotational stiffness comes from both the rotational stiffness of the sheeting itself and the rotational stiffness of the purlin-sheeting connection. The magnitude of the shear and rotational stiffness supplied by the sheeting governs the degree to which lateral displacement and rotation of the purlin about its longitudinal axis are restricted. Both shear and rotational stiffness cause a significant increase in the strength of the attached purlin, and neglect of their existence in design can result in highly over-conservative estimates of the purlin load-carrying capacity (Lucas, Al-Bermani and Kitipornchai (1997a,b)).

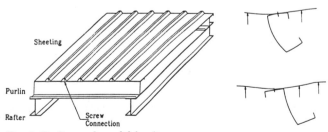

Fig. 1: Purlins under uplift loading

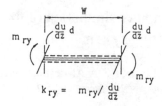

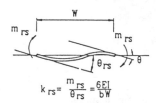

$$k_{ry} = m_{ry}/\frac{du}{dz}$$

(b) Sheeting Shear Stiffness (k_{ry})

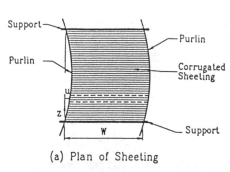

(a) Plan of Sheeting

Fig. 2: Sheeting restraint

$$k_{rs} = \frac{m_{rs}}{\theta_{rs}} = \frac{6EI}{bW}$$

(c) Sheeting Rotational Stiffness (k_{rs})

The finite element model simulates the physical interaction of both the purlin and the sheeting, and allows both the membrane and flexural restraining effects of the sheeting to be accounted for, without the need for either experimental input or overly simplifying assumptions. The model uses a nonlinear elasto-plastic finite element analysis, incorporating a rectangular thin-plate element previously developed by Chin, Al-Bermani and Kitipornchai (1994). The loading is applied across the sheeting, as would occur in the physical system, and is transferred to the purlin at the screw connections and at other points of contact between the purlin flange and the sheeting.

Results from the analysis are compared with experimental results from a test program carried out at the University of Sydney (Hancock et al. (1990,1992), Rousch and Hancock (1994,1995)), in order to show the validity of the model. Single, double and triple span purlins under both uplift and downwards loading are considered.

2.   PREVIOUS RESEARCH

Channel or zed section purlins undergo significant cross-sectional distortion from the onset of loading. The nature of the purlin-sheeting connection makes the shear and rotational stiffness provided by the sheeting to the purlin difficult to quantify. The rotational stiffness varies with sheeting type, purlin type and dimensions, screw spacing, and connection details. In the past, the approach in dealing with these complexities has often been to neglect the effects of purlin cross-sectional distortion and/or the effect of the rotational restraint provided by the sheeting. The assumption that the purlin fails in a mode of flexural-torsional buckling has also frequently been made.

The tested purlins under wind uplift were found to fail suddenly by localised failure at the free flange-web junction, the free flange lip-stiffener or across the full width of the free flange. The lapped continuous purlins under downwards load were all found to fail by a mode of localised failure at the end of the lap. In no cases was flexural-torsional buckling significantly visible.

3.   ANALYTICAL MODEL

3.1 Purlin-sheeting interaction

Fig. 1 shows the behaviour of channel and zed section purlins under uplift loading. The channel section tends to rotate around the middle of the flange where the screw connection is located, while the zed section tends to rotate around the upper flange-web junction. In both cases, the sheeting

604

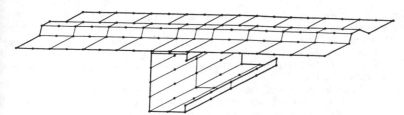

Fig. 3: Finite element modelling of purlin-sheeting system

exerts a retarding effect on the purlin rotation, both at the screw connection and at all other points of contact. These *'points of contact'* can be thought of as nodes with dependent degrees of freedom. The lateral w and vertical v deflections, and in some cases rotation θ_x, of the sheeting and purlin nodes at these contact points can be related, either by a direct equality expression or by some other linear function. An effective way of incorporating these relationships (or constraints) is by use of Lagrange Multipliers (Cook (1981)), a method in which the stiffness matrix of the structure is modified in order to enforce prescribed relationships that couple dependent degrees of freedom.

In a full scale roof system, a series of purlins would run parallel to each other with continuous lapped sheeting spanning the roof. In the analysis, a single purlin is modelled with sheeting the width of one span between the purlins, as shown in Fig. 3. The continuous nature of the sheeting is modelled by incorporating appropriate boundary conditions. Yield stresses for purlin and sheeting are taken as 450 and 550 MPa respectively.

3.2 Failure criterion

Due to the restraint provided by the sheeting to the attached flange of the purlin, purlins may fail by a combination of local plastic collapse, local buckling and yielding, rather than by overall buckling. For the purlins analysed, once plastification was reached (generally in the free flange region near the midspan of the purlin) very small load steps were used until the solution failed to converge in the set number of iterations. This divergence of the solution, occurring shortly after plastification commenced in the purlin, was taken as indicating purlin failure and the load at which this occurred was taken as the predicted failure load. A full description of the nonlinear analytical method may be found elsewhere (Lucas, Al-Bermani and Kitipornchai (1997a,b)).

3.3 Comparison with experimental results

In order to verify the accuracy of the nonlinear analysis, the response predicted by the model is compared with experimental results from the Vacuum Test Rig Program carried out at the University of Sydney. Information regarding the Test Program is given in the papers by Hancock et al. (1990,1992) and Rousch and Hancock (1994,1995). Single, double and triple span purlins were tested and compared with the analytical model in this paper.

3.3.1 Single span purlins

Tests S7T1, S7T2, S7T3 and S7T5 were carried out on simply supported 7m purlins attached to a 1400mm width of Spandek Hi-Ten sheeting (Lysaght Building Industries (1991)) under uplift loading. The details of these tests are briefly outlined below:

Test S7T1	Span:	7m	Section:	Z200-15	No. Rows Bridging:	0
Test S7T2	Span:	7m	Section:	C200-15	No. Rows Bridging:	0
Test S7T3	Span:	7m	Section:	C200-15	No. Rows Bridging:	1
Test S7T5	Span:	7m	Section:	C200-15	No. Rows Bridging:	2

Table 1 presents the comparison of the experimental results with the failure loads predicted by the analysis. The analysis shows very good agreement with the experimental results, the ratio of predicted to test failure loads ranging from 0.96 to 1.00. Tests S7T1 and S7T2, the unbridged purlins, when analysed using the model, commenced yielding in the lower flange elements nearest the web, 350mm from the centre of the purlin. The tested purlins failed by local plastic collapse of the free flange in a similar location.

The analysis of S7T3, the purlin with one row of bridging located at midspan, indicated that yielding started at midspan in both lower flange elements and in the lower lip element. The tested purlin failed by collapse across the whole bottom flange, also at midspan. Test S7T5, the purlin with two rows of bridging, failed by both lip stiffener buckles near the quarter points of the purlin as well as local plastic collapse of the free flange at midspan. The analysis of this purlin showed the onset of yield occurring in the lip element 1925mm from the end of the purlin, very near the location of the lip stiffener buckles.

The analytical model is also able to predict the load-deflection response of the purlins. The vertical and lateral deflections of the unbridged purlins of a typical test S7T2 are shown in Figs 4a and 4b, respectively. The deflections given in these figures were measured at the lower flange-web junction of the purlins at midspan. Gauges were placed on each of the purlins in the test set-up, 1 being the purlin at the bottom of the test frame (curve M1 in the figures), 2 being located in the middle (curve M2) and 3 being the purlin at the top of the frame (curve M3). The lateral deflection of the purlins is taken as positive when it takes the form shown in Fig. 1. Fig. 5 shows the calculated deflected shape of single span purlin cross-sections at midspan at various load levels. They indicate the large degree of cross-sectional distortion experienced by the purlin during uplift loading.

In general, the analysis shows good correlation with the measured deflections. Some discrepancy between the predicted and measured values is to be expected due to initial imperfections in the purlins, follower force due to the vacuum loading (ie., applied load in the analysis remains normal to the initial undeformed plane of the sheeting) and movement in the screw connection during loading. No information was available in order to assess the influence of any of these variables. The agreement between the measured and predicted response is therefore felt to be adequate.

3.3.2 Double span purlins

Continuous zed section purlins consisting of two 10.5m spans were tested under uplift loading in tests S2T1, S2T2 and S2T3. These purlins were arranged at 1200mm centres and attached to either Monoclad (Stramit Industries (1993)) or Trimdek (Lysaght Building Industries (1991)) sheeting. Details of these tests are briefly summarised here:

Test S2T1	Span:	10.5m	Section:	Z300-25	No. Rows Bridging:	0
Test S2T2	Span:	10.5m	Section:	Z300-25	No. Rows Bridging:	1
Test S2T3	Span:	10.5m	Section:	Z300-25	No. Rows Bridging:	2

The three tested purlins failed by local plastic collapse at the flange-web junction near the centre of the purlin, along with inner flange general failure at the end of the lap in the third test. In the analysis, the purlins all commenced yielding in the lower flange element closest to the web. This element was located at the division closest to the end of the lapped region of the purlin. The load at which the tested purlins failed is compared with the failure load predicted by the analysis in Table 1. All predicted loads are within 5% of the experimental result.

3.3.3 Triple span purlins

Four purlins are presented in this section for comparison with the analytical model, two under uplift loading (Test S1T4 and Test S1T5) and two under downwards loading (Test S4T5 and Test S4T6). The details of these tests are briefly given below (where U indicates uplift loading and D indicates downwards loading).

Test S1T4	Span: 7m	Section:	Z200-15	No. Rows Bridging:	0 (U)
Test S1T5	Span: 7m	Section:	Z200-15	No. Rows Bridging:	1 (U)
Test S4T5	Span: 7m	Section:	Z150-19	No. Rows Bridging:	0 (D)
Test S4T6	Span: 7m	Section:	Z150-19	No. Rows Bridging:	1 (D)

In each case the purlin is made up of three spans, lapped over the internal supports and attached to 1200mm spans of Monoclad or Trimdek sheeting.

Comparison of the failure loads of the tested purlins and those predicted by the analytical model are presented in Table 1. For both the uplift and downwards loading cases, the model predicts within 5% of the tested purlin failure load.

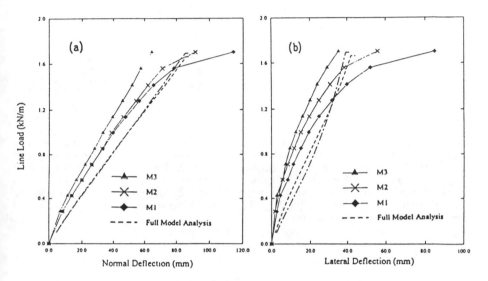

Single Span C200-15 Purlin, Uplift Loading   Single Span C200-15 Purlin, Uplift Loading

Fig. 4: Load-deflection of S7T2

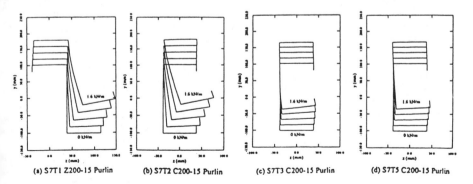

(a) S7T1 Z200-15 Purlin   (b) S7T2 C200-15 Purlin   (c) S7T3 C200-15 Purlin   (d) S7T5 C200-15 Purlin

Fig. 5: Deflected shape of single span purlins at midspan

Table 1: Comparison of Test and Predicted Failure Loads (kN/m)

Test	Test Failure Load	Predicted Load	Predicted/Test
Single Spans			
S7T1	1.85	1.87	1.01
S7T2	1.70	1.65	0.97
S7T3	1.70	1.64	0.96
S7T5	1.95	1.95	1.00
Double Spans			
S2T1	4.33	4.23	0.98
S2T2	4.93	4.87	0.99
S2T3	5.77	5.65	0.98
Triple Spans			
S1T4	2.58	2.60	1.01
S1T5	2.94	2.83	0.96
S4T5	2.92	2.90	0.99
S4T6	2.69	2.81	1.04

4   CONCLUSION

An analytical model for predicting the behaviour of purlin-sheeting systems has been presented in this paper. The model allows for the shear and rotational restraining effects of the sheeting to be incorporated into the analysis, without requiring over simplifying assumptions or experimental input. The use of the thin plate elements allows cross-sectional distortion of the purlin to be included in the analysis and the models are able to determine the ultimate load of a purlin, whether this occurs by yielding or by local buckling. The validity of the models was shown by their good correlation with experimental results.

5.   REFERENCES

Chin, C. K., Al-Bermani, F. G. A. & Kitipornchai, S. (1994), Nonlinear analysis of thin-walled structures using plate elements. *International Journal for Numerical Methods in Engineering*, 37, 1697-1711.

Cook, R. C. (1981), *Concepts and Applications of Finite Element Analysis*. 2nd edn, John Wiley and Sons, New York.

Hancock, G. J., Celeban, M., Healy, C., Georgiou, P. N. & Ings, N. L. (1990), Tests of purlins with screw fastened sheeting under wind uplift. *Proc., Tenth International Specialty Conference on Cold-Formed Steel Structures*, University of Missouri-Rolla, St. Louis, 393-419.

Hancock, G. J., Celeban, M. & Healy, C. (1992), Tests of continuous purlins under downwards loading. *Proc., Eleventh International Specialty Conference on Cold-Formed Steel Structures*, University of Missouri-Rolla, St. Louis, 157-179.

Lucas, R.M., Al-Bermani, F.G.A. and Kitipornchai, S. (1997a), Modelling of cold-formed purlin-sheeting systems, part 1: full model. *Thin-Walled Structures*, to appear.

Lucas, R.M., Al-Bermani, F.G.A. and Kitipornchai, S. (1997b), Modelling of cold-formed purlin-sheeting systems, part 2: simplified model. *Thin-Walled Structures*, to appear.

Lysaght Building Industries (1991), *Design Manual: Steel Roofing and Walling*.

Rousch, C. J. & Hancock, G. J. (1994), A non-linear analysis model for simply-supported and continuous purlins. *Research Report R688, School of Civil and Mining Engineering, The University of Sydney*.

Rousch, C. J. & Hancock, G. J. (1995), Tests of channel and Z-section purlins undergoing non-linear twisting. *Research Report R708, School of Civil and Mining Engineering, The University of Sydney*.

Stramit Industries (1993), *Monoclad: High Strength Roof and Wall Cladding*.

Pull-out capacity of screwed connections in steel cladding systems

M. Mahendran & R. B. Tang
Physical Infrastructure Centre, School of Civil Engineering, Queensland University of Technology, Brisbane, Qld, Australia

ABSTRACT: An investigation was carried out to study the local failure which occurs when a screw fastener pulls out of steel battens, purlins or girts in steel cladding systems. Both two-span cladding tests and small scale tests were conducted. In order to model the pull-out behaviour of steel cladding systems more accurately, an improved design formula was developed in terms of thickness, ultimate tensile strength of steel, thread diameter and pitch of screw. This paper presents the details of this experimental investigation and its results.

1. INTRODUCTION

Damage investigations following storms and cyclones have always shown that disengagement of steel roof and wall cladding systems has occurred due to local failures of their screwed connections under wind uplift or suction loading (see Figure 1). The thin intermittently crest-fixed profiled steel sheeting often pulls-through (pull-over) the screw heads (Figure 1a) due to the large stress concentration around the fastener holes under static and cyclic wind uplift loading. Either static or fatigue type pull-through failures lead to rapid disengagement of roof and wall claddings and the eventual collapse of the entire building. The local pull-through failure phenomenon has been thoroughly investigated (Mahendran, 1994). On the other hand, the steel cladding systems can also suffer from another type of local failure when the screw fasteners pull-out of the steel battens, purlins or girts (Figure 1b). In recent times, very thin high strength steel battens of various shapes have been used in buildings where the local pull-out failure has become critical. Such a pull-out failure also leads to the rapid disengagement of the roof and wall claddings. However, this failure mode has not been well researched for steel cladding systems. Therefore an experimental investigation using both two-span cladding tests

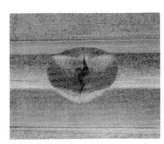
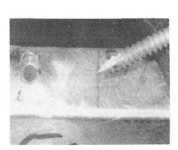
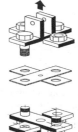

(a) Pull-through Failure   (b) Pull-out Failure
Figure 1. Local Failures   Figure 2. Cross Tension Test

and small scale tests was conducted under static wind uplift/suction load conditions. The applicability of the general design formula for pull-out strength of roof and wall cladding systems was investigated first. An improved formula was then developed in terms of the thickness, ultimate tensile strength of steel, thread diameter and pitch of screw fasteners. This paper presents the details of this investigation and its results.

2. CURRENT DESIGN AND TEST METHODS

The American provisions (AISI, 1989) include design formulae for screw connections in tension as shown by Equation 1. The pull-out capacity of screw fasteners of nominal diameter d, F_{ou}, is calculated as follows with a capacity reduction factor of 0.5:

$$F_{ou} = 0.85 \ t \ d \ f_u \hspace{3cm} \text{(Eq. 1)}$$

where   t = thickness of member and f_u = ultimate tensile strength of steel

In contrast to the American situation, the Australian design for the pull-out failure of screwed connections in tension is mainly based on experiments. However, Equation 1 has been included in the new limit states cold-formed steel structures code AS4600 (SAA, 1996). Since Equation1 was developed for conventional fasteners and thicker mild steel, there is a need to verify the applicability of the formula for thinner, high strength steel that is being commonly used in Australia. At present, the American and Australian codes recommend the use of 75% of the specified minimum strength for high strength steels, such as G550 steel with a yield stress greater than 550 MPa and thickness less than 0.9 mm, to allow for their reduced ductility. As an alternative to the design method, the new code AS4600 (SAA, 1996) also includes a new standard test method for single point fasteners (screws, blind rivets, self-piercing rivets and clinches) in tension (Figure 2). This method was used by Macindoe et al. (1995) to review the applicability of American design formula given by Equation 1 for thin high strength steels such as G550 steels. However, their work is not specific to roof and wall cladding systems.

3. EXPERIMENTAL INVESTIGATIONS

Since the main aim of this investigation was to develop specific design information for the pull-out strength of steel cladding systems, the more general standard cross tension test method was not used, but two-span cladding tests and small scale batten/purlin tests were conducted to better simulate the realistic behaviour of steel roof and wall cladding systems.

3-1. Two-span Cladding Tests

The conventional two-span cladding test method using air bags was conducted first. The test span was 900 mm and the width of trapezoidal sheeting was about 800 mm which involved 5 screws. The average pull-out failure load for each screw fastener in the middle steel batten was 2.0 kN for the No.14-10 HiTeks screw fasteners (IITW, 1995) on 0.8 mm BMT G550 steel batten (Note: BMT = Base Metal Thickness, G550 steel = minimum yield stress of 550 MPa). A different method of simulating the uniform wind uplift pressure by using bricks was also conducted and the average pull-out failure load was 1.9 kN/fastener. This is in reasonable agreement with the results from the air bag method.

3-2. Small Scale Tests

Since pull-out failures are localized around the screw holes on the batten/purlin (Figure 1b), a small scale test method was used to simulate this failure. The batten/purlin is a continuous span system with screw fasteners exerting a tension force when the cladding is subjected to a wind uplift pressure. Therefore attempts were made to model this using a single batten/purlin of 900mm span with four screw fasteners located at their nominal spacing of 190 mm (Figure 3). Equal tension force was applied to screw heads using a distributed loading method.

In order to simplify the multiple screw fastener test method further, a shorter span batten with only one screw fastener was used. Tension force was applied to the head of the fastener (see Figure 4). Since the pull-out failure essentially involves the local deformation around the fastener hole, the test method shown in Figure 4 was expected to produce the same results as other methods. However, in order to decide the best span, the test batten span was varied from 100 to 600 mm to study the effect of the span on pull-out failure load.

Table 1 presents the results of all the small scale tests including the multiple screw fastener and single screw fastener methods for the No.14-10 HiTeks screw fasteners and 0.8 mm BMT G550 steel batten. As seen from these test results, the difference in pull-out failure loads between the multiple screw fastener method (2.03 kN/fastener) and the single screw fastener method (2.14 kN/fastener - average value) is insignificant considering the possible variation in these experimental results. It was also found that changing the test span in the single screw fastener method did not cause any changes to the failure load. Small scale tests produced similar pull-out failure loads to those of two-span cladding tests. Hence it was considered that a single screw fastener method with a span of 300 mm was the most appropriate and adequate method for this investigation. This simple test method would simulate the local flexing of the steel batten around the fastener hole and the appropriate tension loading in the screw fastener to produce the pull-out failure load one would obtain by testing a two-span cladding system.

Table 1. Pull-out Failure Load Using Small Scale Test Methods

	Single Screw Fastener Method						Multiple Screw
Span (mm)	600	500	400	300	200	100	Fastener Method
Failure Load (kN/f)	2.22	2.00	2.27	2.12	2.18	2.05	2.03

Figure 3. Multiple Screw Fastener Tests

Figure 4. Single Screw Fastener Tests

Table 2. Details of Steel Battens and Purlins

Steel Grades	BMT (mm)		Yield Str. fy (MPa)		Ultimate Str. fu (MPa)	
	Nom.	Meas.	Nom.	Meas.	Nom.	Meas.
G250	0.40	0.38		358		415
Battens	0.60	0.54	250	359	320	399
	1.00	0.95		332		390
G550	0.42	0.43		717		721
Battens	0.60	0.61	550	696	550	703
	0.95	0.95		639		655
G500	1.20	1.20	500	635	520	647
Battens	1.60	1.58	450	584	480	604
G450	1.90	1.79		497		560
Purlins	2.40	2.30	450	465	480	587
	3.00	2.93		450		553

Table 3. Details of Screw Fasteners

Screw Type	Gauge	Diameter d (mm)		Thread Form (/inch)	Thread Pitch p (mm)
		Nom.	Meas.		
	10-16	4.87	4.67	16	1.59
	10-24	4.87	4.67	24	1.06
	12-11	5.43	5.52	11	2.31
HiTeks	12-14	5.43	5.47	14	1.81
	12-24	5.43	5.36	24	1.06
	14-10	6.41	6.39	10	2.54
	14-20	6.41	6.22	20	1.27
	10-12	4.87	4.81	12	2.12
Type 17	12-11	5.43	5.53	11	2.31
	14-10	6.41	6.34	10	2.54
Series 500	12-24	5.43	5.49	24	1.06

Note: BMT = Base Metal Thickness,   Nom. = Nominal,   Meas. = Measured,   Str. = Stress

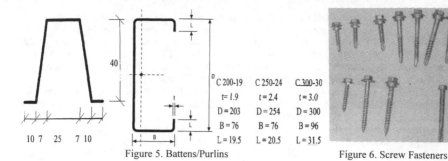

C 200-19	C 250-24	C 300-30
t= 1.9	t = 2.4	t = 3.0
D = 203	D = 254	D = 300
B = 76	B = 76	B = 96
L = 19.5	L = 20.5	L = 31.5

Figure 5. Battens/Purlins   Figure 6. Screw Fasteners

3-3. Experimental Program

A series of pull-out tests was conducted for a range of steel battens, purlins or girts and screw fasteners, which are commonly used in the building industry. The steel battens, purlins/girts covered a range of different thickness from 0.4 mm to 3.0 mm BMT, and steel grades from G250 to G550 (minimum yield stresses from 250 to 550 MPa). The screw fasteners covered a range of different screw gauges from 10 to 14 (nominal thread diameter d from 4.87 to 6.41mm), and thread form from 10 to 24 threads per inch (thread pitch p from 2.54 to 1.06mm). Tables 2 and 3 give the details of steel battens and purlins and screw fasteners used in this investigation. Five tests were conducted for a combination of each batten/purlin in Table 2 and each type of screw fastener in Table 3, resulting in a total of 592 tests. The tensile strength properties of steel used in battens/purlins were also measured (see Table 2).

A preliminary series of tests on battens with different geometry showed that the batten geometry has very little effect on pull-out strength. Hence a batten geometry that is commonly used in the building industry was chosen. For the tests on thicker purlins, three available purlins with different sizes were used. Figure 5 shows the geometry of battens and purlins used in this investigation. As seen in Table 3 and Figure 6, screw fasteners with three different drill points, namely, HiTeks, Type 17 and Series 500, were chosen (IITW, 1995). For each type of screw fastener, the thread diameter and thread form were varied as in Table 3. Using the single screw fastener test set-up shown in Figure 4, the test specimens were loaded until the screw fasteners pulled-out of the battens/purlins.

4. EXPERIMENTAL RESULTS AND DISCUSSIONS

4-1. Results

Table 4 presents typical experimental results for one HiTeks screw fastener and G550 steel batten. Other results are presented in Mahendran and Tang (1996). The results were grouped based on thickness and steel grade, analysed and comparisons made based on these groups.

Table 4. Experimental Results for 10-24×25 HiTeks Screw Fasteners

Thickness	Steel	Failure Load (N/fastener)		
t (mm)	Grade	Experimental Records (N/f)	Mean (N/f)	Std. Dev.
0.42		715, 758, 793, 648, 743, 755, 815	746	58
0.60	G550	930, 918, 1030, 990, 890	952	64
0.95		2100, 1890, 2100, 2100, 2120	2062	109

4-2. Comparison of Test to Predicted Values Based on Current Design Formula

The pull-out failure loads from tests (see Table 4) were compared with the predictions from the current design formula (Equation 1) using both the measured and specified (nominal) values for the properties of the steel and screw fasteners. Tables 5 and 6 present these comparisons in that order for each grade and thickness of steel and groups of screw fasteners (Case 1: All, Case 2: HiTeks + Type 17, Case 3: HiTeks, Case 4: Type 17). In Table 6, 75% of the specified minimum strength of G550 steel (412 MPa) was used for G550 steels of thickness less than 0.90 m according to the American and the new Australian codes (AISI,1989, SAA,1996).

Results in Table 5 show that the mean ratios of Test to Predicted values are less than 1.0 for all cases except for the thicker G450 steel, which reveals the inadequacy of the current design formula in predicting the pull-out failure loads. The current design formula is less conservative for the thinner G500 + G550 steels than for G250 steel for all types of screw fasteners (Cases 1 to 4). However, for the thicker G450 steel, the formula appears to be conservative. The mean ratio of Test to Predicted value is lower for all grades of thinner steel (not only for G550 steel). These observations imply that the current design formula is conservative only for thicker and softer grade steels, and agree well with Macindoe et al.'s (1995) observations. It may be unsafe to use the design formula for thinner steels less than 1.5 mm, in particular for G550 steel.

Table 5. Test to Predicted Values Based on Current Design Formula and Measured Properties

Steel		Case 1		Case 2		Case 3		Case 4	
Grade	Thickness	Mean	COV	Mean	COV	Mean	COV	Mean	COV
G250	t ≤ 1.50	0.88	0.20	0.90	0.18	0.83	0.16	1.07	0.06
G550 + G500	t ≤ 1.50	0.78	0.24	0.80	0.23	0.74	0.24	0.91	0.15
G450	1.5 ≤ t ≤ 3.0	1.21	0.14	1.22	0.14	1.20	0.14	1.28	0.12
G450 + G500 + G550		0.99	0.28	1.00	0.28	0.97	0.30	1.08	0.22
G250 to G550	t ≤ 1.50	0.82	0.23	0.83	0.23	0.78	0.22	0.98	0.14
G250 to G550	t ≤ 3.00	0.96	0.27	1.00	0.26	0.93	0.28	1.08	0.19

By comparing the results in Tables 5 and 6 with Macindoe et al.'s (1995) results obtained using the general test method of cross-tension specimens, it was found that Macindoe et al.'s results gave higher mean ratios of Test to Predicted values in all cases; for example, Macindoe et al.'s results gave a mean value of 1.27 for G250 steels and Case 1 screw fasteners compared with 0.88 in this investigation. This implies that the general test method of using cross-tension specimens could have produced unconservative results compared with the method used here to model the actual pull-out failure in battens and purlins/girts. However, it must be noted that Macindoe et al.'s test results included many thicker steels of 1.2, 1.6, 2.5 and 2.9mm. The use of many thicker steels would have also contributed to the difference between the Test to Predicted values from this investigation and Macindoe et al. (1995). When the results for individual thicknesses were compared, the same trend was observed. These observations indicate that the cross-tension test method may produce unconservative results.

Table 6. Test to Predicted Values Based on Current Design Formula and Specified Properties

Steel		Case 1		n	Cor.	Factor	Case 2		Case 3		Case 4	
Grade	Thickness	Mean	COV		Yield	Φ	Mean	COV	Mean	COV	Mean	COV
G250	t ≤ 1.50	1.02	0.20	165	0.80	0.43	1.04	0.19	0.95	0.17	1.25	0.06
G550 + G500	t ≤ 1.50	1.14	0.17	217	0.68	0.44	1.16	0.17	1.07	0.15	1.34	0.08
G450	1.5 ≤ t ≤ 3.0	1.39	0.14	210	0.83	0.69	1.40	0.14	1.37	0.13	1.48	0.14
G450 + G500 + G550		1.26	0.18	427	0.76	0.53	1.28	0.18	1.22	0.19	1.40	0.12
G250 to G550	t ≤ 1.50	1.09	0.19	382	0.73	0.43	1.11	0.18	1.02	0.17	1.31	0.08
G250 to G550	t ≤ 3.00	1.20	0.21	592	0.77	0.48	1.21	0.20	1.15	0.21	1.36	0.12

Notes: 1. n = Number of Tests   2. Cor. Yield = Correction for Yield

As seen in Table 5, the type of screw fastener has not caused any significant difference in results. The Type 17 screw (Case 4) is the only one which appears to provide slightly higher mean ratios of Test to Predicted values. Therefore in the discussion of results, only the case of all self-drilling screw fasteners (Case 1) was considered. Table 6 results using specified properties reveal that the mean values have increased to more than 1.0 for all cases. The use of

75% of specified tensile strength for G550 steel less than 0.90 mm has caused the mean ratios of Test to Predicted value for G550 + G500 steels (1.14) to be greater than that of G250 steel (1.02). Therefore the use of current design formula with specified properties is preferred, and appears to be capable of predicting the pull-out strengths. These observations are similar to those made by Macindoe et al. (1995).

4-3. Comparison of Test to Predicted Values Based on the Modified Design Formula of Macindoe et al. (1995)

Macindoe et al. (1995) modified the predictive equation for F_{ou} to better model the observed behaviour (Equation 2). By using the term $f_u^{0.5}$, this equation eliminates the need for the use of 75% of the specified minimum strength for G550 steels with a thickness less than 0.9 mm.

$$F_{ou} = 35 \sqrt{(t^{2.2} \ d \ f_u)} \tag{Eq. 2}$$

The pull-out failure load results from tests (see Table 4) were compared with the predictions from Equation 2 using both the measured and specified (nominal) values for the properties of steel and screw fastener in Table 7. For the latter case, 75% of the specified minimum strength of G550 steel (412 MPa) was not used for G550 steels of thickness less than 0.9 mm.

As seen in the results in Table 7, the mean ratios of Test to Predicted values are greater than 1.0 for all cases which reveal the adequacy of the modified design formula. Unlike the current design formula (Equation 1), the modified formula is more conservative for the thinner G500+G550 steels than for G250 steel for all types of screws. The results for the thicker G450 steel indicate similar conservativeness observed with Equation 1. The mean ratios of Test to Predicted values are fairly constant across all thicknesses which means that the modified formula has eliminated the lower Test to Predicted values for thinner steels observed using the current design formula.

Table 7. Test to Predicted Values Based on Macindoe et al.'s Modified Design Formula for Case 1 Screw Fasteners

Steel		Measured Properties		Specified Properties				
Grade	Thickness	Mean	COV	Mean	COV	n	Cor. Yield	Factor Φ
G250	t ≤ 1.50	1.05	0.20	1.09	0.21	165	0.80	0.45
G500 + G550	t ≤ 1.50	1.19	0.21	1.35	0.20	217	0.68	0.49
G450	1.5 ≤ t ≤ 3.0	1.53	0.14	1.61	0.15	210	0.83	0.80
G450 + G500 + G550		1.36	0.21	1.48	0.19	427	0.76	0.61
G250 + G550	t ≤ 1.50	1.13	0.22	1.24	0.23	382	0.73	0.46
G250 to G550	t ≤ 3.00	1.27	0.24	1.37	0.23	592	0.77	0.52

4-4. Comparison of Test to Predicted Values Based on a New Design Formula

Although Macindoe et al.'s (1995) modified formula appears to better model the pull-out strength than the current design formula (Equation 1), further attempts were made in this investigation to develop specific design formulae for the pull-out failure in the battens and purlins/girts commonly used in the building industry. In order to find the more accurate equation for the pull-out strength F_{ou} of steel roof and wall cladding systems, all the parameters on which the strength is dependent were included in the analysis. Therefore the thread diameter d and thread pitch p of the screw fastener and base metal thickness t and tensile strength f_u of the batten/purlin material were all included in the new design formula given by Equation 3. The use of ultimate tensile strength f_u gave a better correlation between the actual and predicted results than the yield strength f_y. Therefore, f_u was used in Equation 3. When compared with Equations 1 and 2, the new equation includes an additional parameter, the thread pitch, as it was often found to affect the pull-out capacity.

$$F_{ou} = k \ d^m \ p^n \ t^v \ f_u^w \tag{Eq. 3}$$

The unknown constants of k, m, n, v and w were determined using a solver in Microsoft Excel based on the method of least squares. Separate equations were derived for different groups as shown in Table 8. Test to Predicted values using Equations 1 and 2 are also included in Table 8 for comparisons with the corresponding values from the new formula. The new formulae with appropriate values for the parameters k, m, n, v and w in Equation 3 appear to provide improved mean (closer to 1.0) and coefficient of variation values (COV less than 0.2) in all cases. However, in order to reduce this to a single equation for all groups, the parameters m, n, v and w were forced to be 1.0, 0.2, 1.3 and 1.0, respectively. The values of k were changed to get the best agreement with test results. Initial k values gave a mean value very close to one, however, they were changed in order to recommend a capacity reduction factor of 0.5 used by the American and Australian codes (AISI, 1989, SAA, 1996). This is considered acceptable as the COV values are still within 0.18 (see Table 9). Table 9 presents the Test to Predicted values based on these changes to the above parameters for Case 1. These equations are much simpler and at the same time they are quite satisfactory as the mean and coefficient of variation values are similar to those in Table 8 and are acceptable. Therefore the formula is simplified with abovementioned parameters of m, n, v, w, and recommended k constant of 0.70 for thinner ($t \le 1.5$ mm) steel battens made of G250, G500 and G550 steel, 0.80 for thicker ($1.5 < t \le 3.0$mm) steel purlins and girts made of G450 steel, and 0.75 for all steel battens ($t \le 3.0$ mm) and purlins/girts made of G250, G450, G500 and G550 steel. It must be noted that in the above equation, d, p and t are in mm and f_u is in MPa.

Table 8. Test to Predicted Values Using the New Design Formula and Measured Properties for Case 1 Screw Fasteners

Steel		Coefficients					Simplified Formula		Current Formula		Modified Formula	
Grade	Thickness	k	m	n	v	w	Mean	COV	Mean	COV	Mean	COV
G250	$t \le 1.50$	1.40	0.6	0.3	1.2	1.0	0.99	0.13	0.88	0.20	1.05	0.20
G550 + G500	$t \le 1.50$	0.95	0.8	0.2	1.4	1.0	1.00	0.17	0.78	0.24	1.19	0.21
G450	$1.5 \le t \le 3.0$	0.90	0.9	0.2	1.3	1.0	0.98	0.10	1.21	0.14	1.53	0.14
G450 + G500 + G550		0.80	0.9	0.2	1.4	1.0	1.02	0.14	0.99	0.28	1.36	0.21
G250 to G550	$t \le 1.50$	1.30	1.0	0.2	1.3	0.9	1.02	0.18	0.82	0.23	1.13	0.21
G250 to G550	$t \le 3.00$	0.80	0.9	0.2	1.4	1.0	1.07	0.16	0.96	0.27	1.27	0.24

Table 9. Test to Predicted Values Using the New Simplified Design Formula

Steel		Coefficients					Measured Properties			Specified Properties		
Grade	Thickness	k	m	n	v	w	Mean	COV	Φ	Mean	COV	Φ
G250	$t \le 1.50$	0.75	1.0	0.2	1.3	1.0	1.04	0.15	0.61	1.19	0.17	0.54
G500 + G550	$t \le 1.50$	0.70	1.0	0.2	1.3	1.0	0.94	0.16	0.53	1.19	0.16	0.54
G450	$1.5 \le t \le 3.0$	0.80	1.0	0.2	1.3	1.0	0.93	0.10	0.59	1.06	0.11	0.55
G450 + G500 + G550		0.75	1.0	0.2	1.3	1.0	0.93	0.15	0.55	1.12	0.13	0.55
G250 to G550	$t \le 1.50$	0.70	1.0	0.2	1.3	1.0	1.02	0.18	0.56	1.22	0.17	0.55
G250 to G550	$t \le 3.00$	0.75	1.0	0.2	1.3	1.0	0.96	0.16	0.56	1.14	0.15	0.54

4-5. Capacity Factors for the Pull-out Failure of Screwed Connections

The proposed equations in this paper can predict average pull-out strengths based on the test data. The actual pull-out strength of a real connection can be considerably less than the value predicted by these equations because of the expected variations in material, fabrication and loading effects. Therefore a capacity reduction factor commonly used in design codes should be recommended. For this purpose the statistical model recommended by AISI (1992) was used. This model is also used in the new Australian Cold-formed Steel Structures code (Macindoe et al., 1995). Based on this model with appropriate material and fabrication factors, the capacity reduction factor ϕ is given by the following equation.

$$\phi = 1.65 \ P_m \ exp. \ (-\beta_0 \ \sqrt{0.0641 + C_p V_p^2}) \qquad \text{(Eq. 4)}$$

In the ϕ factor calculations for the three design formulae (Equations 1, 2 and 4, a value of 3.5 for the target reliability index β_0, Correction factor $C_p = (n-1)/(n-3)$ where n is number of the tests conducted, and the mean and COV values P_m and V_p from experiments, were used. Tables

6, 7 and 9 present the final ϕ factors Specified (Nominal) properties were used in the derivation of ϕ factor, and therefore included a correction factor for yield. However, measured properties were also used for the new design formula without a yield correction factor. As expected, both these approaches produced approximately the same ϕ factors (Table 9).

Comparison of the results in Tables 5 to 9 indicate that the new simplified design formula has less scatter. The mean ratios of Test to Predicted values are more uniform and closer to 1.0 than in other cases. The coefficient of variation is on average less than 0.18 and fairly uniform across different groups whereas the other formulae produced a bigger scatter. Comparison of average and maximum errors for the three formulae confirmed that the new formula produces less errors than other formulae. Based on these observations and previous results, Equation 4 is recommended with a reduction factor ϕ of 0.5. This was possible as the ϕ factors were greater than 0.5 and were of similar magnitude (approximately 0.55) for all groups. Although steel and screw fasteners used in this investigation were obtained from particular manufacturers, results should be equally applicable to other steels and screw fasteners provided they comply with the respective specifications for the grades of steels and fasteners used in this investigation.

5. CONCLUSIONS

An experimental investigation involving 592 small scale tests was conducted to determine the pull-out strength of commonly used steel cladding systems. Analysis of the experimental results showed that the current design formula for the pull-out strength may not be suitable for the screw fasteners and the thin high strength steels considered in this investigation. This design formula gave conservative results only for thicker ($1.5 \leq t \leq 3.0$ mm), softer grade steels. However, a smaller capacity reduction factor of 0.4 may allow the use of current design formula for pull-out strengths. A modified design formula recommended by Macindoe et al. (1995) appears to be more suitable than the current design formula.

A simple design formula that models the pull-out failure more accurately has been developed for the battens, purlins and girts used in the building industry. This formula has been developed in terms of not only the thickness and ultimate tensile strength of steel and the thread diameter of the screw fasteners, but also the pitch of screw fasteners. Although thicker G450 steels gave higher pull-out strengths than other steels, the same formula was used with different constant values. For this improved formula a capacity reduction factor of 0.5 was given in the American and the new Australian Cold-formed Steel Structures codes was found to be acceptable.

This experimental investigation has also shown that the cross-tension test method proposed in the new Australian cold-formed steel structures code may not produce conservative test results as expected. However, further testing is needed to confirm this.

6. ACKNOWLEDGEMENTS

The authors wish to thank BHP Sheet and Coil Products, Stramit Industries, and IITW Construction Products for the donations of experimental materials, and QUT for its financial support for this research project through the ARC Small Grants Scheme.

7. REFERENCES

American Iron and Steel Institute (AISI) 1989, Specification for the Design of Cold-formed Steel Structural Members, Washington, DC.

IITW Construction Products 1995, The Buildex Screw Book, Melbourne.

Macindoe, L. Adams, J. and Pham, L. 1995, Performance of Single Point Fasteners - CSIRO Division of Building, Construction and Engineering, Melbourne.

Mahendran, M. 1994, Behaviour and Design of Profiled Steel Roof Claddings under High Wind Forces, Engng. Struct., Vol.16, No.5.

Mahendran, M. and Tang, R.B. 1996, Pull-out Strength of Steel Roof and Wall Cladding Systems, Research Report 96-38, Queensland University of Technology, Brisbane, Australia

Standards Australia (SAA) 1996, AS/NZS4600 Cold-Formed Steel Structures Code.

The Mechanics of Structures and Materials, Grzebieta, Al-Mahaidi & Wilson (eds)
© *1997 Balkema, Rotterdam, ISBN 90 5410 900 9*

Buckling behaviour of stiffened hollow flange beams

M. Mahendran & P. Avery
Physical Infrastructure Centre, School of Civil Engineering, Queensland University of Technology,
Brisbane, Qld, Australia

ABSTRACT: The Hollow Flange Beam (HFB) is a new cold-formed and resistance welded section developed in Australia. Due to its unique geometry comprising two torsionally rigid triangular flanges and a slender web, the HFB is susceptible to a lateral distortional buckling mode of failure involving web distortion. An investigation using finite element analyses and experiments showed that the use of transverse web plate stiffeners effectively eliminated lateral distortional buckling of HFBs and thus any associated reduction in flexural capacity. This led to the development of a special stiffener that is screw-fastened to the flanges on alternate sides of the web. This paper presents the details of the investigation and the results.

1. INTRODUCTION

A new cold-formed Hollow Flange Beam (HFB) developed in Australia has a unique geometry comprising two torsionally rigid closed triangular flanges and a slender web. The HFB is manufactured from a single strip of high strength steel (G450) using electric resistance welding. Unlike the commonly observed lateral torsional buckling of steel beams, this HFB is susceptible to a lateral distortional buckling mode of failure, characterised by simultaneous lateral deflection, twist and cross-section change due to web distortion as seen in Figure 1. The investigation described in this paper was based on finite element analyses and experiments and concentrated on the use of transverse web plate stiffeners to eliminate lateral distortional buckling of HFBs and improve their performance. This paper presents the details of the finite element analyses and experiments, the results, and the final stiffener arrangement.

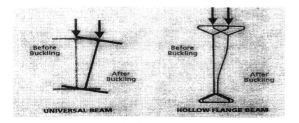

Figure 1. Lateral Torsional and Lateral Distortional Buckling Modes

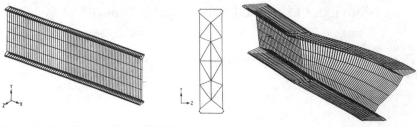

Figure 2. Finite Element Mesh of HFB Members

2. FINITE ELEMENT ANALYSIS

For this investigation, the finite element program NASTRAN (MSC, 1994) was used. Both elastic buckling and non-linear analyses were used to study the lateral distortional buckling behaviour of HFBs. In order to represent membrane and flexural behaviour including web distortion and large displacements, the quadrilateral shell elements were used. Triangular shell elements were used to model the stiffeners. The final mesh details can be seen in Figure 2.

This investigation was confined to constant bending moment and simple supports because this is generally the most conservative loading distribution, and the current design charts were developed using this assumption (Dempsey et al., 1993). This bending moment distribution was generated using two equal point loads located outside the span, as shown in Figure 3. The support constraints included vertical translation, lateral translation and twist. These are the conditions assumed for the derivation of the lateral torsional buckling formula used in both AS1538 and AS4100 (SA, 1990, 1988). The model was designed to transmit only major axis bending from the cantilever to the simple span, with the cantilever fully restrained against lateral deflection to prevent it from buckling. This was achieved by physically separating the cantilever and the simple span with a very small gap, and connecting them only with a linear constraint equation. In order to transmit only major axis bending, adjacent nodes on the cross-sections of the cantilever and the simple span were tied together with equal z-axis rotation (all nodes) and x-axis translation. This arrangement produced ideal boundary conditions at the support: vertical and lateral translational restraints, twist restraint, freedom to rotate about the major and minor axes, and no warping restraint. This model, referred to as the "ideal" model, was used for a large number of parametric studies using elastic buckling analysis.

The experimental set-up used to verify the FEA results used a continuous beam as shown in Figure 3. In order to model the actual experimental set-up, the cantilever was modelled as fully continuous with the main span, and a spring element was used to simulate the partial lateral restraint provided by the jack. The stiffness of this spring element was based on measured

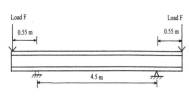

Figure 3. Experimental Set-up

values, and was found to be relatively small due to the arrangement of the hydraulic jacks with spherical seats. This model, referred to as the "experimental" model, was used for the comparison with experimental results. For the spans and sections tested, there was negligible difference between the "ideal" and "experimental" models. This indicates that the quasi warping restraint provided by the cantilever and the partial lateral restraint provided by the jack were insignificant. However, substantial differences between the two models were observed for short spans. Therefore the "ideal" model was used as the basis of all design recommendations.

In the nonlinear strength analysis of the "experimental" model, an initial imperfection of the beam was assumed as a linear variation in lateral displacement, varying from zero at the support to a maximum value of two wall thicknesses at midspan in order to include the effects of manufacturing tolerances and residual stresses (Salmi and Talja, 1992). The nonlinear model also included a stress-strain curve of G450 steel. The accuracy of the final model was validated by comparison of an unstiffened model with existing solutions from a finite strip elastic buckling program (Dempsey, 1993) and with experimental results. Preliminary analyses indicated that the HFB exhibited no post buckling strength for the lateral distortional buckling mode of failure. The elastic buckling moments were found to be only slightly less than the non-linear ultimate moments for both the unstiffened and stiffened cases, therefore elastic buckling was deemed to be the most appropriate method. They showed that significant distortion did occur in HFBs (see Figure 2), and that this distortion did cause a reduction in bending strength.

The following parameters that could influence the lateral buckling behaviour of HFBs stiffened with web stiffeners were investigated using the 'ideal' model: Stiffener types such as transverse web plate or RHS, longitudinal box or cross stiffener, located on one or both sides of the member, Stiffener size (5 to 25 mm plate), Stiffener welding - welded to flange only, welded to web only or welded to both flange and web, Stiffener spacing - a single stiffener location at midspan or two stiffener locations at the third points of the span, HFB section and Span.

2.1 Effect of Stiffener Type and Welding

Table 1 presents the FEA results into the stiffener types for a 4 m span 300 90HFB28 beam with 5 mm transverse web plate stiffeners located at third points of the span.

The rectangular hollow section (RHS) stiffener (Type B) does not provide a significantly higher strength increment than a simple transverse web plate stiffener. This is because the effect of a stiffener is mostly due to the constraints provided, which are independent of the stiffener size. For the same reason the single sided fully welded stiffener (Type C) provides a similar, but slightly less, strength increment compared to the double sided fully welded stiffener (Type A). A transverse web stiffener welded to the flanges only (Type D) is just as effective as a fully welded transverse web stiffener (Type A), while a transverse web stiffener welded to the web only (Type E) has virtually no effect. This is because the majority of the strength increment is provided by the tying together of the rotational degrees of freedom of the flanges, forcing the section to remain undistorted in the vicinity of the stiffener. These findings

Table 1. Effect of Stiffener Type on Elastic Buckling Moment

Stiffener Type	Elastic Buckling Moment (kNm)
No Stiffeners	45.8
Type A - 5 mm Plate on Both Sides, Welded to Web and Flanges	57.2
Type B - RHS on Both Sides, Welded to Web and Flanges	58.9
Type C - 5 mm Plate on One Side, Welded to Web and Flanges	55.6
Type D - 5 mm Plates on Both Sides, Welded to Flanges only	57.2
Type E - 5 mm Plates on Both Sides, Welded to Web only	46.8
Type F - 5 mm Plate on One Side, Welded to Flanges only	55.6

suggest that a single sided transverse web stiffener of nominal size, welded to the flanges only (Type F) may provide a strength increment only slightly less than a fully welded two sided web (Type A) or RHS stiffener. However, 5 mm transverse web plate stiffeners welded to the flanges on both sides of the web (Type D) are recommended subject to experimental verification. Subsequent analyses were restricted to this recommended configuration.

2.2 Effect of Stiffener Thickness

Since the section properties of the stiffener are less significant than the constraint they provide, there is little variation in the strength increment for the members stiffened with 5, 10, 15 and 20 mm plate stiffeners. The effect of stiffener thickness was found to be virtually independent of section but was significantly greater for short spans and almost negligible for long spans. The importance of these results is that stiffener plate thickness is not particularly significant. Fabricators can therefore use any available scrap plate greater than a nominal 5 mm thickness.

2.3 Effect of Location of Number of Stiffeners

Figure 4 illustrates the results obtained from the analysis of 300 90HFB28 with 8 different spans and with Type D stiffeners at midspan (n_s=1), third points (n_s=2), and quarter points (n_s=3) and without stiffeners (n_s=0). Note that n_s represents the number of cross-sections within the span at which stiffeners are located. The lateral torsional buckling (LTB) moment is also provided. Figure 5 illustrates the design bending moments calculated using the procedure outlined in AS1538 (SA, 1988) and the elastic buckling moments from Figure 4. They agree well with the design moments from the HFB Design Manual for unstiffened HFBs (Dempsey et al, 1993). The design moments based on the lateral torsional buckling moment (LTB) are also provided. The increase in strength due to stiffening can be seen in Figure 5.

Figure 4 indicates that as the number of stiffeners increases, the buckling moment tends towards the LTB moment. Increasing the number of stiffeners has a diminishing return, because the LTB moment effectively places a ceiling on the capacity. Hence any subsequent strength increase due to the inclusion of more than 3 stiffeners is due to local increments in the section properties only, and hence will be relatively minor, and not economical. It is clear that the greatest improvement is between the unstiffened and the 1 stiffener cases, with a smaller increment between 1 and 2 stiffeners, and a negligible increment between 2 and 3 stiffeners. In

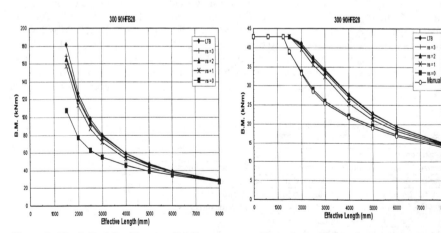

Figure 4. Elastic Buckling Moments (B.M.)   Figure 5. Design Bending Moments (B.M.)

summary, the FEA results indicated that 5 mm transverse web plate stiffeners at third points of the span were adequate to effectively eliminate lateral distortional buckling. They also showed that the stiffener welded to the flanges on one side alone is adequate in achieving this as the difference in buckling capacity was small.

3. EXPERIMENTAL INVESTIGATION

3.1 Experimental Program

The experimental program (Mahendran and Avery, 1996) was designed such that the results from the FEA could be verified adequately before using them in the design of HFBs. A total of ten 6 m long HFB specimens were loaded to failure under a constant bending moment within their span of 4.5 m (see Figure 3 and Table 2). In all the experiments 5 mm transverse web plate stiffeners fabricated from mild (G250) steel plate were used at third points of the span.

Experiments 1 and 2 were first conducted to verify the improvement to the lateral buckling capacity of HFBs with the use of these stiffeners. The FEA indicated that the stiffener welded to the flanges alone is effective in improving the buckling capacity of HFB and that it may be sufficient to weld stiffeners on one side only instead of welding on both sides. Therefore a number of experiments were conducted to verify these predictions (Experiments 2 to 5). Some experiments had stiffeners on both sides of the web at third points of the span (a total of four stiffeners) whereas others had stiffeners which were either on one side only or on alternate sides of the web (a total of two stiffeners). Since the FEA predicted no difference between

Table 2. Elastic Lateral Distortional Buckling and Ultimate Failure Moments

Expt.	HFB Section	Number of Stiffeners and Location	Elastic Buckling Moment – FEA (kNm)	Ult. Failure Moment-Experiment (kNm)
1	300 90HFB28	None	43.3	42.9
2	300 90HFB28	4 - Both sides – WFW	51.8	56.4
3	300 90HFB28	2 - One side only – WF	50.3	48.8
4	300 90HFB28	2 - One side only – WF	50.3	52.3
5	300 90HFB28	2 - Alternate sides – WF	50.3	63.8
6	300 90HFB28	2 - Alternate sides – SF	50.3	57.8
7	300 90HFB28	4 - Both sides – SF	51.8	55.0
8	450 90HFB38	2 - Alternate sides – SF	68.0	71.5
9	250 90HFB28	2 - Alternate sides - SF	51.0	49.5
10	300 90HFB28	None	43.3	42.9

Note: WFW – Welded to Flanges and Web, WF – Welded to Flanges only;
SF – Screw Fastened to Flanges Only; The FEA buckling moments for 450 90HFB38 and 250 90HFB28 with no stiffeners are 57.1 and 44.5 kNm, respectively.

(a) Stiffener Welded only to Flanges   (b) Special Stiffener Screw-fastened to Flanges
Figure 6. Transverse Web Plate Stiffeners

welding to both flanges and web, and welding to flanges only, most experiments (3 to 5) had flange welding only as shown in Figure 6 (a). Since the FEA predicted that welding to flanges alone would be sufficient, a special stiffener was developed that could be screw fastened to the flanges. This stiffener was fabricated by cold-bending a 5 mm plate to fit the inclined flanges of HFB and was easily fastened to the flanges using No.14 screw fasteners (see Figure 6 (b)). Experiments 6 to 9 were therefore conducted with these special stiffeners.

3.2 Experimental Set-up and Procedure

Two hydraulic jacks, located on the overhangs at a distance of 550 mm from each support, were used to produce a constant bending moment over a span of 4500 mm (see Figure 3). Two special loading devices were used to transmit the jack load into the webs of the HFB specimen. This eliminated the load height effects and flange crushing. The hydraulic jacks had spherical seats at its ends and thus were considered to provide little torsional and/or lateral restraint to the cantilever sections of the test beam. However, it is possible that the cantilever may have been subjected to a partial lateral restraint. Therefore the "experimental" model, was used in the comparison of results from experiments. To enable direct comparison with the FEA results and other theoretical solutions, the preferred restraint conditions at the supports were for the cross-section to be restrained from vertical and lateral translation, and prevented from twisting about the longitudinal axis of the member, while being free to rotate about the major and minor axes for both the "ideal" and "experimental" models in the FEA. These conditions were met by using a specially designed, but relatively simple, support configuration (Figure 3). Details of the experimental set-up including the support and loading conditions are given in Mahendran and Avery (1996). Zhao et al. (1995) presents a good discussion of the required support and loading conditions for lateral buckling tests. During testing the load was applied incrementally with reducing load steps as the expected failure load was approached. The magnitude of the applied constant bending moment within the span was obtained by multiplying the jack load by the distance of jack from the support of 550 mm.

3.3 Experimental Results and Discussion

All the experiments and FEA showed that HFBs have very little post-buckling strength beyond lateral buckling. The buckling moment of unstiffened 300 90HFB28 (Experiment 1) was estimated to be 42.6 kNm compared with the ultimate moment of 42.9 kNm. The corresponding results from the FEA were 43.3 and 44.0 kNm. Therefore, no attempt was made to differentiate between the elastic buckling and ultimate moments. Table 2 presents the ultimate failure moments from the experiments, and compares with the elastic lateral distortional buckling moments from the FEA. In general, the results agreed quite well.

Experiments with unstiffened HFBs (Experiments 1 and 10) verified the premature lateral distortional buckling failure of HFBs. Figure 7(a) presents the experimental and non-linear FEA results for the stiffened and unstiffened 300 90HFB28 beams. The non-linear analysis was able to predict the bending moment versus lateral deflection curve, and ultimate moment capacity quite well. The results in Figure 7(a) and Table 2 clearly show the improvement of approx. 20% in the lateral buckling capacity of HFB when stiffeners were used on both sides of the web. The capacity of 56.4 kNm in Experiment 2 is rather high and may have been due to experimental variation. The use of stiffeners appeared to have eliminated the distortion of the HFB section. Figure 3 shows the typical lateral buckling failures of HFBs.

Figure 7(b) presents the bending moment versus lateral deflection results from Experiments 3 to 5 for which only 2 stiffeners were used either on the same side or on alternate sides of the web. The FEA results for the HFB with stiffeners on the same side of the web are also presented in Figure 7(b) and compared with experimental results. It is assumed that the FEA results for stiffeners on alternate sides of the web will be nearly identical to those for stiffeners on the same side of the web. As seen from the ultimate moment results in Figure 7(b) and Table

622

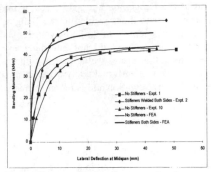
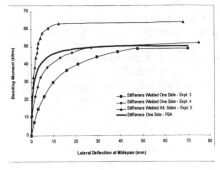

(a) 4 stiffeners welded to flanges and webs   (b) 2 stiffeners welded to flanges only

Figure 7. Bending Moment Vs Lateral Deflection Curves

2, welding stiffeners on the same side (Experiment 3 and 4 - 48.8 and 52.3 kNm) appeared to be detrimental compared to welding stiffeners on alternate sides (Experiment 5 - 63.8 kNm). When the stiffeners were welded to the same side of the beam it was found that the welding process introduced an initial bow in the form of a single half sine wave which was in phase with the expected lateral buckling mode within the span. This had the potential of reducing the ultimate moment capacity of the beams. On the other hand, when the stiffeners were welded to alternate sides of the beam, an initial bow in the form of two half sine waves was introduced in the beam. This initial bow was not in phase with the expected lateral buckling mode and this could have caused the higher buckling capacity of 63.8 kNm. Since the FEA did not include these imperfections due to welding, it could not predict this variation in capacity. For the same reason, the experimental moment versus lateral deflection curves did not agree well with the FEA curve. Despite these results, Experiments 3 to 5 gave confidence in the use of only two stiffeners as the experimental capacities (48.8, 52.3 and 63.8 kNm) were all considerably higher than that of unstiffened HFB (42.9 kNm) and thus confirmed the FEA predictions.

Since the previous experiments with stiffeners welded to the flanges only showed that they were equally effective as those welded to both flanges and webs, the special stiffener shown in Figure 6(b) was used in Experiments 6 to 9. Results were of the same order (57.8 and 55.0 kNm) for Experiments 6 and 7 with stiffeners on one side (2 stiffeners) and both sides of the web (4 stiffeners), respectively. This confirmed that stiffeners on one side were equally effective as stiffeners on both sides of the web for 300 90HFB28 sections. In fact, the experiment with 2 stiffeners produced a higher ultimate moment than that with 4 stiffeners. Both experimental results (57.8 and 55.0 kNm) appeared to be of the same order as the corresponding results (52.3 and 56.4 kNm from experiments 4 and 2) when stiffeners were welded. This implies that screw-fastening the special stiffener to the flanges will be adequate and can eliminate the need for welding the stiffeners to the 300 90HFB28 beams. The lateral deflections prior to buckling were quite small for HFBs with screw-fastened stiffeners compared to the HFBs with welded stiffeners since the former does not introduce any geometrical imperfections or residual stresses. This is a significant advantage.

Experiments 8 and 9 involving other HFB sections, the largest section 450 90HFB38 and one of the smaller sections, 250 90HFB28, also confirmed the analytical predictions and other observations, in particular the adequacy of the new screw-fastened stiffeners. The use of screw-fastened stiffeners improved the buckling moment from 57.1 kNm to 68.0 kNm for the larger HFB section (19% increase) and 44.5 to 51.0 kNm for the smaller section (15% increase). It is to be noted that based on the FEA, up to about 50% increase can be expected in the buckling moment of stiffened HFBs for medium spans in the range of 2 to 4 m.

Since the new stiffeners are simply screw-fastened to the flanges on alternate sides of the beams and improve the buckling capacity in a similar manner to those welded to the flanges, they are recommended rather than welded stiffeners. In Experiments 6 to 9, the new stiffeners were screw fastened to alternate sides of the HFB web. Since they did not introduce any residual stresses or geometric imperfections in the beam as in the case of welding, it is unlikely that screw-fastening to the same side or alternate sides of the web will make any difference to the results. However, the latter method was preferred in this investigation and is recommended.

4. CONCLUSIONS AND RECOMMENDATIONS

This investigation using finite element analyses and experiments showed that the effects of lateral distortional buckling can be effectively and economically eliminated in a Hollow Flange Beam by the use of transverse web plate stiffeners. The minimum stiffener configuration should consist of 5 mm thick mild (G250) steel plate, welded to the flanges, and situated on alternate sides of the section at third points of the span. An 'easy-to-install' special stiffener that is screw-fastened to the flanges on alternate sides of the HFB web was then developed for this purpose. The use of stiffeners provides an increase in the design moment capacity of more than 35%, and cost saving of more than $60 per beam for certain conditions. It is recommended that the design moments for stiffened Hollow Flange Beams shown in Figure 5 be adopted. Similar curves for all other sections are available in Avery and Mahendran (1996). For HFBs with the minimum stiffener configuration, a design flexural buckling strength equal to the design strength based on 95% of the elastic lateral torsional buckling moment can conservatively adopted for members spanning more than 3 metres and subject to constant bending moment.

5. ACKNOWLEDGEMENTS

The authors would like to thank the following people: Mr. Justin Riley and Daren King (QUT), Mr. Ross Dempsey (Palmer Tube Mills Pty. Ltd.), and QUT laboratory staff.

6. REFERENCES

Avery, P. and Mahendran, M.,1996, Finite Element Analysis of Hollow Flange Beams with Web Stiffeners, Research Report 96-16, QUT, Brisbane, Australia.
Dempsey, R.I., 1993, Hollow Flange Beam Member Design Manual, Palmer Tube Mills..
Mahendran, M., and Avery, P., 1996, Buckling Experiments on Hollow Flange Beams with Web Stiffeners, Research Report 96-15, QUT, Brisbane, Australia.
Macneal Schwendler Corporation (MSC) (1994) MSC/NASTRAN Users' Manual, USA
Salmi, P. and Talja, A.,1992, Bending strength of Beams with Non-linear Analysis, Proc. of the 12th Int. Specialty Conf. on Cold-Formed Steel Structures, St Louis, USA, pp.45-63.
Standards Australia (SA), 1988, Cold Formed Steel Structures, AS1538, Sydney
Standards Australia (SA), 1990, Steel Structures, AS4100, Sydney
Zhao, X.L., Hancock, G.J. and Trahair, N.S., 1995, Lateral Buckling Tests of Cold-formed RHS Beams, J. of Structural Engineering, ASCE, Vol.121, No.11, pp.1565-1573.

Timber and masonry (masonry)

The Mechanics of Structures and Materials, Grzebieta, Al-Mahaidi & Wilson (eds)
© 1997 Balkema, Rotterdam, ISBN 90 5410 900 9

Parametric studies on the dynamic characteristics of unreinforced masonry

Y. Zhuge
School of Engineering, University of South Australia, Adelaide, S.A., Australia

D. Thambiratnam & J. Corderoy
Queensland University of Technology, Brisbane, Qld, Australia

ABSTRACT: In order to investigate the seismic behaviour of unreinforced masonry, an analytical model has been developed and is implemented in a nonlinear finite element program. The model is capable of performing either the quasi static or the real time history analysis of masonry structures. In this paper, the influence of some important parameters is studied by using this iterative finite element computer program.

1 INTRODUCTION

Unreinforced masonry (URM) structures are widely used throughout Australia and other regions around the world, even though they seem to have experienced the worst damage during earthquakes. However, it has been found that even when unreinforced, masonry has a substantial deformation capacity after cracking, if it is designed with suitable compressive loads and material properties. Therefore, it is necessary to study the dynamic behaviour of masonry to provide a better understanding of its earthquake response.

The behaviour of URM is much more complex than that of concrete, masonry is a two-phase material and its properties are therefore dependent upon the properties of its constituents, the brick and the mortar. The influence of mortar joint as a plane of weakness is a significant feature which is not present in concrete and this makes the numerical modelling of URM very difficult.

In order to predict the complex behaviour of URM walls under in-plane dynamic loads, including earthquake excitation, a comprehensive analytical model has been developed by the authors and is implemented in a nonlinear finite element program (Zhuge et al. 1995,1996). The model is capable of performing both static and time history analyses of masonry structures and has been calibrated by using results from experimental testing of several masonry panels.

When the structure is subjected to seismic loading, the seismic loads depend mostly upon the mass and fundamental period of the structure. In this paper, a parametric study is carried out to obtain a better understanding of the effects of these factors to URM and provide an optimum design method, so that the seismic strength of the structure against cracking and damage will be maximised.

By varying the frequency and the amplitudes, the component of earthquake excitations has also been investigated as well as the effect of vertical compressive stress to check existing masonry structures to withstand dynamic loads.

It should be noticed that the parameters are selected to investigate the full range of analytical model characteristics and may not necessarily represent realistic characteristics of particular URM walls.

2 ANALYTICAL MODEL FOR MASONRY UNDER DYNAMIC LOADS

An analytical model, based on the finite element method, has been developed to predict the complex behaviour of URM walls under in-plane earthquake excitation, a detailed discussion of the model can be found in (Zhuge 1995). An anisotropic material model which was originally derived for concrete by Darwin and Pecknold (1977), is developed to analyse URM under various states of stresses before and post-failure. The principal of the model is "equivalent uniaxial strain" concept which means the elastic moduli for the constitutive law are the elastic moduli in the two principal stresses directions which can be determined from uniaxial stress-strain curve. A failure criterion has also been developed, which combines both biaxial and Coulomb shear failure models. The resultant failure model is capable of predicting joint failure and this is achieved through the use of the ubiquitous joint model.

The nonlinear finite element program has also been developed based on the analytical model. The nonlinear dynamic analysis is carried out with the modified Newton-Raphson iteration scheme in conjunction with the Newmark time integration algorithm. The model has been validated by comparison with various experimental results, which include URM walls under in-plane monotonic and cyclic loads and under ground excitation. Good agreement is observed in the correlation (Zhuge, et al. 1995, 1996)

3 PARAMETRIC STUDIES

It was found from various experimental results that the vertical compressive stress plays an important role in determining the failure pattern of a wall and that a wall has substantial deformation capacity after cracking if a suitable vertical load is applied. In this section the effect of vertical compressive load is investigated first for walls under dynamic loads.

When a structure is subjected to seismic loading, the response depends primarily upon the fundamental period of the structure. By varying the frequency and the amplitude of the dynamic loading, the components of dynamic excitation are changed and these may influence the response of the structures as well. A parametric study is carried out to obtain a better understanding of the effects of these factors to URM and to optimise the performance of the structures.

For all the dynamic analyses conducted in this section, the ground acceleration is input in the form (sinusoidal base motion):

$$\ddot{u}_g = A \sin \frac{2\pi t}{T} \tag{1}$$

where A is its amplitude, taken as a function of the acceleration due to gravity g and T is its period.

3.1 *Effect of vertical load on the wall behaviour*

The response of a wall with a vertical compressive stress σ_m=0.25MPa is compared with the result of wall 4 (σ_m=0.01MPa) tested by Klopp and Griffith (1993) under the same sinusoidal base motion. The material properties of the wall are summarised in Table 1. The wall has a width/height ratio of 1.58 (1200 x 760 mm) and the period T = 1 sec.

The horizontal displacements at the top of the wall are plotted as a function of time in Figure 1. The effect of vertical compressive stress is evident. With a lower vertical compressive stress of 0.01MPa, the crack initiated at 2.15sec with $\ddot{u}_{g\,max} = 0.783g$ and the wall failed at

3.23sec with $\ddot{u}_{g\,max} = 1.002g$. As σ_m increased to 0.25MPa, both the cracking and the ultimate strength are increased. The crack initiated at 4.15sec with $\ddot{u}_{g\,max} = 1.20g$ and wall did not fail with this maximum ground acceleration.

However, it should be noticed that when the value of the vertical compressive stress is very high, the ultimate strength decreased and the wall fails in a brittle manner. This has been proved by the results of static analysis (Zhuge, 1995).

Table 1. Material properties of the wall.

f_m (MPa)		f_t (MPa)		E_0 (MPa)		ρ (kg/m³)
Brickwork	Mortar	Masonry		Masonry	Mortar	Masonry
8.8	2.9	0.4		1404*	1000	1937

* A dynamic Young's modulus was determined based on the dynamic shear stiffness measured in the laboratory with a lower value equal to 1065MPa.

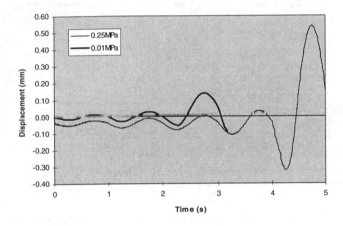

Figure 1. Effect of vertical compressive stress on the time history analysis.

The failure pattern of the wall with σ_m=0.25MPa is shown in Figure 2(b) and compared with the original analysis where σ_m=0.01MPa (Figure 2(a)). When the vertical compressive stress was low, the tensile cracks initiated and propagated along the interface between the concrete base and mortar joint (Figure 2(a)). When σ_m increased, the crack initiated in the middle of the wall and the final failure pattern has a diagonal shape (Figure 2(b)). Therefore, the vertical compressive stress has significant effect on the dynamic analysis for both the ultimate strength and the failure pattern of the wall.

3.2 *Effect of the fundamental period of structure*

The fundamental period of rigid low-rise brittle structures like masonry is normally in the range of 0.1 to 0.5 seconds. In this section, four URM wall structures with typical periods corresponding to those of one to three storey buildings are selected for the parametric study. The material properties used are those determined from the laboratory wall specimens tested by Klopp and Griffith (1993). The geometry and fundamental period of these four walls can be found in Table 2. The vertical compressive stress at the top of the wall is assumed to be equal to 0.01MPa. The walls are subjected to two cycles of increasing ground acceleration and the amplitudes and period of the ground acceleration (Equation 1) are taken as A_1=0.343g, A_2=0.616g and T=0.5sec.

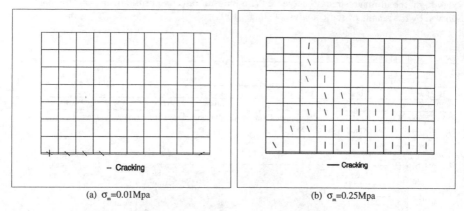

(a) σ_m=0.01Mpa　　　　　　　　　　(b) σ_m=0.25Mpa

Figure 2. Effect of vertical compressive stress on the failure pattern for dynamic analysis.

Table 2. Geometry and fundamental period of the walls.

Wall No.	Height H (m)	Width L (m)	Thickness t (m)	Fundamental Period T_n (sec)
1	3.6	10.0	0.23	0.045
2	7.8	10.0	0.23	0.118
3	7.8	3.9	0.23	0.200
4	12.0	5.0	0.23	0.354

The results of horizontal displacements at the top of the structure vs. the time are plotted in Figure 3. It can be seen that the maximum lateral displacement increases with an increase in the fundamental period.

The times to reach maximum displacement during the first half cycle, cracking and failure times for each wall are compared in Table 3. The maximum ground acceleration for the first half cycle occurred at 0.125sec as the period of this sinusoidal base motion is equal to 0.5sec. However, it can be seen from Table 3, that as the natural period of the wall increased, the wall did not respond as a rigid body anymore. The difference between the ground motion and the response of the wall increases with the increase of the natural period of the wall. There is a phase shift in the response.

It is also found in Table 3 that both the cracking and ultimate strength decrease as the fundamental period of the structure increases. Walls 2 and 3 have the same height, but wall 2 is longer in plan configuration and thus has a higher value of aspect ratio. The difference between the dynamic response of these two walls is evident. Wall 2 did not crack for the first cycle when A_1=0.343g. It cracked during the second cycle of excitation when A_2=0.616g, but did not fail throughout the whole analysis, whereas, wall 3 cracked in the first cycle and failed in the early stage of the second cycle excitation. Therefore, the ultimate strength of the wall increases with an increase in the aspect ratio for dynamic analysis.

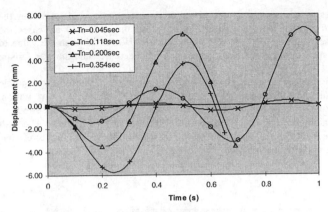

Figure 3. Effect of the natural period of the structures.

Table 3. Comparison of cracking and failure.

Wall No.	Time of Max. Displacement of the first half cycle (sec)	Cracking Time (sec)	Failure time (sec)
1	0.125	_*	_**
2	0.165	0.59	_**
3	0.205	0.14	0.69
4	0.245	0.16	0.66

* No crack occurred.
**No failure occurred.

3.3 Effect of dynamic loading components

In this section, the effect of the frequency of excitation is investigated. The wall geometry and material properties correspond to wall 2 detailed in section 3.2. The walls are subjected to two cycles of increasing ground acceleration and the amplitudes are taken as $A_1=0.4g$, $A_2=0.6g$. The frequency (f) of the ground acceleration varied from 2Hz to 10Hz.

The results of maximum displacements at the top of the wall vs. the relative time (t/T_t), where T_t represents the total running time of the ground acceleration, are shown in Figure 4. The effect of dynamic frequency is significant. It can be seen that the maximum displacement decreases with an increase of the dynamic load frequency. The effect of dynamic load frequency confirms the earlier work by Tercelj et al. (1977). It was found from their tests that the shear strength of masonry wall increased by 12% when loading frequency was increased from 1.0Hz to 5.0Hz, if the fundamental period of the wall is not close to the period of dynamic loading.

4 CONCLUSIONS

In this paper, a parametric study has been carried out by using the nonlinear finite element program developed by the authors for the behaviour of URM under in-plane seismic loads. The effect of some important parameters, such as vertical compressive stress, fundamental period of structures and dynamic loading components is investigated and the major conclusions are summarised as follows:

The vertical compressive stress on the wall plays an important role in determining the ultimate strength, failure pattern and post-failure behaviour of the wall. In all cases the cracking strength increases with the vertical compressive stress. However, when the value of the vertical compressive stress is very high, the ultimate strength decreased and the wall fails in a brittle manner.

The fundamental period of the structure has a significant effect on the nonlinear dynamic response of the URM walls if the fundamental period is not close to the period of excitation. Both the cracking and ultimate strength decrease with an increase in the fundamental period.

The period of the dynamic loading has effect on the behaviour of URM walls. The wall becomes stiffer as the frequency of the dynamic loading increased. This stiffening effect has also been observed in the experimental testing.

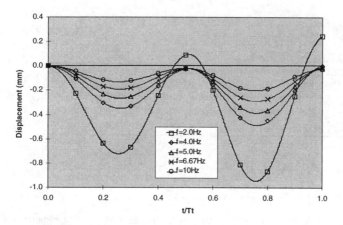

Figure 4. Effect of frequency of dynamic load.

REFERENCES

Darwin, D. and Pecknold, D.A. 1977. Nonlinear Biaxial Stress-Strain Law for Concrete. *J. of the Engg. Mech. Div., ASCE,*(103): 229-241.
Klopp, G. M. and Griffith, M. C. 1993. Earthquake Simulator Tests of Unreinforced Brick Panels. *Proc. 13th Aust. Conf. on the mech. of Struct. & Materials, Australia*: 469-475.
Tercelj, S., Sheppard, P. and Turnsek, V. 1977. The Influence of Frequency on the Shear Strength and Ductility of Masonry Walls in Dynamic Loading Tests. *Proc. 6th World Conf. on Earthq. Engg. New Delhi, India*: 2992-2999.
Zhuge, Y. 1995. Nonlinear Dynamic Response of Unreinforced Masonry Under In-Plane Lateral Loads. *PhD Thesis, Queensland University of Technology.*
Zhuge, Y., Thambiratnam, D. and Corderoy, J. 1995. Numerical Modelling of Unreinforced Masonry Shear Walls. *Proc. of 14th Aust. Conf. on the Mech. of Struct. & Materials, Hobart, Australia*: 78-83.
Zhuge, Y., Thambiratnam, D. and Corderoy, J. 1996. Earthquake Response of Unreinforced Masonry. *AEES Annual Seminar, Adelaide, Australia, Oct*: 7.1-7.6.

Effect of a damp proof course on the shear strength of masonry

M. P. Rajakaruna
School of Engineering, University of South Australia, S.A., Adelaide, Australia

ABSTRACT: Moisture movement in masonry walls is usually prevented by a damp proof course in some mortar joints in the wall. A damp proof course creates a plane of weakness and reduces the shear capacity of a masonry joint. According to the design procedure of the SAA Masonry Code (AS3700) a masonry joint containing a damp proof course is deemed to have zero shear strength unless substantiated by suitable experimental data. This paper describes experimental work carried out to generate data to substantiate the shear strength of masonry containing a damp proof course using material commonly used by the building industry.

INTRODUCTION

Dampness in masonry buildings occurs from moisture movement due to various causes. Moisture infiltration through cracks in mortar or by capillary action has a damaging effect on masonry walls leading to deterioration of bricks and mortar, corrosion of embedded metal items, damage to finishes and may eventually render a building uninhabitable.

Moisture movement in masonry walls is commonly prevented by the inclusion of an impervious barrier known as a damp proof course (DPC). This barrier is formed by placing an impervious membrane either on the brick mortar interface or within the mortar itself. Damp proof courses are commonly located in several layers around the lower courses of brickwork and in exposed positions where water can easily enter the masonry.

A damp proof course creates a plane of weakness in a masonry wall leading to a reduction in the shear capacity. When placed directly on bricks a damp proof course breaks the bond between the bricks and mortar and weakens the joint in shear. Under these conditions frictional forces arising from vertical compressive loads above the plane of the damp proof course provides a significant contribution to shear strength.

The current SAA Masonry code (AS3700) is deficient in the interpretation of shear strength of a joint containing a damp proof course. The current provisions stipulate a bond strength and frictional capacity be taken as zero providing no ability to transfer lateral forces at a damp proof course.

This paper endeavours to contribute test data to quantify the shear capacity of a masonry joint containing a damp proof course made with materials commonly used in South Australia. The results could be used in conjunction with the appropriate capacity reduction factors of AS3700 in the design of masonry shear walls.

SHEAR CAPACITY OF A MASONRY JOINT - AS3700 PROVISIONS

The shear capacity of a masonry joint is considered to be due to two factors. The first due to bond between the mortar and brick called the characteristic shear strength and the second due to friction dependent on precompression loads.

The shear capacity of an unreinforced member is determined from the following expression in Section 5.7 of AS3700:

$$V_d \leq V_0 + K_v f_d A_{dw} \text{ or } 5V_0 \text{ whichever is less}$$

where V_d = the design shear force,

V_o = $C_m f'_{ms} A_{dw}$, the characteristic shear strength of the joint,

$K_v f_d A_{dw}$ = shear strength due to friction.

According to Sections 4.5.4 and 5.7.2 of AS3700, a masonry joint containing a membrane type damp proof course is deemed to have zero shear strength unless substantiated by suitable test data. In most cases a damp proof course nullifies the characteristic shear strength of masonry. However there is a significant amount of shear capacity due to frictional forces arising from vertical loads above the damp proof course except in non-load bearing walls (Page 1994).

MATERALS & METHODS

The investigation was limited to the material most commonly used in South Australia. Test specimens were made of standard clay house bricks with hard burnt centres having negligible suction. Two types of mortar were used, a 1:1:6 (cement : lime : sand) mix used for masonry above the damp proof course known as 'Instant Mortar' and a 1:0:3 'Basic Mortar' mix recommended for areas that require damp proofing. Both mixes were available in premixed bags from Adelaide Brighton Cement.

The most common type of damp proof course material used in South Australia is made of embossed polyethylene having a thickness of 0.5mm and a width of 110mm. It is available in 50 metre rolls under the names 'Dry-Cor' & 'Supercourse 500'. Diamond shaped protrusions arranged in a regular grid pattern are seen on both surfaces of the membrane.

The polyethylene coated aluminium damp proof membrane is not extensively used in South Australia. This material is made by sandwiching a thin layer of aluminium between two layers of polyethylene. The surface is relatively smooth with subtle diamond shaped protrusions on both sides. It is 0.5mm thick and available in 230mm wide, 10 meter lengths under the trade name 'Polyflash'.

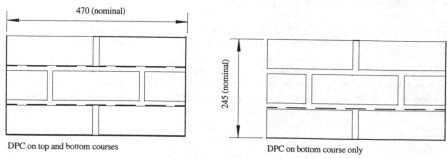

DPC on top and bottom courses   DPC on bottom course only

Figure 1 Shear specimen with damp proof course placed directly on brick

634

The test specimens were two bricks long and three courses high laid in running bond. This configuration returns conservative results with less variability compared to specimens made from a single brick laid in three courses (Greenfield 1990b). The damp proof course was directly placed on the brick as is the practice commonly adopted in South Australia. Although higher strengths are obtained when the damp proof course is sandwiched within the mortar joint (Page 1994) this practice is not prevalent in the local building industry. Test specimens were made with the damp proof course laid on both courses as recommended in AS2904 as well as on the bottom course only to provide a closer representation of the actual situation (Figure 1).

Specimens were cured for seven days under polyethylene sheeting before testing. The loading arrangement is shown in Figures 2 or 3 depending on the direction of shear force being applied. Vertical precompression loads and horizontal shear forces were applied by hydraulic jacks and measured with load cells. Dial gauges were mounted to measure the relative movement among the courses.

Each test specimen was pre loaded to the desired compressive stress and the horizontal shear force was applied to the central course of the specimen at a slow rate (AS2904). Relative movement of the courses was monitored at incremental loads. Slight adjustments were continuously made during the test to maintain a constant vertical load.

Three specimens were tested for each combination of damp proof course type, location and precompression stress. Control samples without the damp proof course were tested under similar conditions for comparison. Precompression levels ranged from 0.0 to 1.5 MPa for samples without a damp proof course. Minimum precompression level was increased to 0.05 MPa for samples containing a damp proof course due to their inherent weakness prior to testing.

Control specimens not containing a damp proof course was often failed suddenly, accompanied with a loud noise and a sharp drop in the horizontal force. Failure of specimens containing a damp proof course was more subtle and was presumed to have occurred when large displacements in the middle course relative to the top and bottom courses were recorded on the dial gauges.

RESULTS

From the test results the mean shear stress and normal stress at failure were determined. The sum of areas of the two bed joints were used for the calculation of shear stress assuming a

Figure 2 In-plane test

Figure 3 Out of plane test

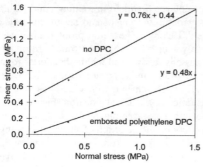

Figure 4 In-plane test, 1:1:6 mortar

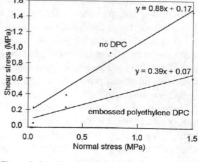

Figure 5 Out of plane test, 1:1:6 mortar

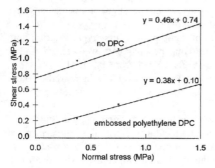

Figure 6 In- plane test, 1:0:3 mortar

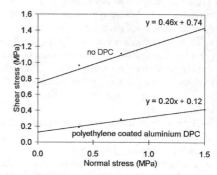

Figure 7 In-plane test 1:0:3 mortar

uniform distribution of shear stress in the specimen loaded in double shear. The results were plotted and a line of best fit was drawn for each set of data as shown in Figures 4 to 7. The results for the samples made without the damp proof course is shown on the same graph for comparison.

The slope of the line is the coefficient of friction also known as the shear factor K_v in AS3700 and the intercept represents the shear strength of the joint. These results showed the an adverse effect membrane type damp proof courses have on the shear strength of masonry joints.

Table 1 Shear strength of joint and shear factor for joints containing a damp proof course

Type of test	Mortar	Damp proof course	Shear strength of joint (MPa)	Shear factor K_v
In-plane DPC on bottom course	1:1:6	embossed polyethylene	0.00	0.48
Out of plane DPC on bottom course	1:1:6	embossed polyethylene	0.07	0.39
In-plane DPC on both courses	1:0:3	embossed polyethylene	0.10	0.38
In-plane DPC on both courses	1:0:3	polyethylene coated aluminium	0.12	0.20

DISCUSSION

The results from a limited number of laboratory tests (Figures 4 to 7) clearly shows the reduction in shear strength of masonry when a damp proof course is included. The techniques used in the determination of shear strength in this investigation are similar to those adopted by Page (1994); hence the results are comparable.

The shear bond strengths of masonry measured (Table 1) are marginally higher than values reported by Page (1994). However this component of shear strength is very small and the zero shear strength prescribed by AS3700 for design purposes appears to be reasonable.

A significant contribution to shear strength at a damp proof course arises from the frictional resistance generated by precompression loads. The shear factors for embossed polyethylene damp proof course obtained from this investigation are considerably lower than those reported by Page (1994). Nevertheless the results exceed the maximum value of 0.30 recommended by AS3700 for normal mortar joints. The frictional shear resistance of polyethylene coated aluminium damp proof course is considerably lower than 0.26 reported by Page (1994) and consequently further testing of this material is recommended.

ACKNOWLEDGEMENTS

The assistance of students P. Millard, B. Stapleton, S. Green and D. Waltham in carrying out the tests at the School of Civil Engineering Laboratories of the University of South Australia is gratefully acknowledged.

REFERENCES

AS2904-1986. *Damp Proof Courses and Flashings*. Standards Association of Australia.

AS3700-1988. *SAA Masonry Code*. Standards Association of Australia.

Greenfield, A. 1990a. Shear Strength of Masonry Containing a Damp Proof Course. *Ceramic Research RP 786:3-7*.

Greenfield, A. 1990b. Shear Strength of Masonry Containing a Damp Proof Course. - Part 2. *Ceramic Research RP 789:3-5*.

Page, A. 1994. A Note on the Shear Capacity of Membrane Damp Proof Courses. *Research Report No 097.05.1994*. Newcastle: The University of Newcastle, Australia.

The Mechanics of Structures and Materials, Grzebieta, Al-Mahaidi & Wilson (eds)
© *1997 Balkema, Rotterdam, ISBN 90 5410 900 9*

Experimental determination of the dynamic modulus of elasticity of masonry units

J.M.Nichols & Y.Z.Totoev
Department of Civil, Surveying and Environmental Engineering, University of Newcastle, Callaghan, N.S.W., Australia

ABSTRACT: One of the parameters that has been identified as influencing the structural response of buildings is the dynamic Modulus of Elasticity of masonry units. The first aim of this paper is to experimentally investigate the use of high frequency sinusoidal loading to determine the dynamic Modulus of Elasticity of masonry units. The Longitudinal Vibration and the Ultrasonic Pulse methods were originally developed for the dynamic testing of concrete specimens. These testing procedures are appropriate as each applies only a minimum stress to the masonry units. This procedure provides results for the masonry units within the elastic range. The second aim is to compare the dynamic Modulus of Elasticity results to the Modulus of Elasticity obtained using quasi-static methods from the same population of masonry units. A test rig has been developed for measuring the elastic properties of masonry units under uniaxial loading.

1 INTRODUCTION

Seismicity within Australian is now recognized as a design issue, as a result of several major earthquakes within the last 40 years in NSW, SA and WA. One of the common building materials used over the last century in Australia is masonry. There has been a growth in the use of unreinforced masonry(URM) in the last 50 years, particularly in light commercial, housing and as infill panels to steel and concrete framed buildings. These types of buildings can pose a significant hazard during seismic events. The main reasons for this performance problem are the relative mass to strength ratio, ductility issues and poor workmanship (Melchers & Page,1992). Unreinforced masonry can however under some circumstances withstand an interplate earthquake and perform well within its design limitations, when constructed to accepted standards.(Tena-Colunga & Abrams, 1992). A reasonable proportion of the larger masonry buildings and dwellings built within in Australia and elsewhere in intraplate regions would have been designed on the basis of static loading design rules and assuming zero or low seismic loads. The design of buildings within Australian must now consider the minimum loading from the Australian Standard(AS) Earthquake Loading Code (AS 1170.4) using either an equivalent static loading, frequency domain or time domain analysis. Material properties are required for these methods of analysis, irrespective of the numerical method.

The first objective of this paper is to experimentally determine the dynamic Modulus of Elasticity for masonry units using two non-destructive techniques. The second objective is to measure the Young's Modulus and Poisson's Ratio using quasi-static methods. The third objective is to compare the quasi-static and dynamic results. The dynamic results are based on the assumption that Poisson's Ratio is invariant and can be measured using the quasi-static proce-

dures. Three pressed clay bricks designated by colour (red, brown, biscuit), one calcium silicate and one concrete brick were used in the experimental work.

2 BACKGROUND

Young's Modulus is an intrinsic property constant for a material. It can be estimated using the
15 to 85 % stress levels in the elastic range from a quasi-static test procedure on a previously
untested population sample (Krajcinovic 1996, LeMaitre 1992). Hookes' law is defined as :

$$\sigma = E \varepsilon \tag{1}$$

where σ defines the stress (MPa), ε is defined by the natural (rather than the engineering) strain
formula (Strains) and E is Young's Modulus (MPa). Young's Modulus and Poisson's Ratio
have been previously measured for a number of Australian brick types. These results were
within the range of 7,000 to 12,000 MPa and 0.12 to 0.29 respectively (Dhanasekar 1985).

Two distinct failure mechanisms can be identified for a "macroscopically homogeneous body
of brittle material". The failure mechanism of the first kind is "quasi-static or stress wave
loading of low intensity" when a single failure plane predominates activated by a "flaw in the
material being stressed to a critical condition". The failure mechanism of the second kind for
masonry or dynamic failure mode occurs "during impulsive loading where multiple fractures on
different planes can be nucleated and they grow to a significant size without arresting each
other." Ceramics and other brittle material have been shown to have a greater fracture resistance to loading that causes the failure mechanism of the second kind (Freund, 1990).

These distinct mechanisms have been observed in masonry testing (Tercelj et.al, 1969, Klopp
1996). The first or quasi-static mechanism is the basis for the development of key static masonry design rules (Page, 1979). The results form part of the dynamic test program on masonry
panels being undertaken at Newcastle. This research on the two failure mechanisms would
suggest that at an equivalent time the following inequality holds, (where ~ denotes dynamic and
- denotes static Modulus of Elasticity):

$$\widetilde{E_t} \geq \overline{E_t} \tag{2}$$

This qualitative observation can be attributed of Tercelj et.al.,(1969). The dynamic Modulus of
Elasticity can be determined from Equation (3) for the Longitudinal Dynamic Test Method
(LDTM) and from Equation (4) for the Ultrasonic Pulse Method (UPM). The LDTM uses a
small audio striker to provide pulses along the longitudonal axis of the brick. The UPM uses
the measurement of the travel time of ultrasonic pulses in the transverse and longitudonal axes.

$$f = \sqrt{\frac{\widetilde{E}}{4L^2 \rho}} \tag{3}$$

$$V = \sqrt{\frac{\widetilde{E}}{\rho} \frac{(1-\upsilon)}{(1+\upsilon)(1-2\upsilon)}} \tag{4}$$

where f (Hz) is the fundamental natural frequency, L (m) is the length of the specimen,
ρ (kg/m^3) is the density, V (m/s) is the pulse velocity and υ is Poisson's Ratio.

The ultrasonic pulse method can be used to determine if the masonry unit samples show any
pattern of anisotrophy, such as may be exhibited as a result of the process of manufacture.

3 EXPERIMENTAL METHOD

The LVTM uses a dynamic test rig that is a modified version of the Electrodynamic's Standard Material Tester EMFCO SCT/5 (EMFCO, n.d.) This test rig is noted in the specification as to complying with British Standard (BS) 1881: 52 (Longitudinal Vibration). Specimens were saw-cut from a standard range of bricks supplied by local manufacturers. Each specimen was microwave dried and then tested in the rig. Each was then allowed to stabilize at room temperature before being retested. A Tektronix Function Generator FG501 with controlled frequency was used to generate the applied sinusoidal loading function. This signal was amplified using a Peavey Electronics Corp. XR400 Amplifier to feed the 3 Ohm coil on the test rig. Each specimen was clamped on the test rig using an 11 mm rad. jaw clamp at the midpoint of the cut brick. A piezoelectric crystal pick-up detects the signal which was monitored on an Tektronix Oscilloscope 7603 for peak amplitude. The frequencies used were in the range from 5 to 9 kHz. A schematic arrangement of the equipment is shown in Figure 1.

The second dynamic method UPM uses a standard measurement system. This method uses the CNS Portable Ultrasonic non destructive tester (CNS 1978). A calibrating specimen is provided with the rig. Testing was at 50 kHz about the longitudonal and transverse axes of the specimen. A schematic arrangement of the Ultrasonic Puls Method is shown in Figure 2.

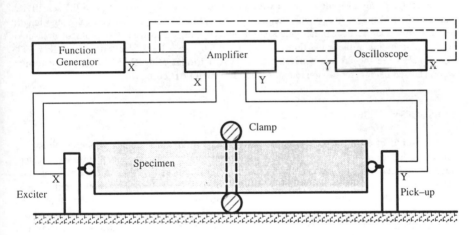

Figure 1 Layout of the dynamic test rig for the Longitudonal Vibration Test Procedure.

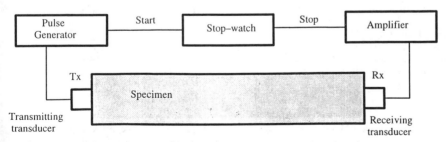

Figure 2 Layout of the dynamic test rig for the Ultrasonic Puls Test Procedure

The quasi-static Test Method uses uniaxial compression applied to the smaller end faces of the brick samples with a Tinius Olsen 1800 kN Universal Testing Machine(UTM). A pressure cell in the UTM generates an analog output signal. Initially this signal was converted to a digital signal using a Gedge Systems(Aus) GS1650P Peak Indicator . The analog to digital signal generator proved to be too coarse (± 2.5kN) for this testing and caused a substantial component of the error in the measurement. Final testing had the analog signal being fed directly into the data logger. Calibration of the signal was undertaken using the UTM dial scale.

Each brick was tested for Young's Modulus about the longest axis to provide a reasonable gauge length for the measurements, to protect the Linear Velocity Displacement Transducers (LVDTs) and to partially negate end effects. The test method was based on the relevant Australian Standard for testing masonry units. Bricks were immersed for a minimum of two hours before being tested, except for the few that were used in both the dynamic and quasi-static testing. Plywood capping was used between the solid platens and the brick.

A rectangular test rig capable of measuring the relative displacements about two axes was designed to provide a repeatable measurement protocol. The rig is similar to the standard cylindrical concrete test rig, except that it is modified to measure Poisson's Ratio. Vertical or longitudinal displacement was measured using two LVDTs Type RDP Electronics D2-200A. These have a total movement of 11 mm. Horizontal displacement was measured using two LVDTs Type Solartron DFG 5.0. These have a total movement of 12 mm. Each of the LVDTs was calibrated using a Mitutoyo gauge with a range of 0 - 25 ± 0.005 mm and the sensitivity tested using a Mitutoyo gauge with range of 0 - 1 ± 0.0002 mm. . The vertical gauge length was 100 mm and the horizontal gauge length was 90 mm. The final test protocol measured both displacements at the same time. Signals were fed into a Data Electronics Datataker 600. The signals were logged and converted to an ASCII format using DASYLab 3. The results were analysed using a regression macro written for MINITAB 10.2.

4 RESULTS

These series of experiments were undertaken to determine the Young's Modulus, Poisson's Ratio and dynamic Modulus of Elasticity of masonry units. Five masonry units were tested, three pressed clay bricks, a concrete and a calcium silicate bricks. All bricks were of local State manufacture and the three pressed bricks were tagged by colour red, brown and biscuit. Young's Modulus and Poisson's Ratio were measured using the quasi-static test method. The test results are based on the gross area of the cross section of each unit. No allowance has been made for the frog. The quasi-static test results for each set of the five brick types are shown in Table 1.

Table 1. Young's Modulus, Poisson's Ratio and typical dimensions for the five brick types.

Brick Type	Number Tested	Young's Modulus MPa	Poisson's Ratio	Length : Width :Depth mm
Pressed Clay Red	6	14,000	0.22	226:111:75
Pressed Clay Biscuit	4	10,000	0.29	230:110:76
Pressed Clay Brown	5	7,000	0.21	227:108:74
Calcium Silicate	3	6,000	0.17	229:108:78
Concrete	4	14,000	0.33	232:109:77

The range of Young's Modulus was from 1 GPa for a pressed clay red brick to 56 GPa for a concrete brick. The stress-strain curves were generally elastic (typical R^2 coefficient ~ 0.95) over the 15 to 85 % range of the peak compressive stress used to determine Young's Modulus.

Table 2. Typical Brick Properties for the specimens for the Longitudinal Vibration Test Method.

Brick Type	Description	Specimen Type 1	Specimen Type 2	Specimen Type 3
	Typical Length mm	180	200	230
	Typical Cross section area mm^2	1100	1800-2020	2250-2500
Pressed Clay Red	Density kg/m^3	2170	2070	2320
Pressed Clay Biscuit	Density	2220	2230	2270
Pressed Clay Brown	Density	2130	2180	2130
Calcium Silicate	Density	1810	1740	1760
Concrete	Density	2010	2100	2190

A few of the pressed bricks exhibited stress-strain curves that would suggest that there is a variation of density within the brick. This variation can probably be attributed to the compaction associated with the creation of the frog. Initial results would point to a greater material density on the frog side of the brick. The non-symmetric shape of the pressed brick requires the use of averaged results between the two sets of displacement for the measurement of Young's Modulus for some bricks. Poisson's Ratio results ranged from 0.1 to 0.4.

The specimens used for the Longitudinal Vibration Test Method were cut from full bricks. Three different specimen sizes were used in the experiments . Typical dimensions for each of the specimen sizes, presented as Length (mm),:Cross Sectional area (mm^2) and the density (kg/m^3) are presented in Table 2.

Three full size brick specimens for each of the five brick types were tested using the Ultrasonic Pulse Method. The Longitudinal Vibration Test Method uses Equation (4) and the results of the frequency measurements, the dimensions and the density results. The Ultrasonic Pulse Method uses Equation (5), Poisson's ratio from the quasi-static testing, the transit time results, the dimensions and the density results. The results for the quasi-static and dynamic testing are shown in Figure 3.

The calcium silicate and the concrete bricks exhibit isotropic behaviour in the Longitudonal Vibration Test as can be seen in Figure 3. The remaining bricks exhibit a slight anistrophic behaviour that is probably attributable to the method of manufacture.

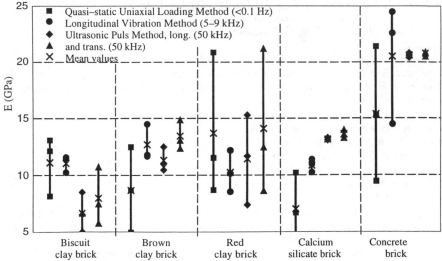

Figure 3 Modulus of Elasticity Results for the three different test procedures.

5 CONCLUSION

This study was designed to compare the quasi-statically measured Young's Modulus to the dynamically measured Modulus of Elasticity. The work forms part of a research program into the response of masonry panels subjected to seismic frequency and intensity of loading. The quantification of the Young's Modulus and Poisson's Ratio was made with three test methods. The first method was quasi-statically in a Universal testing Machine and the second two methods were variations on standard techniques used extensively in concrete research. A set of protocols and a test rig were developed to quasi-statically measure the Young's Modulus of masonry units. This test rig is based on the rig used to test concrete cylindrical specimens.

Two dynamic methods are used to measure the Modulus of Elasticity of concrete cylindrical specimens. These two methods are the Longitudonal Vibration and Ultrasonic Pulse Velocity Test Methods. The limitation for these procedures is the frequency dependence of the results.

The results for the quasi-static measurement of the Young's Modulus and Poisson's ratio for the masonry units is presented in Table 1. The results for the dynamic measurement of the Modulus of Elasticity for the two dynamic methods are shown on Figure 3. These results are within the range of Young's Modulus normally expected for these types of masonry units.

Dynamic measurement of the Modulus of Elasticity is a practicable alternative to quasi-static destructive testing for clay masonry units. There is no evidence of frequency dependence for the clay masonry units within the range of frequencies available with these two methods. There appears to be a strong frequency dependence for the sand based masonry units.

Testing is required using many specimens to quantify the relationship. The frequency dependence of the stiffness of masonry walls was noted by others at lower frequencies than used with these two test methods. Further research is suggested on masonry in the non elastic range.

REFERENCES

CNS Electronics Ltd. 1978. *Pundit Manual for use with Portable Ultrasonic Non-destructive digital indicating tester.* London: CNS.

Dhanasekar, M. 1985.*The performance of brick masonry subjected to in plane loading.* Dissertation No. 990. Newcastle: University of Newcastle, Australia.

EMFCO. n.d. *Specification for the SCT/5 Test Rig.* London: EMFCO.

Freund, L.B. 1990. *Dynamic Fracture Mechanics,* Cambridge :Cambridge University Press.

Klopp, G.M. 1996. *Seismic design of unreinforced masonry structure.,* Adelaide : University of Adelaide.

Krajcinovic, D. 1996. *Damage Mechanics.* New York: Elsevier.

LeMaitre, J. 1992. *A course on Damage Mechanics.* Berlin : Springer-Verlag.

Melchers, R.E. & Page, A.W. 1992. The Newcastle Earthquake. *Building and Structures. 94:143-56.*

Page, A.W. 1979. *The inplane deformation and failure of brickwork.* Dissertation No 636. Newcastle: University of Newcastle, Australia.

Paulson, T.J. & Abrams D.P. 1990. Correlation between static and dynamic response of model masonry structures. *Earthquake Spectra.* 6:573 - 91.

Tena-Colunga, A. & Abrams, D.P. 1992. Response of unreinforced masonry building during the Loma Prieta earthquake. *Proceedings of the Tenth World Conference on Earthquake Engineering.* 3:2292-9.

Tercelj, S. Sheppard, P. & Turnsek, V. 1969. The influence of frequency on the shear strength and ductility of masonry walls in dynamic loading tests. *Proceedings of the Fifth International Conference on Earthquake Engineering.* 3:2292 - 9.

Tomazevic, M. Lutman, M. & Petkovic, L. 1996. Seismic behaviour of masonry walls : experimental simulation. *ASCE Journal of Structural Engineering,* 122:1040-7

The Mechanics of Structures and Materials, Grzebieta, Al-Mahaidi & Wilson (eds)
© *1997 Balkema, Rotterdam, ISBN 90 5410 900 9*

The modelling of masonry veneer under dynamic load

J. Kautto, P.W. Kleeman & A.W. Page
*Department of Civil, Surveying and Environmental Engineering, The University of Newcastle,
N.S.W., Australia*

ABSTRACT: The seismic performance of masonry veneer in Australia is largely unknown. A simple finite element model simulating masonry veneer under both static and dynamic loads has been developed. Dynamic results from time domain analysis are presented together with possible refinements to the original finite element model. Results of static analyses using the model are also briefly presented.

1 INTRODUCTION

Australia has long been regarded as an earthquake free continent. Earthquakes have occurred in the past, but until recently few had been located close enough to populated areas to cause significant damage. The 1989 Newcastle Earthquake caused widespread damage to the city and surrounds and graphically illustrated that even moderate earthquakes can cause significant damage if buildings are not designed for seismic effects. The damage also emphasized possible inadequacies with current design and construction methods for masonry housing, particularly related to the detailing of the structural elements and their connections (Page 1993). With the publication of the new Earthquake Loading Code AS1170.4, and its subsequent adoption by the Building Code of Australia, it is now mandatory to consider seismic effects for all structures including housing.

In the past, housing has been designed for wind load with no consideration of earthquake loading. Details which have been developed for wind effects are not necessarily suitable for earthquake due to the differing nature of the loads, with earthquake load being more dependent on the mass of the structural elements. This therefore makes unreinforced masonry, with its high mass and brittle behaviour particularly susceptible to damage from earthquake loads. A preliminary study of these aspects has been previously reported (Kautto & Page 1995). Some other studies of the performance of housing under seismic loading have been presented (Gad et al 1995), but the emphasis has been on the performance of the back-up frame rather than the veneer. Masonry is widely used in housing in the form of cavity veneer or single skin construction. Veneer construction in particular is very common, particularly in the eastern states. Since unreinforced masonry veneer is so widely used, there is an obvious need for research into its seismic behaviour to ascertain if current details and construction procedures are adequate.

This paper presents an overview of current and future research at The University of Newcastle into the behaviour of masonry housing under seismic loads together with the results of some preliminary analyses. In this research particular emphasis has been placed on the modelling of masonry veneer in an effort to determine its performance under seismic and/or dynamic loads.

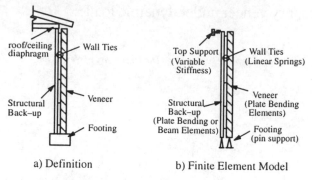

a) Definition   b) Finite Element Model

Figure 1. Masonry Veneer

2 DEFINITION OF MASONRY VENEER

Masonry veneer consists of an outer non-loadbearing leaf of masonry attached to an inner masonry leaf or structural timber or steel back-up frame. The veneer provides an external weather barrier, whilst at the same time adding to the aesthetics of the structure.

Wall ties attach the otherwise free-standing veneer to the structural back-up and therefore play a crucial role in the adequate performance of the masonry veneer. The wall ties are formed from light gauge steel plate or wire, and nailed or secured to the back-up frame and embedded in the mortar joints of the veneer. Unfortunately the wall ties are usually the most neglected component in such walls and often installed incorrectly, thus directly affecting the performance of the assemblage.

The structural back-up serves to laterally support the veneer (via the wall ties) and usually spans vertically from the footing to the ceiling/roof system which acts as a diaphragm. The back-up can be classed as either flexible or stiff, with a flexible back-up having a stiffness far lower than the veneer it is supporting, or a stiff back-up having a stiffness comparable to the veneer. A flexible back-up usually consists of either a timber or steel frame in the form of a stud wall. A stiff back-up usually consists of another leaf of masonry. Masonry veneer with stiff back-up is usually referred to as a "cavity wall", although it is not a true cavity wall in which both leaves are loadbearing. A typical veneer is shown in Figure 1a.

3 FINITE ELEMENT MODEL

A finite element model was used to simulate the behaviour of the masonry veneer subjected to lateral (face) loading. Both static and dynamic analyses were performed. The properties of the model are outlined below.

3.1 *Masonry*

The masonry veneer and rigid back-up walls (if appropriate) were modelled using 4-noded, 12 degree of freedom, orthotropic elastic plate bending elements (although in most cases isotropic behaviour was assumed). Typical elastic properties were assumed for the masonry (E_x=8000MPa, E_y=8000Mpa, Poisson's Ratio = 0.2). The mesh geometry was chosen to suit the layout of the wall ties which are typically located at 600mm centres in the vertical direction and in line with each row of studs. A more refined mesh could have been accommodated but was deemed unnecessary for the problem. Provision for the insertion of both horizontal and vertical cracks in the masonry was also included in the program by de-coupling the appropriate plate bending elements when bending moments exceeded values corresponding to the flexural strength of the masonry in the appropriate direction. Inclined cracks could also be incorporated as appropriate by modifying the relevant plate bending element stiffness matrix. All cracks were inserted into the model manually.

3.2 *Flexible Back-up*

Flexible back-up systems were modelled as simple beam elements. Since only the nodal translations were required, to reduce the total number of degrees of freedom at each node, the beam rotations were eliminated by static condensation. The stiffness properties were determined from typical timber or steel stud wall systems, with the following values being used : elastic modulus = 6.9 GPa (typical for F5 Radiata Pine) , Poisson's Ratio = 0.2.

3.3 *Wall Ties*

In the preliminary analysis wall ties were modelled as simple linear springs, with non-linear behaviour being neglected for the static analysis. However under dynamic loading the non-linear behaviour of the ties may be crucial and will be incorporated into the model when the results of current experimental work evaluating tie stiffness characteristics under cyclic loading is complete.

3.4 *Supports*

The top of the back-up frame in veneer construction is usually supported by the roof/ceiling system which acts as a diaphragm and spans horizontally between the side walls. Since the diaphragm may have some flexibility, the top support for the back-up was modelled using a linear spring to allow this effect to be simulated. Due to the lack of fixity at the base of the veneer and its back-up frame, the relevant degrees of freedom for the points were maintained in the global stiffness matrix so that if a rigid support was required, a large spring stiffness could be inserted. This strategy was adopted in order to keep the boundary conditions as general as possible.

In the analyses reported in this paper, the veneer and the back-up are assumed to span in the vertical direction only with no two way plate action. This is the most critical orientation with regard to the veneer and the structural back-up. In modelling this behaviour, a representative vertically spanning strip encompassing one stud and the corresponding masonry and wall ties was therefore used. The typical arrangement is shown in Figure 1b.

It is common practice to simulate the load effects for both wind and seismic analysis by applying equivalent static forces. Both equivalent static and dynamic analyses were performed for the vertically spanning veneer systems subjected to lateral load effects. This allowed direct comparison of the results. These are outline in the following sections.

4 STATIC ANALYSIS

Static loads were applied as either pressures or suctions to both the veneer and back-up. Detailed results for these analyses have been previously reported (Page et al 1996). Results for a typical one-way analysis indicated that the topmost tie was the most heavily loaded if the veneer remained uncracked, with the veneer spanning from the base to the top tie, with the intermediate ties being only lightly loaded (see Figure 2a.). Stresses in the masonry veneer were greatest at mid-height and of the same order as the flexural strength of the masonry, indicating that cracking might occur in this area. If cracking was allowed to occur at approximately mid-height the tie force distribution changed, with the ties closest to the crack now becoming more heavily loaded along with the topmost tie (see Figure 2b). The results

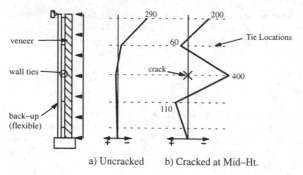

a) Uncracked   b) Cracked at Mid–Ht.

Figure 2. Tie forces(N) under static load (p=0.41KPa) with flexible back–up

from this simple static analysis showed that it is erroneous to calculate the forces in the wall ties based on a local tributary area as is commonly done in practice.

New design procedures have been developed based on this more representative distribution of tie forces in a statically loaded masonry veneer wall (Page et al 1996).

5 DYNAMIC ANALYSIS

The dynamic analysis was performed in the time domain and to keep the model as simple as possible a lumped-mass system was utilized (see Figure 3a.). This allowed static condensation of the global stiffness matrix to remove the rotational degrees of freedom and therefore reduce the overall size of the problem (Clough & Penzien 1975). The seismic effects were applied as ground accelerations obtained from suitable earthquake traces. The input for the dynamic analysis was in the form of an earthquake trace obtained from the 1982 Miramichi earthquake, recorded at Loggie Lodge in Canada (see Figure 3b.). This input was chosen as it was considered to reasonably represent the form of earthquake that occurs in Australia (Melchers 1995), although what constitutes a "typical" Australian earthquake is still open to debate. The modelling of the ties and the structural back-up was the same as for the static analysis. Two different solution techniques were utilized in order to provide a means of checking the consistency of the results; modal superposition, and direct integration.

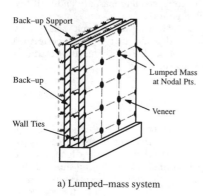

a) Lumped–mass system

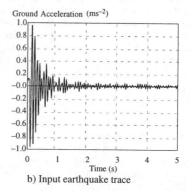

b) Input earthquake trace

Figure 3. Dynamic Model

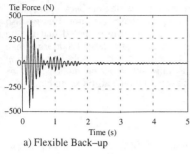
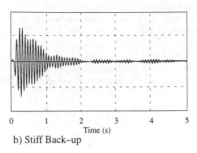

Tie Force (N)

a) Flexible Back–up   b) Stiff Back–up

Figure 4. Dynamic Results: Top Tie Force versus Time

5.1 *Modal Superposition*

The first technique utilized was modal superposition. Mode shapes and fundamental frequencies were determined from the eigenvalues and eigenvectors for a given wall geometry. Damping of 5% critical damping was assumed for all modes.

Since coordinate transformations were required for all input and output (Coates, Coutie & Kong 1988), the calculations were slowed down considerably using the modal superposition technique. A time domain analysis was then performed on each mode using an exact recursion formula based on Duhamel's integral (Irvine 1986). Results were in the form of time histories of displacements, tie forces and bending moments in the back-up and veneer.

5.2 *Direct Integration*

The second solution technique was direct integration. The Newmark integration scheme with constant time stepping was utilized (Bathe & Wilson 1976). Rayleigh damping was assumed in the calculation of the damping matrix based on the first two mode shapes and their respective percentages of critical damping (5%). Unlike modal superposition, no coordinate transformations were required, and therefore input accelerations and output displacements were not modified. The results were also presented as time histories of displacements, tie forces and back-up and masonry bending moments.

Modal superposition analysis, although slow, was considered more accurate than direct integration using a given time step as the recursive formula utilized gave exact results for any time step provided the input was accurately prescribed (Irvine 1986). It was therefore useful in checking the relative accuracy of the direct integration method and in finding the largest time step that yielded similar results. The main benefit of direct integration is that it is much simpler and faster to implement non-linear analysis as no coordinate transformations are required to complete the analysis. This saving in computational time was considered crucial for the later more involved analyses. Analysis of the elastic response for an input earthquake indicated that components such as wall-ties are likely to respond non-linearly and hence modify the response of a masonry veneer to dynamic load.

6 RESULTS OF DYNAMIC ANALYSIS

In the dynamic analysis one-way bending was considered most critical in terms of wall performance and the largest time step which yielded similar results for each of the previously described dynamic analysis techniques was utilized to minimize computational time. Figure 4. shows the variation of force in the topmost tie with time for a strip of wall 1.8m wide by 2.4m high with either a flexible or stiff back-up (since the tie is elastic, it is also representative of the relative displacement between the veneer and back-up). It can be seen that both forms of veneer react almost instantaneously to the dynamic input and that the flexible back-up response remains in step with the excitation whilst the stiff back-up response does not. Energy also dissipates much more quickly with a flexible back-up. It should also be noted that the tie forces cycle rapidly from tension to compression and are initially quite large (and larger than those predicted by the static analysis). For such a load/time history, the simple linear spring model used for the wall ties may be inadequate and a more sophisticated model could be required. This aspect is being clarified in a parallel testing program on wall-tie assemblages.

7 MODELLING OF COMPONENTS

The effect of rapid load reversals on components (such as wall ties) could be significant, as degradation of their elastic properties can occur (due to degradation of the tie itself and/or its attachments to the back-up or embedment in the mortar joints). As a result of the initial dynamic results an experimental testing program is underway to determine the cyclic behaviour of wall ties. Wall ties will be the most critical component as they effectively provide the only means of lateral support to the free-standing veneer wall (provided the back-up itself does not fail). Once the experimental results are available, the non-linear characteristics of the wall ties will be included in the model. Direct integration will be utilized for the non-linear analysis. With the inclusion of non-linear component behaviour the dynamic analysis results will most likely differ considerably from those presented in Figure 4.

8 SUMMARY & CONCLUSIONS

The Newcastle Earthquake highlighted the need for research into the behaviour of masonry structures under dynamic load conditions in Australia. Of particular interest was masonry veneer which forms a large part of domestic housing. A linear elastic finite element model simulating masonry veneer under both static and dynamic loads was developed in order to determine its performance.

Dynamic analysis was in the time domain and two different solution techniques, modal superpositon and direct integration were utilized. It was found that upon review of the initial dynamic results a linear elastic model could be inadequate and a more sophisticated model taking into account the cyclic behaviour of components such as wall ties may be needed.

9 REFERENCES

Bathe, K-J. & Wilson,E.L. 1976. *Numerical methods in Finite Element Analysis*.Englewood Cliffs: Prentice-Hall
Clough, R.W. &. Penzien, J. 1975. *Dynamics of Structures*. New York: McGraw-Hill.
Coates, R.C., Coutie,M.G. & Kong,.F.K. 1988. *Structural Analysis 3rd edition*. Wokingham: Van Nostrand Reinhold (UK) Co. Ltd.
Gad, E.F., Duffield, C.F. ,Stark, G. & Pham, L. 1995. Contribution of Non-Structural Components to the Dynamic Performance of Domestic Steel Framed Structures. *PCEE '95, Proc. Pacific Conf. Earthquake Eng. Melbourne, 20-22 November 1995: 3: 177-186*.
Irvine, H.M. 1986. *Structural Dynamics for the Practising Engineer*. Sydney: Allen & Unwin.
Kautto, J.A. & Page, A.W. 1995. The Impact of the Earthquake Loading Code on Masonry Housing. *Proc. 4th Aust. Masonry Conf., Sydney, 23-24 November 1995: 160-170*
Page, A.W.1993.The Design, Detailing and Construction of Masonry-The Lessons From the Newcastle Earthquake. *Civil Engineering Transactions. CE34: 4: 343-353*.
Page, A.W., Kautto, J. & Kleeman,P.W. 1996. A Design Procedure for cavity and Veneer Wall Ties. *Masonry International: 10: 2: 55-62*.
Melchers, R.E. & Morison, D.W. 1995. Studies of Structural Response to Typical Intra-Plate Ground Shaking. *PCEE '95, Proc. Pacific Conf. Earthquake Eng. Melbourne , 20-22 November 1995: 1: 207-215*.

The Mechanics of Structures and Materials, Grzebieta, Al-Mahaidi & Wilson (eds)
© *1997 Balkema, Rotterdam, ISBN 90 5410 900 9*

Investigations of the bond wrench method of testing masonry

W. Samarasinghe & S. J. Lawrence
Building, Construction and Engineering, CSIRO, Sydney, N.S.W., Australia

A. W. Page
Department of Civil Engineering and Surveying, University of Newcastle, Callaghan, N.S.W., Australia

ABSTRACT: The bond wrench is a simple and practical piece of equipment which can be used to test masonry bond for laboratory as well as site investigations. The bond wrench test method provides more data from a single masonry test specimen relative to other known methods. However, it has produced variable results and there are mixed opinions on its validity.

This paper describes some of the deficiencies associated with the present bond wrench commonly used in Australia. The effect of geometrical configuration of the wrench and the clamping effect of the restraining frame on the stress distribution were investigated by experimental and analytical means using an Aluminium calibration block and finite element analysis. It is shown that the current bond wrench does not produce the assumed stress distribution in the test specimen and as a result it underestimates the masonry strength. Suitable parameters for a new bond wrench are proposed, which eliminate the ill effects of the present apparatus.

1 INTRODUCTION

The primary function of mortar in a masonry wall is to bind the building units firmly together to produce a structurally sound and watertight wall. If the adhesion is poor, rainwater may more easily penetrate the voids formed between the units and mortar, and the resistance of the masonry to eccentric and lateral loads will be reduced. Therefore, the extent of the bond between unit and mortar is a good measure of the quality of masonry.

The Australian Masonry Code AS 3700 (Standards Association of Australia, 1988) recommends assessment of the bond strength by either the bond wrench test method or the bond beam test method. Although the older beam test is still permissible, the preferred method of test is the bond wrench, because the beam test is wasteful of material - giving only one test result from a prism whereas the bond wrench tests every joint. Other advantages of the bond wrench are that it is easy to handle and it can be used to measure flexural bond strength of both in-situ masonry and laboratory specimens.

The bond wrench method of testing was originally developed in Australia and it has now been accepted overseas (in the USA and Canada). In the USA, use of the bond wrench in the laboratory is now covered by ASTM Standard C1072 (ASTM, 1994). A keen interest is also being shown in the UK and Europe towards adopting this method as a standard test procedure for measuring bond.

Despite the popularity of the technique, some concern has been expressed that the bond wrench yields a relatively high variability, and therefore a low characteristic bond strength, in contrast to the beam test. Against this background of uncertainty the CSIRO and the

University of Newcastle, in collaboration with the industry, embarked on research to evaluate the performance of the bond wrench. This work constitutes part of a major research project which is investigating the factors determining tensile bond strength of masonry and the mechanisms of bond formation (Lawrence & Page, 1995).

2 THE BOND WRENCH

In principle, the bond wrench (Figure 1) is simply a long lever, which is clamped to a brick (or concrete block) at one end. The other end is free. An increasing force is gradually applied at the free end until the brick is rotated free from the mortar joint immediately below it. The load at which this occurs is a measure of the strength of the masonry. The bond strength is calculated using the conventional bending formula. It is assumed that the stress distribution through the thickness of the masonry is triangular and that the stress distribution along the length of the brick at any section is uniform.

In laboratory testing the specimen is supported by a reaction frame. The brick below the joint to be tested is clamped along the longitudinal faces ensuring that the bottom interface is between 1 mm and 3 mm above the clamped area.

The bond wrench can be used on existing structures, during or soon after construction or at any time during the subsequent life for trouble-shooting failures. In-situ tests are done by cutting perpends either side of the test unit and attaching the bond wrench.

The most commonly used bond wrench in Australia was developed at Deakin University. It consists of a lever about 1 m long, weighing about 3.4 kg. At one end of the lever are jaws (gripping bars) which can be adjusted to fit the thickness of masonry. The gripping bar on the compression face (19 mm x 19 mm x 200 mm) grips the brick about 40 mm below the top face. The gripping bar on the tension face is 15 mm wide x 19 mm high x 200 mm long. At the other end of the lever is a crossbar handle. Load is applied manually by putting body weight on to the crossbar handle, and the maximum load applied is recorded in a strain-gauge-operated LCD-type display unit mounted on the loading lever.

3 EXPERIMENTAL PROGRAMME

The experimental investigation consisted of two phases. The first phase developed a device for measuring the strains that a bond wrench applies to a typical masonry specimen. The device consisted of a strain-gauged aluminium block with dimensions similar to a typical masonry couplet (two bricks and a mortar joint).

Figure 2 shows the aluminium calibration block and its strain gauge configuration. Altogether, 19 gauges were mounted where the interface between brick and mortar is normally located (Note that strain gauge numbers 10 to 19 were not used). The strain gauges

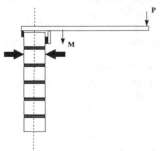

Figure 1. Bond Wrench Schematic

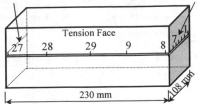

Figure 2. Aluminium Calibration Block Showing the Strain Gauge Locations

were fixed only on one of the longer faces since the presence of the gripping bar of the bond wrench obstructs the other face.

The functionality of the gauges was checked by testing the strain-gauged aluminium block under uniform compression. The elastic modulus of the aluminium was measured as 68,700 MPa.

The second phase of the investigation involved placing the aluminium calibration block into a bond wrench testing apparatus and measuring the strains applied to the block.

The average strain distribution measured on the sides of the calibration block (the average of corresponding gauges between numbers 1-7 and 20-26) is shown in Figure 3. A set of lines is drawn to show the strain distribution through the thickness at different stages.

The initial condition (with the calibration block sitting on a level surface) coincides with the horizontal axis of the graph. The other two curves represent the conditions imposed by the bottom clamps and subsequently by the bond wrench with 495 N load applied at the end of the lever. The expected distribution according to the conventional bending formula is shown for comparison.

The results reveal that the stress distribution through the thickness of the specimen is markedly different from that due to the conventional bending formula. They also show that the clamping by the retaining frame induces a significant tensile strain on the test specimen. An investigation of the ASTM bond wrench has shown similar results (McGinley, 1993).

As a criterion for acceptance of materials and workmanship in masonry construction, AS 3700 has adopted a minimum flexural strength of 0.2 MPa. The maximum strain value recorded due to clamping pressure is equivalent to a stress of 0.17 MPa. Therefore, this unaccounted stress produced in actual testing could lead to the rejection of acceptable masonry. The specified distance between the bottom interface and the clamps in AS 3700

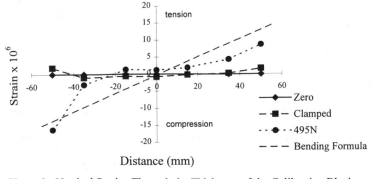

Figure 3. Vertical Strains Through the Thickness of the Calibration Block

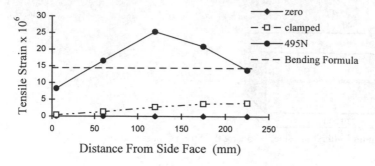

Figure 4. Vertical Strains on the Tensile Face of the Calibration Block

(1 to 3 mm) may be quite critical to this effect. The stress concentration effects would be reduced by increasing this distance.

Figure 4 shows the distribution of strain measured across the tensile face of the calibration block and clearly suggests that the bond wrench produces a distinctly non-uniform stress distribution at the extreme fibres of the tension face of the test specimen. It produces a significantly high, localised stress at the centre of the test specimen, right below where the tensile gripping bar connects to the loading lever. Masonry is a brittle material and therefore failure is likely to be triggered by the peak stress and the test would tend to underestimate the bond strength. According to Figure 4 the predicted bond strength can be as much as 43% less than the actual bond strength.

4 NUMERICAL ANALYSIS

A finite element analysis was performed using three dimensional elastic brick elements to investigate the phenomenon illustrated by the experiments and to design a suitable configuration for the wrench which can eliminate its present defects. The whole assembly of the aluminium block and the bond wrench were analysed using the PAFEC finite element program.

Figure 5 shows the stress distribution at the front face and side faces of the aluminium calibration block, 10 mm away from the bottom clamp (the likely position of the top brick-mortar interface of a couplet). Both experimental and analytical results exhibit a similar pattern of stress distribution with a peak at the centre at the front face. However, the difference between the experimental and analytical values seems to be quite significant. This could be due to unknown clamping pressure applied during the experiments, which could not be simulated in the analytical model.

The stresses on the side faces agree closely with the measured non-linear stress pattern through the thickness of the specimen.

5 A NEW BOND WRENCH

Based on the analytical and test results it is obvious that a new configuration for the bond wrench is required which can produce the assumed linear stress distribution through the thickness at any section of the specimen, and hence uniform tensile stresses along the front face.

Through a series of preliminary analyses the basic configuration of the wrench was established as shown in Figure 6 (dimensions shown are in mm). Both gripping bars were supported at two points, unlike the present wrench. The bottom clamp was 22 mm away from the likely position of the mortar joint.

The calculated stresses through the thickness of the specimen and on the tensile face (Figure 7 & Figure 8) reveal that the ill effects of the wrench on the specimen can be minimised by a proper geometrical design of the wrench. The tensile stresses induced at the front face by the new wrench are much more uniform than those produced by the present wrench and agree well with the bending formula. Analysis revealed that supporting the gripping bars at the fifth points gives the most satisfactory results. Furthermore, it revealed that clamping the specimen just below the bottom interface is not appropriate. A minimum of 25 mm should be maintained between the bottom face of the mortar joint and the top edge of the bottom clamp to avoid undesirable stress concentration effects near the interface.

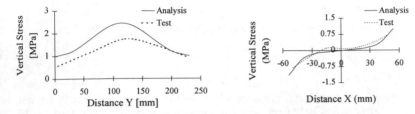

Figure 5. Stress Distribution at the Tensile Face and Side Faces

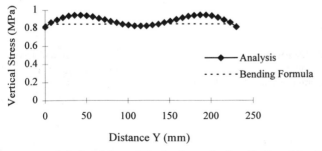

Figure 6. Plan View and Elevation of the New Wrench

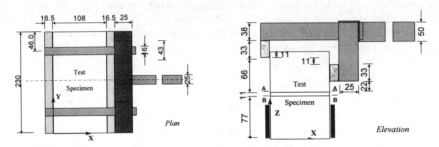

Figure 7. Calculated Stress Distribution at the Tensile Face (New Wrench)

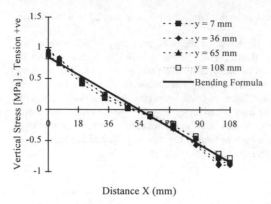

Distance X (mm)

Figure 8. Calculated Stress Distribution Through the Thickness (New Wrench)

6 CONCLUSION

The present bond wrench apparatus produces a non-uniform stress distribution in the test specimen that differs significantly from the assumed linear distribution. Most importantly, due to its geometrical shape, a localised stress peak occurs near the tensile face of the test specimen causing a premature bond failure. The work reported here is limited to a study on an aluminium block similar to the geometrical configuration of a brick couplet. However, subsequent studies (outside the scope of this paper) proved that the elastic moduli of brick and mortar do not significantly influence the stress pattern.

The new configuration proposed for the bond wrench appears to satisfy the requirement of inducing a stress distribution much closer to the assumed conditions.

7 REFERENCES

1. American Society for Testing and Materials, 1994, Standard Test Method for Measurement of Masonry Flexural Bond Strength, ASTM C1072-94.
2. Lawrence, S.J. & Page, A.W., 1995, Mortar Bond – A Major Research Program, Fourth Australasian Masonry Conference, UTS, Sydney, November 1995.
3. McGinley, W M, 1993, Bond Wrench Testing – Calibration Procedures and Proposed Apparatus and Testing Procedure Modifications, Proceedings of the Sixth North American Masonry Conference, June 1993, Philadelphia, Pennsylvania. pp.159-172.
4. Standards Association of Australia, 1988, SAA Masonry Code, AS 3700-1988.

8 ACKNOWLEDGMENT

The work described has benefited from financial support given by the Clay Brick and Paver Institute of Australia, the Concrete Masonry Association of Australia, the Cement and Concrete Association of Australia and the Australian Research Council.

The Mechanics of Structures and Materials, Grzebieta, Al-Mahaidi & Wilson (eds)
© *1997 Balkema, Rotterdam, ISBN 90 5410 900 9*

A study of bond mechanisms in masonry

H.O. Sugo & A.W. Page
Department of Civil, Surveying and Environmental Engineering, The University of Newcastle,
Callaghan, N.S.W., Australia

S.J. Lawrence
Division of Building, Construction and Engineering, CSIRO, Sydney, N.S.W., Australia

ABSTRACT: The bond strength developed by combining three different mortars with a typical dry pressed masonry clay unit was studied. The resultant mortar structures were evaluated using optical and SEM techniques in an attempt to interpret the observed bond strengths and failure modes. In addition the effective transport of mortar fluid was also monitored after the first hour of mortar/unit contact. For the brick/mortar combinations studied this information seems to correlate well with the transport of fines due to brick suction observed in the thin section studies and inturn reflect on the bond strength and resultant mode of failure.

1 INTRODUCTION

One of the main functions of mortar in masonry is to bind the masonry units together to impart structural integrity to the assemblage. The formation of an effective bond between the mortar and masonry units is thus an important aspect of masonry research as it has a direct influence on the behaviour of masonry structures. Despite this, the mechanisms of bond formation in masonry are not yet fully understood. It is known that bond strength is influenced by the effects of workmanship, the proportions of mortar ingredients (cement, lime and sand), and the use of plasticisers such as air entraining agents. These effects have been the subject of many "macro" studies using such techniques as the bond wrench. However, little work has been directed at establishing the basic bond forming mechanisms.

A major collaborative study on bond involving the University of Newcastle and CSIRO is in progress (Lawrence & Page 1995). This project is studying bond at both applied and fundamental levels with the ultimate aim of obtaining a more thorough understanding of the bonding processes. The relationship between bond strength and the corresponding brick/mortar interface properties is being studied for various unit/mortar combinations. As part of the fundamental study a small scale mechanical test has been developed to evaluate bond strength. Optical and scanning electron microscopy techniques are being employed in an attempt to identify the critical elements of bond formation.

The literature on brick/mortar bond has been reviewed by Goodwin & West (1982). Further studies on mortar structure/bond have been published (Chase 1984, Lawrence & Cao 1987, Marusin 1990, Lange et al 1996 and Sugo et al. 1996), and a recent discussion of mortar adhesion mechanisms (Robinson 1996). Significant advances in the understanding of fluid and mass transport effects have been made for when fresh mortar contacts masonry units (Groot 1993).

2 EXPERIMENTAL PROCEDURE

Three mortars were studied: a 1:0:6 (cement:lime:sand by volume) reference mortar; a general purpose 1:1:6 mortar typically used for clay masonry; and a 1:0:6 + 0.005 parts methyl cellulose by weight of cement (water retainer), as recommended for concrete and calcium silicate masonry.

Table-1. Particle Size Distribution of Beach Sand.

Sieve Size	2.36 mm	1.18 mm	600 µm	300 µm	150 µm	75 µm
% Passing	100	100	96	25	1	0

The 1:1:6 mortar for example consisted of 3.0kg Type GP cement, 1.3kg of hydrated lime and 22.9kg of washed and dried beach sand. The sieve analysis for the sand used is given in Table-1. The sand was a sharp dune sand with a poor size distribution. However due to its availability, it is commonly used by bricklayers in the Newcastle area.

A horizontal drum and paddle mixer was used to mix the mortar batches. All batches were mixed dry for two minutes, with water additions over the next three to four minutes. A further five minutes of mixing was allowed after the last water addition. Due to the different workability of each mortar the quantity of water added was left to the bricklayer's judgment, with the total mass of water added to each mix being recorded. Flow, cone penetration and gravimetric air content tests were carried out on the fresh mortars in accordance with AS2701 Sections 6-8.

The masonry units used were solid dry pressed fired clay units. The initial rate of absorption (IRA), 24 hour cold water absorption and 5 hour boil absorption were determined in accordance with AS1226 Sections 8 and 9 to be 3.4kg/m^2/min, 9% and 12% respectively.

For each brick/mortar combination twenty brickwork couplets were constructed. Every second couplet was separated 1 hour after construction and the mortar scraped clean off the bed surfaces. From the mass differences after the mortar had been dried to constant mass at 60°C and the batch proportions used during mixing, the mean water/cement ratio (w/c) within the bed joint were estimated. The 1 hour time period has been selected from previous studies as a reasonable time period to allow suction effects to be negligible (Groot 1993). The remaining ten couplets were covered with black plastic and allowed to cure in the laboratory for 7 days.

At the end of the 7 days the height of the couplets was reduced symmetrically about the bed joint to a net height of 100mm to facilitate coring. Five 25mm diameter cores were then drilled along the centre line of each couplet, perpendicular to the interface, using a custom made thin walled diamond coring drill. A precompression stress of 0.25MPa was applied to each core specimen during drilling to minimise any vibrational effects. Water was used as a coolant and to flush out debris during drilling.

Following coring the specimens were trimmed to an overall length of 75mm. They were air dried overnight at room temperature and then placed in desiccators over fresh silica gel to dry to constant mass. Threaded brass plates were glued at both ends to enable double ended ball joints to be attached, in order to apply a uniaxial load during testing. The length to diameter ratio for the specimens was 3:1 to ensure a reasonable uniform stress distribution across the mortar bed joint.

Uniaxial tension tests were then carried out on each specimen under a controlled rate of cross-head displacement of 0.5 mm/min. The maximum load as well as the failure mode were recorded for each specimen. Overall ten to twelve specimens of each mortar type were tested. After failure the fracture surfaces of the specimens were studied using an Olympus SZ6045 stereo microscope, with further investigations being carried out using a JEOL 840 Scanning Electron Microscope (SEM). Sawn, fractured and polished thin sections across the bed joints were also studied.

3 RESULTS

During mixing both the 1:0:6 and 1:0:6 + methyl cellulose mortars had an oversanded appearance. The three mortars also showed different workabilities. The 1:0:6 mortar was very difficult to work with, it segregated, bled, and slumped very easily making it difficult to obtain fully bedded joints. The 1:1:6 mortar had good coherency and workability was adequate given the nature of sand used. The effect of the water retainer was very noticeable with the workability of the 1:0:6 + methyl cellulose mortar being significantly improved over the reference 1:0:6 batch. It did not segregate or bleed and had similar workability to the 1:1:6, although its coherency was reduced. A summary of the fresh mortar properties together with the initial w/c and 1 hour w/c ratios are presented in Table-2 and the direct tension test results are presented in Table-3.

Table-2 Fresh Mortars Properties, Initial and 1 Hour W/C Ratio

Mortar Type	Flow %	Cone Penetration	Gravimetric Air Content (%)	Bulk Density (kg/m³)	Initial w/c Ratio	1 hour w/c Ratio
1:0:6	120	60	5	1980	1.84	0.17±0.05
1:1:6	85	70	4	1990	1.90	0.46±0.05
1:0:6 + methyl cellulose	100	70	11	1890	1.55	0.22±0.07

Table-3 Observed Bond Strengths and Failure Modes for 25mm ∅ Specimens Tested in Direct Tension

Couplet No.	1:0:6 Mortar		1:1:6 Mortar		1:0:6 + methyl cellulose Mortar	
	Bond Strength (MPa)	Failure Mode	Bond Strength (MPa)	Failure Mode	Bond Strength (MPa)	Failure Mode
1	0.931	M/C CJ	1.82	M/C CJ	0.885	M/C 2mm ↑ BI
	0.740	M/C 2mm ↑ BI	1.86	M/C CJ	0.933	M/C 2mm ↑ BI
	0.795	M/C 2mm ↑ BI	1.72	M/C CJ		
2	0.932	M/C 3mm ↓ TI	1.73	M/C CJ	0.842	M/C 2mm ↑ BI
	0.839	M/C 2mm ↓ TI	1.70	M/C 3mm ↓ TI	0.973	M/C 2mm ↑ BI
	0.955	M/C 4mm ↑ BI	1.44	M/C CJ	0.837	M/C 3mm ↑ BI
3	1.12	M/C 2mm ↑ BI	2.00	M/C CJ	0.672	M/C 2mm ↑ BI
	0.921	M/C 2mm ↓ TI	1.78	M/C CJ	0.786	M/C 2mm ↑ BI
			1.47	M/C 3mm ↓ TI	0.637	M/C 2mm ↑ BI
			1.32	I/A BI,V		
4	1.03	M/C 3mm ↓ TI	1.87	M/C CJ	0.317	M/C 2mm ↑ BI
	0.670	M/C 2mm ↓ TI	1.57	M/C 3mm ↓ TI	0.647	M/C 3mm ↑ BI
	0.290	M/C 2mm ↓ TI, V			0.730	M/C 2mm ↑ BI
Mean	0.84 MPa		1.69 MPa		0.75 MPa	
std dev.	0.22		0.20		0.18	

M/C= Mortar Cohesive Failure   I/A= Interface Adhesive Failure   ↑/↓ = Up/Down from
BI= Bottom Interface   TI= Top Interface   CJ= Centre of Joint   V= Void present

Optical examinations of the failed surfaces of the 1:0:6 and 1:0:6 + methyl cellulose mortars revealed relatively small volumes of paste bridging the sand grains together. The surface of the sand grains had a distinctly clean appearance with little to none cover of cement paste. The 1:1:6 mortar showed comparatively a much greater volume of paste linking the aggregate particles together. A coating of paste around the sand grains was also visible.

Observations of sections produced by fracture, perpendicular to the bed surface, revealed a layer of cementitious material along the interface approximately 1mm thick for both lime free mortars. There was no noticeable distinction in the layer thickness between these two mortars. Adjacent to this cement rich layer was a larger layer extending towards the centre of the joint with a sandy appearance. This migration of cementitious material also occurred in the 1:1:6 mortar but was not as pronounced. The fracture across the mortar joint in this case again revealed the ability of the 1:1:6 mortar to form well developed links between the sand grains.

The volume and distribution of paste within the bed joint can be more clearly observed by studying polished sections across the mortar joint at low magnifications in the SEM. All three mortars showed good contact along the brick interfaces. Figure-1 shows the bottom brick/mortar interface for the 1:0:6 mortar. Note the absence of paste away from the interface and the nature of contact between the sand particles in this region. The polished section of the 1:0:6 + methyl cellulose mortar showed similar features.

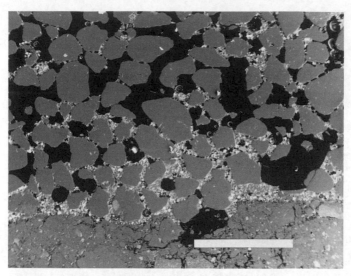

Figure-1 Bottom Brick-1:0:6 Mortar Interface, showing the segregation of paste along interface and adjacent cement depleted region, secondary electron image of polished section, length of bar = 1mm

Figure-2 Bottom Brick-1:1:6 Mortar Interface, showing reduced migration of fines and increased volume of paste, note the paste/aggregate links, secondary electron image of polished section, bar length = 1mm

The greater volume of paste and well developed bridges linking the sand grains together for the 1:1:6 mortar may be observed in Figure-2. The thin coating of paste around the sand grains, typically ranges from 20-100μm, may be observed at higher magnifications.

Energy Dispersive X-ray analyses were also carried on the polished sections of the three mortars to identify particles observed within the paste near the interface and in the bulk of the mortar. The

660

analysis essentially showed only two types of particles: roundish particles of about 10μm in diameter which were identified as C_2S; and smaller very angular particles of C_4AF. No C_3S particles could be identified.

The nature of the hydration products were also studied at higher magnifications using the SEM. This was done along the brick/mortar interfaces and across the mortar joint. Although there were some variations, the hydration products along the interface for the 1:0:6 mortar showed mixed regions of dense Type IV calcium silicate hydrates (CSH), fine platelets of $Ca(OH)_2$, and growths of rod like CSH, some 3-5μm long with diameters less than 0.05μm. Away from the interface the degree of hydration appeared to be reduced with mainly coarse Type I CSH forming.

The 1:0:6 + methyl cellulose showed similar structures to the 1:0:6 mortar along the interface. The presence of Type IV CSH and fine platelets of $Ca(OH)_2$ at the centre of the mortar joint indicated a more uniform level of hydration. Some acicular structures, possibly ettringite, were also observed as well as the relatively clean surface of the sand aggregate.

Some differences in the morphology of the CSH produced by the 1:1:6 mortar were visible. Along the interface the bulk of the CSH consisted of cement grains covered with Type I CSH, this was mixed with regions of Type IV CSH, much larger platelets of $Ca(OH)_2$ and areas of rod like CSH. Away from the interface the cement grains appeared to be less densely packed, regions of large $Ca(OH)_2$ crystals and voids with small $Ca(OH)_2$ crystals could be observed. The coating around the sand particles could also be observed to consist primarily of evenly spaced cement grains with Type I CSH products and mixed in some cases with irregular shaped crystals of $Ca(OH)_2$.

4 DISCUSSION

The properties of the fresh mortars shown in Table-2 reflect the effects of lime and methyl cellulose. For similar w/c ratios the 1:0:6 and 1:1:6 batches produced mortars with similar air contents and bulk densities. The effect of lime was shown by the reduced flow and the improvement in workability observed during brick laying. The methyl cellulose decreases the water demand of the mortar with a minor reduction in bulk density accompanied by an increase in the air content. This increase is probably the result of air entrapment rather than entrainment, as no small discrete spherical bubbles could be observed in the hardened mortar, which would be typical for bubbles produced by an air entraining agent.

The 1 hour w/c ratio is a semi-quantitative value of the retention of moisture by the mortar. This measurement represents the mean value across the mortar bed joint since steep moisture gradients have been shown to occur in the mortar adjacent to the interfaces (Groot 1993). These values confirm the capacity of brick suction to reduce the mortar w/c ratio. Initially brick suction causes the mortar paste to flow to the interface, being a combination of mortar fluids and fine solids. A stage is reached where migration of fines stops but fluid flow continues.

From the 1 hour w/c values it can be seen that the 1:0:6 and 1:0:6 + methyl cellulose mortars had low water retentivity. This is also shown by the amount of cementitious fines which have been transported to the interface as shown in the polished sections. These w/c ratios were lower than the theoretical value of 0.36 necessary for full hydration. Although full hydration would not be possible at 7 days, hydration differences were observed between the interface and the centre of the mortar joint. This was more noticeable in the 1:0:6 mortar and most likely reflects the capacity of the brick to act as a moisture reservoir to aid hydration (Groot 1993). The roles of methyl cellulose and lime in aiding hydration away from the interface were also evident. The fine platelets of $Ca(OH)_2$ observed in the lime free mortars originate from the hydration of the C_3S and C_2S.

The 1:1:6 mortar had a significantly higher 1 hour w/c ratio. This is beneficial not only for hydration purposes but also in minimising the excessive depletion of paste by brick suction and increasing the volume of paste available to provide mortar bond and cohesion. This was shown by the well developed links between the aggregate particles. The increased paste volume is partly due to the additional volume of lime with a more significant component resulting from the increased w/c ratio (2-3 times greater) which has a further multiplying effect (≈ 3) when the relative densities of water and cement are taken into account. Thus it appears that the distribution of cementitious paste across the bed joint and bond strength may be influenced by the rheology of the paste as well as the stage where paste transport stops and fluid only flow continues.

The observed bond strengths in this study were mortar cohesive strengths rather than interface strengths since the predominant failure modes were within the mortar. The observations of the resultant mortar structures allow the bond strengths and failure modes presented in Table-3 to be interpreted. Failure for the 1:0:6 and 1:0:6 + methyl cellulose mortars occurred in cement depleted zones of lower cohesive strength adjacent to the brick interfaces. The 1:1:6 mortar, not having been as severely affected by the brick suction effects, fails more towards the centre of the bed joint. The greater bond strength shown by the 1:1:6 mortar can be attributed to the more substantial paste/aggregate links offered by the lime mortar. The lime mortar also promotes a cementitious cover to form around the sand particles which may also contribute to the strength.

The sharpness of the sand used together with the predominantly single size distribution and the lack of fines directly influenced the workability of the mortars. The 1:0:6 mortar was more sensitive to these sand properties and as a result had a high water demand and produced a mortar with poor workability. This made laying fully bedded joints difficult with a substantial reduction in the bonded area around the perimeter of the couplet. From a practical perspective, this effect could significantly influence the strength obtained by a bond wrench test which tests the entire couplet joint. Using the selective coring technique, where only the central part of the couplet was used, overcame these defects and allowed specimens which were representative of the true behaviour to be tested.

5 CONCLUSIONS

For the masonry units used in this study, brick suction forces were capable of causing transport of mortar fluids and cementitious material to the masonry unit interface. This process increases the cohesive strength of the mortar by reducing the initially high w/c ratio and contributes to the bond forming mechanism by providing good contact of paste along the brick/mortar interfaces. If excessive transport of cementitious material to the interface results, a zone of lower cohesive strength within the mortar adjacent to the interface is formed. This was observed for the 1:0:6 and 1:0:6 + methyl cellulose mortars and was consistent with the failures modes of these mortars. The addition of lime to the mortar increased the water retentivity, promoted a cover of cementitious paste around the sand aggregate and reduced the tendency for a cement depleted layer to be formed. Some minor differences of the microconstituents formed by the three mortars were also observed but these appear to be of secondary importance.

6 REFERENCES

Chase, G.W. Investigations of the Interface Between Brick and Mortar, *The Masonry Society Journal, Vol. 3, No. 2, 1984, pp. T1-T9.*

Goodwin, J.F. & West, W.H. A Review of the Literature on Brick/Mortar Bond. *Proc. Of the British Ceramic Society, Vol. 30, No. 23, 1982, pp. 23-37.*

Groot, C.J.W.P Effects of Water on Mortar-Brick Bond. Delft University of Technology, Delft, The Netherlands, 1993.

Lange, D.A., DeFord, H.D. & Ahmed, A. 1996. Microstructure and Mechanisms of Bond in Masonry. *Proc. 7th North American Masonry Conference, University of Notre Dame, South Bend, Indiana, June, 1996, pp. 167-174.*

Lawrence, S.J. & Cao, H.T. 1987. An Experimental Study of the Interface Between Brick and Mortar. *Proc. of the 4th North American Masonry Conference, Los Angeles, August, 1987.*

Lawrence, S.J. & Page, A.W. 1995. Mortar Bond- A Major Research Program. *Proc. of the 4th Australasian Masonry Conference, University of Technology, Sydney, November, 1995 pp. 31-37.*

Marusin, S.L. Investigations of Shale Brick Interface with Cement-Lime and Polymer Modified Mortars. *J. of the American Ceramic Society, Vol. 73, No. 8, 1990, pp. 2301-2308.*

Robinson, G.C. 1996. Adhesion Mechanisms in Masonry. *The American Ceramic Society Bulletin, Vol. 75, No.2, 1996, pp.81-86.*

Sugo, H.O., Page, A.W. & Lawrence S. J. 1996. Influence of the Macro and Micro Constituents of Air Entrained Mortars on Masonry Bond Strength. *Proc. 7th North American Masonry Conference, University of Notre Dame, South Bend, Indiana, June, 1996, pp. 230-241.*

The Mechanics of Structures and Materials, Grzebieta, Al-Mahaidi & Wilson (eds)
© *1997 Balkema, Rotterdam, ISBN 90 5410 900 9*

Stress-strain relations for grouted concrete masonry under cyclic compression

M. Dhanasekar
Department of Civil Engineering, Central Queensland University, Qld, Australia
R. E. Loov & N. G. Shrive
Department of Civil Engineering, University of Calgary, Canada

ABSTRACT: Theoretical and numerical investigations on the response of masonry to earthquake or cyclic loading are inhibited by the non availability of a suitable material model. In this paper we address this important issue and present an appropriate and easy to implement material model for masonry subjected to cyclic compressive loading. The stress-strain response of grouted concrete masonry with and without lateral confining reinforcing steel to grout fill under cyclic compression was investigated. To confine the grout fill either a fine wire mesh commonly known as chicken mesh or the welded wire mesh fabric was used. Some specimens were tested under monotonic compression and the others were tested under either one, or five cycles of loading for each load level. The stress-strain curves are broken into envelope, common point, unloading and reloading curves. Each component of the curve was fitted with an algebraic equation which are reported in the paper.

1. INTRODUCTION

The response of masonry to earthquake loading has been widely studied by several researchers through expensive experimental investigation. Theoretical and numerical investigations on the response of masonry to earthquake or cyclic loading are inhibited by the non availability of a suitable material model.

Concrete masonry is usually grout filled at selected cores. Vertical reinforcing bars without lateral confining steel are placed within the grouted cores to resist out-of-plane flexure induced by the lateral loading. These bars are regarded ineffective in resisting compression and the compressive forces are resisted by the masonry shell and the grout in a combined mechanism. Under compressive loading the grout cracks and exert outward lateral pressure onto the masonry shell which subsequently fails prematurely without realising its ultimate compression capacity. The compressive strength of grouted concrete masonry is significantly enhanced if confining lateral steel is placed within the grout core. This paper reports the stress-strain response of grouted concrete masonry with and without confining steel under cyclic compression.

2. RESEARCH SIGNIFICANCE AND METHODOLOGY

Lightly reinforced (known as partially reinforced) masonry is popular in countries with moderate levels of seismic activity such as Australia and Canada. Unfortunately, the response of this form of masonry to earthquake is not fully defined in the literature even though uniaxial compressive strength and stiffness of clay block masonry have been reported by Kumar and

Dhanasekar(1995) and the strength properties of grouted concrete masonry has been studied by several investigators including Page et al(1991) and Scrivener and Baker(1988). We have carried out experiments on grouted and ungrouted concrete masonry prisms subjected to cyclic compressive loading. Stress-strain relations capable of defining the response under any combination of loading-unloading-reloading path of ungrouted (hollow) concrete masonry prisms to cyclic loading has already been reported (Dhanasekar et al, 1997). In this paper we report similar stress-strain relations for the grouted concrete masonry.

2.1 Stress-strain response of concrete and masonry

Concrete and masonry are similar materials that exhibit significant nonlinearity in their stress-strain curves. Desayi and Krishnan(1964) proposed an equation for concrete subjected to monotonic compression. The equation relates the stress (σ) in terms of strain (ε), strain at peak stress(ε_0), and a constant (E) as shown in Eqn. 1.

$$\sigma = \frac{E\,\varepsilon}{1 + \left(\dfrac{\varepsilon}{\varepsilon_0}\right)} \qquad \qquad (1)$$

Kent and Park(1971) proposed a model that is easy for numerical implementation. Monotonic and cyclic envelope curves have typically an ascending part (represented by a parabola) and a descending part (represented by a straight line) as given in Eqns. 2a & 2b. The common point curve is related to the envelope curve through a reduction factor of 0.9 for both stresses and strains. The unloading paths are parabolic while the reloading paths are linear up to common points and nonlinear beyond common points.

$$\sigma = \sigma_{peak}\left[1 - Z\left[\varepsilon - 0.002\right]\right] \qquad \qquad (2a)$$

where

$$Z = \frac{0.5}{\dfrac{3 + 0.29\,\sigma_{peak}}{145\,\sigma_{peak} - 1000} - 0.002} \qquad \qquad (2b)$$

An exponential curve was fitted to the envelope and common points data obtained for the solid clay brick masonry by Naraine and Sinha(1989,1991) as shown in Eqn. 3 in which ε and σ are normalised strain and stress; and α and β are constants. The unloading and reloading curves were evaluated using a graphical method by them.

$$\sigma = \frac{\beta \cdot \varepsilon \cdot e^{(1 + \varepsilon/\alpha)}}{\alpha} \qquad \qquad (3)$$

Loov(1991) defined a single curve given in Eqn. 4 that fits the data of several concretes of different strength using only two constants B and n. The same equation (in a modified form) is used to fit the envelope and common point curves for hollow masonry reported in this paper.

$$\sigma = \sigma_{peak}\left(\frac{(1 + B + (1/(n-1)))\varepsilon}{1 + B\varepsilon + (1/(n-1))\varepsilon^n}\right) \qquad \qquad (4)$$

664

3. EXPERIMENTAL INVESTIGATION

Grouted masonry fails due to a combined mechanism caused by the applied vertical compressive loading and internal pressure at grouted cores due to the lateral expansion of the grout. The compressive strength of grouted masonry could be enhanced by minimising the lateral expansion of the grout. To minimise the lateral expansion of the grout Khalaf et al (1994) used rectangular hoops and Kumar and Dhanasekar(1995) used triangular ties. In this research we have used fine wire mesh (FWM) commonly known as chicken mesh and welded wire mesh (WWM) fabric commonly used in reinforced concrete slab construction. FWM (1mm gauge wire on a 50mm X 50mm grid) was rolled into cylindrical shape and inserted into the masonry core prior to grouting. WWM (5mm gauge wires on a 75mm X 75mm grid) was bent into a square prism shape and inserted into the masonry core prior to grouting.

Thirty-three three-high prisms were constructed and tested at the Structures laboratory of the University of Calgary, Canada. Six masonry units and six grout cylinders were also tested. The specimens were constructed from concrete hollow block units of gross dimension 390mm X 190mm X 190mm and ready mix mortar of Type S by a professional bricklayer. They were moist cured under polythene cover for seven days and then grouted. Grout was mixed at site (cement and pea gravel of 350kg/m^3 each and water-cement ratio of 0.79). Specimens were cured for a further 28 days and then tested.

Two LVDTs were placed at 300mm gauge length on opposite sides of the prism to monitor the axial deformation.

An MTS universal testing machine of 2000kN capacity was used for the testing. At least one specimen was tested under monotonic compression while other specimens were tested under cyclic compression. Results of twenty grouted prisms and six grout cylinders are reported in this paper. The specimens were designated using a format of XXXX-Y#n in which "XXX" defines the type of test specimen (GUP for grouted unconfined prisms; GMP for grouted mesh confined prisms; GCP for grouted cage confined prisms and CYL for grout cylinders), "x" stands for the specimen number (1 to 8), "Y" stands for the type of loading (M for monotonic and C for cyclic) and "n" stands for the number of cycles at each level of loading (0 for monotonic loading; and 1 or 5 for cyclic loading).

The prisms failed by uniformly distributed longitudinal cracking that ran parallel to the direction of loading and subsequent outward bulging of the masonry shells. All specimens exhibited nonlinear responses from the beginning of loading as shown in Fig. 1.

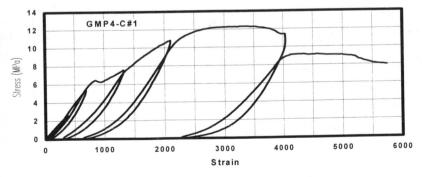

Figure 1. Typical stress-strain curves evaluated by the experiment

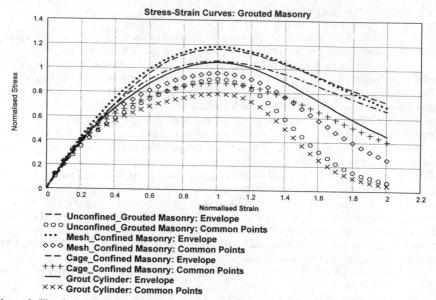

Figure 2 Fitted envelope and common point curves for masonry (non-dimensional plane)

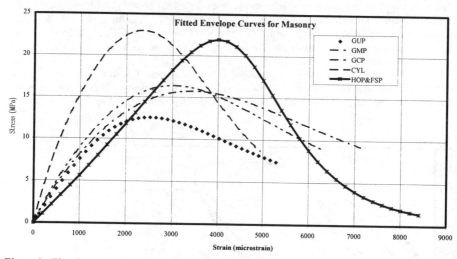

Figure 3. Fitted common points curve in the dimensional stress-strain plane

4.0 STRESS-STRAIN EQUATIONS

Stress-strain curves were normalised using the mean peak stress (σ_0) and mean strain corresponding to peak stress (ε_0). The following σ_0 and ε_0 values were used: for GUP prisms 10.8MPa and 2518 microstrain; for the GMP prisms 13.9MPa and 3024 microstrain; for GCP prisms 14.9MPa and 3433 microstrain; and for CYL cylinders 21.9MPa and 2351 microstrain. The normalised cyclic stress-strain curves were scanned by a computer program and broken into reloading and unloading curves. The program also monitored the normalised strain data to decide on the onset of load reversal. From the unloading and reloading data, the program further evaluated common points which were interpolated from four very closely spaced unloading and reloading data points. The peaks of each reloading data sets were stored as envelope data points. The four data sets (namely, the envelope, common points, unloading and reloading curves) were plotted to ensure their integrity.

Each data set was analysed separately for best fit using MATHCAD 6.0. Generally two constants were fitted using a nonlinear GENFIT function. The "guess' values input initially were adjusted until they matched the GENFIT calculated values.

Envelope and common point curves were fitted with the formulation shown in Eqn. 5 which is an alternate form of Eqn. 4. This equation forces y to be zero but allows (dy/dx) to take on any value at x = 0; and (dy/dx) to be zero and y to be y_{max} at x = 1. In this equation x represents normalised strain ($\varepsilon/\varepsilon_0$) and y represents normalised stress (σ/σ_0). (u_0 and u_1 are constants determine for each curve separately).

$$y = y_{max}\left(\frac{(1 + u_0(1 + u_1))x}{u_0(1 + u_1 x) + x^{(u_0 + 1)}}\right) \quad \dots \dots (5)$$

The envelope and common points equations were then plotted for each group of test specimens (GUP, GMP, GCP, & CYL) separately to assess the consistency in the prediction. All the fitted envelope and common point curves are shown in Fig. 2. As the curves are in normalised form, the effect of grout confinement is not obvious from Fig. 2. These curves are transformed back to the dimensional stress-strain plane to view the practical significance of the grout confinement. The transformed common points curve is shown in Fig. 3. Envelope curves appear similar to the common points curve and, therefore, are not presented separately.

The benefit of lateral reinforcement to the response of grouted concrete masonry is illustrated in Fig. 3. The unconfined masonry exhibits ultimate strain levels very similar to that of the grout cylinder, where as the FWM and WWM confinement increase the ultimate strain significantly which reflect increase in compressive strain ductility. The strength of FWM and WWM masonry is also increased.

The unloading curves were fitted with Eqn. 6. There were two constants (u_0 and u_1) fitted for each of the curve. On an average u_0 was close to 1.0 and u_1 was close to 2.0 suggesting the possibility of fitting a simple parabola, $y = x^2$. The reloading curve was fitted with a cubic equation shown in Eqn. 7. The constants u_0 and u_1 in the equation corresponds to the slopes at x = 0 and x = 1 respectively.

$$y = u_0 x^{u_1} \quad \dots \dots (6)$$

$$y = u_0 x + (3 - 2u_0 - u_1)x^2 + (u_0 + u_1 - 2)x^3 \quad \dots \dots (7)$$

5.0 SUMMARY

Equations for the Stress-Strain curves of grouted concrete masonry under cyclic compression are defined. The envelope and common points curve exhibit the effect of grout confinement. Masonry with confined grout exhibit substantially higher ultimate strain and increased compressive strength. Unloading curves could be approximated by a simple parabolic formulation while the reloading curves required cubic equations for better representation.

6.0 ACKNOWLEDGEMENTS

The support of the James Goldston Faculty of Engineering, Central Queensland University, Australia, the Department of Civil Engineering, University of Calgary, Canada and the Natural Sciences and Engineering Research Council of Canada is gratefully acknowledged.

7.0 REFERENCES

Desayi, P. & Krishnan, S., 1964. Equation for the Stress-Strain Curve of Concrete. *J. ACI.*, 345-350.

Dhanasekar, M., Loov, R.E., McCullough, D., & Shrive, N.G., 1997. Stress-Strain Relations for Hollow Masonry under Cyclic Compression. *Proc. 11th IB²MaC*, Shanghai, (in Press).

Kent, D.C. & Park, R, 1971. Flexural Members with Confined Concrete. *J. of Struct. Dvn., ASCE*, v. 97, 1969-1970.

Khalaf, F.M., Hendry, A.W., & Fairbairn, D.R., 1994. Study of the Compressive Strength of Blockwork Masonry. *ACI Struc. Journal.* 91(4), 367-374.

Kumar, M. & Dhanasekar, M., 1995. Effect of Lateral Reinforcement Detailing on the Strength and Stiffness of Reinforced Clay Block Masonry. Proc. 5th EASEC, GoldCoast, 1005-1010.

Loov, R.E., 1991. A General Stress-Strain Curve for Concrete. *Conf. CSCE*, 302-311.

Naraine, K., & Sinha, S., 1989. Behaviour of Brick Masonry under Cyclic Compressive Loading. *J. of Const. Engg. & Mgmt., ASCE*, 115(2), 1432-1445.

Naraine, K. & Sinha, S., 1991. Model for Cyclic Compressive Behaviour of Brick Masonry. *ACI Struct. Journal*, 88, 603-609.

Page, A.W., Simundic, G., & Xie, H., 1991. A Study on the Relationship between Unit, Prism and Wall Strength for Hollow Masonry in Compression. *Proc. 9th IB²MaC*, Berlin, 236-243.

Paulay, T. & Priestley, M.J.N., 1992. *Seismic Design of Reinforced Concrete and Masonry Buildings*. NY, John Wiley & Sons Inc.

Scrivener, J.C. & Baker, L., 1986. Factors Influencing Grouted Masonry Prism Compressive Strength. *Proc. 8th IB²MaC*, Dublin, 875-883.

Timber and masonry (timber)

The Mechanics of Structures and Materials, Grzebieta, Al-Mahaidi & Wilson (eds)
© 1997 Balkema, Rotterdam, ISBN 90 5410 900 9

Limit states design for timber I-beams with trellis webs

C.C.Rodger & C.Q.Li
Department of Civil Engineering, Monash University, Vic., Australia

ABSTRACT: This paper is concerned with the limit states design of timber hardwood I-beams with trellis webs. Timber I-beams with trellis webs have the potential of achieving substantial market share through cost-effective construction and design provided sufficient strength and serviceability limits are guaranteed. However a design methodology of this kind does not currently exist. In this paper, an overview of structural reliability and its applications to timber structures are presented together with a design procedure for timber I-beams with trellis webs. Theoretical models and experimental results are used to quantify the design parameters and verify the failure modes. An example is given to illustrate the design methodology.

1 INTRODUCTION

The objective of this paper is to present a procedure used in the development of a limit states design for trellis webbed I-beams, based on structural reliability theory. Timber I-beams with trellis webs are floor joists encompassing the use of sawmill off cuts in a load bearing capacity. Desirable aspects of constructing the web out of "shorts" include the increased cross-sectional properties, the availability of larger spanning members and the economic benefit offered over conventional joisting arrangements.

The pending release of a limit states design code for timber structures will provide design procedures for structural members of sawn timber. However the unique configuration of the timber I-beams with trellis requires the use of a different design procedure, taking into account its unique failure modes. Structural reliability is a probabilistic measure of the violation of a limit state, typically strength or serviceability. The major advantage of the implementation of structural reliability is that structural performance can be measured in terms of the probability of failure. In accordance with the Limit States Design Code, AS1793 (SAA, 1975) the limit states design (LSD) criteria may be defined as

$$R^* \geq S^* \tag{1}$$

where R^* and S^* represent the design strength and the design load effect respectively. The design strength and load effect can be determined as a function of the characteristic resistance, R_K, the characteristic load effect, S_K, and the resistance and load factors, γ_R and γ_s, respectively. The design load effect and resistance then become

$$S^* = \gamma_s S_k \tag{2}$$

$$R^* = \gamma_R R_k \tag{3}$$

Typically the characteristic values are taken as extreme, ie. the characteristic resistance is taken as the fifth percentile value while the characteristic load is taken as the ninety-fifth percentile value. The determination of the resistance factor, γ_R, is particularly important in the case of trellis webbed I-beams. γ_R must not only compensate for the orthotropic nature of the timber but also include the duration of load and degradation effects as well as the quality of workmanship apparent in the construction of a composite structure. In timber structure design "k" factors are used to account for non-standard conditions. Typical load models for both sawn timber and composite members including trellis webbed beams are consistent with those applied to steel and concrete structures and are detailed in texts such as Melchers (1987) and Pham (1985).

2 LIMIT STATES DESIGN FOR TIMBER

To provide a "soft" conversion from working stress design (WSD) to limit states design it is important that experienced timber designers can obtain similar results without going through an extensive training course. The introduction of a LSD code for timber, although utilising different concepts in design, can be expressed in a similar way using "k" factors. For instance in AS1720.1 (SAA, 1988), the permissible design stress in bending is determined relative to a basic working stress, F_b' by:

$$F_b = k_1 k_2 k_4 k_5 k_6 k_8 k_{11} k_{12} F_b'$$ (4)

while the LSD code would be represented by the moment as a function of the characteristic strength, R_k', and the section modulus, Z, and a capacity (resistance) factor, ϕ :

$$M_d = \phi k_1 k_2 k_4 k_5 k_8 k_{11} k_{12} Z R_k'$$ (5)

The "k" factors are consistent with those prescribed in AS1720.1 (SAA, 1988) although the duration of load factor, k_1, must be factored in the LSD code. Detailed analysis and application of LSD on sawn timber have been examined by Carson and Leicester (1994) while the implication of LSD on WSD timber design codes is discussed in detail by Webster (1984).

The determination of characteristic values for timber products is encompassed in the timber-in-grade strength and stiffness evaluation standard, AS/NZS 4063 (SAA, 1992) where the characteristic strength of a timber product is given as

$$R_k = \left[1 - \left(\frac{2.7 V_R}{\sqrt{n}} \right) \right] R_{0.05}$$ (6)

where V_R is the coefficient of variation, n is the sample size of the data and $R_{0.05}$ represents the fifth percentile value of the data. The characteristic strength is then converted to LSD format by the normalised characteristic strength, $R_{k,norm}$, which is also defined in the standard as a function of the coefficient of variation and a LSD capacity factor, ϕ:

$$R_{k,norm} = \frac{1.35}{\phi} \frac{R_K}{(1.3 + 0.7 V_R)}$$ (7)

3 STRUCTURAL RELIABILITY

To understand the implications of load and resistance factor design the application of probability and structural reliability must be examined. The reliability of a structural system is expressed in terms of the probability of the structural resistance being exceeded by the loading imposed on the system. Thus the probability of failure is the probability of a limit state violation in accordance with the predefined LSD criteria which can be expressed as follows

$$p_f = P[G(R,S) \leq 0]$$
$$= \Phi(-\beta) \tag{8}$$

where G() is the limit state function of the structure, $\Phi()$ is standard normal distribution function and β is known as safety index, which is commonly defined as

$$\beta = \frac{\mu_G}{\sigma_G} \tag{9a}$$

when R and S are of normal distribution and the limit states function is in the linear form, $G(R,S) = R - S$, the safety index can be obtained as

$$\beta = \frac{\mu_R - \mu_S}{\sqrt{\left(\mu_R V_R\right)^2 + \left(\mu_S V_S\right)^2}} \tag{9b}$$

when R and S are of log-normal distribution the limit state function becomes $G(R,S) = R / S$ and β is obtained by

$$\beta = \frac{\ln\left(\mu_R / \mu_S\right)}{\sqrt{\left(V_R^2 + V_S^2\right)}} \tag{9c}$$

Further discussion on the safety index in relation to log-normal variables is beyond the scope of this paper but is available in Melchers (1987).

4   LOAD AND RESISTANCE FACTOR DESIGN

Reliability based limit states design is generally expressed in a load and resistance factor design format of which the general form is

$$\phi R_n \geq \sum_{k=1}^{i} \gamma_k S_{Kn} \tag{10}$$

where the subscript n denotes nominal values. Compared with equations (2) and (3) ϕ replaces γ_R. In equation (10) i is the number of load combinations while γ_K is the load factor.

The derivation of load and resistance factors, γ_K and ϕ is based on first-order second-moment structural reliability theory. This involves the transformation of design (random) variables to standard normal variates. In the standard normal space, there exists a design point, corresponding to the safety index , which can be expressed as

$$x_i^* = \mu_{xi}\left(1 - \alpha_i \beta_c V_{Xi}\right) \tag{11}$$

where x_i are design parameters (random variables), the directional cosines, α_i can be calculated for a given limit state function with respect to the design variable x_i. The coefficient of variation is determined from statistical studies of the design variables.

By converting the mean to its characteristic value it can be shown

$$x_i^* = \frac{1 - \alpha_i \beta_c V_{Xi}}{1 \pm k_{xi} V_{Xi}} X_{ki} \tag{12}$$

where k_{xi} is a coefficient corresponding to the characteristic fractal of the normal distribution, ("-" is for resistance and "+" is for loads) and X_{ki} represents the basic variables used in design. The expression $(1 \pm k_{Xi}V_{Xi})$ for load and resistance values represents lower and upper percentile values respectively.

For a linear limit state function the load and resistance factors, γ_i and ϕ, can be determined by replacing i with R and S respectively. The following relation is therefore obtained

$$\gamma_s = \frac{1 - \alpha_s \beta_o V_S}{1 + k_s V_S}$$ (13a)

$$\phi = \frac{1 - \alpha_R \beta_o V_R}{1 - k_R V_R}$$ (13b)

These are the load and resistance factors used in LSD.

5   TIMBER I-BEAMS WITH TRELLIS WEBS

Timber I-beams with trellis webs (refer figure 1) utilise timber "shorts" as the web with hardwood timber flanges being used in a similar fashion to Plywood webbed I and box beams. The prescribed use as floor joists neglects the ability to withstand concentrated point loads, thus the maximum force applied to the member in service (bending moment) is resisted by the flanges. The economic benefit offered by trellis webbed beams together with advantages such as (i) increased sectional properties through the use of an I section for a given volume of timber, (ii) larger spanning members, (iii) allowance for the implementation of services makes the trellis webbed I beam a marketable product.

A finite element analysis was completed (Rodger and Li, 1996) with common floor loadings being applied to determine the effect of the alteration of the parameters l_1 and l_2 on the strength and serviceability of trellis webbed I-beams. Correlations for strength (shear stress) and serviceability (mid-span deflection) were developed relative to design parameters. The finite element analysis together with tests currently being undertaken revealed the mode of failure to be the shear stress in the glue line at the end bay of trellis webbed I-beams. Although it is common for sawn timber members to fail due to serviceability the trellis webbed members displayed shear strength failure, hence the design criterion ie., the limit state, is based on the load effect for shear and not serviceability.

5.1   DESIGN FORMAT FOR TRELLIS WEBBED I-BEAMS

The resistance of a trellis webbed I-beam depends on the resistance of its components and connections. As previously discussed the ultimate capacity of the member is deemed to be the shear capacity at the glue line. According to FEA and experiments the shear stress in the glue line is a function of the parameter l_2, the distance between the web sections. The load effect is therefore the shear stress in the web-flange connection at the end bay. Numerical analysis of a coarse meshed finite element model (Rodger and Li, 1996) displayed a linear correlation between the parameter l_2 and the shear stress along the glue line with a coefficient of correlation of 0.98. The load effect can consequently be expressed as

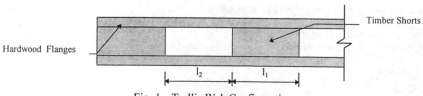

Fig. 1   Trellis Web Configuration

$$\tau_S = S_K = \frac{(l_2 + b)w.s}{d} \tag{14}$$

where τ_S is the shear stress in MPa, the load effect as a function of the load, w in N/mm^2, s the tributary width in mm, and the length parameters b (in mm) and d (in mm^2), to be determined from experimental and/or theoretical modelling. Theoretical analysis for beams of a 3 metre span (Rodger and Li, 1996) revealed these parameters to be 150 and 605x10^3 respectively. For the following example a joist spacing of 1000 mm is adapted together with floor loading conditions prescribed in AS1170.1.

Thus a design limit state of a trellis webbed I-beam, based on a shear failure mode, becomes

$$G = \tau_R - \frac{(l_2 + 150)1000}{605 \times 10^3}.w \tag{15}$$

So that the design format is

$$\phi\tau_R \geq \gamma_S \frac{(l_2 + 150)w}{605} \tag{16}$$

5.2   CHARACTERISTIC VALUES

The shear resistance offered by the structure is to be determined by the results from standard tests being carried out as prescribed in AS/NZS 4063 (1992). The data obtained from these tests, including shear stress, load capacity and MOE, are then statistically evaluated to determine the mean, standard deviation, and fifth percentile values. Due to the lack of test data, values for sawn timber of F17 grade hardwood (from AS1720.1) are employed as the characteristic resistance of trellis webbed I-beams for evaluation purposes. A coefficient of variation similar to that of other composite timber products is also used. Design values for the characteristic strength and coefficient of variation are $R_K=1.70$ and $V_R=0.15$ respectively. By using equation (7) the normalised characteristic strength can be obtained to be $R_{k,norm}=1.81$ MPa.

The load values employed are those prescribed by the Australian loading Code, AS1170.1. For design calculations action of the dead load is negligible compared to that produced under live load action. The nominal value for live load is, $L_{nom}=1.5$ kPa (domestic flooring), with an assumed coefficient of variation of, $V_w=0.25$.

5.3   LOAD AND RESISTANCE FACTORS

To determine the load and resistance factors the safety index has to be known, which can be determined as a function of the parameter l_2. The 'target' safety index, selected by either engineering judgement or code compliance, is used to determine the optimal distance between webs, ie. l_2. Procedures exist (Li, 1996) for the optimisation of a structural system relative to costs including the cost of failure, cost of construction and repair. However due to the lack of data the target safety index is taken as a typical value of 2.5.

By substituting in predetermined values into equation (9) a plot of safety index against l_2 can be developed (refer figure 2). By recommending a safety index of 2.5 a distance between the web sections of approximately 250 mm is recommended.

The load and resistance factors (partial factors) can now be determined in accordance with equations (13a) and (13b) where directional cosines, α_R and α_S of the limit state function expressed in equation (15) are then calculated for a safety index of 2.5. The prescribed numerical example can be evaluated for load and resistance factors

$$\phi = (1 - \alpha_R\beta_o V_R)\left(\frac{\mu_R}{R_n}\right)$$

$$\therefore \phi = \left(1 - 0.568(2.5)(0.15)\right)(1.13) = 0.89$$

$$\gamma_S = \left(1 - \alpha_S \beta_o V_S\right)\left(\frac{\mu_w}{W_n}\right)$$

$$\therefore \gamma_S = \left(1 - (-0.823)(2.5)(0.30)\right)(0.75) = 1.21$$

Thus for a trellis webbed I beam with l_2=250 mm the resistance and load factors are ϕ=0.89 and γ_S=1.21 respectively.

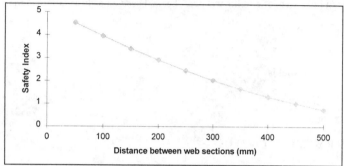

Fig. 2   Safety index for trellis webbed I-beams

6   CONCLUSION

A limit states design procedure has been developed for timber I-beams with trellis webs. The adaptation of the current WSD timber engineering code to LSD has been discussed together with structural reliability concepts used in the development of load and resistance factors for normally distributed loading and resistance models. The failure mode of timber I-beams with trellis webs was analysed using finite element analysis and adapted for the load effect while resistance values for F17 grade timber were employed. A numerical example of the LSD of Timber I-beams with trellis webs has been presented for illustration.

7   BIBLIOGRAPHY

Carson, J. and Leicester, R, 1994, 'Australian Standards for structural timber', *Pacific Timber Engineering Conference 1994*, vol.2, pp. 207-212.

Li, C.Q. (1996), 'Limit States Design for Timber Structures', Proc. 25th Forest Products Research Conference, 18 - 21 Nov., 1996, Melbourne.

Melchers, R.E. (1987) *Structural Reliability Analysis and Prediction*, John Wiley and Sons, Chichester, U.K.

Pham, L (1985), 'Load Combinations and Probabilistic Load Models for Limit State Code', Transactions of The Institution of Engineers, Australia. Civil Engineering, Vol. CE27, No. 1, February 1985.

Rodger, C.C and Li, C.Q. (1996), 'Finite Element Analysis of Hardwood I-Beams with Trellis Webs', Proc. 25th Forest Products Research Conference, 18 - 21 Nov., 1996, Melbourne.

Standards Association of Australia 1975, AS1793.1-1975, *Limit State Design Method*, Standards Australia, Homebush, NSW.

Standards Association of Australia 1988, AS1720.1-1988, *SAA Timber Structures Code, Part 1: Design Methods*, Standards Australia, Homebush, NSW.

Standards Association of Australia 1989, AS1170.1-1989, *SAA Loading Code, Part 1: Dead and live loads and load combinations*, Standards Australia, Homebush, NSW.

Standards Association of Australia 1992, AS/NZS 4063:1992, Timber-Stress-graded-In-grade strength and stiffness, Standards Australia, Homebush, NSW.

Webster, J.A, 1984, 'Some Practical Issues in the Development of a limit states design code for timber structures', *Pacific Timber Engineering Conference 1984*, Volume 2, pp. 923-936.

The Mechanics of Structures and Materials, Grzebieta, Al-Mahaidi & Wilson (eds)
© 1997 Balkema, Rotterdam, ISBN 90 5410 900 9

A review of the design of timber pole houses

Frank Bullen & David J.Wood
School of Civil Engineering, Queensland University of Technology, Brisbane, Qld, Australia

ABSTRACT : As with any structure, design of timber pole frame houses concerns transfer of wind loads through structural elements into the supporting ground. The study described here considered the aspects of pole embedment depth, borehole backfill material, and the development of lateral resistance in a pole house during construction. Good pole embedment can help control both the lateral and rotational response of the pole house to lateral loads. Field studies on pole embedment used four embedment depths and three methods of backfill. Individual pole load/deflection behaviour was compared to existing theoretical methods used for predicting deflection. The response of the house to simulated wind loads was investigated by subjecting the pole house to lateral loading at various stages of construction. As expected the structure became stiffer as the construction progressed.

1.   INTRODUCTION

Although the Australian pole frame building industry is well established, design and construction techniques are generally historically derived and appear to lack good theoretical basis. There has been very little detailed evaluation of pole houses to investigate their response to external loads such as wind forces. A traditional timber framed house with trussed roof and ceiling diaphragm, sheds wind loads to the top plate. These forces are transferred to the end walls via the ceiling diaphragm and then into the foundation/soil. Pole houses possess the same conventional system with the additional advantage of poles transferring load directly from the roof level into the ground.

2   USES AND BENEFITS OF POLE HOUSES

The uses of pole framing are diverse and widespread, including raising houses above flood plains, earthquake resistance and to provide ventilation. Historically in Australia logs and round timbers have been used extensively in farming, the livestock industry, and domestic structures. Pole houses essentially consist of a braced 3D grid of poles which provide support for the floor, walls and roof assembly. A number of benefits can be obtained using pole houses as listed below.

- Pole frame housing is ideal for sloping land as the 3D grid can be easily adapted to even steep slopes and other difficult sites.
- The poles cause minimal disturbance to the ground, there is no disruption to the general landscape and the house is perceived to blend more naturally with its environment.

- The open underside of the house provides a space for parking vehicles, storage, shaded play areas, maintenance access, family recreation, and ventilation.
- In a fire the large timber members in a pole house retain their structural integrity. Poles have a very good area/perimeter ratio and thus provide time for people to move to safety.

3   POLE EMBEDMENT

Good pole installation assists in controlling the response of the structure to lateral forces such as wind loads. In Australia the depth of pole embedment is typically 4 times the diameter of the pole, inclusive of its surrounding backfill. A 230 mm pole would typically be installed in a 450 mm and backfilled with concrete. This forms an equivalent "composite" pile of 450 mm in diameter which would then require an embedment depth of about 1.6 m.

The lateral capacity obtained from shallow depths cannot be analysed by normal pile theories which tend to ignore upper soil layers to a depth of 2-3 diameters. Contribution from these top layers is disregarded due to possible disturbance and to allow full development of soil resistance. Deflection and rotation calculations outlined in AS2159 (Australian Standard Piling Code) (1978) do not apply for pile length to diameter ratios less than 5. Backfilling is also an important issue with some designers and builders support the use of concrete while others suggest that embedment in concrete causes timber degradation.

3.1   *Existing Methods of Determining Pole Embedment*

The common methods used to design pole embedment depth are the Broms Method, the Uniform Building Code (UBC) method , the UBC + Restraining Slab method, and use of Rutledge design charts. All the methods are based on pole rotation and the development of passive soil forces to provide lateral resistance. The different passive stress distributions assumed result in different pole embedment depths to resist an applied lateral force.

Table 1. Summary of calculated pole embedment depths (230 mm diameter)

Method	Depth	Comments
Broms	1.8 m	The top 1.5 diameters of soil is ignored. Lateral passive resistance is then assumed constant at $9c_u$
UBC	2.0 m	Passive resistance increases linearly with depth to a maximum at about 4 m.
UBC + Slab	1.5 m	Slab ensures restraint at ground level, allowing a reduction in depth over the UBC Method
Rutledge	1.2 m	Used for advertising signs etc., and not recommended for habitable structures.

When the different methods are applied to poles of 230 mm diameter, they give a range of embedment depths from 1.2 m (Rutledge) up to 2 m (UBC method as shown in Table 1. The problems in design arise because most of the recognised procedures are insensitive where the depth to diameter ratio becomes less than 4, which is an upper limit for most pole house footings.

3.2   *Field Investigation of Embedment*

Twelve, 230 mm diameter poles were installed in the field using 3 backfill materials and 4 embedment depths of 1200 mm, 1500 mm, 1800 mm and 2100 mm. The backfill methods were concrete, decomposed granite (deco), and half concrete (bottom) and half decomposed granite (top). The deco was classified as a silty, sandy gravel under the UCS System. The pole boreholes were plugged with 200 mm of dry mix concrete which was air cured for 28 days.

The average stiffness for the poles was 12.4 GPa with a coefficient of variation of 12%. Detailed information on pole properties is available elsewhere (Bullen and Wood 1996). The site was fairly uniform in that it contained a stiff, yellow brown, gravely, silty clay with an undrained shear strength of around 80 kPa and a CBR of 15%. All poles were loaded laterally using a horizontal steel cable attached to the pole 2.1 m above ground level, which corresponds to the first floor of a pole house framing system. The poles were each loaded to failure or to a maximum of 30 kN which is approximately 10 times the maximum working load expected in a typical pole frame house.

3.3   Experimental Results of Pole Tests

Experimental results of the study are summarised in Table 2 and Figure 1. More detailed results are available in Bullen and Wood (1996). Pole deflection decreased as the foundation embedment depth increased from 1200 mm to 2400 mm and deflections for the poles backfilled with the compacted decomposed granite were the highest. The mode of pole deflection was a composite of pole flexure and rotation which changed with backfill material.

Table 2. Summary of pole deflections for all material and depths
3 kN and 5 kN (top, middle, bottom)

Pole Depth	Load (kN)	Concrete (mm)	1/2 + 1/2 (mm)	Deco (mm)
1.2 m	3	9.2, 4.7, 0.6	20.2, 10.8, 3.4	39.0, 22.5, 7.1
	5	15.5, 8.5, 1.4	34.9, 19.2, 6.5	77.9, 47.3, 14.3
1.5 m	3	8.1, 3.5, 0.5	14.9, 7.6, 1.6	20.0, 11.8, 3.5
	5	13.5, 6.1, 0.9	26.1, 13.7, 3.1	46.5, 26.2, 8.4
1.8 m	3	6.3, 2.5, 0.2	13.1, 8.0, 1.6	18.5, 10.7, 3.6
	5	10.7, 4.2, 0.4	26.0, 14.4, 3.4	36.5, 20.5, 6.8
2.1 m	3	6.4, 2.4, 0.2	11.9, 6.0, 1.4	11.8, 6.7, 1.3
	5	10.5, 4.0, 0.5	21.0, 10.6, 2.7	22.2, 12.4, 2.7

Top of Pole Deflections for 2100 embedment

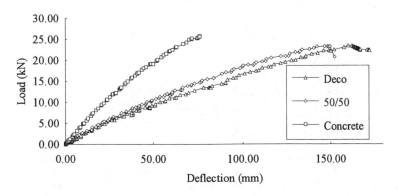

Figure 1. Top of pole deflections for the three-backfill methods (2.1 m embedment)

679

The performance for the poles backfilled with concrete gave the best correlation with the UBC method and possessed the greatest load capacity and stiffness of the 3 materials. The difficulty in compacting the decomposed granite within the confines of the borehole/pole contributes to its poor performance. In such a construction method high quality material such as crushed rock should be used and minimum compaction specifications applied to ensure good load transfer between the pole and the surrounding soil mass. The concrete backfill method is definitely the easiest to carry out, although at a slight increase in material cost.

A widely used "rule of thumb" method for control of deflection is that the deflection should be less than h/300; in this case 2100/300, or 7 mm. The variable h is the height from the soil surface to point of load application. The concrete encased pole had a deflection of 6 mm at the design load of 3 kN. Thus the full depth concrete backfilling method appears to the most suitable for use in pole house construction.

4.   POLE HOUSE INVESTIGATION

The deflection response of a pole frame house to lateral loads during various stages of construction was determined. The house was designed on a grid of nine poles at 3 m spacing with the centre row of poles supporting the ridge beam of the roof. The poles were CCA treated pine, 230 mm in diameter which are commonly used in pole house construction. The design wind speed was 41 m/s, the local design wind velocity for coastal areas in South East Queensland, described as Region B in the wind code AS1170.2 (SAA Loading Code) (1989.

The Timber Research and Development Advisory Council of Queensland (TRADAC),W33N-W41N, Timber Framing Manual, was used to determine the size of timber components. Hardwood bearers and joists were F14 stress grade and weatherboard type cladding and tongue and groove flooring used due to their low racking resistance.

4.1   *Pole House Construction and Testing*

The poles were embedded 1800 mm, in pre-bored 450 mm diameter holes, founded on a 200 mm bed of dry mix concrete, and then backfilled with concrete. Bearers and roof beams were bolted into place and floor joists and rafters installed. As exposed rafter ceilings are a common feature, the ceiling was a plywood diaphragm laid on top of the rafters. The battens were laid on top of the plywood, and the ply nailed through to the rafters below and the battens above. The roof sheeting was then installed.

Table 3. Timber element stiffness

Element	Size (mm)	Mean Stiffness (CV)
Poles (9)	230 round	13200 MPa (0.13)
Bearers (4)	200 x 75	15100 MPa (0.13)
Beams (2)	175 x 75	18300 MPa (0.20)
Rafters (18)	175 x 50	22000 MPa (0.18)
Joists (32)	150 x 50	21600 MPa (0.10)

Each pole, bearer, roof beam, rafter and joist was subjected to 4 point bending to obtain individual stiffness for use in an elastic model. The mean stiffness information for each element type is provided in Table 3. Individual stiffness data are not provided but are available from the authors. The pole house was built and tested in staged construction of poles, floor bearers, roof beams, joists, rafters, addition of the plywood ceiling, roof sheeting and ridge capping, flooring, bracing walls, exterior cladding and finally, interior gyprock cladding.

Lateral loading was applied via cables simultaneously to all poles at the front and back of the house, and a load spreader used to split load equally to the eaves level and the floor level. The poles exhibited load sharing as soon as the framing was connected and poles in the same row deflected a similar amount due to the interconnected framing grid. The addition of the floor did not change the deflection at the eaves but a marked reduction in deflection occurred after the bracing walls were constructed, reducing the deflections from around 30 mm to about 15 mm at each pole, at a total load of 22 kN (equating to 2.4 kN/ pole).

4.2   Interpretation of House Test Data

The elastic frame package SPACEGASS (ITS 1996) was used to model the pole house. The individual timber member stiffness measured in the laboratory was used as input for the analysis. The elastic package is limited to 25 material types and it was necessary to group some element properties. Individual pole, bearer and roof beam stiffness data were used while the rafters and joists were grouped into lots and represented by mean values of the lot. A comparison of data generated using the elastic frame model and experimental data is summarised in Table 4.

Table 4. Comparison of experimental data and elastic model.

Site	Act	SG	Act	SG	Act	SG
Eaves	32.0	23.3 (pole 1)	35.8	24.9 (pole 2)	29.4	22.2 (pole 3)
Floor	8.4	6.1	8.0	6.7	7.1	5.8
Eaves	26.7	23.3 (pole 4)	23.1	24.9 (pole 5)	23.6	22.2 (pole 6)
Floor	5.8	6.0	3.3	6.7	7.6	5.8
Eaves	33.6	23.3 (pole 7)	32.2	24.9 (pole 8)	31.4	22.2 (pole 9)
Floor	7.0	6.1	3.1	6.7	5.8	5.8

Each box represents one pole in the house plan. Deflections are shown for each pole at eaves level(top line) and floor level. Actual deflections denoted as Act, SPACEGASS deflections as SG.

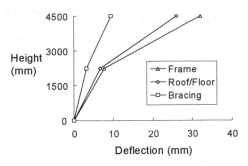

Figure 2 Load deflection plots for various construction stages.

The data in Table 4 are for the design load on the house in the early stages of construction, with roof and floor beams erected, and rafters and joists attached, but no bracing walls. The SPACEGASS model returns more consistent deflections than the field test. During construction the building stiffened considerably, especially after the addition of the bracing walls. Figure 2 shows the gradual increase in overall stiffness, where the deflection profile at the maximum load of 22 kN for each stage of construction is shown. The computer analysis is continuing and will be reported in detail in the future.

4.3 Comments on Full Scale House Testing

The deflections predicted by the elastic model were less than those measured during testing for several reasons.
- The timber elements are anisotropic and possess different stiffness in tension, compression and bending, while the model assumes that each elements is isotropic.
- Each timber element has material properties which vary along the length of the element, while the model treats each element as homogeneous over its full length.
- The main reason is believed to be that the joints in the pole house are nailed and will undergo small internal displacements during loading while the model idealises the joints as pins, with no internal movement

5.   CONCLUSIONS

It was found that even shallow embedded poles possessed reasonable lateral resistance. The deflection the poles reduced as the depth of embedment increased and for the decomposed granite tip deflection reduced from 39 mm to 12 mm. The pole tip deflection for the composite backfill reduced from 20 mm to 12 mm as the depth of embedment increased and for concrete backfilled poles deflection reduced from 9 mm to 6 mm. Pole behaviour was best predicted by using the UBC method, where passive resistance is assumed to increase linearly with depth from the surface (dependent upon the assumed soil characteristics).

The pole house deflection decreased as more elements were added and at the eaves level reduced from 27 mm (frame), to 26 mm (frame plus roof and floor) to 7 mm (frame, floor and bracing). The load-deflection characteristics of the pole house could be approximated using an elastic frame package, if individual and representative element stiffness were applied. Differences between measured and predicted values were attributed to anisotropic and non-homogeneous nature of the timber elements and by the "flexibility" of the nailed joints.

The testing carried out at various construction stages showed that loads transferred from the 3D frame into the cladding. At present the model does not possess the capability to model cladding. The use of high stiffness diagonal cross members in the elastic frame package will be used to represent cladding and to calibrate the model. This will allow the model to more accurately predict the response of pole houses to lateral loads.

6.   REFERENCES

Bullen and Wood (1996). The lateral load capacity of shallow embedded poles. Proc. Int. Wood Engineering Conf. Vol.3, pp.178-184.
Integrated Technical Software Pty Ltd (ITS) (1996). Spacegass Version 7. Three Dimensional Structural Frame Analysis Software.
Standards Australia. (1978). AS 2159. SAA Piling Code. Standards Association of Australia
Standards Australia. AS1170.2 The Australian Standard for Minimum Design Loads on Structures (SAA Loading Code)
Timber Research and Development Advisory Council of Queensland (TRADAC). 1994. The TRADAC W33N-W41N Timber Framing Manual, 3rd Edition. (TRADAC Brisbane).

The Mechanics of Structures and Materials, Grzebieta, Al-Mahaidi & Wilson (eds)
© *1997 Balkema, Rotterdam, ISBN 90 5410 900 9*

Scanning timber for strength predictions

R.H.Leicester
CSIRO Building, Construction and Engineering, Melbourne, Vic., Australia

ABSTRACT: This paper describes the use of neural networks to predict the strength of structural timber on the basis of data obtained from six scanners. These networks were found to be simple to apply, but there are doubts on the robustness of the algorithms for coping with changes in the characteristics of the timber resource. One method suggested for coping with this difficulty is to use a combination of several algorithms, each one set up to cope with a different sub-group of timber.

1 INTRODUCTION

Timber is a highly variable material, with some properties, such as strength, varying by a factor of more than 10:1 within the same tree. Hence, for practical applications timber is sorted into 'grades' of varying quality, as illustrated in Figure 1. Each grade has an assigned design property related to the 5-percentile value of strength within that grade; the commercial value of the grade is related to the design property. Obviously the efficiency of a grading operation increases with the accuracy with which strength can be predicted by nondestructive methods.

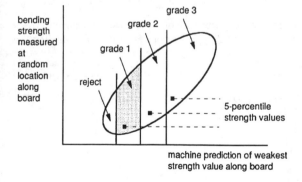

Figure 1: Schematic illustration of a grading operation to sort timber according to strength

A new grading system under development is illustrated in Figure 2. It comprises the following units:
(i) A knot detector, comprising four microwave scanners that are intended to detect the presence of knots as the timber moves past it;

(ii) A microwave scanner used to make a continuous measurement of the slope of grain along a piece of timber; and

(iii) A conventional mechanical stress grader that measures the local stiffness of the timber at discrete intervals along its length.

Details of the grading system have been given in a previous paper (Leicester and Seath, 1996). It is the purpose of this paper to discuss a mathematical procedure used for processing the system signals to predict the strength of a piece of timber at cross-sections along it's length.

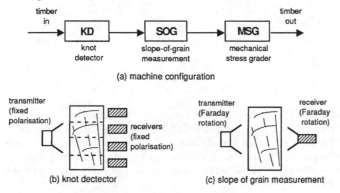

Figure 2: Schematic illustration of the grading system

2 DATA AQUISITION

Data to calibrate the system was obtained by passing some 1000 pieces of timber through the grading machine and recording all readings made by the scanners. Then each piece was broken in bending at one or more locations.

3 DATA-REDUCTION

As a piece of timber is passed through the stress grader each of the six scanner channels takes a reading every 6.5 mm along the length of timber. A typical record for one channel is shown in Figure 3.

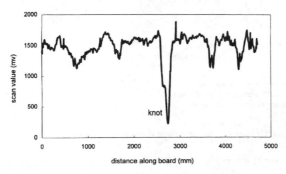

Figure 3: Typical signal from a scanner of the knot detector unit

As a first step in the data processing, the signals of the knot scanner are converted into a simple binary form that indicate either the presence or absence of a knot. This is achieved by using a running average to filter out the longer wave length variations of signals on the knot scanners. An example of the relationship between the use of electronic and visual methods to measure knots is shown in Figure 4. A reasonable match between the two procedures was taken to be an indication that preprocessing was satisfactory.

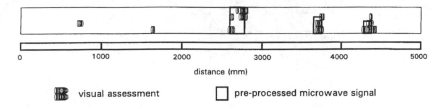

Figure 4: Comparison between electronic and visual assessments of knot presences

The second step in the processing is to select three parameters from each signal to be used as a basis for estimating the strength for each cross-section of the timber under investigation.Thus, 18 parameters were evaluated for each cross-section along a piece of timber. Strength predictions were then based on the use of these input parameters. Several sets of parameters were tried; the choice of parameters was based on the author's past experience in developing grading methods for structural timber. Typically the parameters chosen were of the following type.
(i)   An integrated measurement of intensity in the vicinity of the cross-section;
(ii) A mean value based on an average over a 2000 mm length; and
(iii) A measure of variability, typically the coefficient of variation, based on the whole length of the lumber.

4 NEURAL NETWORK

The first prediction algorithm developed was based on a neural network such as that shown in Figure 5. Within this network, a set of input parameters denoted by X_i are used to predict the output strength R. The input parameters used are the 18 parameters described in the previous section. The network shown comprises two internal layers of neurons and one output neuron, an architecture that is adequate for mapping most functions, even discontinuous functions (Beale and Jackson, 1994).

A sigmoid function is used for each neuron. This function is defined by

$$y = sigm(x) = 1/(1 + e^{-x}) \tag{1}$$

where y denotes the output signal for an input value of x.

This function, illustrated in Figure 6, has the characteristic of an on-off switch for $x < -5$ and $x > +5$. It also has the simple differential

$$dy/dx = y(1 - y) \tag{2}$$

which is useful in writing learning equations for the network.

685

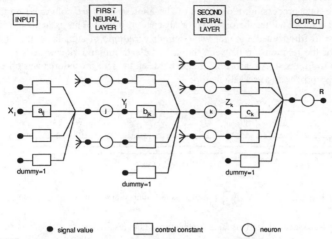

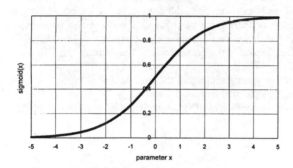

Figure 5: Schematic illustration of the neural network used

Figure 6: Graph of the sigmoid function

The output of the network depends on the selection of the control constants a_{ij}, b_{jk} and c_k. Specifically, the output R may be derived by the following sequential computation.

$$Y_j = \text{sigm} \left[\sum_i a_{ij} X_i \right] \tag{3a}$$

$$Z_k = \text{sigm} \left[\sum_j b_{jk} Y_j \right] \tag{3b}$$

$$R = \text{sigm} \left[\sum_k c_k Z_k \right] \tag{3c}$$

where Yy_j and Z_k denote the output signals from the first and second neural layers respectively.

686

The optimum values of the control constants are derived through a 'learning' process. As each item of data is presented to the network, the error in the prediction of R is used to adjust the value of the control constants. The correction is made proportional to the differentials with respect to R. Making use of equation (2), these differentials are given by

$$\partial R/\partial c_k = R(1 - R) Z_k \qquad (4a)$$

$$\partial R/\partial b_{jk} = R(1 - R) Z_k (1 - Z_k) Y_j c_k \qquad (4b)$$

$$\partial R/\partial a_{ij} = \sum_k R(1 - R) Z_k (1 - Z_k) Y_j (1 - Y_j) X_i c_k b_{jk} \qquad (4c)$$

5 APPLICATION OF THE NETWORK

Because the stress grader is intended for commercial application, there are two important matters in the application of the neural network, or in fact for any other type of algorithm. One is to find a measure of the value of the network and the second is to find some measure of the robustness of the network.

As mentioned earlier with reference to Figure 1, the commercial value of stress-graded timber relates to 5-percentile strength. Hence one measure of value is to specify the 5-percentile value of the lowest acceptable grade and then to determine the quantity or 'recovery' of the total population that qualifies as stress-graded timber. An example of this is shown in Figure 7. The Figure also illustrates the relative benefit obtained through the use of microwave scanners.

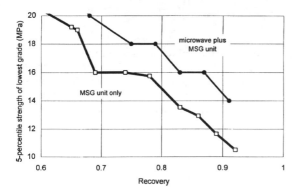

Figure 7:   Example of the computed increase in timber recovery due to the use of microwave scanners

The matter of robustness is more difficult to achieve because of the limited test samples available to assess any algorithm that is developed. To attempt this, two procedures were used. First the test timber was obtained from two quite different resources. About 400 pieces of timber from the first resource were used to develop the algorithms and then they were tested by application to 400 pieces of timber from the second resource. Second, two sets of equipment were applied for scanning the timber. These two sets of equipment were different in terms of their component geometry and hence different in terms of the microwave scanning signals produced. The software criterion used was that any preprocessing algorithms developed should produce the same output with both sets of equipment.

Finally, robustness in the face of varying resource was pursued through the technique of subdividing the timber into various types (related to botanical features) and then to develop a separate algorithm for each type. Using the input parameters mentioned earlier, it was found possible to divide the timber into (i) juvenile wood, (ii) mature wood containing large knots, and (iii) mature wood that is clear.

6 EXPERIENCE WITH THE NEURAL NETWORK

It was found quite simple to program equations (3) and (4) and to set the network into a learning mode. In order to reach an optimum result it was found necessary to run the available data sets through the network several thousand times, and to shuffle the sequence of the data set every dozen or so runs. Numerous networks were trialled. Both one and two neural layer networks were used, and four to 16 neurons per layer were tried. The use of excessive numbers of neurons usually resulted in many neurons being either permanently in the 'on' or the 'off' configuration.

Because a neural network produces a highly nonlinear function of the input parameters, and can in fact quite easily produce discontinuous functions, it is very difficult to find out how the network is operating. This is not a problem when there is unlimited data for use in training and checking the network. However, this complexity causes a serious difficulty when limited data is available and it is desired to make the algorithm sufficiently robust that it will still predict accurately as the timber resource changes.

For this project it was found that a single 4-neuron layer produced a reasonable compromise between efficiency and robustness. This was, however, dependent on using a suitable preprocessing procedure for the microwave signals, and also on subdividing the timber into three types as mentioned earlier.

7 CONCLUDING COMMENT

As a general comment it can be said that the use of neural networks to predict strength on the basis of multiple parameters was found easy to apply but did raise doubts on the robustness of the system. Currently several other procedures are being assessed for their ability to produce robust algorithms, including variance segmentation (Rouger, 1996) and conventional regression analysis. As a long term goal it is intended to try procedures that use the total signal rather than a limited set of parameters derived therefrom.

8 REFERENCES

Beale, R. & T. Jackson (1994). *Neural computing: An introduction.* Institute of Physics Publishing, Bristol, IK, 240 pages.

Leicester, R.H. & C.A. Seath (1996). Application of microwave scanners for stress-grading. *Proc. of International Wood Engineering Conference*, New Orleans, USA, October 28–31, 435–442.

Rouger, F. (1996). Application of a modified statistical segmentation method to timber strength grading. *Proc. of European Workshop on Application of Statistics and Probabilities in Wood Mechanics*, Bordeaux, France, February 22–23, 15 pages.

The Mechanics of Structures and Materials, Grzebieta, Al-Mahaidi & Wilson (eds)
© *1997 Balkema, Rotterdam, ISBN 90 5410 900 9*

Mechano-sorptive deformation in timber

H.R.Milner
Monash University, Melbourne, Vic., Australia

R.H.Leicester
CSIRO Building, Construction and Engineering, Melbourne, Vic., Australia

ABSTRACT: Mechano-sorptive deformation of wood plays a significant role in the deformation and strength of timber construction. A literature review indicates that the available information of mechano-sorptive deformation is probably not adequate for structural design purposes. More reliable quantified information is required on the effects of moisture and stress levels, particularly high stress levels, on the magnitude of mechano-sorptive deformation. More information is also required on the nature of the deformations incurred during several thousand climate cycles.

1 INTRODUCTION

With reference to structural timber, the term 'mechano-sorptive deformation' refers to the deformation of wood under the combined action of stress and moisture change. This deformation occurs in addition to the deformations due to elasticity, shrinkage, swelling and visco-elastic creep. A well known example of mechano-sorptive effects is illustrated in Figure 1. Mechano-sorptive deformations are frequently a significant part (sometimes the greatest part) of deformations in a timber structure and may be wholly or partly irreversible. Mechano-sorptive effects are to be found for timber having a moisture content in the 0–25 per cent range, i.e. in dry or seasoned timber.

Methods for the prediction mechano-sorptive deformations are important in the design of timber structures. Examples of structural behaviour for which mechano-sorptive theory is used include the following:
(a) the long term deformation of a structure,
(b) the long term creep that affects the buckling load capacity of structures, and
(c) the long term reduction in load capacity of structural elements brought about by the redistribution of stresses due to mechano-sorptive creep. An example of this last effect is shown in Figure 2.

In the following, the existing literature is reviewed to assess the adequacy of the information for practical purposes. In recent years, there has been considerable research on this topic, most noticeably by the group of laboratories in Europe taking part in the COST 508 project. In 1991, the COST 508 group held a significant workshop on mechano-sorptive deformation at the Department of Building Materials, Royal Institute of Technology, Sweden; in 1996, at the conclusion of the COST 508 project, an International Conference was held at Stuggart. At both these meetings numerous papers were produced concerning recent studies on the topic of mechano-sorptive deformation.

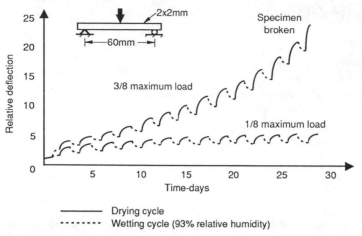

Figure 1. Wooden beam subjected to moisture cycling (after Hearman and Paton 1964).

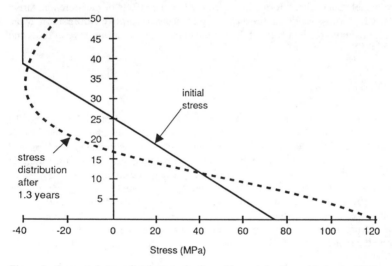

Figure 2. Computed stress distribution of softwood beam (after Lu and Leicester 1994).

2 ELEMENT MODEL

Some of the more comprehensive models are those given by Toratti (1992), Martensson (1992) and Gril (1996). In principle, most of the proposed models can be represented by a set of elastic springs and mechano-sorptive elements such as that shown in Figure 3. The strain ε_m of a mechano-sorptive element is related to the applied stress σ and change in moisture content m by the relationships of the type.

$$d\varepsilon_m = h(\sigma, m) \cdot |dm| \tag{1}$$

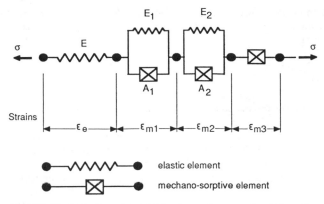

Figure 3. Typical schematic model for the mechano-sorptive deformation of a wood element.

where h(σ, m) is a function of stress type, stress level, moisture content. The function h(σ, m) has a sign depending on whether the moisture content change is positive or negative; usually the strain increment increases only as the moisture content decreases. On unloading, a typical model will exhibit some residual strain; the elements of the model may be chosen so that with the condition $\sigma = 0$, either part or all of the residual strains of the model will gradually disappear with time due to climate effects; in Figure 3 the components ε_{m1} and ε_{m2} will gradually dissipate with time in this way.

3 LIMITATIONS

There are several deficiencies in current knowledge which limit the reliability and thus the practical application of the models. The function h(σ, m) is very nonlinear, particularly at high stresses and high moisture contents; the characteristics of this function at high stress levels are particularly important when attempting to assess the redistribution of stresses that lead to failure. One of the few papers to address this issue is that by Liu $et\ al$. (1992); unfortunately the information given in this report is not very satisfactory.

Another deficiency in the available information relates to the conditions under which there will be an upper limit to the total possible creep. The daily cycles of moisture change comprise some 7000 cycles in a 20 year life of a structural member; however the largest number of moisture cycles undertaken in a reported test are 70 cycles on 10 x 10 mm square beams by Mohager (1987).

With respect to the use of mechano-sorptive theory to predict creep buckling effects, it is necessary to predict deformations under conditions of an incrementally increasing stress. A discussion on this topic, together with some experimental data has been given in a previous paper (Leicester, 1971). However, there have been few studies that examine experimentally the ability of mechano-sorptive models to predict deformation under changing stress conditions. Some information is given in the paper by Navi $et\ al$. (1996).

Finally, comment should be made that for application to engineering design, some assessment of the uncertainty of the predictions of mechano-sorptive deformation must be made. Much of this uncertainty is due to the natural variability of timber and the way in which the variability of natural characteristics, such as slope of grain and wood density, affect creep parameters. A few attempts to assess the uncertainty have been made, such as that by

Martensson and Thelandersson (1992). However, most such attempts do not take into account the greatest source of uncertainty, i.e. the uncertainty in our predictive knowledge of mechano-sorptive effects in real structures.

4 MOISTURE MOVEMENT

The application of mechano-sorptive theory requires predictions of moisture changes within structural elements. This prediction requires two pieces of information. The first is the prediction of the local climate around a structural element. Information for this can be undertaken by theoretical methods (e.g. ASHRAE, 1993) or by direct measurement on existing buildings coupled with external climate data (e.g. Cole 1993). There is a considerable body of research knowledge available for this, but predictions are still associated with considerable uncertainties.

Once the external climate is known, the moisture content changes within a structural timber element can be made with a good degree of accuracy using current moisture transport theories (e.g. Siau, 1995). In fact, for certain idealised climates, closed form solutions are available (Leicester and Lu, 1992; Bajaj and Milner, 1996).

5 FIELD VALIDATION

Because of the current uncertainties associated with predictions of mechano-sorptive effects, most code specifications are based on long duration experiments that simulate practical conditions. Typical examples are the assessment of creep effects by Ranta-Maunus *et al* .(1996) and strength loss effects by Barrett (1996).

6 RESEARCH

Since every structural property of every form of timber is subjected to mechano-sorptive creep, there is a wide range of current research related to the topic. The following are a few examples to illustrate the extent of the subject.

There is continuing fundamental research to examine mechano-sorptive behaviour at the cellular level (Houska, 1996; Gril, 1991, 1993, 1996), the behaviour of juvenile and reaction wood (Houska, 1996), and characteristics perpendicular to the grain (Dill-Langer *et al.*, 1996). Dinwoodie and Bonfield (1995) have reviewed the data for panel products. Leicester and Lu (1996), Ranta-Maunus (1996) and Jensen and Hoffmeyer (1996) have examined the stress redistribution due to mechano-sorptive effects and their influence on the strength of structural timber elements.

7 CONCLUSIONS

The literature survey has indicated that despite the extensive and wide range of research that has already been undertaken on the mechano-sorptive deformations of timber, the subject is so complex that currently it is doubtful whether a quantitative theory can be used to make predictions sufficiently accurate for practical application; the theory however is probably sufficiently robust to make comparative or relative assessments between two sets of conditions.

The most urgent targets for immediate research are the effects of moisture content and high stress levels on mechano-sorptive deformations, the effects of large numbers of moisture cycles, and the effects of changing levels of stress. More accurately quantified models of climate would also be useful.

8 REFERENCES

American Society of Heating, Refrigerating and Air-Conditioning Engineers 1993. ASHRAE Handbook of Fundamentals.

Bajaj, A.S. & H.R. Milner (1996). Time dependent moisture diffusion models for timber. *Proc. of 25th Forest Products Conference*, Melbourne, Australia, November 18–21.

Barrett, J.D. 1996. Duration of load: past, present and future. In *Proc. 1996 International Conference on Wood Mechanics*, EMPA, 14-16 May, Stuttgart, Germany, 119–138.

Cole, I.S. 1993. Wall Cavity Microclimate and material durability parameters– An Australian survey. In S. Nagataki, T. Nireki & F. Tomosawa (eds), *Durability of Building Materials and Components 6*. E & FN Spon, pp.627–636.

COST 508-Wood Mechanics. 1996. *Proc. 1996 International Conference on Wood Mechanics*. 14-16 May, EMPA, Stuttgart, Germany, 675 pages.

Dill-Langer, G. S. Aicher & H.W. Reinhardt 1996. Creep in glulam in tensions perpendicular to the grain at 20°C/65% RH and at sheltered outdoor climate conditions. In *Proc. 1996 International Conference on Wood Mechanics*, EMPA, 14-16 May, Stuttgart, Germany, 155–174.

Dinwoodie, J.M. & P.W. Bonfield 1995. Recent European research on the rheological behaviour of wood based panels. In *Proc. of COST 508 Workshop on Mechanical Properties of Panel Products*, Building Research Establishment, Watford UK.

Gril, J. 1996. Principles of mechano-sorption. In *Proc. 1996 International Conference on Wood Mechanics*, EMPA, 14-16 May, Stuttgart, Germany, 1–18.

Gril, J. & M. Norimoto 1993. Compression of wood at high temperature. In *Proc. of COST 508 Workshop on Wood: Plasticity and Damage*, University of Limerick, Ireland.

Hearmon, R.F.S. & J.M. Paton 1964. Moisture content changes and creep of wood. *Forest Products J*, No.8, 357–359.

Houska, M. 1996. Mechano sorptive creep in adult, juvenile and reaction wood. In *Proc. 1996 International Conference on Wood Mechanics*, EMPA, 14-16 May, Stuttgart, Germany, 47–62.

Jenson, S.K. & P. Hoffmeyer 1996. Mechano sorptive behaviour of notched beams in bending. In *Proc. 1996 International Conference on Wood Mechanics*, EMPA, Stuttgart, Germany, 14-16 May, 203–214.

Liu, T., K. Odeen, & T. Toratti 1992. *Creep in wood under long term loads in constant and varying environments*. The Royale Institute of Technology, Stockholm, Sweden. TRITA-BYMA.

Leicester, R.H. 1971. Lateral deflections of timber beam-columns during drying. *Wood Science and Technology*, Vol. 5, No. 3, pp.221–231.

Leicester, R.H. 1971. A rheological model for mechano-sorptive deflections of beams. *Wood Science and Technology*, Vol. 5, No. 3, pp.211–220.

Leicester, R.H. & J.P. Lu (1992). Effects of shape and size on the mechano-sorptive deformation of beams. *Proc. of IUFRO S.05.02 Timber Engineering Meeting*, Bordeaux, France, August 17–21, 12 pages.

Lu, J.P. & R.H. Leicester 1994 Deformation and strength loss due to mechano-sorptive effects. In *Proc. Pacific Timber Engineering Conference*, Gold Coast, Australia, July, Vol. 1,.140–143.

Martensson, A. 1992. *Mechanical behaviour of wood exposed to humidity variations*. Lund Institute of Technology, Report TVBK-1006, 189 pages.

Martensson, A. & S. Thelandersson 1992. Control of deflections in timber structures with reference to Eurocode 5. In *Proc. of Meeting No. 25, CIB-W18*, Ahus, Sweden.

Mohager, S. 1987. Studier av krypning hos tra (studies of creep of wood, in Swedish) -Doctoral dissertation. Kungliga Tekniska Hogskolan, Stockholm, Sweden, 139p.

Navi, P., V. PiHet & C. Huet 1996. Some new aspects of wood mechanosorptive effects. In *Proc. 1996 International Conference on Wood Mechanics*, EMPA, 14-16 May, Stuttgart, Germany, 19–34.

Ranta-Maunus, A. 1996. The influence of changing state of stress caused by mechano-sorptive creep on the duration of load effect. In *Proc. 1996 International Conference on Wood Mechanics*, EMPA, Stuttgart, Germany, 14-16 May, 187–202.

Ranta-Maunus, A., S. Gowda & M. Kortsmaa 1996. Creep of timber during 4 years in natural environments.

Royal Institute of Technology 1991. *Fundamental aspects on creep in wood.* Stockholm, Sweden.

Siau, J.F. 1995. *Wood: Influence of moisture on physical properties.* Dept of Wood Science and Forest Products, Virginia Polytechnic Institute and State University, USA 227 pages.

Toratti, T. 1992. *Creep of timber beams in a variable environment.* Helsinki University of Technology, 182 pages.

The Mechanics of Structures and Materials, Grzebieta, Al-Mahaidi & Wilson (eds)
© *1997 Balkema, Rotterdam, ISBN 90 5410 900 9*

The effect of moisture content on strength and stiffness of timber

P. De Leo
Barclay Mowlem Ltd, Australia

G. N. Boughton
Curtin University of Technology, Perth, W.A., Australia

ABSTRACT: Current design codes model moisture content effects in timber by reducing the strength where the equilibrium moisture content increases during the life of the structure. However, code models of timber behaviour do not vary the elastic stiffness of timber under the same conditions. Recent in-grade testing shows that the characteristic strength of commercially available timber has very little variation with moisture content, but the characteristic stiffness does vary significantly with moisture content. This paper presents those results and shows how they can be modelled for use in structural design.

1 INTRODUCTION

Limit states design targets consistent levels of reliability or performance for structural elements. With the "soft conversion" of the Australian Timber Structures Code to limit states format this year, the modification factors for timber strength have had to be assessed for their role in delivering consistent performance prediction. All of the modification factors in the strength limit state models of timber performance give a simplified representation of a complex behaviour. This paper examines the strength modification factor for partial seasoning k_4 and its representation in limit states format codes.

It presents some research into the bending strength of timber under different moisture conditions. The effect of moisture on flexural stiffness was also examined.

1.1 *Early models of timber structural behaviour*

Early attempts to characterise the strength of wood were frustrated by the variance in the test data. The variation existed between trees grown in different locations and climates, between timber cut from different locations in a single tree trunk, and with the number and position of knots and other growth characteristics within a single piece of timber. The most consistent characterisation of wood fibre strength was obtained where small specimens were cut out of straight grained wood specifically to avoid growth characteristics. These were known as "small clear specimens" and their strength gave an indication of the potential wood fibre strength for a given species. It was an upper bound on the strength of commercial timber, with the strength being reduced by various growth characteristics such as knots and splits. (Boughton and Crews, 1997)

Timber is graded into various groups which can be expected to have similar strength properties. Where the design codes were underpinned by small clears data, each of these graded groups was given design strengths that were less than the test results for the small clear specimens. The design strengths were found by reducing the small clear specimen strengths in accordance with the grade allocated. The design process then further modified the strength to account for other factors that can affect the strength of the timber. Most of these are a function of the design setting and include: duration of load, moisture content of the timber, temperature, and size factors.

1.2 *Limit states models of structural behaviour*

More recent models of timber strength focus directly on the strength of timber in the market place rather than introducing the steps associated with small clear specimen testing. This is the process of in-grade testing. Large number of specimens are tested to destruction under loading and restraint conditions that model timber use in the current building industry. This gives a real indication of the full distribution of the strength of graded timber in the market place. (Foschi et al, 1989)

The design strength of in-grade tested material is taken from the 5th percentile strength of the tested population with some reductions to give an adequate factor of safety (AS/NZS4063, 1992). Consistent with the behaviour model that uses the 5th percentile strength, modification of that strength now has to reflect the effect of environmental conditions on the 5th percentile strength.

In the Australian context, the new limit states version of AS 1720.1 (1997) - the Timber Design Code has incorporated in-grade test data into the new limit states format and has modified the strengths previously obtained from small clear testing to give compatibility with in-grade data and the limit states format. In line with this philosophy, the modification factors should reflect the effect of changes of the environment on the 5th percentile strength. This includes modification for partial seasoning, or differing moisture contents of the timber in service.

2 MOISTURE IN TIMBER

When trees are growing, the wood is saturated with water. The spaces in the cells are filled with water, and the cell walls contain water in the interstices in the walls. After felling, and certainly after sawing into lengths of rectangular cross section, the moisture starts to escape from the wood. The drying starts with loss of water in the spaces in the cells. This can be accomplished with low energy input and leaves the moisture content at around 25% to 30% for most commercial species in Australia. This is known as "fibre saturation point" (Bootle, 1983). The harder task is the removal of water from the cell walls.

In commercial operations, one of the final stages of drying is the high temperature kiln drying of the timber which removes some water from the cell walls to lower the moisture content from 25% to around 12%. This allows the timber to be classified as seasoned timber.

(Seasoned timber has a moisture content of less than or equal to 15% which is close to the equilibrium moisture content for indoor use in most parts of Australia.) (AS2858, 1986) Seasoned timber is often specified for timber which will be used indoors. Because the moisture in the timber is close to equilibrium with the moisture in the air, there is little movement of moisture into or out of the timber. This gives it stable dimensions and mechanical properties that do not change with time. Seasoned timber used internally is also less susceptible to creep deflections under long term loads.

2.1 *Effect of moisture on wood fibre*

The moisture content of wood affects strength of the wood fibres. The cells of the wood fibre are loosely bound into a structure and have their greatest strength parallel to the axis of the cells. Moisture in the cell walls lubricates the bond between the cells and so the strength of the wood fibre decreases with increases in moisture. This has been found in studies of the affect of moisture on strength of small clear specimens. The curves below show results from such tests for a variety of mechanical properties found by T.R. Wilson(1932) of the US Department of Agriculture in the 1930's. His relationships for variation of strength and stiffness with moisture content are shown in Figure 1 to Figure 3 as "Small clears data". The small clear samples test data shows that as timber is seasoned, an increase in mechanical properties of the wood fibre can be expected. Wilson's work has provided the basis for moisture modification factors used in most timber design codes throughout the world. In the most recent working stress version of the Australian Timber Structures Code (AS1720.1, 1988), Wilson's data was represented by the partial seasoning factors k_4 and k_5.

2.2   Code representations of moisture in wood

Seasoned Timber becoming Partially Seasoned   In some applications (and an extreme example is cooling towers), the timber will be exposed to a moist environment and will draw moisture from the atmosphere. If seasoned timber is used in these applications, the moisture content may rise to above 15%. The timber can no longer be regarded as seasoned and is said to be partially seasoned.

The current version of the limit states timber design code accounts for the effects of partial seasoning using the k_4 factor. For seasoned timber that absorbs moisture from its environment, the k_4 factor is found by decreasing the strength of the timber using equation (1):

$$k_4 = 1 - 0.3\left(\frac{emc - 15}{10}\right) \text{ but } \geq 0.7 \tag{1}$$

This reduces the strength of the seasoned timber in accordance with its expected equilibrium moisture content in service (*emc*).

Unseasoned Timber becoming Partially Seasoned   For unseasoned timber that is used in a dry environment, (for example indoors), moisture will be given up to the atmosphere and the timber will move closer to a moisture content of 15% that would allow it to be classified as seasoned. Initially unseasoned timber that has dried to a moisture content that is less than 25% but more than 15% is also said to be partially seasoned. Here the amount of partial seasoning that can take place is a function of the rate of removal of moisture which in turn is affected by the thickness of the material. k_4 for this case, is a function of the thickness of the material.

2.3   Effects of moisture on the strength of commercial timber

Madsen (1992) tested three commercially available Canadian timber species and species mixes (Douglas Fir, Hem Fir and Spruce Pine Fir) to evaluate the effect that moisture had on mechanical properties. He used an in-grade testing program that reflected a limit state design philosophy - materials loaded in a way that reflected in-service loading configurations.

In summary, Madsen found that:

- Compressive strength is highly sensitive to moisture content
- Tensile strength is not sensitive to moisture content
- Flexural strength is sensitive to moisture content for material with higher strengths but not for material at lower strength.
- Tension and compression perpendicular to grain are sensitive to moisture content
- Modulus of elasticity (*E*) is moderately sensitive to moisture content
- Flexural stiffness (*E I*) is not sensitive to moisture content (While *E* decreases with increasing moisture content, the dimensional changes associated with swelling under increasing moisture gives an increase in *I*. The product *E I* remains largely unchanged.)

A test program to reflect timber in the Australian market place and the Australian design context was devised (De Leo, 1996). A single large sample of seasoned timber consisting of 348 lengths of F5 and 606 lengths of F8 seasoned *radiata pine* was prepared. Each of the F5 and F8 samples was split into three separate groups for different moisture conditioning. Two of the three groups will be detailed in this paper. The other concerned re-drying issues and will be covered in separate publications.

Group A was reconditioned so that the parcel had an average moisture content near 25%. This group simulated seasoned material that had been produced in a normal milling and drying process, but used in an environment where water is absorbed by the timber so that its moisture content had increased. This is close to the lower limit of k_4 (0.7 given in equation 1) in the limit state version of the code.

Group B was tested in its supplied condition (seasoned) so that the moisture content remained less than 15%. This is appropriate for most of the seasoned structural timber used in an indoor environment and gave the conditions for the upper limit of k_4 for seasoned timber (1.0 from equation 1) in the limit state version of the code.

The two parcels of timber represented the range of k4 values for seasoned timber in the current version of the limit states code.

Flexural testing and analysis of each length was performed in accordance with AS/NZS 4063(1992). The load deflection curve, failure load and mode of failure of each specimen was recorded and collated. Moisture content evaluation was performed on a random sample of each of the groups of tested timber in compliance with AS1080.1 (1997). Figures 1, 2 and 3 show the results of the investigation:

3.1   *Effect of moisture content on strength*

Figure 1 shows the complete set of data with bending strength plotted against the average moisture content for each of the two groups of timber. The 5%ile, 50%ile and 95%ile have been found for each of the groups of data produced by the test program. These have been connected by straight lines. The relationship given by the limit states version of AS 1720.1 (and based on k_4 times the given bending strength of F5 and F8 material) is also shown on the plot. The code model gives no indication of the gain in strength for moisture contents below 15%, but shows a decrease in strength for moisture contents between 15% and 25%.

Wilson's work was performed on small clear specimens. The data reported in this work pertains to commercial sized specimens. It is difficult to compare the relationships predicted by Wilson with the in-grade data, but as the timber used in the small clear specimen tests would have been similar in character to the timber at the 95%ile strength, Wilson's curve has been superimposed on the top part of the data plots.

The 95%ile line from the in-grade test data shows a significant decrease in strength as the moisture content increases. The behaviour of these stronger test specimens was quite similar to that expected from small clear specimens with some of the failures being compression failures. The relationship between timber strength and moisture content for these specimens could be seen to follow the trends established by Wilson for small clear specimens.

The 5%ile line from the test data was essentially horizontal. There appeared to be no effect of moisture content on strength of timber when measured at the 5%ile level. The failures of the material near the 5%ile line was generally associated with tension perpendicular to grain at a growth characteristic such as a knot. In this case, the failure was not so much associated with tension failure of wood fibre, but with the separation of the wood fibres.

The strength of wood fibre can be affected by moisture in the wood structure, whereas the failure at a growth characteristic seems to be more a function of the type of feature than the rather subtle lubricating affects of moisture in the fibre structure.

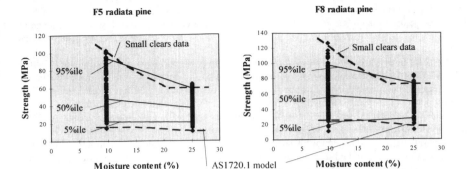

Figure 1 Strength versus moisture content for all strength specimens

The 50%ile line showed a trend that was midway between the two effects detailed for the 5th percentile and the 95th percentile.

Figure 2 shows the strength and moisture content pairs for each of the samples that were tested for moisture content. Here, because the moisture content was known for each and every data pair, the strength could be associated with the moisture content of each piece rather than the average for the batch. This shows the scatter in moisture content values, but has fewer data points as there were fewer moisture content determinations. The 5%ile, 50%ile and 95%ile lines have also been plotted on this figure. The trends seen in Figure 1 are also evident in Figure 2.

The reference lines given by the small clear specimen data, and the model of variation in strength with moisture content predicted by AS1720.1 (1997), are also shown in these plots. The trend given by the small clear specimen data reflects the behaviour of the in-grade specimens at the 95%ile. The trend given by the AS1720.1 (1997) model of moisture content effect on the strength of timber shows a decrease in strength with increasing moisture, and is similar in slope to that of the 50%ile line from the in-grade data. However, the test data at the 5%ile shows little effect of moisture on strength.

k_4 The data shows that $k_4 = 1.0$ for all seasoned specimens

3.2   Effect of moisture content on stiffness

Moisture content also affects the deflection of loaded timber members. All timber design behaviour models allow for time dependent deflection (creep). There is sound evidence for mechano-sorptive creep in which movement of moisture into or out of the timber causes larger

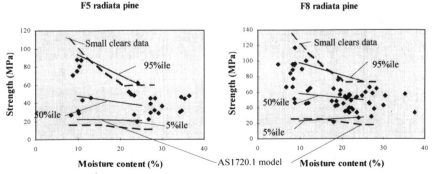

Figure 2 Strength versus moisture content for all moisture content specimens

creep deflections (Toratti 1991). The Australian Standard allows for different duration of load factors to be used for timber with different initial moisture content. This allows for a change in only the long term component of deflection. Allowance for creep has no effect on instantaneous deflection.

The short term MoE was evaluated for each of the test specimens as indicated in AS4063 (1992). In order to calculate the E value, the nominal dimensions of the in-grade specimens were used. Specifically, no allowance for swelling of timber on wetting, or shrinkage of the section during drying was made. This mirrors the dimensions used in the design process. Hence any relationship derived from the test results should be applicable in the design arena. In effect the trend shown in Figure 3 shows the effect of moisture on $E\,I$, or bending stiffness. It shows the flexural stiffness as a function of moisture content of the specimen. The trends observed in this work were more marked than those found by Madsen (1992). This is because of the lower shrinkage for *radiata pine* compared with that for the species used in the Canadian experiments.

Figure 3 also shows that all percentile limits gave the same general trend. Increases in moisture content reduced the 95 percentile, 50 percentile and 5 percentile stiffness of the timber. In this case, the small clear data relationship has been matched against the mean stiffness given by the test data.

The current code relationship has no variation in elastic stiffness with the changes in moisture content.

j_4 An appropriate modeling of instantaneous stiffness of seasoned timber that receives partial seasoning would allow for a decrease in stiffness as equilibrium moisture content rises. Such a relationship is shown in equation (2), and has been plotted in Figure 3 as the proposed behaviour model.

$$E' = j_4\,E \quad \text{with} \quad j_4 = \left[\frac{1}{1 + 0.25\left(\dfrac{emc - 15}{10}\right)} \right] \text{ but } \geq 0.8 \tag{2}$$

where
E = seasoned characteristic Modulus of Elasticity
E' = Modulus of Elasticity corrected for moisture
emc = annual average moisture content

The proposed model for stiffness variation is conservative when compared with the 50%ile stiffness line. (Code stiffness values are derived from average stiffness from in-grade test data.)

In the stiffness data, the trends established by small clear testing were also reflected in the in-grade data. This was because the stiffness determined for a single piece of timber is affected by

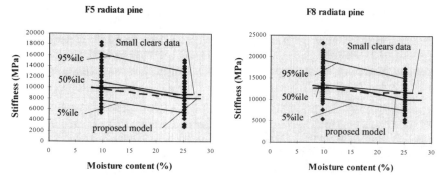

Figure 3 Stiffness versus moisture content

700

the behaviour of all of the timber in the test span. Any growth characteristics will have a relatively minor effect as they will be averaged with all of the other wood fibre in the test span.

4   CONCLUSIONS

For design with the strength limit state, it is important that the effects of the environment on mechanical properties of timber be modelled correctly. The data that support these conclusions were established for *radiata pine* (which constitutes the majority of structural timber used in Australia).

- *Strength limited design* uses a characteristic strength that is based on the 5%ile strength of timber within the grade. The study has shown that there was essentially no variation in the characteristic bending strength with moisture content.
- The partial seasoning factor k_4 should be 1.0. for seasoned timber used where rises in the equilibrium moisture content may cause partial seasoning of the material in service.
- *Stiffness limited design* uses a characteristic stiffness that is based on the mean or average stiffness of the timber within the grade. This study has shown that there was a reduction in instantaneous (elastic) bending stiffness with an increase in moisture content. There is no partial seasoning factor for elastic deformation of timber in the current version of AS1720.1
- A new factor for the effect of partial seasoning on deformation should be introduced. This factor j_4 (equation 2) can be used to reduce the elastic modulus of timber used in design where partial seasoning of an initially seasoned product can be anticipated.
- j_4 would be used to multiply the E used for both long term and short term deflection to give an appropriate value for the higher moisture content of the partially seasoned timber.

5   ACKNOWLEDGMENTS

The authors gratefully acknowledge the financial support of Pine Australia in the execution of the research project reported here. The moisture conditioning of the timber was provided by Wespine Industries and assistance in the laboratory testing phase was provided by the technical staff of the school of Civil Engineering at Curtin University of Technology. Ms Debbie Falck ably assisted in the analysis of the data.

6   REFERENCES

Australian Standards (1986) *AS2858   Timber - Softwood - Visually Stress-graded for Structural Purposes*, Sydney: Standards Australia

Australian Standards (1992) *AS4063   Timber - Stress Graded - In-Grade Strength and Stiffness Evaluation*, Sydney: Standards Australia

Australian Standards (1997) *AS1080.1 Timber - Methods of test Method 1: moisture content*, Sydney: Standards Australia

Australian Standards (1997) *AS1720.1 Timber Structures - Part 1: Design methods*, Sydney: Standards Australia

Bootle, K.R. (1983) *"Wood in Australia"* McGraw Hill Book Company, Sydney.

Boughton, G.N. and Crews, K.I. *"Limit States Timber Design to AS1720.1"* Curtin University Publishing

De Leo P (1996) *Effect of Moisture on In-grade Strength and Stiffness of Radiata pine* Undergraduate Thesis, Perth, Curtin University of Technology.

Foschi, R.O. et al *"Reliability-Based Design of Wood Structures"* Structural Research Series, Report No 34, University of British Columbia, Vancouver BC.

Madsen B (1992) *Structural Behaviour of Timber*, Vancouver: Timber Engineering Limited

Toratti T (1991) *Creep in wood in varying environment humidity - part 1 simulation of creep* Helsinki University of Technology.

Wilson T R C (1932) *Strength - Moisture Relations for Wood* Technical Bulletin No 282, Washington DC: Forest Service, US Department of Agriculture

Design capacity tables for timber

G.N.Boughton & D.J.Falck
Curtin University of Technology, Perth, W.A., Australia

ABSTRACT: Some design capacity tables for timber will be released in 1998. These tables have a format very similar to that of the design capacity tables for steel members. The paper looks at the potential for use of the Design Capacity Tables for Timber (DCTT) and demonstrates that there is a use for them in the offices of practicing engineers and in education as well. The rationale of the tables and the technical justification of them is presented.

1   INTRODUCTION

Timber is an engineered building material that has been experiencing a renaissance in the past few years.

- In 1996, in-grade tested timber products were released in the Australian market place. This means that the manufacturer can publish strength data for design purposes that is underpinned by test results for the product in the market place.
- In 1997, the limit states version of the timber design code was published. This gives designers the ability to use identical loads for the strength limit states for concrete, steel masonry and timber components in the same structure. It also provides a uniform philosophy for the structural design process.

A number of large timber structures continue to be built throughout the world. These included large span dome structures such as the Viking Ship Hall at the Lillehamer Winter Olympics and some large exhibition halls currently under construction for the Sydney 2000 Olympics.

New developments in manufactured timber products which include Laminated Veneer Lumber (LVL) give greater structural reliability because the variance in the properties of these manufactured products has been decreased. Timber must be seen as a legitimate structural engineering material and a full suite of design aids is needed to ensure that design can be accomplished effectively.

1.1   *Limit states timber design code*

The limit states version of the timber design code AS1720.1 (1997) has adopted a similar philosophy to that of the other limit states design codes in Australia, though the structural model to represent timber behaviour bears a great deal of similarity to that used for the working stress version of the code.

The *serviceability limit state* is modelled using elastic deflection formulae, but a correction factor is applied to increase the calculated deflections to allow for the effects of creep of timber under longer-term loads. The timber design code allows the use of equation (1) for prediction of deformation of a simply supported beam under uniformly distributed load.

$$\delta_{total} = \frac{5}{384} \frac{\sum(w_i \, j_2) \, L^4}{E \, I} \tag{1}$$

The correction factor for creep effects is j_2 and is applied to each component of the load combination. It is a function of the duration of each component of the load, and is also affected by the initial moisture content of the timber. The actual value of the j_2 factor is independent of the size of the timber member used.

The *strength limit state* is checked by ensuring that the design capacity of the member is greater than the load effect (factored for the strength limit state). The combined effects of the load factors and the capacity factor ϕ give a level of safety that is uniform across all of the major building materials. The form of the equation for bending members at the strength limit state is given by equation (2). It shows that the behaviour model is based on the elastic modulus of the cross section Z and the ultimate bending strength f'_b. The elastic model gives a satisfactory level of performance when used with a design strength that is based on the 5th percentile of bending strength data. The basic bending strength $\phi f'_b Z$ is modified for the design setting by use of some modification factors (k factors).

$$\phi \, M \geq M^* \tag{2}$$

where $\quad \phi \, M = \phi \, k_1 \, k_4 \, k_6 \, k_9 \, k_{11} \, k_{12} \, f'_b \, Z \tag{3}$

1.2 Modification factors for timber

Classical elastic behaviour models show that the capacity of a structural element is given by the strength that is appropriate for the design action times the relevant geometric property of the cross section. However, other factors can influence the elastic capacity, some increasing it and others decreasing it. Many of these factors have been incorporated into timber design codes and some relate to the way the environment affects the strength of timber materials, while others model buckling phenomena. Bending, tension and compression members all use different combinations of the same k factors, with bending members using the largest number of them. They can be grouped as follows:

modification factors that are not functions of the size of the member
k_1 duration of load factor (strength)   function of load characteristics
k_4 partial seasoning factor   function of environment
k_6 temperature factor   function of environment
k_9 strength sharing factor   function of geometry of structural system

modification factors that are functions of the size of the member
k_{11} size factor   function of cross section dimensions
k_{12} stability factor   function of dimensions and restraint

Timber's behaviour is a function of a quite complex microstructure, and each of the above effects represents a complex relationship between strength and the parameter indicated. The model presented in AS1720.1 (1997) is quite simplified. In many cases, the simplified relationship has been obtained by research based on wood fibre strength rather than the strength of commercial timber as detailed in De Leo and Boughton (1997). The modifications are assumed to be independent, and this is largely the case.

In terms of the design process, the evaluation of k_1 k_4 k_6 and k_9 can be undertaken immediately after the structural scheme has been found and the design loads found. Their values will also be common to many members in the greater structure so can be carried from the design of other members. k_{11} and k_{12} will need to be evaluated each time that a member is sized. (They also happen to be the most difficult to calculate.) The size factor k_{11} is tabulated for some timber materials and must be evaluated from a formula for others. In common with steel member design, the stability factor k_{12} requires quantification of the lateral restraint system, evaluation of a slenderness parameter and then calculation of a stability factor k_{12}.

1.3   *Steps in hand design of timber members*

The steps in the design of timber members are very similar to those in the design of any structural member. They include:
- Selection of a timber species, product and grade.
- Evaluation of design loads for all limit states.
- Choice of critical limit state (strength or serviceability).
- Selection of a size to satisfy the critical limit state. (often an iterative process).
- Checking of performance to other limit states.

The design process required for all materials can prove tedious, and inevitably requires a number of pages of calculations. As well, the selection of bending members and compression members often requires a number of iterations through the design loop to determine an optimal member size. Given these circumstances, designers may limit the range of different grades or products that they consider in the sizing of structural members. This is a limitation that sometimes means that the most cost-effective timber product for an application is not considered.

2   DESIGN CAPACITY TABLES FOR TIMBER (DCTT)

The design dilemma has been faced by structural engineers for as long as structures have been designed. In the past, the design process has been accelerated by production of design charts or tables (AISC 1994) that enable designers to speed their evaluation of the behaviour of different types and sizes of members.

More recently, computer programs that incorporate structural analysis and design modules have been produced to enable a designer to perform almost all of the required calculations from a computer program that incorporates all of the constraints set out in the codes of practice. Spread sheets prepared in-house are also used to good advantage where similar types of structures or structural elements are designed on a regular basis.

However many structural design practices do not have a regular turn over of timber design work that can justify the purchase and investment in training of a design and analysis package, and do not have the repetitive work that justifies the development of spread sheets. These people could make use of Design Capacity Tables for Timber if they were available. Some engineers simply prefer to use tables rather than computers!
- The use of any Design Capacity Tables requires the practitioner to understand the code on which they are based, and to be able to make appropriate decisions about loadings, deflection criteria and restraint types.
- The DCTT provides a tool for determining performance. In common with any design aid, it still requires a designer to show creativity to establish an appropriate structural system to resist all loads on the structure.
- The use of the tables enable a number of different products, grades or species to be explored with little more effort than turning the pages.

Design Capacity Tables must be seen as a tool for applying a material behaviour model to expedite design. They must be used with a clear understanding of the behavioural model and of the design standard.

705

The Design Capacity Tables for Timber (DCTT) (Boughton and Falck, 1998) presents a series of tables that enable simple calculation of design capacities for the most common timber products under a variety of load effects. They include tables for bending members, compression members, tension members and bolted and nailed connections. These can also be used to facilitate the checking of members for combined actions.

The tables do not eliminate calculate them, but they reduce work load by cutting the number of calculations by at least half, and minimise frustration by eliminating iteration in design.

2.1 Technical basis of the Design Capacity Tables

The DCTT contains tables of data for each action and for varying restraint conditions. In all of the equations given below the terms in square brackets [ ] represent the numbers that are presented in the DCTT.

For the serviceability limit state, equation (1) can be rewritten in a design format. This is shown in equation (4) with g_a as a loading and support configuration factor that is tabulated in the DCTT for a number of different loading conditions and spans.

For simply supported members carrying uniformly distributed loads $g_a = \dfrac{5\,L^4}{384}$

For bending

$$[E\,I] = \frac{\sum(g_a\,j_2\,w_i)}{\delta_{\lim}}$$

(4)

EI values can be used to select members. Two different values are tabulated depending on the limits specified:

- $EI_{average}$ for limits appropriate to appearance or comfort
- $EI_{5\%ile}$ for limits appropriate to structural load paths or damage

For strength, the design capacity tables tabulate only those parameters that are a function of the grade, size and species of timber in the member:

For bending $\qquad k_{11}\,k_{12}\,f'_b\,Z \qquad$ equivalent to $\left[\dfrac{\phi\,M}{\phi\,k_1\,k_4\,k_6\,k_9}\right]$ (5)

For compression $\qquad k_{11}\,k_{12}\,f'_c\,A \qquad$ equivalent to $\left[\dfrac{\phi\,N_c}{\phi\,k_1\,k_4\,k_6\,k_9}\right]$ (6)

For tension $\qquad k_{11}\,f'_t\,A \qquad$ equivalent to $\left[\dfrac{\phi\,N_t}{\phi\,k_1\,k_4\,k_6}\right]$ (7)

This enables the parameters that are independent of the size of the members to be evaluated once at the commencement of the design and used to calculate the terms given for bending below in equation (8). These can be compared directly with the values tabulated in the DCTT.

For bending $\qquad \dfrac{M*}{\phi\,k_1\,k_4\,k_6\,k_9}$ used to find $\left[\dfrac{\phi\,M}{\phi\,k_1\,k_4\,k_6\,k_9}\right] \geq \dfrac{M*}{\phi\,k_1\,k_4\,k_6\,k_9}$ (8)

Design capacities can be found by multiplying the tabulated values by the appropriate independent k factors. This is shown only for bending as equation (9). Similar relationships are used in the sizing of compression and tension members.

$$(\phi\,M) = \phi\,k_1\,k_4\,k_6\,k_9\left[\frac{\phi\,M}{\phi\,k_1\,k_4\,k_6\,k_9}\right]$$

(9)

2.2 Example of use of Design Capacity Tables for Timber

A 3.2 m long floor joist is being designed for an office refurbishment in Melbourne.
Unfactored dead load   0.1 kN/m (sheet flooring nailed @ 450 mm centres)
Unfactored design live load   1.8 kN/m
Serviceability limits   span / 350 for short term serviceability live load only (visual)
                               15 mm for total short term serviceability load (partition under)
Design an MGP10 member to suit with spacing between floor joists of 450 mm.

(MGP10 is an new grade of Australian grown seasoned pine timber supported by in-grade tested and verified properties. It is now commonly available throughout the country in dressed sizes. The sizes shown in the DCTT are readily available in most major centres.)
   [ ] shows references from the DCTT. [Table 2.2.1] is a table from the DCTT. Section 2 is bending.
   < > shows references from the Timber Structures Code. <Table 2.5> is a table from AS1720.1, 1997.

Trying serviceability first

Short term servic'ty load $\psi_s Q = 0.7 \times 1.8 = 1.26$ kN/m   limit $= \dfrac{3500}{350} = 10$ mm (visual requ't)

$j_2 = 1$ for short term load <2.4.1.2>   and $g_a = \dfrac{5L^4}{384}$ [Table A1]

$$\frac{\sum\left(g_a \, j_2 \, w_i\right)}{\delta_{lim}} \quad \Rightarrow \quad \frac{5 \, L^4}{384} \frac{j_2 \, \psi_s \, w_Q}{\delta_{lim}} = \frac{5 \times 3200^4 \times 1.0 \times 1.26}{384 \times 10} = 172 \times 10^9 \text{ Nmm}^2 \text{ (10)}$$

use average *EI* value as the limit is given by a visual requirement (no damage)

Total short term servic'ty load   limit = 15 mm (damage to partition)
       $j_2 = 2$ for dead load portion of load combination <2.4.1.2>

$$\frac{\sum\left(g_a \, j_2 \, w_i\right)}{\delta_{lim}} \quad \Rightarrow$$

$$\frac{5 \, L^4}{384} \frac{\left(j_2 \, \psi_s \, w_Q + j_2 \, w_G\right)}{\delta_{lim}} = \frac{5 \times 3200^4 \times \left(1.0 \times 1.26 + 2.0 \times 0.1\right)}{384 \times 15} = 133 \times 10^9 \text{ Nmm}^2 \text{ (11)}$$

use 5%ile *EI* value as the exceeding limit will cause damage to a partition
Using [Table 2.2.1] (exerpt shown in Figure 1) givesEquation (1)gives
(10) => $EI_{avg} > 172 \times 10^9$ Nmm^2 use a 190 x 35 or a 170 x 45   [Table 2.2.1]
(11) => $EI_{5\%ile} > 133 \times 10^9$ Nmm^2 use a 190 x 35 or a 190 x 45   [Table 2.2.1]

Checking for strength

$$M^* = \frac{\left(1.25G + 1.5Q\right) L^2}{8} = \frac{\left(1.25 \times 0.1 + 1.5 \times 1.8\right) 3200^2}{8} = 3.62 \times 10^6 \text{ Nmm}$$

The following parameters can be obtained from AS1720.1(1997)
$\phi = 0.75$ MGP material in bending - primary elements in normal structure <Table 2.5>
$k_1 = 0.94$ Medium duration of strength load (eg crowd)   <2.4.1.1>
$k_4 = 1.0$ Seasoned timber used indoors   <2.4.2>
$k_6 = 1.0$ Melbourne not in the tropics!!   <2.4.3>

$$k_9 = 1 + 0.2\left(1 - 2\,\frac{450}{3500}\right) = 1.15 \quad \text{for discrete load sharing system typical of joists <2.4.5>}$$

equation (8) gives $\dfrac{M*}{\phi\, k_1\, k_4\, k_6\, k_9} = \dfrac{3.62}{0.75 \times 0.94 \times 1.0 \times 1.0 \times 1.15} = 4.46$ kNm

with sheeting nailed to top (compression) edge @ 450 mm centres $L_{ay} = 0.45$ m

[Table2.2.1] (excerpt shown in Figure 1) gives that 450 mm restraint of compression edge gives Full Lateral Restraint (FLR) for all members shown. Hence the major axis bending capacity will be given by this table.

MGP10
Seasoned

Size	Area	major axis		minor axis		major axis		
		$EI_{average}$	$EI_{5\%}$	$EI_{average}$	$EI_{5\%}$	$\dfrac{\phi M}{\phi k_{1-9}}$	FLR_{comp}	FTR_{tens}
	(mm²)	(10⁹Nmm²)	(10⁹Nmm²)	(10⁹Nmm²)	(10⁹Nmm²)	(kNm)	(m)	(m)
70x35	2450	10.0	7.00	2.50	1.75	0.457	2.1	any
90x35	3150	21.3	14.9	3.22	2.25	0.756	1.63	any
120x35	4200	50.4	35.3	4.29	3.00	1.34	1.23	any
140x35	4900	80.0	56.0	5.00	3.50	1.83	1.05	any
170x35	5950	143.	100.	6.07	4.25	2.59	0.901	any
190x35	6650	200.	140.	6.79	4.75	3.13	0.833	any
240x35	8400	403.	282.	8.58	6.00	4.62	0.713	1.86
290x35	10150	711.	498.	10.4	7.25	6.2	0.642	1.40
70x45	3150	12.9	9.00	5.32	3.72	0.588	3.47	any
90x45	4050	27.3	19.1	6.83	4.78	0.972	2.7	any
120x45	5400	64.8	45.4	9.11	6.38	1.73	2.03	any
140x45	6300	103.	72.0	10.6	7.44	2.35	1.74	any
170x45	7650	184.	129.	12.9	9.04	3.33	1.49	any
190x45	8550	257.	180.	14.4	10.1	4.03	1.38	any

Figure 1 Design Capacity Tables - general bending data for F5 seasoned timber

For $\left[\dfrac{\phi M}{\phi\, k_1\, k_4\, k_6\, k_9}\right] = 4.46$ kNm can be used to select a section from either [Table 2.2.1]

(shown above) or [Table 2.2.2]. it shows that a 240x35 MGP10 section would be needed.

The sections previously evaluated for serviceability needed to be increased.

The problem was solved without much more work than the calculation of the load effects for the strength and serviceability limit states. If alternative materials are considered, few further calculations are required - different tables would be used. (In some cases, the use of different materials may give a different ϕ value). The use of the DCTT is simpler than the use of hand calculations for timber design. However, it requires a few more choices and calculations than the use of Design Capacity Tables for steel. This reflects the greater detail required for . behaviour models for timber. The extra calculations are not difficult.

More examples of the use of the tables are given in the Design Capacity Tables for Timber.

2.3   *Educational value of the Design Capacity Tables for Timber*

Timber engineering units are already under pressure in university undergraduate engineering courses. (Boughton, Woodard and Crews 1997). This has generally meant there is less time available in civil/structural units to cover an increasing range of structural timber products.

It is still essential that timber engineering instruction cover the basic behaviour model for timber and the way it is represented in AS1720.1. However, it is generally possible only to illustrate those principles with a few worked examples or assignment problems. This means that in many courses, engineered timber products such as MGP, LVL or glued laminated timber (glulam) are not investigated by the students in detail. However, the basic principles of use of these products are the same as any other timber product, and the DCTT will give opportunity to

708

compare their performance with other timber products without investing a great deal of time in calculations. Rather the principles can be addressed.

For example, the DCTT can be used to show that a 195x38 mm GL12 member and a 170x36 mm LVL member have satisfactory performance for the example in section 2.4. Time can be spent explaining that the better performance comes from the higher stiffness of those two products and the higher ϕ factor that comes with better quality control in manufacture.

Trends in the performance of timber members can also be identified from the tables. It can be shown that as the distance between lateral restraints increases, so the bending capacity of a given member decreases.

3   CONCLUSIONS

The use of design capacity tables enables practising engineers to feel that they have control over the determination of loads. It also allows them to readily identify the restraint offered to the members that they are designing. However, they do take the tedium of hand calculation out of the task of manually selecting member sizes.

The fact that a range of sizes and restraint conditions can be seen at a glance enables a designer to readily identify whether changing the restraint conditions will have a substantial effect on the cost and performance of the structural element.

In order to make effective use of any design aid, including design capacity tables, a designer must have a clear understanding of the behavioural model used for the material. Many of the size-independent k factors are evaluated external to the use of the Design Capacity Tables for Timber, and enable a designer to maintain skills and understanding in that area.

As an educational aid, the Design Capacity Tables allow students to concentrate on the concepts, behavioural models and the relative economics of different structural configurations and products, without being distracted by the tedium of repetitive and iterative design calculations.

4   REFERENCES

Australian Institute of Steel Construction (1994) *Design Capacity Tables vol 1 open sections* Second edition, Sydney, AISC.

Australian Standards (1997) *AS1720.1 Timber Structures - Part 1: Design methods*, Sydney: Standards Australia

De Leo P and Boughton G N (1997) "The Effect of Moisture Content on Strength and Stiffness of Timber" *Proceedings 15th Australasian Conference on the Mechanics of Structures and Materials:* Melbourne, Balkema.

Boughton G N and Falck D J (1998) *Design Capacity Tables for Timber* Perth, Forest and Wood Products Research and Development Corporation. *(In press)*

Boughton G N, Woodard A C, and Crews K I (1997) "Cooperative Timber Education and Research between Industry and Universities" *Proceedings 15th Australasian Conference on the Mechanics of Structures and Materials:* Melbourne, Balkema.

The Mechanics of Structures and Materials, Grzebieta, Al-Mahaidi & Wilson (eds)
© *1997 Balkema, Rotterdam, ISBN 90 5410 900 9*

Cooperative timber education and research between industry and universities

G.N.Boughton
Curtin University of Technology, Perth, W.A., Australia

A.C.Woodard
Timber Promotion Council Victoria, Melbourne, Vic., Australia

K.I.Crews
University of Technology, Sydney, N.S.W., Australia

ABSTRACT: The paper examines collaborative education and research programs that have been operating in Australia.
In spite of the fact that timber is among our oldest structural materials, developments in grading and production technology have brought substantial changes in the behaviour models for structural timber. The paper addresses ways in which cooperative education and research initiatives can be further developed to lessen work loads for academics and at the same time ensure that teaching and research are relevant to the needs of the market place.

1 INTRODUCTION

Timber education in Australia is currently undergoing an exciting and invigorating renaissance as a result of a number of recently established industry-based educational initiatives. The focus is on collaborative development of relevant curricula and educational resources to service the needs of University Engineering, Building, and Architectural courses, as well as the TAFE courses. In all of these areas, there is a threat to reduce timber content in courses in the face of:
- increasing financial pressure on tertiary education forcing reduction of staff numbers,
- pressure to include more material in courses that means students have to be safeguarded from saturation.

Mode of delivery and educational resources must be changed to reflect these pressures.

1.1 *Continuing relevance of timber engineering*

Though timber is still the major material used in domestic framing in Australia, as a primary engineering material, it has had a somewhat cyclic popularity. Timber was the dominant structural material in Australia until the turn of the century (Nolan 1994) when the advent of wrought iron and locally produced concrete made substantial inroads into timber's structural market share. In periods of steel shortage such as those during the two world wars, many impressive long span structures were built (Nolan 1994).

Today, the comparatively limited use of timber in engineering is due to a combination of factors including: a shortage of engineers confident in the design of timber, a perception by some architects that timber is not an environmentally responsible material, and in some cases, by a lack of fabrication capacity (Pellerin et al, 1990). These impediments to the structural use of timber have been slowly changing.

Clients and architects have an appreciation of the warmth and vitality that can be given to a building by conspicuous use of timber. As well, architects are becoming more aware of the lower energy of production for timber products and carbon storage in timber that makes it a sound environmental choice. There is an increase in the number of architects specifying timber as the preferred structural material on the basis of its beneficial effects on the environment.

Engineers must be equipped to design structures using appropriate connections, members and details in timber. Timber design skills remain relevant for structural engineers in the 1990s and will continue into the next century. Timber design in Australia is currently used in the following areas:

- Concrete formwork. - some concrete formwork is fabricated from timber members supporting plywood.
- Multi-residential timber framed construction - most major members in multi-residential construction must be sized by a chartered engineer.
- Domestic construction - some principal members must be designed and detailed by an engineer
- Commercial premises - eg restaurants, small shops or offices. Structural timber members are designed by an engineer.
- Larger structures, swimming pool enclosures, buildings of cultural significance, churches, salt storage sheds, warehouses are all sometimes designed for construction in timber. These are substantial engineered structures in their own right.

1.2   *Pressures on engineering curricula*

All tertiary courses are currently feeling the squeeze that comes from more and more information, skills and concepts that must be fitted into their current curricula. Under other circumstances, an appropriate response may be to lengthen course duration to enable more information to be imparted. However, regular contact with universities across the country has shown that some university curriculum committees maintain that work loads in engineering courses are already too great to enable students to assimilate all of the information that is currently delivered, let alone any more. Reduced funding across the board is forcing tertiary education to look at ways of becoming more efficient at delivering information and developing vocational and life skills.

Information overload is experienced in all field of tertiary education. In the context of timber engineering education, it is driven by two things:

- An increase in the knowledge of our world coming from continued research. This includes better models of the behaviour of many systems which give us better design, analysis and manufacturing tools. This in turn has enabled the design of more ambitious and cost competitive timber structures.
- An increase in the range and variety of commercially available products that people are expected to be able to use. Development of new materials across the building sector and changes in the way traditional materials are used, has meant that people who work within the building industry have to have competence and understanding of a larger range of products and technologies than ever before. New building products and systems that are now commercially available across the country include: light gauge steel framing, tilt-up concrete construction, structural aluminium, reinforced plastics, precast concrete, as well as a host of new timber products such as Laminated Veneer Lumber (LVL), Medium Density and Fibreboard (MDF), composite beams such as "posi-strut", and I shaped beams.

It is not physically possible to fit all of this material into courses, so there is pressure on tertiary education to carefully consider the relevance of the material in courses.

In engineering courses, the time given to timber is variable. Many schools of civil engineering have lost staff experienced in timber engineering and have not replaced them. McDowell (1992) presented a report that showed few engineering schools in Australia included much timber in their curricula.

There have been a number of books of worked examples (Milner and Smith, 1993) and (Maddison, 1989), but no comprehensive Australian text in timber engineering has been available for almost 20 years. There are few resources that can enable a lecturer with little personal expertise in timber to teach in that area. In many cases, the response by universities in this situation has been to drop timber from its curriculum in spite of its continuing relevance for practicing engineers.

2 COOPERATIVE TIMBER EDUCATION PROGRAM

In 1996 a timber industry initiative saw the establishment of the National Timber Education Program. In its current form, it is funded by the timber industries through the Forest and Wood Products Research and Development Corporation (FWPRDC) and managed jointly by the FWPRDC and an industry representative body the National Timber Development Council (NTDC). The program has the following objectives:
- To develop a national educational strategy, program and resources that address educational needs of professions that use or specify timber products (engineering, architecture and building).
- To establish and coordinate national networks of educators and state education officers to maximise the efficient use of resources earmarked for education.
- To initiate professional development and undergraduate programs where needed in cooperation with target group organisations, institutions and timber industry networks.
- To provide resources that will inspire and equip undergraduate students, postgraduate students and practicing professionals towards the appropriate sue of timber in the building industry.

Simply stated, the Timber Education Program aims to offer assistance to already overworked academics as they prepare courses or units that bring students into contact with timber. It does this by facilitating co-operation between universities and with the provision of teaching resources. To date two engineering teaching packages have been released - one focussing on material science units and the other on structural design units. (FWPRDC 1997). A timber engineering text has also been published (Boughton and Crews 1997).

2.1 *Resources from Timber Education Program*

Teaching resources have been prepared and distributed to cover the basic elements of timber engineering in civil/structural engineering undergraduate courses. The resource kits include: slides, lecture outline and commentary, overhead transparencies, student notes and some simple models.

The Timber Education Program also maintains a network among lecturers who have teaching and/or research interest in timber. This enables sharing of ideas and resources and fosters the collaboration in both teaching and research referred to above. Newsletters are published, web pages maintained and an electronic mailing list run for exchange of news and views among the network participants.

2.2 *Institution of Engineers report on engineering education*

The IEAust report on Engineering Education into the Future (IEAust 1996) made a number of recommendations that supported the establishment of strong links with industry.

Recommendation 12 - Engineering schools must be prepared to collaborate to produce innovative courseware.

Recommendation 13 - there must be greater collaboration between the engineering schools and industry.

These two recommendations and the text that supported them confirmed that if a widely based course was to be offered by every engineering school, in spite of decreasing staff numbers, it must rely on collaboration between engineering schools and cooperative arrangements with industry. Few engineering schools will have the required staff numbers to support all specialities across the discipline.

The Timber Education Program addresses both of these areas of co-operation in two ways. It establishes networks among lecturing staff around the nation to foster collaborative efforts, and it devises, distributes and maintains up-to-date teaching resources to support lecturing staff with current information on products and behavioural models of timber performance.

3   TIMBER ENGINEERING EDUCATION

The teaching resource packages developed by industry and distributed by the Timber Education Program aim to provide knowledge of the physical phenomenon of the material response, and develop an intuitive understanding of timber's behaviour under load, so that timber can be appropriately detailed to avoid problems. The philosophy is equally applicable to many other engineering materials.

Timber is a composite material that is naturally produced. It has a well defined grain structure which gives it orthotropic properties. It also has naturally occurring growth characteristics such as knots, which change the orientation of the grain locally. In terms of materials models, timber has properties that make is similar in some respects to steel and similar in other respects to concrete. Some of its characteristics are very similar to fibre-reinforced composites such as FRP. A study of timber mechanics and behaviour serves as a good introduction to many materials with which engineers need familiarity and an intuitive understanding.

Similarity with steel	Similarity with concrete	Similarity with fibre composites
buckling in axial compression	creep under long duration loads	affected by moisture, temperature
lateral torsional buckling	strength a function of size	affected by duration of load
elastic modelling of strength allowed	large variance ratio for properties	properties a function of orientation

3.1   *Mechanics of timber materials*

The teaching resources provided for the timber component of a material science unit aims to impart an appreciation of the nature of timber, and the way in which its behaviour is underpinned by the microstructure of the material.

The behaviour of timber is quite fascinating. The education packages that cover it can be creative and entertaining. For example:

Timber's rather unique structure makes it interesting to study in a materials science context. Timber fibres consist of hollow cells with spiral wound walls that have great tensile strength and stiffness parallel to the axis of the cells (also the grain direction). In contrast, loads applied normal to the grain, tend to separate the cells (if tension) or squash the thin walled cells (if compression) as shown in Figure 1. This microstructure explains why both strength and stiffness parallel to grain are an order of magnitude greater than the strength and stiffness perpendicular to the grain. A good visual model of the microstructure of timber is a group of drinking straws. Such a model is included in the teaching resource package for timber materials.

Clear modelling assists in "de-mystifying" timber behaviour. Knowledge of the microstructure of wood can also help students to understand duration of load and moisture content effects on timber strength. The relationship between the material structure and the mechanical properties must be understood to enable undergraduate students to use any structural material. The study of the structural use of timber allows engineers to develop the materials

714

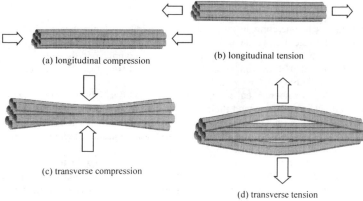

(a) longitudinal compression

(b) longitudinal tension

(c) transverse compression

(d) transverse tension

Figure 1 Hollow tube structure of timber

skills needed to develop an understanding of the use of a wide range of engineering materials. This applies to both static and dynamic properties.

The hollow cell structure and the loose connection between adjacent cells across the grain also gives timber a high damping ratio. This makes it good for attenuating vibrations, and together with the high degree of structural redundancy in light timber framing, makes timber structures appropriate for resisting earthquake forces.

The performance of timber also embodies durability aspects, and again the microstructure of timber can help explain some of the reasons for deterioration of timber. An understanding of these issues will help a designer to make wise decisions about the detailing of timber elements and connections.

3.2   Cooperative delivery of Timber Engineering Education

A number of American Universities have formed cooperatives in which course development is shared amongst a number of courses. High quality course materials can be developed within the cooperative by pooling resources, and then delivered to a much larger number of students than just one university could provide. The industry based model presented here follows a similar logic, in that resources are available in all Australian civil engineering schools.

Currently, computer aided learning tools are under preparation for use in a number of universities across the nation. These include a CDrom based reference work (Heaney and Kneen, 1997) and a self-paced learning package is under development at RMIT. Both of these resources will have the potential to reduce teaching hours without compromising the quality of learning received by the students. Eventually, Internet provision of teaching can be used to extend timber engineering education beyond the lecture room.

The timber industry itself has the capacity to contribute to educational experiences through the sponsorship of research projects, design competitions and provision of specialist lectures. The education model proposed by the IE Aust certainly includes these features. In some other cases, the timber industry has sponsored visits to saw mills and production forests to demonstrate the forest and resource management techniques that have enabled the use of sustainably produced timber products in the Sydney 2000 Olympics site. This is more than just technical education. It provides inspiration that leads to better student motivation and enhances creativity in design and awareness of changing community values.

4 CONCLUSIONS

University education has been operating in an environment of diminishing budgets for some years and current trends indicate that the declining budgets are likely to continue. Educational development must therefore be planned to take into account reducing university based resources. Cooperative education programs will maximise the benefits attained from teaching and learning resources that are developed within the Australian university system.

Industry funded and developed resources have some additional benefits. The resources will be appropriate for professionals in Australia, and their dissemination by industry will foster further cooperation between industry and academia. It gives support and assistance to already over-worked academics while at the same time packaging the information that the academics need to keep the lecture material up-to date and relevant to contemporary practice.

Ultimately, it will be possible to put many of the resources developed in this education program onto the Internet or into self-paced learning packages. These resources will enable distance education to be more easily achieved. Whole options within the course can be run in this mode so that students in courses where timber has not traditionally been part of the syllabus can undertake some level of timber education. Professionals can also take advantage of the distance education facility to lift their own levels of competence in timber design.

The partnership between the timber industry and academia has as its most tangible benefits a collection of education resources. The timber industry has produced some of these as a service to education institutions and enable the effective dissemination of up-to-date information to the students which will be the professionals of tomorrow. The collaboration has the potential to deliver high quality, well co-ordinated research and teaching for many years to come.

5 REFERENCES

Boughton, G.N. and Crews, K.I. (1997) *"Limit states timber design to AS1720.1"* Curtin University Press.

FWPRDC (1997a) *4 hour Timber Materials Teaching Resource Package*, Perth Forest and Wood Products Research and Development Corporation

FWPRDC (1997b) *8 hour Introduction to Timber Engineering Teaching Resource Package*, Perth Forest and Wood Products Research and Development Corporation

Heaney, A and Kneen P (1998) *"Timber structures"* CDrom format reference material published by Forest and Wood Products Research and Development Corporation

Institution of Engineers, Australia (1996) *Changing the culture: Engineering Education into the Future*, Canberra, IEAust

Maddison, B.H. (1989) *"Handbook of Timber Engineering Design"* University of Western Australia Press.

Milner, R.H. and Smith, G.W. (1993) *"Worked Examples for timber Structures"* National Association of Forest Industries ISBN 1 86346 024 1

Nolan G, (1994) "The culture of using timber as a building material in Australia" *Proceedings of the 1994 Pacific timber engineering conference*, Gold Coast, Australia

Pellerin, R.F.; Galligan, W.L.; Barnes, M.A.; Kent, P.M.; Leichti, R.J. (1990) "Developing continuing education in wood engineering for design professionals." Proc 1990 International Timber Engineering Conference, October 1990; Tokyo

Author index